Spectral Geometry and Inverse Scattering Theory

Huaian Diao • Hongyu Liu

Spectral Geometry and Inverse Scattering Theory

Springer

Huaian Diao
School of Mathematics
Jilin University
Changchun, China

Hongyu Liu
Department of Mathematics
City University of Hong Kong
Hong Kong, China

ISBN 978-3-031-34617-0 ISBN 978-3-031-34615-6 (eBook)
https://doi.org/10.1007/978-3-031-34615-6

Mathematics Subject Classification: 35R30, 35P05, 35P25, 78A46, 74J25

This Springer imprint is published by the registered company Springer Nature Switzerland AG
The registered company address is: Gewerbestrasse 11, 6330 Cham, Switzerland

Preface

This book is devoted to the mathematical study of geometrical inverse scattering problems. In the physical setup, one sends incident wave fields to interrogate an unknown/inaccessible object, called a *scatterer*, to generate the scattered wave fields. Then one collects/measures the scattered data, encoded into the so-called far-field pattern, and from which to infer geometrical/medium properties of the scatterer. This type of inverse scattering problems arises in various industrial applications of practical importance, including radar/sonar, geophysical exploration, medical imaging, and nondestructive testing. We are mainly concerned with the determination/identification of the shape of the scatterer, independent of its medium property, by the associated far-field pattern. We shall include the acoustic scattering, electromagnetic scattering, and elastic scattering problems. We shall mainly work within the time-harmonic regime, and the study leads to dealing with the spectral properties of various PDOs (partial differential operators), including the Laplacian, Maxwellian, and Lamé operators. Depending on the concerned geometrical inverse scattering problem, the spectral problem can be the classical Dirichlet/Neumann eigenvalue problem, or can be the so-called transmission eigenvalue problem.

In this book, we shall cover several fundamental aspects of the geometrical inverse scattering problems mentioned above, including the unique identifiability and stability/sensitivity issues, as well as the reconstruction methods. There are some pioneering contributions in the literature in this intriguing field, and hence in this book, we shall only present overviews on these results with new perspectives and novel insights, and instead we shall be mainly concerned with the new developments, especially by the two authors as well as their coauthors. We shall adopt a spectral geometric point of view. Specifically, we shall be mainly concerned with the geometric structures of the underlying eigenfunctions arising in the spectral problems mentioned above as well as their applications to the geometrical inverse scattering problems. Spectral geometry is a classical topic with a long history and rich results in the literature. Nevertheless, we study this topic from the perspective of inverse scattering theory. It leads to many new results that are of both theoretical and practical significance. Recently, there is considerable progress in the literature following this spectral geometric viewpoint due to the fruitful collaborations between the research groups of the two authors of this book. Hence, we think it is a good time to summarize these results in this book. Moreover, the field is still being under extensive development. Hence, the book can serve as a handy reference for researchers who are interested in this intriguing and active field of research. During the writing of this book, many people has helped us including Xiaoxu Fei, Qingle Meng, Yiwen Pei, Baiyi Sun, Longyue Tao, Jiexin Tang, Ruixiang Tang, Li Wang, Ke Yang, and Long Zhang, which we would like to gratefully acknowledge. Huaian Diao would like to acknowledge the research support by the National Natural Science Foundation of

China (grant no. 12371422). Hongyu Liu would like to acknowledge the research support by the Hong Kong RGC General Research Funds (projects 11311122, 12301420 and 11300821), the NSFC/RGC Joint Research Fund (project N_CityU101/21), and the France-Hong Kong ANR/RGC Joint Research Grant, A_CityU203/19.

Changchun, China
Hong Kong, China

Huaian Diao
Hongyu Liu

Contents

Chapter 1
Introduction

1.1 General Study on Inverse Problems

Inverse problems are concerned with the determination of causes by knowing the consequences. In modern language, they are concerned with data assimilation that seeks to optimally combine theory with observations. In its abstract formulation, an inverse problem can be described as follows. Let $(X, \|\cdot\|_X)$ denote the space of target objects (usually consisting of quantities or geometries) and $(Y, \|\cdot\|_Y)$ denote the space of observation data. Let $\mathscr{F}$ denote the model which sends a target object $\Omega \in X$ to a set of data $\boldsymbol{y} \subset Y$ and is determined by the underlying physical process. In its typical form, an inverse problem is given by the following operator equation:

$$\mathscr{F}(\Omega) = \boldsymbol{y}, \quad \Omega \in X, \ \ \boldsymbol{y} \subset Y. \tag{1.1.1}$$

That is, the inverse problem is to determine Ω by knowledge of $\mathscr{F}$ and $\boldsymbol{y}$. In (1.1.1), $\mathscr{F}$ is more than often described by a certain PDE (partial differential equation) system and Ω is a certain missing part of $\mathscr{F}$, say, e.g., the coefficient(s) of the PDE, and hence, the inverse problem is generally nonlinear. Moreover, the inverse problem is usually ill-posed in the sense that a small variation on $\boldsymbol{y}$ may cause a large variation on Ω or even the nonexistence of a true Ω. This shall be more evident in our subsequent discussion. Hence, the mathematical study on inverse problems is intriguing and challenging. We broadly summarize the following six categories.

Modeling. Inverse problems lie at the heart of scientific inquiry and widely occur in many industrial and scientific applications including radar and sonar, medical imaging, geophysical probing, nondestructive testing and cosmological exploration, to name just a few. Depending on the specific applications, the inverse problems can be modeled within certain differential or integral systems, which account for the underlying physical processes. Numerous models arise from many existing industrial and engineering applications and some have become classics in the relevant fields. Nevertheless, there are still many active efforts on refining or extending those models to cater for the new applications and new needs. On the other hand, many frontier and cutting-edge applications lead to many new inverse problems and sophisticated models are a key ingredient to the success of those rapidly growing technologies. It is remarked that even for a given and fixed application, the modeling is usually not necessarily unique. Modeling is not the focus of this book. Nonetheless, the study presented in this book may motivate or serve as an inspiring source or guiding role in the modeling of new or novel industrial and engineering applications. To support this viewpoint, we refer to [4, 5] on the mathematical modeling of geomagnetic anomaly detection and [11, 15, 19] on the mathematical modeling of some novel AI applications.

Uniqueness and Identifiability. For a given inverse problem of the form (1.1.1) with an established model $\mathscr{F}$. The first issue to be asked is whether one can really identify the unknown/inaccessible object Ω by using the measurement data $\boldsymbol{y}$. At the first level, one would need to check

H. Diao, H. Liu, *Spectral Geometry and Inverse Scattering Theory*,
https://doi.org/10.1007/978-3-031-34615-6_1

whether within such a model (which accounts for the underlying probing means), some meaningful observation data y can be generated. Here by meaningful observation data y, we mean that y should indeed carry useful information of the underlying target object Ω. This is the identifiability issue and closely related to the non-identifiability issue that shall be discussed in what follows.

At the second level, one would need to ask whether the dataset is sufficient to guarantee the unique determination of a target object. That is, for two admissible objects $\Omega_1, \Omega_2 \in X$:

$$\Omega_1 = \Omega_2 \quad \text{if and only if } \mathscr{F}(\Omega_1) = \mathscr{F}(\Omega_2). \tag{1.1.2}$$

The uniqueness result in (1.1.2) guarantees that for a given observation dataset y, it determines uniquely a solution Ω to the inverse problem (1.1.1). In general, one needs to impose a certain a priori knowledge on the underlying object in order to establish the desired unique identification result. That is, one usually has an admissible class $\mathscr{C} \subset X$ which is defined by the a priori knowledge of the unknown objects such that

$$\Omega_1, \Omega_2 \in \mathscr{C}, \ \ \Omega_1 = \Omega_2 \quad \text{if and only if } \mathscr{F}(\Omega_1) = \mathscr{F}(\Omega_2). \tag{1.1.3}$$

At the third level, one would ask whether the dataset in deriving the uniqueness result in (1.1.2) is optimal/minimal. A pertinent mathematical notion in determining the optimality/minimality of a given dataset for a specific inverse problem (1.1.1) is the so-called *cardinality* of a quantity. It is defined to be the number of independent variables of the given quantity. In establishing the uniqueness result in (1.1.2) for the inverse problem (1.1.1), it is minimally required that the cardinalities of the unknown Ω and the known y should be at least the same. In such a case, the inverse problem is referred to as *formally-determined.* If the cardinality of Ω, written as $\text{Card}(\Omega)$, is smaller than that of y, i.e., $\text{Card}(y)$, the inverse problem is referred to as *over-determined.* If $\text{Card}(\Omega) > \text{Card}(y)$, the inverse problem is referred to as *under-determined.* In certain under-determined cases, one may still expect the unique identifiability of the inverse problem, say e.g. if strong a priori knowledge is imposed on the unknown, or one is only interested in recovering the unknown to be within a certain fidelity region, which are actually equivalent to reducing the cardinality of the unknown. In general, establishing the uniqueness identifiability result for an inverse problem is more challenging with formally determined data than that with overdetermined data.

Stability and Sensitivity. Stability is a certain quantitative version of the uniqueness result in (1.1.3) for an inverse problem (1.1.1). It usually takes the following form:

$$\|\Omega_1 - \Omega_2\|_X \leq \psi(\|\mathscr{F}(\Omega_1) - \mathscr{F}(\Omega_2)\|_Y), \ \ \Omega_1, \Omega_2 \in \mathscr{C}, \tag{1.1.4}$$

where $\psi : \mathbb{R}_+ \to \mathbb{R}_+$ signifies a stability function and generically fulfils $\lim_{t\to 0}\psi(t) = 0$. Clearly, the stability result in (1.1.4) implies the uniqueness result in (1.1.3).

In many inverse problems, it turns out that the stability estimate takes a logarithmic form, say, e.g., $\psi(t) = (\ln|\ln t|)^{-1}$. This partly evidences that inverse problems are ill-posed. In fact, considering the inverse problem (1.1.1), we let Ω_ε and y_ε be such that

$$\mathscr{F}(\Omega_\varepsilon) = y_\varepsilon, \ \ \|y_\varepsilon - y\|_Y \leq \varepsilon \in \mathbb{R}_+. \tag{1.1.5}$$

If the logarithmic stability mentioned above is sharp, one has $\|\Omega - \Omega_\varepsilon\|_X \sim (\ln|\ln \varepsilon|)^{-1}$, which can be large even if ε is very small, say, e.g., $\varepsilon = e^{-10}$. On the other hand, we would like to remark that stability estimates for inverse problems usually strongly depend on many a priori parameters from the admissible class $\mathscr{C}$, signifying a general principle in inverse problems that “the lack of information cannot be remedied by any mathematical trickeries.”

Another closely related issue is the so-called sensitivity analysis which aims to validate the “well-posedness” of an inverse problem in certain aspects. For this, instead of considering small variation in the observation dataset y, one considers small variation to a certain specific feature of the unknown Ω. If such a small variation causes a significant change of a certain quantity $\mathbf{I}(y)$

that is defined over the corresponding measured data, then we say that the measured quantity $\mathbf{I}(y)$ is very sensitive with respect to the aforementioned specific feature of the unknown Ω. In such a case, it is unobjectionable to expect that the specific feature of Ω mentioned above can be more robustly recovered via using the quantity $\mathbf{I}(y)$.

Non-identifiability and Invisibility. It may occur that for any $\Omega_1, \Omega_2 \in \mathscr{C} \subset X$, one always has $\mathscr{F}(\Omega_1) = \mathscr{F}(\Omega_2)$. In such a case, $\mathscr{C}$ forms an equivalence class (with respect to the measurement dataset y). Hence, the unique identifiability issue to (1.1.1) is whether one can recover the unknown Ω up to such an equivalence class. There is an extreme case that

$$\mathscr{F}(\Omega) = \mathscr{F}(\emptyset), \tag{1.1.6}$$

where we use $\emptyset$ to signify a void object in X. In such a case, we see that *invisibility* occurs, meaning that the probing produces no meaningful data, and the target object Ω is invisible with respect to such probing means. However, it is always meaningful to investigate the non-identifiability for a given inverse problem. In fact, by characterizing the non-identifiability, one can achieve the identifiability of the inverse problem by excluding the non-identifiability scenarios. According to our earlier discussion, though the identifiability result may still not readily imply the unique identification result, it gives the necessary condition for the uniqueness result to be true. In fact, in some inverse problems, such condition is also the necessary condition for the uniqueness result to be held true. This shall become more evident when discussing the inverse scattering problems in what follows.

Moreover, in many practical applications, invisibility is a desirable outcome, in particular when one intends to hide a target object from being detected. Artificially achieving invisibility, a.k.a. invisibility cloaking, can be described as follows. Let $\mathfrak{m}$ denote a certain mechanism that is concatenated with a target object Ω to yield $\Omega \oplus \mathfrak{m}$. $\mathfrak{m}$ is called an invisibility cloaking device if it holds that

$$\mathscr{F}(\Omega \oplus \mathfrak{m}) = \mathscr{F}(\emptyset). \tag{1.1.7}$$

The notion of invisibility cloaking can be generalized by replacing $\emptyset$ in (1.1.7) to be a given object $\Omega_0 \in X$. In such a case, $\mathfrak{m}$ is said to be an illusion device which makes Ω look like Ω_0 with respect to the observation y. Invisibility cloaking constitutes an important field of contemporary research with rich results in the literature. We shall not touch much about invisibility cloaking in this book. We refer to [16–18] and the references cited therein for related mathematical studies in the literature on this intriguing topic.

Reconstruction and Resolution. For a physically and theoretically sound inverse problem, the next crucial issue is to develop effective and efficient numerical reconstruction schemes. The existing numerical strategies can be broadly classified as direct approaches, indirect approaches, and ML (machine learning)-based approaches.

A direct approach for the inverse problem (1.1.1) usually adopts an optimization viewpoint and tackles an optimization problem as follows:

$$\min_{\Omega \in \mathscr{C}} \|\mathscr{F}(\Omega) - y\|_Y + \mathfrak{R}(\Omega), \tag{1.1.8}$$

where $\mathfrak{R}$ signifies a regularization functional. The regularization term is used to tackle the ill-posedness of the inverse problem as well as to incorporate the useful a priori knowledge into the reconstruction process. We refer to [1–3, 9] for the direct treatments of general inverse problems, both linear and nonlinear. It is noted that in the nonlinear case, linearizing strategies are usually indispensable to linearize the nonlinear inverse problems.

Though direct approaches are handy and generally workable, they lose the track of the speciality and particularity of the underlying specific inverse problem, which can enable the development of more effective and efficient reconstruction algorithms. In fact, in this book, we shall show that the geometric patterns of the underlying spectral problems can enable us to develop indirect imaging schemes for inverse scattering problems that can even achieve super-resolution effects. For the reconstruction methods of inverse problems, resolution is also key issue. Indeed, our comprehension of the surrounding world is bounded both below and above due to the resolution limit. Hence, resolution is always a key issue for inverse problems, not only theoretically but also computationally. In this book, we shall present some theoretical and computational discussion on the resolution analysis to inverse scattering problems.

In recent years, the significant increase in computing powers makes machine learning a powerful and widely applicable method for many scientific problems, including inverse problems particularly. The basic idea is train a numerical model of the continuous model $\mathscr{F}$ in (1.1.1) based on the so-called training dataset consisting of a large number of input and output pairs. This is a rapidly growing field with an avalanche of research outputs. It is impossible to give a full account of the contemporary developments, even those for inverse problems. We shall not touch this topic in this book and only refer to [8, 20, 21] for related research along this direction for inverse scattering problems that are within the theme of the current book.

Applications. As discussed earlier, inverse problems are of foundational importance to many scientific and technological applications. The mathematical study and practical applications of inverse problems mutually promote each other and are mutually beneficial to each other. This book is mainly devoted to the mathematical and numerical analysis of inverse problems. As mentioned earlier, we hope that the mathematical treatments presented in this book may motivate or serve as an inspiring source for real applications.

1.2 Inverse Scattering Problems and Spectral Theory

In physics, mathematics, and related fields, a wave is a propagating dynamic disturbance (change from equilibrium) of one or more quantities. Due to its propagating, penetrating, and nondestructive nature, waves are widely used for probing the real world. The associated inverse problems are generally referred to as the inverse scattering problems. The study on inverse scattering problems lies in the core of many areas of science and technology, including radar and sonar, geophysical detection, medical imaging, nondestructive testing, remote sensing, and cosmological exploration. Due to its practical importance, it is still the central theme of many ongoing researches. However, there are various challenges and new inverse scattering problems arising from emerging technologies, and the study on inverse scattering problems is far from complete.

Figure 1.1 provides the schematic illustration of a typical inverse scattering scenario. The system consists of three parts: the input is a certain incident wave field; O is the target object which causes the perturbation/scattering of the incident wave field; $\mathscr{M}$ represents the output that is the measurement of the scattered wave field. The inverse scattering problem is to recover O by $\mathscr{M}$. Following our general discussion in the previous section on inverse problems, we introduce an abstract operator $\mathscr{F}$ which sends the underlying scatterer O to the corresponding measurement $\mathscr{M}$, and then the inverse scattering problem can be formulated as

$$\mathscr{F}(O) = \mathscr{M}. \tag{1.2.1}$$

The operator $\mathscr{F}$ is defined by the physical mechanism (namely, the PDE system) governing the wave propagation. In this book, we consider three types of waves including the acoustic, electromagnetic, and mechanical waves. We are mainly concerned with the scattering of linear waves in the time-harmonic regime. The basic wave equations are, respectively, the Helmholtz equation, Maxwell's equations, and Lamé equations. We shall also consider in our study that more than two waves are involved in a single inverse problem. Depending on the specific situation, the scatterer O and the measurement $\mathscr{M}$ could vary and be complicated. In the inverse scattering theory, the reconstruction could be the recovery of the geometric parameters of the target scatterer O, e.g.,

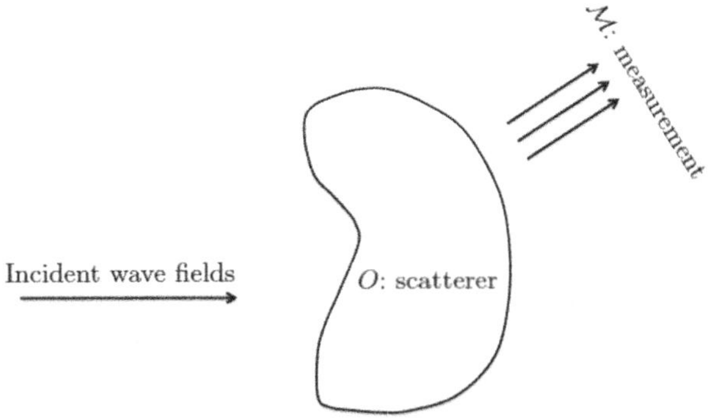

Fig. 1.1 Schematic illustration of a typical inverse scattering scenario

its location or shape, or the recovery of the complicated medium parameters of the scatterer O, or some other local or global features of the scatterer. Two remarks are in order. First, the inverse scattering problem depicted in Fig. 1.1 is a representative one in our study. There are other setups that arise in different applications, say, e.g., instead of generating the scattering by (artificially) sending an incident wave, the scattering can be induced by a certain source located inside or outside the scatterer. In the former case, measurement $\mathscr{M}$ is referred to as *active measurement*, whereas in the latter case, measurement $\mathscr{M}$ is referred to as the *passive measurement*. In this book, we shall stick to the active measurement. Nevertheless, many treatments in this book are of a local feature, namely, they are localized within a certain subregion in the space. Hence, many results can be straightforwardly extended to the case with passive measurement. Second, we would like to refer to [4–7, 10, 12–14] for mathematical study of inverse problems associated with time-dependent or even nonlinear waves.

The key component in the mathematical description of a specific time-harmonic wave scattering is a certain PDO (partial differential operator), which is denoted as $\mathscr{P}(\mathbf{x},\partial)$, $\mathbf{x}\in\mathbb{R}^N$, $N\geq 2$. The forward scattering problem takes the following abstract form:

$$\mathscr{P}(\mathbf{x},\partial)\Psi(\mathbf{x})=k^2\Psi(\mathbf{x}) \quad \text{in } \mathbb{R}^N, \tag{1.2.2}$$

where $k\in\mathbb{R}_+$ and Ψ is an $\mathbb{C}^m$-valued function, $m\geq 1$. In the physical setup, Ψ denotes the wave field and k signifies the (normalized) angular frequency of the wave field in the sense that the time-dependent wave field takes the form $e^{-\mathrm{i}kt}\Psi(\mathbf{x})$. As depicted in Fig. 1.1, we let Ψ^{in} denote the incident wave field which fulfils a certain differential equation as follows:

$$\mathscr{P}_0(\mathbf{x},\partial)\Psi^{in}(\mathbf{x})=k^2\Psi^{in}(\mathbf{x}) \quad \text{in } \mathbb{R}^N, \tag{1.2.3}$$

where $\mathscr{P}_0(\mathbf{x},\partial)$ signifies the PDO that governs the wave propagation in the background space (without the presence of the scattering object). Let $\Psi^{sc}(\mathbf{x})=\Psi(\mathbf{x})-\Psi^{in}(\mathbf{x})$ denote the scattered wave field, which accounts for perturbation/scattering of the wave field Ψ^{in} due to the difference between $\mathscr{P}(\mathbf{x},\partial)$ and $\mathscr{P}_0(\mathbf{x},\partial)$, namely, the scatterer O. It is further required that

$$\Psi^{sc}(\mathbf{x}) \text{ satisfies a certain asymptotic condition as } |\mathbf{x}|\to\infty. \tag{1.2.4}$$

The asymptotic condition in (1.2.4) is usually referred as the *radiation condition*, which is used to characterize the outgoing nature of the scattered wave filed. The outward radiating wave is the wave field that can generate useful information for the inverse problem (1.2.1). The abstract ingredients in (1.2.2)–(1.2.4) shall be made more concrete when we discuss the specific wave scattering in what follows. The measurement dataset $\mathscr{M}$ is usually encoded into the so-called far-field pattern of Ψ^{sc}, written as $\Psi^\infty(\hat{\mathbf{x}})$, where $\hat{\mathbf{x}}:=\mathbf{x}/|\mathbf{x}|\in\mathbb{S}^{N-1}$ with $\mathbb{S}^{N-1}:=\{\mathbf{x}\in\mathbb{R}^N;|\mathbf{x}|=1\}$.

Hence, the inverse scattering problem (1.2.1) can be recast as:

$$\mathscr{F}(O) = \Psi^\infty(\hat{\mathbf{x}}; \Psi^{in}), \ \ \hat{\mathbf{x}} \in \mathbb{S}^{N-1}, \ \Psi^{in} \in \mathscr{W}, \tag{1.2.5}$$

where $\mathscr{F}$ is defined by the forward scattering system (1.2.2)–(1.2.4), and as mentioned earlier, the scatterer O is defined via the difference of $\mathscr{P}(\mathbf{x}, \partial)$ and $\mathscr{P}_0(\mathbf{x}, \partial)$. In (1.2.5), the set $\mathscr{W}$ signifies a certain set consisting of the incident fields used for the inverse scattering problems. Here, we would like to point out that in many inverse scattering problems, the measurement information encoded into the so-called far-field pattern is equivalent to the so-called boundary measurement encoded into a certain boundary operator defined on the boundary of the scatterer.

With the above preparations, most of the discussion in Sect. 1.1 for the general inverse problem (1.1.1) can be directly transplanted to the inverse scattering problem (1.2.1). Specifically, we shall be mainly concerned with the following issues in this book for various inverse scattering problems from acoustic, electromagnetic, and elastic scattering: uniqueness and identifiability, stability and sensitivity, and non-identifiability and invisibility as well as reconstruction algorithms and resolution analysis. Instead of presenting an overview of the content of this book, we choose to present an overview on a specific topic at the beginning of each chapter. As mentioned earlier, we are mainly concerned with the recovery of the geometric features of the scatterer O by minimal/optimal measurement for the inverse scattering problem (1.2.1). This shall lead to several peculiar classes of spectral problems associated with the PDOs $\mathscr{P}(\mathbf{x}, \partial)$ and $\mathscr{P}_0(\mathbf{x}, \partial)$, where the spectral geometric point of view shall play an essential role in our study.

References

1. H. Ammari, *An introduction to mathematics of emerging biomedical imaging*, Springer, Berlin, 2008.
2. H. Ammari and H. Kang, *Reconstruction of small inhomogeneities from boundary measurements*, Springer-Verlag, Berlin, 2004.
3. D. Colton and R. Kress, *Inverse acoustic and electromagnetic scattering theory*, volume 93 of Applied Mathematical Sciences. Springer, Cham, 4th edition, 2019.
4. Y. Deng, J. Li and H. Liu, *On identifying magnetized anomalies using geomagnetic monitoring*, Archive for Rational Mechanics and Analysis, **231** (2019), no. 1, 153–187
5. Y. Deng, J. Li and H. Liu, *On identifying magnetized anomalies using geomagnetic monitoring within a magnetohydrodynamic model*, Archive for Rational Mechanics and Analysis, **235** (2020), 691–721.
6. Y. Deng, H. Liu and W.-Y. Tsui, *Identifying variations of magnetic anomalies using geomagnetic monitoring*, Discrete and Continuous Dynamical Systems, series A, **40** (2020), no. 11, 6411–6440.
7. H. Diao, X. Fei, H. Liu and L. Wang, *Determining anomalies in a semilinear elliptic equation by a minimal number of measurements*, arXiv: 2206.02500, (2022).
8. Y. Gao, H. Liu, X. Wang and K. Zhang, *On an artificial neural network for inverse scattering problems*, J. Comput. Phys., **448** (2022), Paper No. 110771, 15 pp.
9. M.V. Klibanov and J. Li, *Inverse problems and Carleman estimates: global uniqueness, global convergence and experimental data*, De Gruyter, Berlin, 2021.
10. J. Li, H. Liu and S. Ma, *Determining a random Schrodinger operator: both potential and source are random*, Communications in Mathematical Physics, **381** (2021), no. 2, 527–556.
11. J. Li, H. Liu, W.-Y. Tsui and X. Wang, *An inverse scattering approach for geometric body generation: a machine learning perspective*, Math. Eng., **1** (2019), no. 4, 800–823.
12. Y.-H. Lin and H. Liu, *Inverse problems for fractional equations with a minimal number of measurements*, arXiv:2203.03010, (2022).
13. Y.-H. Lin, H. Liu and X. Liu, *Determining a nonlinear hyperbolic system with unknown sources and nonlinearity*, arXiv:2107.10219, (2021).
14. H. Liu and G. Uhlmann, *Determining both sound speed and internal source in thermo- and photo-acoustic tomography*, Inverse Problems, **31** (2015) 105005.
15. H. Liu, Y. Wang and C. Yang, *Mathematical design of a novel gesture-based instruction/input device using wave detection*, SIAM J. Imaging Sci., **9** (2016), no. 2, 822–841.
16. G. Uhlmann, *Inside Out: Inverse Problems and Applications*, Cambridge University Press, Cambridge, (2003).
17. G. Uhlmann, Visibility and invisibility, in *ICIAM 07—6th International Congress on Industrial and Applied Mathematics*, 381–408, Eur. Math. Soc., Zürich, (2009).

18. G. Uhlmann, *Inverse Problems: Visibility and Invisibility*, Journé es équations aux dérivées partielles, (2012), article no. 11, 64pp.
19. X. Wang, Y. Guo, J. Li and H. Liu, *Mathematical design of a novel input/instruction device using a moving acoustic emitter*, Inverse Problems 33 (2017), no. 10, 105009, 19 pp.
20. W. Yin, W. Yang and H. Liu, *A neural network scheme for recovering scattering obstacles with limited phaseless far-field data*, J. Comput. Phys., **417** (2020), 109594, 18 pp.
21. P. Zhang, P. Meng, W. Yin and H. Liu, *A neural network method for time-dependent inverse source problem with limited-aperture data*, J. Comput. Appl. Math., **421** (2023), Paper No. 114842.

Chapter 2
Geometric Structures of Laplacian Eigenfunctions

2.1 Overview

Laplacian eigenvalue problem is arguably the simplest PDE eigenvalue problem, which is stated as finding $u \in L^2(\Omega)$ and $\lambda \in \mathbb{R}$ such that

$$-\Delta u = \lambda u, \tag{2.1.1}$$

where Ω is an open set in $\mathbb{R}^n, n = 2, 3$, under a certain homogeneous boundary condition on $\partial\Omega$, such as the Dirichlet, Neumann or Robin condition. The solution u to (2.1.1) is referred to as a (generalized) Laplacian eigenfunction. There is a long and colourful history on the spectral theory of Laplacian eigenvalues and eigenfunctions; see, e.g. [3, 21, 27, 35, 36, 42, 46, 51, 61, 63, 64, 66]. It still remains an inspiring source for many technical, practical and computational developments [12, 13, 57–59, 66].

In this chapter, we are concerned with the geometric structures of Laplacian eigenfunctions as well as their applications to inverse scattering theory. There is a rich theory on the geometric properties of Laplacian eigenfunctions in the literature; see, e.g. the review papers [32,38,46]. The celebrated Courant's nodal domain theorem states that the first Dirichlet eigenfunction does not change sign in Ω and the nth eigenfunction (counting multiplicity) u_n has at most n nodal domains in [21]. In particular, a famous conjecture concerning the topology of the 2nd Dirichlet eigenfunction states that in $\mathbb{R}^2$, the nodal line of u_2 divides Ω by intersecting its boundary at exactly two points if Ω is convex (cf. [65]). A large amount of literature has been devoted to this conjecture and significant progress has been made in various situations [2,28–31,37,39,41,44,47,52,56,60]. The "hot spots" conjecture formulated by J. Rauch in 1974 says that the maximum of the second Neumann eigenfunction is attained at a boundary point. This conjecture was proved to be true for a class of planar domains [4], but the statement may not be correct in general; see several counterexamples [7, 11, 14, 40]. The hot spots conjecture was proved recently for a new class of domains (possibly non-convex and non-Euclidean) [45]. Another famous longstanding problem in spectral theory is the Schiffer conjecture which states that if a Neumann eigenfunction takes constant value on the boundary, then the domain must be a ball. The Schiffer conjecture is closely related to the Pompeium property in integral geometry (cf. [22, 25, 65]) and has also an interesting connection to invisibility cloaking (cf. [48]). In [33], the nodal set of the second Dirichlet Laplacian eigenfunction was proved to be close to a straight line when the eccentricity of a bounded and convex domain $\Omega \subset \mathbb{R}^2$ is large. On the other hand, one may also have some estimate about the size of eigenfunctions, e.g. the size of the first eigenfunction can be estimated uniformly for all convex domains; see [34]. Other geometrical characteristics may also be analysed, e.g. the volume of a set on which an eigenfunction is positive [54], and lower and upper bounds for the length of the nodal line of an eigenfunction of the Laplace operator in two-dimensional Riemannian manifolds [10, 53].

H. Diao, H. Liu, *Spectral Geometry and Inverse Scattering Theory*,
https://doi.org/10.1007/978-3-031-34615-6_2

As we see from the above, the study of the geometric structures of Laplacian eigenfunctions is intriguing and challenging. We shall present some results on the geometric structures of Laplacian eigenfunctions and their deep relationship to the quantitative behaviours of the eigenfunctions, where we follow the treatment in [15, 16]. Roughly speaking, in Sects. 2.3–2.8, we consider the intersections of certain homogeneous line segments and their implications to the quantitative analytic properties of the underlying eigenfunction. The specific geometric and mathematical setup is heavily motivated by the inverse obstacle scattering problems, which is concerned with the recovery of geometric and physical properties of unknown obstacles from the measurement of the wave pattern due to an impinging field. It is a long-standing and fundamental problem in inverse scattering whether a one-to-one correspondence holds between the geometric shapes of a set of obstacles and their scattering wave patterns due to a single impinging wave field [5, 6, 15–18, 20, 23, 49]. This is also known as the Schiffer problem in the inverse scattering theory [19]. We shall give more detailed introduction on this aspect in Sects. 4.1 and 4.2.

2.2 Motivation and Discussion of Main Results in $\mathbb{R}^2$

We first introduce three definitions related to our main results in $\mathbb{R}^2$.*

Definition 2.2.1. Let Ω be a bounded Lipschitz domain in $\mathbb{R}^2$. For a Laplacian eigenfunction u in (2.1.1), a line segment $\Gamma_h \subset \Omega$ is called a *nodal line* of u if $u = 0$ on Γ_h, where $h \in \mathbb{R}_+$ signifies the length of the line segment. For a given complex-valued function $\eta \in L^\infty(\Gamma_h)$, if it holds that

$$\partial_\nu u(\mathbf{x}) + \eta(\mathbf{x})u(\mathbf{x}) = 0, \quad \mathbf{x} \in \Gamma_h, \tag{2.2.1}$$

then Γ_h is referred to as a *generalized singular line* of u. For the special case that $\eta \equiv 0$ in (2.2.1), a generalized singular line is also called a *singular line* of u in Ω. We use $\mathscr{N}_\Omega^\lambda$, $\mathscr{S}_\Omega^\lambda$ and $\mathscr{M}_\Omega^\lambda$ to denote the sets of nodal, singular and generalized singular lines, respectively, of an eigenfunction u in (2.1.1).

According to Definition 2.2.1, a singular line is obviously a generalized singular line. However, for unambiguity and definiteness, Γ_h in (2.2.1) is called a generalized singular line only if η is not identically zero; otherwise, it is referred to as a singular line. We like to point out that as u is (real) analytic inside Ω, a nodal line or a singular line can be extended by the analytic continuation within Ω (cf. [43]). We are mainly concerned with the local properties of the eigenfunction u around the intersecting point of two lines, and hence, the length h of Γ_h does not play an essential role as long as it is positive. We further emphasize that no any specific boundary condition is specified for u on $\partial\Omega$ in Definition 2.2.1, that is, all our subsequent results hold for a generic Laplacian eigenfunction as long as it satisfies (2.1.1), therefore applicable to the particular Dirichlet, Neumann or Robin eigenfunction. As mentioned earlier, one of the major motivations of our current study comes from attacking the fundamental Schiffer problem in the inverse scattering theory. The localized feature of our results, which is independent of any specific boundary condition of u on $\partial\Omega$, shall play a key role in the study. This shall become more evident in our subsequent analysis.

Definition 2.2.2. Let Γ and Γ' be two line segments in Ω that intersect with each other. Denote the intersecting angle $\theta = \angle(\Gamma, \Gamma') \in (0, 2\pi)$. Set

$$\theta = \alpha \cdot 2\pi, \quad \alpha \in (0, 1).$$

---

* Sections 2.2–2.8 are reproduced from X. Cao, H. Diao H. Liu and J. Zou, *On nodal and generalized singular structures of Laplacian eigenfunctions and applications*, Journal de Mathématiques Pures et Appliquées, 143 (2020), 116–161.

then θ is called an *irrational angle* if α is an irrational number, and it is called a rational angle of degree q if $\alpha = p/q$ with $p,q \in \mathbb{N}$ and irreducible.

Definition 2.2.3. Let u satisfy (2.1.1) and be a nontrivial eigenfunction. For a given point $\mathbf{x}_0 \in \Omega$, if there exists a number $N \in \mathbb{N} \cup \{0\}$ such that

$$\lim_{r\to+0} \frac{1}{r^m} \int_{B(\mathbf{x}_0,r)} |u(\mathbf{x})|\, \mathrm{d}\mathbf{x} = 0 \quad \text{for } m = 0,1,\ldots,N+1, \tag{2.2.2}$$

where $B(\mathbf{x}_0,r)$ is a disk centred at $\mathbf{x}_0$ with radius $r \in \mathbb{R}_+$, we say that u vanishes at $\mathbf{x}_0$ up to the order N. The largest possible N such that (2.2.2) is fulfilled is called the vanishing order of u at $\mathbf{x}_0$, and we write

$$\mathrm{Vani}(u;\mathbf{x}_0) = N.$$

If (2.2.2) holds for any $N \in \mathbb{N}$, then we say that the vanishing order is infinity.

Since u to (2.1.1) is analytic in Ω, it is straightforward to verify that $\mathrm{Vani}(u;\mathbf{x}_0)$ is actually the lowest order of the nontrivial homogeneous polynomial in the Taylor series expansion of u at $\mathbf{x}_0$. Moreover, by the strong unique continuation principle, we know that if the vanishing order of u is infinity at a given point $\mathbf{x}_0 \in \Omega$, then u is identically zero in Ω.

To elucidate the study, we next consider a simple example which connects the vanishing order of an eigenfunction with the intersecting angle of its nodal lines. Set

$$u(\mathbf{x}) = J_n(\sqrt{\lambda}\, r)\sin n\theta, \quad \mathbf{x} = (x_1,x_2) = r(\cos\theta, \sin\theta) \in \Omega,$$

where J_n is the first-kind Bessel function of order $n \in \mathbb{N}$ (cf. Section 3.4 of [19]). We know $u(\mathbf{x})$ is a single spherical wave mode and satisfies (2.1.1), and we can verify that

$$\mathrm{Vani}(u;\mathbf{0}) = n.$$

In particular, it is noted that if one considers u in a central disk B_{r_0} with $\sqrt{\lambda}\, r_0$ being a root of $J_n(t)$ or $J_n'(t)$, then u is actually a Dirichlet or Neumann eigenfunction in $\Omega = B_{r_0}$. However, we are more interested in the nodal lines of u, and it can be easily seen that

$$\Gamma_h^m := \{\mathbf{x} = r(\cos\theta_m, \sin\theta_m); 0 < r < h, \theta_m = \frac{m}{n}\pi\}, \quad m = 0,1,2,\ldots,2n-1. \tag{2.2.3}$$

The nodal lines in (2.2.3) all intersect at the origin and the intersecting angle between any two of them is rational of degree n. Clearly, this simple example reveals an intriguing connection between the intersecting angle of two nodal lines and the vanishing order of the eigenfunction at the intersecting point. The aim of the present section is to provide an accurate and comprehensive characterization of such a relationship in the most general scenario. Roughly speaking, we shall show that the vanishing order is generically infinity if the intersecting angle is *irrational*, and the vanishing order is finite if the intersecting angle is *rational*. In the latter case, the vanishing order is actually the degree of rationality of the intersecting angle. The result is not only discussed for the nodal lines, but also for the generalized singular lines. Hence, the study uncovers a deep relationship between the nodal and generalized singular structures of the Laplacian eigenfunctions and the quantitative behaviours of the eigenfunctions. On the other hand, as it will become more evident in Sects. 4.1 and 4.2, the aforementioned homogeneous line segments of the general Laplacian eigenfunctions occur naturally when dealing with the fundamental Schiffer problem in inverse scattering within the general polygonal geometry.

The rest of the chapter is organized as follows. In Sects. 2.3 and 2.5, we consider the case that the intersecting angle is irrational and show that the vanishing order is infinity. In Sects. 2.4, 2.6 and 2.7, we study the case that the intersecting angle is rational and the vanishing order is finite. Section 2.4 is devoted to the presentation and discussions of the main results, whereas Sects. 2.6

and 2.7 are concentrated on the corresponding rigorous proofs. Section 2.8 discusses a generic condition required in Sects. 2.3–2.7. In Sects. 2.9–2.13, we generalize the geometrical structures of Laplacian eigenfunctions in $\mathbb{R}^2$ developed in Sects. 2.2–2.8 to the three-dimensional cases.

2.3 Irrational Intersection and Infinite Vanishing Order: Two Intersecting Nodal and Singular Lines

In this section, we consider a relatively simple case that two nodal lines or two singular lines intersect at an irrational angle. We show that in such a case, the vanishing order of the eigenfunction at the intersecting point is generically infinite, and hence, it is identically vanishing in Ω.

Theorem 2.3.1. *Let u be a Laplacian eigenfunction to* (2.1.1). *If there are two nodal lines Γ_h^+ and Γ_h^- from $\mathscr{N}_\Omega^\lambda$ such that*

$$\Gamma_h^+ \cap \Gamma_h^- = \mathbf{x}_0 \in \Omega \quad \text{and} \quad \angle(\Gamma_h^+, \Gamma_h^-) = \alpha \cdot 2\pi, \tag{2.3.1}$$

where $\alpha \in (0,1)$ is irrational. Then the vanishing order of u at $\mathbf{x}_0$ is infinite, namely

$$\mathrm{Vani}(u; \mathbf{x}_0) = +\infty.$$

In order to prove Theorem 2.3.1, we need some auxiliary results about reflection principles of nodal and singular lines from the following two lemmas. In what follows, for a line segment $\Gamma \subset \mathbb{R}^2$, we define $\mathscr{R}_\Gamma$ to be the (mirror) reflection in $\mathbb{R}^2$ with respect to the line containing Γ.

Lemma 2.3.1. *Let u be a Laplacian eigenfunction to* (2.1.1). *There holds the following reflection principles:*

1. *Let $\Gamma \in \mathscr{N}_\Omega^\lambda$ (resp. $\Gamma \in \mathscr{S}_\Omega^\lambda$) and $\Gamma' \in \mathscr{N}_\Omega^\lambda \cup \mathscr{S}_\Omega^\lambda$. If $\widetilde{\Gamma} = \mathscr{R}_{\Gamma'}(\Gamma) \subset \Omega$, then $\widetilde{\Gamma} \in \mathscr{N}_\Omega^\lambda$ (resp. $\widetilde{\Gamma} \in \mathscr{S}_\Omega^\lambda$);*
2. *Let $\Gamma \in \mathscr{M}_\Omega^\lambda$ with $\partial_\nu u + \eta u = 0$ on Γ and $\Gamma' \in \mathscr{N}_\Omega^\lambda$. If $\widetilde{\Gamma} = \mathscr{R}_{\Gamma'}(\Gamma) \subset \Omega$, then $\widetilde{\Gamma} \in \mathscr{M}_\Omega^\lambda$ satisfies $\partial_{\widetilde{\nu}} u + \widetilde{\eta} u = 0$ on $\widetilde{\Gamma}$, where $\widetilde{\nu} = \mathscr{R}_{\Gamma'}(\nu)$ and $\widetilde{\eta} = \mathscr{R}_{\Gamma'}(\eta)$.*

The reflection principles are rather standard for the Laplacian [49, 50]. The first reflection principle in Lemma 2.3.1 shall be used in the proof of Theorem 2.3.1, whereas the second one is needed in our subsequent study.

Lemma 2.3.2. *Let $0 < \alpha_1 < 1$ be an irrational number. Define*

$$\alpha_{n+1} = 1 - \left\lfloor \frac{1}{\alpha_n} \right\rfloor \alpha_n, \quad n = 1, 2, \ldots, \tag{2.3.2}$$

where $\lfloor \cdot \rfloor$ is the floor function. Then it holds that

$$\lim_{n \to \infty} \alpha_n = 0.$$

Proof. We prove this lemma by contradiction. First, by induction, it is easy to see that $\{\alpha_n\} \subset \mathbb{R} \backslash \mathbb{Q}$, where $\mathbb{Q}$ is the set of rational numbers and $\alpha_n > 0$. Furthermore, by using

$$0 < \frac{1}{\alpha_n} - \left\lfloor \frac{1}{\alpha_n} \right\rfloor < 1,$$

we know that the sequence $\{\alpha_n\}$ is bounded below and decreasing. Suppose that

$$\lim_{n \to \infty} \alpha_n = \beta_0 > 0. \tag{2.3.3}$$

Since $1/\alpha_n$ is not an integer, we know from [62, p.15, Eq.(2.1.7)] that the Fourier series expansion of $\lfloor 1/\alpha_n \rfloor$ is given as follows

$$\left\lfloor \frac{1}{\alpha_n} \right\rfloor = \frac{1}{\alpha_n} - \frac{1}{2} + \frac{1}{\pi} \sum_{k=1}^{\infty} \frac{\sin(2k\pi/\alpha_n)}{k}. \tag{2.3.4}$$

Next, we show for a sufficiently small $\varepsilon_0 \in \mathbb{R}_+ := \{x \in \mathbb{R} | x > 0\}$, the Fourier series

$$\sum_{k=1}^{\infty} k^{-1} \sin(2k\pi x) \tag{2.3.5}$$

is uniformly convergent with respect to $x \in (\frac{1}{\beta_0} - \varepsilon_0, \frac{1}{\beta_0} + \varepsilon_0)$. To this end, we first prove by absurdity that $\beta_0 \neq \frac{1}{N}$ for any $N \in \mathbb{N}$. It is obvious to see from (2.3.3) that $\beta_0 < 1$, since $\{\alpha_n\}$ is decreasing with β_0 as its infimum. Now assume contrarily that $\beta_0 = \frac{1}{N}$ for some $N \in \mathbb{N} \backslash \{1\}$ and

$$\lim_{n \to \infty} \frac{1}{\alpha_n} = \frac{1}{\beta_0} = N.$$

Since the sequence $\{\frac{1}{\alpha_n}\}$ is increasing and bounded above by $\frac{1}{N}$, we know that there exists a sufficiently large $n_1 \in \mathbb{N}$ such that for any $n \geq n_1$, there holds

$$N - 1 \leq \frac{1}{\alpha_n} \leq N.$$

Hence,

$$\frac{1}{N} \leq \alpha_n \leq \frac{1}{N-1} \quad \text{for any} \quad n \geq n_1, \tag{2.3.6}$$

and

$$\left\lfloor \frac{1}{\alpha_n} \right\rfloor = N - 1.$$

On the other hand, from Definition (2.3.2), we can also deduce that

$$\frac{1}{N} \leq \alpha_{n+1} = 1 - \left\lfloor \frac{1}{\alpha_n} \right\rfloor \alpha_n = 1 - (N-1)\alpha_n \leq \frac{1}{N-1} \quad \text{for } n \geq n_1,$$

which further yields

$$\frac{N-2}{(N-1)^2} \leq \alpha_n \leq \frac{1}{N} \quad \text{for } n \geq n_1. \tag{2.3.7}$$

Combining (2.3.6) and (2.3.7), we can obtain that

$$\alpha_n = \frac{1}{N} \quad \text{for } n \geq n_1,$$

which contradicts to the fact that $\alpha_n \in \mathbb{R} \backslash \mathbb{Q}$ for $n = 1, 2, \ldots$. Therefore we must have $\beta_0 \neq \frac{1}{N}$ for any $N \in \mathbb{N}$.

We proceed to prove the uniform convergence of the Fourier series (2.3.5) with respect to $x \in (\frac{1}{\beta_0} - \varepsilon_0, \frac{1}{\beta_0} + \varepsilon_0)$ by Dirichlet's test [24]. Indeed, since $\{\frac{1}{k}\}$ is decreasing with respect to k and $\lim_{k \to \infty} \frac{1}{k} = 0$ uniformly in $(\frac{1}{\beta_0} - \varepsilon_0, \frac{1}{\beta_0} + \varepsilon_0)$, it suffices to show that the partial sum $\sum_{k=1}^{K_0} \sin(2k\pi x)$ is uniformly bounded in $(\frac{1}{\beta_0} - \varepsilon_0, \frac{1}{\beta_0} + \varepsilon_0)$ for $K_0 = 1, 2, \ldots$. By Edwards [26, p. 110, Eq.(7.1.3)], we actually have

$$\left|\sum_{k=1}^{K_0}\sin(2k\pi x)\right| = \left|\frac{\cos\pi x-\cos(2K_0+1)\pi x}{2\sin\pi x}\right| \le \left|\frac{1}{\sin\pi x}\right| \le \frac{2}{|\sin\frac{\pi}{\beta_0}|}, \tag{2.3.8}$$

which readily gives the desired uniform boundedness. The last inequality in (2.3.8) holds because $\beta_0 \neq \frac{1}{N}$ for any $N\in\mathbb{N}$ and hence $\sin\frac{\pi}{\beta_0}\neq 0$. By the continuity, we know there holds that $|\sin\pi x| \ge \frac{1}{2}|\sin\frac{\pi}{\beta_0}|$ for $x\in(\frac{1}{\beta_0}-\varepsilon_0,\frac{1}{\beta_0}+\varepsilon_0)$.

Taking $x=\alpha_n^{-1}$ in $\sum_{k=1}^{\infty}k^{-1}\sin(2k\pi x)$ and utilizing the uniform convergence result in $\alpha_n\in(\beta_0-\varepsilon_0,\beta_0+\varepsilon_0)$ for a sufficiently small $\varepsilon_0\in\mathbb{R}_+$, we can let $n\to\infty$ on both sides of (2.3.4) to derive that

$$\lim_{n\to\infty}\left\lfloor\frac{1}{\alpha_n}\right\rfloor = \lim_{n\to\infty}\left(\frac{1}{\alpha_n}-\frac{1}{2}+\frac{1}{\pi}\sum_{k=1}^{\infty}\frac{\sin(2k\pi/\alpha_n)}{k}\right) = \left\lfloor\frac{1}{\beta_0}\right\rfloor. \tag{2.3.9}$$

By using (2.3.9) and taking the limits of both sides of (2.3.2), we further have that

$$\beta_0 = 1-\left\lfloor\frac{1}{\beta_0}\right\rfloor\beta_0. \tag{2.3.10}$$

Dividing β_0 on both sides of (2.3.10), we finally arrive at a contradiction

$$1=\frac{1}{\beta_0}-\left\lfloor\frac{1}{\beta_0}\right\rfloor,$$

which completes the proof. □

We are now ready to present the proof of Theorem 2.3.1.

Proof of Theorem 2.3.1. By a rigid motion if necessary, we can assume without loss of generality that $\mathbf{x}_0=\mathbf{0}$ is the origin and Γ_h^- lies in the x_1^+-axis while Γ_h^+ has the angle $2\alpha\pi$ away from Γ_h^- in the anticlockwise direction. Since we are mainly concerned with the local properties, it is assumed that $h\in\mathbb{R}_+$ is sufficiently small such that $\mathscr{R}_{\Gamma_h^+}(\Gamma_h^-)\Subset\Omega$. In what follows, with the help of Lemmas 2.3.1 and 2.3.2, we show that there exists a dense set of nodal lines around the origin.

To begin with, we prove that one can assume $\alpha\in(\mathbb{R}\backslash\mathbb{Q})\cap(0,1/3)$ in (2.3.1). That is, there exist two nodal lines $\Gamma_h^{\pm}$ satisfying (2.3.1) with $\alpha\in(0,1/3)$ being irrational. To that end, we first show that the other two separate cases $\alpha\in(\mathbb{R}\backslash\mathbb{Q})\cap(1/3,1/2)$ and $\alpha\in(\mathbb{R}\backslash\mathbb{Q})\cap(1/2,1)$ can be reduced to the case $\alpha\in(\mathbb{R}\backslash\mathbb{Q})\cap(0,1/3)$. We first consider $\alpha\in(\mathbb{R}\backslash\mathbb{Q})\cap(1/3,1/2)$. By Lemma 2.3.1, we set

$$\Gamma_{1,h}^{(0)}=\mathscr{R}_{\Gamma_h^+}(\Gamma_h^-)\in\mathscr{N}_\Omega^\lambda,\quad \Gamma_{1,h}^{(1)}=\mathscr{R}_{\Gamma_{1,h}^{(0)}}(\Gamma_h^+)\in\mathscr{N}_\Omega^\lambda. \tag{2.3.11}$$

Then it can be directly verified that $\Gamma_{1,h}^{(1)}$ has an angle $\alpha^{(1)}\cdot 2\pi$ away from Γ_h^- in the anticlockwise direction, where

$$\alpha^{(1)} := 1-\left\lfloor\frac{1}{\alpha}\right\rfloor\alpha\in(\mathbb{R}\backslash\mathbb{Q})\cap(0,1/3).$$

Hence, one can replace Γ_h^+ in (2.3.1) by $\Gamma_{1,h}^{(1)}$ in (2.3.11) to obtain the desired result. Next, if $\alpha\in(1/2,1)$, we set

$$\Gamma_{1,h}^{(1)}=\mathscr{R}_{\Gamma_h^-}(\Gamma_h^+)\in\mathscr{N}_\Omega^\lambda. \tag{2.3.12}$$

Then $\Gamma_{1,h}^{(1)}$ has an angle $2(1-\alpha)\pi$ away from Γ_h^- in the anticlockwise direction. It can be directly verified that

$$\alpha^{(2)} := 1-\alpha\in(\mathbb{R}\backslash\mathbb{Q})\cap(0,1/2).$$

If $\alpha^{(2)} \in (0, 1/3)$, then one can replace Γ_h^+ in (2.3.1) by $\Gamma_{1,h}^{(1)}$ in (2.3.12) to obtain the desired result, whereas if $\alpha^{(2)} \in (1/3, 1/2)$, one can follow the same reflection argument as the previous case (cf. (2.3.11)) to obtain a nodal line for replacing Γ_h^+ in (2.3.1) satisfying $\alpha \in (\mathbb{R}\backslash\mathbb{Q}) \cap (0, 1/3)$.

Next, starting from (2.3.1) with $\alpha \in (\mathbb{R}\backslash\mathbb{Q}) \cap (0, 1/3)$, we perform a series of reflections to obtain the dense set of nodal lines around the origin as mentioned at the beginning of the proof; see Fig. 2.1 for a schematic illustration. We write $\Gamma_{0,h}^{(1)} = \Gamma_h^-$ and $\Gamma_{1,h}^{(1)} = \Gamma_h^+$. By Lemma 2.3.1, we

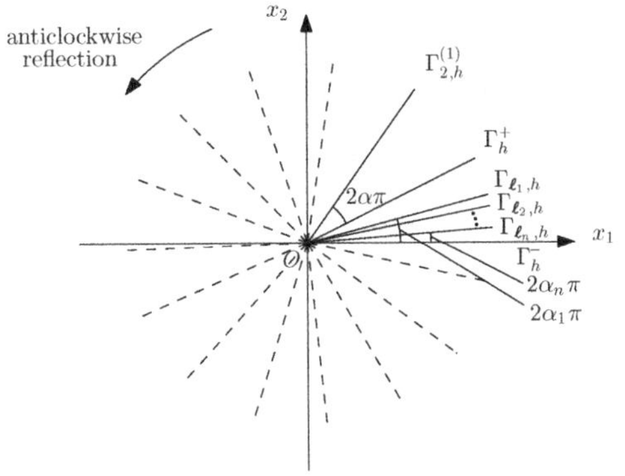

Fig. 2.1 Schematic illustration of the reflection argument in the proof of Theorem 2.3.1

see that

$$\Gamma_{2,h}^{(1)} := \mathscr{R}_{\Gamma_{1,h}^{(1)}}\left(\Gamma_{0,h}^{(1)}\right) \in \mathscr{N}_\Omega^\lambda.$$

Since $\alpha \in (0, 1/3)$, we know that

$$\ell_1 := \left\lfloor \frac{1}{\alpha} \right\rfloor \geq 3.$$

By repeating the reflections, one can find $(\ell_1 - 2)$ nodal lines as follows:

$$\Gamma_{m,h}^{(1)} := \mathscr{R}_{\Gamma_{m-1,h}^{(1)}}\left(\Gamma_{m-2,h}^{(1)}\right) \in \mathscr{N}_\Omega^\lambda, \quad m = 3, \ldots, \ell_1.$$

Furthermore, it is easy to verify that

$$\angle(\Gamma_{\ell_1,h}^{(1)}, \Gamma_h^-) = \alpha_1 \cdot 2\pi, \quad \alpha_1 := 1 - \left\lfloor \frac{1}{\alpha} \right\rfloor \alpha < \alpha.$$

Let

$$\Gamma_{\ell_1,h} = \mathscr{R}_{\Gamma_h^-}\left(\Gamma_{\ell_1,h}^{(1)}\right) \in \mathscr{N}_\Omega^\lambda.$$

Then one can see that $\Gamma_{\ell_1,h}$ has an angle $2\alpha_1\pi$ away from Γ_h^- in the anticlockwise direction.

Next, by replacing Γ_h^+ with $\Gamma_{\ell_1,h}$ and repeating the above reflection argument, one can find a nodal line $\Gamma_{\ell_2,h}$ such that

$$\Gamma_{\ell_2,h} \in \mathscr{N}_\Omega^\lambda, \quad \ell_2 := \left\lfloor \frac{1}{\alpha_1} \right\rfloor, \quad \angle(\Gamma_{\ell_2,h}, \Gamma_h^-) = \alpha_2 \cdot 2\pi \quad \text{and} \quad \alpha_2 := 1 - \left\lfloor \frac{1}{\alpha_1} \right\rfloor \alpha_1 < \alpha_1.$$

Furthermore, we know $\Gamma_{\ell_2,h}$ has an angle $2\alpha_2\pi$ away from Γ_h^- in the anticlockwise direction. Clearly, by further repeating this reflection argument, one can find a series of nodal lines $\Gamma_{\ell_n,h}$ such that

$$\Gamma_{\ell_n,h} \in \mathcal{N}_\Omega^\lambda, \quad \ell_n := \left\lfloor \frac{1}{\alpha_{n-1}} \right\rfloor, \quad \angle(\Gamma_{\ell_n,h}, \Gamma_h^-) = \alpha_n \cdot 2\pi, \quad \text{and } \alpha_n := 1 - \left\lfloor \frac{1}{\alpha_{n-1}} \right\rfloor \alpha_{n-1}. \quad (2.3.13)$$

Moreover, we get that $\Gamma_{\ell_n,h}$ has the angle $2\alpha_n\pi$ away from Γ_h^- in the anticlockwise direction. Then by Lemma 2.3.2 we have

$$\lim_{n\to\infty} \alpha_n = 0. \quad (2.3.14)$$

Combining (2.3.13) and (2.3.14), we see that $\{\Gamma_{\ell_n,h}\}$ forms a dense set of nodal lines around the origin. Hence, by the continuity of u one readily has that u is identically zero. This completes the proof of Theorem 2.3.1. □

The next theorem is concerned with the intersection of two singular lines.

Theorem 2.3.2. *Let u be a Laplacian eigenfunction to (2.1.1). If there are two singular lines Γ_h^+ and Γ_h^- from $\mathcal{S}_\Omega^\lambda$ such that*

$$\Gamma_h^+ \cap \Gamma_h^- = \mathbf{x}_0 \in \Omega \quad \text{and} \quad \angle(\Gamma_h^+, \Gamma_h^-) = \alpha \cdot 2\pi,$$

where $\alpha \in (0,1)$ is irrational, then there holds that

$$\begin{aligned} \mathrm{Vani}(u;\mathbf{x}_0) &= 0, && \text{if } u(\mathbf{x}_0) \neq 0; \\ \mathrm{Vani}(u;\mathbf{x}_0) &= +\infty, && \text{if } u(\mathbf{x}_0) = 0. \end{aligned}$$

Moreover, if $u(\mathbf{x}_0) \neq 0$, we have the following expansion of u in a neighbourhood of $\mathbf{x}_0$:

$$u(\mathbf{x}) = u(\mathbf{x}_0) J_0(\sqrt{\lambda} r), \quad \mathbf{x} = \mathbf{x}_0 + r(\cos\theta, \sin\theta).$$

where $J_0(t)$ is the zeroth Bessel function of the first kind.

In order to prove Theorem 2.3.2, we need some auxiliary results from the following three lemmas, especially about the spherical wave expansion, for which we refer to [19] for more details. In what follows, $\mathrm{i} := \sqrt{-1}$ is used for the imaginary unit.

Lemma 2.3.3 ([19, Section 3.4]). *Suppose that u is an eigenfunction to (2.1.1), then u has the following spherical wave expansion in polar coordinates around the origin:*

$$u(\mathbf{x}) = \sum_{n=0}^{\infty} \left(a_n e^{\mathrm{i}n\theta} + b_n e^{-\mathrm{i}n\theta} \right) J_n\left(\sqrt{\lambda} r\right), \quad (2.3.15)$$

where $\mathbf{x} = (x_1, x_2) = r(\cos\theta, \sin\theta) \in \mathbb{R}^2$, λ is the corresponding eigenvalue of (2.1.1), a_n and b_n are constants, and $J_n(t)$ is the n-th Bessel function of the first kind.

Lemma 2.3.4. *Let Γ be a line segment that can be parameterized in polar coordinates as $\mathbf{x} \in \Gamma$, where $\mathbf{x} = r(\cos\theta, \sin\theta)$ with $0 \le r < \infty$ and θ fixed. Let ν be the unit normal vector to Γ; then it holds that*

$$\frac{\partial u}{\partial \nu} = \pm \frac{1}{r} \frac{\partial u}{\partial \theta}.$$

Proof. Using the polar coordinates and the chain rule, we have

$$\frac{\partial u}{\partial x_1} = \frac{\partial u}{\partial r} \cos\theta - \frac{\sin\theta}{r} \frac{\partial u}{\partial \theta}, \quad \frac{\partial u}{\partial x_2} = \frac{\partial u}{\partial r} \sin\theta + \frac{\cos\theta}{r} \frac{\partial u}{\partial \theta}.$$

Thus

$$\frac{\partial u}{\partial \nu} = \left(\frac{\partial u}{\partial r}\Big|_\Gamma \cos\theta - \frac{\sin\theta}{r} \frac{\partial u}{\partial \theta}\Big|_\Gamma \right) \cos\varphi + \left(\frac{\partial u}{\partial r}\Big|_\Gamma \sin\theta + \frac{\cos\theta}{r} \frac{\partial u}{\partial \theta}\Big|_\Gamma \right) \sin\varphi, \quad (2.3.16)$$

where $\nu = (\cos\varphi, \sin\varphi)$ denotes the unit normal vector to Γ. Using the fact $|\varphi - \theta| = \pi/2$, we complete the proof. □

Lemma 2.3.5. *Suppose that for* $0 < h \ll 1$ *and* $t \in (0,h)$,

$$\sum_{n=0}^{\infty} \alpha_n J_n(t) = 0, \tag{2.3.17}$$

where $J_n(t)$ *is the n-th Bessel function of the first kind. Then*

$$\alpha_n = 0, \quad n = 0,1,2,\ldots.$$

Proof. From [19], we know that

$$J_n(t) = \frac{t^n}{2^n n!}\left(1 + \sum_{p=1}^{\infty} \frac{(-1)^p n!}{p!(n+p)!}\left(\frac{t}{2}\right)^{2p}\right). \tag{2.3.18}$$

Substituting (2.3.18) into (2.3.17) and comparing the coefficient of t^n ($n = 0,1,2,\ldots$), we can prove this lemma. □

Now we are in a position to prove Theorem 2.3.2.

Proof of Theorem 2.3.2. Without loss of generality, we assume that two singular lines Γ_h^+ and Γ_h^- intersect with each other at the origin. Using the reflection principle and a similar argument to the proof of Theorem 2.3.1, for any line segment $\Gamma \Subset \Omega = \{\mathbf{x}; \mathbf{x} = r(\cos\beta, \sin\beta), 0 \le r \le h\}$ pointed out from the origin, we can show that

$$\frac{\partial u}{\partial \nu_\Gamma} \equiv 0 \text{ in } \Omega,$$

where ν_Γ is a unit normal vector to Γ. Recalling the expansion (2.3.15), it is easy to see that

$$\left.\frac{\partial u}{\partial \theta}\right|_\Gamma = \sum_{n=0}^{\infty} \mathrm{i}n\left(a_n e^{\mathrm{i}n\beta} - b_n e^{-\mathrm{i}n\beta}\right) J_n\left(\sqrt{\lambda} r\right) = 0. \tag{2.3.19}$$

Taking $\beta = 0$ in (2.3.19), we derive from Lemma 2.3.5 that

$$\sum_{n=0}^{\infty} \mathrm{i}n(a_n - b_n) J_n\left(\sqrt{\lambda} r\right) = 0,$$

thus

$$\mathrm{i}n(a_n - b_n) = 0, \quad n = 1,2,\ldots.$$

Moreover, evaluating (2.3.19) at $\beta = \alpha\pi$ where $\alpha \in \mathbb{R}\backslash\mathbb{Q}$,, we can deduce that

$$\mathrm{i}n\left(a_n e^{\mathrm{i}n\alpha\pi} - b_n e^{-\mathrm{i}n\alpha\pi}\right) = 0, \quad n = 1,2,\ldots.$$

Hence, a_n and b_n satisfy

$$\begin{bmatrix} 1 & -1 \\ e^{\mathrm{i}n\alpha\pi} & -e^{-\mathrm{i}n\alpha\pi} \end{bmatrix} \begin{bmatrix} a_n \\ b_n \end{bmatrix} = 0, \quad n = 1,2,\ldots,$$

which readily implies that $a_n = b_n = 0$ for $n = 1,2,\ldots$, in view of Lemma 2.3.5. Therefore, we obtain the simplified form of $u(\mathbf{x})$:

$$u(\mathbf{x}) = (a_0 + b_0) J_0(\sqrt{\lambda} r).$$

Finally, from the assumptions in the theorem we can easily see

$$a_0 + b_0 = u(\mathbf{0}) \neq 0, \quad \text{if } u(\mathbf{0}) \neq 0;$$
$$a_0 + b_0 = u(\mathbf{0}) = 0, \quad \text{if } u(\mathbf{0}) = 0,$$

which complete the proof. □

2.4 Rational Intersection and Finite Vanishing Order in $\mathbb{R}^2$

In this section, we consider the general case that two line segments from Definition 2.2.1 intersect at a rational angle. Throughout the present section, we let Γ_h^+ and Γ_h^- signify the two line segments which could be either one of the three types: nodal line, singular line or generalized singular line. It is also assumed that for a generalized singular line of the form (2.2.1), the parameter η is a constant. Nevertheless, we would like to point out that for the case that η is a function in the generalized singular line, we can derive similar conclusions, but through more tedious and subtle calculations. We shall address this point more in Sect. 2.6. Let η_1 and η_2 signify the parameters associated with Γ_h^- and Γ_h^+, respectively, if they are generalized singular lines. Set

$$\angle(\Gamma_h^+, \Gamma_h^-) = \alpha \cdot \pi, \quad \alpha \in (0, 2), \tag{2.4.1}$$

where α is a rational number of the form $\alpha = p/q$ with $p, q \in \mathbb{N}$ and irreducible. Since the Laplace operator $-\Delta$ is invariant under rigid motions, without loss of generality, we assume throughout the rest of this work that

$$\Gamma_h^+ \cap \Gamma_h^- = \mathbf{0} \in \Omega, \tag{2.4.2}$$

and Γ_h^- coincides with the $x_1{}^+$-axis while Γ_h^+ has the angle $\alpha \cdot \pi$ away from Γ_h^- in the anticlockwise direction; see Fig. 2.2 for a schematic illustration.

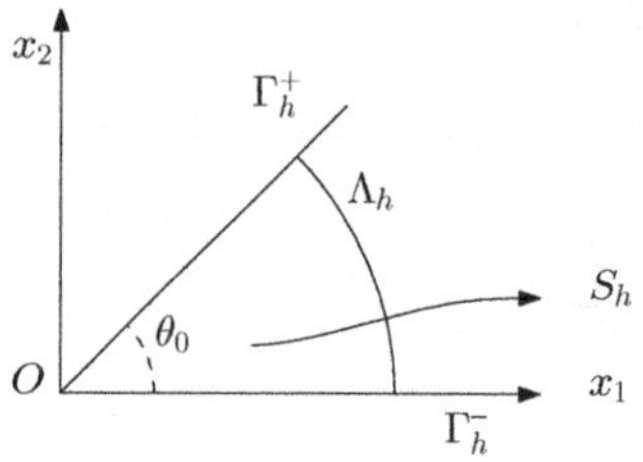

Fig. 2.2 Schematic illustration of the geometry of two intersecting lines with an angle $\theta_0 = \alpha \cdot \pi$ for some $\alpha \in (0, 1) \cap \mathbb{Q}$

Finally, we mainly deal with the case that the two intersecting line segments Γ_h^+ and Γ_h^- form an angle satisfying

$$\angle(\Gamma_h^+, \Gamma_h^-) = \alpha \cdot \pi, \quad \alpha \in (0, 1), \tag{2.4.3}$$

and the other case with $1 < \alpha < 2$ can be reduced to the previous case by a straightforward argument. Indeed, for $1 < \alpha < 2$, we know that Γ_h^+ belongs to the half-plane of $x_2 < 0$ (see Fig. 2.2). Let $\widetilde{\Gamma}_h^+$ be the extended line segment of length h in the half-plane of $x_2 > 0$. Since the Laplacian eigenfunction u is real analytic in the interior of Ω and $\Gamma_h^+ \Subset \Omega$, u is real analytic on Γ_h^+. Moreover, u fulfils a certain homogeneous condition on Γ_h^+. By the analytic continuation (cf. [43]), we know that $\widetilde{\Gamma}_h^+$ fulfils the same homogeneous condition as that on Γ_h^+. That is, $\widetilde{\Gamma}_h^+$ is of the same type of Γ_h^+. Hence, instead of studying the intersection of Γ_h^+ and Γ_h^-, one can study the intersection of $\widetilde{\Gamma}_h^+$ and Γ_h^- and its relations to the vanishing order of the eigenfunction. Clearly, the angle between $\widetilde{\Gamma}_h^+$ and Γ_h^- satisfies (2.4.3).

For a clear exposition, the rest of the section is devoted to the presentation and discussion of our main results, and their proofs shall be postponed to Sects. 2.6 and 2.7. In Sect. 2.6, we consider the case where the vanishing of the eigenfunction is up to the third order, whereas in Sect. 2.7, we consider the case of general vanishing orders.

Theorem 2.4.1. *Let u be a Laplacian eigenfunction to* (2.1.1). *Suppose that there are two generalized singular lines Γ_h^+ and Γ_h^- from $\mathscr{M}_\Omega^\lambda$ such that* (2.4.2) *and* (2.4.3) *hold. Assume that $\eta_1 \equiv C_1$ and $\eta_2 \equiv C_2$, where C_1 and C_2 are two constants. Then the Laplacian eigenfunction u vanishes up to the order N at $\mathbf{0}$:*

$$N \geq n, \quad \text{if } u(\mathbf{0}) = 0 \text{ and } \alpha \neq \frac{q}{p}, \; p = 1, \ldots, n-1, \tag{2.4.4}$$

where $n \in \mathbb{N}$, $n \geq 3$, and for a fixed p, $q = 1, 2, \ldots, p-1$.

In Theorem 2.4.1, we require that $n \geq 3$. That means, we exclude the special case that the intersecting angle is $\pi/2$. Nevertheless, we shall discuss this special case in Remark 2.1 with more details in what follows. In the next two theorems, we consider the case of two intersecting singular and nodal lines, respectively.

Theorem 2.4.2. *Let u be a Laplacian eigenfunction to* (2.1.1). *Suppose that there are two nodal lines Γ_h^+ and Γ_h^- from $\mathscr{N}_\Omega^\lambda$ such that* (2.4.2) *and* (2.4.3) *hold. Then the Laplacian eigenfunction u vanishes up to the order N at $\mathbf{0}$:*

$$N \geq n, \quad \text{if } \alpha \neq \frac{q}{p}, \; p = 1, \ldots, n-1,$$

where $n \in \mathbb{N}$, $n \geq 3$ and for a fixed p, $q = 1, 2, \ldots, p-1$.

Theorem 2.4.3. *Let u be a Laplacian eigenfunction to* (2.1.1). *Suppose that there are two singular lines Γ_h^+ and Γ_h^- from $\mathscr{S}_\Omega^\lambda$ such that* (2.4.2) *and* (2.4.3) *hold. Then the Laplacian eigenfunction u vanishes up to the order N at $\mathbf{0}$:*

$$N \geq n, \quad \text{if } u(\mathbf{0}) = 0 \text{ and } \alpha \neq \frac{q}{p}, \; p = 1, \ldots, n-1,$$

where $n \in \mathbb{N}$, $n \geq 3$ and for a fixed p, $q = 1, 2, \ldots, p-1$.

Example 2.1. Let $\Omega := \{(x_1, x_2) \in \mathbb{R}^2 \mid -2\pi \leq x_1 \leq 2\pi, \, -4\pi \leq x_2 \leq 4\pi\}$ be a rectangle. It is easy to see that

$$u(x_1, x_2) = \sin x_1 \sin 2x_2$$

is an eigenfunction to (2.1.1) with a homogeneous Dirichlet boundary condition on $\partial\Omega$. The corresponding eigenvalue is $\lambda = 5$. One pair of nodal lines of u in Ω are $\{(x_1, x_2) \mid x_2 = 0, -2\pi + h \leq x_1 \leq 2\pi - h\}$ and $\{(x_1, x_2) \mid x_1 = 0, -4\pi + h \leq x_2 \leq 4\pi - h\}$ for a fixed $0 < h < 2\pi$, which are perpendicular to each other at the origin. Therefore, from Theorem 2.4.2, since $\angle(\Gamma_h^+, \Gamma_h^-) = \pi/2$ which implies that $\alpha \neq 1$, the vanishing order N at the origin is 2. In fact, by the explicit expression of u, we know that the order of the lowest nontrivial homogeneous polynomial of the Taylor expansion of u at the origin is 2, which coincides with the conclusion given by Theorem 2.4.3.

We now proceed to consider that a nodal line intersects with a generalized singular line. Without loss of generality, we can assume that Γ_h^- is the generalized singular line, while Γ_h^+ is the nodal line.

Theorem 2.4.4. *Let u be a Laplacian eigenfunction to* (2.1.1). *Suppose that a generalized singular line $\Gamma_h^+ \in \mathscr{M}_\Omega^\lambda$ intersects with a nodal line $\Gamma_h^- \in \mathscr{N}_\Omega^\lambda$ at* $\mathbf{0}$ *with the angle $\angle(\Gamma_h^+, \Gamma_h^-) = \alpha \cdot \pi$. Assume that the boundary parameter $\eta_2 \equiv C_2$ on Γ_h^+ is a constant. Then the Laplacian eigenfunction u vanishes up to the order N at* $\mathbf{0}$*:*

$$N \geq n, \quad \text{if } \alpha \neq \frac{2q+1}{2p}, \; p = 1, \ldots, n-1, \tag{2.4.5}$$

where $n \in \mathbb{N}$, $n \geq 2$ and for a fixed p, $q = 0, 1, \ldots, p-1$.

Next, we consider the intersection of a singular line and a generalized singular line. Similar to Theorem 2.4.4, without loss of generality, we can assume that Γ_h^- is the generalized singular line. Indeed, the vanishing order of the eigenfunction in such a case can be obtained from formally taking η_2 on Γ_h^+ to be zero in Theorem 2.4.1.

Theorem 2.4.5. *Let u be a Laplacian eigenfunction to* (2.1.1). *Suppose that a singular line $\Gamma_h^+ \in \mathscr{S}_\Omega^\lambda$ intersects with a generalized singular line $\Gamma_h^- \in \mathscr{M}_\Omega^\lambda$ at the origin with the angle $\angle(\Gamma_h^+, \Gamma_h^-) = \alpha \cdot \pi$. Assume that the boundary parameter η_1 on Γ_h^- is a non-zero constant, i.e., $\eta_1 \equiv C_1 \neq 0$. Then the Laplacian eigenfunction u vanishes up to the order N at* $\mathbf{0}$*:*

$$N \geq n, \quad \text{if } u(\mathbf{0}) = 0 \text{ and } \alpha \neq \frac{q}{p}, \; p = 1, \ldots, n-1, \tag{2.4.6}$$

where $n \in \mathbb{N}$, $n \geq 3$ and $q = 1, 2, \ldots, p-1$ for a fixed p.

Using a similar proof to Theorem 2.4.4, we can find the relationship between the vanishing order of the Laplacian eigenfunction and the intersecting angle between a singular line and a nodal line.

Theorem 2.4.6. *Let u be a Laplacian eigenfunction to* (2.1.1). *Suppose that a singular line $\Gamma_h^+ \in \mathscr{S}_\Omega^\lambda$ intersects with a nodal line $\Gamma_h^- \in \mathscr{N}_\Omega^\lambda$ at the origin with the angle $\angle(\Gamma_h^+, \Gamma_h^-) = \alpha \cdot \pi$. Then the Laplacian eigenfunction u vanishes up to the order N at* $\mathbf{0}$*:*

$$N \geq n, \quad \text{if } \alpha \neq \frac{2q+1}{2p}, \; p = 1, \ldots, n-1, \tag{2.4.7}$$

where $n \in \mathbb{N}$, $n \geq 2$, and for a fixed p, $q = 0, 1, \ldots, p-1$.

Remark 2.1. As mentioned after Theorem 2.4.1, we exclude the special case that the intersecting angle between two lines is $\pi/2$. In fact, for Theorems 2.4.1–2.4.3 and 2.4.5, one may see from their proofs in Sect. 2.6 that if $\angle(\Gamma_h^+, \Gamma_h^-) = \pi/2$, then there holds that $\nabla u(\mathbf{0}) = 0$ if $u(\mathbf{0}) = 0$. That means, the eigenfunction is vanishing at least to the second order in such a case. For the other two cases in Theorems 2.4.4 and 2.4.6, we can only have that if $\alpha = 1/2$ and $u(\mathbf{0}) = 0$, the eigenfunction is vanishing at least to the first order.

Remark 2.2. It is noted that in Theorems 2.4.4 and 2.4.6, we require that $n \geq 2$, whereas in other theorems, we require that $n \geq 3$. In particular, when $n = 2$, $\alpha \neq 1/2$, one can conclude in Theorems 2.4.4 and 2.4.6 that the eigenfunction is vanishing at least to the second order. This conclusion is different from Theorems 2.4.1–2.4.3 and 2.4.5, where one has that if $\alpha \neq 1/2$, then the eigenfunction is vanishing at least to the third order.

Remark 2.3. We point out that all the vanishing orders of the eigenfunction in Theorems 2.4.1–2.4.6 depend only on the intersecting angle of the two homogeneous line segments and are independent of the underlying eigenvalue (cf. (2.1.1)). This is probably due to the special geometries of the study. We believe when the curved geometries are considered, the underlying eigenvalues should be involved.

2.5 Irrational Intersection and Infinite Vanishing Order: General Cases in $\mathbb{R}^2$

In this section we consider the irrational intersection, namely, α in (2.4.1) is an irrational number. We show that the eigenfunction is generically vanishing to infinity, namely, u is identically zero in Ω. Here, the generic condition is provided by u vanishing or not at the intersecting point. We shall present more discussions on this generic condition in Sect. 2.8. By verifying the conditions in (2.4.4), (2.4.5), (2.4.6) and (2.4.7) for an irrational α, we can readily obtain from Theorems 2.4.1, 2.4.4, 2.4.5 and 2.4.6 the results in the following four theorems.

Theorem 2.5.1. *Let u be a Laplacian eigenfunction to* (2.1.1). *Suppose that there are two generalized singular lines Γ_h^+ and Γ_h^- from $\mathscr{M}_\Omega^\lambda$ such that* (2.4.2) *and* (2.4.3) *hold. Assume that $\eta_1 \equiv C_1$ and $\eta_2 \equiv C_2$, where C_1 and C_2 are two constants. If $\angle(\Gamma_h^+, \Gamma_h^-) = \alpha \cdot \pi$ with $\alpha \in (0,2)$ being irrational, then there holds that*

$$\begin{aligned} \mathrm{Vani}(u;\mathbf{0}) &= 0, \quad && \text{if } u(\mathbf{0}) \neq 0; \\ \mathrm{Vani}(u;\mathbf{0}) &= +\infty, \quad && \text{if } u(\mathbf{0}) = 0. \end{aligned}$$

Theorem 2.5.2. *Let u be a Laplacian eigenfunction to* (2.1.1). *Suppose that a generalized singular line $\Gamma_h^- \in \mathscr{M}_\Omega^\lambda$ intersects with a nodal line $\Gamma_h^+ \in \mathscr{N}_\Omega^\lambda$ at $\mathbf{0}$ with the angle $\angle(\Gamma_h^+, \Gamma_h^-) = \alpha \cdot \pi$. Assume that the boundary parameter $\eta_1 \equiv C_1$ on Γ_h^- is a constant. If $\alpha \in (0,2)$ is irrational, then there holds*

$$\mathrm{Vani}(u;\mathbf{0}) = +\infty.$$

Theorem 2.5.3. *Let u be a Laplacian eigenfunction to* (2.1.1). *Suppose that a singular line $\Gamma_h^+ \in \mathscr{S}_\Omega^\lambda$ intersects with a generalized singular line $\Gamma_h^- \in \mathscr{M}_\Omega^\lambda$ at $\mathbf{0}$ with the angle $\angle(\Gamma_h^+, \Gamma_h^-) = \alpha \cdot \pi$. Assume that the boundary parameter $\eta_1 \equiv C_1$ on Γ_h^- is a constant. If $\alpha \in (0,2)$ is irrational, then there hold that*

$$\begin{aligned} \mathrm{Vani}(u;\mathbf{0}) &= 0, \quad && \text{if } u(\mathbf{0}) \neq 0; \\ \mathrm{Vani}(u;\mathbf{0}) &= +\infty, \quad && \text{if } u(\mathbf{0}) = 0. \end{aligned}$$

Theorem 2.5.4. *Let u be a Laplacian eigenfunction to* (2.1.1). *Suppose that a singular line $\Gamma_h^- \in \mathscr{S}_\Omega^\lambda$ intersects with a nodal line $\Gamma_h^+ \in \mathscr{N}_\Omega^\lambda$ at $\mathbf{0}$ with the angle $\angle(\Gamma_h^+, \Gamma_h^-) = \alpha \cdot \pi$. If $\alpha \in (0,2)$ is irrational, then there holds*

$$\mathrm{Vani}(u;\mathbf{0}) = +\infty.$$

For readers' convenience, we now summarize in Table 2.1 all our main results in Sects. 2.4 and 2.5 about the vanishing order N of the Laplacian eigenfunction u to (2.1.1) at an intersecting point $\mathbf{x}_0$ of two line segments Γ_h^+ and Γ_h^- which form an angle $\angle(\Gamma_h^+, \Gamma_h^-) = \alpha \cdot \pi$ for $\alpha \in (0,1)$. We recall that the sets of nodal, singular and generalized singular lines are denoted by $\mathscr{N}_\Omega^\lambda$, $\mathscr{S}_\Omega^\lambda$ and $\mathscr{M}_\Omega^\lambda$, respectively.

2.6 Proofs of the Theorems in Sect. 2.4 up to the Third Order

In this section, we present the proofs of the theorems shown in Sect. 2.4, but confined to the case that the vanishing order N is at most 3. We develop a mathematical scheme by making use of tools from microlocal analysis that possesses several remarkable properties. Next, we first introduce the so-called complex geometrical optics (CGO) solutions constructed in [8] for the subsequent use. As before, we let (r, θ) denote the polar coordinates in $\mathbb{R}^2$; that is, for $\mathbf{x} = (x_1, x_2) \in \mathbb{R}^2$, one has

	Rational intersection (Finite vanishing order)	Irrational intersection (Infinite vanishing order)
$\Gamma_h^+, \Gamma_h^- \in \mathscr{M}_\Omega^\lambda$	$N \geq n$, if $u(\mathbf{x}_0) = 0$ and $\alpha \neq \frac{q}{p}$, $p = 1, \ldots, n-1$ and $q = 1, \ldots, p-1$	$N = +\infty$, if $u(\mathbf{x}_0) = 0$; $N = 0$, if $u(\mathbf{x}_0) \neq 0$
$\Gamma_h^+, \Gamma_h^- \in \mathscr{N}_\Omega^\lambda$	$N \geq n$, if $\alpha \neq \frac{q}{p}$, $p = 1, \ldots, n-1$ and $q = 1, \ldots, p-1$	$N = +\infty$
$\Gamma_h^+, \Gamma_h^- \in \mathscr{S}_\Omega^\lambda$	$N \geq n$, if $u(\mathbf{x}_0) = 0$ and $\alpha \neq \frac{q}{p}$, $p = 1, \ldots, n-1$ and $q = 1, \ldots, p-1$	$N = +\infty$, if $u(\mathbf{x}_0) = 0$; $N = 0$, if $u(\mathbf{x}_0) \neq 0$
$\Gamma_h^+ \in \mathscr{M}_\Omega^\lambda, \Gamma_h^- \in \mathscr{N}_\Omega^\lambda$	$N \geq n$, if $\alpha \neq \frac{2q+1}{p}$, $p = 1, \ldots, n-1$ and $q = 0, 1, \ldots, p-1$	$N = +\infty$
$\Gamma_h^+ \in \mathscr{S}_\Omega^\lambda, \Gamma_h^- \in \mathscr{N}_\Omega^\lambda$	$N \geq n$, if $\alpha \neq \frac{2q+1}{p}$, $p = 1, \ldots, n-1$ and $q = 0, 1, \ldots, p-1$	$N = +\infty$
$\Gamma_h^+ \in \mathscr{S}_\Omega^\lambda, \Gamma_h^- \in \mathscr{M}_\Omega^\lambda$	$N \geq n$, if $u(\mathbf{x}_0) = 0$ and $\alpha \neq \frac{q}{p}$, $p = 1, \ldots, n-1$ and $q = 1, \ldots, p-1$	$N = +\infty$, if $u(\mathbf{x}_0) = 0$; $N = 0$, if $u(\mathbf{x}_0) \neq 0$

Table 2.1 Vanishing orders of Laplacian eigenfunction at intersecting points

$r = |\mathbf{x}|$ and $\theta = \arg(x_1 + \mathrm{i}x_2)$. Let B_h be the central disk of radius $h \in \mathbb{R}_+$. Let $\Gamma^\pm$ signify the infinite extension of $\Gamma_h^\pm$ in the half-space $x_2 \geq 0$. Let $\theta_m = 0$ and $\theta_M \in (0, \pi)$ be respectively the polar angles of Γ^- and Γ^+. Consider the open sector

$$W = \left\{ \mathbf{x} \in \mathbb{R}^2; \mathbf{x} \neq 0,\ \theta_m < \arg(x_1 + \mathrm{i}x_2) < \theta_M \right\}, \tag{2.6.1}$$

which is formed by the two half-lines Γ^- and Γ^+. We have the following result.

Lemma 2.6.1 ([8, Lemma 2.2 and Proposition 2.3]). *Let*

$$u_0(\mathbf{x}) := \exp\left(\sqrt{r}\left(\cos\left(\frac{\theta}{2} + \pi\right) + \mathrm{i}\sin\left(\frac{\theta}{2} + \pi\right)\right)\right). \tag{2.6.2}$$

Then $\Delta u_0 = 0$ in $\mathbb{R}^2 \backslash \mathbb{R}^2_{0,-}$, where $\mathbb{R}^2_{0,-} := \{\mathbf{x} \in \mathbb{R}^2 | \mathbf{x} = (x_1, x_2); x_1 \leq 0, x_2 = 0\}$, and $s \mapsto u_0(s\mathbf{x})$ decays exponentially in $\mathbb{R}_+$ whenever $\mathbf{x}$ is in the same domain of harmonicity. Furthermore, it holds for $\alpha, s > 0$ that

$$\int_W |u_0(s\mathbf{x})||\mathbf{x}|^\alpha \mathrm{d}\mathbf{x} \leq \frac{2(\theta_M - \theta_m)\Gamma(2\alpha + 4)}{\delta_W^{2\alpha+4}} s^{-\alpha-2}, \tag{2.6.3}$$

where $\delta_W = -\max_{\theta_m < \theta < \theta_M} \cos(\theta/2 + \pi) > 0$, and

$$\int_W u_0(s\mathbf{x}) \mathrm{d}\mathbf{x} = 6\mathrm{i}(e^{-2\theta_M \mathrm{i}} - e^{-2\theta_m \mathrm{i}})s^{-2}, \tag{2.6.4}$$

while for $h \in \mathbb{R}_+$,

$$\int_{W \backslash B_h} |u_0(s\mathbf{x})| \mathrm{d}\mathbf{x} \leq \frac{6(\theta_M - \theta_m)}{\delta_W^4} s^{-2} e^{-\delta_W \sqrt{hs}/2}. \tag{2.6.5}$$

Henceforth, for notational convenience, we set

$$\theta_0 = \alpha \cdot \pi = \angle(\Gamma_h^+, \Gamma_h^-), \tag{2.6.6}$$

for $\alpha \in (0, 1)$. In order to make use of the CGO solution $u_0(s\mathbf{x})$ given in Lemma 2.6.1 as a test function to analyse the vanishing order of u at the origin, we consider the following domain (see Fig. 2.2 for the illustration):

$$S_h = W \cap B_h, \tag{2.6.7}$$

where $\partial S_h = \Gamma_h^+ \cup \Gamma_h^- \cup \Lambda_h$ and

$$\begin{aligned}\Gamma_h^+ &= \{\mathbf{x} \in \mathbb{R}^2;\ 0 \le \sqrt{x_1^2 + x_2^2} \le h,\ \arg(x_1 + \mathrm{i}x_2) = \theta_0\},\\ \Gamma_h^- &= \{\mathbf{x} \in \mathbb{R}^2;\ 0 \le \sqrt{x_1^2 + x_2^2} \le h,\ \arg(x_1 + \mathrm{i}x_2) = 0\},\\ \Lambda_h &= W \cap \partial B_h.\end{aligned} \tag{2.6.8}$$

In the definition of the generalized singular line, we recall that the polar angles of the exterior normal vectors of Γ_h^+ (with respect to the domain W) and Γ_h^- are, respectively,

$$\varphi_M = \theta_0 + \frac{\pi}{2}, \quad \varphi_m = -\frac{\pi}{2}. \tag{2.6.9}$$

In order to investigate the relationship between the vanishing order of u at the origin and the intersecting angle of the generalized singular lines $\Gamma_h^{\pm}$, we consider the following equations

$$\Delta u + \lambda u = 0 \qquad \text{in } B_h, \tag{2.6.10a}$$

$$\frac{\partial u}{\partial \nu} + \eta_1 u = 0 \qquad \text{on } \Gamma_h^-, \tag{2.6.10b}$$

$$\frac{\partial u}{\partial \nu} + \eta_2 u = 0 \qquad \text{on } \Gamma_h^+. \tag{2.6.10c}$$

Next, we derive several crucial auxiliary results regarding the function u satisfying (2.6.10a)–(2.6.10c).

Recall that S_h is defined in (2.6.7). Since u_0 is only smooth in $S_h \backslash B_\varepsilon (0 < \varepsilon < h)$, we cannot use Green's formula for the Laplacian eigenfunction u and the CGO solution u_0 in S_h directly. Instead, we may overcome this difficulty by taking a limit of the volume integral with the integrand $u_0 \Delta u$ over $S_h \backslash B_\varepsilon$ and carefully investigate the boundary integrals on $\partial(S_h \backslash B_\varepsilon)$. We only present the result in the following proposition and its proof is similar to the argument of Lemma 3.2 in [8].

Lemma 2.6.2. *The CGO solution $u_0(sx)$ defined in* (2.6.2) *is harmonic in $S_h \backslash \mathbf{0}$ and decays exponentially as $s \to \infty$ for $0 < \theta < \theta_0$, where θ_0 is the intersecting angle of Γ_h^+ and Γ_h^-. Moreover, for the Laplacian eigenfunction u to* (2.1.1), *Green's formula holds*

$$\int_{S_h} (u_0(s\mathbf{x})\Delta u - u\Delta u_0(s\mathbf{x}))\,\mathrm{d}\mathbf{x} = I_1^+ + I_1^- + I_2, \tag{2.6.11}$$

where

$$\begin{aligned} I_1^+ &= \int_{\Gamma_h^+} \left(u_0(s\mathbf{x}) \frac{\partial u}{\partial \nu} - u(\mathbf{x}) \frac{\partial u_0(s\mathbf{x})}{\partial \nu} \right) \mathrm{d}\sigma,\\ I_1^- &= \int_{\Gamma_h^-} \left(u_0(s\mathbf{x}) \frac{\partial u}{\partial \nu} - u(\mathbf{x}) \frac{\partial u_0(s\mathbf{x})}{\partial \nu} \right) \mathrm{d}\sigma,\\ I_2 &= \int_{\Lambda_h} \left(u_0(s\mathbf{x}) \frac{\partial u}{\partial \nu} - u(\mathbf{x}) \frac{\partial u_0(s\mathbf{x})}{\partial \nu} \right) \mathrm{d}\sigma. \end{aligned} \tag{2.6.12}$$

From Lemma 2.3.4, by direct calculations, we have the following proposition regarding the exterior normal derivative of the CGO solution $u_0(s\mathbf{x})$ on any straight line.

Proposition 2.6.1. *For any straight line Γ, where $\mathbf{x} = r(\cos\theta, \sin\theta) \in \Gamma$, let the exterior unit normal vector to Γ be $\nu = (\cos\varphi, \sin\varphi)$. Then the CGO solution $u_0(s\mathbf{x})$ given in Lemma 2.6.1 fulfils*

$$\left.\frac{\partial u_0(s\mathbf{x})}{\partial \nu}\right|_{\Gamma} = \beta(\theta) e^{\sqrt{sr}\zeta(\theta)} \sqrt{\frac{s}{r}}, \tag{2.6.13}$$

where $\zeta(\theta)$ *and* $\beta(\theta)$ *are given by*

$$\zeta(\theta)=e^{\mathrm{i}(\theta/2+\pi)}=-e^{\mathrm{i}\theta/2},\ \beta(\theta)=\frac{1}{2}\sin(\varphi-\theta)\tilde{\zeta}(\theta),\ \tilde{\zeta}(\theta)=-\mathrm{i}e^{\mathrm{i}\theta/2}. \tag{2.6.14}$$

By induction and straightforward calculations, we can derive the explicit formulas of the following integrals in Lemma 2.6.3, which is essential in showing the relationship between the vanishing order of u and the intersecting angle of the generalized singular lines $\Gamma_h^{\pm}$. The detailed proof of Lemma 2.6.3 is omitted.

Lemma 2.6.3. *For a given* $\zeta(\theta)\in\mathbb{C}$ *and* $\ell=0,1,2,\ldots$, *it holds that*

$$\begin{aligned}\int_0^h r^{\ell}e^{\sqrt{sr}\zeta(\theta)}\mathrm{d}r=&\frac{2}{s^{\ell+1}}\Big\{\frac{(2\ell+1)!}{\zeta(\theta)^{2\ell+2}}+e^{\sqrt{sh}\zeta(\theta)}\sum_{j=0}^{2\ell+1}\frac{(-1)^j(2\ell+1)!}{(2\ell+1-j)!\zeta(\theta)^{j+1}}\\&\times(sh)^{(2\ell+1-j)/2}\Big\},\\\int_0^h r^{\ell}e^{\sqrt{sr}\zeta(\theta)}\sqrt{\frac{s}{r}}\mathrm{d}r=&\frac{2}{s^{\ell}}\Big\{-\frac{(2\ell)!}{\zeta(\theta)^{2\ell+1}}+e^{\sqrt{sh}\zeta(\theta)}\sum_{j=0}^{2\ell}\frac{(-1)^j(2\ell)!}{(2\ell-j)!\zeta(\theta)^{j+1}}\\&\times(sh)^{(2\ell-j)/2}\Big\}.\end{aligned}$$

Furthermore, the following asymptotic expansions are true for $\Re(\zeta(\theta))<0$ *and* $s\to\infty$:

$$\begin{aligned}\int_0^h r^{\ell}e^{\sqrt{sr}\zeta(\theta)}\mathrm{d}r&=\frac{2}{s^{\ell+1}}\cdot\frac{(2\ell+1)!}{\zeta(\theta)^{2\ell+2}}+\mathscr{O}\left(s^{-1/2}e^{\sqrt{sh}\zeta(\theta)}\right),\\\int_0^h r^{\ell}e^{\sqrt{sr}\zeta(\theta)}\sqrt{\frac{s}{r}}\mathrm{d}r&=-\frac{2}{s^{\ell}}\cdot\frac{(2\ell)!}{\zeta(\theta)^{2\ell+1}}+\mathscr{O}\left(e^{\sqrt{sh}\zeta(\theta)}\right).\end{aligned} \tag{2.6.15}$$

In Lemmas 2.6.4 and 2.6.5 below, we will investigate the asymptotic behaviours of the integrals associated with u and the CGO solution $u_0(s\mathbf{x})$ (or their corresponding exterior normal derivatives) with respect to the positive parameter s as $s\to\infty$.

Lemma 2.6.4. *Recall that* Γ_h^- *and* Γ_h^+ *are defined in* (2.6.8). *Denote*

$$I_{11}^+=\int_{\Gamma_h^+}u(\mathbf{x})\frac{\partial u_0(s\mathbf{x})}{\partial\nu}\mathrm{d}\sigma,\quad I_{11}^-=\int_{\Gamma_h^-}u(\mathbf{x})\frac{\partial u_0(s\mathbf{x})}{\partial\nu}\mathrm{d}\sigma. \tag{2.6.16}$$

Then the following asymptotic expansions hold with respect to s *as* $s\to\infty$:

$$\begin{aligned}I_{11}^+&=-\frac{2\beta(\theta_0)}{\zeta(\theta_0)}u(\mathbf{0})-\frac{1}{s}\cdot\frac{4\beta(\theta_0)}{\zeta(\theta_0)^3}c_1(\theta_0)-\frac{1}{s^2}\cdot\frac{48\beta(\theta_0)}{\zeta(\theta_0)^5}c_2(\theta_0)+\mathscr{O}(s^{-3}),\\I_{11}^-&=-\frac{2\beta(0)}{\zeta(0)}u(\mathbf{0})-\frac{1}{s}\cdot\frac{4\beta(0)}{\zeta(0)^3}c_1(0)-\frac{1}{s^2}\cdot\frac{48\beta(0)}{\zeta(0)^5}c_2(0)+\mathscr{O}(s^{-3}),\end{aligned} \tag{2.6.17}$$

where

$$\begin{aligned}c_1(\theta)&=\frac{\partial u}{\partial x_1}\Big|_{\mathbf{x}=\mathbf{0}}\cos\theta+\frac{\partial u}{\partial x_2}\Big|_{\mathbf{x}=\mathbf{0}}\sin\theta,\\c_2(\theta)&=\frac{1}{2}\left(\frac{\partial^2 u}{\partial x_1^2}\Big|_{\mathbf{x}=\mathbf{0}}\cos^2\theta+\frac{\partial^2 u}{\partial x_1x_2}\Big|_{\mathbf{x}=\mathbf{0}}\sin 2\theta+\frac{\partial^2 u}{\partial x_2^2}\Big|_{\mathbf{x}=\mathbf{0}}\sin^2\theta\right).\end{aligned} \tag{2.6.18}$$

Proof. It is easy to see that the exterior unit normal vector to Γ_h^+ is

$$\nu=(\cos\varphi_M,\sin\varphi_M),\quad \varphi_M=\theta_0+\frac{\pi}{2}. \tag{2.6.19}$$

From Proposition 2.6.1, on Γ_h^+, we obtain that

$$\left.\frac{\partial u_0(s\mathbf{x})}{\partial \nu}\right|_{\Gamma_h^+} = \beta(\theta_0)e^{\sqrt{sr}\zeta(\theta_0)}\sqrt{\frac{s}{r}}, \tag{2.6.20}$$

where $\zeta(\theta_0) = -e^{\mathrm{i}\theta_0/2}$, and

$$\beta(\theta_0) = \frac{1}{2}\sin(\varphi_M - \theta_0)\tilde{\zeta}(\theta_0) = -\frac{\mathrm{i}e^{\mathrm{i}\theta_0/2}}{2}, \quad \tilde{\zeta}(\theta_0) = -\mathrm{i}e^{\mathrm{i}\theta_0/2}. \tag{2.6.21}$$

Noting the analyticity of the Laplacian eigenfunction u to (2.1.1) in Ω, we have the expansion near a neighbourhood of the origin:

$$u(\mathbf{x}) = \sum_{\alpha \in \mathbb{N}_0^2, |\alpha| \geq 0} \frac{(\partial^\alpha u)(\mathbf{0})}{\alpha!}\mathbf{x}^\alpha, \tag{2.6.22}$$

where $\mathbb{N}_0^2 = \{(\alpha_1, \alpha_2) \mid \alpha_j \in \mathbb{N} \cup \{0\},\ \ j = 1, 2\}$. Then substituting (2.6.20) and (2.6.22) into I_{11}^+, we deduce that

$$\begin{aligned}
I_{11}^+ &= \int_{\Gamma_h^+} u(\mathbf{x})\frac{\partial u_0(s\mathbf{x})}{\partial \nu}\mathrm{d}\sigma = u(\mathbf{0})\beta(\theta_0)\int_0^h e^{\sqrt{sr}\zeta(\theta_0)}\sqrt{\frac{s}{r}}\mathrm{d}r \\
&+ \left(\left.\frac{\partial u}{\partial x_1}\right|_{\mathbf{x}=\mathbf{0}}\cos\theta_0 + \left.\frac{\partial u}{\partial x_2}\right|_{\mathbf{x}=\mathbf{0}}\sin\theta_0\right)\beta(\theta_0)\int_0^h re^{\sqrt{sr}\zeta(\theta_0)}\sqrt{\frac{s}{r}}\mathrm{d}r \\
&+ \frac{1}{2}\left(\left.\frac{\partial^2 u}{\partial x_1^2}\right|_{\mathbf{x}=\mathbf{0}}\cos^2\theta_0 + 2\left.\frac{\partial^2 u}{\partial x_1 x_2}\right|_{\mathbf{x}=\mathbf{0}}\sin\theta_0\cos\theta_0 + \left.\frac{\partial^2 u}{\partial x_2^2}\right|_{\mathbf{x}=\mathbf{0}}\sin^2\theta_0\right) \\
&\times \beta(\theta_0)\int_0^h r^2 e^{\sqrt{sr}\zeta(\theta_0)}\sqrt{\frac{s}{r}}\mathrm{d}r + r_{I_{11}^+},
\end{aligned} \tag{2.6.23}$$

where

$$r_{I_{11}^+} = \sum_{\alpha \in \mathbb{N}_0^2, |\alpha| \geq 3} \frac{(\partial^\alpha u)(\mathbf{0})}{\alpha!}\int_{\Gamma_h^+}\mathbf{x}^\alpha\frac{\partial u_0(s\mathbf{x})}{\partial \nu}\mathrm{d}\sigma.$$

But this term can be estimated as $s \to \infty$ by means of (2.6.15),

$$\left|r_{I_{11}^+}\right| \leq |\beta(\theta_0)|\int_0^h r^3 \cdot e^{\sqrt{sr}\Re(\zeta(\theta_0))}\sqrt{\frac{s}{r}}\mathrm{d}r \sum_{\substack{\alpha \in \mathbb{N}_0^2, \\ |\alpha| \geq 3}} h^{|\alpha|-3}\left|\frac{(\partial^\alpha u)(\mathbf{0})}{\alpha!}\right| = \mathcal{O}(s^{-3}),$$

where we have used the fact that $\Re(\zeta(\theta_0)) = -\cos(\theta_0/2) < 0$ for $\theta_0 \in (0, \pi)$. Now the asymptotic expansion (2.6.17) for I_{11}^+ follows directly by substituting (2.6.15) into (2.6.23). Similar argument also applies to the asymptotic expansion of I_{11}^- and the detail is omitted.

The proof is complete. □

Lemma 2.6.5. *Recall that Γ_h^- and Γ_h^+ are defined in* (2.6.8)*. Suppose that u satisfies the boundary conditions* (2.6.10b) *and* (2.6.10c) *on Γ_h^- and Γ_h^+, respectively. Moreover, assume that $\eta_1 \in C^\gamma(\Gamma_h^-)$ and $\eta_2 \in C^\gamma(\Gamma_h^+)$ for $\gamma \in (0, 1]$, and*

$$I_{12}^+ = -\int_{\Gamma_h^+} u_0(s\mathbf{x})\frac{\partial u}{\partial \nu}\mathrm{d}\sigma, \quad I_{12}^- = -\int_{\Gamma_h^-} u_0(s\mathbf{x})\frac{\partial u}{\partial \nu}\mathrm{d}\sigma. \tag{2.6.24}$$

Then the following asymptotic expansions hold for $I_{12}^\pm$ with respect to s as $s \to \infty$:

$$
\begin{aligned}
I_{12}^{+} &= \frac{2}{s} \cdot \frac{\eta_2(\mathbf{0})u(\mathbf{0})}{\zeta(\theta_0)^2} + \frac{12}{s^2} \cdot \frac{\eta_2(\mathbf{0})c_1(\theta_0)}{\zeta(\theta_0)^4} + u(\mathbf{0}) \cdot \mathscr{O}(s^{-1-\gamma}) + \mathscr{O}(s^{-2-\gamma}), \\
I_{12}^{-} &= \frac{2}{s} \cdot \frac{\eta_1(\mathbf{0})u(\mathbf{0})}{\zeta(0)^2} + \frac{12}{s^2} \cdot \frac{\eta_1(\mathbf{0})c_1(0)}{\zeta(0)^4} + u(\mathbf{0}) \cdot \mathscr{O}(s^{-1-\gamma}) + \mathscr{O}(s^{-2-\gamma}).
\end{aligned} \tag{2.6.25}
$$

Proof. Since η_1 and η_2 are of C^γ-smooth, we have

$$
\eta_i(x) = \eta_i(\mathbf{0}) + \delta\eta_i(x), \quad |\delta\eta_i| \le \|\eta_i\|_{C^\gamma} \cdot |x|^\gamma. \tag{2.6.26}
$$

Then using (2.6.15), (2.6.22) and (2.6.26), we can deduce in polar coordinates on Γ_h^+ that

$$
\begin{aligned}
I_{12}^{+} = \int_{\Gamma_h^+} u_0(s\mathbf{x})\eta_2 u \mathrm{d}\sigma &= \eta_2(\mathbf{0})u(\mathbf{0}) \int_0^h e^{\sqrt{sr}\zeta(\theta_0)} \mathrm{d}r + u(\mathbf{0}) \int_0^h \delta\eta_2 e^{\sqrt{sr}\zeta(\theta_0)} \mathrm{d}r \\
&+ \eta_2(\mathbf{0})c_1(\theta_0) \int_0^h r e^{\sqrt{sr}\zeta(\theta_0)} \mathrm{d}r + r_{I_{12}^+},
\end{aligned} \tag{2.6.27}
$$

where

$$
\begin{aligned}
r_{I_{12}^+} = \eta_2(\mathbf{0}) &\sum_{\alpha \in \mathbb{N}_0^2, |\alpha| \ge 2} \frac{(\partial^\alpha u)(\mathbf{0})}{\alpha!} \int_{\Gamma_h^+} u_0(s\mathbf{x})\mathbf{x}^\alpha \mathrm{d}\sigma \\
+ &\sum_{\alpha \in \mathbb{N}_0^2, |\alpha| \ge 1} \frac{(\partial^\alpha u)(\mathbf{0})}{\alpha!} \int_{\Gamma_h^+} u_0(s\mathbf{x})\mathbf{x}^\alpha \delta\eta_2 \mathrm{d}\sigma.
\end{aligned} \tag{2.6.28}
$$

For this term, it follows from (2.6.15) that

$$
\begin{aligned}
\left| r_{I_{12}^+} \right| \le |\eta_2(\mathbf{0})| &\sum_{\alpha \in \mathbb{N}_0^2, |\alpha| \ge 2} h^{|\alpha|-2} \left| \frac{(\partial^\alpha u)(\mathbf{0})}{\alpha!} \right| \left| \int_0^h r^2 e^{\sqrt{sr}\Re(\zeta(\theta_0 - \pi))} \mathrm{d}r \right. \\
+ \|\eta_2\|_{C^\gamma} &\sum_{\alpha \in \mathbb{N}_0^2, |\alpha| \ge 1} h^{|\alpha|-1} \left| \frac{(\partial^\alpha u)(\mathbf{0})}{\alpha!} \right| \left| \int_0^h r^{1+\gamma} e^{\sqrt{sr}\Re(\zeta(\theta_0 - \pi))} \mathrm{d}r \right. \\
= \mathscr{O}(s^{-2-\gamma}).
\end{aligned} \tag{2.6.29}
$$

Using this and (2.6.15) again, we can derive (2.6.25) from (2.6.27). Similar derivation can be done for the asymptotic expansion of I_{12}^-. □

The following lemma is about the exterior normal derivative of $\partial_\nu u$ with respect to any singular line of a Laplacian eigenfunction u,

Lemma 2.6.6. *Suppose $\Gamma := \{\mathbf{x} \in \mathbb{R}^2 \mid \mathbf{x} = r(\cos\theta, \sin\theta), r > 0\}$ (θ is fixed) is a singular line of the Laplacian eigenfunction u, and φ is the polar angle of the unit normal vector to Γ, then*

$$
\cos\varphi\cos\theta \left.\frac{\partial^2 u}{\partial x_1^2}\right|_{\mathbf{x}=\mathbf{0}} + \sin\varphi\sin\theta \left.\frac{\partial^2 u}{\partial x_2^2}\right|_{\mathbf{x}=\mathbf{0}} + \sin(\varphi+\theta) \left.\frac{\partial^2 u}{\partial x_1 \partial x_2}\right|_{\mathbf{x}=\mathbf{0}} = 0. \tag{2.6.30}
$$

Proof. Recalling the definition of a singular line, one has

$$
\nabla\left(\frac{\partial u}{\partial \nu}\right) \cdot (\cos\theta, \sin\theta)^\top = 0. \tag{2.6.31}
$$

Then we can derive (2.6.30) by evaluating (2.6.31) in more detail at $\mathbf{x} = \mathbf{0}$. □

For the subsequent analysis, we also need the following lemma.

Lemma 2.6.7. *Suppose that u has the expansion* (2.6.22). *For any straight line segment* $\Gamma_h := \{x \in \mathbb{R}^2 \mid \mathbf{x} = r(\cos\theta, \sin\theta), 0 \le r \le h \ll 1\}$ *(with θ fixed) satisfying* $\Re(\zeta(\theta)) < 0$ *where* $\zeta(\theta)$ *is defined in* (2.6.14), *we assume that* $\nu = (\cos\varphi, \sin\varphi)$ *is the exterior unit normal vector to* Γ_h. *Then it holds as* $s \to \infty$ *that*

$$I_{12} = \int_{\Gamma_h} u_0(s\mathbf{x}) \frac{\partial u}{\partial \nu} \mathrm{d}\sigma = \frac{2}{s} \cdot \left(\frac{\partial u}{\partial x_1}\bigg|_{\mathbf{x}=\mathbf{0}} \cos\varphi + \frac{\partial u}{\partial x_2}\bigg|_{\mathbf{x}=\mathbf{0}} \sin\varphi \right) \cdot \frac{1}{\zeta(\theta)^2} \tag{2.6.32}$$
$$+ \frac{12}{s^2} \cdot \left(\frac{\partial^2 u}{\partial x_1^2}\bigg|_{\mathbf{x}=\mathbf{0}} \cos\varphi\cos\theta + \frac{\partial^2 u}{\partial x_2^2}\bigg|_{\mathbf{x}=\mathbf{0}} \sin\varphi\sin\theta + \frac{\partial^2 u}{\partial x_1 \partial x_2}\bigg|_{\mathbf{x}=\mathbf{0}} \sin(\varphi+\theta) \right)$$
$$\times \frac{1}{\zeta(\theta)^4} + \mathcal{O}(s^{-3}).$$

Proof. Using the polar coordinate on Γ_h, it is easy to see that

$$\frac{\partial u}{\partial \nu} = \frac{\partial u}{\partial x_1}\bigg|_{\mathbf{x}=\mathbf{0}} \cos\varphi + \frac{\partial u}{\partial x_2}\bigg|_{\mathbf{x}=\mathbf{0}} \sin\varphi + r\left(\frac{\partial^2 u}{\partial x_1^2}\bigg|_{\mathbf{x}=\mathbf{0}} \cos\varphi\cos\theta + \frac{\partial^2 u}{\partial x_2^2}\bigg|_{\mathbf{x}=\mathbf{0}} \sin\varphi\sin\theta \right.$$
$$\left. + \frac{\partial^2 u}{\partial x_1 \partial x_2}\bigg|_{\mathbf{x}=\mathbf{0}} \sin(\varphi+\theta) \right) + R(u, r, \varphi, \theta) \tag{2.6.33}$$

where

$$|R(u, r, \varphi, \theta)| \le r^2 \sum_{\alpha \in \mathbb{N}_0^2, |\alpha| \ge 3} h^{|\alpha|-3} \left| \frac{(\partial^\alpha u)(\mathbf{0})}{(\alpha-1)!} \right|.$$

Therefore, we can further write

$$I_{12} = \int_{\Gamma_h} u_0(s\mathbf{x}) \frac{\partial u}{\partial \nu} \mathrm{d}\sigma = \left(\frac{\partial u}{\partial x_1}\bigg|_{\mathbf{x}=\mathbf{0}} \cos\varphi + \frac{\partial u}{\partial x_2}\bigg|_{\mathbf{x}=\mathbf{0}} \sin\varphi \right) \int_0^h e^{\sqrt{sr}\zeta(\theta)} \mathrm{d}r$$
$$+ \left(\frac{\partial^2 u}{\partial x_1^2}\bigg|_{\mathbf{x}=\mathbf{0}} \cos\varphi\cos\theta + \frac{\partial^2 u}{\partial x_2^2}\bigg|_{\mathbf{x}=\mathbf{0}} \sin\varphi\sin\theta + \frac{\partial^2 u}{\partial x_1 \partial x_2}\bigg|_{\mathbf{x}=\mathbf{0}} \sin(\varphi+\theta) \right) \tag{2.6.34}$$
$$\times \int_0^h r e^{\sqrt{sr}\zeta(\theta)} \mathrm{d}r + \int_0^h R(u, r, \varphi, \theta) e^{\sqrt{sr}\zeta(\theta)} \mathrm{d}r.$$

Then the desired result follows from the estimate as $s \to \infty$ by using (2.6.15):

$$\left| \int_0^h R(u, r, \varphi, \theta) e^{\sqrt{sr}\zeta(\theta)} \mathrm{d}r \right| \le \int_0^h r^2 e^{\sqrt{sr}\zeta(\theta)} \mathrm{d}r \cdot \sum_{\alpha \in \mathbb{N}_0^2, |\alpha| \ge 3} h^{|\alpha|-2} \left| \frac{(\partial^\alpha u)(\mathbf{0})}{(\alpha-1)!} \right|$$
$$= \mathcal{O}(s^{-3}).$$

□

Suppose that Γ_h is a nodal line of u. Using the polar coordinate and evaluating (2.6.22) on Γ_h, we can prove the following lemma.

Lemma 2.6.8. *Suppose that u is a Laplacian eigenfunction to* (2.1.1) *and* $u = 0$ *on* Γ_h, *where* $\Gamma_h := \{\mathbf{x} \in \mathbb{R}^2; x = r(\cos\theta, \sin\theta), 0 \le r \le h\}$ *(with θ fixed) is a line segment. Then the functions* $c_1(\theta)$ *and* $c_2(\theta)$ *defined in* (2.6.18) *are both identically zero.*

Now we are in a position to present the proof of the theorems in Sect. 2.4 in the specific case that the vanishing order N is up to 3. Before that, we make two important remarks.

First, throughout the present section, if a generalized singular line of the form (2.2.1) is involved, the parameter η can be a C^1 function other than a constant. Indeed, our argument in the present section can deal with this more general case. In principle, we believe that the theorems in Sect. 2.4 can also be extended to the more general case that η is a function other than a constant. However, when dealing with higher vanishing orders, the corresponding analysis becomes radically more tedious and complicated. Hence, in the next section for the general vanishing order case, we shall stick to the case that η is a constant.

Second, in proving the vanishing order of the eigenfunction (up to 3), we shall only make use of the eigenfunction confined in S_h. This is achieved by means of the auxiliary results given in Lemmas 2.6.2–2.6.7. It is emphasized that this is in sharp contrast to the argument in the next section (for the higher vanishing orders), which applies the spherical wave expansion of the eigenfunction in B_h. Hence, the argument in the present section is "localized". This "localization" property enables one to consider, e.g. the quantitative behaviours of the Dirichlet eigenfunction in Ω up to the boundary $\partial\Omega$. Moreover, combining with our first remark above, the argument can also be used to study the quantitative behaviours of eigenfunctions associated with a general second-order elliptic operator other than the Laplacian.

We first deal with Theorem 2.4.1. According to our discussion above, we actually prove the following more general theorem.

Theorem 2.6.1. *Let u be a Laplacian eigenfunction to* (2.1.1). *Suppose that there are two generalized singular lines Γ_h^+ and Γ_h^- from $\mathcal{M}_\Omega^\lambda$ such that* (2.4.2) *and* (2.4.3) *hold. Assume that $\eta_1 \in C^1(\Gamma_h^-)$ and $\eta_2 \in C^1(\Gamma_h^+)$. If the conditions*

$$u(\mathbf{0}) = 0 \text{ and } \alpha \neq \frac{1}{2} \tag{2.6.35}$$

are satisfied, the Laplacian eigenfunction u vanishes up to the order 3 *at* $\mathbf{0}$.

Proof. Recall Fig. 2.2. Evaluating (2.6.10b) and (2.6.10c) at $\mathbf{0}$, using $u(\mathbf{0}) = 0$ we derive

$$\nabla u\Big|_{\mathbf{x}=\mathbf{0}} \cdot (\cos\varphi_m, \sin\varphi_m) = -\eta_1(\mathbf{0})u(\mathbf{0}) = 0,$$
$$\nabla u\Big|_{\mathbf{x}=\mathbf{0}} \cdot (\cos\varphi_M, \sin\varphi_M) = -\eta_2(\mathbf{0})u(\mathbf{0}) = 0.$$

Since $\theta_0 \in (0,\pi)$, we know that Γ_h^- and Γ_h^+ are non-collinear. Therefore, it is easy to see

$$\nabla u(\mathbf{0}) = 0. \tag{2.6.36}$$

Noting that u is the Laplacian eigenfunction satisfying (2.6.10a), we derive the integral equality from (2.6.11):

$$-\lambda \int_{S_h} u_0(s\mathbf{x})u(\mathbf{x})\mathrm{d}\mathbf{x} = \int_{S_h} (u_0(s\mathbf{x})\Delta u - u\Delta u_0(s\mathbf{x}))\,\mathrm{d}\mathbf{x} = I_1^+ + I_1^- + I_2, \tag{2.6.37}$$

where S_h is defined in (2.6.7), and $I_1^\pm$ and I_2 are defined in (2.6.12).

Since $u \in H^2(B_h)$, which can be embedded into $C^\gamma(B_h)(0<\gamma<1)$, we know that

$$u(\mathbf{x}) = u(\mathbf{0}) + \delta u(\mathbf{x}), \quad |\delta u(\mathbf{x})| \leq \|u\|_{C^\gamma}|\mathbf{x}|^\gamma, \tag{2.6.38}$$

from which it follows that

$$\int_{S_h} u_0(s\mathbf{x})u(\mathbf{x})\mathrm{d}\mathbf{x} = u(\mathbf{0})\int_{S_h} u_0(s\mathbf{x})\mathrm{d}\mathbf{x} + I_3 = u(\mathbf{0})\left(\int_W u_0(s\mathbf{x})\mathrm{d}\mathbf{x} - I_4\right) + I_3, \tag{2.6.39}$$

where

$$I_3 = \int_{S_h} u_0(s\mathbf{x})\delta u(\mathbf{x})\mathrm{d}\mathbf{x}, \quad I_4 = \int_{W\backslash B_h} u_0(s\mathbf{x})\mathrm{d}\mathbf{x}.$$

Substituting (2.6.39) into (2.6.37) and combining with (2.6.4), we derive that

$$-6\lambda \mathrm{i}(e^{-2\mathrm{i}\theta_0}-1)s^{-2}u(\mathbf{0}) = I_1^+ + I_1^- + I_2 + \lambda I_3 - u(\mathbf{0})\lambda I_4. \tag{2.6.40}$$

Using $u(\mathbf{0}) = 0$, we further obtain

$$0 = I_1^+ + I_1^- + I_2 + \lambda I_3. \tag{2.6.41}$$

Since $u \in H^2(B_h)$, we have from [8, Page 6263] and (2.6.3) that for some $c' > 0$,

$$|I_2| \le Ce^{-c'\sqrt{s}}, \quad |I_3| \le \mathcal{O}(s^{-\gamma-2}) \quad \text{as } s \to \infty. \tag{2.6.42}$$

Recalling the definitions of $I_1^\pm, I_2$, $I_{11}^\pm$ and $I_{12}^\pm$ given by (2.6.12), (2.6.16) and (2.6.24), respectively, it is easy to see that

$$I_1^+ = -(I_{11}^+ + I_{12}^+), \quad I_1^- = -(I_{11}^- + I_{12}^-). \tag{2.6.43}$$

Since η_1 and η_2 are C^1 functions on the boundary $\Gamma_h^\pm$, they fulfil the requirement of Lemma 2.6.5. Using (2.6.36) and recalling $c_1(\theta_0)$ and $c_1(0)$ given in (2.6.18), we know that

$$c_1(\theta_0) = c_1(0) = 0. \tag{2.6.44}$$

Substituting $u(\mathbf{0}) = 0$ and (2.6.44) into (2.6.17) and (2.6.25) yields

$$\begin{aligned} &I_{11}^+ = -\frac{1}{s^2}\cdot\frac{48\beta(\theta_0)}{\zeta(\theta_0)^5}c_2(\theta_0) + \mathcal{O}(s^{-3}), \quad I_{11}^- = -\frac{1}{s^2}\cdot\frac{48\beta(0)}{\zeta(0)^5}c_2(0) + \mathcal{O}(s^{-3}), \\ &I_{12}^+ = \mathcal{O}(s^{-2-\gamma}), \quad I_{12}^- = \mathcal{O}(s^{-2-\gamma}). \end{aligned} \tag{2.6.45}$$

But by means of (2.6.43), we deduce by substituting (2.6.45) into (2.6.41) that

$$\frac{1}{s^2}\cdot\frac{48\beta(\theta_0)}{\zeta(\theta_0)^5}c_2(\theta_0) + \frac{1}{s^2}\cdot\frac{48\beta(0)}{\zeta(0)^5}c_2(0) = -\lambda I_3 - I_2 + \mathcal{O}(s^{-2-\gamma}), \tag{2.6.46}$$

where $c_2(\theta)$ is defined in (2.6.18). Multiplying s^2 on both sides of (2.6.46), combining with (2.6.42) and letting $s \to \infty$, we can obtain that

$$\frac{\beta(\theta_0)}{\zeta(\theta_0)^5}c_2(\theta_0) + \frac{\beta(0)}{\zeta(0)^5}c_2(0) = 0. \tag{2.6.47}$$

Using the eigen-equation, $-\Delta u = \lambda u$, we can see

$$\left.\frac{\partial^2 u}{\partial x_1^2}\right|_{\mathbf{x}=\mathbf{0}} + \left.\frac{\partial^2 u}{\partial x_2^2}\right|_{\mathbf{x}=\mathbf{0}} = -\lambda u(\mathbf{0}) = 0. \tag{2.6.48}$$

Substituting this equation into the expression of $c_2(\theta)$, we derive that

$$c_2(0) = \frac{1}{2}\left.\frac{\partial^2 u}{\partial x_1^2}\right|_{\mathbf{x}=\mathbf{0}}, \quad c_2(\theta_0) = \frac{1}{2}\left(\left.\frac{\partial^2 u}{\partial x_1^2}\right|_{\mathbf{x}=\mathbf{0}}\cos 2\theta_0 + \left.\frac{\partial^2 u}{\partial x_1 x_2}\right|_{\mathbf{x}=\mathbf{0}}\sin 2\theta_0\right). \tag{2.6.49}$$

From (2.6.14), it is easy to see that

$$\frac{\beta(\theta_0)}{\zeta(\theta_0)^5} = \frac{\mathrm{i}e^{-2\mathrm{i}\theta_0}}{2}, \quad \frac{\beta(0)}{\zeta(0)^5} = -\frac{\mathrm{i}}{2}. \tag{2.6.50}$$

Then substituting (2.6.49) and (2.6.50) into (2.6.47), we can deduce that

$$\left(1-e^{-2\mathrm{i}\theta_0}\cos 2\theta_0\right)\frac{\partial^2 u}{\partial x_1^2}\bigg|_{\mathbf{x}=\mathbf{0}} - e^{-2\mathrm{i}\theta_0}\sin 2\theta_0\frac{\partial^2 u}{\partial x_1 x_2}\bigg|_{\mathbf{x}=\mathbf{0}} = 0. \tag{2.6.51}$$

Recalling that

$$f(x) := \frac{\partial u}{\partial \nu} + \eta_2 u \equiv 0 \qquad \text{on } \Gamma_h^+,$$

we know the directional derivative of f with respect to the direction $(\cos\theta_0, \sin\theta_0)$ satisfies

$$\nabla\left(\frac{\partial u}{\partial \nu} + \eta_2 u\right)\cdot(\cos\theta_0, \sin\theta_0)^\top = 0. \tag{2.6.52}$$

Since $\eta_1 \in C^1(\Gamma_h^-)$ and $\eta_2 \in C^1(\Gamma_h^+)$, we can use the fact that

$$\nabla\left(\frac{\partial u}{\partial \nu} + \eta_2 u\right) = \begin{bmatrix} \dfrac{\partial^2 u}{\partial x_1^2}\cos\varphi_M + \dfrac{\partial^2 u}{\partial x_1\partial x_2}\sin\varphi_M + \dfrac{\partial \eta_2}{\partial x_1}u + \eta_2\dfrac{\partial u}{\partial x_1} \\ \dfrac{\partial^2 u}{\partial x_1\partial x_2}\cos\varphi_M + \dfrac{\partial^2 u}{\partial x_2^2}\sin\varphi_M + \dfrac{\partial \eta_2}{\partial x_2}u + \eta_2\dfrac{\partial u}{\partial x_2} \end{bmatrix},$$

where $\varphi_M = \theta_0 + \frac{\pi}{2}$, and evaluate (2.6.52) at $\mathbf{x} = \mathbf{0}$, then derive by using $u(\mathbf{0}) = 0$ and (2.6.36) that

$$\cos\varphi_M\cos\theta_0\frac{\partial^2 u}{\partial x_1^2}\bigg|_{\mathbf{x}=\mathbf{0}} + \sin\varphi_M\sin\theta_0\frac{\partial^2 u}{\partial x_2^2}\bigg|_{\mathbf{x}=\mathbf{0}} + \sin(\varphi_M+\theta_0)\frac{\partial^2 u}{\partial x_1\partial x_2}\bigg|_{\mathbf{x}=\mathbf{0}} = 0. \tag{2.6.53}$$

Substituting (2.6.48) into (2.6.53), together with $\varphi_M = \theta_0 + \frac{\pi}{2}$, we can further obtain that

$$\sin 2\theta_0\frac{\partial^2 u}{\partial x_1^2}\bigg|_{\mathbf{x}=\mathbf{0}} - \cos 2\theta_0\frac{\partial^2 u}{\partial x_1\partial x_2}\bigg|_{\mathbf{x}=\mathbf{0}} = 0. \tag{2.6.54}$$

Now combining (2.6.51) with (2.6.54), we can get a system of linear equations with respect to $\frac{\partial^2 u}{\partial x_1^2}(\mathbf{0})$ and $\frac{\partial^2 u}{\partial x_1\partial x_2}(\mathbf{0})$, with the determinant of its coefficient matrix given by

$$\begin{vmatrix} 1-e^{-2\mathrm{i}\theta_0}\cos 2\theta_0 & -e^{-2\mathrm{i}\theta_0}\sin 2\theta_0 \\ \sin 2\theta_0 & -\cos 2\theta_0 \end{vmatrix} = -\cos 2\theta_0 + e^{-2\mathrm{i}\theta_0} = -\mathrm{i}\sin 2\theta_0 \neq 0$$

since $\theta_0 \neq \pi/2$. Therefore, together with (2.6.48) we can conclude that

$$\frac{\partial^2 u}{\partial x_1^2}\bigg|_{x=0} = \frac{\partial^2 u}{\partial x_1\partial x_2}\bigg|_{\mathbf{x}=\mathbf{0}} = \frac{\partial^2 u}{\partial x_2^2}\bigg|_{\mathbf{x}=\mathbf{0}} = 0.$$

Since the order of the lowest nontrivial homogeneous polynomial in Taylor expansion (2.6.22) around the origin is larger than 2, its vanishing order is at least up to 3. This completes the proof. □

We next deal with Theorem 2.4.4, but under a more general situation with $\eta_2 \in C^1(\Gamma_h^+)$.

Theorem 2.6.2. *Let u be a Laplacian eigenfunction to* (2.1.1)*. Suppose that a generalized singular line* $\Gamma_h^+ \in \mathcal{M}_\Omega^\lambda$ *intersects with a nodal line* $\Gamma_h^- \in \mathcal{N}_\Omega^\lambda$ *such that* (2.4.2) *and* (2.4.3) *hold, and* $\eta_2 \in C^1(\Gamma_h^+)$*. If the following condition is fulfilled*

$$\alpha \neq \frac{1}{4}, \frac{1}{2} \text{ and } \frac{3}{4} \tag{2.6.55}$$

then the Laplacian eigenfunction u vanishes up to the order 3 *at* $\mathbf{0}$.

Proof. Since $u = 0$ on Γ_h^-, we know from (2.6.48) that

$$\left.\frac{\partial u}{\partial x_1}\right|_{\mathbf{x}=\mathbf{0}} = \left.\frac{\partial^2 u}{\partial x_1^2}\right|_{\mathbf{x}=\mathbf{0}} = \left.\frac{\partial^2 u}{\partial x_2^2}\right|_{\mathbf{x}=\mathbf{0}} = 0. \tag{2.6.56}$$

Further, we derive by evaluating

$$\frac{\partial u}{\partial \nu} + \eta_2 u = 0$$

on Γ_h^+ at $\mathbf{x} = \mathbf{0}$ that

$$\left.\frac{\partial u}{\partial x_1}\right|_{\mathbf{x}=\mathbf{0}} \cos\varphi_M + \left.\frac{\partial u}{\partial x_2}\right|_{\mathbf{x}=\mathbf{0}} \sin\varphi_M = -\eta_2(\mathbf{0})u(\mathbf{0}) = 0, \tag{2.6.57}$$

where $\varphi_M = \theta_0 + \pi/2$. Then substituting (2.6.56) into (2.6.57), it is easy to see that

$$\cos\theta_0 \cdot \left.\frac{\partial u}{\partial x_2}\right|_{\mathbf{x}=\mathbf{0}} = 0.$$

Hence, if $\theta_0 \neq \pi/2$, we have

$$\left.\frac{\partial u}{\partial x_2}\right|_{\mathbf{x}=\mathbf{0}} = 0. \tag{2.6.58}$$

Now recalling that we have the boundary condition (2.6.10c) on Γ_h^+, with $\eta_2 \in C^1(\Gamma_h^+)$, and the fact that $u = 0$ on Γ_h^-, we can provide the following integral equality by the same argument for deriving (2.6.41)

$$0 = I_1^+ - I_{12}^- + I_2 + \lambda I_3, \tag{2.6.59}$$

where $I_1^+ = -(I_{11}^+ + I_{12}^+)$, I_2, I_{12}^- and I_3 are defined in (2.6.16), (2.6.12), (2.6.24) and (2.6.39) , respectively. For the term I_{11}^+, it follows from (2.6.17) that

$$I_{11}^+ = -\frac{2\beta(\theta_0)}{\zeta(\theta_0)}u(\mathbf{0}) - \frac{1}{s}\cdot\frac{4\beta(\theta_0)}{\zeta(\theta_0)^3}c_1(\theta_0) - \frac{1}{s^2}\cdot\frac{48\beta(\theta_0)}{\zeta(\theta_0)^5}c_2(\theta_0) + \mathscr{O}(s^{-3}),$$

which can be further reduced to

$$I_{11}^+ = -\frac{1}{s^2}\cdot\frac{24\beta(\theta_0)}{\zeta(\theta_0)^5}\cdot\left.\frac{\partial^2 u}{\partial x_1 x_2}\right|_{\mathbf{x}=\mathbf{0}}\sin 2\theta_0 + \mathscr{O}(s^{-3}) \tag{2.6.60}$$

by (2.6.56) and (2.6.58).

To estimate the term I_{12}^+, we recall that η_2 has the expansion (2.6.26). We have $c_1(\theta_0) = 0$ from (2.6.56) and (2.6.58). Then using $u(\mathbf{0}) = c_1(\theta_0) = 0$, we deduce from (2.6.25) that

$$I_{12}^+ = \mathscr{O}(s^{-2-\gamma}). \tag{2.6.61}$$

Next, we estimate I_{12}^-. It follows from (2.6.32), combining with (2.6.56) and (2.6.58), that

$$I_{12}^- = -\int_{\Gamma_h^-} u_0(s\mathbf{x})\frac{\partial u}{\partial \nu}\mathrm{d}\sigma = \frac{12}{s^2}\cdot\left.\frac{\partial^2 u}{\partial x_1 \partial x_2}\right|_{\mathbf{x}=\mathbf{0}}\cdot\frac{1}{\zeta(0)^4} - \mathscr{O}(s^{-3}). \tag{2.6.62}$$

Substituting (2.6.60)–(2.6.62) into (2.6.59), we can get that

$$\frac{1}{s^2}\cdot\frac{24\beta(\theta_0)}{\zeta(\theta_0)^5}\cdot\left.\frac{\partial^2 u}{\partial x_1 x_2}\right|_{\mathbf{x}=\mathbf{0}}\sin 2\theta_0 - \frac{12}{s^2}\cdot\left.\frac{\partial^2 u}{\partial x_1 \partial x_2}\right|_{\mathbf{x}=\mathbf{0}}\cdot\frac{1}{\zeta(0)^4} - \mathscr{O}\left(s^{-2-\gamma}\right) = -(I_2 + \lambda I_3).$$

Multiplying s^2 on both sides of the above equality, we deduce from (2.6.42) that

$$\left(\frac{2\beta(\theta_0)}{\zeta(\theta_0)^5}\cdot\sin 2\theta_0-\frac{1}{\zeta(0)^4}\right)\cdot\frac{\partial^2 u}{\partial x_1 x_2}\bigg|_{\mathbf{x}=\mathbf{0}}=0$$

as $s\to\infty$. From (2.6.50) we see that

$$\frac{2\beta(\theta_0)}{\zeta(\theta_0)^5}\cdot\sin 2\theta_0-\frac{1}{\zeta(0)^4}=\mathrm{i}e^{-2\mathrm{i}\theta_0}\sin 2\theta_0-1=-\cos 2\theta_0 e^{-2\mathrm{i}\theta_0}\neq 0$$

under the conditions $\theta_0\neq\pi/4$ and $\theta_0\neq 3\pi/4$. Hence, we obtain that

$$\frac{\partial^2 u}{\partial x_1\partial x_2}\bigg|_{\mathbf{x}=\mathbf{0}}=0,$$

which completes the proof. □

2.6.1 *Proof of Theorem 2.4.2*

Using $u=0$ on $\Gamma_h^{\pm}$, we have

$$\nabla u\Big|_{\mathbf{x}=\mathbf{0}}\cdot(1,0)^\top=\nabla u\Big|_{\mathbf{x}=\mathbf{0}}\cdot(\cos\theta_0,\sin\theta_0)^\top=0.$$

This implies

$$\nabla u\Big|_{\mathbf{x}=\mathbf{0}}=0. \tag{2.6.63}$$

Now we recall that u has the expansion (2.6.22); then we can derive on Γ_h^- by using (2.6.63) and polar coordinates that

$$\sum_{\substack{\alpha\in\mathbb{N}_0^2,|\alpha|\geq 2\\ \alpha=(\alpha_1,\alpha_2)}}\frac{(\partial^\alpha u)(\mathbf{0})}{\alpha!}r^{|\alpha|}\cos^{\alpha_1}(0)\sin^{\alpha_2}(0)\Big|_{x\in\Gamma_h^-}\equiv 0,\quad 0\leq r\leq h,$$

from which it is not difficult to see that

$$c_2(0)=\frac{1}{2}\frac{\partial^2 u}{\partial x_1^2}\Big|_{\mathbf{x}=\mathbf{0}}\cdot\cos^2 0=0. \tag{2.6.64}$$

This can be used, along with (2.6.48), to deduce that

$$\frac{\partial^2 u}{\partial x_2^2}\Big|_{\mathbf{x}=\mathbf{0}}=0. \tag{2.6.65}$$

By means of the fact $u=0$ on $\Gamma_h^{\pm}$ again, we have the integral identitys

$$0=-I_{12}^+-I_{12}^-+I_2+\lambda I_3, \tag{2.6.66}$$

where I_2, $I_{12}^{\pm}$ and I_3 are defined in (2.6.12), (2.6.24) and (2.6.39), respectively. By Lemma 2.6.7, together with (2.6.63)–(2.6.65), it is easy to see that

$$\begin{aligned} I_{12}^+&=\frac{12}{s^2}\cdot\frac{\partial^2 u}{\partial x_1\partial x_2}\Big|_{\mathbf{x}=\mathbf{0}}\sin(\varphi_M+\theta_0)\cdot\frac{1}{\zeta(\theta_0)^4}+\mathcal{O}(s^{-3}),\\ I_{12}^-&=\frac{12}{s^2}\cdot\frac{\partial^2 u}{\partial x_1\partial x_2}\Big|_{\mathbf{x}=\mathbf{0}}\sin(\varphi_m)\cdot\frac{1}{\zeta(0)^4}+\mathcal{O}(s^{-3}),\end{aligned} \tag{2.6.67}$$

where $\varphi_m = -\pi/2$ and $\varphi_M = \theta_0 + \pi/2$ are the arguments of the exterior unit normal vectors to Γ_h^- and Γ_h^+, respectively. From (2.6.14), we have

$$\zeta(\theta_0)^4 = e^{2\mathrm{i}\theta_0}, \quad \zeta(0)^4 = 1.$$

Substituting (2.6.67) into (2.6.66), we derive that

$$\frac{12}{s^2}\left(1 - \cos 2\theta_0 e^{-2\mathrm{i}\theta_0}\right) \frac{\partial^2 u}{\partial x_1 \partial x_2}\bigg|_{\mathbf{x}=\mathbf{0}} + I_2 + \lambda I_3 - \mathscr{O}(s^{-3}) = 0. \tag{2.6.68}$$

Now multiplying s^2 on both sides of (2.6.68), noting (2.6.42), we have

$$\left(1 - \cos 2\theta_0 e^{-2\mathrm{i}\theta_0}\right) \frac{\partial^2 u}{\partial x_1 \partial x_2}\bigg|_{\mathbf{x}=\mathbf{0}} = 0,$$

from which we can deduce that

$$\frac{\partial^2 u}{\partial x_1 \partial x_2}\bigg|_{\mathbf{x}=\mathbf{0}} = 0$$

if $\theta_0 \neq \pi/2$. This completes our proof. □

2.6.2 Proof of Theorem 2.4.6

Since u satisfies the boundary condition $u = 0$ on Γ_h^-, we know that

$$u(\mathbf{0}) = 0, \quad \nabla u\bigg|_{\mathbf{x}=\mathbf{0}} \cdot (1,0)^\top = 0 \Longrightarrow \frac{\partial u}{\partial x_1}\bigg|_{\mathbf{x}=\mathbf{0}} = 0.$$

Recalling that $\varphi_M = \theta_0 + \pi/2$ is the argument of the exterior unit normal vector of Γ_h^+ and $\partial_\nu u = 0$ on Γ_h^+, we easily see

$$\frac{\partial u}{\partial x_1}\bigg|_{\mathbf{x}=\mathbf{0}}(-\sin\theta_0) + \frac{\partial u}{\partial x_2}\bigg|_{\mathbf{x}=\mathbf{0}}\cos\theta_0 = 0.$$

Therefore, if $\theta_0 \neq \pi/2$, we know

$$\frac{\partial u}{\partial x_2}\bigg|_{\mathbf{x}=\mathbf{0}} = 0.$$

Furthermore, since Γ_h^- is a nodal line, we know by Lemma 2.6.8 that

$$\frac{\partial^2 u}{\partial x_1^2}\bigg|_{\mathbf{x}=\mathbf{0}} = 0.$$

Substituting $u(\mathbf{0}) = 0$ into (2.6.48), we have

$$\frac{\partial^2 u}{\partial x_2^2}\bigg|_{\mathbf{x}=\mathbf{0}} = -\frac{\partial^2 u}{\partial x_1^2}\bigg|_{\mathbf{x}=\mathbf{0}} = 0.$$

Then using the fact that Γ_h^+ is a singular line of u, we derive from Lemma 2.6.6 that

$$\sin(\varphi_M + \theta_0) \cdot \frac{\partial^2 u}{\partial x_1 \partial x_2}\bigg|_{x=0} = \cos 2\theta_0 \cdot \frac{\partial^2 u}{\partial x_1 \partial x_2}\bigg|_{\mathbf{x}=\mathbf{0}} = 0.$$

Hence, if $\theta_0 \neq \pi/4$ and $\theta_0 \neq 3\pi/4$, we can prove

$$\left.\frac{\partial^2 u}{\partial x_1 \partial x_2}\right|_{\mathbf{x}=\mathbf{0}} = 0,$$

which means that the expansion (2.6.22) of u has at least nontrivial homogeneous polynomial with the order of three, hence completing our proof. □

Remark 2.4. In the Sect. 2.6.2 above, we may also use the CGO solution as the test function to study the vanishing property of the second-order partial derivatives of u at the origin with respect to θ_0, which can lead to the same conclusion.

2.7 Proofs of the Theorems in Sect. 2.4 for General Cases

In this section, detailed proofs of the theorems for general vanishing order in Sect. 2.4 are presented, by using the spherical wave expansion of the Laplacian eigenfunction u near the intersecting point between two line segments.

From (2.3.15) and (2.3.16), we have the following lemma regarding the exterior normal derivative of u on $\Gamma_h^\pm$ by using the spherical wave expansion (2.3.15) of u.

Lemma 2.7.1. *Under the polar coordinate, we have the following expansion of the normal derivative of u given by* (2.3.15) *on $\Gamma_h^\pm$ around the origin*

$$\begin{aligned}
\left.\frac{\partial u}{\partial \nu}\right|_{\Gamma_h^+} &= \left.\frac{1}{r}\frac{\partial u}{\partial \theta}\right|_{\theta=\theta_0} = \frac{1}{r}\sum_{n=0}^{\infty} \mathrm{i}n\left(a_n e^{\mathrm{i}n\theta_0} - b_n e^{-\mathrm{i}n\theta_0}\right) J_n\left(\sqrt{\lambda} r\right), \\
\left.\frac{\partial u}{\partial \nu}\right|_{\Gamma_h^-} &= -\left.\frac{1}{r}\frac{\partial u}{\partial \theta}\right|_{\theta=0} = -\frac{1}{r}\sum_{n=0}^{\infty} \mathrm{i}n\left(a_n - b_n\right) J_n\left(\sqrt{\lambda} r\right).
\end{aligned} \tag{2.7.1}$$

In Sect. 2.6, we proved the vanishing order up to three of the Laplacian eigenfunction at the origin. In the next lemma, we clarify the relationship between the coefficients a_n, b_n in (2.3.15) and the vanishing order of u at the origin.

Lemma 2.7.2. *Suppose that u has the spherical wave expansion* (2.3.15) *around the origin, and the coefficients a_n, b_n in* (2.3.15) *satisfy*

$$a_0 + b_0 = 0,\ a_n = b_n = 0,\ n = 1, 2, \ldots, N-1,\ N \geq 2, \tag{2.7.2}$$

if and only if the vanishing order of u at the origin is at least N.

Proof. Substituting (2.7.2) into (2.3.15) yields

$$u(x) = \sum_{n=N}^{\infty} \left(a_n e^{\mathrm{i}n\theta} + b_n e^{-\mathrm{i}n\theta}\right) J_n\left(\sqrt{\lambda} r\right). \tag{2.7.3}$$

The lemma is readily proved by noting from (2.3.18) that the power of the lowest order in (2.7.3) with respect to r is N. □

Recalling the definition of the generalized singular line of the Laplacian eigenfunction u and using the spherical wave expansion (2.3.15) of u and Lemma 2.7.1, we can deduce some equations for the undetermined coefficients $\{a_n, b_n\}$ in (2.3.15). These equations will be used in the proof of Theorem 2.4.1.

Lemma 2.7.3. *Let $\Gamma_h^\pm$ be two generalized singular lines of u with the boundary parameters $\eta_1 \equiv C_1$ and $\eta_2 \equiv C_2$ defined on Γ_h^- and Γ_h^+, respectively, where C_1 and C_2 are two constants. Suppose that $\Gamma_h^\pm$ intersect with each other at the origin and* (2.7.2) *is fulfilled. Then the following equations hold for the coefficients $\{a_N, b_N\}$ and $\{a_{N+1}, b_{N+1}\}$ in* (2.3.15)*:*

$$a_N e^{\mathrm{i}N\theta_0} - b_N e^{-\mathrm{i}N\theta_0} = 0, \ a_N - b_N = 0, \tag{2.7.4}$$

$$2C_2(a_N e^{\mathrm{i}N\theta_0} + b_N e^{-\mathrm{i}N\theta_0}) + \mathrm{i}\sqrt{\lambda}(a_{N+1}e^{\mathrm{i}(N+1)\theta_0} - b_{N+1}e^{-\mathrm{i}(N+1)\theta_0}) = 0, \tag{2.7.5}$$

$$2C_1(a_N + b_N) - \mathrm{i}\sqrt{\lambda}(a_{N+1} - b_{N+1}) = 0. \tag{2.7.6}$$

Proof. Substituting (2.3.15) and (2.7.1) into

$$\frac{\partial u}{\partial \nu} + \eta_2 u = 0 \text{ on } \Gamma_h^+ \quad \text{and} \quad \frac{\partial u}{\partial \nu} + \eta_1 u = 0 \text{ on } \Gamma_h^-,$$

and combining with the following identity (cf. [1])

$$J_n(\sqrt{\lambda}r) = \frac{\sqrt{\lambda}r}{2n}\left(J_{n-1}(\sqrt{\lambda}r) + J_{n+1}(\sqrt{\lambda}r)\right), \quad n \in \mathbb{N},$$

we can obtain

$$\begin{aligned}&\sum_{n=1}^{\infty}\frac{\mathrm{i}\sqrt{\lambda}}{2}(a_n e^{\mathrm{i}n\theta_0} - b_n e^{-\mathrm{i}n\theta_0})\left(J_{n-1}(\sqrt{\lambda}r) + J_{n+1}(\sqrt{\lambda}r)\right)\\ &\quad + C_2\sum_{n=0}^{\infty}(a_n e^{\mathrm{i}n\theta_0} + b_n e^{-\mathrm{i}n\theta_0})J_n(\sqrt{\lambda}r) = 0,\end{aligned} \tag{2.7.7}$$

and

$$\sum_{n=1}^{\infty}\frac{\mathrm{i}\sqrt{\lambda}}{2}(a_n - b_n)\left(J_{n-1}(\sqrt{\lambda}r) + J_{n+1}(\sqrt{\lambda}r)\right) - C_1\sum_{n=0}^{\infty}(a_n + b_n)J_n(\sqrt{\lambda}r) = 0. \tag{2.7.8}$$

Using (2.7.2) and comparing the coefficients of r^{N-1} and r^N on both sides of (2.7.7), one can derive by straightforward calculations the first equation in (2.7.4) and (2.7.5). Similarly, we can derive the second equation in (2.7.4) and (2.7.6) on Γ_h^- by using (2.7.2) and (2.7.8). □

Similar to Lemma 2.7.3, we can obtain the equations for $\{a_n, b_n\}$ by using (2.3.15) and the boundary conditions on $\Gamma_h^{\pm}$ for the singular lines in Lemmas 2.7.4.

Lemma 2.7.4. *Suppose that $\Gamma_h^{\pm}$ are two singular lines of u which intersect with each other at the origin and (2.7.2) is fulfilled. Then the following equations hold for the coefficients $\{a_N, b_N\}$ in (2.3.15):*

$$a_N e^{\mathrm{i}N\theta_0} - b_N e^{-\mathrm{i}N\theta_0} = 0, \tag{2.7.9}$$

$$a_N - b_N = 0. \tag{2.7.10}$$

Proof. Using (2.7.1) and Lemma 2.3.5, together with (2.7.2), we can derive (2.7.9) and (2.7.10). □

In particular, for two intersecting nodal lines $\Gamma_h^{\pm}$, we can derive the equations for $\{a_n, b_n\}$, $n = 0, 1, \ldots$, by Lemma 2.3.5 as follows.

Lemma 2.7.5. *Suppose that $\Gamma_h^{\pm}$ are two nodal lines of u which intersect with each other at the origin, then the following equations hold for the coefficients $\{a_n, b_n\}$ in (2.3.15), $n = 0, 1, \ldots$:*

$$a_n e^{\mathrm{i}n\theta_0} + b_n e^{-\mathrm{i}n\theta_0} = 0,$$

$$a_n + b_n = 0.$$

Proof. Substituting (2.3.15) into $u = 0$ on $\Gamma_h^{\pm}$, we can obtain that

$$0=\sum_{n=0}^{\infty}\left(a_n e^{\mathrm{i}n\theta_0}+b_n e^{-\mathrm{i}n\theta_0}\right)J_n(\sqrt{\lambda}r),\quad 0=\sum_{n=0}^{\infty}(a_n+b_n)J_n(\sqrt{\lambda}r).$$

Then the desired results follow directly from Lemma 2.3.5. □

In the rest of this section, we provide detailed proofs of theorems for general vanishing orders in Sect. 2.4.

2.7.1 Proof of Theorem 2.4.1

According to Theorem 2.6.1, we know that Theorem 2.4.1 holds at least for $N=3$. Hence, by virtue of Lemma 2.7.2, we see that (2.7.2) holds for $N=3$. Now we apply the mathematical induction and assume that (2.7.2) holds for any $N\geq 3$ and $N\in\mathbb{N}$. Then using (2.7.4) we can directly derive by virtue of (2.4.4) that

$$\begin{vmatrix} e^{2\mathrm{i}N\theta_0} & -1 \\ 1 & -1 \end{vmatrix} = 1-e^{2\mathrm{i}N\theta_0}\neq 0$$

for $\theta_0\neq\frac{m\pi}{N}$ $(m=0,1,\ldots,N-1)$. This implies that $a_N=b_N=0$, and completes the proof of Theorem 2.4.1. □

2.7.2 Proof of Theorem 2.4.2

Since $u=0$ on $\Gamma_h^{\pm}$, by Lemma 2.7.5 we know

$$\begin{vmatrix} e^{\mathrm{i}n\theta_0} & e^{-\mathrm{i}n\theta_0} \\ 1 & 1 \end{vmatrix} = e^{-\mathrm{i}n\theta_0}(e^{2\mathrm{i}n\theta_0}-1)\neq 0$$

if $\theta_0\neq\frac{m\pi}{n}$ $(m=0,1,\ldots,n-1)$. This readily implies that $a_n=b_n=0$, $n=1,2,\ldots$. □

2.7.3 Proof of Theorem 2.4.3

In combination with Lemma 2.7.4, Theorem 2.4.3 can be proved by following a completely similar argument to the one for Theorem 2.4.1 in Sect. 2.7.1 by formally taking $\eta_1\equiv 0$ and $\eta_2\equiv 0$.

2.7.4 Proof of Theorem 2.4.4

From Theorem 2.6.2, we know that Theorem 2.4.4 holds at least for $N=3$. Therefore, by virtue of Lemma 2.7.2, one can conclude that (2.7.2) holds for $N=3$. Now we apply the mathematical induction and assume that (2.7.2) holds for any $N\geq 3$ and $N\in\mathbb{N}$. Since $u=0$ on Γ_h^-, from Lemma 2.7.5, one has

$$a_N+b_N=0. \tag{2.7.11}$$

Since $\frac{\partial u}{\partial \nu}+\eta_2 u=0$ on Γ_h^+, by virtue of (2.7.4), it holds that

$$a_N e^{2\mathrm{i}N\theta_0}-b_N=0. \tag{2.7.12}$$

In view of (2.7.11) and (2.7.12), if $\theta_0 \neq \frac{(2m+1)\pi}{2N}$ $(m = 0, 1, \ldots, N-1)$, then

$$\begin{vmatrix} e^{2\mathrm{i}N\theta_0} & -1 \\ 1 & 1 \end{vmatrix} = e^{2\mathrm{i}N\theta_0} + 1 \neq 0,$$

which readily shows that $a_N = b_N = 0$.

The proof is complete. □

2.8 Discussions About the Condition $u(\mathbf{0}) = 0$

We recall a critical condition $u(\mathbf{0}) = 0$ that was used in Theorems 2.4.1, 2.4.3 and 2.4.5 in Sect. 2.4. It is also noted that in other three theorems of the same section, the condition that $u(\mathbf{0}) = 0$ is always fulfilled because one of the two line segments is a nodal line there. In this section, by illustrating with several specific examples, we show that the condition $u(\mathbf{0}) = 0$ can be fulfilled in certain scenarios in Theorems 2.4.1, 2.4.3 and 2.4.5, if one imposes certain generic conditions on the boundary parameters C_i, the intersecting angle $\alpha \cdot \pi$ and the eigenvalue λ.

It is stated in the introduction that one of the main motivations of the study in this work is the unique identifiability in inverse scattering problems. As we will see in the next section, we are able to develop a powerful mathematical strategy so that this condition is always fulfilled by making use of a linear combination of two eigenfunctions.

Proposition 2.8.1. *Let u be a Laplacian eigenfunction to* (2.1.1), *with its Fourier series given by* (2.3.15). *Suppose that there are two generalized singular lines Γ_h^+ and Γ_h^- from $\mathscr{M}_\Omega^\lambda$ such that* (2.4.2) *and* (2.4.3) *hold. Assume that $\eta_1 \equiv C_1$ and $\eta_2 \equiv C_2$ for two constants C_1 and C_2. Then if $\alpha = 1$ and $C_1 \neq C_2$, the Laplacian eigenfunction u fulfils $u(\mathbf{0}) = 0$. If $\alpha \neq 1$, two coefficients a_1 and b_1 in* (2.3.15) *can be expressed explicitly by*

$$a_1 = \frac{1}{\sqrt{\lambda}\sin\theta_0}(C_1 e^{-\mathrm{i}\theta_0} + C_2)u(\mathbf{0}), \quad b_1 = \frac{1}{\sqrt{\lambda}\sin\theta_0}(C_1 e^{\mathrm{i}\theta_0} + C_2)u(\mathbf{0}). \tag{2.8.1}$$

Proof. Recall the Laplacian eigenfunction u has the spherical wave expansion (2.3.15) in polar coordinates around the origin. Then we can obtain from Lemma 2.7.3, and noting that $\alpha = 1$ implies $\theta_0 = \pi$, the following equations

$$\begin{cases} 2C_2(a_0 + b_0) - \mathrm{i}\sqrt{\lambda}(a_1 - b_1) = 0, \\ 2C_1(a_0 + b_0) - \mathrm{i}\sqrt{\lambda}(a_1 - b_1) = 0. \end{cases}$$

Noting that $a_0 + b_0 = u(\mathbf{0})$ and using the assumption $C_1 \neq C_2$, we can derive from the above equations that $u(\mathbf{0}) = 0$. And (2.8.1) follows readily from (2.7.5) and (2.7.6) if $\alpha \neq 1$. □

Proposition 2.8.2. *Let u be a Laplacian eigenfunction to* (2.1.1). *Suppose that there are two generalized singular lines Γ_h^+ and Γ_h^- from $\mathscr{M}_\Omega^\lambda$ such that* (2.4.2) *and* (2.4.3) *hold. Assume that $\eta_1 \equiv C_1$ and $\eta_2 \equiv C_2$ for two constants C_1 and C_2. If $\alpha \neq \frac{1}{2}$, then a_2 and b_2 in* (2.3.15) *can be expressed explicitly as*

$$\begin{aligned} a_2 &= \frac{2u(\mathbf{0})}{\lambda \sin 2\theta_0 \sin\theta_0}(C_1C_2 + C_1C_2 e^{-\mathrm{i}2\theta_0} + C_2^2\cos\theta_0 + C_1^2\cos\theta_0 e^{-\mathrm{i}2\theta_0}), \\ b_2 &= \frac{2u(\mathbf{0})}{\lambda \sin 2\theta_0 \sin\theta_0}(C_1C_2 + C_1C_2 e^{\mathrm{i}2\theta_0} + C_2^2\cos\theta_0 + C_1^2\cos\theta_0 e^{\mathrm{i}2\theta_0}). \end{aligned} \tag{2.8.2}$$

Proof. Substituting (2.8.1) into (2.7.5) and (2.7.6) and taking $n = 2$, we can obtain that

$$
\begin{cases}
2C_2(a_1e^{\mathrm{i}\theta_0}+b_1e^{-\mathrm{i}\theta_0})+\mathrm{i}\sqrt{\lambda}(a_2e^{\mathrm{i}2\theta_0}-b_2e^{-\mathrm{i}2\theta_0})=0,\\
2C_1(a_1+b_1)-\mathrm{i}\sqrt{\lambda}(a_2-b_2)=0.
\end{cases}
\tag{2.8.3}
$$

After rearranging the terms, (2.8.3) can be further simplified as

$$
\begin{cases}
a_2e^{\mathrm{i}2\theta_0}-b_2e^{-\mathrm{i}2\theta_0}=\frac{2C_2\mathrm{i}}{\sqrt{\lambda}}(a_1e^{\mathrm{i}\theta_0}+b_1e^{-\mathrm{i}\theta_0}),\\
a_2-b_2=-\frac{2C_1\mathrm{i}}{\sqrt{\lambda}}(a_1+b_1).
\end{cases}
\tag{2.8.4}
$$

Substituting (2.8.1) into (2.8.4), then we can derive by direct calculations (2.8.2) for the explicit expressions of a_2 and b_2. □

Proposition 2.8.3. *Let u be a Laplacian eigenfunction to (2.1.1). Suppose that there are two generalized singular lines Γ_h^+ and Γ_h^- from $\mathcal{M}_\Omega^\lambda$ such that (2.4.2) and (2.4.3) hold Assume that $\eta_1\equiv C_1$ and $\eta_2\equiv C_2$ for two constants C_1 and C_2. Then if $\alpha=\frac{1}{3}$, $C_1\neq C_2$ and*

$$
1+\frac{4}{3\lambda}\left(C_1^2+C_1C_2+C_2^2\right)\neq 0,
\tag{2.8.5}
$$

the Laplacian eigenfunction u fulfils $u(\mathbf{0})=0$. Furthermore, if $\alpha\neq\frac{1}{3}$, then a_3 and b_3 in (2.3.15) can be expressed explicitly as

$$
a_3=\frac{1}{\sqrt{\lambda}\sin 3\theta_0}(B_1-B_2e^{-\mathrm{i}3\theta_0}),\quad b_3=\frac{1}{\sqrt{\lambda}\sin 3\theta_0}(B_1-B_2e^{\mathrm{i}3\theta_0}),
\tag{2.8.6}
$$

where B_1 and B_2 are given by

$$
B_1=\mathrm{i}\sqrt{\lambda}(a_1e^{\mathrm{i}\theta_0}-b_1e^{-\mathrm{i}\theta_0})+C_2(a_2e^{\mathrm{i}2\theta_0}+b_2e^{-\mathrm{i}2\theta_0})+C_2u(\mathbf{0}),
\tag{2.8.7}
$$

$$
B_2=\mathrm{i}\sqrt{\lambda}(a_1-b_1)-C_1(a_2+b_2)-C_1u(\mathbf{0}),
\tag{2.8.8}
$$

with (a_1,b_1) and (a_2,b_2) defined in (2.8.1) and (2.8.2), respectively.

Proof. Recall (2.7.7) and (2.7.8). Comparing the coefficients of r^2 on both sides of these two equations, we can derive that

$$
\begin{aligned}
&\mathrm{i}\sqrt{\lambda}(a_1e^{\mathrm{i}\theta_0}-b_1e^{-\mathrm{i}\theta_0})+\frac{1}{2}\mathrm{i}\sqrt{\lambda}(a_3e^{\mathrm{i}3\theta_0}-b_3e^{-\mathrm{i}3\theta_0})\\
&+C_2(a_2e^{\mathrm{i}2\theta_0}+b_2e^{-\mathrm{i}2\theta_0})+C_2(a_0+b_0)=0\quad\text{on }\Gamma_h^+,
\end{aligned}
\tag{2.8.9}
$$

and

$$
\mathrm{i}\sqrt{\lambda}(a_1-b_1)+\frac{1}{2}\mathrm{i}\sqrt{\lambda}(a_3-b_3)-C_1(a_2+b_2)-C_1(a_0+b_0)=0\quad\text{on }\Gamma_h^-.
\tag{2.8.10}
$$

Since $a_0+b_0=u(\mathbf{0})$, substituting (2.8.7) and (2.8.8) into (2.8.9) and (2.8.10), we can further obtain that

$$
a_3e^{\mathrm{i}3\theta_0}-b_3e^{-\mathrm{i}3\theta_0}=\frac{2\mathrm{i}B_1}{\sqrt{\lambda}},\quad a_3-b_3=\frac{2\mathrm{i}B_2}{\sqrt{\lambda}}.
\tag{2.8.11}
$$

Since $\alpha=\frac{1}{3}$, taking $\theta_0=\frac{\pi}{3}$ in (2.8.11), we have

$$
a_3-b_3=-\frac{2\mathrm{i}B_1}{\sqrt{\lambda}}=\frac{2\mathrm{i}B_2}{\sqrt{\lambda}},
$$

which indicates that

$$
B_1+B_2=0,
\tag{2.8.12}
$$

where

$$B_1 = \mathrm{i}\sqrt{\lambda}(a_1 e^{\mathrm{i}\frac{\pi}{3}} - b_1 e^{-\mathrm{i}\frac{\pi}{3}}) + C_2(a_2 e^{\mathrm{i}\frac{2\pi}{3}} + b_2 e^{-\mathrm{i}\frac{2\pi}{3}}) + C_2 u(\mathbf{0}), \tag{2.8.13}$$

and B_2 is defined in (2.8.8). Substituting (2.8.1) and (2.8.2) into (2.8.13) and (2.8.8), after straightforward calculations, (2.8.12) can be reduced to

$$(C_1 - C_2)\left[1 + \frac{4}{3\lambda}\left(C_1^2 + C_1 C_2 + C_2^2\right)\right] u(\mathbf{0}) = 0.$$

This implies $u(\mathbf{0}) = 0$ by noting that $C_1 \neq C_2$ and using (2.8.5). ¿Similarly, we can deduce (2.8.6) from (2.8.11) directly for $\alpha \neq 1/3$. □

We end this section with two important remarks.

Remark 2.5. By tracing the proofs of Propositions 2.8.1–2.8.3 and repeating similar arguments, we can find that under some mild assumptions on C_1, C_2, the intersecting angle $\alpha \cdot \pi$ and the eigenvalue λ, the property $u(\mathbf{0}) = 0$ still holds for the rational intersecting angle $\alpha \cdot \pi$ generically except for $\alpha = \pi/2m$, where $m = 1, 2, \ldots$. The detailed arguments are rather tedious and technical, but straightforward.

Remark 2.6. In Propositions 2.8.1–2.8.3, we studied the property $u(\mathbf{0}) = 0$ for two intersected generalized singular lines only for some conditions on C_1, C_2, the intersecting angle $\alpha \cdot \pi$ and λ. Other situations may be analysed similarly, e.g. either $C_1 = 0$ or $C_2 = 0$. But as shown in Theorem 2.3.2, we cannot guarantee $u(\mathbf{0}) = 0$ by imposing some conditions on the intersecting angle between two intersecting singular lines.

2.9 Main Definitions for Laplacian Eigenfunction in $\mathbb{R}^3$

Sections 2.9–2.13 can be regarded as the continuation as well as a significant further development of the study in Sects. 2.2–2.8, where the intersection of two nodal or generalized singular lines is considered. In Sects. 2.9–2.13, we follow the treatment in [16]. Results in Sects. 2.2–2.8 consider a geometrical property that the vanishing order (analytic quantity) of an eigenfunction at an intersecting point is related to the rationality (geometric quantity) of the corresponding intersecting angle. It is very natural to explore if similar results can be discussed in the more important three-dimensional case, about the intersections of nodal or generalized singular planes and their implications to the analytic behaviours of the eigenfunctions. But the geometric setup is much more complicated in three dimensions, so is the corresponding analysis: the intersection of two planes produces an edge corner, whereas the intersection of more than three planes produces a vertex corner; see Fig. 2.3 for a schematic illustration. We aim to derive a comprehensive characterization of the relationship between the analytic behaviours of an eigenfunction at a corner point and the geometric quantities of that corner. More specifically, at the edge corner case, we can show that the vanishing order of the eigenfunction is related to the rationality of the intersecting angle in a similar manner to the two-dimensional case, whereas at the vertex corner case, the vanishing order of the eigenfunction is proved to be related to the intersecting angle in a more complicated and mysterious manner through the roots of the Legendre polynomials. As an important application, these spectral results are applied to provide several fundamental unique identifiability results for the geometrical inverse scattering problem of determining an impenetrable obstacle as well as the (possibly) surface impedance by at most a few far-field measurements in the polyhedral setup. We shall present more relevant discussions in this aspect in Sects. 4.1 and 4.2. We shall investigate in Sects. 2.10, 2.11 and 2.12 the detailed vanishing properties of the Laplacian eigenfunctions in $\mathbb{R}^3$ at places where two or more nodal/singular/generalized singular planes intersect.

The rest of this section is mainly devoted to the introduction of the mathematical setup for the study in $\mathbb{R}^3$. We emphasize that compared with the conventional notion of Laplacian eigenfunctions, we do not prescribe any homogeneous boundary condition for u in (2.1.1). This means the spectral results that we shall discuss in this chapter apply to any function that satisfies (2.1.1) in the interior of Ω, in particular, including all the conventional Laplacian eigenfunctions with various boundary conditions. We next introduce several critical definitions for our subsequent analysis. In what follows, for Π being a flat plane in $\mathbb{R}^3$, any non-empty open connected subset $\Sigma \Subset \Pi$ is called a *cell* of Π. Let $\widetilde{\Pi} = \Pi_\Sigma$ denote the connected component of $\Pi \cap \Omega$ that contains Σ.

Definition 2.9.1. Consider a nontrivial eigenfunction u to (2.1.1). Let $\Sigma \subset \Omega$ be a cell of Π and $\eta \in \mathbb{C}$ be a constant. If $u|_\Sigma = 0$, Σ is said to be a nodal cell of u in Ω. By analytic continuation, it is seen that $u|_{\widetilde{\Pi}} = 0$, and $\widetilde{\Pi}$ is said to be a nodal plane of u. In a similar manner, in the case $(\partial_\nu u + \eta u)\big|_\Sigma = 0$, where ν is a unit one-sided normal direction of Π and $\eta \in \mathbb{C}$ is a constant, Σ and $\widetilde{\Pi}$ are called the generalized singular cell and plane, respectively. In the particular case $\eta \equiv 0$, a generalized singular plane is also called a singular plane. Let $\mathcal{N}_\Omega^\lambda$, $\mathcal{S}_\Omega^\lambda$ and $\mathcal{M}_\Omega^\lambda$ signify the sets of nodal, singular and generalized singular planes, respectively, of u in (2.1.1).

According to Definition 2.9.1, a nodal/generalized singular plane is actually a cell that is fully extended in Ω. Indeed, by the fact that u is analytic in Ω, we know that if the homogeneous condition is satisfied on a cell, then it is also satisfied on the so-called plane in Definition 2.9.1 by the analytic continuation. In what follows, most of the planes are actually the nodal/generalized singular planes in the sense of Definition 2.9.1, which should be clear from the context. Moreover, we would like to emphasize that in defining a generalized singular plane, the parameter η can be replaced to be a complex-valued real-analytic function. Indeed, all of the results obtained in this work hold for the case that η is a variable function as mentioned above. However, in order to ease the exposition, we stick to the case that η is a constant till to Sect. 2.12, and we shall make more relevant remarks in Sect. 2.13.

Let $B_\rho(\mathbf{x})$ denote a ball of radius $\rho \in \mathbb{R}_+$ and centred at $\mathbf{x} \in \mathbb{R}^3$.

Definition 2.9.2. Let Π_1 and Π_2 be two adjacent faces of a polyhedron $\mathcal{P}$ in Ω. Let $\boldsymbol{l}$ be a connected open portion of the edge formed by Π_1 and Π_2 such that $\boldsymbol{l} \Subset \Omega$. Then any $\mathbf{x} \in \boldsymbol{l}$ is said to be an edge corner point; see Fig. 2.3 for a schematic illustration. For notational convenience, we also let $\mathcal{E}(\Pi_1, \Pi_2, \boldsymbol{l})$ denote the edge corner as illustrated in Fig. 2.3.

Definition 2.9.3. Let $\{\Pi_\ell\}_{\ell=1}^n$ $(n \geq 3)$ be n planes in Ω such that they form a polyhedral cone $\mathcal{K}$ with the vertex $\mathbf{x}_0 \in \Omega$. Let $\rho \in \mathbb{R}_+$ be sufficiently small such that $B_\rho(\mathbf{x}_0) \subset \Omega$; then $\mathcal{K} \cap B_\rho(\mathbf{x}_0)$ is called a vertex corner associated with Π_1, Π_2, ..., Π_n and denoted by $\mathcal{V}(\{\Pi_\ell\}_{\ell=1}^n, \mathbf{x}_0)$; see Fig. 2.3 for a schematic illustration.

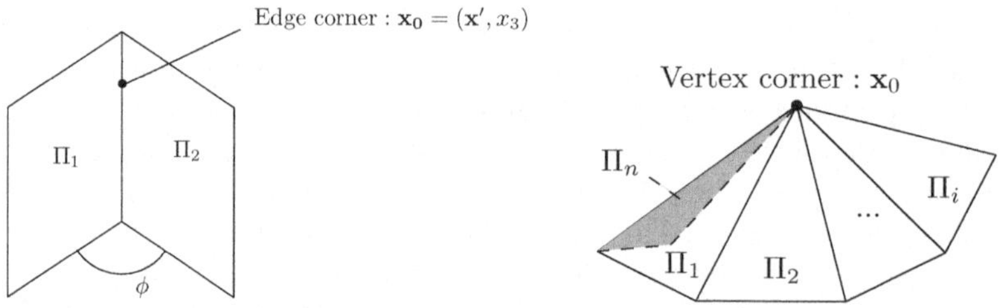

Fig. 2.3 Schematic illustrations of edge corner and vertex corner

It is obvious that a vertex corner $\mathcal{V}(\{\Pi_\ell\}_{\ell=1}^n, \mathbf{x}_0)$ is composed of finite many-edge corners, which are intersected by any two adjacent planes. Moreover, a vertex corner must be an edge corner. Definitions 2.9.1–2.9.3 describe some geometric notions. Next, we introduce several analytic notions for the Laplacian eigenfunction.

Definition 2.9.4. Let u be a nontrivial eigenfunction in (2.1.1). For a given point $\mathbf{x}_0 \in \Omega$, if there exists a number $N \in \mathbb{N} \cup \{0\}$ such that

$$\lim_{\rho \to +0} \frac{1}{\rho^m} \int_{B_\rho(\mathbf{x}_0)} |u(\mathbf{x})| \, \mathrm{d}\mathbf{x} = 0 \quad \text{for } m = 0, 1, \ldots, N+2, \tag{2.9.1}$$

we say that u vanishes at $\mathbf{x}_0$ up to the order N. The largest possible N such that (2.9.1) is fulfilled is called the vanishing order of u at $\mathbf{x}_0$, and we write

$$\mathrm{Vani}(u; \mathbf{x}_0) = N.$$

If (2.9.1) holds for any $N \in \mathbb{N}$, then we say that the vanishing order is infinity.

By the strong UCP, if the vanishing order of u at $\mathbf{x}_0 \in \Omega$ is infinite, we know that $u \equiv 0$ in Ω. Similarly, we can introduce the definition of the vanishing order of u at an edge or vertex corner.

Definition 2.9.5. Let u be a nontrivial eigenfunction to (2.1.1). Consider an edge corner $\mathscr{E}(\Pi_1, \Pi_2, \boldsymbol{l}) \Subset \Omega$. For any given $\mathbf{x}_0 \in \boldsymbol{l}$, if

$$\mathrm{Vani}(u; \mathbf{x}_0) = N,$$

we say that u vanishes at $\mathbf{x}_0$ associated with the edge corner $\mathscr{E}(\Pi_1, \Pi_2, \boldsymbol{l}) \Subset \Omega$ up to order N, denoted by

$$\mathrm{Vani}(u; \mathbf{x}_0, \Pi_1, \Pi_2) = N.$$

For a vertex corner $\mathbf{x}_0 \in \Omega$ which is intersected by Π_i, $i = 1, 2, \ldots n$, the vanishing order of u at $\mathbf{x}_0$ is defined by

$$\mathrm{Vani}(u; \mathbf{x}_0) := \max\Big\{ \max_{i=1,2,\ldots n-1} \mathrm{Vani}(u; \mathbf{x}_0, \Pi_i, \Pi_{i+1}), \mathrm{Vani}(u; \mathbf{x}_0, \Pi_n, \Pi_1) \Big\}.$$

2.10 Vanishing Orders at Edge Corners

In this section, we study the vanishing property of the Laplacian eigenfunction at an edge corner $\mathbf{x}_0 \in \boldsymbol{l}$ associated with $\mathscr{E}(\Pi_1, \Pi_2, \boldsymbol{l})$. The two planes Π_ℓ ($\ell = 1, 2$) could be either one of the following three types: nodal, singular or generalized singular planes. First, we give a definition of the irrational or rational dihedral angle of two intersecting planes.

Definition 2.10.1. Let Π_1 and Π_2 be two planes in $\mathbb{R}^3$ that intersect with each other. Let $\phi \in (0, \pi)$ be one of the associated intersecting dihedral angle of Π_1 and Π_2 satisfying

$$\phi = \alpha \cdot \pi, \quad \alpha \in (0, 1).$$

Then ϕ is said to be an *irrational dihedral angle* if α is an irrational number; and it is said to be a rational dihedral angle of degree q if $\alpha = p/q$ with $p, q \in \mathbb{N}$ and is irreducible.

Since $-\Delta$ is invariant under rigid motions, throughout the rest of this chapter, we assume that the edge corner $\mathscr{E}(\Pi_1, \Pi_2, \boldsymbol{l})$ satisfies

$$\boldsymbol{l} = \Big\{ \mathbf{x} = (\mathbf{x}', x_3) \in \mathbb{R}^3; \mathbf{x}' = 0, \ x_3 \in (-H, H) \Big\} \Subset \Omega, \tag{2.10.1}$$

where $2H$ is the length of $\boldsymbol{l}$. That is, $\boldsymbol{l}$ coincides with the x_3-axis. We further assume that Π_1 coincides with the (x_1, x_3)-plane while Π_2 possesses a dihedral angle $\alpha\pi$ away from Π_1 in the anticlockwise direction; see Fig. 2.4 for a schematic illustration. Clearly, we can assume that $\alpha \in (0, 1)$. Moreover, when we consider the vanishing order at an edge corner of $\mathscr{E}(\Pi_1, \Pi_2, \boldsymbol{l})$, we assume throughout this section that the edge corner under consideration is the origin $\mathbf{0} \in \boldsymbol{l}$.

In the next subsection, we first study a relatively simpler case that at least one of the intersecting planes of $\mathscr{E}(\Pi_1,\Pi_2,\boldsymbol{l})$ is a nodal plane. Without loss of generality, we assume $u|_{\Pi_1}\equiv 0$ throughout this subsection.

2.10.1 Vanishing Orders at an Edge Corner with at Least One Plane Being Nodal

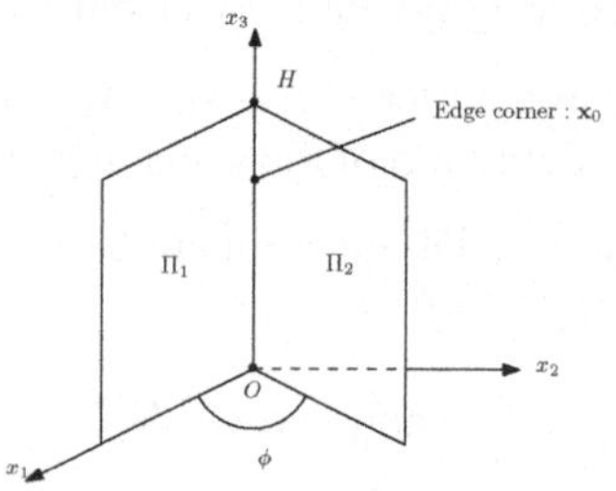

Fig. 2.4 Schematic illustration of two intersecting planes with an edge corner and the dihedral angle ϕ

We first derive several important auxiliary results for the subsequent analysis, for which we will often use the spherical coordinate of any point $\mathbf{x}$ in $\mathbb{R}^3$:

$$\mathbf{x}=(r\sin\theta\cos\phi, r\sin\theta\sin\phi, r\cos\theta):=(r,\theta,\phi),\ r\geq 0,\ \theta\in[0,\pi),\ \phi\in[0,2\pi). \tag{2.10.2}$$

Then the following proposition is a consequence of direct computing using spherical coordinates.

Proposition 2.10.1. *Let Π be any of the two planes associated with $\mathscr{E}(\Pi_1,\Pi_2,\boldsymbol{l})$. For any point $\mathbf{x}\in\Pi$, we know that ϕ defined in* (2.10.2) *is fixed; see Fig. 2.4. Let ν be the unit normal vector that is perpendicular to Π. Then*

$$\frac{\partial u}{\partial \nu}=\pm\frac{1}{r\sin\theta}\frac{\partial u}{\partial\phi}.$$

Lemma 2.10.1 ([19, Section 3.3]). *The solution u to* (2.1.1) *has the spherical wave expansion in spherical coordinates around the origin:*

$$u(\mathbf{x})=4\pi\sum_{n=0}^{\infty}\sum_{m=-n}^{n}\mathrm{i}^n a_n^m j_n(\sqrt{\lambda}r)Y_n^m(\theta,\phi), \tag{2.10.3}$$

where $j_n(t)$ is the spherical Bessel function of order n, and $Y_n^m(\theta,\phi)$ is the spherical harmonics given by

$$Y_n^m(\theta,\phi)=\sqrt{\frac{2n+1}{4\pi}\frac{(n-|m|)!}{(n+|m|)!}}P_n^{|m|}(\cos\theta)e^{\mathrm{i}m\phi}$$

with $P_n^m(t)$ being the associated Legendre functions.

Lemma 2.10.2 ([55, Theorem 2.4.4]). *In the spherical coordinate system, the associated Legendre functions fulfil the following orthogonality condition for any fixed $n\in\mathbb{N}$, and any two integers $m\geq 0$ and $l\leq n$:*

$$\int_0^{\pi}\frac{P_n^m(\cos\theta)P_n^l(\cos\theta)}{\sin\theta}d\theta=\begin{cases}0 & \text{if}\quad l\neq m,\\ \frac{(n+m)!}{m(n-m)!} & \text{if}\quad l=m.\end{cases} \tag{2.10.4}$$

Lemma 2.10.3. *Suppose that for $t \in (0,h)$, $h \in \mathbb{R}_+$,*

$$\sum_{n=0}^{\infty} \alpha_n j_n(t) = 0, \tag{2.10.5}$$

where $j_n(t)$ is the n-th spherical Bessel function. Then

$$\alpha_n = 0, \quad n = 0,1,2,\ldots. \tag{2.10.6}$$

Proof. By Colton and Kress [19, Section 2.4] we know that

$$j_n(t) := \sum_{p=0}^{\infty} \frac{(-1)^p t^{n+2p}}{2^p p! 1\cdot 3 \cdots (2n+2p+1)} = \frac{t^n}{(2n+1)!!}\left(1 + \sum_{p=1}^{\infty} \frac{(-1)^p t^{2p}}{2^p p! N_{l,n}}\right), \tag{2.10.7}$$

where $N_{l,n} = (2n+3)\cdot(2n+5)\cdots(2n+2p+1)$. Substituting (2.10.7) into (2.10.5) and comparing the coefficient of t^n ($n = 1,2,\ldots$), we can deduce (2.10.6). □

We are now in a position to study the general vanishing orders with the help of the spherical wave expansion of the Laplacian eigenfunction u to (2.1.1) around an intersecting edge corner.

Lemma 2.10.4. *Let u be a Laplacian eigenfunction to* (2.1.1). *Suppose that there exists an edge corner $\mathscr{E}(\Pi_1, \Pi_2, \boldsymbol{l})$ such that*

$$\mathscr{E}(\Pi_1, \Pi_2, \boldsymbol{l}) \Subset \Omega,$$

where Π_1 and Π_2 are from either of $\mathscr{N}_{\Omega}^{\lambda}$, $\mathscr{S}_{\Omega}^{\lambda}$ or $\mathscr{M}_{\Omega}^{\lambda}$. If there exists a sufficiently small $\varepsilon \in \mathbb{R}_+$ such that

$$u|_{B_\varepsilon(\mathbf{0})\cap \boldsymbol{l}} = 0, \tag{2.10.8}$$

then it holds for the coefficients in (2.10.3) *that*

$$a_n^0 = 0, \quad n \in \mathbb{N}\cup\{0\}. \tag{2.10.9}$$

Proof. Since the line segment $\boldsymbol{l}$ associated with $\mathscr{E}(\Pi_1, \Pi_2, \boldsymbol{l})$ coincides with the x_3-axis, we know $\theta = 0$ or π for $\mathbf{x} \in \boldsymbol{l}$ in the spherical coordinate system (2.10.2). Combining with Lemma 2.10.1, we know under the condition (2.10.8) that

$$u|_{B_\varepsilon(\mathbf{0})\cap \boldsymbol{l}} = 4\pi \sum_{n=0}^{\infty}\sum_{m=-n}^{n} \mathrm{i}^n a_n^m j_n(\sqrt{\lambda} r)\sqrt{\frac{2n+1}{4\pi}}\sqrt{\frac{(n-|m|)!}{(n+|m|)!}} P_n^{|m|}(\pm 1) e^{\mathrm{i}m\phi} = 0. \tag{2.10.10}$$

On the other hand, we have that for $m \in \mathbb{N}$ (cf. [9]),

$$P_n^{-m} = (-1)^m \frac{(n-m)!}{(n+m)!} P_n^m,\ P_n^m(\pm 1) = 0,\ P_n^0(+1) = 1,\ P_n^0(-1) = (-1)^n. \tag{2.10.11}$$

Substituting (2.10.11) into (2.10.10), it is easy to see that

$$\sum_{n=0}^{\infty} \mathrm{i}^n \sqrt{\frac{2n+1}{4\pi}} a_n^0 j_n(\sqrt{\lambda} r) = 0.$$

By virtue of Lemma 2.10.3, we readily see

$$\mathrm{i}^n \sqrt{\frac{2n+1}{4\pi}} a_n^0 = 0 \quad \text{for } n = 0,1,2,\ldots,$$

which completes the proof of Lemma 2.10.4. □

First, we consider the case that two nodal planes intersect with each other to yield the edge corner.

Theorem 2.10.1. *Let u be a Laplacian eigenfunction to (2.1.1). Consider an edge corner $\mathscr{E}(\Pi_1,\Pi_2,\boldsymbol{l}) \Subset \Omega$ where the two planes Π_ℓ, $\ell=1,2$ are assumed to be nodal, namely, $\Pi_\ell \in \mathscr{N}_\Omega^\lambda$ ($\ell=1,2$). If the corresponding dihedral angle can be written as*

$$\angle(\Pi_1,\Pi_2)=\phi=\alpha\cdot\pi,\quad \alpha\in(0,1),$$

where α satisfies for an $N\in\mathbb{N}$, $N\geq 3$,

$$\alpha\neq\frac{q}{p},\quad p=1,2,\ldots,N-1,\ q=1,2,\ldots,p-1, \tag{2.10.12}$$

then u vanishes up to order at least N at the edge corner $\mathbf{0}$.

Proof. Since $u|_{\Pi_i}\equiv 0$, $i=1,2$, it follows from Lemma 2.10.1 that

$$u|_{\Pi_1}=4\pi\sum_{n=0}^{\infty}\sum_{m=-n}^{n}\mathrm{i}^n a_n^m j_n(\sqrt{\lambda}r)\sqrt{\frac{2n+1}{4\pi}}\sqrt{\frac{(n-|m|)!}{(n+|m|)!}}P_n^{|m|}(\cos\theta)=0, \tag{2.10.13}$$

$$u|_{\Pi_2}=4\pi\sum_{n=0}^{\infty}\sum_{m=-n}^{n}\mathrm{i}^n a_n^m j_n(\sqrt{\lambda}r)\sqrt{\frac{2n+1}{4\pi}}\sqrt{\frac{(n-|m|)!}{(n+|m|)!}}P_n^{|m|}(\cos\theta)e^{\mathrm{i}m\alpha\cdot\pi}=0, \tag{2.10.14}$$

where $\phi=0$ on Π_1 and $\phi=\alpha\cdot\pi,\alpha\in(0,1)$ on Π_2. It is obvious that $u|_{\boldsymbol{l}}=0$, and then we have (2.10.9) from Lemma 2.10.4. Thus, comparing the coefficient of r and substituting $a_n^0=0$ for $n=0,1$ into (2.10.13) and (2.10.14), we obtain

$$(a_1^1+a_1^{-1})P_1^1(\cos\theta)=0,\quad (a_1^1 e^{\mathrm{i}\alpha\cdot\pi}+a_1^{-1}e^{-\mathrm{i}\alpha\cdot\pi})P_1^1(\cos\theta)=0.$$

Since $\theta\in(0,\pi)$ is arbitrary, utilizing the orthogonality condition (2.10.4), we can deduce

$$a_1^1+a_1^{-1}=0,\quad a_1^1 e^{\mathrm{i}\alpha\cdot\pi}+a_1^{-1}e^{-\mathrm{i}\alpha\cdot\pi}=0.$$

Therefore, if $\alpha\neq 0,1$, we derive that $a_1^{\pm 1}=0$.

Assume that $a_{n-1}^m=0$, $m=\pm1,\pm2,\ldots,\pm(n-1)$. We next show by induction that $a_n^m=0$, $m=\pm1,\pm2,\ldots,\pm n$. Indeed, comparing the coefficients of r^n, we obtain

$$\sum_{m=-n}^{n}\mathrm{i}^n a_n^m\frac{\sqrt{\lambda}^n}{(2n+1)!!}\sqrt{\frac{2n+1}{4\pi}}\sqrt{\frac{(n-|m|)!}{(n+|m|)!}}P_n^{|m|}(\cos\theta)=0, \tag{2.10.15}$$

$$\sum_{m=-n}^{n}\mathrm{i}^n a_n^m\frac{\sqrt{\lambda}^n}{(2n+1)!!}\sqrt{\frac{2n+1}{4\pi}}\sqrt{\frac{(n-|m|)!}{(n+|m|)!}}P_n^{|m|}(\cos\theta)e^{\mathrm{i}m\alpha\cdot\pi}=0. \tag{2.10.16}$$

Similarly, substituting $a_n^0=0$ into (2.10.15) and (2.10.16), noting that θ is arbitrary and utilizing the orthogonality condition (2.10.4) again, we can derive for $m=1,2,\ldots$ that

$$a_n^m+a_n^{-m}=0,\quad a_n^m e^{\mathrm{i}m\alpha\cdot\pi}+a_n^{-m}e^{-\mathrm{i}m\alpha\cdot\pi}=0. \tag{2.10.17}$$

Hence, if $\alpha\neq\frac{k}{m}$, $k=1,2,\ldots,m-1$, the coefficient matrix fulfils

$$\begin{vmatrix}1 & 1\\ e^{\mathrm{i}m\alpha\cdot\pi} & e^{-\mathrm{i}m\alpha\cdot\pi}\end{vmatrix}=-2\mathrm{i}\sin m\alpha\cdot\pi\neq 0,$$

which yields that $a_n^m = 0$ for $m = \pm 1, \pm 2, \ldots, \pm n$, hence completing the proof of Theorem 2.10.1. □

Remark 2.10.1. In the proof of Theorem 2.10.1, we make use of the boundary conditions of u on Π_1 and Π_2 as well as the orthogonality property (2.10.4) to arrive at the homogeneous linear system (2.10.17), which in turn proves that $a_n^{\pm m} = 0$ provided a certain condition on the dihedral angle $\alpha \cdot \pi$ is fulfilled. This type of argument shall be frequently used in the proofs of Theorems 2.10.2, 2.10.3 and 2.11.1 in what follows.

We now proceed to consider the case that a nodal plane $\Pi_1 \in \mathscr{N}_\Omega^\lambda$ intersects with a generalized singular plane $\Pi_2 \in \mathscr{M}_\Omega^\lambda$.

Theorem 2.10.2. *Let u be a Laplacian eigenfunction to (2.1.1). Consider an edge corner $\mathscr{E}(\Pi_1, \Pi_2, \boldsymbol{l}) \Subset \Omega$ such that*

$$\Pi_1 \in \mathscr{N}_\Omega^\lambda, \quad \Pi_2 \in \mathscr{M}_\Omega^\lambda \quad and \quad \angle(\Pi_1, \Pi_2) = \phi = \alpha \cdot \pi, \quad \alpha \in (0,1).$$

If for an $N \in \mathbb{N}$, $N \geq 2$, there holds

$$\alpha \neq \frac{2q+1}{2p}, \quad p = 1, 2, \ldots, N-1, \; q = 1, 2, \ldots, p-1,$$

and then u vanishes up to order at least N at the edge corner $\mathbf{0}$.

Proof. Since $u|_{\Pi_1} \equiv 0$, it is direct to know that $u|_{\boldsymbol{l}} \equiv 0$, which indicates that $a_n^0 = 0$ for $n = 0, 1, 2, \ldots$ from Lemma 2.10.4. Furthermore, by Lemma 2.10.1 we have

$$u|_{\Pi_1} = 4\pi \sum_{n=0}^{\infty} \sum_{m=-n}^{n} \mathrm{i}^n a_n^m j_n(\sqrt{\lambda} r) \sqrt{\frac{2n+1}{4\pi}} \sqrt{\frac{(n-|m|)!}{(n+|m|)!}} P_n^{|m|}(\cos\theta) = 0. \tag{2.10.18}$$

Combining with Proposition 2.10.1, we derive the following expression on Π_2:

$$\begin{aligned} \frac{\partial u}{\partial \nu} + \eta u \Big|_{\Pi_2} &= \frac{1}{r\sin\theta} \frac{\partial u}{\partial \phi} + \eta u \Big|_{\phi = \alpha\cdot\pi} \\ &= \frac{1}{r\sin\theta} 4\pi \sum_{n=0}^{\infty} \sum_{m=-n}^{n} \mathrm{i}^{n+1} m a_n^m j_n(\sqrt{\lambda} r) \sqrt{\frac{2n+1}{4\pi}} \sqrt{\frac{(n-|m|)!}{(n+|m|)!}} P_n^{|m|}(\cos\theta) e^{\mathrm{i} m\alpha\cdot\pi} \\ &+ \eta \cdot 4\pi \sum_{n=0}^{\infty} \sum_{m=-n}^{n} \mathrm{i}^n a_n^m j_n(\sqrt{\lambda} r) \sqrt{\frac{2n+1}{4\pi}} \sqrt{\frac{(n-|m|)!}{(n+|m|)!}} P_n^{|m|}(\cos\theta) e^{\mathrm{i} m\alpha\cdot\pi} = 0. \end{aligned} \tag{2.10.19}$$

Since $\theta \in (0, \pi)$ and $r > 0$, multiplying $r\sin\theta$ on both sides of (2.10.19), we can obtain that

$$\begin{aligned} &\sum_{n=0}^{\infty} \sum_{m=-n}^{n} \mathrm{i}^{n+1} m a_n^m j_n(\sqrt{\lambda} r) \sqrt{\frac{2n+1}{4\pi}} \sqrt{\frac{(n-|m|)!}{(n+|m|)!}} P_n^{|m|}(\cos\theta) e^{\mathrm{i} m\alpha\cdot\pi} \\ &+ \eta \cdot r\sin\theta \sum_{n=0}^{\infty} \sum_{m=-n}^{n} \mathrm{i}^n a_n^m j_n(\sqrt{\lambda} r) \sqrt{\frac{2n+1}{4\pi}} \sqrt{\frac{(n-|m|)!}{(n+|m|)!}} P_n^{|m|}(\cos\theta) e^{\mathrm{i} m\alpha\cdot\pi} = 0. \end{aligned} \tag{2.10.20}$$

Following a similar argument to Theorem 2.10.1, we may compare the coefficients of r in (2.10.18) and (2.10.20), respectively. First, for (2.10.18) we have

$$\sum_{m=-1}^{1} \mathrm{i} a_1^m \frac{\sqrt{\lambda}}{3!!} \sqrt{\frac{3}{4\pi}} \sqrt{\frac{(1-|m|)!}{(1+|m|)!}} P_1^{|m|}(\cos\theta) = 0.$$

Since $a_1^0 = 0$, using (2.10.4), we can deduce that

$$a_1^1 + a_1^{-1} = 0. \tag{2.10.21}$$

Then for (2.10.20), we have

$$\sum_{m=-1}^{1} m a_1^m \frac{\sqrt{\lambda}}{3!!} \sqrt{\frac{3}{4\pi}} \sqrt{\frac{(1-|m|)!}{(1+|m|)!}} P_1^m(\cos\theta) e^{\mathrm{i}m\alpha\cdot\pi} = 0, \tag{2.10.22}$$

since $a_0^0 = 0$. By the orthogonality condition of P_1^m for arbitrary $\theta \in (0,\pi)$ and the fact that $a_1^0 = 0$, we can simplify (2.10.22) to get

$$a_1^1 e^{\mathrm{i}\alpha\cdot\pi} - a_1^{-1} e^{-\mathrm{i}\alpha\cdot\pi} = 0.$$

Combining (2.10.21) with (2.10.22), we can obtain that if $\alpha \neq \frac{1}{2}$, then $a_1^{\pm 1} = 0$. By induction, we assume that $a_{n-1}^m = 0$, $m = \pm 1, \pm 2, \ldots, \pm(n-1)$. Considering the coefficients of r^n in (2.10.18), we have

$$\sum_{m=-n}^{n} \mathrm{i}^n a_n^m \frac{\sqrt{\lambda}^n}{(2n+1)!!} \sqrt{\frac{2n+1}{4\pi}} \sqrt{\frac{(n-|m|)!}{(n+|m|)!}} P_n^{|m|}(\cos\theta) = 0,$$

from which we can derive

$$a_n^m + a_n^{-m} = 0 \quad \text{for } m = 1, 2, \ldots \tag{2.10.23}$$

by virtue of the fact that $a_n^0 = 0$ and (2.10.4). Similarly, for (2.10.20), we know the coefficients of r^n fulfil that

$$\begin{aligned} &\sum_{m=-n}^{n} \mathrm{i}^{n+1} m a_n^m \frac{\sqrt{\lambda}^n}{(2n+1)!!} \sqrt{\frac{2n+1}{4\pi}} \sqrt{\frac{(n-|m|)!}{(n+|m|)!}} P_n^{|m|}(\cos\theta) e^{\mathrm{i}m\alpha\cdot\pi} \\ &+ \eta \cdot \sin\theta \sum_{m=-(n-1)}^{n-1} \mathrm{i}^{n-1} a_{n-1}^m \frac{\sqrt{\lambda}^{n-1}}{(2n-1)!!} \sqrt{\frac{2n-1}{4\pi}} \sqrt{\frac{(n-1-|m|)!}{(n-1+|m|)!}} P_{n-1}^{|m|}(\cos\theta) e^{\mathrm{i}m\alpha\cdot\pi} = 0. \end{aligned} \tag{2.10.24}$$

Substituting $a_{n-1}^m = 0$, $m = \pm 1, \pm 2, \ldots, \pm(n-1)$, and $a_n^0 = 0$ into (2.10.24), utilizing (2.10.4) again we derive

$$a_n^m e^{\mathrm{i}m\alpha\cdot\pi} - a_n^{-m} e^{-\mathrm{i}m\alpha\cdot\pi} = 0. \tag{2.10.25}$$

Therefore, by the virtue of Remark 2.10.1, we can deduce from (2.10.23) and (2.10.25) that if $\alpha \neq \frac{2k+1}{2m}$ $(k = 0, 1, \ldots, m-1)$, $a_n^m = 0$, $m = \pm 1, \pm 2, \ldots, \pm n$, hence completing the proof of Theorem 2.10.2. □

It is straightforward to verify from the proof of Theorem 2.10.2 that η can be 0. In such a case, we have the following result.

Corollary 2.10.1. *Let u be a Laplacian eigenfunction to (2.1.1). Consider an edge corner $\mathscr{E}(\Pi_1, \Pi_2, \boldsymbol{l}) \Subset \Omega$ such that*

$$\Pi_1 \in \mathscr{N}_\Omega^\lambda, \quad \Pi_2 \in \mathscr{S}_\Omega^\lambda \quad \text{and} \quad \angle(\Pi_1, \Pi_2) = \phi = \alpha \cdot \pi, \quad \alpha \in (0,1).$$

If for an $N \in \mathbb{N}$, $N \geq 2$, there holds

$$\alpha \neq \frac{2q+1}{2p}, \quad p = 1, 2, \ldots, N-1, \; q = 1, 2, \ldots, p-1,$$

then u vanishes up to order at least N at the edge corner $\mathbf{0}$.

2.10.2 Vanishing Orders at an Edge Corner Intersected by Generalized Singular Planes

In this subsection, we consider the case that an edge corner $\mathscr{E}(\Pi_1, \Pi_2, \boldsymbol{l})$ is intersected by two generalized singular planes, namely, $\Pi_\ell \in \mathscr{M}_\Omega^\lambda$, $\ell = 1, 2$. In what follows, we signify the boundary parameters on Π_ℓ to be η_ℓ, $\ell = 1, 2$. Then we can derive the following three theorems.

Theorem 2.10.3. *Let u be a Laplacian eigenfunction to (2.1.1). Consider an edge corner $\mathscr{E}(\Pi_1, \Pi_2, \boldsymbol{l}) \Subset \Omega$ with $\Pi_\ell \in \mathscr{M}_\Omega^\lambda, \ell = 1, 2$ and $\angle(\Pi_1, \Pi_2) = \phi = \alpha \cdot \pi$ for $\alpha \in (0, 1)$. If there exists a sufficiently small radius $\varepsilon \in \mathbb{R}_+$ such that*

$$u|_{B_\varepsilon(\mathbf{0}) \cap \boldsymbol{l}} \equiv 0, \tag{2.10.26}$$

and for an $N \in \mathbb{N}$, $N \geq 3$,

$$\alpha \neq \frac{q}{p}, \quad p = 1, 2, \ldots, N-1, \; q = 1, 2, \ldots, p-1,$$

then u vanishes up to the order at least N at the edge corner $\mathbf{0}$.

Proof. Since $u|_{\Pi_i} = \frac{\partial u}{\partial \nu} + \eta_i u = 0$, $i = 1, 2$, we have by using Proposition 2.10.1 that

$$\begin{aligned} \frac{\partial u}{\partial \nu} + \eta_1 u \Big|_{\Pi_1} &= -\frac{1}{r \sin\theta} \frac{\partial u}{\partial \phi} + \eta_1 u \Big|_{\phi=0} = 0, \\ \frac{\partial u}{\partial \nu} + \eta_2 u \Big|_{\Pi_2} &= \frac{1}{r \sin\theta} \frac{\partial u}{\partial \phi} + \eta_2 u \Big|_{\phi=\alpha\cdot\pi} = 0, \end{aligned}$$

which can be written more explicitly in spherical coordinate system by Lemma 2.10.1 as

$$\begin{aligned} &-\sum_{n=0}^{\infty} \sum_{m=-n}^{n} \mathrm{i}^{n+1} m a_n^m j_n(\sqrt{\lambda} r) \sqrt{\frac{2n+1}{4\pi}} \sqrt{\frac{(n-|m|)!}{(n+|m|)!}} P_n^{|m|}(\cos\theta) \\ &+ \eta_1 r \sin\theta \sum_{n=0}^{\infty} \sum_{m=-n}^{n} \mathrm{i}^n a_n^m j_n(\sqrt{\lambda} r) \sqrt{\frac{2n+1}{4\pi}} \sqrt{\frac{(n-|m|)!}{(n+|m|)!}} P_n^{|m|}(\cos\theta) = 0, \end{aligned} \tag{2.10.27}$$

and

$$\begin{aligned} &\sum_{n=0}^{\infty} \sum_{m=-n}^{n} \mathrm{i}^{n+1} m a_n^m j_n(\sqrt{\lambda} r) \sqrt{\frac{2n+1}{4\pi}} \sqrt{\frac{(n-|m|)!}{(n+|m|)!}} P_n^{|m|}(\cos\theta) e^{\mathrm{i} m \alpha\cdot\pi} \\ &+ \eta_2 r \sin\theta \sum_{n=0}^{\infty} \sum_{m=-n}^{n} \mathrm{i}^n a_n^m j_n(\sqrt{\lambda} r) \sqrt{\frac{2n+1}{4\pi}} \sqrt{\frac{(n-|m|)!}{(n+|m|)!}} P_n^{|m|}(\cos\theta) e^{\mathrm{i} m \alpha\cdot\pi} = 0. \end{aligned} \tag{2.10.28}$$

Under the condition (2.10.26), we know from Lemma 2.10.4 that

$$a_n^0 = 0, \qquad \text{for } n = 0, 1, 2, \ldots. \tag{2.10.29}$$

Comparing the coefficients of r^1 in (2.10.27) and (2.10.28), respectively, we have

$$\begin{aligned} \sum_{m=-1}^{1} m a_1^m \frac{\sqrt{\lambda}}{3!!} \sqrt{\frac{3}{4\pi}} \sqrt{\frac{(1-|m|)!}{(1+|m|)!}} P_1^{|m|}(\cos\theta) = 0, \\ -\sum_{m=-1}^{1} m a_1^m \frac{\sqrt{\lambda}}{3!!} \sqrt{\frac{3}{4\pi}} \sqrt{\frac{(1-|m|)!}{(1+|m|)!}} P_1^{|m|}(\cos\theta) e^{\mathrm{i} m \alpha\cdot\pi} = 0. \end{aligned}$$

Utilizing the orthogonality condition (2.10.4) and the fact that $a_1^0 = 0$, we can obtain the linear system with respect to $a_1^{\pm 1}$ as

$$a_1^1 - a_1^{-1} = 0, \quad a_1^1 e^{\mathrm{i}\alpha\cdot\pi} - a_1^{-1} e^{-\mathrm{i}\alpha\cdot\pi} = 0.$$

Since $\alpha \in (0,1)$, which indicates that $\phi \neq 0, \pi$, it is easy to see that $a_1^{\pm 1} = 0$. Using the same argument, by induction, we assume that

$$a_{n-1}^m = 0, \quad m = \pm 1, \pm 2, \ldots, \pm(n-1). \tag{2.10.30}$$

Then by considering the coefficients of r^n in (2.10.27) and (2.10.28), we have

$$\begin{aligned} & -\sum_{m=-n}^{n} \mathrm{i}^{n+1} m a_n^m \frac{\sqrt{\lambda}^n}{(2n+1)!!} \sqrt{\frac{2n+1}{4\pi}} \sqrt{\frac{(n-|m|)!}{(n+|m|)!}} P_n^{|m|}(\cos\theta) \\ & + \eta_1 \sin\theta \sum_{m=-(n-1)}^{n-1} \mathrm{i}^{n-1} a_{n-1}^m \frac{\sqrt{\lambda}^{n-1}}{(2n-1)!!} \sqrt{\frac{2n-1}{4\pi}} \sqrt{\frac{(n-1-|m|)!}{(n-1+|m|)!}} P_{n-1}^{|m|}(\cos\theta) = 0, \end{aligned} \tag{2.10.31}$$

and

$$\begin{aligned} & \sum_{m=-n}^{n} \mathrm{i}^{n+1} m a_n^m \frac{\sqrt{\lambda}^n}{(2n+1)!!} \sqrt{\frac{2n+1}{4\pi}} \sqrt{\frac{(n-|m|)!}{(n+|m|)!}} P_n^{|m|}(\cos\theta) e^{\mathrm{i}m\alpha\cdot\pi} \\ & + \eta_2 \sin\theta \sum_{m=-(n-1)}^{n-1} \mathrm{i}^{n-1} a_{n-1}^m \frac{\sqrt{\lambda}^{n-1}}{(2n-1)!!} \sqrt{\frac{2n-1}{4\pi}} \sqrt{\frac{(n-1-|m|)!}{(n-1+|m|)!}} P_{n-1}^{|m|}(\cos\theta) e^{\mathrm{i}m\alpha\cdot\pi} = 0. \end{aligned} \tag{2.10.32}$$

By induction, substituting (2.10.29) and (2.10.30) into (2.10.31) and (2.10.32), using (2.10.4) we can deduce that for $m \in \mathbb{N}_+$,

$$\begin{cases} a_n^m - a_n^{-m} = 0, \\ a_n^m e^{\mathrm{i}m\alpha\cdot\pi} - a_n^{-m} e^{-\mathrm{i}m\alpha\cdot\pi} = 0. \end{cases} \tag{2.10.33}$$

Hence, if $\alpha \neq \frac{k}{m}$, $k = 1, 2, \ldots, m-1$, by virtue of Remark 2.10.1, we can deduce that $a_n^m = 0$, $m = \pm 1, \pm 2, \ldots, \pm n$, which completes the proof of Theorem 2.10.3. □

Remark 2.10.2. It is important and necessary to assume $u \equiv 0$ on $B_\varepsilon(\mathbf{0}) \cap \boldsymbol{l}$ in Theorem 2.10.3. Otherwise, we cannot derive the recursive equations with respect to a_n^m from (2.10.27) and (2.10.28) to ensure the desired vanishing results.

Remark 2.10.3. It is straightforward to verify in the proof of Theorem 2.10.3 that η_1 and/or η_2 can be taken to be zero. That is, Theorem 2.10.3 also includes the cases that at least one of the two planes Π_ℓ is a singular plane.

2.11 Vanishing Orders at Vertex Corners

In this section, we study the vanishing property of the Laplacian eigenfunction to (2.1.1) at a vertex corner $\mathscr{V}(\{\Pi_\ell\}_{\ell=1}^n, \mathbf{x}_0) \Subset \Omega$, where Π_ℓ could be either a nodal plane, a singular plane or a generalized singular plane. It is known that an edge corner $\mathscr{E}(\Pi_1, \Pi_2, \boldsymbol{l})$ can be regarded as part of a vertex corner $\mathscr{V}(\{\Pi_\ell\}_{\ell=1}^n, \mathbf{x}_0)$. In Sect. 2.10, we have unveiled that the vanishing order of the eigenfunction u at an edge corner can be determined by the intersecting dihedral angle of $\mathscr{E}(\Pi_1,\Pi_2,\boldsymbol{l})$ under a generic condition (cf. (2.10.26)). In this section, we concentrate on the following condition

$$u(\mathbf{x}_0)=0, \tag{2.11.1}$$

to study the vanishing property of u at $\mathbf{x}_0$. We should point out that (2.11.1) is much more relaxed compared with (2.10.26), and it can be easily fulfilled in certain generic case, e.g. superpositions of two eigenfunctions at the point $\mathbf{x}_0$. In particular, such a condition like (2.11.1) can be used to show the unique determination of some polyhedral obstacles in $\mathbb{R}^3$ by finitely many measurements in the inverse obstacle scattering problem; see more details in Sect. 4.1.1.

Similar to Sect. 2.10, without loss of generality, we assume that the vertex corner $\mathbf{x}_0$ of $\mathscr{V}(\{\Pi_\ell\}_{\ell=1}^n, \mathbf{x}_0)$ coincides with the origin. We first focus on the case that $n=3$, which implies that the vertex corner $\mathscr{V}(\{\Pi_\ell\}_{\ell=1}^3, \mathbf{x}_0)$ is formed by three planes; see Fig. 2.5 for a schematic illustration. For $n>3$, the related results can be derived in a similar way; see Theorems 2.11.3–2.11.4. It is obvious that $\mathscr{V}(\{\Pi_\ell\}_{\ell=1}^3, \mathbf{x}_0)$ is formed by three edge corners $\mathscr{E}(\Pi_1,\Pi_2,\boldsymbol{l}_1)$, $\mathscr{E}(\Pi_2,\Pi_3,\boldsymbol{l}_2)$ and $\mathscr{E}(\Pi_3,\Pi_1,\boldsymbol{l}_3)$ where $\boldsymbol{l}_1$, $\boldsymbol{l}_2$ and $\boldsymbol{l}_3$ are three line segments of $\Pi_1\cap\Pi_2$, $\Pi_2\cap\Pi_3$ and $\Pi_3\cap\Pi_1$, respectively. Hence, if either of the three planes Π_ℓ is nodal, say Π_3, then one can apply the results in Sect. 2.10 to the edge corners $\mathscr{E}(\Pi_2,\Pi_3,\boldsymbol{l}_2)$ and $\mathscr{E}(\Pi_3,\Pi_1,\boldsymbol{l}_3)$ to derive a certain vanishing order at the vertex corner, by regarding it as an edge corner associated with $\mathscr{E}(\Pi_2,\Pi_3,\boldsymbol{l}_2)$ and $\mathscr{E}(\Pi_3,\Pi_1,\boldsymbol{l}_3)$, respectively. Hence, we shall mainly focus on the vanishing order generated through the intersection of the two planes Π_1 and Π_2, both of which are assumed not to be nodal.

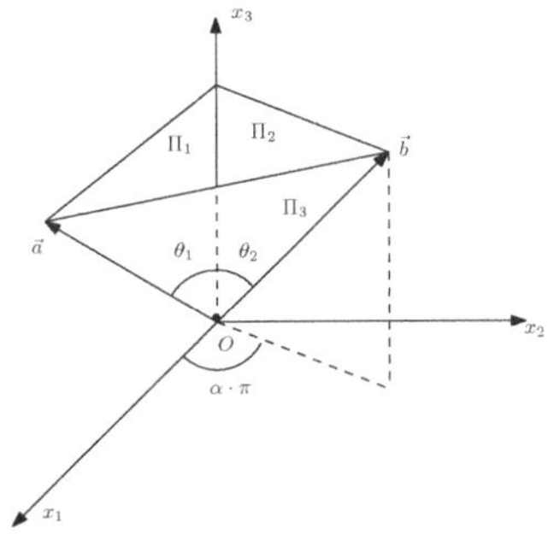

Fig. 2.5 Schematic illustration of a vertex corner that is intersected by Π_1, Π_2 and Π_3

Theorem 2.11.1. *Let u be a Laplacian eigenfunction to (2.1.1). Consider a vertex corner $\mathscr{V}(\{\Pi_\ell\}_{\ell=1}^3,\mathbf{0})\Subset\Omega$ with $\Pi_\ell\in\mathscr{M}_\Omega^\lambda$, $\ell=1,2$, $\angle(\Pi_1,\Pi_2)=\phi=\alpha\cdot\pi$, $\alpha\in(0,1)$ and $\Pi_3\in\mathscr{N}_\Omega^\lambda$. Assume that $\Pi_3=\mathrm{span}\{\vec{a},\vec{b}\}$, where $\vec{a}=(r,\theta_1,0)\in\Pi_1\cap\Pi_3$ and $\vec{b}=(r,\theta_2,\alpha\cdot\pi)\in\Pi_2\cap\Pi_3$ for $r>0$, $\alpha\in(0,1)$, and fixed θ_1 and θ_2 in the spherical coordinate system. If for an $N\in\mathbb{N}$, $N\geq 3$, it holds that*

$$P_p^0(\cos\theta_i)\neq 0, i=1 \text{ or } 2, \text{ and } \alpha\neq\frac{q}{p}, p=1,2,\ldots,N-1,\ q=1,2,\ldots,p-1, \tag{2.11.2}$$

where P_p^0 is the associated Legendre polynomial; then the vanishing order of u at $\mathbf{0}$ generated by the intersection of the two planes Π_1 and Π_2 is at least order N.

Proof. Since Π_1 and Π_2 are two generalized singular planes, we have

$$\frac{\partial u}{\partial\nu}+\eta_1 u\Big|_{\Pi_1}=0 \quad\text{and}\quad \frac{\partial u}{\partial\nu}+\eta_2 u\Big|_{\Pi_2}=0. \tag{2.11.3}$$

By Proposition 2.10.1 and Lemma 2.10.1, we can write (2.11.3) explicitly as

$$
\begin{aligned}
&-\frac{1}{r\sin\theta}\frac{\partial u}{\partial\phi}+\eta_1 u\Big|_{\phi=0}\\
&=-\frac{1}{r\sin\theta}4\pi\sum_{n=0}^{\infty}\sum_{m=-n}^{n}\mathrm{i}^{n+1}ma_n^m j_n(\sqrt{\lambda}r)\sqrt{\frac{2n+1}{4\pi}}\sqrt{\frac{(n-|m|)!}{(n+|m|)!}}P_n^{|m|}(\cos\theta)\\
&+\eta_1 4\pi\sum_{n=0}^{\infty}\sum_{m=-n}^{n}\mathrm{i}^n a_n^m j_n(\sqrt{\lambda}r)\sqrt{\frac{2n+1}{4\pi}}\sqrt{\frac{(n-|m|)!}{(n+|m|)!}}P_n^{|m|}(\cos\theta)=0, \qquad (2.11.4)
\end{aligned}
$$

and

$$
\begin{aligned}
&\frac{1}{r\sin\theta}\frac{\partial u}{\partial\phi}+\eta_2 u\Big|_{\phi=\alpha\cdot\pi}\\
&=\frac{1}{r\sin\theta}4\pi\sum_{n=0}^{\infty}\sum_{m=-n}^{n}\mathrm{i}^{n+1}ma_n^m j_n(\sqrt{\lambda}r)\sqrt{\frac{2n+1}{4\pi}}\sqrt{\frac{(n-|m|)!}{(n+|m|)!}}P_n^{|m|}(\cos\theta)e^{\mathrm{i}m\alpha\cdot\pi}\\
&+\eta_2 4\pi\sum_{n=0}^{\infty}\sum_{m=-n}^{n}\mathrm{i}^n a_n^m j_n(\sqrt{\lambda}r)\sqrt{\frac{2n+1}{4\pi}}\sqrt{\frac{(n-|m|)!}{(n+|m|)!}}P_n^{|m|}(\cos\theta)e^{\mathrm{i}m\alpha\cdot\pi}=0. \qquad (2.11.5)
\end{aligned}
$$

Since $\Pi_3=\mathrm{span}\{\vec{a},\vec{b}\}$, where $\vec{a}=(r,\theta_1,0)\in\Pi_1\cap\Pi_3$ and $\vec{b}=(r,\theta_2,\alpha\cdot\pi)\in\Pi_2\cap\Pi_3$ for fixed θ_1, θ_2 and $u|_{\Pi_3}\equiv 0$. It is direct to see $u|_{\vec{a}}=u|_{\vec{b}}=0$, which further indicates that

$$
u|_{\vec{a}}=4\pi\sum_{n=0}^{\infty}\sum_{m=-n}^{n}\mathrm{i}^n a_n^m j_n(\sqrt{\lambda}r)\sqrt{\frac{2n+1}{4\pi}}\sqrt{\frac{(n-|m|)!}{(n+|m|)!}}P_n^{|m|}(\cos\theta_1)=0, \qquad (2.11.6)
$$

and

$$
u|_{\vec{b}}=4\pi\sum_{n=0}^{\infty}\sum_{m=-n}^{n}\mathrm{i}^n a_n^m j_n(\sqrt{\lambda}r)\sqrt{\frac{2n+1}{4\pi}}\sqrt{\frac{(n-|m|)!}{(n+|m|)!}}P_n^{|m|}(\cos\theta_2)e^{\mathrm{i}m\alpha\cdot\pi}=0. \qquad (2.11.7)
$$

Combining with (2.11.4) and (2.11.5), it suffices to use (2.11.6) or (2.11.7) to study the coefficients of r^n, $n\in\mathbb{N}$. In what follows, without loss of generality, we discuss only (2.11.6). Since $u|_{\vec{a}}\equiv 0$, the coefficient of r^0 fulfils that

$$
4\pi a_0^0\sqrt{\frac{1}{4\pi}}P_0^0(\cos\theta_1)=0,
$$

where we can know that $a_0^0=0$ since $P_0^0\equiv 1$. Considering the coefficients of r, from (2.11.4), (2.11.5) and (2.11.6), we can, respectively, see that

$$
\sum_{m=-1}^{1}ma_1^m\frac{\sqrt{\lambda}}{3!!}\sqrt{\frac{3}{4\pi}}\sqrt{\frac{(1-|m|)!}{(1+|m|)!}}P_1^{|m|}(\cos\theta)+\eta_1\sin\theta a_0^0\sqrt{\frac{1}{4\pi}}P_0^0(\cos\theta)=0, \qquad (2.11.8)
$$

$$
\sum_{m=-1}^{1}ma_1^m\frac{\sqrt{\lambda}}{3!!}\sqrt{\frac{3}{4\pi}}\sqrt{\frac{(1-|m|)!}{(1+|m|)!}}P_1^{|m|}(\cos\theta)e^{\mathrm{i}m\alpha\cdot\pi}-\eta_2\sin\theta\sum_{m=-1}^{1}a_0^0\sqrt{\frac{1}{4\pi}}P_0^0(\cos\theta)e^{\mathrm{i}m\alpha\cdot\pi}=0, \qquad (2.11.9)
$$

$$
\sum_{m=-1}^{1}\mathrm{i}a_1^m\frac{\sqrt{\lambda}}{3!!}\sqrt{\frac{3}{4\pi}}\sqrt{\frac{(1-|m|)!}{(1+|m|)!}}P_1^{|m|}(\cos\theta_1)=0. \qquad (2.11.10)
$$

Substituting $a_0^0=0$ into (2.11.8) and (2.11.9), combining with (2.10.4), we can directly derive the following linear system with respect to $a_1^{\pm 1}$:

$$a_1^1 - a_1^{-1} = 0, \ a_1^1 e^{\mathrm{i}\alpha\cdot\pi} - a_1^{-1} e^{-\mathrm{i}\alpha\cdot\pi} = 0. \tag{2.11.11}$$

Thus, we know that $a_1^{\pm 1} = 0$ since $\alpha \in (0,1)$. As a consequence, if $P_1^0(\cos\theta_1) \neq 0$ in (2.11.10), we can deduce that $a_1^0 = 0$ easily.

By induction, we assume that $a_{n-1}^m = 0$ for $m = 0, \pm 1, \pm 2, \ldots, \pm(n-1)$. Then considering the coefficients of r^n, by (2.11.4), (2.11.5) and (2.11.6), we have

$$\begin{aligned} &-\sum_{m=-n}^{n} \mathrm{i}^{n+1} m a_n^m \frac{\sqrt{\lambda}^n}{(2n+1)!!}\sqrt{\frac{2n+1}{4\pi}}\sqrt{\frac{(n-|m|)!}{(n+|m|)!}}P_n^{|m|}(\cos\theta) \\ &+\eta_1 \sin\theta \sum_{m=-(n-1)}^{n-1} \mathrm{i}^{n-1} a_{n-1}^m \frac{\sqrt{\lambda}^{n-1}}{(2n-1)!!}\sqrt{\frac{2n-1}{4\pi}}\sqrt{\frac{(n-1-|m|)!}{(n-1+|m|)!}}P_{n-1}^{|m|}(\cos\theta) = 0, \end{aligned} \tag{2.11.12}$$

$$\begin{aligned} &\sum_{m=-n}^{n} \mathrm{i}^{n+1} m a_n^m \frac{\sqrt{\lambda}^n}{(2n+1)!!}\sqrt{\frac{2n+1}{4\pi}}\sqrt{\frac{(n-|m|)!}{(n+|m|)!}}P_n^{|m|}(\cos\theta) e^{\mathrm{i}m\alpha\cdot\pi} \\ &+\eta_2 \sin\theta \sum_{m=-(n-1)}^{n-1} \mathrm{i}^{n-1} a_{n-1}^m \frac{\sqrt{\lambda}^{n-1}}{(2n-1)!!}\sqrt{\frac{2n-1}{4\pi}}\sqrt{\frac{(n-1-|m|)!}{(n-1+|m|)!}}P_{n-1}^{|m|}(\cos\theta) e^{\mathrm{i}m\alpha\cdot\pi} = 0, \end{aligned} \tag{2.11.13}$$

and

$$\sum_{m=-n}^{n} \mathrm{i}^{n} a_n^m \frac{\sqrt{\lambda}^n}{(2n+1)!!}\sqrt{\frac{2n+1}{4\pi}}\sqrt{\frac{(n-|m|)!}{(n+|m|)!}}P_n^{|m|}(\cos\theta_1) = 0. \tag{2.11.14}$$

Utilizing the assumption $a_{n-1}^m = 0$ for $n = 0, \pm 1, \pm 2, \ldots, \pm(n-1)$ in (2.11.12) and (2.11.13), we know from the orthogonality condition (2.10.4) that for $m \in \mathbb{N}_+$, a_n^m satisfies

$$a_n^m - a_n^{-m} = 0, \ a_n^m e^{\mathrm{i}m\alpha\cdot\pi} - a_n^{-m} e^{-\mathrm{i}m\alpha\cdot\pi} = 0. \tag{2.11.15}$$

Therefore, if $\alpha \neq \frac{k}{m}$, $k = 1, 2, \ldots, m-1$, and by virtue of Remark 2.10.1 we can derive that $a_n^m = 0$ for $m = \pm 1, \pm 2, \ldots, \pm n$. Now we are in a position to show that $a_n^0 = 0$. Indeed, substituting $a_n^m = 0$, $m = \pm 1, \pm 2, \ldots, \pm n$ into (2.11.14), we can obtain that if $P_n^0(\cos\theta_1) \neq 0$, then $a_n^0 = 0$, which completes the proof of Theorem 2.11.1. □

In the above proof of Theorem 2.11.1, we have analysed only the condition $u|_{\vec{a}} = 0$ for illustration. For the condition $u|_{\vec{b}} = 0$, we give the discussion in the following remark.

Remark 2.11.1. In the proof of Theorem 2.11.1, if we use (2.11.7) instead of (2.11.6), combining with (2.11.4) and (2.11.5), to consider the coefficients of r^n, $n \in \mathbb{N}$, then (2.11.10) becomes

$$\sum_{m=-1}^{1} \mathrm{i} a_1^m \frac{\sqrt{\lambda}}{3!!}\sqrt{\frac{3}{4\pi}}\sqrt{\frac{(1-|m|)!}{(1+|m|)!}}P_1^{|m|}(\cos\theta_2) e^{\mathrm{i}m\alpha\cdot\pi} = 0. \tag{2.11.16}$$

Since we know $a_1^{\pm 1} = 0$ by (2.11.8) and (2.11.9), we can obtain from (2.11.16) that if $P_1^0(\cos\theta_2) \neq 0$, then $a_1^0 = 0$. By induction, in order to study a_n^0, we replace (2.11.14) by

$$\sum_{m=-n}^{n} \mathrm{i}^{n} a_n^m \frac{\sqrt{\lambda}^n}{(2n+1)!!}\sqrt{\frac{2n+1}{4\pi}}\sqrt{\frac{(n-|m|)!}{(n+|m|)!}}P_n^{|m|}(\cos\theta_2) e^{\mathrm{i}m\alpha\cdot\pi} = 0. \tag{2.11.17}$$

Substituting $a_n^m = 0$, $m = \pm 1, \pm 2, \ldots, \pm n$, which is derived from (2.11.15) to (2.11.17), we can deduce that if $P_n^0(\cos\theta_2) \neq 0$, then $a_n^0 = 0$.

Hence, from the above discussions we know that it is actually equivalent to consider $u|_{\vec{a}} = 0$ or $u|_{\vec{b}} = 0$ in the proof of Theorem 2.11.1. Therefore, in our subsequent study, we shall only prove under the condition with respect to $\vec{a}$.

In Theorem 2.11.1, we have considered the case that $\Pi_3 \in \mathscr{N}_\Omega^\lambda$ is a nodal plane. Next, we study a more complicated case that $\Pi_3 \in \mathscr{M}_\Omega^\lambda$ is a generalized singular plane.

Theorem 2.11.2. *Let u be a Laplacian eigenfunction to (2.1.1). Consider a vertex corner $\mathscr{V}(\{\Pi_\ell\}_{\ell=1}^3, \mathbf{0}) \Subset \Omega$ with $\Pi_\ell \in \mathscr{M}_\Omega^\lambda$, $\ell = 1,2,3$ and $\angle(\Pi_1, \Pi_2) = \phi = \alpha \cdot \pi$, $\alpha \in (0,1)$. Assume that $\Pi_3 = \operatorname{span}\{\vec{a}, \vec{b}\}$, where $\vec{a} = (r, \theta_1, 0) \in \Pi_1 \cap \Pi_3$ and $\vec{b} = (r, \theta_2, \alpha \cdot \pi) \in \Pi_2 \cap \Pi_3$ for $r > 0$, $\alpha \in (0,1)$, and fixed $\theta_1 \in (0,\pi)$ and $\theta_2 \in (0,\pi)$ in the spherical coordinate system. If for an $N \in \mathbb{N}$, $N \geq 3$, it holds that*

$$u(\mathbf{0}) = 0,\ P_p^1(\cos\theta_i) \neq 0, i = 1 \text{ or } 2, \text{ and } \alpha \neq \frac{q}{p}, p = 1,2,\ldots,N-1,\ q = 1,2,\ldots,p-1, \tag{2.11.18}$$

where P_p^1 is the associated Legendre polynomial, then the vanishing order of u at $\mathbf{0}$ generated by the intersection of the two planes Π_1 and Π_2 is at least order N.

Proof. Since Π_i, $i = 1,2,3$ are three generalized singular planes, we have

$$\frac{\partial u}{\partial \nu} + \eta_1 u\Big|_{\Pi_1} = 0, \quad \frac{\partial u}{\partial \nu} + \eta_2 u\Big|_{\Pi_2} = 0 \quad \text{and} \quad \frac{\partial u}{\partial \nu} + \eta_3 u\Big|_{\Pi_3} = 0.$$

From Theorem 2.11.1, we have already known that u satisfies (2.11.4) and (2.11.5) on Π_1 and Π_2, respectively. Besides, by Remark 2.11.1, we can obtain that

$$\frac{\partial u}{\partial \nu} + \eta_3 u\Big|_{\vec{a}} = 0. \tag{2.11.19}$$

Since $\Pi_3 = \operatorname{span}\{\vec{a}, \vec{b}\}$, which implies that $\nu = \vec{b} \times \vec{a} = (\sin\theta_2 \sin(\alpha\cdot\pi)\cos\theta_1, \sin\theta_1\cos\theta_2 - \sin\theta_2\cos(\alpha\cdot\pi)\cos\theta_1, -\sin\theta_1\sin\theta_2\sin(\alpha\cdot\pi))^{\mathrm{T}}$, we know that (2.11.19) can be written as

$$\begin{aligned}\frac{\partial u}{\partial \nu} + \eta_3 u\Big|_{\vec{a}} &= \frac{1}{r}\frac{\partial u}{\partial \theta}\sin\theta_2\sin(\alpha\cdot\pi) + \frac{1}{r\sin\theta_1}\frac{\partial u}{\partial \phi} \\ &\cdot (\sin\theta_1\cos\theta_2 - \sin\theta_2\cos\theta_1\cos\alpha\cdot\pi) + \eta_3 u\Big|_{\theta=\theta_1,\phi=0} = 0. \end{aligned} \tag{2.11.20}$$

By Lemma 2.10.1, multiplying $r\sin\theta_1$ on both sides of (2.11.20), the equation can be simplified to

$$\begin{aligned} &\sin\theta_1\sin\theta_2\sin(\alpha\cdot\pi)\sum_{n=0}^{\infty}\sum_{m=-n}^{n} \mathrm{i}^n a_n^m j_n(\sqrt{\lambda}r)\sqrt{\frac{2n+1}{4\pi}}\sqrt{\frac{(n-|m|)!}{(n+|m|)!}}\frac{dP_n^{|m|}(\cos\theta)}{d\theta}\Big|_{\theta=\theta_1} \\ &+ (\sin\theta_1\cos\theta_2 - \sin\theta_2\cos\theta_1\cos\alpha\cdot\pi)\sum_{n=0}^{\infty}\sum_{m=-n}^{n} \mathrm{i}^{n+1} m a_n^m j_n(\sqrt{\lambda}r)\sqrt{\frac{2n+1}{4\pi}}\sqrt{\frac{(n-|m|)!}{(n+|m|)!}} \\ &\cdot P_n^{|m|}(\cos\theta_1) + \eta_3\sin\theta_1 r\sum_{n=0}^{\infty}\sum_{m=-n}^{n} \mathrm{i}^n a_n^m j_n(\sqrt{\lambda}r)\sqrt{\frac{2n+1}{4\pi}}\sqrt{\frac{(n-|m|)!}{(n+|m|)!}}P_n^{|m|}(\cos\theta_1) = 0. \end{aligned} \tag{2.11.21}$$

Since $u(\mathbf{0}) = 0$, we know that $a_0^0 = 0$. Combining (2.11.4), (2.11.5) with (2.11.21), the corresponding coefficients of r respectively fulfil that

$$\sum_{m=-1}^{1} m a_1^m \frac{\sqrt{\lambda}}{3!!}\sqrt{\frac{3}{4\pi}}\sqrt{\frac{(1-|m|)!}{(1+|m|)!}}P_1^{|m|}(\cos\theta) + \eta_1\sin\theta a_0^0\sqrt{\frac{1}{4\pi}}P_0^0(\cos\theta) = 0, \tag{2.11.22}$$

$$\sum_{m=-1}^{1} ma_1^m \frac{\sqrt{\lambda}}{3!!}\sqrt{\frac{3}{4\pi}}\sqrt{\frac{(1-|m|)!}{(1+|m|)!}}P_1^{|m|}(\cos\theta)e^{im\alpha\cdot\pi} - \eta_2 \sin\theta \sum_{m=-1}^{1} a_0^0\sqrt{\frac{1}{4\pi}}P_0^0(\cos\theta)e^{im\alpha\cdot\pi} = 0, \tag{2.11.23}$$

and

$$\begin{aligned}&\sin\theta_1\sin\theta_2\sin(\alpha\cdot\pi)\sum_{m=-1}^{1} ia_1^m\frac{\sqrt{\lambda}}{3!!}\sqrt{\frac{3}{4\pi}}\sqrt{\frac{(1-|m|)!}{(1+|m|)!}}\frac{dP_1^{|m|}(\cos\theta)}{d\theta}\Big|_{\theta=\theta_1}\\&-(\sin\theta_1\cos\theta_2-\sin\theta_2\cos\theta_1\cos(\alpha\cdot\pi))\sum_{m=-1}^{1} ma_1^m\frac{\sqrt{\lambda}}{3!!}\sqrt{\frac{3}{4\pi}}\sqrt{\frac{(1-|m|)!}{(1+|m|)!}}P_1^{|m|}(\cos\theta_1)\\&+\eta_3\sin\theta_1 a_0^0\sqrt{\frac{1}{4\pi}}P_0^0(\cos\theta_1)=0.\end{aligned} \tag{2.11.24}$$

Substituting $a_0^0 = 0$ into (2.11.22) and (2.11.23), utilizing the orthogonality condition we can derive

$$a_1^1 - a_1^{-1} = 0, \ \ a_1^1 e^{i\alpha\cdot\pi} - a_1^{-1}e^{-i\alpha\cdot\pi} = 0, \tag{2.11.25}$$

which yields $a_1^{\pm 1} = 0$ from the fact that $\alpha \in (0,1)$. In addition, taking $a_0^0 = a_1^{\pm 1} = 0$ in (2.11.24), we have

$$\sin\theta_1\sin\theta_2\sin(\alpha\cdot\pi)ia_1^0\frac{\sqrt{\lambda}}{3!!}\sqrt{\frac{3}{4\pi}}(-P_1^1(\cos\theta_1)) = 0. \tag{2.11.26}$$

Hence, by the assumptions on θ_1, θ_2 and α, we can obtain that $a_1^0 = 0$ if $P_1^1(\cos\theta_1) \neq 0$.

Proving by induction, we assume that $a_{n-1}^m = 0$ for $m = 0, \pm 1, \pm 2, \ldots \pm (n-1)$. Then considering the coefficients of r^n in (2.11.4), (2.11.5) and (2.11.21); accordingly, we know that there holds (2.11.12), (2.11.13) and also

$$\begin{aligned}&\sin\theta_1\sin\theta_2\sin(\alpha\cdot\pi)\sum_{m=-n}^{n} i^n a_n^m\frac{\sqrt{\lambda}^n}{(2n+1)!!}\sqrt{\frac{2n+1}{4\pi}}\sqrt{\frac{(n-|m|)!}{(n+|m|)!}}\frac{dP_n^{|m|}(\cos\theta)}{d\theta}\Big|_{\theta=\theta_1}\\&+(\sin\theta_1\cos\theta_2-\sin\theta_2\cos\theta_1\cos\alpha\cdot\pi)\sum_{m=-n}^{n} i^{n+1}ma_n^m\frac{\sqrt{\lambda}^n}{(2n+1)!!}\sqrt{\frac{2n+1}{4\pi}}\sqrt{\frac{(n-|m|)!}{(n+|m|)!}}\\&\cdot P_n^{|m|}(\cos\theta_1)+\eta_3\sin\theta_1\sum_{m=-(n-1)}^{n-1} i^{n-1}a_{n-1}^m\frac{\sqrt{\lambda}^{n-1}}{(2n-1)!!}\sqrt{\frac{2n-1}{4\pi}}\sqrt{\frac{(n-1-|m|)!}{(n-1+|m|)!}}P_{n-1}^{|m|}(\cos\theta_1)=0.\end{aligned}$$

Using the assumption that $a_{n-1}^m = 0$, $m = 0, \pm 1, \pm 2, \ldots \pm (n-1)$ in (2.11.12) and (2.11.13), similar to Theorem 2.11.3, we can obtain that if $\alpha \neq \frac{k}{m}$, $k = 1, 2, \ldots, m-1$, then $a_n^m = 0$ for $m = \pm 1, \pm 2, \ldots \pm n$. Therefore, we can deduce from the last relation above that

$$\sin\theta_1\sin\theta_2\sin(\alpha\cdot\pi)i^n a_n^0\frac{\sqrt{\lambda}^n}{(2n+1)!!}\sqrt{\frac{2n+1}{4\pi}}(-P_n^1(\cos\theta_1)) = 0, \tag{2.11.27}$$

which indicates that $a_n^0 = 0$ if $P_n^1(\cos\theta_1) \neq 0$, hence completing the proof of Theorem 2.11.2. □

Remark 2.11.2. Following a similar argument in Theorem 2.11.2, if we take into account the condition $\frac{\partial u}{\partial \nu} + \eta_3 u\Big|_{\vec{b}} \equiv 0$ on Π_3, then we can derive similar results with respect to θ_2 instead of θ_1.

Remark 2.11.3. By direct verifications in the proof of Theorem 2.11.2, one can show that either of the boundary parameters, η_ℓ or $\ell = 1, 2, 3$, can be taken to be zero. That means, the generalized singular planes in Theorem 2.11.1 can be replaced by singular planes, and the vanishing results still hold.

In Theorems 2.11.1 and 2.11.2, we have considered the vanishing properties at a vertex corner that is intersected by three planes ($n = 3$). In fact, the similar arguments work for the case that $n > 3$, in which the third plane no longer intersects with Π_1 or Π_2. Without loss of generality, we denote the third plane to be discussed by $\Pi_j = \text{span}\{\overrightarrow{OA_j}, \overrightarrow{OA_{j+1}}\}$, where $3 \le j \le n$, and if $j = n$, we assume that $A_{n+1} := A_1$. Let Π_1 coincide with the (x_1,x_3)-plane, Π_2 possesses a dihedral angle $\alpha \cdot \pi$ away from Π_1 in the anticlockwise direction, and $\overrightarrow{OA_2}$ lies on the x_3-axis; see Fig. 2.6 for a schematic illustration.

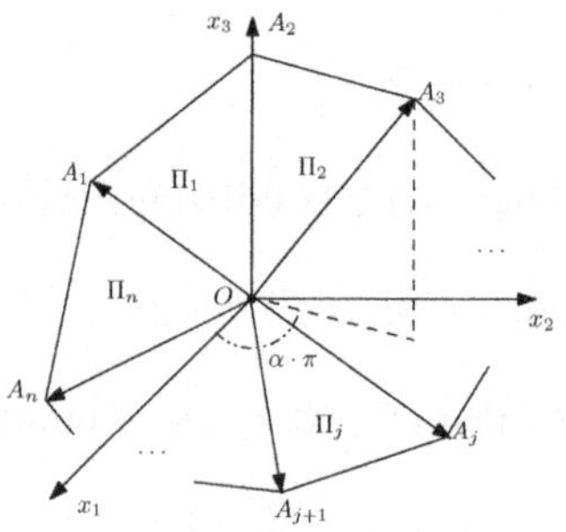

Fig. 2.6 Schematic illustration of a vertex corner that is intersected by $\Pi_1, \Pi_2, \ldots, \Pi_n$ with $n > 3$

Theorem 2.11.3. *Let u be a Laplacian eigenfunction to (2.1.1). Consider a vertex corner $\mathscr{V}(\{\Pi_\ell\}_{\ell=1}^n, \mathbf{0}) \Subset \Omega$ as described above with $\Pi_\ell \in \mathscr{M}_\Omega^\lambda$, $\ell = 1,2$, $\angle(\Pi_1,\Pi_2) = \phi = \alpha\cdot\pi$, $\alpha \in (0,1)$ and $\Pi_j \in \mathscr{N}_\Omega^\lambda$. Assume that $\Pi_j = \text{span}\{\overrightarrow{OA_j}, \overrightarrow{OA_{j+1}}\}$, where $\overrightarrow{OA_j} = (r,\theta_j,\phi_j)$ and $\overrightarrow{OA_{j+1}} = (r,\theta_{j+1},\phi_{j+1})$ for $r > 0$, $\theta_j,\theta_{j+1} \in (0,\pi)$ and $\phi_j,\phi_{j+1} \in (0,2\pi)$ such that $0 < \phi_{j+1} - \phi_j < \pi$ in the spherical coordinate system. If for an $N \in \mathbb{N}$, $N \ge 3$, it holds that*

$$P_p^0(\cos\theta_\tau) \neq 0, \tau = j \text{ or } j+1, \text{ and } \alpha \neq \frac{q}{p}, p = 1,2,\ldots,N-1,\ q = 1,2,\ldots,p-1, \tag{2.11.28}$$

where P_p^0 is the associated Legendre polynomial, then the vanishing order of u at $\mathbf{0}$ generated by the intersection of the two planes Π_1 and Π_2 is at least order N.

Proof. Since Π_1 and Π_2 are two generalized singular planes, we can derive (2.11.4) and (2.11.5) immediately. Considering Π_j, we know that $u|_{\Pi_j} = 0$, which indicates that $u|_{\overrightarrow{OA_j}} \equiv 0$ and $u|_{\overrightarrow{OA_{j+1}}} \equiv 0$. By Remark 2.11.1, it suffices to analyse $u|_{\overrightarrow{OA_j}} \equiv 0$ as follows:

$$u|_{\overrightarrow{OA_j}} = 4\pi \sum_{n=0}^{\infty}\sum_{m=-n}^{n} \mathrm{i}^n a_n^m j_n(\sqrt{\lambda}r)\sqrt{\frac{2n+1}{4\pi}}\sqrt{\frac{(n-|m|)!}{(n+|m|)!}}P_n^{|m|}(\cos\theta_j)e^{\mathrm{i}m\phi_j} = 0. \tag{2.11.29}$$

Taking $m = n = 0$ in (2.11.29), we have $4\pi a_0^0\sqrt{\frac{1}{4\pi}}P_0^0(\cos\theta_j) = 0$, where we can derive $a_0^0 = 0$ since $P_0^0 \equiv 1$. Thus, from (2.11.4), (2.11.5) and (2.11.29), we know that the coefficients of r satisfies (2.11.11) and thus $a_1^{\pm 1} = 0$. Moreover, we have

$$\sum_{m=-1}^{1} \mathrm{i}a_1^m \frac{\sqrt{\lambda}}{3!!}\sqrt{\frac{3}{4\pi}}\sqrt{\frac{(1-|m|)!}{(1+|m|)!}}P_1^{|m|}(\cos\theta_j)e^{\mathrm{i}m\phi_j} = 0, \tag{2.11.30}$$

which can be further simplified as $a_1^0 P_1^0(\cos\theta_j) = 0$ after substituting $a_1^{\pm 1} = 0$ into (2.11.30). Hence, it is easy to see that $a_1^0 = 0$ if $P_1^0(\cos\theta_j) \neq 0$.

By induction, we assume that $a_{n-1}^m = 0$ for $m = 0, \pm 1, \pm 2, \ldots, \pm(n-1)$. Considering the coefficients of r^n, we can obtain (2.11.12) and (2.11.13) which induce (2.11.15) as well as the equation

$$\sum_{m=-n}^{n} \mathrm{i}^n a_n^m \frac{\sqrt{\lambda}^n}{(2n+1)!!} \sqrt{\frac{2n+1}{4\pi}} \sqrt{\frac{(n-|m|)!}{(n+|m|)!}} P_n^{|m|}(\cos\theta_j) e^{\mathrm{i}m\phi_j} = 0. \tag{2.11.31}$$

Since we have already known that if $\alpha \neq \frac{k}{m}$, $k = 1, 2, \ldots, m-1$, then $a_n^m = 0$ for $m = \pm 1, \pm 2, \ldots, \pm n$ from (2.11.15). Substituting this result into (2.11.31), we can deduce $a_n^0 P_n^0(\cos\theta_j) = 0$. Therefore, we know that $a_n^0 = 0$ if $P_n^0(\cos\theta_j) \neq 0$. Similarly, if we utilize the condition $u|_{\overrightarrow{OA_{j+1}}} \equiv 0$, then the same argument and results work for θ_{j+1}, which completes our proof. □

We proceed to consider the case that Π_j is a generalized singular plane instead of a nodal plane as in Theorem 2.11.3.

Theorem 2.11.4. *Let u be a Laplacian eigenfunction to (2.1.1). Consider a vertex corner $\mathscr{V}(\{\Pi_\ell\}_{\ell=1}^n, \mathbf{0}) \Subset \Omega$ with $\Pi_\ell \in \mathscr{M}_\Omega^\lambda$, $\ell = 1, 2$, $\angle(\Pi_1, \Pi_2) = \phi = \alpha \cdot \pi$, $\alpha \in (0,1)$ and $\Pi_j \in \mathscr{M}_\Omega^\lambda$. Assume that $\Pi_j = \mathrm{span}\{\overrightarrow{OA_j}, \overrightarrow{OA_{j+1}}\}$, where $\overrightarrow{OA_j} = (r, \theta_j, \phi_j)$ and $\overrightarrow{OA_{j+1}} = (r, \theta_{j+1}, \phi_{j+1})$ for $r > 0$, $\theta_j, \theta_{j+1} \in (0, \pi)$ and $\phi_j, \phi_{j+1} \in (0, 2\pi)$ such that $0 < \phi_{j+1} - \phi_j < \pi$ in the spherical coordinate system. If for an $N \in \mathbb{N}$, $N \geq 3$, there holds*

$$u(\mathbf{0}) = 0, \quad P_p^1(\cos\theta_\tau) \neq 0, \tau = j \text{ or } j+1, \text{ and } \alpha \neq \frac{q}{p}, \tag{2.11.32}$$

where $p = 1, 2, \ldots, N-1$, $q = 1, 2, \ldots, p-1$ and P_p^1 is the associated Legendre polynomial, then the vanishing order of u at $\mathbf{0}$ generated by the intersection of the two planes Π_1 and Π_2 is at least order N.

Proof. From Theorem 2.11.2 and the fact that Π_1 and Π_2 are two generalized singular planes, we know u fulfils (2.11.4) and (2.11.5). Now considering Π_j, there holds $\frac{\partial u}{\partial \nu} + \eta_j u = 0$ on Π_j. Since $\Pi_j = \mathrm{span}\{\overrightarrow{OA_j}, \overrightarrow{OA_{j+1}}\}$, we have $\frac{\partial u}{\partial \nu} + \eta_j u\Big|_{\overrightarrow{OA_j}} = 0$ and

$$\nu = \overrightarrow{OA_j} \times \overrightarrow{OA_{j+1}} = \begin{pmatrix} \sin\theta_j \sin\phi_j \cos\theta_{j+1} - \sin\theta_{j+1} \sin\phi_{j+1} \cos\theta_j \\ -\sin\theta_j \cos\phi_j \cos\theta_{j+1} + \sin\theta_{j+1} \cos\phi_{j+1} \cos\theta_j \\ \sin\theta_j \cos\phi_j \sin\theta_{j+1} \sin\phi_{j+1} - \sin\theta_{j+1} \cos\phi_{j+1} \sin\theta_j \sin\phi_j \end{pmatrix}.$$

Combining with Lemma 2.10.1, we can obtain by direct computations that

$$\begin{aligned} \frac{\partial u}{\partial \nu} + \eta_j u\Big|_{\overrightarrow{OA_j}} &= \frac{1}{r\sin\theta_j} \frac{\partial u}{\partial \phi} (\sin\theta_{j+1} \cos\theta_j \cos(\phi_j - \phi_{j+1}) - \sin\theta_j \cos\theta_{j+1}) \\ &+ \frac{1}{r} \frac{\partial u}{\partial \theta} \sin\theta_{j+1} \sin(\phi_j - \phi_{j+1}) + \eta_j u\Big|_{\theta = \theta_j, \phi = \phi_j} = 0. \end{aligned} \tag{2.11.33}$$

Since $\theta_j \in (0, \pi)$, multiplying $r\sin\theta_j$ on both sides of (2.11.33), we can deduce that

$$\begin{aligned} &(\sin\theta_{j+1} \cos\theta_j \cos(\phi_j - \phi_{j+1}) - \sin\theta_j \cos\theta_{j+1}) \sum_{n=0}^{\infty} \sum_{m=-n}^{n} \mathrm{i}^{n+1} m a_n^m j_n(\sqrt{\lambda} r) \sqrt{\frac{2n+1}{4\pi}} \\ &\cdot \sqrt{\frac{(n-|m|)!}{(n+|m|)!}} P_n^{|m|}(\cos\theta_j) e^{\mathrm{i}m\phi_j} + \sin\theta_j \sin\theta_{j+1} \sin(\phi_j - \phi_{j+1}) \sum_{n=0}^{\infty} \sum_{m=-n}^{n} \mathrm{i}^n a_n^m j_n(\sqrt{\lambda} r) \\ &\cdot \sqrt{\frac{2n+1}{4\pi}} \sqrt{\frac{(n-|m|)!}{(n+|m|)!}} \frac{dP_n^{|m|}(\cos\theta)}{d\theta}\Big|_{\theta=\theta_j} e^{\mathrm{i}m\phi_j} + \eta_j \sin\theta_j r \sum_{n=0}^{\infty} \sum_{m=-n}^{n} \mathrm{i}^n a_n^m j_n(\sqrt{\lambda} r) \end{aligned}$$

$$\cdot \sqrt{\frac{2n+1}{4\pi}} \sqrt{\frac{(n-|m|)!}{(n+|m|)!}} P_n^{|m|}(\cos\theta_j) e^{\mathrm{i}m\phi_j} = 0. \tag{2.11.34}$$

Since $u(\mathbf{0}) = 0$, we have $a_0^0 = 0$. Considering the coefficients with respect to r in (2.11.4), (2.11.5) and (2.11.34), we know that $a_1^{\pm 1}$ fulfils (2.11.11) which induces that $a_1^{\pm 1} = 0$ since $\alpha \in (0, 1)$. Moreover, it is easy to see from (2.11.34) that

$$\sin\theta_j \sin\theta_{j+1} \sin(\phi_j - \phi_{j+1}) \mathrm{i} a_1^0 \frac{\sqrt{\lambda}}{3!!} \sqrt{\frac{3}{4\pi}} (-P_1^1(\cos\theta_j)) = 0.$$

Since $\theta_j, \theta_{j+1} \in (0, \pi)$ and $0 < \phi_j - \phi_{j+1} < \pi$, we know $a_1^0 = 0$ if $P_1^1(\cos\theta_j) \neq 0$.

Similarly, we assume that $a_{n-1}^m = 0$, $m = 0, \pm 1, \pm 2, \ldots, \pm(n-1)$. Then combining with Theorem 2.11.2, we know that a_n^m satisfies (2.11.12), (2.11.13) and

$$(\sin\theta_{j+1}\cos\theta_j\cos(\phi_j - \phi_{j+1}) - \sin\theta_j\cos\theta_{j+1}) \sum_{m=-n}^{n} \mathrm{i}^{n+1} m a_n^m \frac{\sqrt{\lambda}^n}{(2n+1)!!} \sqrt{\frac{2n+1}{4\pi}}$$

$$\cdot \sqrt{\frac{(n+|m|)!}{(n-|m|)!}} P_n^{|m|}(\cos\theta_j) e^{\mathrm{i}m\phi_j} + \sin\theta_j \sin\theta_{j+1} \sin(\phi_j - \phi_{j+1}) \sum_{m=-n}^{n} \mathrm{i}^n a_n^m \frac{\sqrt{\lambda}^n}{(2n+1)!!}$$

$$\cdot \sqrt{\frac{2n+1}{4\pi}} \sqrt{\frac{(n-|m|)!}{(n+|m|)!}} \frac{dP_n^{|m|}(\cos\theta)}{d\theta}\bigg|_{\theta=\theta_j} e^{\mathrm{i}m\phi_j} + \eta_j \sin\theta_j \sum_{m=-(n-1)}^{n-1} \mathrm{i}^{n-1} a_{n-1}^m \frac{\sqrt{\lambda}^{n-1}}{(2n-1)!!}$$

$$\cdot \sqrt{\frac{2n-1}{4\pi}} \sqrt{\frac{(n-1-|m|)!}{(n-1+|m|)!}} P_{n-1}^{|m|}(\cos\theta_j) e^{\mathrm{i}m\phi_j} = 0. \tag{2.11.35}$$

In (2.11.12) and (2.11.13), utilizing the assumption $a_{n-1}^m = 0$ for $m = 0, \pm 1, \pm 2, \ldots \pm (n-1)$, we know that if $\alpha \neq \frac{k}{m}$, $k = 1, 2, \ldots, m$, then $a_n^m = 0$, $\pm 1, \pm 2, \ldots, \pm n$. Hence, (2.11.35) can be simplified to

$$\sin\theta_j \sin\theta_{j+1} \sin(\phi_j - \phi_{j+1}) \mathrm{i}^n a_n^0 \frac{\sqrt{\lambda}^n}{(2n+1)!!} \sqrt{\frac{2n+1}{4\pi}} (-P_n^1(\cos\theta_j)) = 0.$$

Since $\theta_j, \theta_{j+1} \in (0, \pi)$ and $0 < \phi_j - \phi_{j+1} < \pi$, we can derive that $a_n^0 = 0$ if $P_n^1(\cos\theta_j) \neq 0$. The same results work for θ_{j+1} if we take into account that $\frac{\partial u}{\partial \nu} + \eta_j u\big|_{\overrightarrow{OA_{j+1}}} = 0$. This completes the proof of Theorem 2.11.4. □

Remark 2.11.4. Similarly, to Remark 2.11.3, one can have by direct verifications that the vanishing results in Theorem 2.11.4 still hold if any of the generalized singular planes involved is replaced by a singular plane.

2.12 Irrational Intersections and Infinite Vanishing Orders in $\mathbb{R}^3$

From the results derived in Sects. 2.10 and 2.11, one can identify that the vanishing order of the eigenfunction u at an edge or a vertex corner relies on the degree of the dihedral angle of the underlying corner. In the following two definitions, we first introduce the irrational and rational edge or vertex corner. Then, based on the results in Sects. 2.10 and 2.11, we show that the vanishing order of the eigenfunction at an irrational edge or vertex corner is generically infinity, and hence, it vanishes identically in Ω.

Definition 2.12.1. Let $\mathscr{E}(\Pi_1, \Pi_2, \boldsymbol{l})$ be an edge corner defined in Definition 2.9.2 and the corresponding dihedral angle of Π_1 and Π_2 is denoted by $\phi = \alpha \cdot \pi$, $\alpha \in (0,1)$. If ϕ is an irrational dihedral angle, namely, α is an irrational number, then $\mathscr{E}(\Pi_1, \Pi_2, \boldsymbol{l})$ is said to be an *irrational* edge corner. Otherwise, it is said to be a *rational* edge corner. For a rational edge corner $\mathscr{E}(\Pi_1, \Pi_2, \boldsymbol{l})$, the dihedral angle between Π_1 and Π_2 is called the *rational degree* of $\mathscr{E}(\Pi_1, \Pi_2, \boldsymbol{l})$.

Definition 2.12.2. Let $\mathscr{V}(\{\Pi_\ell\}_{\ell=1}^n, \mathbf{x}_0)$ be a vertex corner defined in Definition 2.9.3, where $n \in \mathbb{N}$ and $n \geq 3$. It is clear that $\mathscr{V}(\{\Pi_\ell\}_{\ell=1}^n, \mathbf{x}_0)$ is composed of the following n edge corners

$$\mathscr{E}_\ell := \mathscr{E}(\Pi_\ell, \Pi_{\ell+1}, \boldsymbol{l}_\ell), \quad \mathscr{E}_n := \mathscr{E}(\Pi_n, \Pi_1, \boldsymbol{l}_n), \quad \Pi_{n+1} := \Pi_1, \quad \ell = 1, 2, \ldots, n-1,$$

where $\boldsymbol{l}_\ell$ is the line segment of $\Pi_\ell \cap \Pi_{\ell+1}$ and $\boldsymbol{l}_n$ is a line segment of $\Pi_n \cap \Pi_1$, respectively. Denote

$$\begin{aligned} I_{\mathsf{IR}} &= \{\ell \in \mathbb{N} \mid 1 \leq \ell \leq n, \quad \mathscr{E}_\ell \text{ is an irrational edge corner}\}, \\ I_{\mathsf{R}} &= \{\ell \in \mathbb{N} \mid 1 \leq \ell \leq n, \quad \mathscr{E}_\ell \text{ is a rational edge corner}\}. \end{aligned} \tag{2.12.1}$$

If $\#I_{\mathsf{IR}} \geq 1$, then $\mathscr{V}(\{\Pi_\ell\}_{\ell=1}^n, \mathbf{x}_0)$ is said to be an *irrational* vertex corner. If $\#I_{\mathsf{IR}} \equiv 0$, then $\mathscr{V}(\{\Pi_\ell\}_{\ell=1}^n, \mathbf{x}_0)$ is said to be a *rational* vertex corner. For a rational vertex corner $\mathscr{V}(\{\Pi_\ell\}_{\ell=1}^n, \mathbf{x}_0)$ composed of edge corners $\mathscr{E}_\ell := \mathscr{E}(\Pi_\ell, \Pi_{\ell+1}, \boldsymbol{l}_\ell)$, the largest degree of $\mathscr{E}_\ell$ $(\ell = 1, \ldots, n)$ is referred to as the *rational degree* of $\mathscr{V}(\{\Pi_\ell\}_{\ell=1}^n, \mathbf{x}_0)$.

When an irrational edge corner $\mathscr{E}(\Pi_1, \Pi_2, \boldsymbol{l})$ is intersected by two nodal planes of u, we can derive the following result from Theorem 2.10.1.

Theorem 2.12.1. *Let u be a Laplacian eigenfunction to* (2.1.1). *Suppose that $\mathscr{E}(\Pi_1, \Pi_2, \boldsymbol{l}) \Subset \Omega$ is an irrational edge corner with $\Pi_1, \Pi_2 \in \mathscr{N}_\Omega^\lambda$. Then it holds that*

$$\mathrm{Vani}(u; \mathbf{0}, \Pi_1, \Pi_2) = +\infty, \quad \mathbf{0} \in \boldsymbol{l}.$$

If the intersecting two planes of the irrational edge corner are either of the three types: a nodal plane, a singular plane or a generalized singular plane, namely, for the general case, we have the irrational intersection results as shown below.

Theorem 2.12.2. *Let u be a Laplacian eigenfunction to* (2.1.1). *Suppose that $\mathscr{E}(\Pi_1, \Pi_2, \boldsymbol{l}) \Subset \Omega$ is an irrational edge corner with $\Pi_1 \in \mathscr{N}_\Omega^\lambda$ and $\Pi_2 \in \mathscr{M}_\Omega^\lambda$. Then it holds that*

$$\mathrm{Vani}(u; \mathbf{0}, \Pi_1, \Pi_2) = +\infty, \quad \mathbf{0} \in \boldsymbol{l}.$$

The same result can be derived for the case that $\eta \equiv 0$, which indicates that Π_2 is a singular plane. The detailed discussion can be found in Theorem 2.10.2.

The next theorem is concerned with the intersection of two generalized singular planes, which is a direct corollary of Theorem 2.10.3.

Theorem 2.12.3. *Let u be a Laplacian eigenfunction to* (2.1.1). *Suppose that $\mathscr{E}(\Pi_1, \Pi_2, \boldsymbol{l}) \Subset \Omega$ is an irrational edge corner with $\Pi_\ell \in \mathscr{M}_\Omega^\lambda$ $(\ell = 1, 2)$. If there exists a sufficiently small $\varepsilon > 0$ such that*

$$u|_{B_\varepsilon(\mathbf{0}) \cap \boldsymbol{l}} \equiv 0, \tag{2.12.2}$$

then it holds that

$$\mathrm{Vani}(u; \mathbf{0}, \Pi_1, \Pi_2) = +\infty, \quad \mathbf{0} \in \boldsymbol{l}.$$

If $\eta_1 = 0$ or $\eta_2 = 0$, which indicates that either Π_1 or Π_2 becomes a singular plane, we can deduce the same vanishing property as Theorem 2.12.3. Moreover, if $\eta_1 = \eta_2 = 0$, for the intersection of two singular planes, we can further obtain the explicit form of u as below.

Theorem 2.12.4. *Let u be a Laplacian eigenfunction to* (2.1.1). *Suppose that $\mathscr{E}(\Pi_1, \Pi_2, \boldsymbol{l}) \Subset \Omega$ is an irrational edge corner and $\Pi_\ell \in \mathscr{S}_\Omega^\lambda$ $(\ell = 1, 2)$. If* (2.12.2) *is satisfied, then it holds that*

$$\mathrm{Vani}(u;\mathbf{0},\Pi_1,\Pi_2)=+\infty,\quad \mathbf{0}\in\boldsymbol{l}. \tag{2.12.3}$$

Moreover, if $u|_{B_\varepsilon(\mathbf{0})\cap\boldsymbol{l}}\not\equiv 0$, then we have the following expansion of u in a neighbourhood of the edge corner $\mathbf{0}$ in the polar coordinate system:

$$u(\mathbf{x})=4\pi\sum_{n=0}^{\infty}\mathrm{i}^n a_n^0 j_n(\sqrt{\lambda}r)Y_n^0(\theta,\phi), \tag{2.12.4}$$

where $Y_n^0(\theta,\phi)$ is the spherical harmonics and $j_n(t)$ is the n-th Bessel function.

Proof. By Theorem 2.10.3 and Remark 2.10.3, it is easy to verify that (2.12.3) holds under the generic condition (2.12.2). However, if (2.12.2) fails to be fulfilled, then we cannot derive $a_n^0=0$ for $n=0,1,2,\ldots$.

Since $\frac{\partial u}{\partial \nu}\Big|_{\Pi_\ell}\equiv 0$, $\ell=1,2$, we can obtain by direct computation that

$$-\sum_{n=0}^{\infty}\sum_{m=-n}^{n}\mathrm{i}^{n+1}ma_n^m j_n(\sqrt{\lambda}r)\sqrt{\frac{2n+1}{4\pi}}\sqrt{\frac{(n-|m|)!}{(n+|m|)!}}P_n^{|m|}(\cos\theta)=0, \tag{2.12.5}$$

and

$$\sum_{n=0}^{\infty}\sum_{m=-n}^{n}\mathrm{i}^{n+1}ma_n^m j_n(\sqrt{\lambda}r)\sqrt{\frac{2n+1}{4\pi}}\sqrt{\frac{(n-|m|)!}{(n+|m|)!}}P_n^{|m|}(\cos\theta)e^{\mathrm{i}m\alpha\cdot\pi}=0, \tag{2.12.6}$$

on Π_1 and Π_2, respectively. By comparing the coefficients of r in (2.12.5) and (2.12.6), with the help of the orthogonality condition, we can still obtain that $a_1^{\pm 1}=0$ since $\alpha\in(0,1)$ for the dihedral angle $\phi=\alpha\cdot\pi$. By induction, following a similar argument to the proof of Theorem 2.10.3, we can deduce that

$$-\sum_{m=-n}^{n}\mathrm{i}^{n+1}ma_n^m\frac{\sqrt{\lambda}^n}{(2n+1)!!}\sqrt{\frac{2n+1}{4\pi}}\sqrt{\frac{(n-|m|)!}{(n+|m|)!}}P_n^{|m|}(\cos\theta)=0, \tag{2.12.7}$$

and

$$\sum_{m=-n}^{n}\mathrm{i}^{n+1}ma_n^m\frac{\sqrt{\lambda}^n}{(2n+1)!!}\sqrt{\frac{2n+1}{4\pi}}\sqrt{\frac{(n-|m|)!}{(n+|m|)!}}P_n^{|m|}(\cos\theta)e^{\mathrm{i}m\alpha\cdot\pi}=0. \tag{2.12.8}$$

Therefore, by (2.10.4), we see that $a_n^m=0$ $(m=\pm1,\pm2,\ldots,\pm n)$ since the corresponding dihedral angle is irrational. Hence, we are able to obtain the explicit expression (2.12.4) around the edge corner $\mathbf{0}$. □

Based on the irrational intersection at an edge corner by two planes, we next consider the corresponding properties at a vertex corner which is intersected by n planes where $n\geq 3$.

Using Theorem 2.11.3 and Remark 2.11.4, we have the following results for an irrational vertex corner.

Theorem 2.12.5. *Let u be a Laplacian eigenfunction to (2.1.1). Consider an irrational vertex corner $\mathscr{V}(\{\Pi_\ell\}_{\ell=1}^n,\mathbf{0})\Subset\Omega$, where the intersecting n planes $\Pi_1,\Pi_2,\ldots,\Pi_n$ could be either of the three types: a nodal plane, a singular plane or a generalized singular plane, $n\in\mathbb{N}$ and $n\geq 3$. Assume that for $i=1,2,\ldots,n$, $\Pi_i=\mathrm{span}\{\overrightarrow{OA_i},\overrightarrow{OA_{i+1}}\}$, where $\overrightarrow{OA_i}=(r,\theta_i,\phi_i)$, $\overrightarrow{OA_{i+1}}=(r,\theta_{i+1},\phi_{i+1})$ for $r>0$, $\theta_i,\theta_{i+1}\in(0,\pi)$ and $\phi_i,\phi_{i+1}\in(0,2\pi)$ such that $0<\phi_{i+1}-\phi_i<\pi$ in the spherical coordinate system. Particularly when $i=n$, we denote $\Pi_{n+1}:=\Pi_1$. Recall that I_{R} and I_{IR} are defined in* (2.12.1). *If one of the following conditions is fulfilled that*

1. *there exists an index $\ell_0\in I_{\mathrm{IR}}$ such that $\Pi_{\ell_0}\in\mathscr{N}_\Omega^\lambda$ or $\Pi_{\ell_0+1}\in\mathscr{N}_\Omega^\lambda$;*

2. *for any* $\ell \in I_{\mathrm{IR}}$, *if* $\Pi_\ell, \Pi_{\ell+1} \in \{\mathscr{S}^\lambda_\Omega \cup \mathscr{M}^\lambda_\Omega\}$, $u(\mathbf{0}) = 0$ *and for a fixed* $\ell_0 \in I_{\mathrm{IR}}$ *there exists an index* $j \in \{1, \ldots, n\}$ *such that the corresponding plane* $\Pi_j = \mathrm{span}\{\overrightarrow{OA_j}, \overrightarrow{OA_{j+1}}\}$ *satisfies* $P^0_p(\cos\theta_\tau) \neq 0$ *and* $P^1_p(\cos\theta_\tau) \neq 0$ *for all* $p \in \mathbb{N}$, $\tau = j, j+1$, *where* P^0_p *and* P^1_p *are the associated Legendre polynomials;*

then there holds that $\mathrm{Vani}(u; \mathbf{0}) = +\infty$.

2.13 Vanishing at Edge and Vertex Corners Involving Generalized Singular Planes with Variable Parameters

In Sects. 2.10–3.5, whenever a generalized singular plane Π is concerned, the parameter η (cf. Definition 2.9.2) was assumed to be a constant. In this section, we remark that with some straightforward modifications, all the results derived in Sects. 2.10–3.5 equally hold for the case that η is a (variable) analytic function on Π. To that end, we make the following crucial observation. In the sequel, for an analytic function f on Π with the following series representation,

$$f(x) = \sum_{\ell=0}^{\infty} a_\ell(\theta, \phi) r^\ell, \quad \mathbf{x} = (r, \theta, \phi) \in \Pi, \tag{2.13.1}$$

we define $\deg_{\mathrm{low}}(f) = N$ if $a_\ell = 0$ for $\ell = 0, 1, \ldots, N$ while $a_{N+1} \neq 0$. Next, we first assume that η is an analytic function of the form (2.13.1) on Π. Recall that u has the expansion (2.10.3) and assume that

$$a^m_n = 0 \ \text{ for } \ n = 0, 1, \ldots, N \text{ and } m = \pm n, \pm(n-1), \ldots, \pm 1, 0. \tag{2.13.2}$$

Then by straightforward calculations, one has

$$\begin{aligned} &\frac{\partial u}{\partial \nu} + \eta u \\ =& \frac{1}{r \sin\theta} 4\pi \sum_{n=N}^{\infty} \sum_{m=-n}^{n} \mathrm{i}^{n+1} m a^m_n j_n(\sqrt{\lambda} r) \sqrt{\frac{2n+1}{4\pi}} \sqrt{\frac{(n-|m|)!}{(n+|m|)!}} P^{|m|}_n(\cos\theta) e^{\mathrm{i} m \alpha \cdot \pi} \\ &+ \sum_{\ell=0}^{\infty} a_\ell(\theta, \phi) r^\ell \cdot 4\pi \sum_{n=N}^{\infty} \sum_{m=-n}^{n} \mathrm{i}^n a^m_n j_n(\sqrt{\lambda} r) \sqrt{\frac{2n+1}{4\pi}} \sqrt{\frac{(n-|m|)!}{(n+|m|)!}} P^{|m|}_n(\cos\theta) e^{\mathrm{i} m \alpha \cdot \pi}. \end{aligned} \tag{2.13.3}$$

Using (2.10.7) and (2.13.3), it is straightforward to verify that $\deg_{\mathrm{low}}(\eta u) \geq N$, while the leading-order term for $\partial u / \partial \nu + \eta u$ is the r^{N-1}-term, completely determined by $\partial u / \partial \nu$. With such an observation, it is straightforward to show that all the results in Sects. 2.10–3.5 hold for the case that η is of the form (2.13.1) on Π. For the general case that η is analytic on Π, since all of our mathematical arguments can actually be localized around the corner points, one can complete the proofs by using the fact that η has the series expansion (2.13.1) locally around the corner points.

References

1. M. Abramowitz and I. A. Stegun, *Handbook of mathematical functions: with formulas, graphs, and mathematical tables*, vol. 55, Courier Corporation, (1964).
2. G. Alessandrini, *Nodal lines of eigenfunctions of the fixed membrane problem in general convex domains*, Comm. Math. Helv., **69** (1994), 142–154.
3. M. S. Ashbaugh and R. D. Benguria, *Isoperimetric inequalities for eigenvalues of the Laplacian,* Proc. Symp. Pure Math., **76** (2007), 105–139.
4. R. Bãnuelos and K. Burdzy, *On the "hot spots" conjecture of J. Rauch,* J. Funct. Anal., **164** (1999), 1–33.

5. G. Bao, H. Zhang and J. Zou, *Unique determination of periodic polyhedral structures by scattered electromagnetic fields*, Trans. Amer. Math. Soc. **363** (2011), 4527–4551.
6. G. Bao, H. Zhang and J. Zou, *Unique determination of periodic polyhedral structures by scattered electromagnetic fields II: the resonance case*, Trans. Amer. Math. Soc. **366** (2014), 1333–1361.
7. R. F. Bass and K. Burdzy, *Fiber Brownian motion and the "hot spots" problem,* Duke Math. J., **105** (2000), 25–58.
8. E. Blåsten, *Nonradiating sources and transmission eigenfunctions vanish at corners and edges*, SIAM J. Math. Anal., **50(6)** (2018), 6255–6270.
9. W. Bosch, *On the computation of derivatives of Legendre functions*, Phys. Chem. Earth, **25**(9–11), 655–659.
10. J. Brüning, Uber Knoten von Eigenfunktionen des Laplace-Beltrami-Operators, Math. Z., **158** (1978), 15–21.
11. K. Burdzy, *The hot spots problem in planar domains with one hole*, Duke Math. J., **129** (2005), 481–502.
12. K. Burdzy, R. Holyst, D. Ingerman, and P. March, *Configurational transition in a Fleming-Viot-type model and probabilistic interpretation of Laplacian eigenfunctions*, J. Phys. A **29** (1996), 2633–2642.
13. K. Burdzy, R. Holyst and P. March, *A Fleming-Viot Particle Representation of the Dirichlet Laplacian*, Comm. Math. Phys., **214** (2000), 679–703.
14. K. Burdzy and W. Werner, *A counterexample to the "hot spots" conjecture,* Ann. Math., **149** (1999), 309–317.
15. X. Cao, H. Diao, H. Liu and J. Zou, *On nodal and generalized singular structures of Laplacian eigenfunctions and applications*, Journal de Mathématiques Pures et Appliquées, 143 (2020), 116–161.
16. X. Cao, H. Diao, H. Liu and J. Zou, *On novel geometric structures of Laplacian eigenfunctions in* $\mathbb{R}^3$ *and applications to inverse problems*, SIAM Journal on Mathematics Analysis, 53(2021), no. 2, 1263–1294.
17. J. Cheng and M. Yamamoto, *Uniqueness in an inverse scattering problem within non-trapping polygonal obstacles with at most two incoming waves*, Inverse Problems, **19** (2003), 1361–1384.
18. D. Colton and R. Kress, *Looking back on inverse scattering theory*, SIAM Review, **60(40)** (2018), 779–807.
19. D. Colton and R. Kress, *Inverse acoustic and electromagnetic scattering theory*, volume 93 of Applied Mathematical Sciences. Springer, Cham, 4th edition, 2019.
20. D. Colton and B. Sleeman, *Uniqueness theorems for the inverse problem of acoustic scattering*, IMA J. Appl. Math., **31** (1983), 253–259.
21. R. Courant and D. Hilbert, *Methods of Mathematical Physics*, Vol. I, Interscience Publishers, New York, (1953).
22. J. Deng, *Some results on the Schiffer conjecture in* $\mathbb{R}^2$, J. Differential Equations, **253** (2012), 2515–2526.
23. H. Diao, H. Liu, L. Zhang and J. Zou, *Unique continuation from a generalized impedance edge-corner for Maxwell's system and applications to inverse problems*, Inverse Problems, 37 (2021), 035004.
24. Peter Gustav Lejeune Dirichlet, *Démonstration d'un théorème d'Abel*, Journal de Mathématiques Pures et Appliquées, 2nd series, tome **7** (1862), 253–255.
25. P. Ebenfelt, *Propagation of singularities from singular and infinite points in certain complex analytic Cauchy problems and an application to the Pompeiu problem*, Duke Math. J., **73** (1994), 561–582.
26. R. E. Edwards, *Fourier series: A modern introduction*, 2nd edition, Springer, (1979).
27. G. Faber, *Beweis, dass unter allen homogenen Membranen von gleicher Flache und gleicher Spannung die kreisformige den tiefste, Grundton gibt,* Sitzungsber. Bayer. Akad. Wiss. Muchen, Math. Phys. Kl., (1923), 169–172.
28. S. Fournais, *The nodal surface of the second eigenfunction of the Laplacian in* $\mathbb{R}^D$ *can be closed,* J. Differential Equation, **173** (2001), 145–159.
29. P. Freitas, *Closed nodal lines and interior hot spots of the second eigenfunction of the Laplacian on surfaces*, Indiana Univ. Math. J., **51** (2002), 305–316.
30. P. Freitas and D. Krejčiřík, *Unbounded planar domains whose second nodal line does not touch the boundary*, Math. Res. Lett., **14** (2007), 107–111.
31. P. Freitas and D. Krejčiřík, *Location of the nodal set for thin curved tubes*, Indiana Univ. Math. J., **57** (2008), 343–375.
32. D. S. Grebenkov and B. T. Nguyen, *Geometrical structure of Laplacian eigenfunctions*, SIAM Rev., **55(4)** (2013), 601–667.
33. D. Grieser and D. Jerison, *Asymptotics of the first nodal line of a convex domain*, Invent. Math., **125** (1996), 197–219.
34. D. Grieser and D. Jerison, *The size of the first eigenfunction of a convex planar domain*, J. Am. Math. Soc., **11** (1998), 41–72.
35. A. Hassell, L. Hillairet, J. Marzuola, *Eigenfunction concentration for polygonal billiards*, Comm. Partial Differential Equations, **34(4–6)** (2009), 475–485.
36. W. K. Hayman, *Some bounds for principal frequency,* Appl. Anal., **7** (1978), 247–254.
37. M. Hoffmann-Ostenhof, T. Hoffmann-Ostenhof, and N. Nadirashvili, *The nodal line of the second eigenfunction of the Laplacian in* $\mathbb{R}^2$ *can be closed,* Duke Math. J., **90** (1997), 631–640.
38. D. Jakobson, N. Nadirashvili and J. Toth, *Geometric properties of eigenfunctions*, Russ. Math. Surv., **56** (2001), 1085–1105.
39. D. Jerison, *The first nodal set of a convex domain*, in: Essays on Fourier analysis in honor of Elias M. Stein (Princeton, NJ, 1991), Princeton Math. Ser. **42** (1995), Princeton University Press, Princeton, NJ, 225–249.

40. D. Jerison and N. Nadirashvili, *The "hot spots" conjecture for domains with two axes of symmetry*, J. Amer. Math. Soc., **13** (2000), 741–772.
41. J. B. Kennedy, *The nodal line of the second eigenfunction of the Robin Laplacian in* $\mathbb{R}^2$ *can be closed*, J. Differential Equations, **251(12)** (2011), 3606–3624.
42. E. Krahn, *Uber eine von Rayleigh formulierte Minimaleigenschaft des Kreises,* Math. Ann., **94** (1925), 97–100.
43. S. G. Krantz and H. R. Parks, *A primer of real analytic functions*, 2nd edition, Birkhäuser Boston, Inc., Boston, MA, (2002).
44. D. Krejčiřík and M. Tušek, *Nodal sets of thin curved layers*, J. Differential Equations **258** (2015), 281–301.
45. D. Krejčiřík and M. Tušek, *Location of hot spots in thin curved strips*, J. Differential Equations, **266** (2019), 2953–2977.
46. J. R. Kuttler and V. G. Sigillito, *Eigenvalues of the Laplacian in two Dimensions,* SIAM Rev., **26** (1984), 163–193.
47. C. S. Lin, *On the second eigenfunction of the Laplacian in* $\mathbb{R}^2$, Comm. Math. Phys., **111** (1987), 161–166.
48. H. Liu, *Schiffer's conjecture, interior transmission eigenvalues and invisibility cloaking: singular problem vs. nonsingular problem*, Contemporary Mathematics, American Math. Soc., **598** (2013).
49. H. Liu and J. Zou, *Uniqueness in an inverse acoustic obstacle scattering problem for both sound-hard and sound-soft polyhedral scatterers*, Inverse Problems, **22** (2006), 515–524.
50. H. Liu and J. Zou, *On unique determination of partially coated polyhedral scatterers with far-field measurements*, Inverse Problems, **23** (2007), 297–308.
51. E. Makai, *A lower estimation of simply connected membranes,* Act. Math. Acad. Sci. Hungary, **16** (1965), 319–327.
52. A. D. Melas, *On the nodal line of the second eigenfunction of the Laplacian in* $\mathbb{R}^2$, J. Diff. Geom., **35** (1992), 255–263.
53. N. Nadirashvili, *On the length of the nodal curve of an eigenfunction of the Laplace operator*, Russ. Math. Surv., **43** (1988), 227–228.
54. N. S. Nadirashvili, *Metric properties of eigenfunctions of the Laplace operator on manifolds*, Ann. Inst. Fourier, **41** (1991), 259–265.
55. J.-C. Nédélec, *Acoustic and Electromagnetic Equations*, Springer-Verlag, New York, (2001).
56. L. E. Payne, *On two conjectures in the fixed membrane eigenvalue problem*, Z. Angew. Math. Phys., **24** (1973), 721–729.
57. J. W. S. Rayleigh, *The Theory of Sound*, 2nd Ed., Dover Publications, New York, **1,2**(1945).
58. B. Sapoval and T. Gobron, *Vibrations of strongly irregular or fractal resonators*, Phys. Rev. E, **47** (1993), 3013–3024.
59. B. Sapoval, T. Gobron and A. Margolina, *Vibrations of fractal drums*, Phys. Rev. Lett., **67** (1991), 2974–2977.
60. R. Schoen and S.-T. Yau, *Lectures on Differential Geometry*, Conference Proceedings and Lecture Notes in Geometry and Topology. **1** (1994), International Press, Boston.
61. A. Shnirelman, *Ergodic properties of eigenfunctions*, Uspechi Math. Nauk, **29** (1974),181–182.
62. E. C. Titchmarsh and D. R. Heath-Brown, *The theory of the Riemann zeta-function*, Oxford University Press, Oxford, (1986).
63. H. Weyl, *Über die asymptotische verteilung der Eigenwerte*, Gott. Nach., 110–117, (1911).
64. H. Weyl, *Das asymptotische Verteilungsgesetz der Eigenwerte linearer partieller Differentialgleichungen*, Math. Ann., **71** (1912), 441–479.
65. S. T. Yau, *Problem section, Seminar on differential geometry*, Ann. of Math. Studies **102** (1982), Princeton Univ. Press, Princeton NJ, 669–706.
66. S. Zelditch, *Eigenfunctions of the Laplacian of Riemannian Manifolds*, book in progress, 2017.

Chapter 3
Geometric Structures of Maxwellian Eigenfunctions

3.1 Overview

We are concerned with the unique continuation property (UCP) of the time-harmonic Maxwell system

$$\nabla \wedge \mathbf{E} - \mathrm{i}k\mathbf{H} = \mathbf{0}, \quad \nabla \wedge \mathbf{H} + \mathrm{i}k\mathbf{E} = \mathbf{0} \quad \text{in } \Omega, \tag{3.1.1}$$

where $\mathrm{i} := \sqrt{-1}$ and $k \in \mathbb{R}_+$, in a particular scenario that is strongly motivated by our study of the inverse electromagnetic scattering problem. In this chapter, we follow the treatment in [6]. Further discussion on the inverse electromagnetic scattering problems can be found in Sect. 4.1.2. We start with the necessary mathematical setup. We consider the domain Ω to be an open set in $\mathbb{R}^3$, bounded or unbounded, and the solution $(\mathbf{E}, \mathbf{H})$ to the system (3.1.1) in the space $H_{loc}(\mathrm{curl}, \Omega)$ defined by

$$\begin{aligned} H_{loc}(\mathrm{curl}, \Omega) &= \{U|_B \in H(\mathrm{curl}, B); B \text{ is any bounded subdomain of } \Omega\}, \\ H(\mathrm{curl}, B) &= \{U \in L^2(B)^3;\ \nabla \wedge U \in L^2(B)^3\}. \end{aligned}$$

We will often use $B_\rho(\mathbf{x})$ for a ball of radius $\rho \in \mathbb{R}_+$ and centered at $\mathbf{x} \in \mathbb{R}^3$. For a set $K \subset \mathbb{R}^3$, $B_\rho(K) := \{\mathbf{x}; \mathbf{x} \in B_\rho(\mathbf{y}) \text{ for any } \mathbf{y} \in K\}$. Let Π_1 and Π_2 be two planes in $\mathbb{R}^3$ such that $\Pi_1 \cap \Pi_2 = \boldsymbol{L}$, where $\boldsymbol{L}$ is a straight line. We suppose that there exists an open line segment $\boldsymbol{l} \Subset \boldsymbol{L}$ and $\rho \in \mathbb{R}_+$ such that $B_\rho(\boldsymbol{l}) \Subset \Omega$. Let $\mathscr{W}(\Pi_1, \Pi_2)$ denote one of the wedge domains formed by Π_1 and Π_2; then $\partial\mathscr{W}(\Pi_1, \Pi_2) \cap B_\rho(\boldsymbol{l})$ is called an edge corner associated with Π_1 and Π_2; see Fig. 3.1 for a schematic illustration. In the sequel, we let $\widetilde{\Pi}_j$, $j = 1, 2$, denote the two flat faces of the edge corner lying on Π_j, respectively, and write the edge corner as $\mathscr{E}(\widetilde{\Pi}_1, \widetilde{\Pi}_2, \boldsymbol{l})$. Any $\mathbf{x} \in \boldsymbol{l}$ is said to be an edge-corner point of $\mathscr{E}(\widetilde{\Pi}_1, \widetilde{\Pi}_2, \boldsymbol{l})$.

Let $\boldsymbol{\eta}_j$ denote a generalized impedance parameter on $\widetilde{\Pi}_j$, which can be one of the three cases:

$$\text{(i) } \boldsymbol{\eta}_j \equiv 0; \quad \text{(ii) } \boldsymbol{\eta}_j \equiv \infty; \quad \text{(iii) } \boldsymbol{\eta}_j \in L^\infty(\widetilde{\Pi}_j). \tag{3.1.2}$$

Let $\nu_j \in \mathbb{S}^2$ be the unit normal vector to Π_j, pointing to the exterior of $\mathscr{W}(\Pi_1, \Pi_2)$. We are interested in the *generalized impedance condition* on $\widetilde{\Pi}_j$ associated with $(\mathbf{E}, \mathbf{H})$ to the Maxwell system (3.1.1), i.e.,

$$\nu_j \wedge (\nabla \wedge \mathbf{E}) + \boldsymbol{\eta}_j (\nu_j \wedge \mathbf{E}) \wedge \nu_j \big|_{\widetilde{\Pi}_j} = 0. \tag{3.1.3}$$

In the case $\boldsymbol{\eta}_j \equiv \infty$, (3.1.3) is understood as

$$(\nu_j \wedge \mathbf{E}) \wedge \nu_j \big|_{\widetilde{\Pi}_j} = 0. \tag{3.1.4}$$

H. Diao, H. Liu, *Spectral Geometry and Inverse Scattering Theory*,
https://doi.org/10.1007/978-3-031-34615-6_3

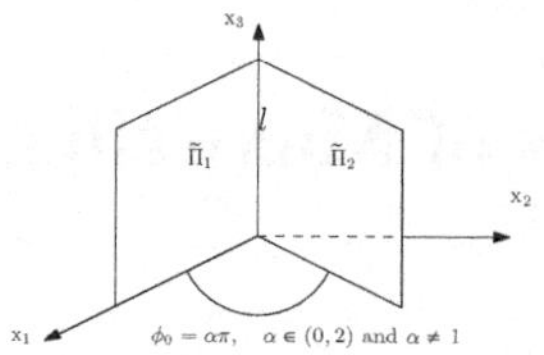

Fig. 3.1 Schematic illustration of two intersecting planes with an edge corner $\mathscr{E}(\widetilde{\Pi}_1, \widetilde{\Pi}_2, \boldsymbol{l})$ and the dihedral angle ϕ_0

An edge corner $\mathscr{E}(\widetilde{\Pi}_1, \widetilde{\Pi}_2, \boldsymbol{l})$ with the generalized impedance condition (3.1.3) imposed on $\widetilde{\Pi}_j$, $j = 1,2$, is called a generalized impedance edge corner associated with the Maxwell system (3.1.1). In this chapter we present the unique continuation property of the solution $(\mathbf{E}, \mathbf{H})$ to (3.1.1) with the presence of a generalized impedance edge corner.

The rest of the chapter is organized as follows. Section 3.2 is devoted to some preliminary knowledge and auxiliary results. In Sects. 3.3 and 3.4, we discuss the UCP results from a generalized impedance edge corner for the Maxwell equations (3.1.1) in two different scenarios.

3.2 Preliminaries and Auxiliary Lemmas

In this section, we collect some preliminary knowledge for the Maxwell system (3.1.1) as well as derive several auxiliary lemmas for our subsequent use.

First, we note that the Maxwell system (3.1.1) is invariant under rigid motions (cf. [3, 12]). Hence, throughout the rest of this chapter and without loss of generality, we can assume that the edge corner $\mathscr{E}(\widetilde{\Pi}_1, \widetilde{\Pi}_2, \boldsymbol{l}) \Subset \Omega$ of our interest is of the following form:

$$\boldsymbol{l} = \left\{ \mathbf{x} = (\mathbf{x}', x_3) \in \mathbb{R}^3; \mathbf{x}' := (x_1, x_2) = \mathbf{0},\ x_3 \in (-h, h) \right\} \Subset \Omega,$$

where $2h \in \mathbb{R}_+$ is the length of $\boldsymbol{l}$, and furthermore, Π_1 coincides with the (x_1, x_3)-plane, while Π_2 possesses a dihedral angle $\phi_0 = \alpha\pi$ away from Π_1 in the anticlockwise direction; see Fig. 3.1 for a schematic illustration. Throughout the chapter, we assume that

$$\alpha \in (0,2) \quad \text{but} \quad \alpha \neq 1. \tag{3.2.1}$$

It can be directly verified that the exterior unit normal vectors ν_j to Π_j, $j = 1,2$ are given by

$$\nu_1 = (0,-1,0)^\top, \quad \nu_2 = (-\sin\phi_0, \cos\phi_0, 0)^\top. \tag{3.2.2}$$

As specified earlier, we have the generalized impedance condition (3.1.3) imposed on $\widetilde{\Pi}_j$, where the boundary impedance parameter η_j fulfils (3.1.2). In order to consider the unique continuation from the edge corner as described above, we introduce the following definition.

Definition 3.2.1. Let $\mathbf{E} \in H_{loc}(\mathrm{curl}, \Omega)$ be a solution to (3.1.1) and suppose there exists an edge corner $\mathscr{E}(\widetilde{\Pi}_1, \widetilde{\Pi}_2, \boldsymbol{l}) \Subset \Omega$ as described above. For a given point $\mathbf{x}_0 \in \boldsymbol{l}$, if there exists a number $N \in \mathbb{N} \cup \{0\}$ such that

$$\lim_{\rho \to +0} \frac{1}{\rho^m} \int_{B_\rho(\mathbf{x}_0)} |\mathbf{E}(\mathbf{x})| \,\mathrm{d}\mathbf{x} = 0 \ \text{ for } m = 0, 1, \ldots, N+2, \tag{3.2.3}$$

we say that $\mathbf{E}$ vanishes at $\mathbf{x}_0$ up to the order N. The largest possible N such that (3.2.3) is fulfilled is called the vanishing order of $\mathbf{E}$ at $\mathbf{x}_0$, and we write

$$\mathrm{Vani}(\mathbf{E};\mathbf{x}_0)=N.$$

If (3.2.3) holds for any $N\in\mathbb{N}$, we say that the vanishing order is infinity.

Since $\mathbf{E}$ is (real) analytic in Ω, we immediately see that if the vanishing order of $\mathbf{E}$ at any point $\mathbf{x}_0\in\boldsymbol{l}$ is infinity, then $\mathbf{E}\equiv 0$ in Ω, namely, the strong unique continuation property holds. In what follows, it is sufficient to consider the UCP at the origin $\mathbf{0}\in\boldsymbol{l}$. Moreover, due to the symmetry role between $(\mathbf{E},\mathbf{H})$ and $(-\mathbf{H},\mathbf{E})$, namely, both of them satisfy the same Maxwell system (3.1.1), we only consider the vanishing order of $\mathbf{E}$, and the same result equally holds for $\mathbf{H}$. It turns out that the vanishing order of $\mathbf{E}$ is related to the *rationality* of the edge-corner angle, i.e., $\alpha\pi$, and we shall make it more rigorous in the sequel.

In the subsequent analysis, we will often use the spherical coordinates of a point $\mathbf{x}$ in $\mathbb{R}^3$:

$$\mathbf{x}=(r\sin\theta\cos\phi, r\sin\theta\sin\phi, r\cos\theta):=(r,\theta,\phi),\ r\geq 0,\ \theta\in[0,\pi),\ \phi\in[0,2\pi). \tag{3.2.4}$$

Let $\hat{\mathbf{x}}=(1,0,0)^\top, \hat{\mathbf{y}}=(0,1,0)^\top, \hat{\mathbf{z}}=(0,0,1)^\top$. Then we know

$$\begin{aligned}
\hat{r}&=\sin\theta\cos\phi\cdot\hat{\mathbf{x}}+\sin\theta\sin\phi\cdot\hat{\mathbf{y}}+\cos\theta\cdot\hat{\mathbf{z}}\\
\hat{\theta}&=\cos\theta\cos\phi\cdot\hat{\mathbf{x}}+\cos\theta\sin\phi\cdot\hat{\mathbf{y}}-\sin\theta\cdot\hat{\mathbf{z}}\\
\hat{\phi}&=-\sin\phi\cdot\hat{\mathbf{x}}+\cos\phi\cdot\hat{\mathbf{y}}
\end{aligned} \tag{3.2.5}$$

constitutes an orthonormal basis in the spherical coordinate system.

Definition 3.2.2. Suppose that $\psi(r,\theta)$ is a complex-valued function for $(r,\theta)\in\Sigma:=[0,r_0]\times[-\theta_0,\theta_0]$, where $r_0,\theta_0\in\mathbb{R}_+$. ψ is said to belong to class $\mathscr{A}$ in Σ if it allows an absolutely convergent series representation as follows:

$$\psi(r,\theta)=a_0+\sum_{j=1}^{\infty}a_j(\theta)r^j, \tag{3.2.6}$$

where $a_0\in\mathbb{C}\backslash\{0\}$ and $a_j(\theta)\in C[-\theta_0,\theta_0]$.

Here are two simple scenarios for $\psi(r,\theta)$ to belong to the class $\mathscr{A}$: first, ψ is a nonzero constant; second, $\psi(r,\theta)$ is real analytic in Σ with r_0,θ_0 being sufficiently small and $\psi(0,\theta)$ independent of θ. For an impedance parameter η_j in (3.1.3) in the third case, namely, $\eta_j\in L^\infty(\widetilde{\Pi}_j)$, we readily see that in the (r,θ,ϕ)-coordinate, $\phi|_{\widetilde{\Pi}_1}=0$ and $\phi|_{\widetilde{\Pi}_2}=\phi_0$. In what follows, if for any $\mathbf{x}_0\in\boldsymbol{l}$ there exists a neighborhood $\Sigma_{\mathbf{x}_0}$ of $\mathbf{x}_0$ which is of the form in Definition 3.2.2 and is contained in $\overline{\widetilde{\Pi}_j}$ such that $\psi_{\mathbf{x}_0}(r,\theta):=\eta_j(\mathbf{x}-\mathbf{x}_0)$ belongs to the class $\mathscr{A}$ in $\Sigma_{\mathbf{x}_0}$, then we say that η_j belongs to the class $\mathscr{A}(\boldsymbol{l})$. It is emphasized that η_j belonging to the class $\mathscr{A}(\boldsymbol{l})$ is a local property, which is localized around a neighborhood of $\boldsymbol{l}$ on $\widetilde{\Pi}_j$. In fact, our subsequent analysis of the UCP from the edge corner $\mathscr{E}(\widetilde{\Pi}_1,\widetilde{\Pi}_2,\boldsymbol{l})$ is confined locally around a neighborhood of $\boldsymbol{l}$ and, indeed, around a neighborhood of the origin $\mathbf{0}$ according to our earlier discussion.

Next, we consider the Fourier representations of the solutions to (3.1.1) in terms of the spherical waves. Throughout the rest of the chapter, for a fixed $l\in\mathbb{N}$, we adopt the notation

$$[l]_0:=\{0,\pm 1,\ldots,\pm l\},\quad [l]_1:=\{\pm 1,\ldots,\pm l\}. \tag{3.2.7}$$

Recall that the spherical harmonics $Y_l^m(\theta,\phi)$ are given by

$$Y_l^m(\theta,\phi)=c_l^m P_l^{|m|}(\cos\theta)e^{\mathrm{i}m\phi},\quad c_l^m=\sqrt{\frac{2l+1}{4\pi}\frac{(l-|m|)!}{(l+|m|)!}}, \tag{3.2.8}$$

where $P_l^m(t)$ is the Legendre function. For simplicity, we shall often write the notation Y_l^m for $Y_l^m(\theta,\phi)$. For our subsequent use, the following lemma presents some important properties of the associated Legendre functions [1].

Lemma 3.2.1. *The following recursive relationships of Legendre functions hold for $l \in \mathbb{N}$ and $m \in [l]_0$,*

$$\begin{aligned} \frac{\mathrm{d}P_l^{|m|}(\cos\theta)}{\mathrm{d}\theta} &= \frac{1}{2}\left[(l+|m|)(l-|m|+1)P_l^{|m|-1}(\cos\theta) - P_l^{|m|+1}(\cos\theta)\right], \\ \frac{|m|}{\sin\theta}P_l^{|m|}(\cos\theta) &= -\frac{1}{2}\left[P_{l-1}^{|m|+1}(\cos\theta) + (l+|m|-1)(l+|m|)P_{l-1}^{|m|-1}(\cos\theta)\right], \end{aligned} \tag{3.2.9}$$

If $P_l^m(\cos\theta)$ is evaluated at $\theta = 0$, for $l \in \mathbb{N}\cup\{0\}$, we have

$$P_l^m(1) = 0, \quad m \in [l]_1; \quad P_l^0(1) = 1. \tag{3.2.10}$$

For a fixed $n \in \mathbb{N}\cup\{0\}$ and $m \in \mathbb{N}$ with $m \le n$, it holds that

$$P_n^{-m}(\cos\theta) = (-1)^m \frac{(n-m)!}{(n+m)!} P_n^m(\cos\theta). \tag{3.2.11}$$

Recall that the spherical Bessel function $j_\ell(t)$ of the order ℓ is defined by

$$j_\ell(t) = \frac{t^\ell}{(2\ell+1)!!}\left(1 - \sum_{l=1}^{\infty} \frac{(-1)^l t^{2l}}{2^l l!(2\ell+3)\dots(2\ell+2l+1)}\right) = \frac{t^\ell}{(2\ell+1)!!} + \mathscr{O}(t^{\ell+2}). \tag{3.2.12}$$

There holds the following recursive relationships [1]:

$$\frac{j_\ell(t)}{t} = \frac{j_{\ell-1}(t) + j_{\ell+1}(t)}{2\ell+1}, \quad j_\ell'(t) = \frac{\ell j_{\ell-1}(t) - (\ell+1)j_{\ell+1}(t)}{2\ell+1}, \quad \ell \in \mathbb{N}. \tag{3.2.13}$$

Lemma 3.2.2 (**[5]**). *Recall that $\hat{\mathbf{r}}, \hat{\boldsymbol{\theta}}$ and $\hat{\boldsymbol{\phi}}$ are defined in (3.2.5). Denote*

$$\mathbf{M}_l^m(\mathbf{x}) = j_l(kr)\cdot \mathbf{X}_l^m, \quad \mathbf{N}_l^m(\mathbf{x}) = \mathrm{i}\left(\frac{j_l(kr)}{kr} + j_l'(kr)\right)\mathbf{Z}_l^m - \frac{\sqrt{l(l+1)}}{kr}\cdot j_l(kr)Y_l^m \cdot \hat{\mathbf{r}}, \tag{3.2.14}$$

where $k \in \mathbb{R}_+$, $j_l'(kr)$ is the derivative of $j_l(kr)$ with respect to kr, and

$$\mathbf{X}_l^m = \frac{\mathrm{i}}{\sqrt{l(l+1)}}\left(\frac{\mathrm{i}\cdot m}{\sin\theta}Y_l^m\hat{\boldsymbol{\theta}} - \frac{\partial Y_l^m}{\partial\theta}\cdot\hat{\boldsymbol{\phi}}\right), \quad \mathbf{Z}_l^m = \frac{\mathrm{i}}{\sqrt{l(l+1)}}\left(\frac{\partial Y_l^m}{\partial\theta}\hat{\boldsymbol{\theta}} + \frac{\mathrm{i}\cdot m}{\sin\theta}Y_l^m\hat{\boldsymbol{\phi}}\right).$$

The solution $\mathbf{E}(\mathbf{x})$ to (3.1.1) has the following Fourier expansion around $\mathbf{0}$,

$$\mathbf{E}(\mathbf{x}) = \sum_{l=1}^{\infty}\sum_{m=-l}^{l}\left(a_l^m\cdot\mathbf{M}_l^m(\mathbf{x}) + b_l^m\cdot\mathbf{N}_l^m(\mathbf{x})\right), \quad a_l^m, b_l^m \in \mathbb{C},$$

which (along with its derivatives) converges uniformly in $B_{\rho_0}(\mathbf{0})$ for a sufficiently small $\rho_0 \in \mathbb{R}_+$.

In this section, we adopt the following notation:

$$\sum_{l,m,N} = \sum_{l=N}^{+\infty}\sum_{m=-l}^{l}, \quad N \in \mathbb{N}.$$

Using (3.2.13), from Lemma 3.2.2, we can derive that

$$\mathbf{E}(\mathbf{x}) = -\Sigma_{l,m,1} \frac{1}{\sqrt{l(l+1)}} \Bigg\{ b_l^m \cdot l(l+1) p_l(kr) \cdot Y_l^m \cdot \hat{\boldsymbol{r}} + \left[a_l^m \cdot j_l(kr) \frac{m}{\sin\theta} Y_l^m \right. \tag{3.2.15}$$

$$\left. + b_l^m \cdot q_l(kr) \cdot \frac{\partial Y_l^m}{\partial \theta} \right] \cdot \hat{\boldsymbol{\theta}} + \mathrm{i} \left[a_l^m \cdot j_l(kr) \frac{\partial Y_l^m}{\partial \theta} + b_l^m \cdot q_l(kr) \frac{m}{\sin\theta} Y_l^m \right] \cdot \hat{\boldsymbol{\phi}} \Bigg\},$$

where

$$p_l(kr) = \frac{j_{l-1}(kr) + j_{l+1}(kr)}{2l+1}, \quad q_l(kr) = \frac{(l+1) j_{l-1}(kr) - l j_{l+1}(kr)}{2l+1}. \tag{3.2.16}$$

Remark 3.2.1. In view of (3.2.12), we know the lowest-order terms of $p_l(kr)$ and $q_l(kr)$ with respect to the power of r are given, respectively, by

$$\frac{k^{l-1}}{(2l+1)(2l-1)!!} r^{l-1} \text{ and } \frac{(l+1)k^{l-1}}{(2l+1)(2l-1)!!} r^{l-1}.$$

Lemma 3.2.3 ([7, Proposition 2.1.7]). *If the power series* $\sum_\mu a_\mu \mathbf{x}^\mu$ *converges at a point* $\mathbf{x}_0$, *then it converges uniformly and absolutely on compact subsets of* $U(\mathbf{x_0})$, *where*

$$U(\mathbf{x}_0) = \{(r_1 x_{0,1}, \ldots, r_n x_{0,n}) : -1 < r_j < 1, j = 1, \ldots, n\}, \mathbf{x}_0 = (x_{0,1}, \ldots, x_{0,n}) \in \mathbb{R}^n.$$

Using Definition 3.2.1 and (3.2.15), we can obtain the following lemma.

Lemma 3.2.4. *Let* $\mathbf{E}$ *be a solution to* (3.1.1). *Recall that* $\mathbf{E}$ *has the radial wave expansion* (3.2.15) *in* $B_{\rho_0}(\mathbf{0})$. *For a fixed* $N \in \mathbb{N}$, *if*

$$a_l^m = b_l^m = 0, \quad m \in [l]_0, \quad l = 1, 2, \ldots, N, \tag{3.2.17}$$

where $[l]_0$ *is defined in* (3.2.7), *then*

$$\mathrm{Vani}(\mathbf{E}; \mathbf{0}) \geq N. \tag{3.2.18}$$

Conversely, if there exists $N \in \mathbb{N}$ *such that* (3.2.18) *holds, then we have* (3.2.17).

Proof. From Lemma 3.2.3, we know that (3.2.15) converges uniformly and absolutely in $B_{\rho_1}(\mathbf{0})$, where $0 < \rho_1 < \rho_0$. Substituting (3.2.17) into (3.2.15), we have

$$\mathbf{E}(\mathbf{x}) = -\sum_{l=N+1}^{\infty} \sum_{m=-l}^{l} \frac{1}{\sqrt{l(l+1)}} \Bigg\{ b_l^m \cdot l(l+1) p_l(kr) \cdot Y_l^m \cdot \hat{\boldsymbol{r}} \tag{3.2.19}$$

$$+ \left[a_l^m \cdot j_l(kr) \frac{m}{\sin\theta} Y_l^m + b_l^m \cdot q_l(kr) \cdot \frac{\partial Y_l^m}{\partial \theta} \right] \cdot \hat{\boldsymbol{\theta}}$$

$$+ \mathrm{i} \left[a_l^m \cdot j_l(kr) \frac{\partial Y_l^m}{\partial \theta} + b_l^m \cdot q_l(kr) \frac{m}{\sin\theta} Y_l^m \right] \cdot \hat{\boldsymbol{\phi}} \Bigg\}.$$

From Remark 3.2.1, the lowest order of r with respect to the power of r in (3.2.19) is N. Therefore,

$$\frac{\mathbf{E}(\mathbf{x})}{r^N} = -\sum_{l=N+1}^{\infty} \sum_{m=-l}^{l} \frac{1}{\sqrt{l(l+1)}} \Bigg\{ b_l^m \cdot l(l+1) \frac{p_l(kr)}{r^N} \cdot Y_l^m \cdot \hat{\boldsymbol{r}}$$

$$+ \left[a_l^m \cdot \frac{j_l(kr)}{r^N} \frac{m}{\sin\theta} Y_l^m + b_l^m \cdot \frac{q_l(kr)}{r^N} \cdot \frac{\partial Y_l^m}{\partial \theta} \right] \cdot \hat{\boldsymbol{\theta}}$$

$$+ \mathrm{i} \left[a_l^m \cdot \frac{j_l(kr)}{r^N} \frac{\partial Y_l^m}{\partial \theta} + b_l^m \cdot \frac{q_l(kr)}{r^N} \frac{m}{\sin\theta} Y_l^m \right] \cdot \hat{\boldsymbol{\phi}} \Bigg\}$$

converges uniformly and absolutely in $B_{\rho_1}(\mathbf{0})$, which implies

$$\left|\frac{\mathbf{E}(\mathbf{x})}{r^N}\right| = \mathscr{O}(1), \quad \text{as } r \to +0. \tag{3.2.20}$$

In view of Definition 3.2.1, by virtue of (3.2.20), we have

$$\lim_{\rho\to+0}\frac{1}{\rho^m}\int_{B_\rho(\mathbf{0})}|\mathbf{E}(\mathbf{x})|\,\mathrm{d}\mathbf{x} \le \lim_{\rho\to+0}\frac{\rho^{N+2}}{\rho^m}\int_0^\rho\int_0^\pi\int_0^{2\pi}\left|\frac{\mathbf{E}(\mathbf{x})}{r^N}\right|\mathrm{d}r\mathrm{d}\theta\mathrm{d}\phi = 0,$$

which holds for $m = 0, 1, \ldots, N+2$, and this proves (3.2.18). The other direction of the conclusion can be proved by using similar arguments. □

Lemma 3.2.5. *Let* $\mathbf{E}$ *be a solution to* (3.1.1)*. Recall that* $\mathbf{E}$ *has the radial wave expansion* (3.2.15) *in* $B_{\rho_0}(\mathbf{0})$*. Consider an edge corner* $\mathscr{E}(\widetilde{\Pi}_1, \widetilde{\Pi}_2, \boldsymbol{l}) \Subset \Omega$ *associated with* $\mathbf{E}$*, with* ν_i *defined in* (3.2.2) *being the outward unit normal vectors to* Π_i*,* $i = 1, 2$*. Then it holds that*

$$\begin{aligned}
\nu_1 \wedge \mathbf{E}|_{\widetilde{\Pi}_1} = {} & \Sigma_{l,m,1} - \frac{1}{\sqrt{l(l+1)}}\Bigg\{ b_l^m l(l+1)p_l(kr)Y_l^m\Big|_{\phi=0}\mathbf{e_1}(\theta,0) + \Bigg(a_l^m j_l(kr)\frac{m}{\sin\theta}Y_l^m\Big|_{\phi=0} \\
& + b_l^m \cdot q_l(kr)\frac{\partial Y_l^m}{\partial\theta}\Big|_{\phi=0}\Bigg)\mathbf{e_2}(\theta,0)\Bigg\}, \\
\nu_2 \wedge \mathbf{E}|_{\widetilde{\Pi}_2} = {} & \Sigma_{l,m,1} - \frac{1}{\sqrt{l(l+1)}}\Bigg\{ b_l^m l(l+1)p_l(kr)Y_l^m\Big|_{\phi=\phi_0}\mathbf{e_1}(\theta,\phi_0) + \Bigg(a_l^m j_l(kr)\frac{m}{\sin\theta}Y_l^m\Big|_{\phi=\phi_0} \\
& + b_l^m \cdot q_l(kr)\frac{\partial Y_l^m}{\partial\theta}\Big|_{\phi=\phi_0}\Bigg)\mathbf{e_2}(\theta,\phi_0)\Bigg\},
\end{aligned} \tag{3.2.21}$$

where

$$e_1(\theta,\phi) = [\cos\phi\cos\theta, \sin\phi\cos\theta, -\sin\theta]^\top, \; e_2(\theta,\phi) = -[\cos\phi\sin\theta, \sin\phi\sin\theta, \cos\theta]^\top, \tag{3.2.22}$$

are linearly independent for any θ *and* ϕ*. Furthermore, we have*

$$\begin{aligned}
\nu_1 \wedge (\nabla \wedge \mathbf{E}|_{\widetilde{\Pi}_1}) = {} & \mathrm{i}k\sum_{l,m,1}\frac{1}{\sqrt{l(l+1)}}\Bigg\{ a_l^m l(l+1)p_l(kr)Y_l^m\Big|_{\phi=0}\cdot\mathbf{e_1}(\theta,0) \\
& + \Bigg(-b_l^m j_l(kr)\cdot\frac{m}{\sin\theta}Y_l^m + a_l^m\cdot q_l(kr)\cdot\frac{\partial Y_l^m}{\partial\theta}\Big|_{\phi=0}\Bigg)\cdot\mathbf{e_2}(\theta,0)\Bigg\}, \\
\nu_2 \wedge (\nabla \wedge \mathbf{E}|_{\widetilde{\Pi}_2}) = {} & \mathrm{i}k\sum_{l,m,1}\frac{1}{\sqrt{l(l+1)}}\Bigg\{ a_l^m l(l+1)p_l(kr)Y_l^m\Big|_{\phi=\phi_0}\cdot\mathbf{e_1}(\theta,\phi_0) \\
& + \Bigg(-b_l^m j_l(kr)\cdot\frac{m}{\sin\theta}Y_l^m\Big|_{\phi=\phi_0} + a_l^m q_l(kr)\cdot\frac{\partial Y_l^m}{\partial\theta}\Big|_{\phi=\phi_0}\Bigg)\cdot\mathbf{e_2}(\theta,\phi_0)\Bigg\}.
\end{aligned} \tag{3.2.23}$$

Proof. Using the fact that $\phi = \phi_0$ for $\mathbf{x} = (r, \theta, \phi) \in \Pi_2$, it is easy to see that

$$\nu_2 \wedge (\hat{\boldsymbol{r}}|_{\phi=\phi_0}) = \begin{bmatrix}\cos\phi_0\cos\theta \\ \sin\phi_0\sin\theta \\ -\sin\theta\end{bmatrix}, \; \nu_2 \wedge (\hat{\boldsymbol{\theta}}|_{\phi=\phi_0}) = \begin{bmatrix}-\cos\phi_0\sin\theta \\ -\sin\phi_0\sin\theta \\ -\cos\theta\end{bmatrix}, \; \nu_2 \wedge (\hat{\boldsymbol{\phi}}|_{\phi=\phi_0}) = \mathbf{0}, \tag{3.2.24}$$

from which we can derive the second equation of (3.2.21). The first equation of (3.2.21) can be obtained in a similar way.

Recall that $\mathbf{M}_l^m(\mathbf{x})$ and $\mathbf{N}_l^m(\mathbf{x})$ are defined in (3.2.14). Using the identity $\nabla\wedge\mathbf{M}_l^m(\mathbf{x}) = -\mathrm{i}k\mathbf{N}_l^m(\mathbf{x})$ and $\nabla\wedge\mathbf{N}_l^m(\mathbf{x}) = \mathrm{i}k\mathbf{M}_l^m(\mathbf{x})$ (cf. [5]) we can obtain that

$$
\begin{aligned}
\nabla\wedge\mathbf{E}|_{\tilde{\Pi}_1} = \mathrm{i}k\sum_{l=1}^{\infty}\sum_{m=-l}^{l}\frac{1}{\sqrt{l(l+1)}}\Bigg\{ & a_l^m\cdot l(l+1)p_l(kr)Y_l^m\cdot\nu_1\wedge\hat{\boldsymbol{r}}|_{\phi=0} \\
& +\left(-b_l^m j_l(kr)\frac{m}{\sin\theta}Y_l^m + a_l^m q_l(kr)\frac{\partial Y_l^m}{\partial\theta}\right)\cdot\nu_1\wedge\hat{\boldsymbol{\theta}}|_{\phi=0} \\
& +\left(-b_l^m j_l(kr)\mathrm{i}\frac{\partial Y_l^m}{\partial\theta} + a_l^m\cdot q_l(kr)\frac{\mathrm{i}m}{\sin\theta}Y_l^m\right)\cdot\nu_1\wedge\hat{\boldsymbol{\phi}}|_{\phi=0}\Bigg\}, \\
\nabla\wedge\mathbf{E}|_{\tilde{\Pi}_2} = \mathrm{i}k\sum_{l=1}^{\infty}\sum_{m=-l}^{l}\frac{1}{\sqrt{l(l+1)}}\Bigg\{ & a_l^m\cdot l(l+1)p_l(kr)Y_l^m\cdot\nu_2\wedge\hat{\boldsymbol{r}}|_{\phi=\phi_0} \\
& +\left(-b_l^m j_l(kr)\frac{m}{\sin\theta}Y_l^m + a_l^m q_l(kr)\frac{\partial Y_l^m}{\partial\theta}\right)\nu_2\wedge\hat{\boldsymbol{\theta}}|_{\phi=\phi_0} \\
& +\left(-b_l^m j_l(kr)\mathrm{i}\frac{\partial Y_l^m}{\partial\theta} + a_l^m\cdot q_l(kr)\frac{\mathrm{i}m}{\sin\theta}Y_l^m\right)\nu_2\wedge\hat{\boldsymbol{\phi}}|_{\phi=\phi_0}\Bigg\}.
\end{aligned}
\tag{3.2.25}
$$

Combing (3.2.25) with (3.2.24), together with straightforward though a bit tedious calculations, one can deduce the second equation of (3.2.23). The first equation of (3.2.23) can be shown in a similar manner.

The proof is complete. □

Lemma 3.2.6. *Under the same notations and conditions as the ones in Lemma 3.2.5, we further assume that η_1, η_2 belong to the class $\mathscr{A}(\boldsymbol{l})$. Then we have*

$$
\begin{aligned}
& \nu_1\wedge(\nabla\wedge\mathbf{E}|_{\tilde{\Pi}_1}) + \eta_1(\nu_1\wedge\mathbf{E}|_{\tilde{\Pi}_1})\wedge\nu_1 \qquad (3.2.26) \\
= & \sum_{l,m,1}\frac{1}{\sqrt{l(l+1)}}\Bigg\{\left(\mathrm{i}ka_l^m l(l+1)p_l(kr)Y_l^m - \eta_1 a_l^m j_l(kr)\tfrac{m}{\sin\theta}Y_l^m - \eta_1 b_l^m q_l(kr)\tfrac{\partial Y_l^m}{\partial\theta}\right)\boldsymbol{e_1}(\theta,0) \\
& +\left(-\mathrm{i}kb_l^m j_l(kr)\tfrac{m}{\sin\theta}Y_l^m + \mathrm{i}ka_l^m q_l(kr)\tfrac{\partial Y_l^m}{\partial\theta} + \eta_1 b_l^m l(l+1)p_l(kr)Y_l^m\right)\cdot\boldsymbol{e_2}(\theta,0)\Bigg\},
\end{aligned}
$$

and

$$
\begin{aligned}
& \nu_2\wedge(\nabla\wedge\mathbf{E}|_{\tilde{\Pi}_2}) + \eta_2(\nu_2\wedge\mathbf{E}|_{\tilde{\Pi}_2})\wedge\nu_2 \qquad (3.2.27) \\
= & \sum_{l,m,1}\frac{1}{\sqrt{l(l+1)}}\Bigg\{\left(\mathrm{i}ka_l^m l(l+1)p_l(kr)Y_l^m - \eta_2 a_l^m j_l(kr)\frac{m}{\sin\theta}Y_l^m - \eta_2 b_l^m q_l(kr)\frac{\partial Y_l^m}{\partial\theta}\right)\boldsymbol{e_1}(\theta,\phi_0) \\
& +\left(-\mathrm{i}kb_l^m j_l(kr)\frac{m}{\sin\theta}Y_l^m + \mathrm{i}ka_l^m q_l(kr)\frac{\partial Y_l^m}{\partial\theta} + \eta_2 b_l^m l(l+1)p_l(kr)Y_l^m\right)\cdot\boldsymbol{e_2}(\theta,\phi_0)\Bigg\},
\end{aligned}
$$

where $\boldsymbol{e_1}(\theta,0)$, $\boldsymbol{e_2}(\theta,0)$ $\boldsymbol{e_1}(\theta,\phi_0)$, and $\boldsymbol{e_2}(\theta,\phi_0)$ are defined in (3.2.22).

Proof. Recall that ν_2 is defined in (3.2.2); $\hat{\boldsymbol{r}}, \hat{\boldsymbol{\theta}}$, and $\hat{\boldsymbol{\phi}}$ are given by (3.2.5). Then it is easy to see that

$$(\nu_2 \wedge \hat{\mathbf{r}}) \wedge \nu_2 = (\cos\phi_0 \sin\theta, \sin\phi_0 \sin\theta, \cos\theta)^\top, \tag{3.2.28}$$

$$(\nu_2 \wedge \hat{\boldsymbol{\theta}}) \wedge \nu_2 = (\cos\phi_0 \cos\theta, \sin\phi_0 \cos\theta, -\sin\theta)^\top, \quad (\nu_2 \wedge \hat{\boldsymbol{\phi}}) \wedge \nu_2 = \mathbf{0}. \tag{3.2.29}$$

Using (3.2.15) and (3.2.23), we can derive that

$$\begin{aligned} &\nu_2 \wedge (\nabla \wedge \mathbf{E}|_{\widetilde{\Pi}_2}) + \eta_2 (\nu_2 \wedge \mathbf{E}|_{\widetilde{\Pi}_2}) \wedge \nu_2 \\ &= \sum_{l=1}^{\infty} \sum_{m=-l}^{l} \frac{1}{\sqrt{l(l+1)}} \Bigg\{ \mathrm{i}k \Bigg(a_l^m l(l+1) p_l(kr) Y_l^m \cdot \nu_2 \wedge \hat{\mathbf{r}} \\ &+ \Big(-b_l^m j_l(kr) \cdot \frac{m}{\sin\theta} Y_l^m + a_l^m q_l(kr) \frac{\partial Y_l^m}{\partial\theta} \Big) \cdot \nu_2 \wedge \hat{\boldsymbol{\theta}} \\ &+ \Big(-b_l^m j_l(kr) \mathrm{i} \frac{\partial Y_l^m}{\partial\theta} + a_l^m q_l(kr) \frac{\mathrm{i}m}{\sin\theta} Y_l^m \Big) \nu_2 \wedge \hat{\boldsymbol{\phi}} \Bigg) \\ &- \eta_2 \Bigg(\Big(b_l^m \cdot l(l+1) p_l(kr) \cdot Y_l^m \Big) \cdot (\nu_2 \wedge \hat{\mathbf{r}}) \wedge \nu_2 \\ &+ \Big(a_l^m j_l(kr) \frac{m}{\sin\theta} Y_l^m + b_l^m q_l(kr) \frac{\partial Y_l^m}{\partial\theta} \Big) (\nu_2 \wedge \hat{\boldsymbol{\theta}}) \wedge \nu_2 \\ &+ \mathrm{i} \Big(a_l^m j_l(kr) \frac{\partial Y_l^m}{\partial\theta} + \frac{m b_l^m q_l(kr)}{\sin\theta} Y_l^m \Big) (\nu_2 \wedge \hat{\boldsymbol{\phi}}) \wedge \nu_2 \Bigg) \Bigg\}. \end{aligned} \tag{3.2.30}$$

Substituting (3.2.24) and (3.2.28) into (3.2.30), together with straightforward calculations, we can obtain (3.2.27). Equation (3.2.27) can be derived in a similar manner. □

3.3 Vanishing Orders for an Edge Corner $\mathscr{E}(\widetilde{\Pi}_1, \widetilde{\Pi}_2, \boldsymbol{l})$ with $\eta_j \in \mathscr{A}(\boldsymbol{l})$

In this section, we consider the case that $\mathscr{E}(\widetilde{\Pi}_1, \widetilde{\Pi}_2, \boldsymbol{l})$ is an edge corner, with both η_1 and η_2 belonging to the class $\mathscr{A}(\boldsymbol{l})$. We shall derive the vanishing order of $\mathbf{E}$ to (3.1.1) at the origin $\mathbf{0} \in \boldsymbol{l}$. The major idea is to make use of the radial wave expansion (3.2.15) of $\mathbf{E}$ in $B_{\rho_0}(\mathbf{0})$ and to investigate the relationships between $a_n^{\pm 1}$, a_n^0 and $b_n^{\pm 1}$, b_n^0. Henceforth, according to Definition 3.2.2, we assume that η_j, $j = 1, 2$, are given by the following absolutely convergent series at $\mathbf{0} \in \boldsymbol{l}$:

$$\eta_1 = \eta_1 + \sum_{j=1}^{\infty} \eta_{1,j}(\theta) r^j, \tag{3.3.1}$$

$$\eta_2 = \eta_2 + \sum_{j=1}^{\infty} \eta_{2,j}(\theta) r^j \tag{3.3.2}$$

where $\eta_\ell \in \mathbb{C} \backslash \{0\}$, $\eta_{\ell,j}(\theta) \in C[-\pi, \pi]$, and $r \in [-h, h]$, $\ell = 1, 2$.

Lemma 3.3.1. *Let* $\mathbf{E}$ *be a solution to* (3.1.1)*, whose radial wave expansion in* $B_{\rho_0}(\mathbf{0})$ *is given by* (3.2.15)*. Consider a generalized impedance edge corner* $\mathscr{E}(\widetilde{\Pi}_1, \widetilde{\Pi}_2, \boldsymbol{l}) \Subset \Omega$*, with* $\angle(\Pi_1, \Pi_2) = \phi_0 = \alpha\pi$ *for* $\alpha \in (0, 2)$ *but* $\alpha \neq 1$*. Suppose that the generalized impedance parameters* η_j *on* $\widetilde{\Pi}_j$*,* $j = 1, 2$*, are given by* (3.3.1) *and* (3.3.2)*, then it holds that*

$$0 = \frac{4\mathrm{i}kc_1^1 \sin^2\phi_0}{6\sqrt{2}} (a_1^1 + a_1^{-1}) - \frac{4kc_1^1 \sin\phi_0 \cos\phi_0}{6\sqrt{2}} (a_1^1 - a_1^{-1}) - \frac{(\eta_2 \cos\phi_0 + \eta_1)\sqrt{2} c_1^0}{3} b_1^0, \tag{3.3.3a}$$

$$0 = -\frac{4ikc_1^1 \sin\phi_0 \cos\phi_0}{6\sqrt{2}}(a_1^1 + a_1^{-1}) - \frac{4kc_1^1 \sin^2\phi_0}{6\sqrt{2}}(a_1^1 - a_1^{-1}) - \frac{\eta_2\sqrt{2}c_1^0 \sin\phi_0}{3} b_1^0, \tag{3.3.3b}$$

$$0 = -\frac{4c_1^1(-\eta_1 + \eta_2\cos\phi_0)}{6\sqrt{2}}(b_1^1 + b_1^{-1}) + \frac{4\eta_2 c_1^1 \sin\phi_0 \mathrm{i}}{6\sqrt{2}}(b_1^1 - b_1^{-1}). \tag{3.3.3c}$$

Assume that there exists $n \in \mathbb{N}\backslash\{1\}$ such that

$$a_l^0 = b_l^0 = a_l^{\pm 1} = b_l^{\pm 1} = 0, \quad l = 1, \ldots, n-1. \tag{3.3.4}$$

Then it holds that

$$\begin{aligned}\frac{\eta_1\sqrt{n(n+1)}c_n^0}{2n+1} b_n^0 = &\frac{\mathrm{i}kn(n+1)^2 c_n^1 \sin^2\phi_0}{2(2n+1)\sqrt{n(n+1)}}(a_n^1 + a_n^{-1}) \\ &- \frac{kn(n+1)^2 c_n^1 \sin\phi_0\cos\phi_0}{2(2n+1)\sqrt{n(n+1)}}(a_n^1 - a_n^{-1}) - \frac{\eta_2\sqrt{n(n+1)}c_n^0\cos\phi_0}{2n+1} b_n^0,\end{aligned} \tag{3.3.5a}$$

$$\begin{aligned}\frac{kn(n+1)^2 c_n^1}{2(2n+1)\sqrt{n(n+1)}}(a_n^1 - a_n^{-1}) = &-\frac{\mathrm{i}kn(n+1)^2 c_n^1 \sin\phi_0\cos\phi_0}{2(2n+1)\sqrt{n(n+1)}}(a_n^1 + a_n^{-1}) \\ &+ \frac{kn(n+1)^2 c_n^1 \cos^2\phi_0}{2(2n+1)\sqrt{n(n+1)}}(a_n^1 - a_n^{-1}) - \frac{\eta_2\sqrt{n(n+1)}c_n^0\sin\phi_0}{2n+1} b_n^0,\end{aligned} \tag{3.3.5b}$$

$$\begin{aligned}-\frac{\eta_1 n(n+1)^2 c_n^1}{2(2n+1)\sqrt{n(n+1)}}(b_n^1 + b_n^{-1}) = &\frac{\eta_2 n(n+1)^2 c_n^1 \cos\phi_0}{2(2n+1)\sqrt{n(n+1)}}(b_n^1 + b_n^{-1}) \\ &- \frac{n(n+1)^2 \eta_2 \sin\phi_0 \mathrm{i}}{2(2n+1)\sqrt{n(n+1)}}(b_n^1 - b_n^{-1}).\end{aligned} \tag{3.3.5c}$$

Proof. We shall first derive (3.3.5a), (3.3.5b), and (3.3.5c). Equation (3.3.3a), (3.3.3b), and (3.3.3c) can be obtained in a similar way and we shall sketch the corresponding derivations at the end of the proof.

We first note that

$$(\nu_2 \wedge \mathbf{E}) \wedge \nu_2 = -\nu_2 \wedge (\nu_2 \wedge \mathbf{E}) = -\big(\nu_2 \cdot (\nu_2 \cdot \mathbf{E}) - \mathbf{E}(\nu_2 \cdot \nu_2)\big) = \mathbf{E} - (\nu_2 \cdot \mathbf{E}) \cdot \nu_2. \tag{3.3.6}$$

Hence, we have from

$$\nu_2 \wedge (\nabla \wedge \mathbf{E}|_{\widetilde{\Pi}_2}) + \eta_2(\nu_2 \wedge \mathbf{E}|_{\widetilde{\Pi}_2}) \wedge \nu_2 = \mathbf{0}, \tag{3.3.7}$$

that

$$\nu_2 \wedge (\nabla \wedge \mathbf{E})|_{\widetilde{\Pi}_2} + \eta_2\big(\mathbf{E}|_{\widetilde{\Pi}_2} - (\nu_2 \cdot \mathbf{E}|_{\widetilde{\Pi}_2}) \cdot \nu_2\big) = \mathbf{0}. \tag{3.3.8}$$

Multiplying the cross product with ν_2 from the left on both sides (3.3.8), by using the fact that

$$\nu_2 \wedge \big(\nu_2 \wedge (\nabla \wedge \mathbf{E})|_{\widetilde{\Pi}_2}\big) = \nu_2 \cdot \big(\nu_2 \cdot (\nabla \wedge \mathbf{E})|_{\widetilde{\Pi}_2}\big) - (\nu_2 \cdot \nu_2)(\nabla \wedge \mathbf{E})|_{\widetilde{\Pi}_2},$$

we can obtain that

$$\big(\nu_2 \cdot (\nabla \wedge \mathbf{E})|_{\widetilde{\Pi}_2}\big)\nu_2 + \eta_2(\nu_2 \wedge \mathbf{E}|_{\widetilde{\Pi}_2}) = \nabla \wedge \mathbf{E}|_{\widetilde{\Pi}_2}. \tag{3.3.9}$$

Similarly, since the generalized impedance condition (3.1.3) associated with η_1 is imposed on $\widetilde{\Pi}_1$, using the above argument, we can deduce that

$$\big(\nu_1 \cdot (\nabla \wedge \mathbf{E})|_{\widetilde{\Pi}_1}\big)\nu_1 + \eta_1(\nu_1 \wedge \mathbf{E}|_{\widetilde{\Pi}_1}) = \nabla \wedge \mathbf{E}|_{\widetilde{\Pi}_1}. \tag{3.3.10}$$

Since $\boldsymbol{l} \in \widetilde{\Pi}_1 \cap \widetilde{\Pi}_2$, combining (3.3.9) with (3.3.10), it yields that

$$\big(\nu_1 \cdot (\nabla \wedge \mathbf{E}|_{\boldsymbol{l}})\big)\nu_1 + \eta_1(\nu_1 \wedge \mathbf{E}|_{\boldsymbol{l}}) = \big(\nu_2 \cdot (\nabla \wedge \mathbf{E}|_{\boldsymbol{l}})\big)\nu_2 + \eta_2(\nu_2 \wedge \mathbf{E}|_{\boldsymbol{l}}). \tag{3.3.11}$$

Due to (3.3.4), using (3.2.21) and (3.2.25), by virtue of (3.2.9), it yields that

$$\begin{aligned}
\nabla \wedge \mathbf{E}|_{\widetilde{\Pi}_2} = &\sum_{l=n}^{\infty}\sum_{m=-l}^{l} \frac{\mathrm{i}k}{\sqrt{l(l+1)}}\Bigg\{ a_l^m \cdot l(l+1)p_l(kr)c_l^m P_l^{|m|} \cdot \hat{\boldsymbol{r}}|_{\widetilde{\Pi}_2} \\
&+ \Bigg(b_l^m j_l(kr)c_l^m \frac{\mathrm{sgn}(m)}{2}\,[P_{l-1}^{|m|+1}(\cos\theta) + (l+|m|-1)(l+|m|)P_{l-1}^{|m|-1}(\cos\theta)] \\
&+ a_l^m q_l(kr)c_l^m \frac{1}{2}[(l+|m|)(l-|m|+1)P_l^{|m|-1}(\cos\theta) - P_l^{|m|+1}(\cos\theta)]\Bigg)\hat{\boldsymbol{\theta}}|_{\widetilde{\Pi}_2} \\
&+ \Bigg(- b_l^m j_l(kr)\mathrm{i}c_l^m \frac{1}{2}[(l+|m|)(l-|m|+1)P_l^{|m|-1}(\cos\theta) - P_l^{|m|+1}(\cos\theta)] \\
&- a_l^m \cdot q_l(kr)\mathrm{i}c_l^m \frac{\mathrm{sgn}(m)}{2}\,[P_{l-1}^{|m|+1}(\cos\theta) \\
&+ (l+|m|-1)(l+|m|)P_{l-1}^{|m|-1}(\cos\theta)]\Bigg)\hat{\boldsymbol{\phi}}|_{\Pi_2}\Bigg\}
\end{aligned} \tag{3.3.12}$$

and

$$\begin{aligned}
\nu_2 \wedge \mathbf{E}|_{\widetilde{\Pi}_2} = &\sum_{l=n}^{\infty}\sum_{m=-l}^{l}\Bigg\{\Bigg\{ - b_l^m \cdot \sqrt{l(l+1)}p_l(kr)\cdot c_l^m P_l^{|m|}\Bigg\}\cdot \nu_2 \wedge \hat{\boldsymbol{r}}|_{\widetilde{\Pi}_2} \\
&+ \Bigg\{ a_l^m \cdot j_l(kr)\frac{1}{\sqrt{l(l+1)}}c_l^m \frac{\mathrm{sgn}(m)}{2}\,[P_{l-1}^{|m|+1}(\cos\theta) + (l+|m|-1)(l+|m|) \\
&\times P_{l-1}^{|m|-1}(\cos\theta)] - b_l^m \cdot q_l(kr)\cdot \frac{1}{\sqrt{l(l+1)}}c_l^m \frac{1}{2}[(l+|m|)(l-|m|+1) \\
&\times P_l^{|m|-1}(\cos\theta) - P_l^{|m|+1}(\cos\theta)]\Bigg\}\cdot \nu_2 \wedge \hat{\boldsymbol{\theta}}|_{\widetilde{\Pi}_2} + \Bigg\{ - a_l^m \cdot j_l(kr)\frac{\mathrm{i}}{\sqrt{l(l+1)}}\frac{c_l^m}{2} \\
&\times [(l+|m|)(l-|m|+1)P_l^{|m|-1}(\cos\theta) - P_l^{|m|+1}(\cos\theta)] + b_l^m q_l(kr)\frac{\mathrm{i}}{\sqrt{l(l+1)}} \\
&\times c_l^m \frac{\mathrm{sgn}(m)}{2}\,[P_{l-1}^{|m|+1}(\cos\theta) + (l+|m|-1)(l+|m|)P_{l-1}^{|m|-1}(\cos\theta)]\Bigg\} \times \nu_2 \wedge \hat{\boldsymbol{\phi}}|_{\widetilde{\Pi}_2}\Bigg\},
\end{aligned} \tag{3.3.13}$$

where

$$\mathrm{sgn}(m) = 1 \text{ when } m > 0;\ 0 \text{ when } m = 0;\ -1 \text{ when } m < 0.$$

Recall that if $\mathbf{x} \in \boldsymbol{l}$ one has

$$\theta = \phi = 0, \quad 0 \le r \le h, \tag{3.3.14}$$

where r, θ, and ϕ are the spherical coordinates of $\mathbf{x} \in \boldsymbol{l}$ defined in (3.2.4). It is straightforward to calculate that

$$\begin{aligned}
&\nu_2 \wedge \hat{\boldsymbol{r}}|_{\theta=\phi=0} = \begin{bmatrix}\cos\phi_0\\ \sin\phi_0\\ 0\end{bmatrix},\ \nu_2 \wedge \hat{\boldsymbol{\theta}}|_{\theta=\phi=0} = -\begin{bmatrix}0\\ 0\\ \cos\phi_0\end{bmatrix},\ \nu_2 \wedge \hat{\boldsymbol{\phi}}|_{\theta=\phi=0} = -\begin{bmatrix}0\\ 0\\ \sin\phi_0\end{bmatrix}, \\
&\nu_2 \cdot \hat{\boldsymbol{r}}|_{\theta=\phi=0} = 0, \quad \nu_2 \cdot \hat{\boldsymbol{\theta}}|_{\theta=\phi=0} = -\sin\phi_0, \quad \nu_2 \cdot \hat{\boldsymbol{\phi}}|_{\theta=\phi=0} = \cos\phi_0,
\end{aligned} \tag{3.3.15}$$

where $\hat{\boldsymbol{r}}$, $\hat{\boldsymbol{\theta}}$ and $\hat{\boldsymbol{\phi}}$ are defined in (3.2.4).

Evaluating (3.3.12) and (3.3.13) at $\boldsymbol{l}$, by virtue of (3.2.10) and (3.3.15), we can derive that

$$\begin{aligned}\nabla\wedge\mathbf{E}|_{\boldsymbol{l}} = \sum_{l=n}^{+\infty}\frac{\mathrm{i}k}{\sqrt{l(l+1)}}\Bigg\{ & a_l^0 l(l+1)p_l(kr)c_l^0\cdot\hat{\boldsymbol{r}}|_{\theta=\phi=0}\\ & +\frac{(l+1)l}{2}\Big((b_l^1-b_l^{-1})\cdot j_l(kr)\cdot c_l^1+(a_l^1+a_l^{-1})q_l(kr)c_l^1\Big)\cdot\hat{\boldsymbol{\theta}}|_{\theta=\phi=0}\\ & +\mathrm{i}\frac{(l+1)l}{2}\Big(-(b_l^1+b_l^{-1})\cdot j_l(kr)c_l^1-(a_l^1-a_l^{-1})q_l(kr)c_l^1\Big)\cdot\hat{\boldsymbol{\phi}}|_{\theta=\phi=0}\Bigg\},\\ \nu_2\wedge\mathbf{E}|_{\boldsymbol{l}} = \sum_{l=n}^{+\infty}\frac{1}{\sqrt{l(l+1)}}\Bigg\{\Bigg(& -b_l^0 l(l+1)p_l(kr)c_l^0\cdot\nu_2\wedge\hat{\boldsymbol{r}}|_{\theta=\phi=0}\\ & -\frac{(l+1)l}{2}\Big(-(a_l^1-a_l^{-1})j_l(kr)c_l^1+(b_l^1+b_l^{-1})q_l(kr)c_l^1\Big)\cdot\nu_2\wedge\hat{\boldsymbol{\theta}}|_{\theta=\phi=0}\\ & -\mathrm{i}\frac{(l+1)l}{2}\Big((a_l^1+a_l^{-1})j_l(kr)c_l^1-(b_l^1-b_l^{-1})q_l(kr)c_l^1\Big)\cdot\nu_2\cdot\hat{\boldsymbol{\phi}}|_{\theta=\phi=0}\Bigg)\Bigg\}.\end{aligned}\tag{3.3.16}$$

Therefore, from (3.3.16), we obtain that

$$\begin{aligned}&\nu_2^\top(\nabla\wedge\mathbf{E}|_{\boldsymbol{l}})\nu_2+\boldsymbol{\eta}_2(\nu_2\wedge\mathbf{E}|_{\boldsymbol{l}})\\ &=\sum_{l=n}^{+\infty}\frac{1}{\sqrt{l(l+1)}}\Bigg\{\mathrm{i}k\Bigg[-\sin\phi_0\frac{c_l^1(l+1)l}{2}\Big((b_l^1-b_l^{-1})j_l(kr)+(a_l^1+a_l^{-1})q_l(kr)c_l^1\Big)\\ &-\cos\phi_0\frac{\mathrm{i}(l+1)l}{2}\times\Big((b_l^1+b_l^{-1})j_l(kr)c_l^1+(a_l^1-a_l^{-1})q_l(kr)c_l^1\Big)\Bigg]\\ &\times\begin{bmatrix}-\sin\phi_0\\ \cos\phi_0\\ 0\end{bmatrix}+\boldsymbol{\eta}_2\Bigg[-b_l^0 l(l+1)p_l(kr)c_l^0\begin{bmatrix}\cos\phi_0\\ \sin\phi_0\\ 0\end{bmatrix}\\ &-\frac{(l+1)l}{2}\Big(-(a_l^1-a_l^{-1})j_l(kr)c_l^1+(b_l^1+b_l^{-1})q_l(kr)c_l^1\Big)\begin{bmatrix}0\\ 0\\ -\cos\phi_0\end{bmatrix}\\ &-\frac{\mathrm{i}(l+1)l}{2}\Big((a_l^1+a_l^{-1})j_l(kr)c_l^1-(b_l^1-b_l^{-1})q_l(kr)c_l^1\Big)\cdot\begin{bmatrix}0\\ 0\\ -\sin\phi_0\end{bmatrix}\Bigg]\Bigg\}.\end{aligned}\tag{3.3.17}$$

Using a similar argument for deriving (3.3.17), we have

$$\begin{aligned}&\nu_1^\top(\nabla\wedge\mathbf{E}|_{\boldsymbol{l}})\nu_1+\eta_1(\nu\wedge\mathbf{E}|_{\boldsymbol{l}})\\ &=-\sum_{l=n}^{+\infty}\Bigg\{\frac{kl(l+1)}{2\sqrt{l(l+1)}}\Big[(b_l^1+b_l^{-1})\cdot j_l(kr)\cdot c_l^1+(a_l^1-a_l^{-1})q_l(kr)c_l^1\Big]\begin{bmatrix}0\\ -1\\ 0\end{bmatrix}\\ &+\boldsymbol{\eta}_1\Bigg(-b_l^0\sqrt{l(l+1)}p_l(kr)c_l^0\begin{bmatrix}-1\\ 0\\ 0\end{bmatrix}+\frac{l(l+1)}{2\sqrt{l(l+1)}}\Big((a_l^1-a_l^{-1})j_l(kr)c_l^1\\ &-(b_l^1+b_l^{-1})q_l(kr)c_l^1\Big)\begin{bmatrix}0\\ 0\\ 1\end{bmatrix}\Bigg)\Bigg\}.\end{aligned}\tag{3.3.18}$$

Note that $\eta_\ell, \ell = 1,2$ have the expansions (3.3.1) and (3.3.2) respectively, where the coefficients of r^0 in (3.3.1) and (3.3.2) are the nonzero numbers η_1 and η_2. From Remark 3.2.1, it is easy to see that the lowest order of (3.3.17) and (3.3.18) with respect to the power of r is $n-1$, which is contributed by $p_n(kr)$ and $q_n(kr)$ in (3.3.17) and (3.3.18). Substituting (3.3.17) and (3.3.18) into (3.3.11) and comparing the coefficients of r^{n-1} on both sides of the first, second, and third component of (3.3.11), we can derive (3.3.5a), (3.3.5b), and (3.3.5c), respectively.

We can derive (3.3.3a), (3.3.3b), and (3.3.3c) by similar arguments for (3.3.5a), (3.3.5b), and (3.3.5c). Indeed, the Fourier expansions of (3.3.17) and (3.3.18) can be rewritten with the starting summation index $n=1$. Hence, we can obtain (3.3.3a), (3.3.3b), and (3.3.3c) by comparing the coefficients of r^0 on both sides of (3.3.11) by virtue of (3.3.17) and (3.3.18).

The proof is complete. □

Lemma 3.3.2. *Under the same setup in Lemma 3.3.1, it holds that*

$$0 = -\frac{4\eta_2 c_1^1 \cos^2\phi_0}{6\sqrt{2}}(b_1^1 + b_1^{-1}) + \frac{4\mathrm{i}\eta_2 c_1^1 \sin\phi_0\cos\phi_0}{6\sqrt{2}}(b_1^1 - b_1^{-1}) + \frac{\mathrm{i}k\sqrt{2}c_1^0\cos\phi_0}{3}a_1^0, \tag{3.3.19}$$

$$0 = \frac{4\eta_2 c_1^1 \sin\phi_0\cos\phi_0}{6\sqrt{2}}(b_1^1 + b_1^{-1}) + \frac{4\mathrm{i}\eta_2 c_1^1 \sin^2\phi_0}{6\sqrt{2}}(b_1^1 - b_1^{-1}) + \frac{\mathrm{i}k\sqrt{2}c_1^0\sin\phi_0}{3}a_1^0, \tag{3.3.20}$$

$$0 = \frac{4\mathrm{i}kc_1^1\cos\phi_0}{6\sqrt{2}}(a_1^1 + a_1^{-1}) + \frac{4kc_1^1\sin\phi_0}{6\sqrt{2}}(a_1^1 - a_1^{-1}) + \frac{\eta_2\sqrt{2}c_1^0}{3}b_1^0. \tag{3.3.21}$$

Furthermore, if we assume there exists $n \in \mathbb{N}\backslash\{1\}$ such that (3.3.4) *is fulfilled, then it holds that*

$$\begin{aligned} 0 =& \frac{\mathrm{i}k\sqrt{n(n+1)}c_n^0\cos\phi_0}{2n+1}a_n^0 - \frac{\eta_2 n(n+1)^2 c_n^1\cos^2\phi_0}{2(2n+1)\sqrt{n(n+1)}}(b_n^1 + b_n^{-1}) \\ &+ \frac{\mathrm{i}\eta_2 n(n+1)^2 c_n^1\sin\phi_0\cos\phi_0}{2(2n+1)\sqrt{n(n+1)}}(b_n^1 - b_n^{-1}), \end{aligned} \tag{3.3.22a}$$

$$\begin{aligned} 0 =& \frac{\mathrm{i}kc_n^0\sqrt{n(n+1)}\sin\phi_0}{2n+1}a_n^0 + \frac{\eta_2 n(n+1)^2 c_n^1\sin\phi_0\cos\phi_0}{2(2n+1)\sqrt{n(n+1)}}(b_n^1 + b_n^{-1}) \\ &+ \frac{\mathrm{i}\eta_2 n(n+1)^2 c_n^1\sin^2\phi_0}{2(2n+1)\sqrt{n(n+1)}}(b_n^1 - b_n^{-1}), \end{aligned} \tag{3.3.22b}$$

$$\begin{aligned} 0 =& \frac{\mathrm{i}kn(n+1)^2 c_n^1\cos\phi_0}{2(2n+1)\sqrt{n(n+1)}}(a_n^1 + a_n^{-1}) + \frac{kn(n+1)^2 c_n^1\sin\phi_0}{2(2n+1)\sqrt{n(n+1)}}(a_n^1 - a_n^{-1}) \\ &+ \frac{\eta_2\sqrt{n(n+1)}c_n^0}{2n+1}b_n^0. \end{aligned} \tag{3.3.22c}$$

Proof. We first prove (3.3.22a), (3.3.22b), and (3.3.22c). Since the generalized impedance condition (3.1.3) associated with η_2 is imposed on $\widetilde{\Pi}_2$, we have

$$\nu_2 \wedge (\nabla \wedge \mathbf{E}|_\iota) + \eta_2(\nu_2 \wedge \mathbf{E}|_\iota) \wedge \nu_2 = \mathbf{0}. \tag{3.3.23}$$

Here, we recall (3.3.14). Under the assumption (3.3.4), using (3.2.9), (3.2.10) and (3.2.27), we can obtain that

$$\begin{aligned}
&\nu_2 \wedge (\nabla \wedge \mathbf{E}|_l) + \eta_2(\nu_2 \wedge \mathbf{E}|_l) \wedge \nu_2 = \Sigma_{l=n}^{+\infty} \Bigg\{ \mathrm{i}k \Bigg\{ a_l^0 \sqrt{l(l+1)} p_l(kr) c_l^0 \cdot (\cos\phi_0, \sin\phi_0, 0)^\top \\
&+ \tfrac{l(l+1)}{2\sqrt{l(l+1)}} \Big((b_l^1 - b_l^{-1}) j_l(kr) c_l^1 + (a_l^1 + a_l^{-1}) q_l(kr) c_l^1 \Big) \begin{bmatrix} 0 \\ 0 \\ -\cos\phi_0 \end{bmatrix} - \tfrac{l(l+1)\mathrm{i}}{2\sqrt{l(l+1)}} \Big((b_l^1 + b_l^{-1}) \\
&\times j_l(kr) c_l^1 + (a_l^1 - a_l^{-1}) q_l(kr) c_l^1 \Big) \begin{bmatrix} 0 \\ 0 \\ -\sin\phi_0 \end{bmatrix} \Bigg\} + \eta_2 \Bigg\{ -b_l^0 \sqrt{l(l+1)} p_l(kr) c_l^0 \begin{bmatrix} 0 \\ 0 \\ 1 \end{bmatrix} + \tfrac{l(l+1)}{2\sqrt{l(l+1)}} \\
&\times \Big((a_l^1 - a_l^{-1}) j_l(kr) c_l^1 + (b_l^1 + b_l^{-1}) q_l(kr) c_l^1 \Big) \times \begin{bmatrix} \cos^2\phi_0 \\ -\sin\phi_0\cos\phi_0 \\ 0 \end{bmatrix} + \tfrac{l(l+1)\mathrm{i}}{2\sqrt{l(l+1)}} \Big((a_l^1 + a_l^{-1}) \\
&\times j_l(kr) c_l^1 - (b_l^1 - b_l^{-1}) q_l(kr) c_l^1 \Big) \times \begin{bmatrix} \sin\phi_0\cos\phi_0 \\ \sin^2\phi_0 \\ 0 \end{bmatrix} \Bigg\} \Bigg\}.
\end{aligned} \tag{3.3.24}$$

Note that η_ℓ, $\ell = 1, 2$, have the expansions (3.3.1) and (3.3.2), respectively, where the coefficients of r^0 in (3.3.1) and (3.3.2) are nonzero number η_1 and η_2. In view of Remark 3.2.1, we know that the lowest order of (3.3.24) with respect to the power of r is $n-1$, which is contributed by $p_n(kr)$ and $q_n(kr)$ in (3.3.24). Substituting (3.3.17) and (3.3.24) into (3.3.23), comparing the coefficients of r^{n-1} on both sides of the first, the second, and the third component of (3.3.23), we derive that (3.3.22a), (3.3.22b), and (3.3.22c), respectively.

We can derive (3.3.19), (3.3.20), and (3.3.21) by following similar arguments in deriving (3.3.22a), (3.3.22b), and (3.3.22c). Indeed, the Fourier expansions of (3.3.24) can be rewritten with the starting summation index $n = 1$. Hence, we can obtain (3.3.19), (3.3.20), and (3.3.21) by comparing the coefficients of r^0 on both sides of (3.3.23) by virtue of (3.3.24). □

Lemma 3.3.3. *Under the same setup in Lemma 3.3.1, one has the following linear relations:*

$$\begin{cases} \beta_{11}^1 (b_1^1 + b_1^{-1}) + \beta_{12}^1 (b_1^1 - b_1^{-1}) + \beta_{13}^1 a_1^0 = 0, \\ \beta_{21}^1 (b_1^1 + b_1^{-1}) + \beta_{22}^1 (b_1^1 - b_1^{-1}) + \beta_{23}^1 a_1^0 = 0, \\ \beta_{31}^1 (b_1^1 + b_1^{-1}) + \beta_{32}^1 (b_1^1 - b_1^{-1}) + \beta_{33}^1 a_1^0 = 0, \end{cases} \tag{3.3.25}$$

where

$$\begin{aligned}
\beta_{11}^1 &= -\frac{4\eta_2 c_1^1 \cos^2\phi_0}{6\sqrt{2}}, \quad \beta_{12}^1 = \frac{4\mathrm{i}\eta_2 c_1^1 \sin\phi_0 \cos\phi_0}{6\sqrt{2}}, \quad \beta_{13}^1 = \frac{\mathrm{i}k\sqrt{2} c_1^0 \cos\phi_0}{3}, \\
\beta_{21}^1 &= \frac{4\eta_2 c_1^1 \sin\phi_0 \cos\phi_0}{6\sqrt{2}}, \quad \beta_{22}^1 = \frac{4\mathrm{i}\eta_2 c_1^1 \sin^2\phi_0}{6\sqrt{2}}, \quad \beta_{23}^1 = \frac{\mathrm{i}k\sqrt{2} c_1^0 \sin\phi_0}{3}, \\
\beta_{31}^1 &= -\frac{4c_1^1(-\eta_1 + \eta_2\cos\phi_0)}{6\sqrt{2}}, \quad \beta_{32}^1 = \frac{4\eta_2 c_1^1 \sin\phi_0 \mathrm{i}}{6\sqrt{2}}, \quad \beta_{33}^1 = 0.
\end{aligned}$$

If we assume that there exists $n \in \mathbb{N} \backslash \{1\}$ such that (3.3.4) *is fulfilled, then one has that*

$$\begin{cases} \beta_{11}^n (b_n^1 + b_n^{-1}) + \beta_{12}^n (b_n^1 - b_n^{-1}) + \beta_{13}^n a_n^0 = 0, \\ \beta_{21}^n (b_n^1 + b_n^{-1}) + \beta_{22}^n (b_n^1 - b_n^{-1}) + \beta_{23}^n a_n^0 = 0, \\ \beta_{31}^n (b_n^1 + b_n^{-1}) + \beta_{32}^n (b_n^1 - b_n^{-1}) + \beta_{33}^n a_n^0 = 0, \end{cases} \tag{3.3.26}$$

where

$$\beta_{11}^n = -\frac{\eta_2 n(n+1)^2 c_n^1 \cos^2\phi_0}{2(2n+1)\sqrt{n(n+1)}}, \quad \beta_{12}^n = \frac{\mathrm{i}\eta_2 n(n+1)^2 c_n^1 \sin\phi_0\cos\phi_0}{2(2n+1)\sqrt{n(n+1)}}, \quad \beta_{13}^n = \frac{\mathrm{i}k\sqrt{n(n+1)}c_n^0\cos\phi_0}{2n+1},$$

$$\beta_{21}^n = \frac{\eta_2 n(n+1)^2 c_n^1 \sin\phi_0\cos\phi_0}{2(2n+1)\sqrt{n(n+1)}}, \quad \beta_{22}^n = \frac{\mathrm{i}\eta_2 n(n+1)^2 c_n^1 \sin^2\phi_0}{2(2n+1)\sqrt{n(n+1)}}, \quad \beta_{23}^n = \frac{\mathrm{i}k\sqrt{n(n+1)}c_n^0\sin\phi_0}{2n+1},$$

$$\beta_{31}^n = -\frac{n(n+1)^2 c_n^1(-\eta_1+\eta_2\cos\phi_0)}{2(2n+1)\sqrt{n(n+1)}}, \quad \beta_{32}^n = \frac{\mathrm{i}\eta_2 n(n+1)^2 c_n^1 \sin\phi_0}{2(2n+1)\sqrt{n(n+1)}}, \quad \beta_{33}^n = 0.$$

Furthermore, if $\alpha \neq \frac{1}{2}$ and $\alpha \neq \frac{3}{2}$, then it holds that

$$a_n^0 = b_n^{\pm 1} = 0. \tag{3.3.27}$$

Proof. Combining (3.3.19), (3.3.20) with (3.3.3c), we can obtain (3.3.25). Similarly, by virtue of (3.3.5c), (3.3.22a), and (3.3.22b), we can derive (3.3.26). After straightforward calculations, it can be verified that the determinant of coefficient matrices (3.3.26) is given by

$$|\mathscr{B}_n| = -k\eta_2^2\left(\frac{n+1}{2n+1}\right)^3 \frac{n\sqrt{n(n+1)}}{2}\left(c_n^1\right)^2 c_n^0 \sin^2\phi_0\cos^2\phi_0, \tag{3.3.28}$$

where c_n^0, c_n^1 are nonzero constants defined in (3.2.8). Recall that η_ℓ, $\ell = 1,2$ have the expansions (3.3.1) and (3.3.2), respectively, where the coefficients of r^0 in (3.3.1) and (3.3.2) are nonzero number η_1 and η_2. Since $\phi_0 = \alpha\pi \neq \pi/2$, $\phi_0 = \alpha\pi \neq 3\pi/2$, $\eta_2 \neq 0$, and $k \in \mathbb{R}_+$, we conclude that $\mathscr{B}_n$ are nonsingular, which readily implies (3.3.27). □

The following two important lemmas provide the recursive relationships for $a_n^{\pm m}$ and $b_n^{\pm m}$, $m = 0,1,\ldots,n$, which will be used to characterize the vanishing order of $\mathbf{E}$ with respect to the corresponding dihedral angle of the edge corner $\mathscr{E}(\widetilde{\Pi}_1,\widetilde{\Pi}_2,\boldsymbol{l}) \Subset \Omega$ in Theorem 3.3.1.

Lemma 3.3.4. *Under the same setup as in Lemma 3.3.1, we assume further an $n \in \mathbb{N}\backslash\{1\}$ such that*

$$a_l^m = b_l^m = 0, \quad l = 1,\ldots,n-1, \text{ and } m \in [l]_0. \tag{3.3.29}$$

Then we have the following recursive linear equations:

$$\begin{aligned}
0 =& \mathrm{i}k\frac{\sqrt{n(n+1)}}{2n+1}c_n^0 a_n^0 - \frac{\eta_1(n+1)}{2(2n+1)}\frac{c_n^1(n+1)n}{\sqrt{n(n+1)}}(b_n^1+b_n^{-1}),\\
0 =& \mathrm{i}k\frac{\sqrt{n(n+1)}}{2n+1}c_n^1(a_n^1+a_n^{-1}) - \frac{\eta_1(n+1)}{2(2n+1)}\frac{c_n^2}{\sqrt{n(n+1)}}(n+2)(n-1)(b_n^2+b_n^{-2})\\
&+\frac{\eta_1(n+1)}{2n+1}\frac{c_n^0}{\sqrt{n(n+1)}}b_n^0,\\
0 =& \mathrm{i}k\frac{\sqrt{n(n+1)}}{2n+1}c_n^m(a_n^m+a_n^{-m}) - \frac{\eta_1(n+1)}{2(2n+1)}\frac{c_n^{m+1}}{\sqrt{n(n+1)}}(n+m+1)(n-m)\\
&\times(b_n^{m+1}+b_n^{-(m+1)}) + \frac{\eta_1(n+1)}{2(2n+1)}\frac{c_n^{m-1}}{\sqrt{n(n+1)}}(b_n^{m-1}+b_n^{-(m-1)}), \; m = 2,3,\ldots,n-1,\\
0 =& \mathrm{i}k\frac{\sqrt{n(n+1)}}{2n+1}c_n^n(a_n^n+a_n^{-n}) + \frac{\eta_1(n+1)}{2(2n+1)}\frac{c_n^{n-1}}{\sqrt{n(n+1)}}(b_n^{n-1}+b_n^{-(n-1)}),
\end{aligned} \tag{3.3.30}$$

and

$$0 = \mathrm{i}k\frac{n+1}{2(2n+1)}\frac{c_n^1}{\sqrt{n(n+1)}}(n+1)n(a_n^1+a_n^{-1}) + \eta_1\frac{c_n^0\sqrt{n(n+1)}}{2n+1}b_n^0,$$

$$
\begin{aligned}
0 =& \mathrm{i}k \frac{n+1}{2(2n+1)} \frac{c_n^2}{\sqrt{n(n+1)}} (n+2)(n-1)(a_n^2 + a_n^{-2}) - \mathrm{i}k \frac{n+1}{2n+1} \frac{c_n^0}{\sqrt{n(n+1)}} a_n^0 \\
&+ \eta_1 \frac{c_n^1 \sqrt{n(n+1)}}{2n+1} (b_n^1 + b_n^{-1}), \qquad (3.3.31)\\
0 =& \mathrm{i}k \frac{n+1}{2(2n+1)} \frac{c_n^m}{\sqrt{n(n+1)}} (n+m)(n-m+1)(a_n^m + a_n^{-m}) - \mathrm{i}k \cdot \frac{n+1}{2(2n+1)} \frac{c_n^{m-2}}{\sqrt{n(n+1)}} \\
&\times (a_n^{m-2} + a_n^{-(m-2)}) + \eta_1 \frac{c_n^{m-1} \sqrt{n(n+1)}}{2n+1} (b_n^{m-1} + b_n^{-(m-1)}), \quad m = 3,4,\ldots,n, \\
0 =& -\mathrm{i}k \frac{n+1}{2(2n+1)} \frac{c_n^{n-1}}{\sqrt{n(n+1)}} (a_n^{n-1} + a_n^{-(n-1)}) + \eta_1 \frac{c_n^n \sqrt{n(n+1)}}{2n+1} (b_n^n + b_n^{-n}) = 0.
\end{aligned}
$$

Proof. Since the generalized impedance condition (3.1.3) associated with η_1 is imposed on $\widetilde{\Pi}_1$, substituting (3.3.29) into (3.2.27), by virtue of (3.2.9), we derive that

$$
\begin{aligned}
\mathbf{0} =& \sum_{l=n}^{\infty} \sum_{m=-l}^{l} \frac{1}{\sqrt{l(l+1)}} \Bigg\{ \Bigg(\mathrm{i}k a_l^m l(l+1) p_l(kr) c_l^m P_l^m + \eta_1 a_l^m j_l(kr) c_l^m \frac{\operatorname{sgn}(m)}{2} [P_{l-1}^{|m|+1}(\cos\theta) \\
&+ (l+|m|-1)(l+|m|) P_{l-1}^{|m|-1}(\cos\theta)] - \eta_1 b_l^m q_l(kr) \frac{c_l^m}{2} [(l+|m|)(l-|m|+1) \\
&\times P_l^{|m|-1}(\cos\theta) - P_l^{|m|+1}(\cos\theta)] \Bigg) \mathbf{e_1}(\theta,0) + \Bigg(\mathrm{i}k b_l^m j_l(kr) c_l^m \frac{\operatorname{sgn}(m)}{2} [P_{l-1}^{|m|+1}(\cos\theta) \\
&+ (l+|m|-1)(l+|m|) P_{l-1}^{|m|-1}(\cos\theta)] + \mathrm{i}k a_l^m q_l(kr) \frac{c_l^m}{2} [(l+|m|)(l-|m|+1) \\
&\times P_l^{|m|-1}(\cos\theta) - P_l^{|m|+1}(\cos\theta)] + \eta_1 b_l^m l(l+1) p_l(kr) c_l^m P_l^m \Bigg) \mathbf{e_2}(\theta,0) \Bigg\}, \qquad (3.3.32)
\end{aligned}
$$

where $\mathbf{e_1}(\theta,0)$, $\mathbf{e_2}(\theta,0)$ $\mathbf{e_1}(\theta,\phi_0)$, and $\mathbf{e_2}(\theta,\phi_0)$ are defined in (3.2.22).

Recall that η_ℓ, $\ell = 1,2$ have the expansion (3.3.1) and (3.3.2), respectively, where the coefficients of r^0 in (3.3.1) and (3.3.2) are nonzero number η_1 and η_2. The lowest-order term in (3.3.32) with respect to the power of r is r^{n-1}, which is contributed by $p_n(kr)$ and $q_n(kr)$ from Remark 3.2.1. Furthermore, it is noted that coefficients of r^{n-1} in $p_n(kr)$ and $q_n(kr)$ are $\frac{k^{n-1}}{(2n+1)(2n-1)!!}$ and $\frac{(n+1)k^{n-1}}{(2n+1)(2n-1)!!}$ respectively. Due to the fact that $\mathbf{e}_1(\theta,\phi)$ and $\mathbf{e}_2(\theta,\phi)$ are linearly independent for any θ and ϕ, from Lemma 2.10.3, comparing the coefficient of r^{n-1} on both sides of (3.3.32) associated with $\mathbf{e}_1(\theta,\phi)$ for $\phi = 0$, we have

$$
\begin{aligned}
0 =& \mathrm{i}k \sum_{\substack{m=-n \\ m\neq 0}}^{n} a_n^m \frac{\sqrt{n(n+1)}}{2n+1} c_n^m P_n^m(\cos\theta) - \eta_1 \sum_{m=-n}^{n} b_n^m \cdot \frac{n+1}{2n+1} \frac{1}{\sqrt{n(n+1)}} \frac{c_n^m}{2} \cdot \Bigg((n+m) \\
&\times (n-m+1) P_n^{m-1}(\cos\theta) - P_n^{m+1}(\cos\theta) \Bigg) + \eta_1 b_n^0 \frac{n+1}{2n+1} \frac{c_n^0}{\sqrt{n(n+1)}} P_n^1(\cos\theta), \quad (3.3.33)
\end{aligned}
$$

where for the index $m = 0$ in (3.3.32) we use the property (3.2.11), and c_n^m, $m = 0,1,\ldots,n$ are nonzero constants defined in (3.2.8). Utilizing the orthogonality condition (2.10.4), from (3.3.33) we can deduce (3.3.30).

Similarly, comparing the coefficient of r^{n-1} on both sides of (3.3.32) associated with $\mathbf{e}_2(\theta,\phi)$ for $\phi = 0$ in (3.3.32), from Lemma 2.10.3, we obtain the following $n+1$ equations:

$$0 = \mathrm{i}k \sum_{\substack{m=-n \\ m\neq 0}}^{n} a_n^m \cdot \frac{n+1}{2n+1} \frac{1}{\sqrt{n(n+1)}} c_n^m \Big((n+m)(n-m+1) P_n^{m-1}(\cos\theta) - P_n^{m+1}(\cos\theta) \Big)$$

$$- \mathrm{i}k a_n^0 \cdot \frac{n+1}{2n+1} \frac{1}{\sqrt{n(n+1)}} c_n^0 P_n^1(\cos\theta) + \eta_1 \sum_{m=-n}^{n} b_n^m \frac{\sqrt{n(n+1)}}{2n+1} c_n^m P_n^m(\cos\theta), \qquad (3.3.34)$$

where for the index $m = 0$ in (3.3.32) we use the property (3.2.11). By virtue of (3.3.34), utilizing the orthogonality condition (2.10.4), we can obtain (3.3.31). □

Lemma 3.3.5. *Under the same setup to Lemma 3.3.4 and assuming that there exists $n \in \mathbb{N}\backslash\{1\}$ such that* (3.3.29) *is fulfilled, we have the following recursive linear equations:*

$$0 = \mathrm{i}k \frac{\sqrt{n(n+1)}}{2n+1} c_n^0 a_n^0 - \frac{\eta_2(n+1)}{2(2n+1)} \frac{c_n^1 (n+1)n}{\sqrt{n(n+1)}} (b_n^1 e^{\mathrm{i}\alpha\cdot\pi} + b_n^{-1} e^{-\mathrm{i}\alpha\cdot\pi}),$$

$$0 = \mathrm{i}k \frac{\sqrt{n(n+1)}}{2n+1} c_n^1 (a_n^1 e^{\mathrm{i}\alpha\cdot\pi} + a_n^{-1} e^{-\mathrm{i}\alpha\cdot\pi}) - \frac{\eta_2(n+1)}{2(2n+1)} \frac{c_n^2 (n+2)(n-1)}{\sqrt{n(n+1)}}$$

$$\times (b_n^2 e^{\mathrm{i}2\alpha\cdot\pi} + b_n^{-2} e^{-\mathrm{i}2\alpha\cdot\pi}) + \frac{\eta_2(n+1)}{2n+1} \frac{c_n^0}{\sqrt{n(n+1)}} b_n^0,$$

$$0 = \mathrm{i}k \frac{\sqrt{n(n+1)}}{2n+1} c_n^m (a_n^m e^{\mathrm{i}m\alpha\cdot\pi} + a_n^{-m} e^{-\mathrm{i}m\alpha\cdot\pi}) - \frac{\eta_2(n+1)}{2(2n+1)} \frac{c_n^{m+1}}{\sqrt{n(n+1)}} (n+m+1)$$

$$\times (n-m)(b_n^{m+1} e^{\mathrm{i}(m+1)\alpha\cdot\pi} + b_n^{-(m+1)} e^{-\mathrm{i}(m+1)\alpha\cdot\pi}) + \frac{\eta_2(n+1)}{2(2n+1)} \frac{c_n^{m-1}}{\sqrt{n(n+1)}}$$

$$\times (b_n^{m-1} e^{\mathrm{i}(m-1)\alpha\cdot\pi} + b_n^{-(m-1)} e^{-\mathrm{i}(m-1)\alpha\cdot\pi}), \quad m = 2, 3, \ldots, n-1, \qquad (3.3.35)$$

$$0 = \mathrm{i}k \frac{\sqrt{n(n+1)}}{2n+1} c_n^n (a_n^n e^{\mathrm{i}n\alpha\cdot\pi} + a_n^{-n} e^{-\mathrm{i}n\alpha\cdot\pi}) + \frac{\eta_2(n+1)}{2(2n+1)} \frac{c_n^{n-1}}{\sqrt{n(n+1)}}$$

$$\times (b_n^{n-1} e^{\mathrm{i}(n-1)\alpha\pi} + b_n^{-(n-1)} e^{-\mathrm{i}(n-1)\alpha\pi}),$$

and

$$0 = \mathrm{i}k \frac{n+1}{2(2n+1)} \frac{c_n^1}{\sqrt{n(n+1)}} (n+1)n (a_n^1 e^{\mathrm{i}\alpha\cdot\pi} + a_n^{-1} e^{-\mathrm{i}\alpha\cdot\pi}) + \eta_2 \frac{c_n^0 \sqrt{n(n+1)}}{2n+1} b_n^0,$$

$$0 = \mathrm{i}k \frac{n+1}{2(2n+1)} \frac{c_n^2}{\sqrt{n(n+1)}} (n+2)(n-1)(a_n^2 e^{2\mathrm{i}\alpha\cdot\pi} + a_n^{-2} e^{-2\mathrm{i}\alpha\cdot\pi}) - \mathrm{i}k \frac{n+1}{2n+1}$$

$$\times \frac{c_n^0}{\sqrt{n(n+1)}} a_n^0 + \eta_2 \frac{c_n^1 \sqrt{n(n+1)}}{2n+1} (b_n^1 e^{\mathrm{i}\alpha\cdot\pi} + b_n^{-1} e^{-\mathrm{i}\alpha\cdot\pi}), \qquad (3.3.36)$$

$$0 = \mathrm{i}k \frac{n+1}{2(2n+1)} \frac{c_n^m (n+m)(n-m+1)}{\sqrt{n(n+1)}} (a_n^m e^{\mathrm{i}m\alpha\cdot\pi} + a_n^{-m} e^{-\mathrm{i}m\alpha\cdot\pi}) - \mathrm{i}k \cdot \frac{n+1}{2(2n+1)}$$

$$\times \frac{c_n^{m-2}}{\sqrt{n(n+1)}}(a_n^{m-2}e^{\mathrm{i}(m-2)\alpha\cdot\pi} + a_n^{-(m-2)}e^{-\mathrm{i}(m-2)\alpha\cdot\pi}) + \eta_2\frac{c_n^{m-1}\sqrt{n(n+1)}}{2n+1}$$

$$\times (b_n^{m-1}e^{\mathrm{i}(m-1)\alpha\cdot\pi} + b_n^{-(m-1)}e^{-\mathrm{i}(m-1)\alpha\cdot\pi}), \quad m = 3,4,\ldots,n,$$

$$0 = -\mathrm{i}k\frac{n+1}{2(2n+1)}\frac{c_n^{n-1}}{\sqrt{n(n+1)}}(a_n^{n-1}e^{\mathrm{i}(n-1)\alpha\cdot\pi} + a_n^{-(n-1)}e^{-\mathrm{i}(n-1)\alpha\cdot\pi}) + \eta_2\frac{c_n^n\sqrt{n(n+1)}}{2n+1}$$

$$\times (b_n^n e^{\mathrm{i}n\alpha\cdot\pi} + b_n^{-n}e^{-\mathrm{i}n\alpha\cdot\pi}) = 0.$$

Proof. Since the generalized impedance condition (3.1.3) associated with $\boldsymbol{\eta}_2$ is imposed on $\widetilde{\Pi}_2$, substituting (3.3.29) into (3.2.27), by virtue of (3.2.9), we derive that

$$\begin{aligned}\mathbf{0} = \sum_{l=n}^{\infty}\sum_{m=-l}^{l}\frac{e^{\mathrm{i}m\alpha\cdot\pi}}{\sqrt{l(l+1)}}\Bigg\{\Bigg(&\mathrm{i}ka_l^m l(l+1)p_l(kr)c_l^m P_l^m + \boldsymbol{\eta}_2 a_l^m j_l(kr)c_l^m\frac{\operatorname{sgn}(m)}{2}[P_{l-1}^{|m|+1}(\cos\theta)\\ &+(l+|m|-1)(l+|m|)P_{l-1}^{|m|-1}(\cos\theta)] - \boldsymbol{\eta}_2 b_l^m q_l(kr)\frac{c_l^m}{2}[(l+|m|)(l-|m|+1)\\ &\times P_l^{|m|-1}(\cos\theta) - P_l^{|m|+1}(\cos\theta)]\Bigg)\mathbf{e_1}(\theta,\phi_0) + \Bigg(\mathrm{i}kb_l^m j_l(kr)c_l^m\frac{\operatorname{sgn}(m)}{2}[P_{l-1}^{|m|+1}(\cos\theta)\\ &+(l+|m|-1)(l+|m|)P_{l-1}^{|m|-1}(\cos\theta)] + \mathrm{i}ka_l^m q_l(kr)\frac{c_l^m}{2}[(l+|m|)(l-|m|+1)\\ &\times P_l^{|m|-1}(\cos\theta) - P_l^{|m|+1}(\cos\theta)] + \boldsymbol{\eta}_2 b_l^m l(l+1)p_l(kr)c_l^m P_l^m\Bigg)\mathbf{e_2}(\theta,\phi_0)\Bigg\},\end{aligned} \tag{3.3.37}$$

where $\mathbf{e_1}(\theta,0)$, $\mathbf{e_2}(\theta,0)$ $\mathbf{e_1}(\theta,\phi_0)$, and $\mathbf{e_2}(\theta,\phi_0)$ are defined in (3.2.22).

Recall that η_ℓ, $\ell = 1,2$, have the expansion (3.3.1) and (3.3.2), respectively, where the coefficients of r^0 in (3.3.1) and (3.3.2) are nonzero number η_1 and η_2. Comparing the coefficient of r^{n-1} on both sides of (3.3.37) associated with $\mathrm{e}_1(\theta,\phi)$ and $\mathrm{e}_2(\theta,\phi)$ for $\phi = \phi_0$, respectively, utilizing the orthogonality condition (2.10.4), we can derive (3.3.35) and (3.3.36). □

The next theorem characterizes the vanishing order of $\mathbf{E}$ to (3.1.1) at $\mathbf{0} \in \mathscr{E}(\widetilde{\Pi}_1, \widetilde{\Pi}_2, \boldsymbol{l})$ with $\boldsymbol{\eta}_j \in \mathscr{A}(\boldsymbol{l})$.

Theorem 3.3.1. *Let* $\mathbf{E}$ *be a solution to* (3.1.1)*, whose radial wave expansion in* $B_{\rho_0}(\mathbf{0})$ *is given by* (3.2.15)*. Consider a generalized impedance edge corner* $\mathscr{E}(\widetilde{\Pi}_1, \widetilde{\Pi}_2, \boldsymbol{l}) \Subset \Omega$*, with* $\angle(\Pi_1, \Pi_2) = \phi_0 = \alpha\pi$ *for* $\alpha \in (0,2)$ *but* $\alpha \neq 1$*. Suppose that the generalized impedance parameters* $\boldsymbol{\eta}_j$ *on* $\widetilde{\Pi}_j$*,* $j = 1,2$*, are given by* (3.3.2)*. Then it holds that* $\mathbf{E}$ *vanishes up to the order* N *at* $\mathbf{0}$*:*

$$\operatorname{Vani}(\mathbf{E};\mathbf{0}) \geq \begin{cases} 1, & \text{if } \alpha \neq \frac{1}{2} \text{ and } \alpha \neq \frac{3}{2},\\ N \in \mathbb{N}\backslash\{1\}, & \text{if } \alpha \neq \frac{q}{p},\ p = 1,\ldots,N, \text{ and for a fixed } p,q = 1,\ldots,2p-1. \end{cases}$$

Proof. We prove this theorem by induction. Assume that $\alpha \neq \frac{1}{2}$ and $\alpha \neq \frac{3}{2}$. Since the generalized impedance condition (3.1.3) associated with $\boldsymbol{\eta}_1$ is imposed on $\widetilde{\Pi}_1$, by virtue of (3.2.9) and (3.2.27), we derive that

$$\begin{aligned}\mathbf{0} = \sum\nolimits_{l,m,1}\frac{1}{\sqrt{l(l+1)}}\Bigg\{\Bigg(&\mathrm{i}ka_l^m l(l+1)p_l(kr)c_l^m P_l^m + \boldsymbol{\eta}_1 a_l^m j_l(kr)c_l^m\tfrac{\operatorname{sgn}(m)}{2}[P_{l-1}^{|m|+1}(\cos\theta)\\ &+(l+|m|-1)(l+|m|)P_{l-1}^{|m|-1}(\cos\theta)] - \boldsymbol{\eta}_1 b_l^m q_l(kr)\tfrac{c_l^m}{2}[(l+|m|)(l-|m|+1)\end{aligned}$$

$$
\times P_l^{|m|-1}(\cos\theta) - P_l^{|m|+1}(\cos\theta)]\Big)\boldsymbol{e_1}(\theta,0) + \Big(ikb_l^m j_l(kr)c_l^m \tfrac{\operatorname{sgn}(m)}{2}\big[P_{l-1}^{|m|+1}(\cos\theta)
+(l+|m|-1)(l+|m|)P_{l-1}^{|m|-1}(\cos\theta)\big] + ika_l^m q_l(kr)\tfrac{c_l^m}{2}\big[(l+|m|)(l-|m|+1)
\times P_l^{|m|-1}(\cos\theta) - P_l^{|m|+1}(\cos\theta)\big] + \eta_1 b_l^m l(l+1)p_l(kr)c_l^m P_l^m\Big)\boldsymbol{e_2}(\theta,0)\Big\}, \tag{3.3.38}
$$

where $\boldsymbol{e_1}(\theta,0)$, $\boldsymbol{e_2}(\theta,0)$ $\boldsymbol{e_1}(\theta,\phi_0)$ and $\boldsymbol{e_2}(\theta,\phi_0)$ are defined in (3.2.22). Comparing the coefficient of r^0 associated with $\boldsymbol{e_2}(\theta,0)$ on both sides of (3.3.38), in view of (3.3.2), we can obtain from Lemma 3.2.1 that

$$
0 = \mathrm{i}k\frac{4c_1^1}{6\sqrt{2}}(a_1^1 + a_1^{-1}) + \eta_1\frac{\sqrt{2}c_1^0}{3}b_1^0. \tag{3.3.39}
$$

Combining (3.3.39) with (3.3.3a) and (3.3.3b) from Lemma 3.3.3, we derive that

$$
\mathscr{A}_1\begin{bmatrix} a_1^1 + a_1^{-1} \\ a_1^1 - a_1^{-1} \\ b_1^0 \end{bmatrix} = \mathbf{0}, \quad \mathscr{A}_1 = (\alpha_{ij}^1)_{i,j=1}^3, \tag{3.3.40}
$$

where

$$
\alpha_{11} = \frac{4ikc_1^1\sin^2\phi_0}{6\sqrt{2}}, \quad \alpha_{12} = -\frac{4kc_1^1\sin\phi_0\cos\phi_0}{6\sqrt{2}}, \quad \alpha_{13} = \frac{(-\eta_2\cos\phi_0 - \eta_1)\sqrt{2}c_1^0}{3}
$$

$$
\alpha_{21} = -\frac{4ikc_1^1\sin\phi_0\cos\phi_0}{6\sqrt{2}}, \quad \alpha_{22} = -\frac{4kc_1^1\sin^2\phi_0}{6\sqrt{2}}, \quad \alpha_{23} = -\frac{\eta_2\sqrt{2}c_1^0\sin\phi_0}{3}
$$

$$
\alpha_{31} = \frac{4ikc_1^1}{6\sqrt{2}}, \quad \alpha_{32} = 0, \quad \alpha_{33} = \frac{\sqrt{2}c_1^0\eta_1}{3}.
$$

By direct calculations, it yields that

$$
|\mathscr{A}_1| = -\mathrm{i}k^2\eta_1\left(\frac{2}{3}\right)^3\frac{\sqrt{2}}{2}(c_1^1)^2 c_1^0\sin^2(\alpha\pi).
$$

Since $\alpha \neq \frac{1}{2}$, $\alpha \neq \frac{3}{2}$ and $\eta_1 \neq 0$, by virtue of (3.3.40) and $k \in \mathbb{R}_+$, it can be derived that $a_1^{\pm 1} = b_1^0 = 0$. Recall that (3.3.25) is given by Lemma 3.3.3. In view of $\alpha \neq \frac{1}{2}$ and $\alpha \neq \frac{3}{2}$, $\alpha \in (0,1)$, $k \in \mathbb{R}_+$ and $\eta_2 \neq 0$, using the fact that

$$
|\mathscr{B}_1| = -k\eta_2^2\left(\frac{2}{3}\right)^3\frac{\sqrt{2}}{2}(c_1^1)^2 c_1^0\sin^2(\alpha\pi)\cos^2(\alpha\pi) \neq 0,
$$

where $\mathscr{B}_1$ is defined in (3.3.25), we can obtain that $b_1^{\pm 1} = a_1^0 = 0$. Therefore, from Lemma 3.2.4, we prove that $\mathrm{Vani}(\mathbf{E};\mathbf{0}) \geq 1$ under conditions $\alpha \neq \frac{1}{2}$, $\alpha \neq \frac{3}{2}$, and $\eta_\ell \neq 0$, $\ell = 1,2$.

By mathematical induction, suppose that $\alpha \neq \frac{q}{p}$, $p = 1,\ldots,n-1$ and $q = 1,2,\ldots,2p-1$ for a fixed p; then we have

$$
\mathrm{Vani}(\mathbf{E};\mathbf{0}) \geq n-1.
$$

From Lemma 3.2.4, we know that

$$
a_l^m = b_l^m = 0, \quad m \in [l]_0, \quad l = 1,2,\ldots,n. \tag{3.3.41}
$$

Therefore, we see that (3.3.30), (3.3.31), (3.3.35), and (3.3.36) hold from Lemmas 3.3.4 and 3.3.5. In the following under the assumption

$$\eta_\ell \neq 0 \text{ for } \ell = 1,2 \text{ and } \alpha \neq \frac{q}{p},\ p = 1,\ldots,n, \tag{3.3.42}$$

where $q = 1,2,\ldots,2p-1$ for a fixed p, we shall show that

$$a_n^m = b_n^m = 0, \quad \forall m \in [n]_0 \tag{3.3.43}$$

by utilizing the recursive equations of (3.3.30), (3.3.31), (3.3.35), and (3.3.36). Indeed, following a similar argument as we did for (3.3.40), we come from (3.3.3a) and (3.3.3b) to derive (3.3.45) as follows

$$\mathscr{A}_n \begin{bmatrix} a_n^1 + a_n^{-1} \\ a_n^1 - a_n^{-1} \\ b_n^0 \end{bmatrix} = \mathbf{0}, \quad \mathscr{A}_n = \left(\alpha_{ij}^n\right)_{i,j=1}^3, \tag{3.3.44}$$

where

$$\begin{aligned}
\alpha_{11}^n &= \frac{\mathrm{i}kn(n+1)^2 c_n^1 \sin^2\phi_0}{2(2n+1)\sqrt{n(n+1)}}, \quad \alpha_{12}^n = -\frac{kn(n+1)^2 c_n^1 \sin\phi_0\cos\phi_0}{2(2n+1)\sqrt{n(n+1)}},\\
\alpha_{13}^n &= \frac{(-\eta_2\cos\phi_0 - \eta_1)\sqrt{n(n+1)}c_n^0}{2n+1}, \quad \alpha_{21}^n = -\frac{\mathrm{i}kn(n+1)^2 c_n^1 \sin\phi_0\cos\phi_0}{2(2n+1)\sqrt{n(n+1)}},\\
\alpha_{22}^n &= -\frac{kn(n+1)^2 c_n^1 \sin^2\phi_0}{2(2n+1)\sqrt{n(n+1)}}, \quad \alpha_{23}^n = -\frac{\eta_2\sqrt{n(n+1)}c_n^0\sin\phi_0}{2n+1},\\
\alpha_{31}^n &= \mathrm{i}k\frac{n(n+1)^2 c_n^1}{2(2n+1)\sqrt{n(n+1)}}, \quad \alpha_{32}^n = 0, \quad \alpha_{33}^n = \eta_1\frac{\sqrt{n(n+1)}c_n^0}{2n+1}.
\end{aligned}$$

It can be derived that

$$|\mathscr{A}_n| = -\mathrm{i}k^2\eta_1\left(\frac{n+1}{2n+1}\right)^3 \frac{n\sqrt{n(n+1)}}{2}\left(c_n^1\right)^2 c_n^0 \sin^2(\alpha\pi). \tag{3.3.45}$$

Since $\alpha \in (0,2)$, $\alpha \neq 1$, $\alpha \neq \frac{1}{2}$, $\alpha \neq \frac{3}{2}$, and $\eta_\ell \neq 0$, $\ell = 1,2$, by virtue of (3.3.44), (3.3.45), and Lemma 3.3.3, we have

$$a_n^{\pm1} = a_n^0 = b_n^{\pm1} = b_n^0 = 0. \tag{3.3.46}$$

Substituting (3.3.46) into the second equation of (3.3.30), (3.3.31), (3.3.35), and (3.3.36), since $k \in \mathbb{R}_+$, $\eta_\ell \neq 0$ for $\ell = 1,2$, and $c_n^2 \neq 0$, we obtain that

$$\begin{cases} a_n^2 + a_n^{-2} = 0, \\ a_n^2 e^{2\mathrm{i}\alpha\cdot\pi} + a_n^{-2}e^{-2\mathrm{i}\alpha\cdot\pi} = 0, \end{cases} \quad \begin{cases} b_n^2 + b_n^{-2} = 0, \\ b_n^2 e^{2\mathrm{i}\alpha\cdot\pi} + b_n^{-2}e^{-2\mathrm{i}\alpha\cdot\pi} = 0, \end{cases}$$

which can be shown to prove that $a_n^{\pm2} = b_n^{\pm2} = 0$, since

$$\begin{vmatrix} 1 & 1 \\ e^{\mathrm{i}2\alpha\cdot\pi} & e^{-\mathrm{i}2\alpha\cdot\pi} \end{vmatrix} = -2\mathrm{i}\sin(2\alpha\pi) \neq 0$$

under (3.3.42). Substituting

$$a_n^{\pm1} = b_n^{\pm1} = a_n^{\pm2} = b_n^{\pm2} = 0$$

into the third equation of (3.3.30), (3.3.31), (3.3.35), and (3.3.36), since $k \in \mathbb{R}_+$, $\eta_\ell \neq 0$ for $\ell = 1,2$, and $c_n^3 \neq 0$, we get that

$$\begin{cases} a_n^3 + a_n^{-3} = 0, \\ a_n^3 e^{3\mathrm{i}\alpha\cdot\pi} + a_n^{-3}e^{-3\mathrm{i}\alpha\cdot\pi} - 0, \end{cases} \quad \begin{cases} b_n^3 + b_n^{-3} = 0, \\ b_n^3 e^{3\mathrm{i}\alpha\cdot\pi} + b_n^{-3}e^{-3\mathrm{i}\alpha\cdot\pi} = 0, \end{cases}$$

which can be shown to prove that $a_n^{\pm 3} = b_n^{\pm 3} = 0$, since

$$\begin{vmatrix} 1 & 1 \\ e^{\mathrm{i}3\alpha\cdot\pi} & e^{-\mathrm{i}3\alpha\cdot\pi} \end{vmatrix} = -2\mathrm{i}\sin(3\alpha\pi) \neq 0$$

under (3.3.42). Repeating the above procedures step by step, utilizing the recursive property of (3.3.30), (3.3.31), (3.3.35), and (3.3.36), we can prove (3.3.43). Generally, assume that we have proved that

$$a_n^{\pm m} = b_n^{\pm m} = 0 \text{ for } m = 0, 1, \ldots, \ell.$$

Substituting $a_n^{\pm(\ell-1)} = b_n^{\pm(\ell-2)} = 0$ into the ℓ-th equation of (3.3.30) and (3.3.35), we can obtain that

$$\begin{cases} b_n^{\ell} + b_n^{-\ell} = 0, \\ b_n^{\ell} e^{\mathrm{i}\ell\alpha\cdot\pi} + b_n^{-\ell} e^{-\mathrm{i}\ell\alpha\cdot\pi} = 0, \end{cases} \tag{3.3.47}$$

under the assumption $\eta_1 \neq 0$ and $\eta_2 \neq 0$. Substituting $a_n^{\pm(\ell-2)} = b_n^{\pm(\ell-1)} = 0$ into the ℓ-th equation of (3.3.31) and (3.3.36), we can get that

$$\begin{cases} a_n^{\ell} + a_n^{-\ell} = 0, \\ a_n^{\ell} e^{\mathrm{i}\ell\alpha\cdot\pi} + a_n^{-2} e^{-\mathrm{i}\ell\alpha\cdot\pi} = 0, \end{cases} \tag{3.3.48}$$

Hence, from (3.3.47) and (3.3.48), under (3.3.42) it yields that $a_n^{\pm\ell} = b_n^{\pm\ell} = 0$.

Now we can directly see from (3.3.43) and Lemma 3.2.4 that

$$\mathrm{Vani}(\mathbf{E}; \mathbf{0}) \geq n,$$

which completes the proof of Theorem 3.3.1. □

3.4 Vanishing Orders for an Edge Corner $\mathscr{E}(\widetilde{\Pi}_1, \widetilde{\Pi}_2, \boldsymbol{l})$ with $\eta_j \in \mathscr{A}(\boldsymbol{l})$ or $\eta_j = 0, \infty$

In this section, we investigate the vanishing order of the solution $\mathbf{E}$ to (3.1.1) at an edge-corner point $\mathbf{0} \in \mathscr{E}(\widetilde{\Pi}_1, \widetilde{\Pi}_2, \boldsymbol{l}) \Subset \Omega$, with $\angle(\Pi_1, \Pi_2) = \phi_0 = \alpha\pi$, $\alpha \in (0,2)$, and $\alpha \neq 1$ and two different generalized impedance conditions (3.1.3) on $\widetilde{\Pi}_1$ and $\widetilde{\Pi}_2$. More specifically, we shall first consider the associated generalized impedance parameters of the generalized impedance edge corner $\mathscr{E}(\Pi_1, \Pi_2, \boldsymbol{l})$ (Theorem 3.4.1) to be, respectively, $\eta_1 \equiv \infty$ and $\eta_2 \equiv 0$ and utilize Lemma 3.4.1 to provide the vanishing order of $\mathbf{E}$ at $\mathbf{0}$. Then we will consider (Theorems 3.4.2 and 3.4.3) in the case that $\eta_2 \in \mathscr{A}(\boldsymbol{l})$ has the expansion (3.3.2), whereas η_1 could be either ∞ or 0. The reflection principle established in [10, 11]

is adopted to transform the corresponding generalized impedance edge corner to be the generalized impedance edge corner intersected by two plane cells with the generalized impedance condition (3.1.3) and two associated generalized impedance parameters belonging to $\mathscr{A}(\boldsymbol{l})$.

Lemma 3.4.1. *Let* $\mathbf{E}$ *be a solution to* (3.1.1)*, whose radial wave expansion in* $B_{\rho_0}(\mathbf{0})$ *is given by* (3.2.15)*. Consider a generalized impedance edge corner* $\mathscr{E}(\widetilde{\Pi}_1, \widetilde{\Pi}_2, \boldsymbol{l}) \Subset \Omega$*, with* $\angle(\Pi_1, \Pi_2) = \phi_0 = \alpha\pi$ *for* $\alpha \in (0,2)$ *but* $\alpha \neq 1$*. Suppose that the generalized impedance parameters* η_1 *on* $\widetilde{\Pi}_1$ *and* η_2 *on* $\widetilde{\Pi}_2$ *satisfy (ii) and (i) in* (3.1.2)*, respectively.*

Then it holds that

$$b_1^1 + b_1^{-1} = 0, \quad b_1^0 = 0, \quad a_1^1 - a_1^{-1} = 0, \tag{3.4.1a}$$

$$b_2^m + b_2^{-m} = 0, \quad m = 1, 2, \text{ and } b_2^0 = 0, \tag{3.4.1b}$$

and

$$a_1^1 e^{\mathrm{i}\alpha\cdot\pi} + a_1^{-1} e^{-\mathrm{i}\alpha\cdot\pi} = 0, \quad a_1^0 = 0, \quad b_1^1 e^{\alpha\cdot\pi} - b_1^{-1} e^{-\mathrm{i}\alpha\cdot\pi} = 0, \tag{3.4.2a}$$

$$a_2^m e^{\mathrm{i}m\alpha\cdot\pi} + a_2^{-m} e^{-\mathrm{i}m\alpha\cdot\pi} = 0, \quad m = 1,2, \text{ and } a_2^0 = 0. \tag{3.4.2b}$$

Assume that there exists an $n \in \mathbb{N}$ *such that*

$$a_l^m = b_l^m = 0, \quad l = 1,2,\ldots,n-1, \quad m \in [l]_0, \tag{3.4.3}$$

then we have

$$b_n^m + b_n^{-m} = 0, \quad m = 1,\ldots,n, \text{ and } b_n^0 = 0, \tag{3.4.4a}$$

$$a_n^m e^{\mathrm{i}m\alpha\cdot\pi} + a_n^{-m} e^{-\mathrm{i}m\alpha\cdot\pi} = 0, \quad m = 1,\ldots,n, \text{ and } a_n^0 = 0, \tag{3.4.4b}$$

and

$$\sum_{m=1}^{n} m c_n^m (a_n^m - a_n^{-m}) \frac{P_n^m(\cos\theta)}{\sin\theta} + \sum_{m=-(n+1)}^{n+1} \frac{c_{n+1}^m (n+2)}{2n+3} b_{n+1}^m \left.\frac{\partial Y_{n+1}^m}{\partial\theta}\right|_{\phi=0} = 0, \tag{3.4.5a}$$

$$\sum_{m=1}^{n} m c_n^m (b_n^m e^{\mathrm{i}m\alpha\cdot\pi} - b_n^{-m} e^{-\mathrm{i}m\alpha\cdot\pi}) \frac{P_n^m(\cos\theta)}{\sin\theta} + \sum_{m=-(n+1)}^{n+1} \frac{c_{n+1}^m (n+2)}{2n+3} a_{n+1}^m \left.\frac{\partial Y_{n+1}^m}{\partial\theta}\right|_{\phi=\phi_0} = 0, \tag{3.4.5b}$$

where c_n^m *are nonzero constants defined in* (3.2.8) *for* $m = 0,1,\ldots,n$. *Furthermore, we have*

$$b_{n+1}^m + b_{n+1}^{-m} = 0, \quad m = 1,\ldots,n+1, \text{ and } b_{n+1}^0 = 0, \tag{3.4.6a}$$

$$a_{n+1}^m e^{\mathrm{i}m\alpha\cdot\pi} + a_{n+1}^{-m} e^{-\mathrm{i}m\alpha\cdot\pi} = 0, \quad m = 1,\ldots,n+1, \text{ and } a_{n+1}^0 = 0, \tag{3.4.6b}$$

where c_{n+1}^m *are nonzero constants defined in* (3.2.8) *for* $m = 0,1,\ldots,n+1$.

Proof. We first derive (3.4.1a) and (3.4.1b). Since the generalized impedance condition (3.1.3) associated with η_1 is imposed on $\widetilde{\Pi}_1$ where $\eta_1 \equiv \infty$, using (3.2.21), we have

$$\begin{aligned} \mathbf{0} = \sum_{l=1}^{\infty}\sum_{m=-l}^{l} -\frac{1}{\sqrt{l(l+1)}} \Bigg\{ & b_l^m l(l+1) p_l(kr) Y_l^m \Big|_{\phi=0} \mathbf{e_1}(\theta,0) \\ & + \left(a_l^m j_l(kr) \frac{m}{\sin\theta} Y_l^m \Big|_{\phi=0} + b_l^m \cdot q_l(kr) \frac{\partial Y_l^m}{\partial\theta}\Big|_{\phi=0} \right) \mathbf{e_2}(\theta,0) \Bigg\}, \end{aligned} \tag{3.4.7}$$

where $\mathbf{e}_1(\theta,0)$ and $\mathbf{e}_2(\theta,0)$ are defined in (3.2.22). From Remark 3.2.1, the lowest order of (3.4.7) with respect to the power of r is r^0, which is contributed by $p_1(kr)$ and $q_1(kr)$. Similarly, the second lowest order of (3.4.7) with respect to the power of r is r^1, which is contributed by $j_1(kr)$, $p_2(kr)$, and $q_2(kr)$. Comparing the coefficient of r^0 and r^1 associated with $\mathbf{e}_1(\theta,0)$ on both sides of (3.4.7), utilizing the orthogonality property (2.10.4), we can obtain (3.4.1a) and (3.4.1b). Substituting (3.4.1b) into (3.4.7), comparing the coefficient of r^1 in the resulting equation (3.4.7) associated with $\mathbf{e}_2(\theta,0)$, using Lemma 2.10.3, we deduce that

$$(a_1^1 c_1^1 - a_1^{-1} c_1^{-1}) P_1^1(\cos\theta) = 0, \tag{3.4.8}$$

where $c_1^{\pm 1}$ are nonzero constants defined in (3.2.8). In view of (3.4.8), from (2.10.4) and $c_1^1 = c_1^{-1} \neq 0$, it yields that the last equation of (3.4.2a).

Since the generalized impedance condition (3.1.3) associated with η_2 is imposed on $\widetilde{\Pi}_2$ where $\eta_2 \equiv 0$, by virtue of (3.2.23) it yields that

$$
\begin{aligned}
\mathbf{0} = ik\sum_{l=1}^{\infty}\sum_{m=-l}^{l}\frac{1}{\sqrt{l(l+1)}}\Bigg\{ & a_l^m l(l+1)p_l(kr)Y_l^m\Big|_{\phi=\phi_0}\cdot \boldsymbol{e_1}(\theta,\phi_0) \\
& + \left(-b_l^m j_l(kr)\cdot\frac{m}{\sin\theta}Y_l^m\Big|_{\phi=\phi_0} + a_l^m q_l(kr)\cdot\frac{\partial Y_l^m}{\partial\theta}\Big|_{\phi=\phi_0}\right)\cdot\boldsymbol{e_2}(\theta,\phi_0)\Bigg\},
\end{aligned} \tag{3.4.9}
$$

where $\boldsymbol{e}_1(\theta,\phi_0)$ and $\boldsymbol{e}_2(\theta,\phi_0)$ are defined in (3.2.22). From Remark 3.2.1, the lowest order of (3.4.9) with respect to the power of r is r^0, which is contributed by $p_1(kr)$ and $q_1(kr)$. Similarly, the second lowest order of (3.4.9) with respect to the power of r is r^1, which is contributed by $j_1(kr)$, $p_2(kr)$, and $q_2(kr)$. Comparing the coefficient of r^0 and r^1 associated with $\boldsymbol{e}_1(\theta,\phi_0)$ on both sides of (3.4.9), utilizing the orthogonality property (2.10.4), we can obtain (3.4.2a) and (3.4.2b).

Comparing the coefficient of r^1 in (3.4.9) associated with $\boldsymbol{e}_2(\theta,\phi_0)$, using Lemma 2.10.3, we deduce that

$$
(b_1^1 c_1^1 e^{i\alpha\cdot\pi} - b_1^{-1}c_1^{-1}e^{-i\alpha\cdot\pi})P_1^1(\cos\theta) = 0, \tag{3.4.10}
$$

where $c_1^{\pm 1}$ are nonzero constants defined in (3.2.8). In view of (3.4.10), from (2.10.4) and $c_1^1 = c_1^{-1} \neq 0$, it yields that the last equation of (3.4.2a).

Now we are in the position to prove (3.4.4a), (3.4.5a), and (3.4.6a) under the assumption (3.4.3). Since the generalized impedance condition (3.1.3) associated with η_1 is imposed on $\widetilde{\Pi}_1$ where $\eta_1 \equiv \infty$, substituting (3.4.3) into (3.2.21), it yields that

$$
\begin{aligned}
\mathbf{0} = \sum_{l=n}^{\infty}\sum_{m=-l}^{l} -\frac{1}{\sqrt{l(l+1)}}\Bigg\{ & b_l^m l(l+1)p_l(kr)Y_l^m\Big|_{\phi=0}\boldsymbol{e_1}(\theta,0) \\
& + \left(a_l^m j_l(kr)\frac{m}{\sin\theta}Y_l^m\Big|_{\phi=0} + b_l^m\cdot q_l(kr)\frac{\partial Y_l^m}{\partial\theta}\Big|_{\phi=0}\right)\boldsymbol{e_2}(\theta,0)\Bigg\},
\end{aligned} \tag{3.4.11}
$$

where $\boldsymbol{e}_1(\theta,0)$ and $\boldsymbol{e}_2(\theta,0)$ are defined in (3.2.22).

The lowest-order term in (3.4.11) with respect to the power of r is r^{n-1}, which is contributed by $p_n(kr)$ and $q_n(kr)$ from Remark 3.2.1. Since $\boldsymbol{e}_1(\theta,\phi)$ and $\boldsymbol{e}_2(\theta,\phi)$ are linearly independent for any θ and ϕ, where $\boldsymbol{e}_i(\theta,\phi)$ are defined in (3.2.5), from Lemma 2.10.3, comparing the coefficient of r^{n-1} on both sides of (3.4.11) associated with $\boldsymbol{e}_1(\theta,0)$, we can obtain

$$
\sum_{m=0}^{n} c_n^m(b_n^m + b_n^{-m})P_n^m(\cos\theta) = 0.
$$

Utilizing the orthogonality property (2.10.4), since $c_n^m \neq 0$ for $m \in [n]_0$, (3.4.4a) holds.

From Remark 3.2.1 we know that the second lowest-order term in (3.4.11) with respect to the power of r is r^n, which is related to $j_n(kr)$, $p_{n+1}(kr)$, and $q_{n+1}(kr)$. Since $\boldsymbol{e}_1(\theta,\phi)$ and $\boldsymbol{e}_2(\theta,\phi)$ are linearly independent for any θ and ϕ, comparing the coefficient of r^n on both sides of (3.4.11) associated with $\boldsymbol{e}_1(\theta,0)$, we can obtain

$$
\sum_{m=0}^{n+1} c_{n+1}^m(b_{n+1}^m + b_{n+1}^{-m})P_{n+1}^m(\cos\theta) = 0.
$$

Using the orthogonality property (2.10.4), together with the fact that $c_{n+1}^m \neq 0$ for $m \in [n+1]_0$, we see that (3.4.6a) holds.

Similarly, in view of Remark 3.2.1, comparing the coefficient of r^n on both sides of (3.4.11) associated with $\boldsymbol{e}_2(\theta,0)$, we know that (3.4.5a) holds.

We proceed to derive (3.4.4b), (3.4.5b), and (3.4.6b) under the assumption (3.4.3). Since the generalized impedance condition (3.1.3) associated with η_2 is imposed on $\widetilde{\Pi}_2$ where $\eta_2 \equiv 0$, substituting (3.4.3) into (3.2.23), it yields that

$$\mathbf{0} = \mathrm{i}k \sum_{l=n}^{\infty} \sum_{m=-l}^{l} \frac{1}{\sqrt{l(l+1)}} \left\{ a_l^m l(l+1) p_l(kr) Y_l^m \Big|_{\phi=\phi_0} \cdot \boldsymbol{e_1}(\theta, \phi_0) \right. \\ \left. + \left(- b_l^m j_l(kr) \cdot \frac{m}{\sin\theta} Y_l^m \Big|_{\phi=\phi_0} + a_l^m q_l(kr) \cdot \frac{\partial Y_l^m}{\partial \theta} \Big|_{\phi=\phi_0} \right) \cdot \boldsymbol{e_2}(\theta, \phi_0) \right\}. \tag{3.4.12}$$

where $\boldsymbol{e}_1(\theta, \phi_0)$ and $\boldsymbol{e}_2(\theta, \phi_0)$ are defined in (3.2.22).

The lowest-order term in (3.4.12) with respect to the power of r is r^{n-1}, which is contributed by $p_n(kr)$ and $q_n(kr)$ from Remark 3.2.1. Since $\boldsymbol{e}_1(\theta, \phi)$ and $\boldsymbol{e}_2(\theta, \phi)$ are linearly independent for any θ and ϕ, where $\boldsymbol{e}_i(\theta, \phi)$ are defined in (3.2.22), from Lemma 2.10.3, comparing the coefficient of r^{n-1} on both sides of (3.4.12) associated with $\boldsymbol{e}_1(\theta, \phi_0)$, we can obtain

$$\sum_{m=0}^{n} c_n^m (a_n^m e^{\mathrm{i}m\alpha\cdot\pi} + a_n^{-m} e^{-\mathrm{i}m\alpha\cdot\pi}) P_n^m(\cos\theta) = 0. \tag{3.4.13}$$

Using the orthogonality property (2.10.4), together with the fact that $c_n^m \neq 0$ for $m \in [n]_0$, we can obtain (3.4.4b).

From Remark 3.2.1 we know that the second lowest-order term in (3.4.12) with respect to the power of r is r^n, which is related to $j_n(kr)$, $p_{n+1}(kr)$, and $q_{n+1}(kr)$. Since $\boldsymbol{e}_1(\theta, \phi)$ and $\boldsymbol{e}_2(\theta, \phi)$ are linearly independent for any θ and ϕ, comparing the coefficient of r^n on both sides of (3.4.12) associated with $\boldsymbol{e}_1(\theta, \phi_0)$, we can get

$$\sum_{m=0}^{n+1} c_{n+1}^m (a_{n+1}^m e^{\mathrm{i}m\alpha\cdot\pi} + a_{n+1}^{-m} e^{-\mathrm{i}m\alpha\cdot\pi}) P_{n+1}^m(\cos\theta) = 0.$$

Utilizing the orthogonality property (2.10.4), since $c_{n+1}^m \neq 0$ for $m \in [n+1]_0$, we derive (3.4.6b).

Similarly, in view of Remark 3.2.1, comparing the coefficient of r^n on both sides of (3.4.12) associated with $\boldsymbol{e}_2(\theta, \phi_0)$, we know that (3.4.5b) holds.

The proof is complete. □

Theorem 3.4.1. *Under the same setup in Lemma 3.4.1, we have that*

$$\mathrm{Vani}(\mathbf{E}; \mathbf{0}) \geq N, \quad \text{if } \alpha \neq \frac{q}{2p}, \; p = 1, \ldots, N,$$

where $N \in \mathbb{N}$ and $q = 1, 2, \ldots, 4p-1$ for a fixed p.

Proof. We prove this theorem by induction. We first assume that

$$\alpha \neq \frac{1}{2} \text{ and } \alpha \neq \frac{3}{2}, \tag{3.4.14}$$

and prove that $\mathrm{Vani}(\mathbf{E}; \mathbf{0}) \geq 1$. Since the generalized impedance condition (3.1.3) associated with η_1 is imposed on $\widetilde{\Pi}_1$ with $\eta_1 \equiv \infty$, we know from Lemma 3.4.1 that (3.4.1a) holds. Similarly, since the generalized impedance condition (3.1.3) associated with η_2 is imposed on $\widetilde{\Pi}_2$ with $\eta_2 \equiv 0$, we know (3.4.2a) from Lemma 3.4.1.

Combining (3.4.1a) and (3.4.2a), it yields that

$$\begin{cases} a_1^1 - a_1^{-1} = 0, \\ a_1^1 e^{\mathrm{i}\alpha\cdot\pi} + a_1^{-1} e^{-\mathrm{i}\alpha\cdot\pi} = 0, \end{cases} \quad \begin{cases} b_1^1 + b_1^{-1} = 0, \\ b_1^1 e^{\mathrm{i}\alpha\cdot\pi} - b_1^{-1} e^{-\mathrm{i}\alpha\cdot\pi} = 0. \end{cases} \tag{3.4.15}$$

Under (3.4.14) we have

$$\begin{vmatrix} 1 & -1 \\ e^{i\alpha\cdot\pi} & e^{-i\alpha\cdot\pi} \end{vmatrix} = 2\cos(\alpha\cdot\pi) \neq 0,$$

which implies that $a_1^{\pm 1} = b_1^{\pm 1} = 0$ from (3.4.15). Since $a_1^0 = b_1^0 = 0$, we have proved $\mathrm{Vani}(\mathbf{E};\mathbf{0}) \geq 1$ from Lemma 3.2.4 under the assumption (3.4.14).

Now, by the induction we assume that $\mathrm{Vani}(\mathbf{E};\mathbf{0}) \geq n-1$, under the condition that

$$\alpha \neq \frac{2q+1}{2p}, p = 1,\ldots,n, \text{ and } q = 0,1,\ldots,2p-1 \text{ for a fixed } p. \tag{3.4.16}$$

Then we know from Lemma 3.2.4 that

$$a_l^m = 0 \text{ for } l = 1,\ldots,n-1 \text{ and } m \in [l]_0. \tag{3.4.17}$$

Due to (3.4.17) and the fact that the generalized impedance condition (3.1.3) associated with η_1 is imposed on $\widetilde{\Pi}_1$ with $\eta_1 \equiv \infty$, we have from Lemma 3.4.1, (3.4.4a) and (3.4.6a) that

$$b_n^m + b_n^{-m} = 0, \quad m = 1,\ldots,n, \text{ and } b_n^0 = 0, \tag{3.4.18}$$

$$b_{n+1}^m + b_{n+1}^{-m} = 0, \quad m = 1,\ldots,n+1, \text{ and } b_{n+1}^0 = 0. \tag{3.4.19}$$

Substituting (3.4.19) into the first equation of (3.4.5a) yields that

$$a_2^m - a_2^{-m} = 0, \quad m = 1,\ldots,n, \tag{3.4.20}$$

by noting that $c_{n+1}^m = c_{n+1}^{-m} \neq 0$ for $m = 1,\ldots,n$, where c_{n+1}^m and c_{n+1}^{-m} are defined in (3.2.8).

Similarly, due to (3.4.17) and the fact that the generalized impedance condition (3.1.3) associated with η_2 is imposed on $\widetilde{\Pi}_2$ with $\eta_2 \equiv 0$, using Lemma 3.4.1, we get that

$$a_n^m e^{im\alpha\pi} + a_n^{-m} e^{-im\alpha\pi} = 0, \quad m = 1,\ldots,n, \text{ and } a_n^0 = 0 \tag{3.4.21}$$

by (3.4.4b) and

$$a_{n+1}^m e^{im\alpha\pi} + a_{n+1}^{-m} e^{-im\alpha\pi} = 0, \quad m = 1,\ldots,n+1, \text{ and } a_{n+1}^0 = 0 \tag{3.4.22}$$

by (3.4.6b). Substituting (3.4.22) into the second equation of (3.4.5b) yields that

$$b_n^m e^{im\alpha\cdot\pi} - b_n^{-m} e^{-im\alpha\cdot\pi} = 0, \quad m = 1,\ldots,n \tag{3.4.23}$$

by using the fact that $c_{n+1}^m = c_{n+1}^{-m} \neq 0$ for $m = 1,\ldots,n$ and the definition of $Y_{n+1}^m(\theta,\phi)$, where c_{n+1}^m and c_{n+1}^{-m} are defined in (3.2.8).

Combing (3.4.18), (3.4.20), and (3.4.21) with (3.4.23), we obtain that

$$\begin{cases} a_n^m - a_n^{-m} = 0, \\ a_n^m e^{im\alpha\cdot\pi} + a_n^{-m} e^{-im\alpha\cdot\pi} = 0, \end{cases} \begin{cases} b_n^m + b_n^{-m} = 0, \\ b_n^m e^{im\alpha\cdot\pi} - b_n^{-m} e^{-im\alpha\cdot\pi} = 0, \end{cases} \quad m = 1,\ldots,n. \tag{3.4.24}$$

Under the assumption (3.4.16), it is not difficult to see that

$$\begin{vmatrix} 1 & -1 \\ e^{im\alpha\cdot\pi} & e^{-im\alpha\cdot\pi} \end{vmatrix} = 2\cos(m\alpha\cdot\pi) \neq 0,$$

which imply that $a_n^{\pm m} = b_n^{\pm m} = 0$ in view of (3.4.24). Due to (3.4.18) and (3.4.21), we have $a_n^0 = b_n^0 = 0$; hence, we have proved $\mathrm{Vani}(\mathbf{E};\mathbf{0}) \geq n$ from Lemma 3.2.4 under the assumption (3.4.16). The proof is now completed. □

In the following two theorems, we study the generalized impedance edge corner $\mathscr{E}(\widetilde{\Pi}_1,\widetilde{\Pi}_2,\boldsymbol{l})$ where the generalized impedance parameter η_2 on $\widetilde{\Pi}_2$ satisfies (iii) in (3.1.2) and has the expansion (3.3.2), whereas the generalized impedance parameter η_1 on $\widetilde{\Pi}_1$ satisfies either (i) or (ii) in (3.1.2). For this purpose, we will make use of the reflection principles for the Maxwell equations established in [10, 11], as stated below in Lemma 3.4.2, where for a two-dimensional plane $\Pi \in \mathbb{R}^3$, ν_Π and $\mathscr{R}_\Pi$ are, respectively, the unit normal to Π and the reflection with respect to Π.

Lemma 3.4.2 ([11, Theorems 2.1–2.2]). *Consider a generalized impedance edge corner $\mathscr{E}(\widetilde{\Pi}_1, \widetilde{\Pi}_2,\boldsymbol{l}) \Subset \Omega$, with $\angle(\Pi_1,\Pi_2) = \phi_0 = \alpha\pi$ for $\alpha \in (0,1)$. Assume that the generalized impedance parameter η_2 on $\widetilde{\Pi}_2$ satisfies (iii) in (3.1.2) and has the expansion (3.3.2) while the generalized impedance parameter η_1 on $\widetilde{\Pi}_1$ satisfies (ii) in (3.1.2) (i.e., $\eta_1 \equiv \infty$). Recall that Π_1 be a plane containing $\widetilde{\Pi}_1$. Let $\widetilde{\Pi}_2' = \mathscr{R}_{\Pi_1}(\widetilde{\Pi}_2)$. Then*

$$\nu_{\widetilde{\Pi}_2'} \wedge (\nabla \wedge \mathbf{E}) + \widetilde{\eta}_2(\nu_{\widetilde{\Pi}_2'} \wedge \mathbf{E}) \wedge \nu_{\widetilde{\Pi}_2'} = \mathbf{0} \text{ on } \widetilde{\Pi}_2', \tag{3.4.25}$$

where $\nu_{\Pi_2'}$ points to the interior of $\mathscr{E}(\widetilde{\Pi}_1,\widetilde{\Pi}_2',\boldsymbol{l})$ and $\widetilde{\eta}_2(\mathbf{x}) = \eta_2(\mathscr{R}_{\Pi_1}(\mathbf{x}))$ for $\mathbf{x} \in \widetilde{\Pi}_2'$.

Similarly, consider a generalized impedance edge corner $\mathscr{E}(\widetilde{\Pi}_1,\widetilde{\Pi}_2,\boldsymbol{l}) \Subset \Omega$, with $\angle(\Pi_1,\Pi_2) = \phi_0 = \alpha\pi$ for $\alpha \in (0,1)$. Assume that the generalized impedance parameter η_2 on $\widetilde{\Pi}_2$ satisfies (iii) in (3.1.2) and has the expansion (3.3.2) while the generalized impedance parameter η_1 on $\widetilde{\Pi}_1$ satisfies (i) in (3.1.2) (i.e., $\eta_1 \equiv 0$). Then

$$\nu_{\widetilde{\Pi}_2'} \wedge (\nabla \wedge \mathbf{E}) + \widetilde{\eta}_2(\nu_{\widetilde{\Pi}_2'} \wedge \mathbf{E}) \wedge \nu_{\widetilde{\Pi}_2'} = \mathbf{0} \text{ on } \widetilde{\Pi}_2',$$

where $\nu_{\Pi_2'}$ points to the interior of $\mathscr{E}(\widetilde{\Pi}_1,\widetilde{\Pi}_2',\boldsymbol{l})$ and $\widetilde{\eta}_2(\mathbf{x}) = \eta_2(\mathscr{R}_{\Pi_1}(\mathbf{x}))$ for $\mathbf{x} \in \widetilde{\Pi}_2'$.

Theorem 3.4.2. *Let $\mathbf{E}$ be a solution to (3.1.1). Consider a generalized impedance edge corner $\mathscr{E}(\widetilde{\Pi}_1,\widetilde{\Pi}_2,\boldsymbol{l}) \Subset \Omega$, with $\angle(\Pi_1,\Pi_2) = \phi_0 = \alpha\pi$ for $\alpha \in (0,2)$ but $\alpha \neq 1$. Assume that the generalized impedance parameter η_2 on $\widetilde{\Pi}_2$ satisfies (iii) in (3.1.2) and has the expansion (3.3.2) while the generalized impedance parameter η_1 on $\widetilde{\Pi}_1$ satisfies (ii) in (3.1.2) (i.e., $\eta_1 \equiv \infty$). Then*

$$\mathrm{Vani}(\mathbf{E};\mathbf{0}) \geq N, \quad \text{if } \alpha \neq \frac{q}{2p},\ p = 1,\ldots,N, \tag{3.4.26}$$

where $N \in \mathbb{N}$ and $q = 1,2,\ldots,4p-1$ for a fixed p.

Proof. Let $\widetilde{\Pi}_2' = \mathscr{R}_{\Pi_1}(\widetilde{\Pi}_2)$, where Π_1 is a plane containing $\widetilde{\Pi}_1$. With the help of Lemma 3.4.2, we know that $\mathbf{E}$ satisfies the generalized impedance boundary condition (3.4.25) on $\widetilde{\Pi}_2'$. Since $\mathbf{x} \in \widetilde{\Pi}_2$, we have the spherical coordinate of $\mathbf{x} = (r,\theta,\phi_0)$, where $0 \leq r \leq h$, $\theta \in [-\pi,\pi]$, and $\phi_0 = \alpha\pi$. It is clear that the spherical coordinate of $\mathscr{R}_{\Pi_1}(\mathbf{x}))$ for $\mathbf{x} \in \widetilde{\Pi}_2$ is given by

$$(r,\theta,\phi_1), \text{ with } \phi_1 = 2-\alpha \in (0,2).$$

Recall that η_2 has the expansion (3.3.2). Although $\mathbf{x} \in \widetilde{\Pi}_2$ and $\mathscr{R}_{\Pi_1}(\mathbf{x})) \in \widetilde{\Pi}_2'$ have different azimuthal angles but they have the same polar angle θ, hence we know from Definition 3.2.2 that $\widetilde{\eta}_2$ has the same expansion (3.3.2) as η_2.

We can easily see the following cases for the dihedral angle between $\widetilde{\Pi}_2$ and $\widetilde{\Pi}_2'$:

$$\angle(\widetilde{\Pi}_2,\widetilde{\Pi}_2') = \begin{cases} 2\alpha\pi \in (0,\pi], & \alpha \in (0,1/2), \\ 2(1-\alpha)\pi \in (0,\pi], & \alpha \in [1/2,1), \\ 2(\alpha-1)\pi \in (0,\pi], & \alpha \in (1,3/2), \\ 2(2-\alpha)\pi \in (0,\pi], & \alpha \in [3/2,2). \end{cases}$$

Based on these, we divide our proof into four cases. As we stated earlier, due to the invariance of the Maxwell system (3.1.1) under rigid motions, we may assume that the generalized impedance edge corner $\mathscr{E}(\widetilde{\Pi}_2, \widetilde{\Pi}_2', \boldsymbol{l}) \Subset \Omega$ is placed as in Fig. 3.1.

Case 1. If $\alpha \in (0, 1/2)$, then $2\alpha \in (0, 1)$. By virtue of Theorem 3.3.1, if

$$2\alpha \neq \frac{q}{p}, \quad p = 1, \ldots, N, \text{ and } q = 1, \ldots, p-1 \text{ for a fixed } p, \tag{3.4.27}$$

we have $\mathrm{Vani}(\mathbf{E}; \mathbf{0}) \geq N$. It is easy to see that (3.4.27) is equivalent to

$$\alpha \neq \frac{q}{2p}, \quad p = 1, \ldots, N, \text{ and } q = 1, \ldots, p-1 \text{ for a fixed } p. \tag{3.4.28}$$

Case 2. If $\alpha \in [1/2, 1)$, then $2(1-\alpha) \in (0, 1]$. By virtue of Theorem 3.3.1, if

$$2(1-\alpha) \neq \frac{q}{p}, \quad p = 1, \ldots, N, \text{ and } q = 1, \ldots, p \text{ for a fixed } p, \tag{3.4.29}$$

we have $\mathrm{Vani}(\mathbf{E}; \mathbf{0}) \geq N$. It is easy to see that (3.4.29) is equivalent to

$$\alpha \neq \frac{q}{2p}, \quad p = 1, \ldots, N, \text{ and } q = p, \ldots, 2p-1 \text{ for a fixed } p. \tag{3.4.30}$$

Case 3. If $\alpha \in (1, 3/2)$, then $2(\alpha - 1) \in (0, 1)$. By virtue of Theorem 3.3.1, if

$$2(\alpha-1) \neq \frac{q}{p}, \quad p = 1, \ldots, N, \text{ and } q = 1, \ldots, p-1 \text{ for a fixed } p, \tag{3.4.31}$$

we have $\mathrm{Vani}(\mathbf{E}; \mathbf{0}) \geq N$. It is easy to see that (3.4.31) is equivalent to

$$\alpha \neq \frac{q}{2p}, \quad p = 1, \ldots, N, \text{ and } q = 2p+1, \ldots, 3p-1 \text{ for a fixed } p. \tag{3.4.32}$$

Case 4. If $\alpha \in [3/2, 2)$, then $2(2-\alpha) \in (0, 1]$. By virtue of Theorem 3.3.1, if

$$2(2-\alpha) \neq \frac{q}{p}, \quad p = 1, \ldots, N, \text{ and } q = 1, \ldots, p-1 \text{ for a fixed } p, \tag{3.4.33}$$

we have $\mathrm{Vani}(\mathbf{E}; \mathbf{0}) \geq N$. It is easy to see that (3.4.33) is equivalent to

$$\alpha \neq \frac{q}{2p}, \quad p = 1, \ldots, N, \text{ and } q = 3p, \ldots, 4p-1 \text{ for a fixed } p. \tag{3.4.34}$$

Now Theorem 3.4.2 is a direct consequence of (3.4.28), (3.4.30), (3.4.32), and (3.4.34). □

With the help of Lemma 3.4.2 and similar argument to the one for Theorem 3.4.2, we can demonstrate the following theorem, whose detailed proof is omitted.

Theorem 3.4.3. *Let* $\mathbf{E}$ *be a solution to* (3.1.1)*. Consider a generalized impedance edge corner* $\mathscr{E}(\widetilde{\Pi}_1, \widetilde{\Pi}_2, \boldsymbol{l}) \Subset \Omega$*, with* $\angle(\Pi_1, \Pi_2) = \phi_0 = \alpha\pi$ *for* $\alpha \in (0, 1)$*. Assume that the generalized impedance parameter* η_2 *on* $\widetilde{\Pi}_2$ *satisfies (iii) in* (3.1.2) *and has the expansion* (3.3.2) *while the generalized impedance parameter* η_1 *on* $\widetilde{\Pi}_1$ *satisfies (i) in* (3.1.2) *(i.e.,* $\eta_1 \equiv 0$*). Then*

$$\mathrm{Vani}(\mathbf{E}; \mathbf{0}) \geq N, \quad \text{if } \alpha \neq \frac{q}{2p}, \, p = 1, \ldots, N,$$

where $N \in \mathbb{N}$ *and* $q = 1, 2, \ldots, 4p-1$ *for a fixed* p*.*

3.5 Irrational Intersections and Infinite Vanishing Orders

From the results we provided in Sects. 3.3 and 3.4, we see that the vanishing order of the eigenfunction $\mathbf{E}$ at a generalized impedance edge corner relies on the degree of the dihedral angle of the underlying corner. Next, we introduce the irrational and rational edge corners, and based on the results in Sects. 3.3 and 3.4, we then show that the vanishing order of the eigenfunction at an irrational edge corner is generically infinity, and hence, it vanishes identically in Ω, leading to the strong uniqueness continuation principle in such cases.

Definition 3.5.1. Let $\mathscr{E}(\widetilde{\Pi}_1, \widetilde{\Pi}_2, \boldsymbol{l})$ be an edge corner defined in Sect. 3.1, with the corresponding dihedral angle of $\widetilde{\Pi}_1$ and $\widetilde{\Pi}_2$ given by $\phi_0 = \alpha\pi$ for $\alpha \in (0,2)$ but $\alpha \neq 1$. If α is an irrational number, then the edge corner is called irrational. If α is a rational number of the form q/p with $p, q \in \mathbb{N}$ being irreducible, the edge corner is called rational and p is referred to as its rational degree.

We readily have the following conclusions from Theorems 3.3.1, 3.4.1, 3.4.2, and 3.4.3.

Theorem 3.5.1. *Let* $\mathbf{E}$ *be a solution to* (3.1.1). *Consider an irrational generalized impedance edge corner* $\mathscr{E}(\widetilde{\Pi}_1, \widetilde{\Pi}_2, \boldsymbol{l}) \Subset \Omega$, *with* $\angle(\Pi_1, \Pi_2) = \phi_0 = \alpha\pi$ *for* $\alpha \in (0,2)$ *but* $\alpha \neq 1$. *Under the same requirements on* η_1 *and* η_2 *as the ones for either of Theorems 3.3.1, 3.4.1, 3.4.2, and 3.4.3, it holds that*

$$\mathrm{Vani}(\mathbf{E}; \mathbf{0}) = +\infty, \quad \mathbf{0} \in \boldsymbol{l}.$$

Remark 3.5.1. As remarked at the beginning of Sect. 3.2, the statement of Theorem 2.12.1 actually holds for any edge-corner point rather than just the origin.

References

1. M. Abramowitz and I. A. Stegun, *Handbook of mathematical functions: with formulas, graphs, and mathematical tables*, vol. 55, Courier Corporation, (1964).
2. L. Alessandra and V. Felli, *Unique continuation from the edge of a crack*, arXiv:2004.11177
3. G. Bao, H. Liu and J. Zou, *Nearly cloaking the full Maxwell equations: cloaking active contents with general conducting layers*, J. Math. Pures Appl. (9), **101** (2014), 716–733.
4. D. Colton and R. Kress, *Looking back on inverse scattering theory*, SIAM Review, **60(40)** (2018), 779–807.
5. D. Colton and R. Kress, *Inverse acoustic and electromagnetic scattering theory*, volume 93 of Applied Mathematical Sciences. Springer, Cham, 4th edition, 2019.
6. H. Diao, H. Liu, L. Zhang and J. Zou, *Unique continuation from a generalized impedance edge-corner for Maxwell's system and applications to inverse problems*, Inverse Problems, 37 (2021), 035004.
7. S. G. Krantz and H. R. Parks, *A primer of real analytic functions*, 2nd edition, Birkhäuser Boston, Inc., Boston, MA, (2002).
8. H. Liu, *A global uniqueness for formally determined inverse electromagnetic obstacle scattering*, Inverse Problems, **24** (2008), 035018 (13pp).
9. H. Liu, L. Rondi and J. Xiao, Mosco convergence for $H(curl)$ spaces, higher integrability for Maxwell's equations, and stability in direct and inverse EM scattering problems, *J. Eur. Math. Soc.*, *J. Eur. Math. Soc.*, **21** (2019), no. 10, 2945–2993.
10. H. Liu, M. Yamamoto and J. Zou, *Reflection principle for the Maxwell equations and its application to inverse electromagnetic scattering*, Inverse Problems, **23** (2007), 2357–2366.
11. H. Liu, M. Yamamoto and J. Zou, *New reflection principles for Maxwell's equations and their applications*, Numer. Math. Theor. Meth. Appl., **2(1)** (2009), 1–17.
12. H. Liu and T. Zhou, *On approximate electromagnetic cloaking by transformation media*, SIAM J. Appl. Math., **71** (2011), 218–241.
13. H. Liu and J. Zou, *Uniqueness in an inverse acoustic obstacle scattering problem for both sound-hard and sound-soft polyhedral scatterers*, Inverse Problems, **22** (2006), 515–524.
14. F. Treves, *Introduction to Pseudodifferential and Fourier Integral Operators*, Vol. 1, Plenum Press, New York, (1980).

Chapter 4
Inverse Obstacle and Diffraction Grating Scattering Problems

4.1 Unique Recovery Results for Inverse Obstacle Scattering Problems

4.1.1 Unique Recovery Results for the Inverse Obstacle Problem with the Helmholtz Equation

We first consider the inverse acoustic obstacle problem. Let $k = \omega/c \in \mathbb{R}_+$ be the wavenumber of a time harmonic wave with $\omega \in \mathbb{R}_+$ and $c \in \mathbb{R}_+$ signifying the frequency and sound speed, respectively. Let $\Omega \subset \mathbb{R}^n$, $n = 2,3$, be a bounded domain with a Lipschitz boundary $\partial\Omega$ and a connected complement $\mathbb{R}^n\backslash\overline{\Omega}$. Furthermore, let the incident field u^i be a plane wave of the form

$$u^i := u^i(\mathbf{x};k,\mathbf{d}) = e^{ik\mathbf{x}\cdot\mathbf{d}}, \quad \mathbf{x} \in \mathbb{R}^n, \tag{4.1.1}$$

where $\mathbf{d} \in \mathbb{S}^{n-1}$ denotes the incident direction of the impinging wave and $\mathbb{S}^{n-1} := \{\mathbf{x} \in \mathbb{R}^n : |\mathbf{x}| = 1\}$. Physically, Ω is an impenetrable obstacle that is unknown or inaccessible, and u^i signifies the detecting wave field that is used for probing the obstacle. The presence of the obstacle interrupts the propagation of the incident wave and generates the so-called scattered wave field u^s. Let $u := u^i + u^s$ be the resulting total wave field; then the forward scattering problem can be described by the following Helmholtz system:

$$\begin{cases} \Delta u + k^2 u = 0 & \text{in } \mathbb{R}^n\backslash\overline{\Omega}, \\ u = u^i + u^s & \text{in } \mathbb{R}^n\backslash\overline{\Omega}, \\ \mathscr{B}(u) = 0 & \text{on } \partial\Omega, \\ \lim\limits_{r\to\infty} r^{\frac{n-1}{2}} \left(\dfrac{\partial u^s}{\partial r} - iku^s \right) = 0. \end{cases} \tag{4.1.2}$$

The limiting equation above is known as the Sommerfeld radiation condition which holds uniformly in $\hat{\mathbf{x}} := \mathbf{x}/|\mathbf{x}| \in \mathbb{S}^{n-1}$ and characterizes the outgoing nature of the scattered wave field u^s. The boundary operator $\mathscr{B}$ could be Dirichlet type, $\mathscr{B}(u) = u$; or Neumann type, $\mathscr{B}(u) = \partial_\nu u$; or Robin type, $\mathscr{B}(u) = \partial_\nu u + \eta u$, corresponding to that Ω is a sound-soft, sound-hard or impedance obstacle, respectively. Here ν denotes the exterior unit normal vector to $\partial\Omega$ and $\eta \in L^\infty(\partial\Omega)$ signifies a boundary impedance parameter. It is required that $\Re\eta \geq 0$ and $\Im\eta \geq 0$. In what follows, we formally take $u = 0$ on $\partial\Omega$ as $\partial_\nu u + \eta u = 0$ on $\partial\Omega$ with $\eta = +\infty$. In doing so, we can unify all three boundary conditions as the generalized impedance boundary condition:

$$\mathscr{B}(u) = \partial_\nu u + \eta u = 0 \quad \text{on } \partial\Omega, \tag{4.1.3}$$

H. Diao, H. Liu, *Spectral Geometry and Inverse Scattering Theory*,
https://doi.org/10.1007/978-3-031-34615-6_4

where η could be ∞, corresponding to a sound-soft obstacle. The forward scattering problem (4.1.2) is well understood [16, 47] and there exists a unique solution $u \in H^1_{loc}(\mathbb{R}^2\backslash\overline{\Omega})$ that admits the following asymptotic expansion:

$$u^s(\mathbf{x};k,\mathbf{d}) = \frac{e^{ikr}}{r^{(n-1)/2}} u_\infty(\hat{\mathbf{x}};k,\mathbf{d}) + \mathscr{O}\left(\frac{1}{r^{(n+1)/2}}\right) \quad \text{as } r \to \infty \tag{4.1.4}$$

which holds uniformly with respect to all directions $\hat{\mathbf{x}} := \mathbf{x}/|\mathbf{x}| \in \mathbb{S}^1$. The complex valued function u_∞ in (4.1.4) defined over the unit sphere $\mathbb{S}^1$ is known as the far-field pattern with $\hat{\mathbf{x}} \in \mathbb{S}^1$ signifying the observation direction. *The inverse obstacle scattering problem* is to recover Ω by using the knowledge of the far-field pattern $u_\infty(\hat{\mathbf{x}};k,\mathbf{d})$. By introducing an operator $\mathscr{F}$ which sends the obstacle to the corresponding far-field pattern through the Helmholtz system (4.1.2), the inverse obstacle problem can be formulated as the following abstract operator equation:

$$\mathscr{F}(\Omega,\eta) = u_\infty(\hat{\mathbf{x}};k,\mathbf{d}), \tag{4.1.5}$$

where $\mathscr{F}$ is defined by the forward obstacle scattering system and is non-linear. That is, one intends to determine (Ω,η) from the knowledge of $u_\infty(\hat{\mathbf{x}};k,\mathbf{d})$.

A primary issue for the inverse obstacle problem (4.1.5) is the unique identifiability, which is concerned with the sufficient conditions such that the correspondence between Ω and u_∞ is one-to-one. There is a widespread belief that one can establish uniqueness for (4.1.5) by a single or at most finitely many far-field patterns. We remark that by a single far-field pattern we mean that $u_\infty(\hat{\mathbf{x}};k,\mathbf{d})$ is collected for all $\hat{\mathbf{x}} \in \mathbb{S}^1$, but is associated with a fixed incident $e^{ik\mathbf{x}\cdot\mathbf{d}}$. Phrased in the geometric term, it states that the analytic function u_∞ on the unit sphere associated with at most finitely many k and $\mathbf{d}$ can supply a global parameterization of a generic domain Ω. This problem is known as the *Schiffer problem* in the inverse scattering community. It is named after M. Schiffer for his pioneering contribution around 1960 which actually appeared as a private communication in the monograph by Lax and Phillips [35]. There is a long and colourful history on the study of the Schiffer problem, and we refer to a recent survey paper by Colton and Kress [15] which contains an excellent account of the historical developments of this problem. Uniqueness results with a fixed wave number and all incident directions for the Neumann and transmission boundary conditions by means of boundary integral equation methods can be found in [31], which simplified Isakovs approach [26] to prove uniqueness in inverse scattering for penetrable obstacles.

Recent progress on the Schiffer problem is made on general polyhedral obstacles in $\mathbb{R}^n$, $n \geq 2$. Uniqueness and stability results by using a finite number of far-field patterns can be found in [2, 14, 21, 23, 37, 38, 43–46]. The major idea is to make use of the reflection principle for the Laplacian eigenfunction to propagate the so-called Dirichlet or Neumann hyperplanes. In the two-dimensional case, the Dirichlet and Neumann hyperplanes are actually the nodal and singular lines introduced in the present chapter. In [44], the authors made an effort to answer the unique determination issue for impedance-type obstacles but gave only a partial solution to this fundamental problem. In this subsection, we provide a solution to this inverse obstacle problem in two dimensions, and the approach is uniform to sound-soft, sound-hard and impedance-type obstacles. This approach is completely local and enables us to show in a rather general scenario that one can determine an impedance obstacle as well as its surface impedance by at most two far-field patterns. We follow the treatment in [10]. The corresponding unique determination for a general polyhedral acoustic obstacle in $\mathbb{R}^3$ by finite many far-field measurements shall be considered later by using the local geometrical structure of Laplacian eigenfunctions developed in Sects. 2.9–2.13.

Consider an obstacle Ω associated with the generalized impedance boundary condition (4.1.3). It is called an admissible polygonal obstacle if $\Omega \subset \mathbb{R}^2$ is an open polygon, and on each edge of $\partial\Omega$, η is either a constant (possibly zero) or ∞. That is, each edge $\mathscr{K}$ of an admissible polygonal obstacle is either sound-soft ($\eta \equiv \infty$ on $\mathscr{K}$) or sound-hard ($\eta \equiv 0$ on $\mathscr{K}$), or impedance-type (η

is a constant on $\mathscr{K}$). It is emphasized that η may take different values on different edges of $\partial\Omega$. We write (Ω,η) to signify an admissible polygonal obstacle.

Definition 4.1.1. Let (Ω,η) be an admissible polygonal obstacle. If all the angles of its corners are irrational, then it is said to be an *irrational obstacle*. If a corner angle of Ω is rational, then it is called a *rational obstacle*. The smallest degree of the rational corner angles of Ω is referred to as the *rational degree* of Ω.

It is easy to see that for a rational polygonal obstacle Ω in Definition 4.1.1, the rational degree of Ω is at least 2.

Definition 4.1.2. Ω is said to be an admissible complex polygonal obstacle if it consists of finitely many admissible polygonal obstacles. That is,

$$(\Omega,\eta)=\bigcup_{j=1}^{l}(\Omega_j,\eta_j), \tag{4.1.6}$$

where $l\in\mathbb{N}$ and each (Ω_j,η_j) is an admissible polygonal obstacle. Here, we define

$$\eta=\sum_{j=1}^{l}\eta_j\chi_{\partial\Omega_j}. \tag{4.1.7}$$

Moreover, Ω is said to be irrational if all of its component polygonal obstacles are irrational; otherwise, it is said to be rational. For the latter case, the smallest degree among all the degrees of its rational components is defined to be the degree of the complex obstacle Ω.

Next, we first consider the determination of an admissible complex irrational polygonal obstacle by at most two far-field patterns. We have the following local uniqueness result.

Theorem 4.1. *Let (Ω,η) and $(\widetilde{\Omega},\widetilde{\eta})$ be two admissible complex irrational obstacles. Let $k\in\mathbb{R}_+$ be fixed and $\mathbf{d}_\ell$, $\ell=1,2$ be two distinct incident directions from $\mathbb{S}^1$. Let $\mathbf{G}$ denote the unbounded connected component of $\mathbb{R}^2\backslash\overline{(\Omega\cup\widetilde{\Omega})}$. Let u_∞ and $\widetilde{u}_\infty$ be the far-field patterns associated with (Ω,η) and $(\widetilde{\Omega},\widetilde{\eta})$, respectively. If*

$$u_\infty(\hat{\mathbf{x}},\mathbf{d}_\ell)=\widetilde{u}_\infty(\hat{\mathbf{x}},\mathbf{d}_\ell),\quad \hat{\mathbf{x}}\in\mathbb{S}^1, \ell=1,2, \tag{4.1.8}$$

then one has that

$$\left(\partial\Omega\backslash\partial\overline{\widetilde{\Omega}}\right)\cup\left(\partial\widetilde{\Omega}\backslash\partial\overline{\Omega}\right)$$

cannot have a corner on $\partial\mathbf{G}$.

Proof. We prove the theorem by contradiction. Assume (4.1.8) holds but $\left(\partial\Omega\backslash\partial\overline{\widetilde{\Omega}}\right)\cup\left(\partial\widetilde{\Omega}\backslash\partial\overline{\Omega}\right)$ has a corner $\mathbf{x}_c$ on $\partial\mathbf{G}$. Clearly, $\mathbf{x}_c$ is either a vertex of Ω or a vertex of $\widetilde{\Omega}$. Without loss of generality, we assume that $\mathbf{x}_c$ is a vertex of $\widetilde{\Omega}$. Moreover, we see that $\mathbf{x}_c$ lies outside Ω. Let $h\in\mathbb{R}_+$ be sufficiently small such that $B_h(\mathbf{x}_c)\Subset\mathbb{R}^2\backslash\overline{\Omega}$. Moreover, since $\mathbf{x}_c$ is a vertex of $\widetilde{\Omega}$, we can assume that

$$B_h(\mathbf{x}_c)\cap\partial\widetilde{\Omega}=\Gamma_h^{\pm}, \tag{4.1.9}$$

where $\Gamma_h^{\pm}$ are the two line segments lying on the two edges of $\widetilde{\Omega}$ that intersect at $\mathbf{x}_c$.

Recall that $\mathbf{G}$ denotes the unbounded connected component of $\mathbb{R}^2\backslash\overline{(\Omega\cup\widetilde{\Omega})}$. By (4.1.8) and the Rellich theorem (cf. [16]), we know that

$$u(\mathbf{x};k,\mathbf{d}_\ell)=\widetilde{u}(\mathbf{x};k,\mathbf{d}_\ell),\quad \mathbf{x}\in\mathbf{G},\ \ell=1,2. \tag{4.1.10}$$

It is clear that $\Gamma_h^{\pm} \subset \partial \mathbf{G}$. Hence, by using (4.1.10) as well as the generalized boundary condition (4.1.3) on $\partial\widetilde{\Omega}$, we readily have

$$\partial_\nu u + \widetilde{\eta} u = \partial_\nu \widetilde{u} + \widetilde{\eta}\widetilde{u} = 0 \quad \text{on } \Gamma_h^{\pm}. \tag{4.1.11}$$

It is also noted that in $B_h(\mathbf{x}_c)$, $-\Delta u = k^2 u$. Next, we consider two separate cases.

Case 1. Suppose that either $u(\mathbf{x}_c;k,\mathbf{d}_1)$ or $u(\mathbf{x}_c;k,\mathbf{d}_2)$ is zero. Without loss of generality, we assume that $u(\mathbf{x}_c;k,\mathbf{d}_1) = 0$. By the assumption of the theorem that $\widetilde{\Omega}$ is an irrational obstacle, we see that Γ_h^+ and Γ_h^- intersect with an irrational angle. Hence, by our results in Sects. 2.3 and 2.5, one immediately has that

$$u(\mathbf{x};k,\mathbf{d}_1) = 0 \quad \text{in } B_h(\mathbf{x}_c), \tag{4.1.12}$$

which in turn yields by the analytic continuation that

$$u(\mathbf{x};k,\mathbf{d}_1) = 0 \quad \text{in } \mathbb{R}^2\backslash\overline{\Omega}. \tag{4.1.13}$$

In particular, one has from (4.1.13) that

$$\lim_{|\mathbf{x}|\to\infty} |u(\mathbf{x};k,\mathbf{d}_1)| = 0. \tag{4.1.14}$$

But this contradicts to the fact that follows from (4.1.4):

$$\lim_{|\mathbf{x}|\to\infty} |u(\mathbf{x};k,\mathbf{d}_1)| = \lim_{|\mathbf{x}|\to\infty} \left| e^{ik\mathbf{x}\cdot\mathbf{d}_1} + u^s(\mathbf{x};k,\mathbf{d}_1)\right| = 1. \tag{4.1.15}$$

Case 2. Suppose that both $u(\mathbf{x}_c;k,\mathbf{d}_1) \neq 0$ and $u(\mathbf{x}_c;k,\mathbf{d}_2) \neq 0$. Set

$$\alpha_1 = u(\mathbf{x}_c;k,\mathbf{d}_2) \quad \text{and} \quad \alpha_2 = -u(\mathbf{x}_c;k,\mathbf{d}_1), \tag{4.1.16}$$

and

$$v(\mathbf{x}) = \alpha_1 u(\mathbf{x};k,\mathbf{d}_1) + \alpha_2 u(\mathbf{x};k,\mathbf{d}_2), \quad \mathbf{x} \in B_h(\mathbf{x}_c). \tag{4.1.17}$$

Clearly, there hold

$$-\Delta v = k^2 v \quad \text{in } B_h(\mathbf{x}_c); \quad \partial_\nu v + \widetilde{\eta} v = 0 \quad \text{on } \Gamma_h^{\pm}. \tag{4.1.18}$$

Moreover, by the choice of α_1, α_2 in (4.1.16), one obviously has that $v(\mathbf{x}_c) = 0$. Hence, by our results in Sects. 2.3 and 2.5, one immediately has that

$$v = 0 \quad \text{in } B_h(\mathbf{x}_c), \tag{4.1.19}$$

which in turn yields by the analytic continuation that

$$\alpha_1 u(\mathbf{x};k,\mathbf{d}_1) + \alpha_2 u(\mathbf{x};k,\mathbf{d}_2) = 0 \quad \text{in } \mathbb{R}^2\backslash\overline{\Omega}. \tag{4.1.20}$$

However, since $\mathbf{d}_1$ and $\mathbf{d}_2$ are distinct, we know from [16, Chapter 5] that $u(x;k,\mathbf{d}_1)$ and $u(\mathbf{x};k,\mathbf{d}_2)$ are linearly independent in $\mathbb{R}^2\backslash\overline{\Omega}$. Therefore, one has from (4.1.20) that $\alpha_1 = \alpha_2 = 0$, which contracts to the assumption at the beginning that both α_1 and α_2 are nonzero. □

It is recalled that the convex hull of Ω, denoted by $\mathscr{CH}(\Omega)$, is the smallest convex set that contains Ω. As a direct consequence of Theorem 4.1, we next show that the convex hull of a complex irrational obstacle can be uniquely determined by at most two far-field measurements. Furthermore, the boundary impedance parameter η can be partially identified as well. In fact we have

Corollary 4.1. *Let (Ω,η) and $(\widetilde{\Omega},\widetilde{\eta})$ be two admissible complex irrational obstacles. Let $k\in\mathbb{R}_+$ be fixed and $\mathbf{d}_\ell$, $\ell=1,2$ be two distinct incident directions from $\mathbb{S}^1$. Let $\mathbf{G}$ denote the unbounded connected component of $\mathbb{R}^2\backslash\overline{(\Omega\cup\widetilde{\Omega})}$. Let u_∞ and $\widetilde{u}_\infty$ be, respectively, the far-field patterns associated with (Ω,η) and $(\widetilde{\Omega},\widetilde{\eta})$. If*

$$u_\infty(\hat{\mathbf{x}},\mathbf{d}_\ell)=\widetilde{u}_\infty(\hat{\mathbf{x}},\mathbf{d}_\ell),\ \ \hat{\mathbf{x}}\in\mathbb{S}^1,\ \ell=1,2, \tag{4.1.21}$$

then one has that

$$\mathscr{CH}(\Omega)=\mathscr{CH}(\widetilde{\Omega}):=\Sigma, \tag{4.1.22}$$

and

$$\eta=\widetilde{\eta}\ \ \text{on}\ \ \partial\Omega\cap\partial\widetilde{\Omega}\cap\partial\Sigma. \tag{4.1.23}$$

Proof. From Theorem 4.1, we can immediately conclude (4.1.22). Next, we prove (4.1.23). Let $\mathscr{E}\subset\partial\Omega\cap\partial\widetilde{\Omega}\cap\partial\Sigma$ be an open subset such that $\eta\neq\widetilde{\eta}$ on $\mathscr{E}$. By taking a smaller subset of $\mathscr{E}$ if necessary, we can assume that η ($\widetilde{\eta}$, respectively) is either a fixed constant or ∞ on $\mathscr{E}$. Clearly, one has $u=\widetilde{u}$ in $\mathbb{R}^2\backslash\overline{\Sigma}$. Hence, there holds that

$$\partial_\nu u+\eta u=0,\ \ \partial_\nu\widetilde{u}+\widetilde{\eta}\widetilde{u}=0,\ \ u=\widetilde{u},\ \ \partial_\nu u=\partial_\nu\widetilde{u}\quad\text{on}\ \mathscr{E}. \tag{4.1.24}$$

By direct verification, one can show that

$$u=\partial_\nu u=0\quad\text{on}\ \mathscr{E}, \tag{4.1.25}$$

which in turn yields by the Holmgren uniqueness result (cf. [43]) that $u=0$ in $\mathbb{R}^2\backslash\overline{\Omega}$. Hence, we arrive at the same contradiction as that in (4.1.15), which implies (4.1.23). □

Remark 4.1. Let $\mathscr{V}(\Omega)$ and $\mathscr{V}(\mathscr{CH}(\Omega))$ denote the sets of vertices of Ω and $\mathscr{CH}(\Omega)$, respectively. It is known that $\mathscr{V}(\mathscr{CH}(\Omega))\subset\mathscr{V}(\Omega)$. Theorem 4.1 states that if the corner angle of the polygon Ω at any vertex in $\mathscr{V}(\Omega)$ is irrational, then $\mathscr{CH}(\Omega)$ can be uniquely determined by two far-field patterns. Indeed, from the proof of Theorem 4.1, we see that this requirement can be relaxed to that the corner angle of the polygon Ω at any vertex in $\mathscr{V}(\mathscr{CH}(\Omega))$ is irrational.

We proceed now to consider the unique determination of rational obstacles. Let Ω be a polygon in $\mathbb{R}^2$ and $\mathbf{x}_c$ be a vertex of Ω. In what follows, we define

$$\Omega_r(\mathbf{x}_c)=B_r(\mathbf{x}_c)\cap\mathbb{R}^2\backslash\overline{\Omega},\ \ r\in\mathbb{R}_+. \tag{4.1.26}$$

For a function $f\in L^2_{loc}(\mathbb{R}^2\backslash\overline{\Omega})$, we define

$$\mathscr{L}(f)(\mathbf{x}_c):=\lim_{r\to+0}\frac{1}{|\Omega_r(\mathbf{x}_c)|}\int_{\Omega_r(\mathbf{x}_c)}f(\mathbf{x})\,\mathrm{d}\mathbf{x} \tag{4.1.27}$$

if the limit exists. It is easy to see that if $f(\mathbf{x})$ is continuous in $\overline{\Omega_{\tau_0}(\mathbf{x}_c)}$ for a sufficiently small $\tau_0\in\mathbb{R}_+$, then $\mathscr{L}(f)(\mathbf{x}_c)=f(\mathbf{x}_c)$.

Theorem 4.2. *Let (Ω,η) be an admissible complex rational obstacle of degree $p\geq3$. Let $k\in\mathbb{R}_+$ be fixed and $\mathbf{d}_\ell$, $\ell=1,2$ be two distinct incident directions from $\mathbb{S}^1$. Set $u_\ell(\mathbf{x})=u(\mathbf{x};k,\mathbf{d}_\ell)$ to be the total wave fields associated with (Ω,η) and $e^{\mathrm{i}k\mathbf{x}\cdot\mathbf{d}_\ell}$, $\ell=1,2$, respectively. Recall that $\mathbf{G}$ denotes the unbounded connected component of $\mathbb{R}^2\backslash\overline{(\Omega\cup\widetilde{\Omega})}$. If the following condition is fulfilled,*

$$\mathscr{L}\left(u_2\cdot\nabla u_1-u_1\cdot\nabla u_2\right)(\mathbf{x}_c)\neq0, \tag{4.1.28}$$

where $\mathbf{x}_c$ is any vertex of Ω, then then one has that

$$\left(\partial\Omega\backslash\partial\overline{\widetilde{\Omega}}\right)\cup\left(\partial\widetilde{\Omega}\backslash\partial\overline{\Omega}\right)$$

cannot have a corner on $\partial\mathbf{G}$.

Proof. We prove the theorem by contradiction. Assume that there exists an admissible complex rational obstacle of degree $p \geq 3$, $(\widetilde{\Omega}, \widetilde{\eta})$, such that (4.1.8) holds but $\left(\partial\Omega\backslash\partial\overline{\widetilde{\Omega}}\right) \cup \left(\partial\widetilde{\Omega}\backslash\partial\overline{\Omega}\right)$ has a corner on $\partial\mathbf{G}$. In what follows, we adopt the same notation as those introduced in the proof of Theorem 4.1. Note that the total wave fields $\widetilde{u}_\ell$, $\ell = 1,2$, associated with $(\widetilde{\Omega}, \widetilde{\eta})$, are also assumed to fulfil the condition (4.1.28).

By following a similar argument to the proof of Theorem 4.1, one can show that there exist two line segments $\Gamma_h^\pm$ in $\mathbb{R}^2\backslash\overline{\Omega}$ such that $\partial_\nu u + \widetilde{\eta} u = 0$ on $\Gamma_h^\pm$, and Γ_h^+ and Γ_h^- intersect at a point $\mathbf{x}_c$ which is a vertex of $\widetilde{\Omega}$. Using the fact that $u = \widetilde{u}$ near $\mathbf{x}_c$ and the condition (4.1.28) on $(\widetilde{\Omega}, \widetilde{\eta})$, we actually have

$$u(\mathbf{x}_c;\mathbf{d}_2) \cdot \nabla u(\mathbf{x}_c;\mathbf{d}_1) - u(\mathbf{x}_c;\mathbf{d}_1) \cdot \nabla u(\mathbf{x}_c;\mathbf{d}_2) \neq \mathbf{0}. \tag{4.1.29}$$

Clearly, (4.1.29) implies that $\alpha_1 := u(\mathbf{x}_c;\mathbf{d}_2)$ and $\alpha_2 := -u(\mathbf{x}_c;\mathbf{d}_1)$ cannot be identically zero. Set v to be the one introduced in (4.1.17); then it clearly satisfies

$$v(\mathbf{x}_c) = 0 \quad \text{and} \quad \nabla v(\mathbf{x}_c) \neq \mathbf{0}. \tag{4.1.30}$$

Since $\widetilde{\Omega}$ is rational of degree $p \geq 3$, we know that Γ_h^+ and Γ_h^- intersect either at an irrational angle or at a rational angle of degree $p \geq 3$. In either case, by our results in Sects. 2.3, 2.4 and 2.5, we can see that v is vanishing at least to the second order at $\mathbf{x}_c$. Hence, there holds $\nabla v(\mathbf{x}_c) = 0$, which is a contradiction to (4.1.30). □

Similar to Corollary 4.1, as a direct consequence of Theorem 4.2, under the condition (4.1.28), we next show that the convex hull of a complex rational obstacle of degree $p \geq 3$ can be uniquely determined by at most two far-field measurements. Indeed we have

Corollary 4.2. *Let* (Ω, η) *be an admissible complex rational obstacle of degree* $p \geq 3$. *Let* $k \in \mathbb{R}_+$ *be fixed and* $\mathbf{d}_\ell$, $\ell = 1,2$ *be two distinct incident directions from* $\mathbb{S}^1$. *Set* $u_\ell(\mathbf{x}) = u(\mathbf{x};k,\mathbf{d}_\ell)$ *to be the total wave fields associated with* (Ω, η) *and* $e^{\mathrm{i}k\mathbf{x}\cdot\mathbf{d}_\ell}$, $\ell = 1,2$, *respectively. If* (4.1.28) *is fulfilled, then* $\mathscr{CH}(\Omega)$ *is uniquely determined by* $u_\infty(\hat{\mathbf{x}}, \mathbf{d}_\ell)$, $\ell = 1,2$. *Similar to Corollary 4.1, the boundary impedance parameter* η *can be partially identified as well.*

Remark 4.2. As mentioned earlier, a general rational obstacle is at least of order 2. By Remark 2.1, we can easily extend the proof of Theorem 4.2 to cover the general case that $p = 2$. However, as discussed in Remark 2.1, we need to exclude the case that $\eta \equiv \infty$ and η is a finite number (possibly being zero) on the two intersecting line segments $\Gamma_h^\pm$, respectively (as appeared in the proof of Theorem 4.2).

Remark 4.3. Similar to Remark 4.1, the condition (4.1.28) can be relaxed to hold only at any vertex in $\mathscr{V}(\mathscr{CH}(\Omega))$ in Theorem 4.2 and Corollary 4.2. Furthermore, since in the proof of Theorem 4.2, we only make use of the vanishing up to the second order. By our results in Sect. 2.6, we know that Theorem 4.2 actually holds for a more general case where the surface impedance η can be a C^1 function.

It would be interesting to investigate the sufficient conditions for (4.1.28) to hold. From a practical point of view, the condition (4.1.28) depends on the a priori knowledge of the underlying obstacle as well as the choice of the incident waves. As an illustrating scenario, suppose that the obstacle Ω is sufficiently small compared with the wavelength, namely, $k \cdot \mathrm{diam}(\Omega) \ll 1$. Then from a physical viewpoint, the scattered wave field due to the obstacle is of a much smaller magnitude than that of the incident field, and the incident plane wave dominates in the total wave field $u = u^i + u^s$. In such a case, one can verify that the condition (4.1.28) is fulfilled in the setup described in Theorem 4.2 (thanks to the fact that the condition is actually satisfied by two incident plane waves). However, we shall not explore more about this point. Finally, we also would like to point out that our arguments for the uniqueness results in Theorems 4.1 and 4.2 are "localized" around the corner point $\mathbf{x}_c$. Therefore, one may consider other different types of wave incidences from the incident plane wave (4.1.1), e.g. the point source of the form,

$$u^i(\mathbf{x};\mathbf{z}_0) = H_0^1(k|\mathbf{x}-\mathbf{z}_0|), \ \mathbf{x}, \mathbf{z}_0 \in \mathbb{R}^2,$$

where H_0^1 is the zeroth-order Hankel function of the first kind, and $\mathbf{z}_0$ signifies the location of the source $u^i(\mathbf{x},\mathbf{z}_0)$. $u^i(\mathbf{x};\mathbf{z}_0)$ blows up at the point $\mathbf{z}_0$. By direct verifications, we can show that both the uniqueness results in Theorem 4.1 and 4.2 still hold for this point source incidence.

In the remainder of this subsection, we apply the results we have obtained in Sects. 2.9–2.13 about the vanishing properties of an eigenfunction at a vertex corner to study a fundamental mathematical issue in inverse scattering problems, namely, the unique identifiability of the inverse problem recovering the shape of some unknown objects by certain wave probing data. We follow the treatment in [11]. We shall first obtain some local uniqueness results for the inverse problem since the method developed here is completely local. Nevertheless, if the underlying obstacles are further assumed to be convex, the local uniqueness results imply the global uniqueness results. Moreover, we would like to point out that in deriving those unique determination results, we need to introduce certain restrictive conditions on the underlying polyhedral obstacles. One of such conditions is contained in the following admissibility definition. Very recently, in [12] the unique determination of an impedance polyhedral obstacle or grating in $\mathbb{R}^3$ by a single far-field measurement is established by utilizing the local geometry structure of Laplacian eigenfunctions.

Definition 4.1. Let $\Omega \subset \mathbb{R}^3$ be an open polyhedron associated with the generalized impedance boundary condition (4.1.3). Then Ω is said to be an *admissible polyhedral obstacle* if the following conditions are fulfilled:

- On each face of $\partial\Omega$, the surface impedance η is either a constant (possibly zero) or ∞.
- For any vertex of Ω that is intersected by n planes: $\Pi_1, \Pi_2, \ldots, \Pi_n$, $n \geq 3$, there exists a plane $\Pi_j := \{\overrightarrow{OA_j}, \overrightarrow{OA_{j+1}}\}$, where O denotes the vertex locating at the origin. $\overrightarrow{OA_\tau} = (r, \theta_\tau, \phi_\tau)$, for $r > 0$, $\theta_\tau \in (0,\pi)$ and $\phi_\tau \in (0,2\pi)$ in the spherical coordinate system such that $P_n^0(\cos\theta_\tau) \neq 0$ and $P_n^1(\cos\theta_\tau) \neq 0$, where $\tau = j, j+1$, $n \in \mathbb{N}$, P_n^0 and P_n^1 are the associated Legendre polynomials.

Remark 4.4. By noting the fact that $P_n^0(1) \equiv 1$, $P_n^1(1) \equiv 0$ when $\theta = 0$ for all $n \in \mathbb{N}$ and the continuity of the associated Legendre polynomials, one easily knows that there exists $\delta_0 > 0$ such that for any $\varepsilon > 0$ and $\theta \in (0, \delta_0)$, $P_n^0(\cos\theta) \in (1-\varepsilon, 1)$ and $P_n^1(\cos\theta) \in (0,\varepsilon)$, which in turn imply the existence of θ_τ in Definition 4.1. Therefore, the definition of the aforementioned admissible polyhedral obstacle is well defined.

Remark 4.5. In view of Sect. 2.13, in Definition 4.1, the surface impedance η can be relaxed to that it is either an analytic function (possibly zero) or ∞; see also Remark 4.9 in what follows.

Throughout this section, we signify an admissible polyhedral obstacle as (Ω, η). Then we define the rational and irrational obstacle in $\mathbb{R}^3$ based on Definition 2.12.2 for the rational and irrational vertex corner of Ω.

Definition 4.2. Let (Ω, η) be an admissible polyhedral obstacle. If there exists a rational vertex corner, then it is said to be a *rational obstacle*. If all the vertex corners of Ω are irrational, then it is called an *irrational obstacle*. The smallest degree of the rational corner of Ω is referred to as the *rational degree* of Ω.

Definition 4.3. Ω is said to be an admissible complex polyhedral obstacle if it consists of finitely many admissible polyhedral obstacles. That is,

$$(\Omega,\eta)=\bigcup_{j=1}^{l}(\Omega_j,\eta_j),$$

where $l\in\mathbb{N}$ and each (Ω_j,η_j) is an admissible polyhedral obstacle. Here, we define

$$\eta=\sum_{j=1}^{l}\eta_j\chi_{\partial\Omega_j}.$$

Moreover, Ω is said to be irrational if all of its component polyhedral obstacles are irrational; otherwise, it is said to be rational. For the latter case, the smallest degree among all the degrees of its rational components is defined to be the degree of the complex obstacle Ω.

Next, we give the unique determination result for an admissible complex irrational polyhedral obstacle by at most two far-field patterns.

Theorem 4.3. *Consider a fixed $k\in\mathbb{R}_+$ and two distinct incident directions $\mathbf{d}_1$ and $\mathbf{d}_2$ from $\mathbb{S}^2$. Let (Ω,η) and $(\widetilde{\Omega},\widetilde{\eta})$ be two admissible complex irrational obstacles, with u_∞ and $\widetilde{u}_\infty$ being their corresponding far-field patterns and $\mathbf{G}$ being the unbounded connected component of $\mathbb{R}^3\backslash\overline{(\Omega\cup\widetilde{\Omega})}$. If u_∞ and $\widetilde{u}_\infty$ are the same in the sense that*

$$u_\infty(\hat{\mathbf{x}};k,\mathbf{d}_\ell)=\widetilde{u}_\infty(\hat{\mathbf{x}};k,\mathbf{d}_\ell),\ \ for\ \ \ell=1,2\ \ and\ all\ \ \hat{\mathbf{x}}\in\mathbb{S}^2, \tag{4.1.31}$$

then $(\partial\Omega\backslash\partial\overline{\widetilde{\Omega}})\bigcup(\partial\widetilde{\Omega}\backslash\partial\overline{\Omega})$ cannot possess a vertex corner on $\partial\mathbf{G}$. Moreover,

$$\eta=\widetilde{\eta}\quad on\quad \partial\Omega\cap\partial\widetilde{\Omega}. \tag{4.1.32}$$

Proof. We prove the theorem by contradiction. Assume that $(\partial\Omega\backslash\partial\overline{\widetilde{\Omega}})\bigcup(\partial\widetilde{\Omega}\backslash\partial\overline{\Omega})$ has a vertex corner $\mathbf{x}_c$ on $\partial\mathbf{G}$. Then, $\mathbf{x}_c$ is located either at Ω or $\widetilde{\Omega}$. Without loss of generality, we assume that $\mathbf{x}_c$ is a 3D vertex corner of $\widetilde{\Omega}$, which also indicates that $\mathbf{x}_c$ lies outside Ω. Let $h\in\mathbb{R}_+$ be sufficiently small such that $B_h(x_c)\Subset\mathbb{R}^2\backslash\Omega$, then we can suppose for $n\geq 3$ that

$$B_h(\mathbf{x}_c)\cap\partial\widetilde{\Omega}=\Pi_i,\quad i=1,2,\ldots,n,$$

where Π_i are the n planes lying on the n faces of $\widetilde{\Omega}$ that intersect at $\mathbf{x}_c$.

Recall that $\mathbf{G}$ is the unbounded connected component of $\mathbb{R}^3\backslash\overline{(\Omega\cup\widetilde{\Omega})}$. By (4.1.31) and the Rellich theorem (cf. [16]), we know that

$$u(\mathbf{x};k,\mathbf{d}_\ell)=\widetilde{u}(\mathbf{x};k,\mathbf{d}_\ell),\quad x\in\mathbf{G},\ \ell=1,2. \tag{4.1.33}$$

Since $\Pi_i\subset\partial\mathbf{G}$, $i=1,2,\ldots,n$, combining (4.1.33) with the generalized boundary condition (4.1.3) on $\partial\widetilde{\Omega}$, it is easy to obtain for $n\geq 3$ that

$$\partial_\nu u+\widetilde{\eta}u=\partial_\nu\widetilde{u}+\widetilde{\eta}\widetilde{u}=0\quad\text{on}\ \ \Pi_i,\quad i=1,2,\ldots,n.$$

Furthermore, since $B_h(x_c)\Subset\mathbb{R}^2\backslash\overline{\Omega}$, we have $-\Delta u=k^2u$ in $B_h(\mathbf{x}_c)$. We divide our remaining proof into two separate cases.

Case 1. Suppose that either $u(\mathbf{x}_c;k,\mathbf{d}_1)$ or $u(\mathbf{x}_c;k,\mathbf{d}_2)$ is zero. Without loss of generality, we assume that $u(\mathbf{x}_c;k,\mathbf{d}_1)=0$. By the assumption of the theorem that $\widetilde{\Omega}$ is an admissible irrational obstacle, we know that $\mathbf{x}_c$ is an irrational vertex corner of $\widetilde{\Omega}$, which also implies that there exist Π_{i_0} and Π_{i_0+1} such that the corresponding intersecting dihedral angle is irrational. Hence, by our results in Sects. 2.11 and 2.12, we can immediately derive that

$$u(\mathbf{x};k,\mathbf{d}_1)=0\quad\text{in}\ \ B_h(\mathbf{x}_c),$$

which in turn yields by the analytic continuation that

$$u(\mathbf{x};k,\mathbf{d}_1)=0 \quad \text{in } \mathbb{R}^3\backslash\overline{\Omega}. \tag{4.1.34}$$

In particular, one has from (4.1.34) that

$$\lim_{|\mathbf{x}|\to\infty}|u(\mathbf{x};k,\mathbf{d}_1)|=0. \tag{4.1.35}$$

But this contradicts to the fact that follows from (4.1.4):

$$\lim_{|\mathbf{x}|\to\infty}|u(\mathbf{x};k,\mathbf{d}_1)|=\lim_{|\mathbf{x}|\to\infty}\left|e^{ik\mathbf{x}\cdot\mathbf{d}_1}+u^s(\mathbf{x};k,\mathbf{d}_1)\right|=1. \tag{4.1.36}$$

Case 2. Suppose that both $u(\mathbf{x}_c;k,\mathbf{d}_1)\neq 0$ and $u(\mathbf{x}_c;k,\mathbf{d}_2)\neq 0$. Set

$$\alpha_1=u(\mathbf{x}_c;k,\mathbf{d}_2) \quad \text{and} \quad \alpha_2=-u(\mathbf{x}_c;k,\mathbf{d}_1), \tag{4.1.37}$$

and

$$v(\mathbf{x})=\alpha_1 u(\mathbf{x};k,\mathbf{d}_1)+\alpha_2 u(\mathbf{x};k,\mathbf{d}_2) \quad \forall x\in B_h(\mathbf{x}_c). \tag{4.1.38}$$

It is easy to verify for $n\geq 3$ that v fulfils

$$-\Delta v=k^2 v \quad \text{in } B_h(\mathbf{x}_c) \quad \text{and} \quad \partial_\nu v+\widetilde{\eta}v=0 \quad \text{on } \Pi_i,\ i=1,2,\ldots,n. \tag{4.1.39}$$

Moreover, by the choice of α_1 and α_2 in (4.1.37), it is obvious to see that $v(\mathbf{x}_c)=0$. Hence, by our results in Sects. 2.11 and 2.12, we deduce that

$$v=0 \quad \text{in } B_h(\mathbf{x}_c),$$

and thus

$$\alpha_1 u(\mathbf{x};k,\mathbf{d}_1)+\alpha_2 u(\mathbf{x};k,\mathbf{d}_2)=0 \quad \text{in } \mathbb{R}^3\backslash\overline{\Omega} \tag{4.1.40}$$

by the analytic continuation. However, since $\mathbf{d}_1$ and $\mathbf{d}_2$ are distinct, we know from [16, Chap. 5] that $u(\mathbf{x};k,\mathbf{d}_1)$ and $u(\mathbf{x};k,\mathbf{d}_2)$ are linearly independent in $\mathbb{R}^3\backslash\overline{\Omega}$. Therefore, from (4.1.40) we can obtain that

$$\alpha_1=\alpha_2=0, \tag{4.1.41}$$

which contracts to the assumption at the beginning that both α_1 and α_2 are nonzero.

It remains to prove (4.1.32), and we do it by contradiction. Let $\mathscr{E}\subset\partial\Omega\cap\partial\widetilde{\Omega}$ be an open subset such that $\eta\neq\widetilde{\eta}$ on $\mathscr{E}$. By taking a smaller subset of $\mathscr{E}$ if necessary, we can assume that η (respectively $\widetilde{\eta}$) is either a fixed constant or ∞ on $\mathscr{E}$. Clearly, one has $u=\widetilde{u}$ in $\mathbb{R}^3\backslash\overline{(\Omega\cup\widetilde{\Omega})}$. Hence, it holds that

$$\partial_\nu u+\eta u=0,\ \partial_\nu\widetilde{u}+\widetilde{\eta}\widetilde{u}=0,\ u=\widetilde{u},\ \partial_\nu u=\partial_\nu\widetilde{u} \quad \text{on } \mathscr{E}. \tag{4.1.42}$$

Combining with the assumption that $\eta\neq\widetilde{\eta}$ on $\mathscr{E}$, we can deduce by direct computing that

$$u=\partial_\nu u=0 \quad \text{on } \mathscr{E}, \tag{4.1.43}$$

which in turn yields by Holmgren's uniqueness result (cf. [43]) that $u=0$ in $\mathbb{R}^3\backslash\Omega$. Therefore, we arrive at the same contradiction as that in (4.1.35), leading to the conclusion (4.1.32). □

Theorem 4.3 presents a local uniqueness result by showing the lack of an irrational vertex corner. If the underlying admissible complex irrational obstacles are convex, we can obtain a global unique identifiability result by two far-field patterns as follows.

Corollary 4.3. *Consider a fixed $k \in \mathbb{R}_+$ and two distinct incident directions $\mathbf{d}_1$ and $\mathbf{d}_2$ from $\mathbb{S}^2$. Let (Ω, η) and $(\widetilde{\Omega}, \widetilde{\eta})$ be two convex admissible complex irrational obstacles, with u_∞ and $\widetilde{u}_\infty$ being their corresponding far-field patterns and $\mathbf{G}$ being the unbounded connected component of $\mathbb{R}^3\backslash\overline{(\Omega \cup \widetilde{\Omega})}$. If u_∞ and $\widetilde{u}_\infty$ are the same in the sense that*

$$u_\infty(\hat{\mathbf{x}}; k, \mathbf{d}_\ell) = \widetilde{u}_\infty(\hat{\mathbf{x}}; k, \mathbf{d}_\ell), \text{ for } \ell = 1, 2 \text{ and all } \hat{\mathbf{x}} \in \mathbb{S}^2, \tag{4.1.44}$$

then

$$\Omega = \widetilde{\Omega}, \quad \eta = \widetilde{\eta}.$$

Proof. Since Ω and $\widetilde{\Omega}$ are both convex, due to Krein-Milman theorem [50], we know that a convex polyhedron is fully determined by the set of its vertices. Now we prove by absurdity. If $\Omega \neq \widetilde{\Omega}$, it is obvious that there exists a vertex corner $\mathbf{x}_c$ on $\partial\mathbf{G}$, where $\mathbf{G}$ is the unbounded connected component of $\mathbb{R}^3\backslash\overline{(\Omega \cup \widetilde{\Omega})}$. Following the similar argument in Theorem 4.3, by the condition (4.1.44), we can arrive at the contradictions (4.1.36) or (4.1.41). Therefore, we can prove that $\Omega = \widetilde{\Omega}$. The conclusion $\eta = \widetilde{\eta}$ can be obtained by using the similar argument in Theorem 4.3. □

We proceed to consider the unique determination of rational obstacles. By Definitions 2.12.2 and 4.2, we know that a rational obstacle contains at least one rational vertex corner. Recalling the results in Sects. 2.10 and 2.11, for a fixed rational vertex corner $\mathbf{x}_c$ which is intersected by Π_i, where $\Pi_i = \text{span}\{\overrightarrow{OA_i}, \overrightarrow{OA_{i+1}}\}$ with the n dihedral angles $\angle(\Pi_i, \Pi_{i+1}) = \alpha_i \cdot \pi$, $i = 1, 2, \ldots, n$ ($n \geq 3$), it is direct to verify that the eigenfunction u to $-\Delta u = k^2 u$ in $B_h(\mathbf{x}_c)$ of the form

$$u(\mathbf{x}) = 4\pi \sum_{n=0}^{\infty} \sum_{m=-n}^{n} \mathrm{i}^n a_n^m j_n(\sqrt{\lambda} r) \sqrt{\frac{2n+1}{4\pi} \frac{(n-|m|)!}{(n+|m|)!}} P_n^{|m|}(\cos\theta) e^{\mathrm{i}m\phi}$$

satisfies that $a_0^0 = 0$ if $u(\mathbf{x}_c) = 0$, where $\mathbf{x} = (x_1, x_2, x_3) = r(\sin\theta\cos\phi, \sin\theta\sin\phi, \cos\theta) \in \mathbb{R}^3$, λ is the corresponding eigenvalue, $P_n^m(t)$ denotes the associated Legendre function and $j_n(t)$ signifies the nth spherical Bessel function. Since $\alpha_i \in (0,1)$ for any $i = 1, 2, \ldots, n$, one can immediately obtain that $a_1^{\pm 1} = 0$; see Theorems 2.10.2, 2.11.1 and 2.11.2 for detailed discussions. Moreover, if we denote

$$\overrightarrow{OA_i} = (r, \theta_i, \phi_i) \text{ for } r > 0, \theta_i \in (0, \pi) \text{ and } \phi_i \in (0, 2\pi) \tag{4.1.45}$$

in the spherical coordinate system, then there always holds that $P_1^1(\cos\theta_i) = -\sin\theta_i \neq 0$. However, since $P_1^0(\cos\theta_i) = \cos\theta_i$, we know that $P_1^0(\cos\theta_i) \neq 0$ is only true for $\theta_i \neq \frac{\pi}{2}$, and thus, $a_1^0 = 0$. That is, the eigenfunction u vanishes at least to the second order when $\theta_i \neq \frac{\pi}{2}$, and u vanishes at least to the first order otherwise.

Let Ω be a polyhedron in $\mathbb{R}^3$ and $\mathbf{x}_c$ be a vertex corner of Ω. Then we introduce for $r \in \mathbb{R}_+$ that $\Omega_r(\mathbf{x}_c) = B_r(\mathbf{x}_c) \cap \mathbb{R}^3\backslash\overline{\Omega}$, and define for any $f \in L^2_{loc}(\mathbb{R}^3\backslash\overline{\Omega})$ that

$$\mathscr{L}(f)(\mathbf{x}_c) := \lim_{r \to +0} \frac{1}{|\Omega_r(\mathbf{x}_c)|} \int_{\Omega_r(\mathbf{x}_c)} f(\mathbf{x})\, \mathrm{d}\mathbf{x}$$

if the limit exists. It is easy to see that if $f(\mathbf{x})$ is continuous in $\overline{\Omega_{\varepsilon_0}(\mathbf{x}_c)}$ for a sufficiently small $\varepsilon_0 \in \mathbb{R}_+$, then $\mathscr{L}(f)(\mathbf{x}_c) = f(\mathbf{x}_c)$.

Now we are ready to study the unique determination of rational obstacles.

Theorem 4.4. *Consider a fixed $k \in \mathbb{R}_+$ and two distinct incident directions $\mathbf{d}_1$ and $\mathbf{d}_2$ from $\mathbb{S}^2$. Let (Ω, η) and $(\widetilde{\Omega}, \widetilde{\eta})$ be two admissible complex rational obstacles of degree $p \geq 3$, with $u_\ell(\mathbf{x}) := u(\mathbf{x}; k, \mathbf{d}_\ell)$ and $\widetilde{u}_\ell := \widetilde{u}(\mathbf{x}; k, \mathbf{d}_\ell)$ being their corresponding total wave fields associated with the incident field $e^{\mathrm{i}k\mathbf{x}\cdot\mathbf{d}_\ell}$ and $u_\infty(\hat{\mathbf{x}}; k, \mathbf{d}_\ell)$ and $\widetilde{u}_\infty(\hat{\mathbf{x}}; , k, \mathbf{d}_\ell)$ being their corresponding far-field patterns for $\ell = 1, 2$. We further write $\mathbf{G}$ for the unbounded connected component of $\mathbb{R}^3\backslash\overline{(\Omega \cup \widetilde{\Omega})}$. Then the set $(\partial\Omega\backslash\partial\overline{\widetilde{\Omega}}) \cup (\partial\widetilde{\Omega}\backslash\partial\overline{\Omega})$ cannot possess a vertex corner on $\partial\mathbf{G}$, if the following conditions are* *satisfied:*

$$u_{\ell,\infty}(\hat{\mathbf{x}};k,\mathbf{d}_\ell)=\widetilde{u}_{\ell,\infty}(\hat{\mathbf{x}};k,\mathbf{d}_\ell),\ \ \hat{\mathbf{x}}\in\mathbb{S}^2,\ell=1,2, \tag{4.1.46}$$

$$\mathscr{L}\left(u_2\cdot\nabla u_1-u_1\cdot\nabla u_2\right)(\mathbf{x}_c)\neq 0 \quad \text{and} \quad \mathscr{L}\left(\widetilde{u}_2\cdot\nabla\widetilde{u}_1-\widetilde{u}_1\cdot\nabla\widetilde{u}_2\right)(\mathbf{x}_c)\neq 0 \tag{4.1.47}$$

for all vertices $\mathbf{x}_c$ of Ω. Moreover,

$$\eta=\widetilde{\eta} \quad \text{on} \quad \partial\Omega\cap\partial\widetilde{\Omega}. \tag{4.1.48}$$

Proof. We prove the theorem by contradiction. Assume that (4.1.46) holds but $(\partial\Omega\backslash\partial\overline{\widetilde{\Omega}})\cup(\partial\widetilde{\Omega}\backslash\partial\overline{\Omega})$ has a vertex corner $\mathbf{x}_c$ on $\partial\mathbf{G}$. Without loss of generality, we still assume that $\mathbf{x}_c$ is a vertex corner of $\widetilde{\Omega}$. In what follows, we adopt the same notations as those introduced in the proof of Theorem 4.3.

By following a similar argument to the proof of Theorem 4.3, one can show that there exist n pieces of planes $\Pi_i\subset\partial\mathbf{G}$ intersecting at $\mathbf{x}_c$, such that $\partial_\nu u+\widetilde{\eta}u=0$ on Π_i, $i=1,2,\ldots,n$. Using the fact that $u=\widetilde{u}$ near $\mathbf{x}_c$, we derive by Rellich Lemma and the condition (4.1.47) on $(\widetilde{\Omega},\widetilde{\eta})$ that

$$u(\mathbf{x}_c;k,\mathbf{d}_2)\cdot\nabla u(\mathbf{x}_c;k,\mathbf{d}_1)-u(\mathbf{x}_c;k,\mathbf{d}_1)\cdot\nabla u(\mathbf{x}_c;k,\mathbf{d}_2)\neq 0. \tag{4.1.49}$$

Clearly, this implies that $\alpha_1:=u(\mathbf{x}_c;k,\mathbf{d}_2)$ and $\alpha_2=-u(\mathbf{x}_c;k,\mathbf{d}_1)$ cannot be identically zero. Let v be the same combination as introduced in (4.1.38); then we can directly verify that v fulfils (4.1.39) and

$$v(\mathbf{x}_c)=0 \quad \text{and} \quad \nabla v(\mathbf{x}_c)\neq 0. \tag{4.1.50}$$

Noting that $\widetilde{\Omega}$ is a rational obstacle of degree $p\geq 3$, we know that $\Pi_i, i=1,2,\ldots,n$ intersect either at an irrational vertex corner or a rational vertex corner of degree $p\geq 3$. In either case, we see by our results in Sects. 2.10, 2.11 and 2.12 that v vanishes at least to second order at $\mathbf{x}_c$ if $\theta_i\neq\frac{\pi}{2}$ in (4.1.45) for $i=1,2,\ldots,n$. Hence, there holds that $\nabla v(\mathbf{x}_c)=0$, which is a contradiction to (4.1.50). Following a similar argument in the proof of Theorem 4.3, if $\eta\neq\widetilde{\eta}$, one can directly verify that (4.1.42) and (4.1.43) still hold. Therefore, (4.1.48) can be derived directly by using Holmgren's uniqueness principle. □

Remark 4.6. In the proof of Theorem 4.4, we may illustrate the vanishing order of u by the normal derivatives in Taylor expansion. That is, the conditions that $v(\mathbf{x}_c)=\nabla v(\mathbf{x}_c)=0$ imply that v vanishes at $\mathbf{x}_c$ at least up to the second order. Indeed, this is equivalent to the vanishing condition that $a_0^0=a_1^{\pm 1}=a_1^0=0$ for the coefficients of u in the spherical wave expansion, which follows readily from the main theorems of Sects. 2.10 and 2.11 under the condition that $\theta_i\neq\frac{\pi}{2}$ for $i=1,2,\ldots,n$.

Remark 4.7. The uniqueness results and the corresponding argument in Theorems 4.3 and 4.4 are "localized" around the corner $\mathbf{x}_c$ based on the spectral results in Sects. 2.10 and 2.11. This provides a very effective analytical approach to study inverse scattering problems. Similar to Corollary 4.3, if the underlying admissible complex rational obstacle is convex, then it can be uniquely determined by two far-field patterns under the same generic conditions as those in Theorem 4.4.

Remark 4.8. We would like to point out that the condition (4.1.47) can be fulfilled under certain generic conditions on Ω. For instance, if the obstacle Ω is sufficiently small compared with the wavelength, namely, $k\cdot\mathrm{diam}(\Omega)\ll 1$, then it is known from the physical point of view that the scattered wave field due to the obstacle is of a much smaller magnitude than the incident field, and thus, the incident plane wave dominates in the total wave field $u=u^i+u^s$. Under this circumstance, (4.1.47) can be verified directly. Condition (4.1.47) can hold in more general scenarios, but we shall not explore this technical issue further in this book.

Remark 4.9. We would like to point out that by using the results in Sect. 2.13, Theorems 4.3 and 4.4 equally hold for the case that the surface impedance η is a (variable) analytic function; see also Remark 4.5.

4.1.2 Unique Recovery Results for the Inverse Obstacle Problem with the Maxwell System

Let Ω be a bounded Lipschitz domain in $\mathbb{R}^3$ such that $\mathbb{R}^3\backslash\bar{\Omega}$ is connected and the incident electric and magnetic fields be of the form

$$\mathbf{E}^i(\mathbf{x}) := \mathbf{p}e^{\mathrm{i}k\mathbf{x}\cdot\mathbf{d}}, \quad \mathbf{H}^i(\mathbf{x}) := \frac{1}{\mathrm{i}k}\nabla\wedge\mathbf{p}e^{\mathrm{i}k\mathbf{x}\cdot\mathbf{d}} = \mathbf{d}\wedge\mathbf{p}e^{\mathrm{i}k\mathbf{x}\cdot\mathbf{d}}, \tag{4.1.51}$$

which are known as the time harmonic electromagnetic plane waves, with $\mathbf{p}\in\mathbb{R}^3\backslash\{\mathbf{0}\}, k\in\mathbb{R}_+$ and $\mathbf{d}\in\mathbb{S}^2 := \{\mathbf{x}\in\mathbb{R}^3; |\mathbf{x}|=1\}$ representing the polarization, wave number and direction of propagation, respectively. And it holds that $\mathbf{p}\perp\mathbf{d}$. The associated forward scattering problem can be described by the following time harmonic Maxwell equations (cf. [16]):

$$\begin{cases} \nabla\wedge\mathbf{E}-\mathrm{i}k\mathbf{H}=0 \quad \text{in} \quad \mathbb{R}^3\backslash\overline{\Omega}, \\ \nabla\wedge\mathbf{H}+\mathrm{i}k\mathbf{E}=0 \quad \text{in} \quad \mathbb{R}^3\backslash\overline{\Omega}, \\ \mathbf{E}(\mathbf{x})=\mathbf{E}^i(\mathbf{x})+\mathbf{E}^s(\mathbf{x}), \\ \mathbf{H}(\mathbf{x})=\mathbf{H}^i(\mathbf{x})+\mathbf{H}^s(\mathbf{x}), \\ \mathscr{B}(\mathbf{E})=\mathbf{0} \qquad\qquad \text{on } \partial\Omega, \\ \lim_{|\mathbf{x}|\to\infty}(\mathbf{H}^s\wedge\mathbf{x}-|\mathbf{x}|\mathbf{E}^s)=\mathbf{0}, \end{cases} \tag{4.1.52}$$

where $\mathbf{E}=(E_1,E_2,E_3)$ and $\mathbf{H}=(H_1,H_2,H_3)$ are the total electric and magnetic fields, respectively, formed by the incident fields $\mathbf{E}^i(\mathbf{x})$ and $\mathbf{H}^i(\mathbf{x})$ and scattered fields $\mathbf{E}^s(\mathbf{x})$ and $\mathbf{H}^s(\mathbf{x})$. The last equation of (4.1.52) is the Silver-Müller radiation condition. The boundary condition $\mathscr{B}(\mathbf{E})$ on $\partial\Omega$ could be either of the following three conditions:

1. the Dirichlet condition (i.e., Ω is a perfectly electric conducting (PEC) obstacle):

$$\mathscr{B}(\mathbf{E})=\nu\wedge\mathbf{E};$$

2. the Neumann condition (i.e., Ω is a perfectly magnetic conducting (PMC) obstacle):

$$\mathscr{B}(\mathbf{E})=\nu\wedge(\nabla\wedge\mathbf{E});$$

3. the impedance condition (i.e. Ω is an impedance obstacle):

$$\mathscr{B}(\mathbf{E})=\nu\wedge(\nabla\wedge\mathbf{E})+\eta(\nu\wedge\mathbf{E})\wedge\nu,\ \Re(\eta)\geq 0 \text{ and } \Im(\eta)<0,$$

where ν denotes the exterior unit normal vector to $\partial\Omega$ and $\eta\in L^\infty(\Omega)$. The conditions $\Re(\eta)\geq 0$ and $\Im(\eta)<0$ are the physical requirements.

To ease the exposition, similar to our notation in (3.1.2)–(3.1.3), we unify the aforementioned three types of boundary conditions as

$$\mathscr{B}(\mathbf{E})=\nu\wedge(\nabla\wedge\mathbf{E})+\eta(\nu\wedge\mathbf{E})\wedge\nu \quad \text{on } \partial\Omega, \tag{4.1.53}$$

which yields the Dirichlet and Neumann boundary conditions, respectively, with $\eta=\infty$ and 0.

For the forward scattering problem (4.1.52), it is known that there exists a unique pair of solutions $(\mathbf{E},\mathbf{H})\in H_{\mathrm{loc}}(\mathrm{curl},\mathbb{R}^3\backslash\overline{\Omega})\times H_{\mathrm{loc}}(\mathrm{curl},\mathbb{R}^3\backslash\overline{\Omega})$ (cf. [51]). Furthermore, the radiating fields $\mathbf{E}^s$ and $\mathbf{H}^s$ to (4.1.52) possess the following asymptotic expansions:

$$\begin{aligned}\mathbf{E}^s(\mathbf{x};\Omega,k,\mathbf{d},\mathbf{p}) &= \frac{e^{\mathrm{i}k\mathbf{x}\cdot\mathbf{d}}}{|\mathbf{x}|}\left\{\mathbf{E}_\infty(\hat{\mathbf{x}};\Omega,k,\mathbf{d},\mathbf{p})+\mathscr{O}\left(\frac{1}{|\mathbf{x}|}\right)\right\} \quad \text{as} \quad |\mathbf{x}|\to\infty,\\ \mathbf{H}^s(\mathbf{x};\Omega,k,\mathbf{d},\mathbf{p}) &= \frac{e^{\mathrm{i}k\mathbf{x}\cdot\mathbf{d}}}{|\mathbf{x}|}\left\{\mathbf{H}_\infty(\hat{\mathbf{x}};\Omega,k,\mathbf{d},\mathbf{p})+\mathscr{O}\left(\frac{1}{|\mathbf{x}|}\right)\right\} \quad \text{as} \quad |\mathbf{x}|\to\infty,\end{aligned} \tag{4.1.54}$$

which hold uniformly in the angular variable $\hat{\mathbf{x}} = \mathbf{x}/|\mathbf{x}| \in \mathbb{S}^2$. The functions $\mathbf{E}_\infty(\hat{\mathbf{x}})$ and $\mathbf{H}_\infty(\hat{\mathbf{x}})$ in (4.1.54) are called the electric and magnetic far field patterns, respectively, and both are analytic on the entire unit sphere $\mathbb{S}^2$. As above and also in what follows, the notation $\mathbf{U}(\mathbf{x};\Omega,\mathbf{p},k,\mathbf{d})$ will be frequently used to specify the dependence of a given function $\mathbf{U}$ on the scatterer Ω, the polarization $\mathbf{p}$, the wave number k and the incident direction $\mathbf{d}$.

The inverse electromagnetic obstacle scattering problem corresponding to (4.1.52) is to recover the scatterer Ω (and η as well in the impedance case) by the knowledge of the far-field pattern $\mathbf{E}_\infty(\hat{\mathbf{x}};\Omega,\mathbf{p},k,\mathbf{d})$ (or equivalently $\mathbf{H}_\infty(\hat{\mathbf{x}};\Omega,\mathbf{p},k,\mathbf{d})$). By introducing an operator $\mathscr{F}$ which sends the obstacle to the corresponding far-field pattern, defined by the forward scattering system (4.1.52), the aforementioned inverse problem can be formulated as

$$\mathscr{F}(\Omega,\eta) = \mathbf{E}_\infty(\hat{\mathbf{x}};\Omega,k,\mathbf{d},\mathbf{p}). \tag{4.1.55}$$

This inverse problem (4.1.55) is highly non-linear and severely ill-posed (cf. [16]). It is a long-standing problem that one can establish the one-to-one correspondence for (4.1.55) by a single far-field pattern or a finite number of far-field patterns (namely, with a fixed triplet of k, $\mathbf{d}$ and $\mathbf{p}$ or a finite number of triplets of k, $\mathbf{d}$ and $\mathbf{p}$); see the recent survey paper [15] by Colton and Kress for more discussions about the historical developments of this fundamental problem. Kress in [32] showed that a ball and its boundary condition (for constant impedance η) can be uniquely determined by a single far-field pattern. Also, the far-field patterns associated with all incident directions $\mathbf{d}$, all polarizations $\mathbf{p}\perp\mathbf{d}$ and all observation directions $\hat{\mathbf{x}}$ can uniquely determine the shape of Ω and its boundary condition; see [32] for more details.

Under the assumption that Ω is a polyhedral obstacle associated with $\eta\equiv 0$ or $\eta\equiv\infty$, the unique correspondence, aka unique identifiability, for the inverse problem (4.1.55) by a single far-field measurement was established in the literature; see [36, 40–43]. The unique identifiability for an impedance obstacle of the polyhedral shape for the case that the generalized impedance parameter η, which can be 0, ∞ or a variable function, is established in [19]. To be more specific about the generalized impedance obstacle, we introduce the following definition.

Definition 4.1.3. Let Ω be an open and bounded polyhedron in $\mathbb{R}^3$. Hence, $\partial\Omega$ possesses finitely many edge corners that are formed by the intersections of any two adjacent faces of $\partial\Omega$. Ω is said to be irrational if all of its edge corners are irrational; otherwise, it is called rational, and the smallest degree among the rational degrees of all of its rational corners is referred to the degree of the polyhedron.

Definition 4.1.4. (Ω,η) is said to be an admissible polyhedral obstacle if Ω is an open-bounded polyhedron and η fulfils the following requirements:

(1) For each face of $\partial\Omega$, say $\widetilde{\Pi}$, and each edge of $\widetilde{\Pi}$, say $\boldsymbol{l}$, there exists a neighbourhood $\Sigma_{\boldsymbol{l}} := B_\rho(\boldsymbol{l})\cap\widetilde{\Pi}$ with $\rho\in\mathbb{R}_+$ and $B_\rho(\boldsymbol{l}) := \{\mathbf{x}\in\mathbb{R}^3; |\mathbf{x}-\mathbf{x}'|<\rho, \forall\mathbf{x}'\in\boldsymbol{l}\}$, such that either $\eta|_{\Sigma_{\boldsymbol{l}}} = 0$, or $\eta|_{\Sigma_{\boldsymbol{l}}} = \infty$, or $\eta|_{\Sigma_{\boldsymbol{l}}} \in \mathscr{A}(\boldsymbol{l})$.
(2) On any open subset of the other part of $\partial\Omega$ other than the neighbourhood of each edge of $\partial\Omega$ introduced in (1), η can be 0, or ∞ or $\eta\in L^\infty$.
(3) In the case $\eta\in L^\infty$, one has that $\Re(\eta)\geq 0$ and $\Im(\eta)<0$.

Definition 4.1.5. Ω is said to be an admissible complex polyhedral obstacle if it consists of finitely many admissible polyhedral obstacles. That is,

$$(\Omega,\eta) = \bigcup_{j=1}^{l}(\Omega_j,\eta_j),$$

where $l \in \mathbb{N}$ and each (Ω_j, η_j) is an admissible polyhedral obstacle. Here, we define

$$\eta = \sum_{j=1}^{l} \eta_j \chi_{\partial\Omega_j}.$$

Moreover, Ω is said to be irrational if all of its component polyhedral obstacles are irrational; otherwise, it is said to be rational. For the latter case, the smallest degree among all the degrees of its rational components is defined to be the degree of the complex obstacle Ω.

Next, we first derive a local unique identifiability result in determining an admissible complex irrational polyhedral obstacle by a single far-field pattern.

Theorem 4.1.1. *Consider a fixed triplet of $k \in \mathbb{R}_+$, $\mathbf{d} \in \mathbb{S}^2$ and $\mathbf{p} \in \mathbb{R}^3\backslash\{\mathbf{0}\}$. Let (Ω, η) and $(\widetilde{\Omega}, \widetilde{\eta})$ be two admissible complex irrational obstacles, with $\mathbf{E}_\infty$ and $\widetilde{\mathbf{E}}_\infty$ being their corresponding far-field patterns and $\mathbf{G}$ being the unbounded connected component of $\overline{\mathbb{R}^3\backslash(\Omega \cup \widetilde{\Omega})}$. If $\mathbf{E}_\infty$ and $\widetilde{\mathbf{E}}_\infty$ are the same in the sense that*

$$\mathbf{E}_\infty(\hat{\mathbf{x}}; \Omega, k, \mathbf{d}, \mathbf{p}) = \widetilde{\mathbf{E}}_\infty(\hat{\mathbf{x}}; \widetilde{\Omega}, k, \mathbf{d}, \mathbf{p}), \quad for \quad all \ \hat{\mathbf{x}} \in \mathbb{S}^2, \tag{4.1.56}$$

then $(\partial\Omega\backslash\partial\overline{\widetilde{\Omega}}) \bigcup (\partial\widetilde{\Omega}\backslash\partial\overline{\Omega})$ cannot possess an edge corner on $\partial\mathbf{G}$. Moreover,

$$\eta = \widetilde{\eta} \quad on \quad \partial\Omega \cap \partial\widetilde{\Omega} \cap \partial\mathbf{G}. \tag{4.1.57}$$

Proof. We prove the theorem by contradiction. Assume that $(\partial\Omega\backslash\partial\overline{\widetilde{\Omega}}) \bigcup (\partial\widetilde{\Omega}\backslash\partial\overline{\Omega})$ has an edge corner $\mathbf{x}_c$ on $\partial\mathbf{G}$. Then, $\mathbf{x}_c$ is either located at $\partial\Omega$ or $\partial\widetilde{\Omega}$. Without loss of generality, we assume that $\mathbf{x}_c$ is an edge corner of $\partial\widetilde{\Omega}$, which also indicates that $\mathbf{x}_c$ lies outside Ω. Let $h \in \mathbb{R}_+$ be sufficiently small such that $B_h(\mathbf{x}_c) \Subset \mathbb{R}^2\backslash\overline{\Omega}$, then we have

$$B_h(\mathbf{x}_c) \cap \partial\widetilde{\Omega} = \widetilde{\Pi}_\ell, \quad \ell = 1, 2,$$

where $\widetilde{\Pi}_\ell$ are two flat subsets lying on the faces of $\widetilde{\Omega}$ that intersect at $\mathbf{x}_c$. Moreover, for the subsequent use, we let h be smaller than ρ, where ρ is the parameter in Definition 4.1.4. Hence, we have an edge corner $\mathscr{E}(\widetilde{\Pi}_1, \widetilde{\Pi}_2, \boldsymbol{l}) \in \partial\mathbf{G}$ with $\mathbf{x}_c \in \boldsymbol{l}$, where $\mathbf{G}$ is the unbounded connected component of $\overline{\mathbb{R}^3\backslash(\Omega \cup \widetilde{\Omega})}$. By (4.1.56) and the Rellich theorem (cf. [16]), we know that

$$\mathbf{E}(\mathbf{x}; k, \mathbf{d}, \mathbf{p}) = \widetilde{\mathbf{E}}(\mathbf{x}; k, \mathbf{d}, \mathbf{p}), \quad \mathbf{x} \in \mathbf{G}. \tag{4.1.58}$$

Since $\widetilde{\Pi}_\ell \subset \partial\mathbf{G}$, $\ell = 1, 2$, combining (4.1.58) with the generalized boundary condition (4.1.53) on $\partial\widetilde{\Omega}$, it is easy to obtain that

$$\nu_\ell \wedge (\nabla \wedge \mathbf{E}) + \widetilde{\eta}(\nu_\ell \wedge \mathbf{E}) \wedge \nu_\ell = \nu_\ell \wedge (\nabla \wedge \widetilde{\mathbf{E}}) + \widetilde{\eta}(\nu_\ell \wedge \widetilde{\mathbf{E}}) \wedge \nu_\ell = \mathbf{0} \text{ on } \widetilde{\Pi}_\ell. \tag{4.1.59}$$

We consider the following two separate cases, depending on the values of $\widetilde{\eta}$ on $\widetilde{\Pi}_\ell$ associated with the edge corner $\mathscr{E}(\widetilde{\Pi}_1, \widetilde{\Pi}_2, \boldsymbol{l})$

Case 1. $\widetilde{\eta}\big|_{\widetilde{\Pi}_\ell} = 0$ or $\widetilde{\eta}\big|_{\widetilde{\Pi}_\ell} = \infty$, $\ell = 1, 2$. We only consider the case $\widetilde{\eta}\big|_{\widetilde{\Pi}_\ell} = \infty$ and the other case can be treated in a similar manner. First, we note that one has from (4.1.59),

$$(\nu_\ell \wedge \mathbf{E}) \wedge \nu_\ell = 0 \quad \text{on } \widetilde{\Pi}_\ell, \ \ell = 1, 2. \tag{4.1.60}$$

Let $\widehat{\Pi}_\ell$ denote the full flat extension of $\widetilde{\Pi}_\ell$ within $\mathbb{R}^3\backslash\overline{\Omega}$. We claim that at least one of $\widehat{\Pi}_\ell$ is bounded. In fact, if on the contrary both $\widehat{\Pi}_1$ and $\widehat{\Pi}_2$ are unbounded, then one has from analytic continuation (noting that $\mathbf{E}$ is real analytic in $\mathbb{R}^3\backslash\overline{\Omega}$) and (4.1.60) that

$$\lim_{|\mathbf{x}|\to\infty, \mathbf{x}\in\widehat{\Pi}_\ell} |(\nu_\ell \wedge \mathbf{E}) \wedge \nu_\ell| = 0, \ \ell = 1,2. \tag{4.1.61}$$

Using (4.1.54), we note that $\mathbf{E}^s(\mathbf{x}) \to \mathbf{0}$ as $|\mathbf{x}| \to \infty$, and hence, we further have from (4.1.61) that

$$\lim_{|\mathbf{x}|\to\infty, \mathbf{x}\in\widehat{\Pi}_\ell} \left|(\nu_\ell \wedge \mathbf{E}^i) \wedge \nu_\ell\right| = 0, \ \ell = 1,2, \tag{4.1.62}$$

which together with (4.1.51) readily implies that $|(\nu_\ell \wedge \mathbf{E}) \wedge \nu_\ell| = 0$, $\ell = 1,2$. But this is impossible since ν_1 and ν_2 are linearly independent. Without loss of generality, we can assume that $\widehat{\Pi}_1$ is bounded. Clearly, $\widehat{\Pi}_1$ and part of $\partial\Omega$ form a bounded domain in $\mathbb{R}^3\backslash\overline{\Omega}$, and we denote it as Ω_1. It is noted from (4.1.59) that one has

$$\nu \wedge (\nabla \wedge \mathbf{E}) + \widetilde{\eta}(\nu \wedge \mathbf{E}) \wedge \nu = \mathbf{0} \text{ on } \partial\Omega_1\backslash\widehat{\Pi}_1 \text{ and } \nu \wedge (\nabla \wedge \mathbf{E}) = 0 \text{ on } \widehat{\Pi}_1. \tag{4.1.63}$$

We next show that $\widetilde{\eta}$ can only take 0 or ∞ on $\partial\Omega_1\backslash\widehat{\Pi}_1$. Indeed, we assume on the contrary that there exists a non-empty open subset $\Lambda_1 \subset \partial\Omega_1\backslash\widehat{\Pi}_1$ such that $\widetilde{\eta} \in L^\infty(\Lambda_1)$ with $\Re(\widetilde{\eta}) \geq 0$ and $\Im(\widetilde{\eta}) < 0$, and on $(\partial\Omega_1\backslash\widehat{\Pi}_1)\backslash\overline{\Lambda_1}$, $\widetilde{\eta}$ takes either 0 or ∞. Noting that the Maxwell equations, namely, the first two equations in (4.1.52), are satisfied in Ω_1, we have from Green's formula that

$$\begin{aligned} \mathbf{i}k\int_{\Omega_1} |\mathbf{H}|^2 &= \int_{\Omega_1} (\nabla \wedge \mathbf{E}) \cdot \overline{\mathbf{H}} = \int_{\Omega_1} \mathbf{E} \cdot (\nabla \wedge \overline{\mathbf{H}}) + \int_{\Omega_1} (\overline{\mathbf{H}} \wedge \nu) \cdot \mathbf{E} \\ &= \mathbf{i}k\int_{\Omega_1} |\mathbf{E}|^2 + \int_{\partial\Omega_1} (\overline{\mathbf{H}} \wedge \nu) \cdot \mathbf{E} = \mathbf{i}k\int_{\Omega_1} |\mathbf{E}|^2 + \int_{\Lambda_1} (\overline{\mathbf{H}} \wedge \nu) \cdot \mathbf{E}, \end{aligned} \tag{4.1.64}$$

where in deriving the last equality, we have made use of the fact that $(\overline{\mathbf{H}} \wedge \nu) \cdot \mathbf{E} = 0$ on $\partial\Omega_1\backslash\overline{\Lambda_1}$. Using the fact that $\Im(\widetilde{\eta}) < 0$ on Λ_1, one can readily infer from (4.1.64) that $\nu \wedge \mathbf{E}|_{\Lambda_1} = \mathbf{0}$, which together with (4.1.63) further implies that $\nu \wedge \mathbf{H}|_{\Lambda_1} = 0$. Hence, by Holmgren's uniqueness principle (cf. [16]), one has that

$$\mathbf{E}(\mathbf{x}; k, \mathbf{d}, \mathbf{p}) = \mathbf{0} \text{ in } \mathbb{R}^3\backslash\overline{\Omega}, \tag{4.1.65}$$

which in particular yields that

$$\lim_{|\mathbf{x}|\to\infty} |\mathbf{E}(\mathbf{x}; k, \mathbf{d}, \mathbf{p})| = \mathbf{0}. \tag{4.1.66}$$

But this contradicts to the fact that follows from (4.1.54):

$$\lim_{|\mathbf{x}|\to\infty} |\mathbf{E}(\mathbf{x}; k, \mathbf{d}, \mathbf{p})| = \lim_{|\mathbf{x}|\to\infty} \left|\mathbf{p}e^{\mathbf{i}k\mathbf{x}\cdot\mathbf{d}} + \mathbf{E}^s(\mathbf{x}; k, \mathbf{d}, \mathbf{p})\right| = |\mathbf{p}| \neq 0. \tag{4.1.67}$$

Hence, we actually can find a polyhedral domain $\Omega_1 \subset \mathbb{R}^3\backslash\Omega$ such that one has on $\partial\Omega_1$, either $\nu \wedge \mathbf{E} = 0$ or $\nu \wedge \mathbf{H} = 0$. The situation is reduced to that considered in [36] and [41]. It is noted that in [41], two far-field patterns are used to handle the above situation. However, the pair of incident fields $(\mathbf{E}^i, \mathbf{H}^i)$ in (4.1.51) in our current case is chosen slightly different from that in [41], which enables one to apply the path argument from [36] to arrive at a contradiction by starting from Ω_1.

Case 2. $\widetilde{\eta}\big|_{\widetilde{\Pi}_\ell} \in \mathscr{A}(\boldsymbol{l})$, $\ell = 1,2$, or one of $\widetilde{\eta}\big|_{\widetilde{\Pi}_\ell}$ belongs to $\mathscr{A}(\boldsymbol{l})$, and the other one takes 0 or ∞, or one of $\widetilde{\eta}\big|_{\widetilde{\Pi}_\ell}$ is 0 and the other one is ∞. This falls exactly to the situation considered in Theorem 3.5.1. By the irrationality of the edge corner as well as the strong uniqueness continuation principle in Theorem 3.5.1, we readily have (4.1.65), which again leads to the contradiction (4.1.67).

It remains to prove (4.1.57), and we establish it by contradiction. Let $\Gamma \subset \partial\Omega \cap \partial\widetilde{\Omega} \cap \partial\mathbf{G}$ be an open subset such that $\eta \neq \widetilde{\eta}$ on Γ. By taking a smaller subset of Γ if necessary, we can assume that η (respectively, η) is either L^∞ or 0 or ∞ on Γ. Clearly, $\mathbf{E} = \widetilde{\mathbf{E}}$ in $\mathbf{G}$. Hence, it holds that

$$(\nu \wedge \mathbf{E}) \wedge \nu = (\nu \wedge \widetilde{\mathbf{E}}) \wedge \nu \text{ and } \nu \wedge (\nabla \wedge \mathbf{E}) = \nu \wedge (\nabla \wedge \widetilde{\mathbf{E}}) \quad \text{on } \Gamma.$$

and

$$\nu\wedge(\nabla\wedge\mathbf{E})+\eta(\nu\wedge\mathbf{E})\wedge\nu=\mathbf{0},\quad \nu\wedge(\nabla\wedge\widetilde{\mathbf{E}})+\widetilde{\eta}(\nu\wedge\widetilde{\mathbf{E}})\wedge\nu=\mathbf{0}\quad \text{on } \Gamma.$$

Combining with the assumption that $\eta\neq\widetilde{\eta}$ on $\mathscr{E}$, we can directly deduce that

$$\nu\wedge\mathbf{E}=\nu\wedge\mathbf{H}=0\quad \text{on } \Gamma,$$

which in turn yields by Holmgren's uniqueness principle (cf. [16]) that $\mathbf{E}=\mathbf{0}$ in $\mathbb{R}^3\backslash\Omega$. Therefore, we arrive at the same contradiction as that in (4.1.66) and (4.1.67), which readily proves (4.1.57).

The proof is now completed. □

It is recalled that the convex hull of Ω, denoted by $\mathscr{CH}(\Omega)$, is the smallest convex set that contains Ω. As a direct consequence of Theorem 4.1.1, we show in the next corollary that the convex hull of a complex irrational obstacle can be uniquely determined by one far-field measurement, and the boundary impedance parameter η can be partially identified as well. From this conclusion, we know if the underlying polyhedral obstacle is convex, both the obstacle and its boundary impedance are uniquely determined by a single far-field pattern.

Corollary 4.1.1. *Consider a fixed triplet of $k\in\mathbb{R}_+$, $\mathbf{d}\in\mathbb{S}^2$ and $\mathbf{p}\in\mathbb{R}^3\backslash\{\mathbf{0}\}$. Let (Ω,η) and $(\widetilde{\Omega},\widetilde{\eta})$ be two admissible complex irrational obstacles, with $\mathbf{E}_\infty$ and $\widetilde{\mathbf{E}}_\infty$ being their corresponding far-field patterns. If $\mathbf{E}_\infty$ and $\widetilde{\mathbf{E}}_\infty$ satisfy* (4.1.56), *then one has that*

$$\mathscr{CH}(\Omega)=\mathscr{CH}(\widetilde{\Omega}):=\Sigma,\quad \eta=\widetilde{\eta}\ \text{on}\ \partial\Omega\cap\partial\widetilde{\Omega}\cap\partial\Sigma.$$

As a further application of the UCP results provided in Chap. 3, we consider the unique determination of a rather general class of non-convex obstacles. To that end, we first introduce the aforementioned class of non-convex obstacles.

In the sequel, we denote by $P_S(\mathbf{x})$ the projection of a point $\mathbf{x}\in\mathbb{R}^3$ onto a set S. Let $\partial(\mathscr{CH}(\Omega))=\{\Sigma_\ell\,|\,\ell=1,\ldots,N\}$, where Σ_ℓ, $\ell=1,\ldots,N$ are the finitely many faces of $\mathscr{CH}(\Omega)$. Let $\mathscr{V}(\Omega)$ and $\mathscr{V}(\mathscr{CH}(\Omega))$ denote the sets of vertices of Ω and $\mathscr{CH}(\Omega)$, respectively. It is known that $\mathscr{V}(\mathscr{CH}(\Omega))\subset\mathscr{V}(\Omega)$. For any vertex $\mathbf{v}\in\mathscr{V}(\Omega)\backslash\mathscr{V}(\mathscr{CH}(\Omega))$, we consider the projection, $P_{\Sigma_j}(\mathbf{v})$, where $\Sigma_j\subset\partial(\mathscr{CH}(\Omega))$ is a face. It is assumed that there exists at least one Σ_j such that $\mathbf{v}-P_{\Sigma_j}(\mathbf{v})\subset\mathbb{R}^3\backslash\Omega$. Then for a face $\Sigma_l\subset\partial(\mathscr{CH}(\Omega))$, we say that $\mathbf{v}\vdash\Sigma_l$ if

$$\mathbf{v}-P_{\Sigma_\ell}(\mathbf{v})=\arg\min_{\mathbf{w}-P_{\Sigma_j}(\mathbf{w})\in\mathbb{R}^3\backslash\Omega,\forall\Sigma_j\subset\partial(\mathscr{CH}(\Omega))}\left|\mathbf{w}-P_{\Sigma_j}(\mathbf{w})\right|. \tag{4.1.68}$$

Definition 4.1.6. Let Ω be an admissible polyhedral obstacle and Σ_l be a given face of $\partial(\mathscr{CH}(\Omega))$ and $\mathscr{V}_{\mathscr{C}}$ be a given set of finitely many, discrete and distinct points on Σ_l. Ω is said to be uniformly concave with respect to $\mathscr{V}_{\mathscr{C}}$ if for any $\mathbf{v}\in\mathscr{V}(\Omega)\backslash\mathscr{V}(\mathscr{CH}(\Omega))$, $\mathbf{v}\vdash\Sigma_l$ and

$$\{P_{\Sigma_l}(\mathbf{v})\mid\mathbf{v}\in\mathscr{V}(\Omega)\backslash\mathscr{V}(\mathscr{CH}(\Omega))\}=\mathscr{V}_{\mathscr{C}}.$$

As a simple illustrating example of Definition 4.1.6, we consider two different uniformly concave hexahedrons $\Omega_1:=ABCDE_1$ and $\Omega_2:=ABCDE_2$; see Fig. 4.1. It is easy to see that Ω_1 and Ω_2 have the same convex hull, which is the tetrahedron $ABCD$. The vertexes E_1 and E_2 corresponding to Ω_1 and Ω_2 have the same projecting point on the face $\Sigma:=BCD$ of the convex hull $ABCD$. It is pointed out that the vertex corner $\mathscr{V}(BE_2C,CE_2D,BE_2D,E_2)\in\partial\mathbf{G}$, where BE_2C,CE_2D,BE_2D are faces of Ω_2 and $\mathbf{G}=\mathbb{R}^3\backslash(\Omega_1\cup\Omega_2)$.

Theorem 4.1.2. *Consider a fixed triplet of $k\in\mathbb{R}_+$, $\mathbf{d}\in\mathbb{S}^2$ and $\mathbf{p}\in\mathbb{R}^3\backslash\{\mathbf{0}\}$. Let (Ω,η) and $(\widetilde{\Omega},\widetilde{\eta})$ be two uniformly concave irrational admissible polyhedral obstacles with respect to the set $\mathscr{V}_{\mathscr{C}}$, with $\mathbf{E}_\infty$ and $\widetilde{\mathbf{E}}_\infty$ being their corresponding far-field patterns. If $\mathbf{E}_\infty$ and $\widetilde{\mathbf{E}}_\infty$ satisfy* (4.1.56), *then*

$$\Omega=\widetilde{\Omega}\ \text{and}\ \eta=\widetilde{\eta}.$$

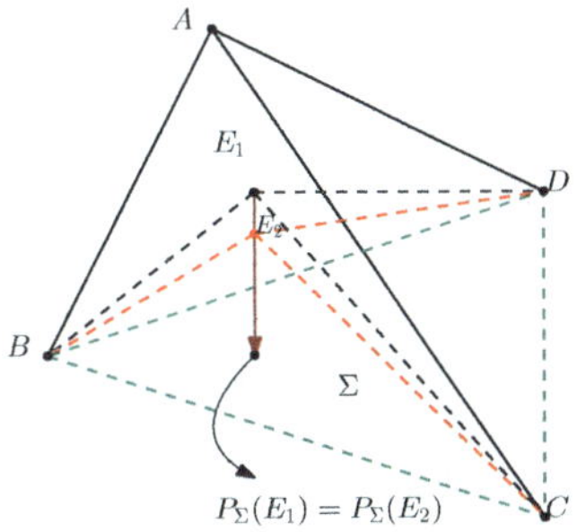

Fig. 4.1 Schematic illustration of two different uniformly concave hexahedrons $ABCDE_1$ and $ABCDE_2$ with $\mathscr{CH}(ABCDE_1)$=$\mathscr{CH}(ABCDE_2) = ABCD$

Proof. We prove this theorem by contradiction. Assume that $\Omega \neq \widetilde{\Omega}$ but (4.1.56) is still fulfilled. From Corollary 4.1.1, we have $\mathscr{CH}(\Omega) = \mathscr{CH}(\widetilde{\Omega})$, which implies that the vertices of Ω contributing to $\mathscr{CH}(\Omega)$ are the same as the corresponding vertices of $\widetilde{\Omega}$ contributing to $\mathscr{CH}(\widetilde{\Omega})$. We shall prove that there must exist an edge corner $\mathscr{E}(\Pi_1, \Pi_2, \mathbf{x}_c) \in \partial\mathbf{G}$, where $\mathbf{G}$ is the unbounded connected component of $\mathbb{R}^3 \backslash \overline{(\Omega \cup \widetilde{\Omega})}$. Since $\Omega \neq \widetilde{\Omega}$, there exists an edge $\boldsymbol{l} \subset \partial\Omega \backslash \partial\widetilde{\Omega}$ or $\boldsymbol{l} \subset \partial\widetilde{\Omega} \backslash \partial\Omega$. Without loss of generality, we assume that $\boldsymbol{l} \subset \partial\widetilde{\Omega} \backslash \partial\Omega$. In the sequel, we let $\mathbf{a}_l$ and $\mathbf{b}_l$ denote the two vertices of the line segment $\boldsymbol{l}$. We divide our remaining proof into two separate cases.

Case 1. Suppose that $\mathbf{a}_l \in \mathscr{V}(\mathscr{CH}(\widetilde{\Omega}))$ and $\mathbf{b}_l \in \mathscr{V}(\mathscr{CH}(\widetilde{\Omega}))$. Therefore, $\boldsymbol{l} \subset \partial\mathbf{G} \cap \partial\widetilde{\Omega}$. There exists a point $\mathbf{x}_c \in \boldsymbol{l}$ and a sufficiently small $h \in \mathbb{R}_+$ such that

$$B_h(\mathbf{x}_c) \cap \partial\widetilde{\Omega} = \widetilde{\Pi}_\ell, \quad \ell = 1, 2,$$

where $\widetilde{\Pi}_\ell$ are two flat subsets lying on the faces of $\widetilde{\Omega}$ that intersect at $\mathbf{x}_c$. Clearly, $\mathbf{x}_c \in \boldsymbol{l}$ is an edge-corner point.

Case 2. Suppose that there exists at least one of $\mathbf{a}_l$ and $\mathbf{b}_l$ belonging to $\mathscr{V}(\widetilde{\Omega}) \backslash \mathscr{V}(\mathscr{CH}(\widetilde{\Omega}))$; namely, $\mathbf{x}_c \in \mathscr{V}(\widetilde{\Omega}) \backslash \mathscr{V}(\mathscr{CH}(\widetilde{\Omega}))$, where $\mathbf{x}_c$ could be either $\mathbf{a}_l$ or $\mathbf{b}_l$. Since Ω and $\widetilde{\Omega}$ are uniformly concave admissible polyhedral obstacles with respect to the set $\mathscr{V}_\mathscr{C}$, there exists a face $\Sigma_\ell \Subset \partial(\mathscr{CH}(\Omega))$ such that $\mathbf{x}_c \vdash \Sigma_\ell$ and $\mathscr{V}_\mathscr{C} \Subset \Sigma_\ell$. Furthermore, we know that there exists a vertex $\mathbf{x}_{c,\Omega} \in \mathscr{V}(\Omega) \backslash \mathscr{V}(\mathscr{CH}(\Omega)$ such that

$$\mathbf{x}_{c,\Omega} \vdash \Sigma_\ell, \quad \boldsymbol{P}_{\Sigma_\ell}(\mathbf{x}_{c,\Omega}) = \boldsymbol{P}_{\Sigma_\ell}(\mathbf{x}_c) \in \mathscr{V}_\mathscr{C}.$$

Since $\mathbf{x}_{c,\Omega}$ and $\mathbf{x}_c$ are distinct, it holds that

$$\mathrm{d}(\mathbf{x}_c, \Sigma_\ell) \neq \mathrm{d}(\mathbf{x}_{c,\Omega}, \Sigma_\ell),$$

where $\mathrm{d}(\mathbf{x}_c, \Sigma_\ell)$ is the distance between $\mathbf{x}_c$ and Σ_ℓ. Without loss of generality, we may assume that $\mathrm{d}(\mathbf{x}_c, \Sigma_\ell) < \mathrm{d}(\mathbf{x}_{c,\Omega}, \Sigma_\ell)$. Hence, one can conclude that $\mathbf{x}_c \in \partial\mathbf{G}$, which also indicates that $\mathbf{x}_c$ lies outside Ω. Let $h \in \mathbb{R}_+$ be sufficiently small such that $B_h(\mathbf{x}_c) \Subset \mathbb{R}^2 \backslash \overline{\Omega}$; then due to the fact that $\mathscr{V}_\mathscr{C}$ is discrete and distinct, we can conclude that

$$B_h(\mathbf{x}_c) \cap \partial\widetilde{\Omega} = \widetilde{\Pi}_\ell, \quad \ell = 1, 2,$$

where $\widetilde{\Pi}_\ell$ are two plane cells lying on the faces of $\widetilde{\Omega}$ that intersect at $\mathbf{x}_c$.

The remaining proof is similar to that of Theorem 4.1.1, which is omitted. □

Finally, we remark that in this section, we only consider the case that the underlying obstacle is irrational in order to make use of the strong unique continuation principle in Theorem 3.5.1. That is, in the contradiction argument in proving Theorems 4.1.1 and 4.1.2, one can find an edge corner that can lead to the vanishing of the total wave field outside the obstacle by the strong unique continuation principle in Theorem 3.5.1. However, we would like to emphasize that the same argument would work for the case that the underlying obstacle is of a general polyhedral shape, subject to some slight modification. In fact, in such a case, it may happen that the edge corner in the contradiction argument is rational, and hence instead of Theorem 3.5.1, one would need to make use of the finite vanishing order results in Theorems 3.3.1, 3.4.1, 3.4.2 and 3.4.3 to obtain that the total wave field is "small" around the edge corner (compared to the total vanishing in the irrational case). Hence, a contradiction can be obtained if one requires that the total wave field outside the obstacle is everywhere "big", which can be fulfilled in certain scenarios of practical interest, see, e.g. [11]. Nevertheless, we shall not explore this direction any further in this subsection.

4.2 Unique Recovery for the Inverse Diffraction Grating Problem with the Helmholtz Equation in $\mathbb{R}^2$

In this section, we consider the unique recovery for the inverse acoustic diffraction grating problem. First, we give a brief review of the basic mathematical model for this inverse problem. Let the profile of a diffraction grating be described by the curve

$$\Lambda_f = \{(x_1, x_2) \in \mathbb{R}^2;\ x_2 = f(x_1)\}, \tag{4.2.1}$$

where f is a periodic Lipschitz function with period 2π. Let

$$\Omega_f = \{\mathbf{x} \in \mathbb{R}^2; x_2 > f(x_1), x_1 \in \mathbb{R}\}$$

be filled with a material whose index of refraction (or wave number) k is a positive constant. Suppose further that the incident wave given by

$$u^i(\mathbf{x}; k, \mathbf{d}) = e^{\mathrm{i}k\mathbf{d}\cdot\mathbf{x}}, \quad \mathbf{d} = (\sin\theta, -\cos\theta)^\top, \quad \theta \in \left(-\frac{\pi}{2}, \frac{\pi}{2}\right),$$

propagates to Λ_f from the top. Then the total wave satisfies the following Helmholtz system:

$$\Delta u + k^2 u = 0 \quad \text{in } \Omega_f; \quad \mathscr{B}(u)\big|_{\Lambda_f} = 0 \quad \text{on } \Lambda_f, \tag{4.2.2}$$

with the generalized impedance boundary condition

$$\mathscr{B}(u) = \partial_\nu u + \eta u = 0 \quad \text{on } \partial\Omega, \tag{4.2.3}$$

where η can be ∞ or 0, corresponding to a sound-soft or sound-hard grating, respectively.

To achieve the uniqueness of (4.2.2), the total wave field u should be α-quasiperiodic in the x_1-direction, with $\alpha = k\sin\theta$, which means that

$$u(x_1 + 2\pi, x_2) = e^{2\mathrm{i}\alpha\pi} \cdot u(x_1, x_2),$$

and the scattered field u^s satisfies the Rayleigh expansion (cf. [48, 49]):

$$u^s(\mathbf{x}; k, \mathbf{d}) = \sum_{n=-\infty}^{+\infty} u_n e^{\mathrm{i}\xi_n(\theta)\cdot\mathbf{x}} \quad \text{for} \quad x_2 > \max_{x_1 \in [0, 2\pi]} f(x_1), \tag{4.2.4}$$

where $u_n \in \mathbb{C} (n \in \mathbb{Z})$ are called the Rayleigh coefficient of u^s, and

$$\xi_n(\theta) = (\alpha_n(\theta), \beta_n(\theta))^\top, \quad \alpha_n(\theta) = n + k\sin\theta,$$
$$\beta_n(\theta) = \begin{cases} \sqrt{k^2 - \alpha_n^2(\theta)}, & \text{if } |\alpha_n(\theta)| \le k \\ \mathrm{i}\sqrt{\alpha_n^2(\theta) - k^2}, & \text{if } |\alpha_n(\theta)| > k \end{cases}. \tag{4.2.5}$$

The existence and uniqueness of the α-quasiperiodic solution to (4.2.2) for the sound-soft or impedance boundary condition with $\eta \in \mathbb{C}$ being a constant satisfying $\Im(\eta) > 0$ can be found in [1,8,28,29]. It should be pointed out that the uniqueness of the direct scattering problem associated with the sound-hard condition is not always true (see [27]). In our subsequent study, we assume the well-posedness of the forward scattering problem and focus on the study of the inverse grating problem.

Introduce a measurement boundary as

$$\Gamma_b := \{(x_1, b) \in \mathbb{R}^2;\ 0 \le x_1 \le 2\pi,\ b > \max_{x_1 \in [0,2\pi]} |f(x_1)|\}.$$

The inverse diffraction grating problem is to determine (Λ_f, η) from the knowledge of $u(\mathbf{x}|_{\Gamma_b}; k, \mathbf{d})$ and can be formulated as the operator equation:

$$\mathscr{F}(\Lambda_f, \eta) = u(\mathbf{x}; k, \mathbf{d}), \quad \mathbf{x} \in \Gamma_b,$$

where $\mathscr{F}$ is defined by the forward diffraction scattering system and is non-linear.

The unique recovery result on the inverse diffraction grating problem with the sound-soft or sound-hard boundary condition by a finite number of incident plane waves can be found in [20,22,29,30]. Recently, a generalized impedance polyhedral grating profile in $\mathbb{R}^3$ and its variable boundary parameter can be uniquely determined by the measurement $u(\mathbf{x}|_{\Gamma_b}; k, \mathbf{d})$ associated with a fixed wave k number and a fixed incident direction $\mathbf{d}$ under generic a priori assumptions; see [12] for more detailed discussions on this aspect. Furthermore, uniqueness results to the shape determination for an electromagnetic doubly periodic structure can be found in [3]. Similarly, the corresponding uniqueness for determining a polyhedral grating in electromagnetic scattering was studied in [5, 6]. In order to study unique identifiability for the impedance or generalized impedance grating problem in $\mathbb{R}^2$, we propose the following admissible polygonal gratings associated with the inverse diffraction grating problem.

Definition 4.2.1. Let (Λ_f, η) be a periodic grating as described in (4.2.1). Suppose there is a partition, $[0, 2\pi] = \cup_{i=1}^{\ell} [a_i, a_{i+1}]$ with $a_i < a_{i+1}$, $a_1 = 0$ and $a_{\ell+1} = 2\pi$. If on each piece $[a_i, a_{i+1}]$, $1 \le i \le \ell$, f is a linear polynomial and η is either a constant (possibly zero) or ∞, then (Λ_f, η) is said to be an admissible polygonal grating.

Definition 4.2.2. Let (Λ_f, η) be an admissible polygonal grating. Let Γ^+ and Γ^- be two adjacent pieces of Λ_f. The intersecting point of Γ^+ and Γ^- is called a corner point of Λ_f, and $\angle(\Gamma^+, \Gamma^-)$ is called a corner angle. If all the corner angles of Λ_f are irrational, then it is said to be an *irrational polygonal grating*. If a corner angle of Λ_f is rational, it is called a *rational polygonal grating*. The smallest degree of the rational corner angles of Λ_f is referred to as the *rational degree* of Λ_f.

Clearly, for a rational polygonal grating Λ_f in Definition 4.2.2, the rational degree of Λ_f is at least 2. Next, we consider our uniqueness result in determining an admissible irrational polygonal grating by at most two incident waves. We first present a useful lemma, whose proof follows from a completely similar argument to that of [16, Theorem 5.1].

Lemma 4.2.1. *Let $\xi_\ell \in \mathbb{R}^2$, $\ell = 1, \ldots, n$, be n vectors which are distinct from each other and D be an open set in $\mathbb{R}^2$. Then all the functions in the following set are linearly independent:*

$$\{e^{\mathrm{i}\xi_\ell \cdot \mathbf{x}};\ \mathbf{x} \in D,\ \ell = 1, 2, \ldots, n\}.$$

Theorem 4.2.1. *Let (Λ_f,η) and $(\Lambda_{\tilde f},\tilde\eta)$ be two admissible irrational polygonal gratings, and* $\mathbf{G}$ *be the unbounded connected component of $\Omega_f\cap\Omega_{\tilde f}$. Let $k\in\mathbb{R}_+$ be fixed and $\mathbf{d}_\ell$, $\ell=1,2$ be two distinct incident directions from $\mathbb{S}^1$, with*

$$\mathbf{d}_\ell=(\sin\theta_\ell,-\cos\theta_\ell)^\top,\quad \theta_\ell\in\left(-\frac{\pi}{2},\frac{\pi}{2}\right).$$

Let $u(\mathbf{x};k,\mathbf{d}_\ell)$ and $\tilde u(\mathbf{x};k,\mathbf{d}_\ell)$ denote the total fields associated with (Λ_f,η) and $(\Lambda_{\tilde f},\tilde\eta)$, respectively, and let Γ_b be a measurement boundary given by

$$\Gamma_b:=\left\{(x_1,b)\in\mathbb{R}^2;\ 0\le x_1\le 2\pi,\ b>\max\left\{\max_{x_1\in[0,2\pi]}|f(x_1)|,\ \max_{x_1\in[0,2\pi]}|\tilde f(x_1)|\right\}\right\},$$

If it holds that

$$u(\mathbf{x};k,\mathbf{d}_\ell)=\tilde u(\mathbf{x};k,\mathbf{d}_\ell),\quad \ell=1,2,\quad \mathbf{x}=(x_1,b)\in\Gamma_b, \tag{4.2.6}$$

then it cannot be true that there exists a corner point of Λ_f lying on $\partial\mathbf{G}\backslash\partial\Lambda_{\tilde f}$, or a corner point of $\Lambda_{\tilde f}$ lying on $\partial\mathbf{G}\backslash\partial\Lambda_f$.

Proof. The proof follows from a similar argument to that for Theorem 4.1.1, and we only sketch the necessary modifications in this new setup. By contradiction and without loss of generality, we assume that there exists a corner point $\mathbf{x}_c$ of Λ_f which lies on $\partial\mathbf{G}\backslash\Lambda_{\tilde f}$.

First, by the well-posedness of the diffraction grating problem (4.2.2)-(4.2.4) as well as the unique continuation, we show that $u(\mathbf{x};k,\mathbf{d}_\ell)=\tilde u(\mathbf{x};k,\mathbf{d}_\ell)$ for $\mathbf{x}\in\mathbf{G}$. In fact, by introducing $w(\mathbf{x};k,\mathbf{d}_\ell):=u(\mathbf{x};k,\mathbf{d}_\ell)-\tilde u(\mathbf{x};k,\mathbf{d}_\ell)$, $\ell=1,2$, we see from (4.2.6) that w fulfils

$$\Delta w+k^2w=0 \text{ in } \mathbf{U};\ w=0 \text{ on } \Gamma_b \text{ and } w \text{ satisfies the Rayleigh expansion (4.2.4)},$$

where $\mathbf{U}:=\mathbf{G}\cap\{\mathbf{x}\in\mathbb{R}^2; x_2>b, x_1\in\mathbb{R}\}$ with $\partial\mathbf{U}=\Gamma_b$. Hence, by the uniqueness of the solution to the diffraction grating problem, we readily know $w=0$ in $\mathbf{U}$. On the other hand, since $u(\mathbf{x};k,\mathbf{d}_\ell)$ and $\tilde u(\mathbf{x};k,\mathbf{d}_\ell)$ are analytic in $\mathbf{G}$, we know $w(\mathbf{x};k,\mathbf{d}_\ell)=0$ in $\mathbf{G}$ by means of the analytic continuation, that is, $u(\mathbf{x};k,\mathbf{d}_\ell)=\tilde u(\mathbf{x};k,\mathbf{d}_\ell)$ for $\mathbf{x}\in\mathbf{G}$. Furthermore, we can prove that

$$u(\mathbf{x};k,\mathbf{d}_\ell)=0\quad\text{or}\quad v(\mathbf{x})=0\quad\text{for}\quad x_2>\max_{x_1\in[0,2\pi]}f(x_1),$$

where v is similarly defined to (4.1.17) and (4.1.16). Next, when $x_2>\max_{x_1\in[0,2\pi]}|f(x_1)|$, $u(\mathbf{x};k,\mathbf{d}_\ell)$ has the Rayleigh expansion (cf. [48, 49]):

$$u(\mathbf{x};k,\mathbf{d}_\ell)=e^{\mathrm{i}k\mathbf{d}_\ell\cdot\mathbf{x}}+\sum_{n=-\infty}^{+\infty}u_ne^{\mathrm{i}\xi_n(\theta_\ell)\cdot\mathbf{x}}\quad\text{for}\quad x_2>\max_{x_1\in[0,2\pi]}f(x_1), \tag{4.2.7}$$

where $\xi_n(\theta_\ell)$, $\alpha_n(\theta_\ell)$, $\beta_n(\theta_\ell)$ are defined in (4.2.5). Using the definition of $\alpha_0(\theta_\ell)$ and $\beta_0(\theta_\ell)$ in (4.2.5), we can easily show that

$$k\mathbf{d}_\ell=(\alpha_0(\theta_\ell),-\beta_0(\theta_\ell))^\top. \tag{4.2.8}$$

We proceed to consider two separate cases.

Case 1. Suppose that either $u(\mathbf{x}_c;k,\mathbf{d}_1)$ or $u(\mathbf{x}_c;k,\mathbf{d}_2)$ is zero. Without loss of generality, we assume the former case. Then

$$u(\mathbf{x};k,\mathbf{d}_1)=0\quad\text{for}\quad x_2>\max_{x_1\in[0,2\pi]}f(x_1).$$

Clearly, any two vectors of $\{\xi_n(\theta_1) \mid n \in \mathbb{Z}\}$ are distinct from each other. Moreover, in view of (4.2.8), $k\mathbf{d}_1 \notin \{\xi_n(\theta_1) \mid n \in \mathbb{Z}\}$ since $|\theta_1| < \pi/2$. In view of (4.2.7), from Lemma 4.2.1 we can arrive at a contradiction.

Case 2. Suppose that both $u(\mathbf{x}_c; k, \mathbf{d}_1) \neq 0$ and $u(\mathbf{x}_c; k, \mathbf{d}_2) \neq 0$. Then it holds that

$$\alpha_1 u(\mathbf{x}; k, \mathbf{d}_1) + \alpha_2 u(\mathbf{x}; k, \mathbf{d}_2) = 0 \quad \text{for} \quad x_2 > \max_{x_1 \in [0,2\pi]} f(x_1), \tag{4.2.9}$$

where $\alpha_\ell \neq 0$, $\ell = 1, 2$ are defined in (4.1.16). Substituting (4.2.7) into (4.2.9), we derive that

$$\sum_{\ell=1}^{2} \alpha_\ell e^{\mathrm{i}k\mathbf{d}_\ell \cdot \mathbf{x}} + \sum_{n=-\infty}^{+\infty} \sum_{\ell=1}^{2} u_n(\theta_\ell) \alpha_\ell e^{\mathrm{i}\xi_n(\theta_\ell)\cdot \mathbf{x}} = 0 \quad \text{for} \quad x_2 > \max_{x_1 \in [0,2\pi]} f(x_1), \tag{4.2.10}$$

where $u_n(\theta_\ell) \in \mathbb{C}(n \in \mathbb{Z})$ are the Rayleigh coefficients of $u^s(\mathbf{x}; k, \mathbf{d}_\ell)$ associated with the incident wave $e^{\mathrm{i}k\mathbf{d}_\ell \cdot \mathbf{x}}$. Clearly, any two vectors of the set

$$\{k\mathbf{d}_1\} \bigcup \{k\mathbf{d}_2\} \bigcup \{\xi_n(\theta_1) \mid n \in \mathbb{Z}\} \bigcup \{\xi_n(\theta_2) \mid n \in \mathbb{Z}\}$$

are distinct since $|\theta_\ell| < \pi/2$ and (4.2.8). Using Lemma 4.2.1 and (4.2.10), we can see $\alpha_\ell = 0$ for $\ell = 1, 2$, which is a contradiction to $\alpha_\ell \neq 0$, $\ell = 1, 2$. □

References

1. H.-D. Alber, *A quasi-periodic boundary value problem for the Laplacian and the continuation of its resolvent*, Proc. Roy. Soc. Edinburgh Sect. A, **82(3-4)** (1978/79), 251–272.
2. G. Alessandrini and L. Rondi, *Determining a sound-soft polyhedral scatterer by a single far-field measurement*, Proc. Amer. Math. Soc., **35** (2005), 1685–1691.
3. H. Ammari, *Uniqueness theorems for an inverse problem in a doubly periodic structure*, Inverse Problems **11** (1995), 823–833.
4. T. Arens, *The scattering of plane elastic waves by a one-dimensional periodic surface,* Math. Meth. Appl. Sci., **22** (1999), 55–72.
5. G. Bao, H. Zhang and J. Zou, *Unique determination of periodic polyhedral structures by scattered electromagnetic fields*, Trans. Amer. Math. Soc. **363** (2011), 4527–4551.
6. G. Bao, H. Zhang and J. Zou, *Unique determination of periodic polyhedral structures by scattered electromagnetic fields II: the resonance case*, Trans. Amer. Math. Soc. **366** (2014), 1333–1361.
7. A. Böttcher, B. Silbermann and A.Y. Karlovich, *Analysis of Toeplitz Operators*, 2nd Edition, Springer-Verlag, Berlin, 2006.
8. M Cadilhac, *Some mathematical aspects of the grating theory*, Electromagnetic Theory of Gratings, Springer, 1980, 53–62.
9. X. Cao, H. Diao and H. Liu, *Determining a piecewise conductive medium body by a single far-field measurement*, CSIAM Trans. Appl. Math., **1** (2020), 740–765.
10. X. Cao, H. Diao, H. Liu and J. Zou, *On nodal and generalized singular structures of Laplacian eigenfunctions and applications*, Journal de Mathématiques Pures et Appliquées, 143 (2020), 116–161.
11. X. Cao, H. Diao, H. Liu and J. Zou, *On novel geometric structures of Laplacian eigenfunctions in* $\mathbb{R}^3$ *and applications to inverse problems*, SIAM Journal on Mathematics Analysis, 53(2021), no. 2, 1263–1294.
12. X. Cao, H. Diao, H. Liu and J. Zou, *Two single-measurement uniqueness results for inverse scattering problems within polyhedral geometries*, Inverse Problems and Imaging, **16** (2022), no. 6, 1501–1528.
13. A. Charalambopoulos, D. Gintides and K. Kiriaki, *On the uniqueness of the inverse elastic scattering problem for periodic structures, Inverse Problems*, **17** (2001), 1923–1935.
14. J. Cheng and M. Yamamoto, *Uniqueness in an inverse scattering problem within non-trapping polygonal obstacles with at most two incoming waves*, Inverse Problems, **19** (2003), 1361–1384.
15. D. Colton and R. Kress, *Looking back on inverse scattering theory*, SIAM Review, **60(40)** (2018), 779–807.
16. D. Colton and R. Kress, *Inverse acoustic and electromagnetic scattering theory*, volume 93 of Applied Mathematical Sciences. Springer, Cham, 4th edition, 2019.
17. H. Diao, H. Liu and L. Wang, *On generalized Holmgren's principle to the Lamé operator with applications to inverse elastic problems*, Calculus of Variations and Partial Differential Equations, **59** (2020), 179.

18. H. Diao, H. Liu and L. Wang, *Further results on generalized Holmgren's principle to the Lamé operator and applications*, Journal of Differential Equations, **309** (2022), 841–882.
19. H. Diao, H. Liu, L. Zhang and J. Zou, *Unique continuation from a generalized impedance edge-corner for Maxwell's system and applications to inverse problems*, Inverse Problems, 37 (2021), 035004.
20. J. Elschner, G. Schmidt and M. Yamamoto, *An inverse problem in periodic diffractive optics: Global uniqueness with a single wavenumber*, Inverse Problems **19** (2003), 779–787.
21. J. Elschner and M. Yamamoto, *Uniqueness in determining polygonal sound-hard obstacles with a single incoming wave*, Inverse Problems, **22** (2006), no. 1, 355–364.
22. J. Elschner and M. Yamamoto, *Uniqueness in determining polygonal periodic structures*, Z. Anal. Anwend. **26** (2007), 165–177.
23. J. Elschner and M. Yamamoto, *Uniqueness in determining polygonal sound-hard obstacles with a single incoming wave*, Inverse Problems, **24**(2008), no. 3, 035004, 7 pp.
24. J. Elschner and M. Yamamoto, *Uniqueness in inverse elastic scattering with finitely many incident waves*, Inverse Problems, **26** (2010), 045005.
25. P. Hähner P and G. C. Hsiao, *Uniqueness theorems in inverse obstacle scattering of elastic waves Inverse Problem*, **9** (1993), 525–534.
26. V. Isakov, *On uniqueness in the inverse transmission scattering problem*, Commun. Part. Diff. Eqns., **15** (1990), 1565–1587.
27. I. V. Kamotski and S. A. Nazarov, *The augmented scattering matrix and exponentially decaying solutions of an elliptic problem in a cylindrical domain*, Journal of Mathematical Sciences, **111** (2002), 3657–3666.
28. A. Kirsch, *Diffraction by periodic structures*, Inverse problems in mathematical physics, Lecture Notes in Phys., **422** (1993), Springer, Berlin, 87–102.
29. A. Kirsch, *Uniqueness theorems in inverse scattering theory for periodic structures*, Inverse Problems, **10(1)** (1994), 145.
30. A. Kirsch and F. Hettlich, *Schiffer's theorem in inverse scattering theory for periodic structures*, Inverse Problems, **13** (1997), 351–361.
31. A. Kirsch and R. Kress, *Uniqueness in inverse obstacle scattering*, Inverse Problems, **9** (1993), 285–299.
32. R. Kress, *Uniqueness in inverse obstacle scattering for eletromagnetic waves*, Proc. URSI General Assembly, Maastricht, (2002).
33. V. D. Kupradze, *Three-dimensional Problems of the Mathematical Theory of Elasticity and Thermoelasticity*, Amsterdam, North-Holland, (1979).
34. J. Lai, H. Liu, J. Xiao and Y. Xu, *The decoupling of elastic waves from a weak formulation perspective*, East Asian Journal on Applied Mathematics, **9(2)** (2019), 241–251.
35. P. Lax and R. Phillips, *Scattering Theory*, Academic Press, New York and London, (1967).
36. H. Liu, *A global uniqueness for formally determined inverse electromagnetic obstacle scattering*, Inverse Problems, **24** (2008), 035018 (13pp).
37. H. Liu, M. Petrini, L. Rondi and J. Xiao, Stable determination of sound-hard polyhedral scatterers by a minimal number of scattering measurements, *J. Differential Equations*, **262(3)** (2017), 1631–1670.
38. H. Liu, L. Rondi and J. Xiao, Mosco convergence for $H(curl)$ spaces, higher integrability for Maxwell's equations, and stability in direct and inverse EM scattering problems, *J. Eur. Math. Soc.*, **21** (2019), no. 10, 2945–2993.
39. H. Liu and J. Xiao, Decoupling elastic waves and its applications, *J. Differential Equations*, **265(8)** (2017), 4442–4480.
40. H. Liu, M. Yamamoto and J. Zou, *Reflection principle for the Maxwell equations and its application to inverse electromagnetic scattering*, Inverse Problems, **23** (2007), 2357–2366.
41. H. Liu, M. Yamamoto and J. Zou, *New reflection principles for Maxwell's equations and their applications*, Numer. Math. Theor. Meth. Appl., **2(1)** (2009), 1–17.
42. H. Liu and T. Zhou, *On approximate electromagnetic cloaking by transformation media*, SIAM J. Appl. Math., **71** (2011), 218–241.
43. H. Liu and J. Zou, *Uniqueness in an inverse acoustic obstacle scattering problem for both sound-hard and sound-soft polyhedral scatterers*, Inverse Problems, **22** (2006), 515–524.
44. H. Liu and J. Zou, *On unique determination of partially coated polyhedral scatterers with far field measurements*, Inverse Problems, **23** (2007), 297–308.
45. H. Liu and J. Zou, *Zeros of the Bessel and spherical Bessel functions and their applications for uniqueness in inverse acoustic obstacle scattering*, IMA J. Appl. Math., **72(6)** (2007), 817–831.
46. H. Liu and J. Zou, *Uniqueness in determining multiple polygonal or polyhedral scatterers of mixed type*, Discrete Continu. Dyn. Syst. B 9 (2008), 375–396.
47. W. Mclean, *Strongly Elliptic Systems and Boundary Integral Equation*, Cambridge University Press, Cambridge, (2000).
48. R. F. Millar, *On the Rayleigh assumption in scattering by a periodic surface*, Proc. Cambridge Philos. Soc., **65** (1969), 773–791.
49. R. F. Millar, *On the Rayleigh assumption in scattering by a periodic surface: Part II*, Proc. Cambridge Philos. Soc., **69** (1971), 217–225.
50. H. Minkowski, *Gesammelte Abhandlungen, 2*, Leipzig: Teubner, 157–161, (1911).
51. J.-C. Nédélec, *Acoustic and Electromagnetic Equations*, Springer-Verlag, New York, (2001).

Chapter 5
Path Argument for Inverse Acoustic and Electromagnetic Obstacle Scattering Problems

5.1 Uniqueness in an Inverse Acoustic Obstacle Scattering Problem for Both Sound-Hard and Sound-Soft Polyhedral Scatterers

5.1.1 Overview

Let us recall the acoustic obstacle scattering problem from sound-hard or sound-soft obstacles. To describe the scattering system, we shall use u^i, u^s and u to represent the incident, scattered and total field, respectively, where $u = u^i + u^s$ and $u^i(x) = \exp\{ik\mathbf{x}\cdot\mathbf{d}\}$ with $\mathrm{i} = \sqrt{-1}$, $\mathbf{d} \in \mathbb{S}^{N-1}$ being the incident direction and $k > 0$ being the wave number. Let D be an impenetrable obstacle, where D is a general compact set in $\mathbb{R}^N (N \geq 2)$ with an open connect complement $G = \mathbb{R}^N \setminus D$. Consider the direct acoustic obstacle scattering problem as follows:

$$\Delta u + k^2 u = 0 \quad \text{in} \quad G = \mathbb{R}^n \setminus D, \tag{5.1.1a}$$

$$\lim_{r\to\infty} r^{(N-1)/2}\left(\frac{\partial u^s}{\partial r} - \mathrm{i}ku^d\right) = 0, \tag{5.1.1b}$$

with $r = |\mathbf{x}|$ for $\mathbf{x} \in \mathbb{R}^N$. The boundary condition on D can be either of the following boundary conditions:

$$u = 0 \quad \text{on} \quad \partial G \quad \text{(the sound-soft obstacle)}, \tag{5.1.2}$$

$$\frac{\partial u}{\partial \nu} = 0 \quad \text{on} \quad \partial G \quad \text{(the sound-soft obstacle)}, \tag{5.1.3}$$

where ν is the unit normal to ∂G pointing to the interior of G.

It is known (cf [18]) that there exists a unique solution $u = u(D;k,\mathbf{d}) \in H^1_{loc}(G)$ to (5.1.1a)–(5.1.2) or (5.1.1a), (5.1.1b) and (5.1.3) if ∂G is Lipschitz continuous, and u is analytic in any compact set in G. The Sommerfeld radiation (5.1.1b) characterizes the outgoing wave and enables us to have the following asymptotic behaviour for the scattered wave u^s:

$$u^s(\mathbf{x}) = \frac{e^{\mathrm{i}k|\mathbf{x}|}}{|\mathbf{x}|^{(N-1)/2}}\left\{u_\infty(\hat{\mathbf{x}}) + \mathscr{O}\left(\frac{1}{|\mathbf{x}|}\right)\right\} \quad \text{as} \quad |\mathbf{x}| \to \infty, \tag{5.1.4}$$

where $\hat{\mathbf{x}} = \frac{\mathbf{x}}{|\mathbf{x}|} \in \mathbb{S}^{N-1}$ and $u_\infty(\hat{\mathbf{x}})$ is defined in the unit sphere $\mathbb{S}^{N-1}$, known as the far-field pattern (cf [5]). We shall also write $u_\infty(\hat{\mathbf{x}}; D, k, \mathbf{d})$ to specify its dependence on the obstacle D, the wave number k and the incident direction $\mathbf{d}$.

The inverse acoustic obstacle scattering problem (IAOSP) is to determine ∂G from the far-field pattern $u_\infty(\hat{\mathbf{x}}; D, k, \mathbf{d})$ which can be observed. This section shall consider the uniqueness for

H. Diao, H. Liu, *Spectral Geometry and Inverse Scattering Theory*,
https://doi.org/10.1007/978-3-031-34615-6_5

inverse acoustic obstacle scattering problems with polyhedral scatterers. An obstacle D is said to be a polyhedral scatterer if it is a compact subset of $\mathbb{R}^N$ with connected complement $G = \mathbb{R}^N \setminus D$, and the boundary of G is composed of a finite union of cells. A cell, as defined in [1], is the closure of an open subset of an $(N-1)$-dimension hyperplane. Based on this definition, we can write a two-dimensional polyhedral scatterer D as

$$D = \left(\bigcup_{i=1}^{m} S_i \right) \cap \left(\bigcap_{l=1}^{n} L_l \right),$$

where each S_i is a polygon (screen) and each L_l is a line segment (crack), and write a three-dimensional polyhedral scatterer D as

$$D = \left(\bigcup_{i=1}^{m} P_i \right) \cap \left(\bigcap_{l=1}^{n} S_l \right),$$

where each P_i is a polyhedron (real body) and each S_l is a cell (screen). We emphasize that a cell need not be an $(N-1)$-dimension polyhedron. Clearly, such a polyhedral scatterer is very general and it admits the simultaneous presence of finitely many solid- and crack-type obstacles.

In Sect. 4.1, we present the uniqueness results in determining **G** by knowledge of a single far-field pattern via studying the geometric structures of the Laplacian eigenfunctions. Though the method is of a local feature, it requires a priori geometric restrictions on the underlying obstacle. In this chapter, we present a method of a global feature that can deal with the unique identifiability issue with more general obstacles. We follow the treatment in [16] and the method is known as the path argument in the literature.

5.1.2 Uniqueness for the Sound-Hard Case

We first introduce some notation and definitions for the subsequent use. Let $u_l(\mathbf{x}), l = 2, \ldots, N$, be the total fields of (5.1.1a), (5.1.1b) and (5.1.3) corresponding to the incident waves $\exp\{ik\mathbf{x}\cdot\mathbf{d}\}$, where $\{\mathbf{d}_l\}_{l=1}^N$, with each $\mathbf{d}_l \in \mathbb{S}^{N-1}$, are assumed to be linearly independent. We shall write $\mathscr{U} = \{u_1, u_2, \ldots, u_N\}$ and the operations on $\mathscr{U}$ are always understood to be elementwise. For example, for any $\nu \in \mathbb{S}^{N-1}$,

$$\frac{\partial \mathscr{U}}{\partial \nu} = \left\{ \frac{\partial u_1}{\partial \nu}, \frac{\partial u_2}{\partial \nu}, \ldots, \frac{\partial u_N}{\partial \nu} \right\},$$

and

$$\frac{\partial \mathscr{U}}{\partial \nu} = 0, \text{ on } S \text{ implies } \frac{\partial u_l}{\partial \nu} = 0 \text{ on } S \text{ for } l =, 2, \ldots, N,$$

where S can be any hypersurface in G and ν is its outward normal. Throughout, we will denote an open ball in $\mathbb{R}^N$ with centre $\mathbf{x}$ and radius r by $B_r(\mathbf{x})$, the closure of $B_r(\mathbf{x})$ by $\bar{B}_r(\mathbf{x})$ and the boundary of $B_r(\mathbf{x})$ by $S_r(\mathbf{x})$. Based on the earlier definition of a general polyhedral scatterer D of our interest, we can write the boundary of $G = \mathbb{R}^N \setminus D$ as

$$\partial G = \bigcup_{l=1}^{n} C_l, \tag{5.1.5}$$

where each C_l is a cell in $\mathbb{R}^N$.

Definition 5.1.1. The set $\mathscr{L}_{\mathscr{U}}$ is called a Neumann set of $\mathscr{U}$ in G if

$$\mathscr{L}_{\mathscr{U}} = \left\{ \mathbf{x} \in G; \left.\frac{\partial \mathscr{U}}{\partial \nu}\right|_{\Pi \cap B_r(\mathbf{x}) \cap G} = 0 \text{ for some } r > 0 \text{ and hyperplane } \Pi \text{ passing through } \mathbf{x} \right\}.$$

We have the following useful result:

Lemma 5.1.1. *For any* $\mathbf{x} \in \mathscr{L}_{\mathscr{U}}$*, let* Π *be the corresponding hyperplane involved in the definition of* $\mathscr{L}_{\mathscr{U}}$ *and* $\widetilde{\Pi}$ *be the open connected component of* $\Pi \setminus D$ *containing* $\mathbf{x}$*; then*

$$\left.\frac{\partial \mathscr{U}}{\partial \nu}\right|_{\widetilde{\Pi}} = 0. \tag{5.1.6}$$

Proof. By Definition 5.1.1, we know that $\frac{\partial \mathscr{U}}{\partial \nu} = 0$ on $\Pi \cap B_r(\mathbf{x}) \cap G$. Since $\mathscr{U}$ is analytic in G(cf[5]), then $\frac{\partial \mathscr{U}}{\partial \nu}$ is analytic in $N-1$ variables on $\Pi \setminus D$, which clearly lies in G. Now observing that $\Pi \cap B_r(\mathbf{x}) \cap G$ is an open set on $\Pi \cap D$, we have $\frac{\partial \mathscr{U}}{\partial \nu} = 0$ on $\widetilde{\Pi}$ by analytic continuation. □

We will refer to $\widetilde{\Pi}$ in the above lemma as the *Neumann hyperplane* in what follows, and obviously, it must be an open connected subset of a hyperplane and its boundary lies on ∂G.

Now, we derive some important properties of the Neumann set $\mathscr{L}_{\mathscr{U}}$.

Lemma 5.1.2. *The Neumann set* $\mathscr{L}_{\mathscr{U}}$ *and all Neumann hyperplanes are bounded. And* $\mathscr{L}_{\mathscr{U}}$ *is closed in the sense that for any sequence* $\{\mathbf{x}_n\}_{n=1}^{\infty} \subset \mathscr{L}_{\mathscr{U}}$*, which converges to a point* $\mathbf{x}_0 \in G$*, we must have* $\mathbf{x}_0 \in \mathscr{L}_{\mathscr{U}}$*, i.e. there exists a Neumann hyperplane* $\widetilde{\Pi}_0$ *passing through* $\mathbf{x}_0$*.*

Proof. We first show the boundedness of $\mathscr{L}_{\mathscr{U}}$. Set $\mathscr{U}^s(\mathbf{x}) = \mathscr{U} - \exp\{ik\mathbf{x}\cdot\mathbf{d}\}$, with $d = \{d_1, d_2, \ldots, d_N\}$ being the scattered fields; then we have

$$\lim_{|\mathbf{x}|\to\infty} |\nabla \mathscr{U}^s(\mathbf{x})| = 0 \tag{5.1.7}$$

i.e. $\lim_{|\mathbf{x}|\to\infty} |\nabla u_l^s(\mathbf{x})| = 0 (l = 1, 2, \ldots, N)$. The limit (5.1.7) can be shown following the proof of Lemma 9 in [3]. Now we demonstrate the boundedness of $\mathscr{L}_{\mathscr{U}}$ by contradiction. If $\mathscr{L}_{\mathscr{U}}$ is unbounded, then there must exist a Neumann hyperplane $\widetilde{\Pi}$ which connects to infinity. To see this, we first note that D is bounded, so one can bound D by a ball $B_R(\mathbf{0})$ with sufficiently large radius R. By the unboundedness of $\mathscr{L}_{\mathscr{U}}$, we know there must exist a point $\mathbf{y} \in \mathscr{L}_{\mathscr{U}} \cap (\mathbb{R}^N \setminus \bar{B}_R(\mathbf{0}))$; then the corresponding Neumann hyperplane $\widetilde{\Pi}_{\mathbf{y}}$ containing $\mathbf{y}$ must connect to infinity. Next, using (5.1.6) and (5.1.7), we have

$$\lim_{\mathbf{x}\in\widetilde{\Pi}:|\mathbf{x}|\to\infty} |\partial_\nu \exp\{ik\mathbf{x}\cdot\mathbf{d}\}| = 0,$$

where $\nu \in \mathbb{S}^{N-1}$ is the unit normal to $\widetilde{\Pi}$. Hence,

$$\lim_{\mathbf{x}\in\widetilde{\Pi}:|\mathbf{x}|\to\infty} |ik(\mathbf{d}\cdot\nu)\exp\{ik\mathbf{x}\cdot\mathbf{d}\}| = 0.$$

Noting that $k \neq 0$, we have $\mathbf{d}\cdot\nu = 0$, or equivalently,

$$\nu\cdot\mathbf{d}_l = 0, \quad 1, 2, \ldots, N.$$

But this is impossible since $\nu \in \mathbb{S}^{N-1}$ and $\{\mathbf{d}_l\}_{l=1}^N$ are linearly independent. Therefore, the set $\mathscr{L}_{\mathscr{U}}$ must be bounded. Clearly, the above proof has also demonstrated that all Neumann hyperplanes must be bounded.

Next, we shall show the closeness of $\mathscr{L}_{\mathscr{U}}$. Let $\{\mathbf{x}_n\}_{n=1}^{\infty}$ be a sequence in $\mathscr{L}_{\mathscr{U}}$ and $\mathbf{x}_0 \in G$, such that $\lim_{n\to\infty} \mathbf{x}_n = \mathbf{x}_0$. Take a sufficiently small hypercube $T_r(\mathbf{x}_0)$ of edge length r and centred at

$\mathbf{x}_0$ such that the closure of $T_r\mathbf{x}_0$ lies entirely in G. Without loss of generality, we may assume that $\{\mathbf{x}_n\}_{n=1}^{\infty} \subset T_r(\mathbf{x}_0)$. Let $\widetilde{\Pi}_n$ be the Neumann hyperplane through $\mathbf{x}_n$ such that

$$\left.\frac{\partial \mathscr{U}}{\partial \nu_n}\right|_{\widetilde{\Pi}_n},$$

where ν_n is the unit normal to $\widetilde{\Pi}_n$. Let us write $\nu(\mathbf{x}_n) = \nu_n$; then by possible extraction of a subsequence, we may assume that $\nu(\mathbf{x}_n) \to \nu_0$ as $n \to \infty$ and write $\nu(\mathbf{x}_0) = \nu_0 \in \mathbb{S}^{N-1}$. Let Π_0 be a hyperplane through $\mathbf{x}_0$ and have ν_0 as its normal, and then we can show that for any $P_0 \in \Pi_0 \cap T_r(\mathbf{x}_0)$, there exists a sequence of points $\{P_n\}_{n=1}^{\infty}$ such that $P_n \in \Pi_n$ for each n, where Π_n is the hyperplane in $\mathbb{R}^N$ containing the Neumann hyperplane $\widetilde{\Pi}_n$ and $\lim_{n\to\infty} P_n = P_0$. To see this, let L be the straight line through P_0 with direction ν_0; then any point $P \in L$ is given by

$$P = P_0 + t\nu_0 \quad \text{for some } t \in \mathbb{R}.$$

Note that the equation for hyperplane Π_n is given by

$$(P - \mathbf{x}_n) \cdot \nu_n = 0 \quad \text{for any} \quad P \in \Pi_n.$$

Since $\nu_n \to \nu_0$ as $n \to \infty$, we can assume that $\nu \cdot \nu_0 \neq 0$ for $n \in \mathbb{N}$. Then by straightforward calculations, we can show that L intersects with each Π_n, and the intersection point is given by

$$P_n = P_0 + t_n \nu_0 \quad \text{with} \quad t_n = \frac{(\mathbf{x}_n - P_0)\cdot \nu_n}{\nu_0 \cdot \nu_n}, n = 1, 2, \ldots.$$

Using the facts that

$$(P_0 - \mathbf{x}_0)\cdot \nu_0 = 0, \quad \lim_{n\to\infty} \mathbf{x}_n = \mathbf{x}_0,$$

we see

$$\lim_{n\to\infty} t_n = \lim_{n\to\infty} \frac{(\mathbf{x}_n - P_0)\cdot \nu_n}{\nu_0 \cdot \nu_n} = \lim_{n\to\infty} \frac{(\mathbf{x}_n - \mathbf{x}_0)\cdot \nu_n}{\nu_0 \cdot \nu_n} + \lim_{n\to\infty} \frac{(\mathbf{x}_0 - P_0)\cdot \nu_n}{\nu_0 \cdot \nu_n} = 0,$$

this implies

$$\lim_{n\to\infty} P_n = P_0.$$

Since P_n converges to $P_0 \in \Pi_0 \cap T_r(\mathbf{x}_0)$ along the ν_0-direction, we may assume that for all n, $P_n \in T_r(\mathbf{x}_0)$, i.e.$P_n \in \widetilde{\Pi}_n \cap T_r(\mathbf{x}_0)$. Noting that $\nabla \mathscr{U}$ is continuous in the closure of $T_r(\mathbf{x}_0)$, we have

$$\frac{\partial \mathscr{U}}{\partial \nu}(P_0) = \nabla \mathscr{U}(P_0)\cdot \nu_0 = \lim_{n\to\infty} \nabla \mathscr{U}(P_n)\cdot \nu_n = 0.$$

Thus, we have $\mathbf{x}_0 \in \mathscr{L}_{\mathscr{U}}$. The proof is completed. □

Next, we recall a fundamental property for a connected set (see Theorem 3.19.9 in [7]), which will be used in our subsequent arguments.

Lemma 5.1.3. *Let E be a metric space, $A \subset E$ be a subset and $B \subset E$ be a connected set such that $A \cap B \neq \emptyset$ and $(E\backslash A) \cap B \neq \emptyset$; then $\partial A \cap B \neq \emptyset$.*

Now, we are ready to present our main uniqueness result for a sound-hard polyhedral scatterer.

Theorem 5.1.1. *Let $\mathbf{d}_l \in \mathbb{S}^{N-1}, l = 1, 2, \ldots, N$, be N linearly independent directions and $k > 0$ be fixed. A polyhedral scatterer D described in 6 is uniquely determined by the far-field patterns $\mathscr{U} = \{u_{1,\infty}, u_{2,\infty}, \ldots, u_{N,\infty}\}$.*

Proof. We shall prove the theorem by contradiction. First, we follow [1] and [8] to show that if Theorem 5.1.1 does not hold, then we can assume that there exists a Neumann hyperplane $\widetilde{\Pi}_1$ in $G = \mathbb{R}^N \setminus D$. To see this, let D' be a polyhedral scatterer different from D and u' be the solution to (5.1.1a), (5.1.1b) and (5.1.3) when D is replaced by D'. And similarly, we write $\mathscr{U}' = \{u'_1, u'_2, \dots, u'_N\}$ and $\mathscr{U}'_\infty = \{u'_{1,\infty}, u'_{2,\infty}, \dots, u'_{N,\infty}\}$ for the total fields and far-field patterns corresponding to the incident wave $\exp\{ik\mathbf{x}\cdot\mathbf{d}_l\}$, $l = 1,2,\dots,N$.

If the theorem is not true, then we can assume $\mathscr{U}_\infty = \mathscr{U}'_\infty$ for N given linearly independent $\mathbf{d}_l \in \mathbb{S}^{N-1}, l = 1,2,\dots,N$, and fixed $k > 0$. Letting Ω be the unbounded connected component of $\mathbb{R}^N \setminus (D \cup D')$, then by [5, Theorem 2.13], we infer that $\mathscr{U} = \mathscr{U}'$ over Ω.

First, we can see $\partial\Omega \not\subset D \cap D'$ from the connectedness of both G and $G' = \mathbb{R}^N \setminus D'$. Indeed, if $\partial\Omega \subset D \cap D'$, then we must have $\Omega = \mathbb{R}^N \setminus D = \mathbb{R}^N \setminus D'$. To see this, we first observe that $\Omega \subset \mathbb{R}^N \setminus D$ and $\Omega \subset \mathbb{R}^N \setminus D'$ by noting $\Omega \subset \mathbb{R}^N \setminus (D \cup D')$. On the other hand, if there exist $\mathbf{x} \in \mathbb{R}^N \setminus D$ and $\mathbf{x}' \in \mathbb{R}^N \setminus D'$ such that $\mathbf{x} \notin \Omega$ and $\mathbf{x}' \notin \Omega$, we obtain from Lemma 5.1.3 (with $A = \Omega$ and $B = \mathbb{R}^N \setminus D$ or $B = \mathbb{R}^N \setminus D'$) that $\partial\Omega \cap (\mathbb{R}^N \setminus D) \neq \emptyset$ and $\partial\Omega \cap (\mathbb{R}^N \setminus D') \neq \emptyset$, which contradicts the assumption that $\partial\Omega \subset D \cap D'$, thus leading to $\mathbb{R}^N \setminus D \subset \Omega$ and $\mathbb{R}^N \setminus D' \subset \Omega$. Therefore, $\Omega = \mathbb{R}^N \setminus D = \mathbb{R}^N \setminus D'$, which implies $D = D' = \mathbb{R}^N \setminus \Omega$. But this contradicts the fact that D and D' are two different polyhedral scatterers.

Using the previous conclusion that $\partial\Omega \not\subset D \cap D'$, we must have $(\partial G' \setminus D) \cap \partial\Omega \neq \emptyset$ or $(\partial G \setminus D') \cap \partial\Omega \neq \emptyset$. Without loss of generality, we may assume the first case held, and therefore, there exists a point $\tilde{\mathbf{x}}' \in (\partial G' \setminus D) \cap \partial\Omega$. We can also assume that $\tilde{\mathbf{x}}'$ belongs to the interior of one of the cells composing $\partial G'$, and so there exists a hyperplane Π_1 and $r > 0$ such that $\tilde{\mathbf{x}}' \in S_1 = \Pi_1 \cap B_r(\tilde{\mathbf{x}}') \subset (\partial G' \setminus D) \cap \partial\Omega$. Since $\mathscr{U} = \mathscr{U}'$ in Ω, by noting $\frac{\partial \mathscr{U}'}{\partial \nu} = 0$ on $S_1 \subset \partial G'$, we have $\frac{\partial \mathscr{U}}{\partial \nu} = 0$ on S_1. Hence, $\tilde{\mathbf{x}}'$ is contained in the Neumann set of $\mathscr{U}$ in G and S_1 is contained in a Neumann hyperplane of $\mathscr{U}$ in G, which we denote by $\widetilde{\Pi}_1$.

Next, we start from this Neumann hyperplane $\widetilde{\Pi}_l$ to build up a contradiction.

In the following, a curve $\gamma = \gamma(t)(t \geq 0)$ is said to be regular if it is C^1-smooth and $\frac{d}{dt}\gamma(t) \neq 0$. And the notation Π_l, with l being an integer, shall always represent a hyperplane in $\mathbb{R}^N$, which contains a Neumann hyperplane $\widetilde{\Pi}_l$. Since G is an unbounded open connected set, hence the open set $G \setminus \widetilde{\Pi}_1$ must contain an open connected component, denoted as $\widetilde{G}$, which connects to the infinity. In fact, $\widetilde{G}$ is unique because $\widetilde{\Pi}_1$ is bounded by Lemma 5.1.2 and G cannot be divided by $\widetilde{\Pi}_l$ into more than one unbounded open component; otherwise, ∂G is unbounded. Thus, $\widetilde{\Pi}_1$ lies on $\partial\widetilde{G}$, due to the fact that every point on $\widetilde{\Pi}_1$ is in G and so can be connected to the infinity. Next, we fix an arbitrary point $\mathbf{x}_1 \in \widetilde{\Pi}_1$. Let $\gamma = \gamma(t)(t \geq 0)$ be a regular curve such that $\gamma(0) = \mathbf{x}_1, \gamma(t)(t > 0)$ lies entirely in $\widetilde{G}$ and $\lim_{t\to\infty} |\gamma(t)| = +\infty$. Clearly, the curve γ lies on one side of Π_1, that is, $\gamma(t) \in \Pi_1$ if $t = 0$, and we set $t_1 = 0$ (we refer to [13, 19] for the properties of open connected set). For convenience, we choose $\gamma(t)$ to be as 'straight' as possible in the sense that there are as few snakelike portions as possible. For example, we may let $\gamma(t)$ be given consecutively connected line segments in $\widetilde{G}$ but with C^1-smooth functions to connect two neighbouring line segments and let it be a straight line outside a sufficiently large ball containing D.

Next, define the distance between two sets A and B in $\mathbb{R}^N$ as usual:

$$d(A,B) = \inf_{\mathbf{x}\in A, \mathbf{y}\in B} |\mathbf{x} - \mathbf{y}|.$$

Let

$$d_l = d(\gamma, C_l), \quad l = 1,2,\dots,n, \tag{5.1.8}$$

and

$$r_0 = \frac{1}{2} \min_{1\leq l\leq n} d_l. \tag{5.1.9}$$

Noting that γ is a closed set in $\mathbb{R}^n$ and $\{C_l\}_{l=1}^n$, which form ∂G, are compact sets, it can be readily seen that $\mathbf{d}_l > 0, l = 1,2,\dots,n$ are attainable. Hence, $r_0 > 0$, and for any point $\mathbf{x} \in \gamma(t)$, we have $\bar{B}_{r_0}(\mathbf{x}) \subset G$.

Let $\tilde{\mathbf{x}}_2^+ = \gamma(\tilde{t}_2) \in S_{r_0}(\mathbf{x}_1) \cap \gamma$ and $\tilde{\mathbf{x}}_2^- \in S_{r_0}(\mathbf{x}_1)$ be the symmetric point of $\tilde{\mathbf{x}}_2^+$ with respect to Π_1. We remark that by Lemma 5.1.3, the set γ must intersect $S_{r_0}(\mathbf{x}_1)$, but the intersection need not necessarily be a unique point. For definiteness, we take $\tilde{t}_2 = \max\left\{t > 0; \gamma(t) \in S_{r_0}(\mathbf{x}_1)\right\}$. Now, let G_1^+ be the connected component of $G\setminus\tilde{\Pi}_1$ containing $\tilde{\mathbf{x}}_2^+$ and G_1^- be the connected component of $G\setminus\tilde{\Pi}_1$ containing $\tilde{\mathbf{x}}_2^-$. It remarked that it may happen that $G_1^+ = G_1^-$. We denote by R_1 the reflection with respect to Π_1, and then let E_1^+ be the connected component of $G_1^+\cap R_1(G_1^-)$ containing $\tilde{\mathbf{x}}_2^+$ and E_1^- be the connected component of $G_1^-\cap R_1(G_1^+)$ containing $\tilde{\mathbf{x}}_2^-$. Observe that $E_1^+ = R_1(E_1^-)$, and if we set $E_1 = E_1^+\cup\tilde{\Pi}_1\cup E_1^-$, then E_1 contains the closed ball $\bar{B}_{r_0}(\mathbf{x}_1)$. Moreover, E_1 is a connected open set with the boundary composed of subsets of the cells $\{C_l\}_{l=1}^n$ and $\{R_1(C_l)\}_{l=1}^n$. One can easily verify that $\mathscr{U}(\mathbf{x}) - R_1\mathscr{U}(\mathbf{x})$, where $R_1\mathscr{U}(\mathbf{x}) = \mathscr{U}(R_1(\mathbf{x}))$ is a solution to the Helmholtz equation in E_1 with zero Dirichlet and Neumann data on $\tilde{\Pi}_1\cap\bar{B}_{r_0}(\mathbf{x}_1)$; therefore, $\mathscr{U}(\mathbf{x}) = R_1\mathscr{U}(\mathbf{x})$ in E_1 by Holmgren's theorem (cf. Theorem 6.12 in [4]), i.e. $\mathscr{U}$ is even symmetric in E_1 with respect to the hyperplane Π_1. This indicates $\left.\frac{\partial\mathscr{U}}{\partial\nu_1}\right|_{E_1}\cap\Pi_1 = 0$, where ν_1 is the unit normal to Π_1. Next, we show that E_1 is bounded. Clearly, we first see $\partial E_1, \partial G_1^\pm$ and $R_1(\partial G_1^\pm)$ are bounded by our construction. If E_1 is unbounded, then E_1 would contain $\mathbb{R}^n\setminus B_r(\mathbf{x}_1)$ for some sufficiently large $r > 0$. Then using $\left.\frac{\partial\mathscr{U}}{\partial\nu_1}\right|_{E_1}\cap\Pi_1 = 0$ and analytic continuation, $\Pi_1\setminus B_r(\mathbf{x}_1)$ are parts of some Neumann hyperplanes. This contradicts Lemma 5.1.2, and so E_1 is bounded. Now by the unboundedness of γ, there must exist a $t_2 > \tilde{t}_2$, such that $\mathbf{x}_2 = \gamma(t_2) \in \partial E_1$. Noting that ∂E_1 is composed of subsets of the cells $\{C_l\}_{l=1}^n$ and $\{R_1(C_l)\}_{l=1}^n$, the set $\mathscr{U}$ takes zero Neumann data on ∂E_1 by using the fact that $R_1\mathscr{U}(\mathbf{x}) = \mathscr{U}(\mathbf{x})$ in E_1. Thus, by analytic continuation, $\mathbf{x}_2 \in \partial E_1$ implies the existence of a Neumann hyperplane passing through $\mathbf{x}_2$, which we denote by $\tilde{\Pi}_2$, and we have $\mathbf{x}_2 = \gamma(t_2) \in \mathscr{L}_{\mathscr{U}}$. Furthermore, we may assume that $\gamma(t_2)$ is the 'last' point on γ to intersect $\tilde{\Pi}_2$, that is

$$t_2 = \max\{t > 0; \gamma(t) \in \tilde{\Pi}_2\} < \infty.$$

The following two facts shall be crucial: $\tilde{\Pi}_2$ is different from $\tilde{\Pi}_1$, since $\tilde{\Pi}_1$ intersects γ only at $\mathbf{x}_1$; the length of $\gamma(t)$ from t_1 to t_2 is larger than r_0, i.e.

$$|\gamma(t_1 \le t \le t_2)| \ge |\gamma(t_1 \le t \le \tilde{t}_2)| \ge r_0.$$

Next, let $\tilde{\mathbf{x}}_3^+ = \gamma(\tilde{t}_3) \in S_{r_0}(\mathbf{x}_2)\cap\gamma$, and $\tilde{\mathbf{x}}_3^- \in S_{r_0}(\mathbf{x}_2)$ be the symmetric point of $\tilde{\mathbf{x}}_3^+$ with respect to Π_2, and then let G_2^+ be the connected component of $G\setminus\tilde{\Pi}_2$ containing $\tilde{\mathbf{x}}_3^+$ and G_2^- be the connected component of $G\setminus\tilde{\Pi}_2$ containing $\tilde{\mathbf{x}}_3^-$. Denote by R_2 the reflection with respect to Π_2, and let E_2^+ be the connected component of $G_2^+\cap R_2(G_2^-)$ containing $\tilde{\mathbf{x}}_3^+$ and E_2^- be the connected component of $G_2^-\cap R_2(G_2^+)$ containing $\tilde{\mathbf{x}}_3^-$. Set $E_2 = E_2^+\cup\tilde{\Pi}_2\cup E_2^-$; then we see that E_2 contains the closed ball $\bar{B}_{r_0}(\mathbf{x}_2)$ and its boundary is composed of sunsets of the cells $\{C_l\}_{l=1}^n$ and $\{R_2(C_l)\}_{l=1}^n$. By a similar argument as used earlier for deriving $\mathbf{x}_2 = \gamma(t_2)$ and $\tilde{\Pi}_2$, there exists a points $\tilde{\mathbf{x}}_3^+ = \gamma(t_3)(t_3 > t_2)$ and a Neumann hyperplane $\tilde{\Pi}_3$ passing through $\mathbf{x}_3$. Again, we may assume that $\mathbf{x}_3$ is the 'last' point to pass through Π_3. We see that $\tilde{\Pi}_3$ is different from $\tilde{\Pi}_1$ and $\tilde{\Pi}_2$, since $\mathbf{x}_1 = \gamma(t_1)$ and $\mathbf{x}_2 = \gamma(t_2)$ are the last point to pass through $\tilde{\Pi}_1$ and $\tilde{\Pi}_2$, respectively, and the length of $\gamma(t)$ from t_2 to t_3 is larger than r_0, i.e.

$$|\gamma(t_2 \le t \le t_3)| \ge r_0.$$

Continuing with the above procedure, we can construct a strictly increasing sequence $\{t_n\}_{n=1}^\infty$ such that for any n, $\mathbf{x}_n = \gamma(t_n) \in \mathscr{L}_{\mathscr{U}}$ and $\tilde{\Pi}_n$ is a Neumann hyperplane passing through $\mathbf{x}_n$. Moreover, those Neumann hyperplanes are different from each other, and the length of $\gamma(t)$ from t_n to t_{n+1} is not less than r_0, i.e.

$$|\gamma(t_n \le t \le t_{n+1})| \ge r_0. \tag{5.1.10}$$

Since $\mathscr{L}_{\mathscr{U}}$ is bounded and $\lim_{t\to\infty}|\gamma(t)| = +\infty$, so we must have $\lim_{n\to\infty} t_n = t_0$ for some finite t_0. Otherwise, we would have $\lim_{n\to\infty} t_n = +\infty$ due to the fact that t_n is strictly increasing and this further implies $\lim_{n\to\infty}|\gamma(t_n)| = +\infty$, contradicting that $\gamma(t_n) = \mathbf{x}_n \in \mathscr{X}_{\mathscr{U}}$ for each n and the boundedness of $\mathscr{X}_{\mathscr{U}}$. Finally, because $\gamma(t)$ is a C^1-smooth curve, we must have that

$$\lim_{n\to\infty}|\gamma(t_n \le t \le t_{n+1})| = \lim_{n\to\infty}\int_{t_n}^{t_{n+1}}|\gamma'(t)|\mathrm{d}t = 0, \tag{5.1.11}$$

which contradicts the inequality (5.1.10), thus completing the proof of Theorem 5.1.1. □

5.1.3 Uniqueness for the Sound-Soft Case

In this section, we extend the arguments in the previous section to the sound-soft case, but with some adaptations, which, we think, might provide some alternative thinking for the further study of the uniqueness issues for IAOSP.

In correspondence with *Neumann set* and *Neumann hyperplane* for the sound-hard case, we introduce the *Dirichlet set* and *Dirichlet hyperplane* for the current sound-soft case. Let $u(\mathbf{x})$ be the total field to (5.1.1a)–(5.1.2) associated with a single incident wave $u^i(\mathbf{x}) = \exp\{ik\mathbf{x}\cdot d\}$ with fixed k and $\mathbf{d}$.

Definition 5.1.2. The set $\mathscr{D}_n$ is called a Dirichlet set of u in G, if

$$\mathscr{D}_u = \{\mathbf{x}\in G; u|_{\Pi\cap B_r(\mathbf{x})\cap G} \text{ for some } r>0 \text{ and hyperplane } \Pi \text{ passing through } \mathbf{x}\}.$$

Similar to Lemma 5.1.1, we have

Lemma 5.1.4. *For any* $\mathbf{x}\in\mathscr{D}_n$*, let* $\widetilde{\Pi}$ *be the corresponding open connected component of* $\Pi\setminus D$ *containing* $\mathbf{x}$*; then the following holds:*

$$u|_{\widetilde{\Pi}} = 0. \tag{5.1.12}$$

We will refer to $\widetilde{\Pi}$ in the above lemma as the Dirichlet hyperplane in the following. Obviously, the same as the Neumann hyperplane, a Dirichlet hyperplane must be an open connected subset of a hyperplane and its boundary lies on ∂G.

The following lemma is a counterpart of Lemma 5.1.2 for the sound-hard case.

Lemma 5.1.5. *The Dirichlet set* $\mathscr{D}_u$ *and all Dirichlet hyperplanes are bounded. And* $\mathscr{D}_u$ *is closed in the sense that for any sequence* $\{\mathbf{x}_n\}_{n=1}^{\infty}\subset\mathscr{D}_u$*, which converges to a point* $\mathbf{x}_0\in G$*, and then we have* $\mathbf{x}_0\in\mathscr{D}_u$*, i.e. there exists a Dirichlet hyperplane* Π_0 *passing through* $\mathbf{x}_0$*.*

Proof. For the boundedness of $\mathscr{D}_u$, we refer to Lemma 3.1 in [1]. And the closeness of $\mathscr{D}_u$ can be proved in a similar way to the proof Lemma 5.1.2. □

Now, the uniqueness result is stated in the following theorem.

Theorem 5.1.2. *The polyhedral scatterer D described in* (5.1.5) *is uniquely determined by a single far-field pattern* u_∞ *corresponding to an incident wave* $\exp\{ik\mathbf{x}\cdot\mathbf{d}\}$ *with* $k>0$ *and* $\mathbf{d}\in\mathbb{S}^{N-1}$ *fixed.*

Proof. By contradiction, similar to the proof of Theorem 5.1.1, we assume that there exists a Dirichlet hyperplane $\widetilde{\Pi}_1$ in $G=\mathbb{R}^N\setminus D$. Then the rest of the proof can be carried out in the same way as for the sound-hard in Theorem 5.1.1. But to provide a possible alterative thinking for the uniqueness for the inverse acoustic scattering problem, we shall present a different analysis below to prove Theorem 5.1.2. As done in Theorem 5.1.1, with the help of the reflection principle for the Dirichlet problem and the auxiliary function $u(\mathbf{x})+Ru(\mathbf{x})$, we can find a countable set of distinct Dirichlet hyperplane $\{\widetilde{\Pi}_n\}_{n=1}^{\infty}$ and a sequence of points $\mathbf{x}_n=\gamma(t_n)$ with $\{t_n\}_{n=1}^{\infty}$ being a strictly increasing sequence and $\mathbf{x}_n$ lying on $\widetilde{\Pi}_n$. Like in the proof of Theorem 5.1.1, we can choose a uniform radius r_0 for the balls $B_{r_0}(\mathbf{x}_n)(n=1,2\ldots)$. But for our purpose, we specially choose a sequence of balls B_{r_0} lying entirely in G with distinct radii r_n. Also, we can find a finite t_0 such that $t_n \to t_0$ as $n \to \infty$ and a Dirichlet hyperplane $\widetilde{\Pi}_0$ passing through $\mathbf{x}_0 = \gamma(t_0)$. Further, there exists a sufficiently small hypercube $\widetilde{\Pi}_n \cap T_r(\mathbf{x}_0) \to \widetilde{\Pi}_0 \cap T_r(\mathbf{x}_0)$ as $n \to \infty$, in the sense that for any $P_0 \in \widetilde{\Pi}_0 \cap T_r(\mathbf{x}_0)$, there exists in the unit normal ν_0-direction to $\widetilde{\Pi}_0$ a sequence $\{P_n\}_{n=1}^{\infty}$, with $P_n \in \Pi_n$ for each $n \in \mathbb{N}$, given by

$$P_n = P_0 + t_n \nu_0, \quad t_n = \frac{(\mathbf{x}_n - P_0)\cdot \nu_n}{\nu_0 \cdot \nu_n}, \; n = 1,2,\ldots, \tag{5.1.13}$$

and P_n converges to P_0. Now, if there are infinitely many P_ns which are different from each other, then using $P_n \to P_0$ as $n \to \infty$, we may assume that $P_n \in \widetilde{\Pi}_n \cap T_r(\mathbf{x}_0)$ for all $n \in \mathbb{N}$. This implies

$$\frac{\partial u}{\partial \nu_0}(P_0) = \lim_{n\to\infty} \frac{u(P_n) - u(P_0)}{t_n} = 0.$$

□

Due to our construction, all Dirichlet hyperplanes $\tilde{\Pi}_n$ are different from each other. Next, we claim that $\frac{\partial u}{\partial \nu_0} = 0$ a.e. on $\widetilde{\Pi}_0 \cap T_r(\mathbf{x}_0)$. If this is not true, we would have a sufficiently small ball $B_{\tilde{r}}(Q_1) \subset T_{r_0}(\mathbf{x}_0)$, where $Q_1 \in \widetilde{\Pi}_0$, such that for every $P_0 \in B_{\tilde{r}}(Q_1) \cap \widetilde{\Pi}_0$, only finitely many out of the sequence $\{P_n\}_{n=1}^{\infty}$ given in (5.1.13) are different from each other. Thus, there exists an $N_0 \in \mathbb{N}$ such that $t_n = 0$ for $n > N_0$, namely, $P_0 \in \Pi_n$ for $n > N_0$. Now, we choose N points $Q_l \in B_{\tilde{r}}(Q_1) \cap \widetilde{\Pi}_0 (l = 1,2,\ldots,N)$ such that the vectors $Q_1Q_l (2 \le l \le N)$ are linearly independent. Let $N_l \in \mathbb{N}$ be the integer such that $Q_l \in \Pi_n$ for $n > N_l, l = 1,2,\ldots,N$ and $M = \max\{N_1,\ldots,N_l\}$; then for $n > M$, we have $Q_1Q_l \subset \Pi_n$, for all $l \ge 2$. Since $Q_1Q_l, l = 2,\ldots,N$ are linearly independent and all lie on both Π_n and Π_0, then Π_n must coincide with Π_0 when $n > M$. This contradicts our construction that $\{\widetilde{\Pi}_n\}_{n=1}^{\infty}$ are countable different Dirichlet hyperplanes. So, we have demonstrated that $\frac{\partial u}{\partial \nu_0} = 0$ a.e. on $\widetilde{\Pi}_0 \cap T_r(\mathbf{x}_0)$, which implies $\frac{\partial u}{\partial \nu_0} = 0$ in $\widetilde{\Pi}_0 \cap T_r(\mathbf{x}_0)$ by the analyticity of u. Noting that we also have $u = 0$ on $\widetilde{\Pi}_0 \cap T_r(\mathbf{x}_0)$, then by Holmgren's theorem we must have $u = 0$ over G. But this contradicts Lemma 5.1.5, so completes the proof of Theorem 5.1.2.

5.2 A Global Uniqueness for Formally Determined Inverse Electromagnetic Obstacle Scattering

5.2.1 Introduction and Statement of the Result

In this section, we shall be mainly concerned with the path argument for the inverse electromagnetic obstacle scattering, where one utilizes the time-harmonic electromagnetic far-field measurements to identify the inaccessible/unknown objects. Here, we follow the treatment in [14].

For a brief description of the forward scattering problem, we let a perfect conducting obstacle $\mathbf{D} \subset \mathbb{R}^3$ be a compact set with connected Lipschitz complement $\mathbf{G} := \mathbb{R}^3 \backslash \mathbf{D}$, and

$$\mathbf{E}^i(\mathbf{x}) := \frac{\mathrm{i}}{k} \nabla \wedge \nabla \wedge \ \mathbf{p}\, e^{ik\mathbf{x}\cdot\mathbf{d}} = ik(\mathbf{d}\times\mathbf{p})\times\mathbf{d}\, e^{ik\mathbf{x}\cdot\mathbf{d}}, \tag{5.2.1}$$

$$\mathbf{H}^i(\mathbf{x}) := \nabla \wedge \ \mathbf{p}\, e^{ik\mathbf{x}\cdot\mathbf{d}} = ik\mathbf{d}\times\mathbf{p}\, e^{ik\mathbf{x}\cdot\mathbf{d}}, \tag{5.2.2}$$

be the incident electric and magnetic fields, where $\mathbf{p} \in \mathbb{R}^3, k > 0$ and $\mathbf{d} \in \mathbb{S}^2 := \{\mathbf{x} \in \mathbb{R}^3; |\mathbf{x}| = 1\}$ represents the polarization, wave number and direction of propagation, respectively. The incident wave propagating in the homogeneous background medium will be perturbed when it encounters an obstacle and produces a scattered field. We denote by $\mathbf{E}^s$ and $\mathbf{H}^s$ the scattered electric and magnetic fields, respectively, and define the total electric and magnetic fields to be

$$\mathbf{E}(\mathbf{x}) = \mathbf{E}^i(\mathbf{x}) + \mathbf{E}^s(\mathbf{x}), \quad \mathbf{H}(\mathbf{x}) = \mathbf{H}^i(\mathbf{x}) + \mathbf{H}^s(\mathbf{x}), \quad \mathbf{x} \in \mathbb{R}^3. \tag{5.2.3}$$

Then the direct scattering problem consists of finding a solution $(\mathbf{E},\mathbf{H}) \in H^1_{loc}(\mathrm{curl}\,;\mathbf{G}) \times H^1_{loc}(\mathrm{curl}\,;\mathbf{G})$ that satisfies the following time-harmonic Maxwell equations:

$$\nabla \wedge \mathbf{E} - ik\,\mathbf{H} = 0, \quad \nabla \wedge \mathbf{H} + ik\,\mathbf{E} = 0 \quad \text{in } \mathbf{G} := \mathbb{R}^3 \backslash \mathbf{D}, \tag{5.2.4}$$

$$\nu \wedge \mathbf{E} = 0 \quad \text{on} \quad \partial\mathbf{G}, \tag{5.2.5}$$

$$\lim_{|\mathbf{x}|\to\infty} (\mathbf{H}^s \wedge \mathbf{x} - |\mathbf{x}|\mathbf{E}^s) = 0, \tag{5.2.6}$$

where the last limit corresponds to the so-called Silver-Müller radiation condition characterizing the fact the scattered wave is radiating.

The forward scattering problem (5.2.1)–(5.2.6) is well understood (see, e.g. [2]). Particularly, $\mathbf{E}$ and $\mathbf{H}$ are (real) analytic in $\mathbf{G}$ in the sense that both the real and imaginary parts of $\mathbf{E}$ and $\mathbf{H}$ are real analytic functions in $\mathbf{G}$; see [5] and we also refer to Chap. 3 [10] for a formal definition of real analytic functions. The asymptotic behaviour of the scattered field $(\mathbf{E}^s, \mathbf{H}^s)$ is governed by (see [5])

$$\mathbf{E}^s(\mathbf{x};\mathbf{D},\mathbf{p},k,\mathbf{d}) = \frac{e^{ik\mathbf{x}\cdot\mathbf{d}}}{|\mathbf{x}|}\left\{\mathbf{E}_\infty(\hat{\mathbf{x}};\mathbf{D},\mathbf{p},k,\mathbf{d}) + \mathscr{O}\left(\frac{1}{|\mathbf{x}|}\right)\right\} \quad \text{as } |\mathbf{x}| \to \infty, \tag{5.2.7}$$

$$\mathbf{H}^s(\mathbf{x};\mathbf{D},\mathbf{p},k,\mathbf{d}) = \frac{e^{ik\mathbf{x}\cdot\mathbf{d}}}{|\mathbf{x}|}\left\{\mathbf{H}_\infty(\hat{\mathbf{x}};\mathbf{D},\mathbf{p},k,\mathbf{d}) + \mathscr{O}\left(\frac{1}{|\mathbf{x}|}\right)\right\} \quad \text{as } |\mathbf{x}| \to \infty, \tag{5.2.8}$$

uniformly for all $\hat{\mathbf{x}} = \mathbf{x}/|\mathbf{x}| \in \mathbb{S}^2$. The functions $\mathbf{E}_\infty(\hat{\mathbf{x}})$ and $\mathbf{H}_\infty(\hat{\mathbf{x}})$ in (5.2.7) and (5.2.8) are called the electric and magnetic far-field patterns, respectively, and both are analytic on the unit sphere $\mathbb{S}^2$. As is noted above, $\mathbf{E}^s(\mathbf{x};\mathbf{D},\mathbf{p},k,\mathbf{d}), \mathbf{E}_\infty(\hat{\mathbf{x}};\mathbf{D},\mathbf{p},k,\mathbf{d})$, etc. will be frequently used to specify their dependence on the observation direction $\hat{\mathbf{x}}$, the polarization $\mathbf{p}$, the wavenumber k and the incident direction $\mathbf{d}$.

In this section we are concerned with the inverse electromagnetic obstacle scattering problem. Namely, through the far-field measurement of the electric far-field pattern (or, equivalently, the magnetic far-field pattern) of the wave which is scattered by $\mathbf{D}$ corresponding to a given incident wave, for one or more choices of its polarization $\mathbf{p}$ or of its wavenumber k, or of its propagation direction $\mathbf{d}$, we would like to recover the position, shape and boundary type of the underlying obstacle D. Let us define the map sending the obstacle to the corresponding far-field pattern as follows:

$$\mathscr{F}_e(\partial\mathbf{G}) = \mathbf{E}_\infty(\hat{\mathbf{x}};\mathbf{D},\mathbf{p},k,\mathbf{d}) \quad \text{for} \quad (\hat{\mathbf{x}};\mathbf{D},\mathbf{p},k,\mathbf{d}) \in \mathbb{S}^2_0 \times \mathbb{U} \times \mathbb{K} \times \widetilde{\mathbb{S}^2_0}, \tag{5.2.9}$$

where $\mathbb{S}^2_0, \widetilde{\mathbb{S}^2_0} \subset \mathbb{S}^2$, $\mathbb{U} \subset \mathbb{R}^3$, $\mathbb{K} \subset \mathbb{R}_+ := \{\mathbf{x} \in \mathbb{R}; \mathbf{x} > 0\}$ and the non-linear operator $\mathscr{F}_e$ is defined by the forward scattering system (5.2.4)–(5.2.6). Clearly, the inverse problem is to recover $\partial\mathbf{G}$ by knowledge of $\mathbf{E}_\infty$; namely, we need to invert the non-linear operator equations (5.2.9). This problem is non-linear and ill-posed, which has many applications in medical imaging, geophysical exploration and nondestructive testing, etc.

It is first recalled that a compact polyhedron in $\mathbb{R}^3$ is a simply connected compact set whose boundary is composed of (open) *faces*, *edges* and *vertices*. In the following, we call $\mathbf{D}$ a *polyhedral obstacle* if it is composed of finitely many (but unknown *a priori*) pairwise disjoint compact polyhedra. That is,

$$\mathbf{D} = \bigcup_{l=1}^{m} D_l, \tag{5.2.10}$$

where m is an unknown integer but must be finite and each $D_l, 1 \leq l \leq m$ is a compact polyhedron such that

$$D_i \cap D_j = \emptyset \quad \text{if} \quad i \leq j \quad \text{and} \quad 1 \neq i, j \leq m.$$

Clearly, the forward scattering problem (5.2.4)–(5.2.6) with such a polyhedral obstacle $\mathbf{D}$ is well-posed. Moreover, we know that the singular behaviours of the weak solution only attach to the edges and vertices, that is, $(\mathbf{E}, \mathbf{H})$ satisfies (5.2.4) in the classical sense in any subdomain of $\mathbf{G}$, which does not meet any corner or edge of $\mathbf{D}$ (see [6]). By the regularity of the strong solution for the forward scattering problem, we know that $\mathbf{E}$ and $\mathbf{H}$ are at least $C^{0,\alpha}$ continuous $(0 < \alpha < 1)$ up to the regular points, namely, points lying in the interior of the open faces of $\mathbf{D}$.

The main result of this chapter is the following.

Theorem 5.2.1. *Let $\mathbf{D}$ and $\widetilde{\mathbf{D}}$ be two perfect polyhedral obstacles. For any fixed $k_0 > 0$, $\mathbf{d}_0 \in \mathbb{S}^2$ and $\mathbf{p}_0 \in \mathbb{R}^3$ such that $\mathbf{d}_0$ and $\mathbf{p}_0$ are linearly independent, we have $\mathbf{D} = \widetilde{\mathbf{D}}$ as long as*

$$\mathbf{E}_\infty(\hat{\mathbf{x}}; \mathbf{D}, \mathbf{p}_0, k_0, \mathbf{d}_0) = \mathbf{E}_\infty(\hat{\mathbf{x}}; \widetilde{\mathbf{D}}, \mathbf{p}_0, k_0, \mathbf{d}_0) \quad \textit{for} \quad \hat{\mathbf{x}} \in \mathbb{S}^2. \tag{5.2.11}$$

The remainder of the section is organized as follows. In Subsect. 5.2.2, we introduce the perfect set and perfect planes and then show several crucial properties of them which shall play a key role in proving Theorem 5.2.1. Subsection 5.2.3 is devoted to proof of Theorem 5.2.1.

5.2.2 Perfect Set and Perfect Planes

First, we fix some notations which shall be used throughout in the remainder of the section. We denote an open ball on $\mathbb{R}^3$ with centre $\mathbf{x}$ and radius r by $B_r(\mathbf{x})$, the closure of $B_r(\mathbf{x})$ by $\bar{B}_r(\mathbf{x})$ and the boundary of $B_r(\mathbf{x})$ by $S_r(\mathbf{x})$. The notation $T_r(\mathbf{x})$ is defined to be an open cube of edge length r, centred at $\mathbf{x}$, while $\bar{T}_r(\mathbf{x})$ is its corresponding closure. Unless specified otherwise, ν shall always denote the inward normal to the concerned domain, or the normal to a two-dimensional plane in $\mathbb{R}^3$. The distance between two sets $\mathscr{A}$ and $\mathscr{B}$ in $\mathbb{R}^3$ is understood as usual to be $\mathbf{d}(\mathscr{A}, \mathscr{B}) = \inf_{\mathbf{x}\in\mathscr{A}, \mathbf{y}\in\mathscr{B}} |\mathbf{x} - \mathbf{y}|$. Finally, a curve $\gamma = \gamma(t)(t \geq 0)$ is said to be regular if it is C^1 smooth and $\frac{d}{dt}(\gamma(t)) \neq 0$.

Henceforth, we let $k_0 > 0$, $\mathbf{d}_0 \in \mathbb{S}^2$ and $\mathbf{p}_0 \in \mathbb{R}^3$ be fixed such that $\mathbf{d}_0$ and $\mathbf{p}_0$ are linearly independent and denote by $\mathbf{E} := \mathbf{E}(\mathbf{x}; \mathbf{D}, \mathbf{p}_0, k_0, \mathbf{d}_0)$ the total electric field in (5.2.4)–(5.2.6) corresponding to a polyhedral perfect conducting obstacle $\mathbf{D}$ as described in (5.2.11). The following definition of a *perfect set* is modified from that in [14] to fit the problem under investigation.

Definition 5.2.1. The set $\mathscr{P}_\mathbf{E}$ is called a perfect set of $\mathbf{E}$ in $\mathbf{G} := \mathbb{R}^3 \backslash \mathbf{D}$ if

$$\mathscr{P}_\mathbf{E} = \{\mathbf{x} \in \mathbf{G}; \nu \wedge \mathbf{E}|_{\Pi \cap B_r(\mathbf{x}) \cap \mathbf{G}} = 0 \text{ for some } r > 0 \text{ and hyperplane } \Pi \text{ passing through} \mathbf{x}\},$$

where ν is the unit normal to the plane Π.

For any $\mathbf{x} \in \mathscr{P}_\mathbf{E}$, we let Π be the plane involved in the definition of $\mathscr{P}_\mathbf{E}$. Furthermore, we let $\widetilde{\Pi}$ be the connected component of $\Pi \backslash \mathbf{D}$ containing $\mathbf{x}$, then by the analyticity of $\mathbf{E}$ in $\mathbf{G}$, we see $\nu \wedge \mathbf{E} = 0$ on $\widetilde{\Pi}$ by classical continuation. In the following, such $\widetilde{\Pi}$ will be referred to as a *perfect plane*. The introduction of the perfect set and perfect plane is motivated by the observation that, when proving Theorem 5.2.1 by contradiction, if two different obstacles produce the same far-field pattern, then outside one obstacle there exists a perfect plane which is extended from an open face of the other obstacle. From now on, $\widetilde{\Pi}_l$, with an integer l, shall always represent a perfect plane in $\mathbf{G}$ which lies on the plane Π_l in $\mathbb{R}^3$.

A very fine property of perfect planes is the so-called *reflection principle*, which constitutes an indispensable ingredient in the path arguments for proving the uniqueness results in [14, 15]. We formulate the principle in the following theorem. Subsequently, we use $\mathscr{R}_\Pi$ to denote the reflection in $\mathbb{R}^3$ with respect to a plane Π.

Theorem 5.2.2. *For a connected polyhedral domain Ω in $\mathbf{G} := \mathbb{R}^3 \backslash \mathbf{D}$, let $\widetilde{\Pi}$ be one of its faces that lies on some perfect plane. Furthermore, let Π be the plane in $\mathbb{R}^3$ containing $\widetilde{\Pi}$ and $\Omega \cup \mathscr{R}_\Pi \Omega \subset \mathbf{G}$. We have two consequences:*
(i) $\nu_\Pi \wedge \mathbf{E} = 0$ on $\Pi \cap (\Omega \cup \mathscr{R}_\Pi \Omega)$.
(ii) Suppose that $\Sigma \subset \partial\Omega$ is a subset of one face of Ω other than $\widetilde{\Pi}$, and the following condition holds

$$\nu_\Sigma \wedge \mathbf{E} = 0 \quad on\ \Sigma, \tag{5.2.12}$$

where ν_Σ is the unit normal to Σ directed to the interior of Ω. Then we have

$$\nu_{\Sigma'} \wedge \mathbf{E} = 0 \quad on\ \Sigma', \tag{5.2.13}$$

where $\Sigma' = \mathscr{R}_\Pi$ and Σ' is the unit normal to Σ' directed to the interior of $\mathscr{R}_\Pi \Omega$.

Proof. The verification for (i) can be found in the proof of Theorem 3.3 in [14], while for (ii), it is given in Theorem 2.3 in [14]. See also Theorem 2.9 and proof of Theorem 3.2 in [15]. □

The reflection principle in item (i) of Theorem 5.2.2 is particularly useful when $(\Pi \cap (\Omega \cup \mathscr{R}_\Pi \Omega) \backslash \widetilde{\Pi}) \neq \emptyset$. Clearly, in such a case, we can find a perfect plane also lying on the plane Π, but different from $\widetilde{\Pi}$.

Next, we would classify all those perfect planes in $\mathbf{G}$ into two sets in $\mathbb{R}^3$: one is bounded and the other is unbounded. In fact, it is verified directly that there might exist unbounded perfect planes. This constitutes one of the major differences from those perfect planes introduced in [14, 15]. All the perfect planes defined there are bounded due to the use of two different incident waves; see Lemma 3.2 in [14]. In our subsequent path argument for proving Theorem 5.2.1, the procedure of continuation of perfect planes along an existing path might be broken down with the presence of an unbounded perfect plane, since one may not be able to find another perfect plane with an unbounded perfect plane by using the reflection principle in Theorem 5.2.2. In the remainder of this subsection, we shall show some critical properties on the unbounded perfect planes.

Lemma 5.2.1. *All the unbounded perfect planes associated with $\mathbf{E}$ in $\mathbf{G}$ are conplane.*

Obviously, Lemma 5.2.1 is divided into the following two lemmas:

Lemma 5.2.2. *There cannot exist two unbounded perfect planes $\widetilde{\Pi}_1$ and $\widetilde{\Pi}_2$ such that $\Pi_1 \nparallel \Pi_2$.*

Lemma 5.2.3. *There cannot exist two different unbounded perfect planes $\widetilde{\Pi}_1$ and $\widetilde{\Pi}_2$ such that $\Pi_1 \parallel \Pi_2$.*

Proof of Lemma 5.2.2. Assume contrarily that $\widetilde{\Pi}_1$ and $\widetilde{\Pi}_2$ are two unbounded perfect planes in $\mathbf{G}$ such that $\Pi_1 \nparallel \Pi_2$. Let ν_1 and ν_2 be the unit normals to Π_1 and Π_2, respectively. Noting that $\mathbf{E}^s(\mathbf{x}) = \mathscr{O}(1/|\mathbf{x}|)$ as $|\mathbf{x}| \to \infty$, we have from

$$\nu_l \wedge \mathbf{E}(\mathbf{x}) = 0 \quad on\ \widetilde{\Pi}_l\ for\ l = 1, 2,$$

that

$$\lim_{\mathbf{x} \in \widetilde{\Pi}_l; |x| \to \infty} |\nu_l \wedge \mathbf{E}^{\mathrm{i}}(\mathbf{x})| = 0 \quad on\ \widetilde{\Pi}_l\ for\ l = 1, 2.$$

Using (5.2.1), we further deduce

$$\nu_l \wedge ((\mathbf{d}_0 \wedge \mathbf{p}_0) \wedge \mathbf{d}_0) = 0 \quad for\ l = 1, 2.$$

That is, $\nu_1 \parallel \nu_2$ since they are both parallel to a fixed vector $(\mathbf{d}_0 \wedge \mathbf{p}_0) \wedge \mathbf{d}_0$, contradiction to our assumption that $\Pi_1 \nparallel \Pi_2$ and completing the proof. □

Proof of Lemma 5.2.3. By contradiction, we assume that there exist two different perfect planes $\widetilde{\Pi}_1$ and $\widetilde{\Pi}_2$ such that $\Pi_1 \parallel \Pi_2$. Let $\mathscr{J} := T_r(0)$ be a sufficiently large cube such that $\mathbf{D} \subset \mathscr{J}$, and by suitable rotation, we may without loss of generality assume that both Π_1 and Π_2 are perpendicular to one face of $\mathscr{J}$. Next, with a little bit abuse of notations, we still denote by $\widetilde{\Pi}_1$ and $\widetilde{\Pi}_2$, those parts of $\widetilde{\Pi}_1$ and $\widetilde{\Pi}_2$ lying outside of $\mathscr{J}$, namely, $\widetilde{\Pi}_1 \backslash \mathscr{J}$ and $\widetilde{\Pi}_2 \backslash \mathscr{J}$, and the same rule applies to $\widetilde{\Pi}_l, l \in \mathbb{Z}$ appearing in the remainder of the proof. Now, in the (unbounded) polyhedral domain $\mathbb{R}^3 \backslash \mathscr{J}$, we can make use of the reflection as stated in (ii) of Theorem 5.2.2 and from $\widetilde{\Pi}_1$ and $\widetilde{\Pi}_2$ to find that

$$\nu \wedge \mathbf{E} = 0 \quad on\ \widetilde{\Pi}_3 := \mathscr{R}_{\Pi_2}(\widetilde{\Pi}_1).$$

Continuing with such argument, from $\widetilde{\Pi}_2$ and $\widetilde{\Pi}_3$ we have

$$\nu \wedge \mathbf{E} = 0 \quad on\ \widetilde{\Pi}_4 := \mathscr{R}_{\Pi_3}(\widetilde{\Pi}_2).$$

By repeating this reflection, we eventually find a sequence of perfect planes $\widetilde{\Pi}_l, l = 1,2,3\ldots$ such that all $\widetilde{\Pi}_l$'s are parallel to each other. Clearly, $\mathbf{d}(\widetilde{\Pi}_l, \widetilde{\Pi}_{l+1}) = \mathbf{d}(\widetilde{\Pi}_1, \widetilde{\Pi}_2) > 0$ being fixed for $l = 1,2,3\ldots$. Hence, there must exist some $l_0 < \infty$ such that $\mathscr{J}$ lies entirely at one side of Π_{l_0}. Using again the reflection principle in Theorem 5.2.2 (ii), we see $\nu \wedge \mathbf{E} = 0$ on $\mathscr{R}_{\Pi_{l_0}}(\partial \mathbf{D})$. Finally, let Σ_1 and Σ_2 be two adjacent faces of $\mathscr{R}_{\Pi_{l_0}}(\partial \mathbf{D})$ and we have from the extension of Σ_1 and Σ_2 two non-parallel unbounded perfect planes, which contradicts to Lemma 5.2.2. The proof is completed. □

We proceed to make an important observation of the reflection principle (i) in Theorem 5.2.2, when $\Omega \cup \mathscr{R}_\Pi \Omega$ is unbounded while $\widetilde{\Pi}$ is bounded. In this case, it is clear that the extension of some part of $(\Pi \cap (\Omega \cup \mathscr{R}_\Pi \Omega)) \backslash \widetilde{\Pi}$ gives at least one unbounded perfect plane. That is, some bounded perfect plane might imply the existence of some correspondingly unbounded perfect plane. Next, we study carefully such special bounded perfect plane $\widetilde{\Pi}_0$, which can be regarded as 'unbounded'. To localize our investigation, we fix an arbitrary point $\mathbf{x}_0 \in \widetilde{\Pi}_0 \cap \mathbf{G}$ and take a sufficiently small ball $B_0 := B_r(\mathbf{x}_0)$ such that $B_0 \subset \mathbf{G}$. B_0 is divided by $\widetilde{\Pi}_0$ into two half balls, which we denote by B_0^+ and B_0^-, respectively. Let $\mathbf{G}_0^+$ be the connected component of $\mathbf{G} \backslash \widetilde{\Pi}_0$ containing B_0^+ and $\mathbf{G}_0^-$ be the connected component of $\mathbf{G} \backslash \widetilde{\Pi}_0$ containing B_0^-. We remark that it may happen that $\mathbf{G}_0^+ = \mathbf{G}_0^-$. Next, let Λ_0^+ be the connected component of $\mathbf{G}_0^+ \cap \mathscr{R}_{\Pi_0}(\mathbf{G}_0^-)$ containing B_0^+ and Λ_0^- be the connected component of $\mathbf{G}_0^- \cap \mathscr{R}_{\Pi_0}(\mathbf{G}_0^+)$ containing B_0^-. Finally, set $\Lambda_0 = \Lambda_0^+ \cup \widetilde{\Pi}_0 \cup \Lambda_0^-$ and we see that Λ_0 is a polyhedral domain which is symmetric with respect to Π_0. Moreover, $B_0 \subset \Lambda_0$. One can easily see that the construction of Λ_0 is only dependent on the perfect plane $\widetilde{\Pi}_0$. Since $\partial \Lambda_0$ is composed of subsets lying either on $\partial \Lambda_0$ or on $\mathscr{R}_{\Pi_0}(\partial \mathbf{D})$, by the reflection principle in (ii) of Theorem 5.2.2, we have $\nu \times \mathbf{E} = 0$ on $\partial \Lambda_0$.

From now on, we shall denote by $\Lambda_{\widetilde{\Pi}_l}$ the symmetric set constructed as above corresponding to a bounded perfect plane $\widetilde{\Pi}_l$, namely, in the above, $\Lambda_{\widetilde{\Pi}_0} := \Lambda_0$. Clearly, in case $\Lambda_{\widetilde{\Pi}_l}$ is unbounded, we see from our earlier discussion that there must exist an unbounded perfect plane which is extended from some part of $(\Pi_l \cap \Lambda_{\widetilde{\Pi}_l}) \backslash \widetilde{\Pi}_l$. Such observation in combination with the result in Lemma 5.2.1 gives the following.

Lemma 5.2.4. *All the bounded planes $\widetilde{\Pi}_l$ with unbounded $\Lambda_{\widetilde{\Pi}_l}$ and all the unbounded perfect planes are conplane.*

Based on Lemma 5.2.4, we introduce the following set consisting of all the 'unbounded' perfect planes

$$\begin{aligned}\mathscr{Q}_{\mathbf{E}} := \{\widetilde{\Pi}; \widetilde{\Pi} &\text{ is an unbounded perfect plane} \\ &\text{or } \widetilde{\Pi} \text{ is a bounded perfect plane but with unbounded } \Lambda_{\widetilde{\Pi}_l}\}.\end{aligned} \tag{5.2.14}$$

Since all the members in $\mathscr{Q}_{\mathbf{E}}$ are conplane, one verifies directly that $\mathscr{Q}_{\mathbf{E}}$ consists of at most finitely many perfect planes by noting the fact that $\mathbf{D}$ is composed of finitely many pairwise disjoint compact polyhedra. We further define $\breve{\mathscr{Q}}_{\mathbf{E}}$ to be the subset of $\mathscr{Q}_{\mathbf{E}}$ consisting of those bounded perfect planes in $\mathscr{Q}_{\mathbf{E}}$. Next, we show some topological properties of the sets $\mathscr{Q}_{\mathbf{E}}$ and $\breve{\mathscr{Q}}_{\mathbf{E}}$.

Lemma 5.2.5. *Let* $\mathbf{G} := \mathbb{R}^3 \backslash \mathbf{D}$*; then*
(i) $\mathbf{G}\backslash\overline{\breve{\mathscr{Q}}_{\mathbf{E}}}$ *is connected.*
(ii) $\mathbf{G}\backslash\overline{\mathscr{Q}_{\mathbf{E}}}$ *has no bounded connected component.*

Proof. We first observe that $\breve{\mathscr{Q}}_{\mathbf{E}}$ is bounded since $\breve{\mathscr{Q}}_{\mathbf{E}} \subset \overline{ch(\mathbf{D})}$, where $ch(\mathbf{D})$ is the convex hull of $\mathbf{D}$. By further noting that $\partial\mathbf{G}$ is bounded, we know that $\mathbf{G}\backslash\overline{\breve{\mathscr{Q}}_{\mathbf{E}}}$ has exactly one unbounded connected component. Hence, if $\mathbf{G}\backslash\overline{\breve{\mathscr{Q}}_{\mathbf{E}}}$ is not connected, it must have some bounded connected component, say $\mathscr{C}_0 \subset \mathbf{G}$. Clearly, there must be one face of the polyhedral domain $\mathscr{C}_0$ that comes from exactly a perfect plane in $\breve{\mathscr{Q}}_{\mathbf{E}}$, say $\widetilde{\Pi}_0$. Now, one can verify directly that $\Lambda_{\widetilde{\Pi}_0} \subset \mathscr{C}_0 \cup \mathscr{R}_{\Pi_0}\mathscr{C}_0$, which is bounded since $\mathscr{C}_0$ is bounded. But this contradicts to the assumption that $\widetilde{\Pi}_0 \in \mathscr{Q}_{\mathbf{E}}$, thus proving assertion (i). Next, assertion (ii) is readily seen from (i). In fact, if $\mathbf{G}\backslash\overline{\mathscr{Q}_{\mathbf{E}}}$ has a bounded connected component, say $\mathscr{D}_0$, then one must have $\mathscr{D}_0 \subset \mathbf{G}\backslash\overline{\breve{\mathscr{Q}}_{\mathbf{E}}}$, which is certainly not true. The proof is completed. □

Correspondingly, we set

$$\mathscr{S}_{\mathbf{E}} := \{\widetilde{\Pi}; \widetilde{\Pi} \text{ is a bounded perfect plane with bounded } \Lambda_{\widetilde{\Pi}}\}. \tag{5.2.15}$$

Finally, we give a lemma concerning the fundamental property of a connected set (see, e.g. Theorem 3.19.9 in [7]), which shall be needed in the following subsection on proving Theorem 5.2.1.

Lemma 5.2.6. *Let* $\mathbb{E}$ *be a metric space,* $\mathscr{A} \subset \mathbb{E}$ *be a subset and* $\mathscr{B} \subset \mathbb{E}$ *be a connected set such that* $\mathscr{A} \cap \mathscr{B} \neq \emptyset$ *and* $(\mathbb{E}\backslash\mathscr{A}) \cap \mathscr{B} \neq \emptyset$*; then* $\partial\mathscr{A} \cap \mathscr{B} \neq \emptyset$.

5.2.3 Proof of Theorem 5.2.1

The entire subsection is devoted to the proof of Theorem 5.2.1 by contradiction. Assume that $\mathbf{D} \neq \widetilde{\mathbf{D}}$ and

$$\mathbf{E}_\infty(\hat{\mathbf{x}}; \mathbf{D}, \mathbf{p}_0, k_0, \mathbf{d}_0) = \mathbf{E}_\infty(\hat{\mathbf{x}}; \widetilde{\mathbf{D}}, \mathbf{p}_0, k_0, \mathbf{d}_0) \quad for\, \hat{\mathbf{x}} \in \mathbb{S}^2. \tag{5.2.16}$$

Let Ω be the (unique) unbounded connected component of $\mathbb{R}^3$ $(\mathbf{D} \cup \widetilde{\mathbf{D}})$. By Rellich's theorem (see Theorem 6.9, [5]); we refer from (5.2.16) that

$$\mathbf{E}(\mathbf{x}; \mathbf{D}) = \mathbf{E}(\mathbf{x}; \widetilde{\mathbf{D}}) \quad for\, \mathbf{x} \in \Omega, \tag{5.2.17}$$

where $\mathbf{E}(\mathbf{x}; \mathbf{D})$ and $\mathbf{E}(\mathbf{x}; \widetilde{\mathbf{D}})$ are abbreviations $\mathbf{E}(\mathbf{x}; \mathbf{D}, \mathbf{p}_0, k_0, \mathbf{d}_0)$ and $\mathbf{E}(\mathbf{x}; \widetilde{\mathbf{D}}, \mathbf{p}_0, k_0, \mathbf{d}_0)$, respectively. Next, noting that $\mathbf{D} \neq \widetilde{\mathbf{D}}$, we see that either $(\mathbb{R}^3\backslash\bar{\Omega})$ $\mathbf{D} \neq \emptyset$ or $(\mathbb{R}^3\backslash\bar{\Omega}) \backslash \widetilde{\mathbf{D}} \neq \emptyset$. Without loss of generality, we assume the former case and let $D^* := (\mathbb{R}\backslash\bar{\Omega})$ $\mathbf{D} \neq \emptyset$. It is easily seen that $D^* \subset \widetilde{\mathbf{D}}$, so D^* is bounded. Moreover, by choosing the connected component if necessary, we assume that D^* is connected. Clearly, D^* is a bounded polyhedral domain in $\mathbf{G} = \mathbb{R}^3\backslash\mathbf{D}$ and $\mathbf{E}(\mathbf{x}; \mathbf{D})$ is defined over D^*. Noting $\partial D^* \subset \partial\Omega \cup \partial\mathbf{D} \subset \partial\mathbf{D} \cup \partial\widetilde{\mathbf{D}}$ and using (5.2.17), we have from the perfect boundary conditions of $\mathbf{E}(\mathbf{x}; \mathbf{D})$ and $\mathbf{E}(\mathbf{x}; \widetilde{\mathbf{D}})$ on $\partial\mathbf{D}$ and $\partial\widetilde{\mathbf{D}}$ that

$$\nu \wedge \mathbf{E}(\mathbf{x}; \mathbf{D}) = 0 \quad \text{on } \partial D^*. \tag{5.2.18}$$

In the following, in order to simply notations, we use as those introduced in Subsect. 5.2.2, e.g. we write $\mathbf{E}(\mathbf{x})$ to denote $\mathbf{E}(\mathbf{x}, \mathbf{D})$, etc. The remainder of the proof will be proceeded into three steps and a brief outline is as follows. In the first step, we will find a perfect plane $\widetilde{\Pi}_1 \in \mathscr{S}_{\mathbf{E}}$, and this is the starting point of the subsequent path argument. In the second step, we would construct implicitly a path, which is a regular curve lying entirely in the exterior of **D** and connecting to infinity. As we mentioned earlier that the path argument might be broken down with the presence of some 'unbounded' perfect planes (namely, perfect planes in $\mathscr{Q}_{\mathbf{E}}$), in order to avoid our subsequent argument being trapped at such 'unbounded' perfect planes, the curve is required to have at most one intersection with $\mathscr{Q}_{\mathbf{E}}$. Fortunately, this can be done by using Lemma 5.2.5. Finally, using the reflection principle in Theorem 5.2.2, we make a continuation of (bounded) perfect planes along the exit path to find a sequence of perfect planes. Then a contradiction is constructed by showing that the continuation must follow the exit path to infinity since we always step a length larger than a fixed positive constant when making such continuation, but on the other hand, all the bounded perfect planes are contained in the convex hull of **D** being bounded. In this final step, we must be carefully treating the possible presence of 'unbounded' perfect planes and this is the main difference of the present path argument from those implemented in Sect. 5.1.

Step I: Existence of a Bounded Perfect Plane $\widetilde{\Pi}_1$ with Bounded $\Lambda_{\widetilde{\Pi}_1}$

We first note that $\partial D^*\backslash\partial\mathbf{D} \neq \emptyset$. Hence, there must be an open face say Σ_0 on ∂D^* that can be extended in **G** to form a perfect plane and it is denoted by $\widetilde{\Pi}_0$. Since $\widetilde{\Pi}_0$ is extended from a face of the bounded polyhedral domain D^* in **G**, we infer from Lemma 5.2.7 that $\widetilde{\Pi}_0$ is bounded. Now, if the symmetric set $\Lambda_{\widetilde{\Pi}_0}$ corresponding to $\widetilde{\Pi}_0$ is bounded, then we are done since we can take $\widetilde{\Pi}_0$ as $\widetilde{\Pi}_1$. So, without loss of generality, we assume that $\Lambda_{\widetilde{\Pi}_0}$ is unbounded. Next, based on Σ_0, we construct a bounded polyhedral domain in **G** which is symmetric with the respect based on Σ_0 but different from $\Lambda_{\widetilde{\Pi}_0}$. The construction procedure is similar to that for $\Lambda_{\widetilde{\Pi}_0}$, and we nonetheless present it here for clearness.

Fix an arbitrary point $\mathbf{x}^* \in \Sigma_0$ and let $B^* := B_\varepsilon(\mathbf{x}^*)$ with $\varepsilon > 0$ being sufficiently small such that B^* is divided by Σ_0 into two (open) half ball B_*^+ and B_*^- satisfying $B_*^+ \subset D^*$ and $B_*^- \subset \mathbf{G}\backslash D^*$. Next, let Θ_*^+ be the connected component of $\mathscr{R}_{\Pi_0}(\mathbf{G}\backslash\overline{D^*}) \cap D^*$ containing B_*^+ and Θ_*^- be the connected component of $\mathscr{R}_{\Pi_0}D^* \cap (\mathbf{G}\backslash\overline{D^*})$ containing B_*^-. Set $\Theta^* = \Theta_*^+ \cup \Sigma_0 \cup \Theta_*^-$. Clearly, the connected component Θ^* is a non-empty bounded polyhedral domain in **G** since $B^* \subset \Theta^* \subset D^* \cup \mathscr{R}_{\Pi_0}D^*$. We remark that Θ_0 is in fact the connected component of $(D^* \cup \mathscr{R}_{\Pi_0}D^*) \cap \Lambda_{\widetilde{\Pi}_0}$ containing Σ_0. By the reflection principle (ii) of Theorem 5.2.2, $\nu \times \mathbf{E}(x) = 0$ on $\partial\Theta^*$. It is obvious that $\partial\Theta^*\backslash\mathbf{D} \neq \emptyset$. Let $\Sigma_1 \subset \partial\Theta^*\backslash\mathbf{D}$ be an open face. By analytic continuation, Σ_1 is extended in **G** to give a perfect plane $\widetilde{\Pi}_1$. Since Θ^* is symmetric with respect to Π_0, we know $\Sigma_1 \not\subset \Pi_0$, and therefore, $\widetilde{\Pi}_1 \notin \mathscr{Q}_{\mathbf{E}}$ by Lemma 5.2.4, i.e. $\widetilde{\Pi}_1$ is bounded with bounded $\Lambda_{\widetilde{\Pi}_1}$.

Step II: Construction of the Exit Path γ

Since both $\partial\mathbf{G}$ and $\widetilde{\Pi}_1$ are bounded, we see that $\mathbf{G}\backslash\widetilde{\Pi}_1$ has a unique unbounded connected component, which is denoted by $\mathscr{U}$. It readily has that $\widetilde{\Pi}_1 \subset \partial\mathscr{U}$ and $\mathscr{U}$ contains the exterior of a sufficiently large ball containing **D**. Next, we fix an arbitrary point $\mathbf{x}_1 \in \widetilde{\Pi}_1$. Let $\gamma := \gamma(t)(t \geq 0)$ be a regular curve such that $\gamma(t_1) = \mathbf{x}_1$ with $\gamma(t)(t > 0)$ lies entirely in $\mathscr{U}$. Furthermore, the curve γ connects to infinity, i.e. $\lim_{t\to\infty}|\gamma(t)| = \infty$. The exit path γ constructed in this way might have a non-empty intersection with $\mathscr{Q}_{\mathbf{E}}$. In this case, we require that $\gamma(t)(t > 0)$ has only one intersection point with $\mathscr{Q}_{\mathbf{E}}$. In fact, in case $\gamma(t) \cap \mathscr{Q}_{\mathbf{E}} \neq \emptyset$, we would modify the curve γ as follows to satisfy such requirement. Let $\mathbf{x}_T := \gamma(T)$ be the 'first' intersection point of $\gamma(t)(t > 0)$ and $\mathscr{Q}_{\mathbf{E}}$; that is,

$$T = \min\{t > 0; \gamma(t) \in \mathscr{Q}_{\mathbf{E}}\} < \infty.$$

Then, let $\mathscr{V}$ be the connected component of $\mathbf{G} \setminus \mathscr{Q}_{\mathbf{E}}$ such that $\mathbf{x}_T \in \partial\mathscr{V}$. Let $\mathscr{W} := \mathscr{U} \cap \mathscr{V}$. It can be verified that $\mathscr{W}$ is an unbounded connected open set such that $\mathbf{x}_T \in \partial\mathscr{W}$. Indeed, the connectedness of $\mathscr{W}$ is obvious by noting that both $\mathscr{U}$ and $\mathscr{V}$ are connected. Whereas the unboundedness of $\mathscr{W}$ is due to the facts that $\mathscr{V}$ is unbounded by Lemma 5.2.5 and $\mathscr{U}$ contains the exterior of a sufficiently large ball containing **D** as mentioned earlier. Next, let $\eta(t)(t \geq T)$ be a regular curve such that $\eta(T) = \mathbf{x}_T, \eta(t)(t > T)$ lies entirely in $\mathscr{W}$ and connects to infinity (i.e. $\lim_{t\to\infty}|\eta(t)| = \infty$). Furthermore, it is trivially required that $\eta(t)$ has C^1 connection with $\gamma(t)(0 \leq t \leq T)$ at $\mathbf{x}_T$. Now, set

$$\tilde{\gamma}(t) = \begin{cases} \gamma(t) & 0 \leq t \leq T, \\ \eta(t) & t > T, \end{cases}$$

then $\tilde{\gamma}(t)(t \geq 0)$ satisfies all our requirements of an exit path.

Step III: Continuation of Bounded Perfect Planes Along γ

Let $d_0 = \mathbf{d}(\gamma, \mathbf{D}) > 0$, which is attainable since $\mathbf{D}$ is compact, and $r_0 = d_0/2$. Clearly, $\bar{B}_{r_0}(\gamma(t)) \subset \mathbf{G}$ for any $t \geq 0$. Let $\tilde{\mathbf{x}}_2^+ = \gamma(\tilde{t}_2) \in S_{r_0}(\mathbf{x}_1) \cap \gamma$, where $\tilde{t}_2$ is taken to be $\tilde{t}_2 = \max\{t > 0; \gamma(t) \in S_{r_0}(\mathbf{x}_1)\}$, and let $\tilde{\mathbf{x}}_2^-$ be the symmetric point of $\tilde{\mathbf{x}}_2^+$ with respect to Π_1. Next, let $\mathbf{G}_1^+$ be the connected component of $\mathbf{G}\backslash\widetilde{\Pi}_1$ containing $\tilde{\mathbf{x}}_2^+$, and $\mathbf{G}_1^-$ be the connected component of $\mathbf{G}\backslash\widetilde{\Pi}_1$ containing $\tilde{\mathbf{x}}_2^-$. Then let Λ_1^+ be the connected component of $\mathbf{G}_1^+ \cap R_{\Pi_1}(\mathbf{G}_1^-)$ containing $\tilde{\mathbf{x}}_2^+$, and Λ_1^- be the connected component of $\mathbf{G}_1^- \cap R_{\Pi_1}(\mathbf{G}_1^+)$ containing $\tilde{\mathbf{x}}_2^-$. Set $\Lambda_1 = \Lambda_1^+ \cup \widetilde{\Pi}_1 \cup \Lambda_1^-$. In fact, Λ_1 is the symmetric set $\Lambda_{\widetilde{\Pi}_1}$ corresponding to the perfect plane $\widetilde{\Pi}_1$ and we present its construction again for convenience of the subsequent argument. Since $\widetilde{\Pi}_1 \in \mathscr{S}_{\mathbf{E}}, \Lambda_1$ is bounded. By Lemma 5.2.6, it is easy to deduce that $\gamma \cap \partial\Lambda_1 \neq \emptyset$. We let $\mathbf{x}_2 = \gamma(t_2)$ be the 'last' intersection point of γ and $\partial\Lambda_1$; namely, $t_2 = \max\{t > 0; \gamma(t) \in \partial\Lambda_1\} < \infty$. This then implies the existence of a perfect plane passing through $\mathbf{x}_2$ which is extended from an open face of $\partial\Lambda_1$ whose closure contains $\mathbf{x}_2$. We denote the perfect plane by $\widetilde{\Pi}_2$. Without loss of generality, we further assume that $\mathbf{x}_2$ is the 'last' intersection point of γ with $\widetilde{\Pi}_2$. By the following result, we know $\widetilde{\Pi}_2$ is bounded. We shall prove at the end of this subsection.

Lemma 5.2.7. *Suppose that $\Lambda \subset \mathbf{G}$ is a bounded polyhedral domain such that*

$$\nu \wedge \mathbf{E} = 0 \quad on\ \partial\Lambda.$$

Then every open face lying on $\partial\Lambda\backslash\mathbf{D}$ cannot be connectedly extended to an unbounded planar domain in $\mathbf{G}$.

Now, we still need to distinguish between two cases of $\widetilde{\Pi}_2 \in \check{\mathscr{Q}}_{\mathbf{E}}$ and $\widetilde{\Pi}_2 \in \mathscr{S}_{\mathbf{E}}$. But for the end of a more general discussion, we next give the introduction procedure for the above reflection argument of finding a different perfect plane with a known one. Suppose that $\widetilde{\Pi}_n \in \mathscr{S}_{\mathbf{E}}, n \in \mathbb{N}$ and $\mathbf{x}_n := \gamma(t_n) \in \gamma \cap \widetilde{\Pi}_n$ is the 'last' intersection point between γ and $\widetilde{\Pi}_n$.

Let $\tilde{\mathbf{x}}_{n+1}^+ = \gamma(\tilde{t}_{n+1}) \in S_{r_0}(\mathbf{x}_n) \cap \gamma$, where $\tilde{t}_{n+1}$ is taken to be $\tilde{t}_{n+1} = \max\{t > 0; \gamma(t) \in S_{r_0}(\mathbf{x}_n)\}$, and let $\tilde{\mathbf{x}}_{n+1}^-$ be the symmetric point of $\tilde{\mathbf{x}}_{n+1}^+$ with respect to Π_1. Next, let $\mathbf{G}_n^+$ be the connected component of $\mathbf{G}\backslash\widetilde{\Pi}_n$ containing $\tilde{\mathbf{x}}_{n+1}^+$ and $\mathbf{G}_n^-$ be the connected component of $\mathbf{G}\backslash\widetilde{\Pi}_n$ containing $\tilde{\mathbf{x}}_{n+1}^-$. Then let Λ_n^+ be the connected component of $\mathbf{G}_n^+ \cap R_{\Pi_n}(\mathbf{G}_n^-)$ containing $\tilde{\mathbf{x}}_{n+1}^+$ and Λ_n^- be the connected component of $\mathbf{G}_n^- \cap R_{\Pi_n}(\mathbf{G}_n^+)$ containing $\tilde{\mathbf{x}}_{n+1}^-$. Set $\Lambda_n = \Lambda_n^+ \cup \widetilde{\Pi}_n \cup \Lambda_n^-$. By our earlier discussion, $\Lambda_n = \Lambda_{\widetilde{\Pi}_n}$, and it is bounded since $\widetilde{\Pi}_n \in \mathscr{S}_{\mathbf{E}}$. By Lemma 5.2.6, we can obtain $\gamma \cap \partial\Lambda_n \neq \emptyset$. We let $\mathbf{x}_{n+1} := \gamma(t_{n+1})$ with $t_{n+1} = \max\{t > 0; \gamma(t) \in \partial\Lambda_n\} < \infty$. This again implies the existence of a perfect plane $\widetilde{\Pi}_{n+1}$ passing through $\mathbf{x}_{n+1}$, and $\widetilde{\Pi}_{n+1}$ is bounded by Lemma 5.2.7. Furthermore, we can also assume that x_{n+1} is the 'last' intersection point of γ with $\widetilde{\Pi}_{n+1}$. In the following, we list several important results that have been achieved:

(i)

$$\mathbf{x}_n, \mathbf{x}_{n+1} \in \widehat{\mathscr{P}_{\mathbf{E}}} := \{\mathbf{x} \in \mathscr{P}_{\mathbf{E}}; \mathbf{x} \in \widetilde{\Pi}\ with\ \widetilde{\Pi} \in \mathscr{S}_{\mathbf{E}} \cup \check{\mathscr{Q}}_{\mathbf{E}}\}; \tag{5.2.19}$$

(ii) $\widetilde{\Pi}_{n+1}$ is different from $\widetilde{\Pi}_n$, since t_n and t_{n+1} with $t_{n+1} > t_n$ are respectively the 'last' intersection points between γ and $\widetilde{\Pi}_n$ and $\widetilde{\Pi}_{n+1}$;

(iii) Both $\widetilde{\Pi}_n$ and $\widetilde{\Pi}_{n+1}$ are bounded;

(iv) Since $B_{r_0}(\mathbf{x}_n) \subset \Lambda_n$, the length of $\gamma(t)$ from t_n to t_{n+1} is not less than r_0, i.e.

$$|\gamma(t_n \leq t \leq t_{n+1})| \geq |\gamma(t_n \leq t \leq \tilde{t}_{n+1})| \geq r_0. \tag{5.2.20}$$

If $\widetilde{\Pi}_{n+1} \in \mathscr{S}_{\mathbf{E}}$, by repeating the above reflection argument, we can find another bounded perfect plane $\widetilde{\Pi}_{n+2}$, and also $\mathbf{x}_{n+2} := \gamma(t_{n+2})$, the 'last' intersection point between γ and $\widetilde{\Pi}_{n+2}$, such that

$$|\gamma(t_{n+1} \leq t \leq x_{n+2})| \geq r_0.$$

In case $\widetilde{\Pi}_{n+1} \in \breve{\mathscr{Q}}_{\mathbf{E}}$, we can no longer guarantee that $\gamma \cap \partial\Lambda_{n+1} \neq \emptyset$ since $\Lambda_{n+1} = \Lambda_{\widetilde{\Pi}_{n+1}}$ is unbounded. Let $B_0 := B_{\varepsilon_0}(\mathbf{x}_{n+1})$ with $\varepsilon_0 > 0$ be sufficiently small such that $B_0 \subset B_{r_0}(\mathbf{x}_{n+1})$ and one of the half ball of B_0 divided by $\widetilde{\Pi}_{n+1}$ is contained entirely in Λ_n. Here, we recall that $\widetilde{\Pi}_{n+1}$ is extended from an open face of Λ_n. Then, let Λ^*_{n+1} be the connected component of $(\Lambda_n \cup \mathscr{R}_{\Pi_{n+1}}\Lambda_n) \cap \Lambda_{n+1}$ containing B_0. It is remarked that this is similar to the construction of Θ^* from D^* in step I. Since Λ_n is bounded, we know Λ^*_{n+1} is bounded. Moreover, by the reflection principle in (ii) of Theorem 5.2.2, $\nu \wedge \mathbf{E}(x) = 0$ on $\partial\Lambda^*_{n+1}$. Now, by Lemma 5.2.6, it is verified directly that $\gamma \cap \Lambda^*_{n+1} \neq \emptyset$. Also, we let $\mathbf{x}_{n+2} := \gamma(t_{n+2})$ be the 'last' intersection point between γ and $\partial\Lambda^*_{n+2}$. By analytic continuation, this implies the existence of a perfect plane $\widetilde{\Pi}_{n+2}$ passing through $\mathbf{x}_{n+2}$, which must be bounded by Lemma 5.2.7. More importantly, noting that $\widetilde{\Pi}_{n+2}$ is not conplane to $\widetilde{\Pi}_{n+1}$, we know by Lemma 5.2.4 that $\widetilde{\Pi}_{n+2} \in \mathscr{S}_{\mathbf{E}}$. As what has been frequently done before, we can further assume that $\mathbf{x}_{n+2}$ is the 'last' intersection point of γ with $\widetilde{\Pi}_{n+2}$. Finally, it is easy to show

$$|\gamma(t_{n+1} \leq t \leq t_{n+2})| \geq \varepsilon_0.$$

By induction and also by noting that $\gamma(t)(t > 0)$ has at most one intersection point with $\mathscr{Q}_{\mathbf{E}}$, we have constructed a sequence of different perfect planes $\widetilde{\Pi}_n, n = 1, 2, 3, \ldots$, all belonging to $\mathscr{S}_{\mathbf{E}}$ except possibly only one belonging to $\breve{\mathscr{Q}}_{\mathbf{E}}$. Moreover, there is a strictly increasing sequence $\{t_n\}_{n=1}^{\infty}$ together with a sequence of points $\mathbf{x}_n = \gamma(t_n) \in \gamma \cap \widetilde{\Pi}_n, n = 1, 2, 3, \ldots$, such that

$$|\gamma(t_n \leq t \leq t_{n+1})| \geq r_0 \quad \textit{when } n > n_0, \tag{5.2.21}$$

where n_0 is the index such that $\widetilde{\Pi}_{n_0} \in \mathscr{Q}_{\mathbf{E}}$ and it might be 0.

Now, we can conclude our proof of Theorem 5.2.1 by a contradiction as follows. Since $\gamma(t_n) \in \widehat{\mathscr{P}}_{\mathbf{E}} \subset \overline{ch(\mathbf{D})}$ being bounded and $\lim_{t\to\infty} |\gamma(t)| = \infty$, we know there must exist some $T_0 < \infty$ such that $\lim_{n\to\infty} t_n = T_0$. Then,

$$\lim_{n\to\infty} |\gamma(t_n \leq t \leq t_{n+1})| = \lim_{n\to\infty} \int_{t_n}^{t_{n+1}} |\gamma'(t)| \mathrm{d}t = 0. \tag{5.2.22}$$

A contradiction to (5.2.21).

Proof of Lemma 5.2.7. Assume contrarily that there is an open face Γ_0 on $\partial\Lambda \backslash \mathbf{D}$ which can be connectedly extended in $\mathbf{G}$ to give an unbounded planar domain. By analytic continuation, this gives an unbounded perfect plane $\widetilde{\Pi}_0$. Since Λ is a bounded polyhedron in $\mathbf{G}$, $\widetilde{\Pi}_0$ must be separated from Λ at some of its edge. Hence, there is another open face Γ_1 on $\partial\Lambda \setminus \mathbf{D}$, such that Γ_0 and Γ_1 have a common edge in $\mathbf{G}$. Again by analytic continuation, we have a perfect plane $\widetilde{\Pi}_1$ from the connected extension of Γ_1 in $\mathbf{G}$. Noting $\widetilde{\Pi}_0 \in \mathscr{Q}_{\mathbf{E}}$, we see $\widetilde{\Pi}_1 \in \mathscr{S}_{\mathbf{E}}$ by Lemma 5.2.4. Next, the argument follows a similar manner as that of step III in proof of Theorem 5.2.1.

Fix an arbitrary point $\mathbf{x}_1 \in \Gamma_0 \cap \Gamma_1$. Let $\gamma := \gamma(t)(t \geq 0)$ be a regular curve such that $\gamma(t_1) = \mathbf{x}_1$ with $t_1 = 0$ and $\gamma(t)(t > 0)$ lies entirely in the unbounded connected component of $\widetilde{\Pi}_0 \backslash \Lambda$ and $\lim_{t\to\infty} |\gamma(t)| = \infty$. Set $\tau_0 = \mathbf{d}(\gamma, \mathbf{D}) > 0$. From our earlier discussion in step III of proof of Theorem 5.2.1, we know $\gamma \cap \partial\Lambda_{\widetilde{\Pi}_1} \neq \emptyset$. Furthermore, letting $\mathbf{x}_2 := \gamma(t_2)$ be the 'last' intersection point of γ with $\partial\Lambda_{\widetilde{\Pi}_1}$, there is another perfect plane $\widetilde{\Pi}_2$ extended from an open face on $\partial\Lambda_{\widetilde{\Pi}_1}$ such that $\widetilde{\Pi}_2$ passing through $\mathbf{x}_2$ and

$$|\gamma(t_1 \leq t \leq t_2)| \geq \tau_0.$$

A crucial observation is that $\gamma \subset \widetilde{\Pi}_0$, we can without loss of generality assume that $\widetilde{\Pi}_2$ is non-parallel to $\widetilde{\Pi}_0$; therefore, $\widetilde{\Pi}_2 \in \mathscr{S}_{\mathbf{E}}$ by Lemma 5.2.4. By repeating the above procedure, we can construct countably many different perfect planes $\widetilde{\Pi}_n \in \mathscr{S}_{\mathbf{E}}, n = 1,2,3,\ldots$, together with a sequence of points $\mathbf{x}_n := \gamma(t_n) \in \gamma \cap \widetilde{\Pi}_n$ satisfying

$$|\gamma(t_n \leq t \leq t_{n+1})| \geq \tau_0.$$

Finally, a similar contradiction is established as that in (5.2.22), thus completing the proof. □

References

1. G. Alessandrini and L. Rondi, *Determining a sound-soft polyhedral scatterer by a single far-field measurement*, Proc. Amer. Math. Soc., **35** (2005), 1685–1691.
2. M. Cessenat, *Mathematical Methods in Electromagnetism: Linear Theory and Applications,* (Singapore: World Scientific)(1996).
3. J. Cheng and M. Yamamoto, *Uniqueness in an inverse scattering problem within non-trapping polygonal obstacles with at most two incoming waves*, Inverse Problems, **19** (2003), 1361–1384.
4. D. Colton and R. Kress, *Integral Equation Method in Scattering Theory*, New York: Wiely, (1983).
5. D. Colton and R. Kress, *Inverse acoustic and electromagnetic scattering theory*, volume 93 of Applied Mathematical Sciences. Springer, Cham, 4th edition, 2019.
6. M. Costabel and M. Dauge, *Singularities of electromagnetic Scattering Theory*, 2nd edn (Berlin: Springer) (2000).
7. J. Dieudonné, *Foundations of modern analysis*, New York:Academic, (1969)
8. J. Elschner and M. Yamamoto, *Uniqueness in determining polygonal sound-hard obstacles with a single incoming wave*, Inverse Problems, **22** (2006), 355–364.
9. J. Elschner and M. Yamamoto, *Uniqueness in determining polygonal sound-hard obstacles with a single incoming wave*, Inverse Problems, **24**(2008), no. 3, 035004, 7 pp.
10. J. Fritz, *Partial Differential Equations*, 3rd edn (New York: Springer)(1978).
11. V. Isakov, *Inverse Problems for Partial Differential Equations*, 2nd edition, Springer-Verlag, New York, (2006).
12. R. Kress, *Uniqueness in inverse obstacle scattering for electromagnetic waves*, Proc. URSI General Assembly, Maastricht, (2002).
13. S. Lang, *Complex Analysis*, 3rd edn, New York: Springer, (1993).
14. H. Liu, M. Yamamoto and J. Zou, *Reflection principle for the Maxwell equations and its application to inverse electromagnetic scattering*, Inverse Problems, **23** (2007), 2357–2366.
15. H. Liu, M. Yamamoto and J. Zou, *New reflection principles for Maxwell's equations and their applications*, Numer. Math. Theor. Meth. Appl., **2(1)** (2009), 1–17.
16. H. Liu and J. Zou, *Uniqueness in an inverse acoustic obstacle scattering problem for both sound-hard and sound-soft polyhedral scatterers*, Inverse Problems, **22** (2006), 515–524.
17. H. Liu and J. Zou, *Uniqueness in determining multiple polygonal or polyhedral scatterers of mixed type*, Discrete Continu. Dyn. Syst. B 9 (2008), 375–396.
18. W. Mclean, *Strongly Elliptic Systems and Boundary Integral Equation*, Cambridge University Press, Cambridge, (2000).
19. E. M. Stein and R. Shakarchi, *Complex analysis II*, Princeton. (2003).

A crucial observation is that [illegible] we can without loss of generality assume that [illegible] is non-[illegible] parallel to [illegible], therefore [illegible] by Lemma 3.2.4. By repeating the above procedure, we can construct countably many different perfect planes [illegible] together with a sequence of points [illegible] satisfying

[illegible]

Finally, a similar contradiction is established as that in (3.2.2), thus completing the proof. □

References

1. [illegible]
2. [illegible]
3. [illegible]
4. [illegible]
5. [illegible]
6. [illegible]
7. [illegible]
8. [illegible]
9. [illegible]
10. [illegible]
11. [illegible]
12. [illegible]
13. [illegible]
14. [illegible]
15. [illegible]
16. [illegible]
17. [illegible]
18. [illegible]
19. [illegible]

Chapter 6
Stability for Inverse Acoustic Obstacle Scattering Problems

6.1 Overview

In this chapter, we present the quantitative path argument in deriving sharp stability estimates for the inverse acoustic obstacle problems with a single or a few far-field measurements. We follow the treatment in [15].* In what follows, a set $\Sigma \subset \mathbb{R}^N$, $N \geq 2$, is called a *scatterer* if it is a compact set such that $\mathbb{R}^N \backslash \Sigma$ is connected. A scatterer is said to be an obstacle if it is the closure of an open set and it is said to be a screen if its interior is empty.

For clarity, we present the forward scattering problem considered in this section again. If an incident time-harmonic acoustic wave encounters a scatterer, then it is perturbed through the creation of a scattered or reflected wave. The total wave is given by the superposition of the incident and the scattered wave and it is characterized by the totdal field u, solution to the following exterior boundary value problem:

$$\begin{cases} \Delta u + k^2 u = 0 & \text{in } \mathbb{R}^N \backslash \Sigma \\ u = u^i + u^s & \text{in } \mathbb{R}^N \backslash \Sigma \\ B.C. & \text{on } \partial \Sigma \\ \lim\limits_{r \to \infty} r^{(N-1)/2} \left(\dfrac{\partial \mathbf{u}^s}{\partial r} - \mathrm{i} k u^s \right) = 0 \; r = \|\mathbf{x}\|. \end{cases}$$

Here $k > 0$ in the reduced wave equation, or Helmholtz equation, is the wavenumber and u^i is the incident field, that is, the field of the incident wave. The incident field is usually an entire solution of the Helmholtz equation; here we shall always assume that the incident wave is a time-harmonic plane wave with direction of propagation $\mathbf{v} \in \mathbb{S}^{N-1}$, that is, $u^i(\mathbf{x}) = \mathrm{e}^{\mathrm{i}k\mathbf{x}\cdot\mathbf{v}}$, $\mathbf{x} \in \mathbb{R}^N$. Instead, u^s is the scattered field, that is, the field of the scattered wave. The last limit is the Sommerfeld radiation condition and corresponds to the fact that the scattered wave is radiating. Moreover, it implies that the scattered field has the following asymptotic behavior:

$$u^s(\mathbf{x}) = \frac{\mathrm{e}^{\mathrm{i}k\|\mathbf{x}\|}}{\|\mathbf{x}\|^{(N-1)/2}} \left\{ u_\infty(\hat{\mathbf{x}}) + O\left(\frac{1}{\|\mathbf{x}\|}\right) \right\},$$

where $\hat{\mathbf{x}} = \mathbf{x}/\|\mathbf{x}\| \in \mathbb{S}^{N-1}$ and u_∞ is the so-called *far-field pattern* of u^s. We shall also write $u_\infty(\hat{\mathbf{x}}; \Sigma, k, \mathbf{v})$ to specify its dependence on the observation direction $\hat{\mathbf{x}} \in \mathbb{S}^{N-1}$, the scatterer Σ, the wavenumber $k > 0$, and the direction of propagation of the incident field $\mathbf{v} \in \mathbb{S}^{N-1}$.

Finally, the boundary condition on the boundary of Σ depends on the physical properties of the scatterer Σ. If Σ is *sound-soft*, then u satisfies a homogeneous Dirichlet condition, whereas if Σ is

H. Diao, H. Liu, *Spectral Geometry and Inverse Scattering Theory*,
https://doi.org/10.1007/978-3-031-34615-6_6

sound-hard, we have a homogeneous Neumann condition. We remark that other conditions such as the impedance boundary condition or transmission conditions for penetrable scatterers may be of interest for the applications.

The plan of the chapter is as follows. In Sect. 6.2 we discuss a few preliminaries. In particular we define and study suitable classes of admissible scatterers and we present a few basic properties of the solutions to the corresponding scattering problems. In Sect. 6.3 the main stability results are stated. In Sect. 6.4 we present the main geometric construction. Finally, in Sect. 6.5 we conclude the proofs of our stability results.

6.2 Classes of Admissible Scatterers and Preliminaries

The integer $N \geq 2$ shall always denote the space dimension. We notice that we always omit the dependence of constants on the space dimension N.

For any $\mathbf{x} \in \mathbb{R}^N$, $N \geq 2$, we denote $\mathbf{x} = (\mathbf{x}', x_N) \in \mathbb{R}^{N-1} \times \mathbb{R}$ and $\mathbf{x} = (\mathbf{x}'', x_{N-1}, x_N) \in \mathbb{R}^{N-2} \times \mathbb{R} \times \mathbb{R}$. For any $s > 0$ and any $\mathbf{x} \in \mathbb{R}^N$, $B_s(\mathbf{x})$ denotes the ball contained in $\mathbb{R}^N$ with radius s and center $\mathbf{x}$, whereas $B'_s(\mathbf{x}')$ denotes the ball contained in $\mathbb{R}^{N-1}$ with radius s and center $\mathbf{x}'$. Moreover, $B_s = B_s(\mathbf{0})$ and $B'_s = B'_s(\mathbf{0})$. For any ball B centered at zero, we denote $B^{\pm} = B \cap \{\mathbf{y} \in \mathbb{R}^N : y_N \gtrless 0\}$. Analogously, for any hyperplane Π in $\mathbb{R}^N$, we use the following notation. If, with respect to a suitable Cartesian coordinate system, we have $\Pi = \{\mathbf{y} \in \mathbb{R}^N : y_N = 0\}$, then for any $\mathbf{x} \in \Pi$ and any $r > 0$ we denote $B_r^{\pm}(\mathbf{x}) = B_r(\mathbf{x}) \cap \{y_N \gtrless 0\}$. Furthermore, we denote with T_Π the reflection in Π, namely, in this case for any $\mathbf{y} = (y_1, \ldots, y_{N-1}, y_N) \in \mathbb{R}^N$, we have $T_\Pi(\mathbf{y}) = (y_1, \ldots, y_{N-1}, -y_N)$. Finally, for any $E \subset \mathbb{R}^N$, we denote $B_s(E) = \bigcup_{\mathbf{x} \in E} B_s(\mathbf{x})$.

Given a point $\mathbf{x} \in \mathbb{R}^N$, a vector $\mathbf{v} \in \mathbb{S}^{N-1}$, and constants $r > 0$ and θ, $0 < \theta \leq \pi/2$, we call $\mathscr{C}(\mathbf{x}, \mathbf{v}, r, \theta)$ the open cone with vertex in $\mathbf{x}$, bisecting vector given by $\mathbf{v}$, radius r, and amplitude given by θ, that is

$$\mathscr{C}(\mathbf{x}, \mathbf{v}, r, \theta) = \left\{ \mathbf{y} \in \mathbb{R}^N : 0 < \|\mathbf{y} - \mathbf{x}\| < r \text{ and } \cos(\theta) < \frac{\mathbf{y} - \mathbf{x}}{\|\mathbf{y} - \mathbf{x}\|} \cdot \mathbf{v} \leq 1 \right\}.$$

We remark that by a cone we always mean a bounded not empty open cone of the kind defined above.

By $\mathscr{H}^s$, $0 \leq s \leq N$, we denote the s-dimensional Hausdorff measure in $\mathbb{R}^N$. We recall that $\mathscr{H}^N$ coincides with the Lebesgue measure.

6.2.1 Classes of Admissible Scatterers and Obstacles

We recall that by a *scatterer* in $\mathbb{R}^N$ we mean a compact set Σ contained in $\mathbb{R}^N$ such that $\mathbb{R}^N \backslash \Sigma$ is connected. We say that a scatterer Σ is an *obstacle* if $\Sigma = \overline{\Omega}$ where Ω is an open set. If the interior of Σ is empty, then we usually call it a *screen*. If Σ is a scatterer in $\mathbb{R}^N$, we shall denote $G = \mathbb{R}^N \backslash \Sigma$, which is then a connected open set containing the exterior of a ball.

A more quantitative assumption on the connectedness of $G = \mathbb{R}^N \backslash \Sigma$ is the following. Let $\delta : (0, +\infty) \to (0, +\infty)$ be a nondecreasing left-continuous function. Let Σ be a compact set contained in $\mathbb{R}^N$. We say that Σ satisfies the *uniform exterior connectedness* with function δ if for any $t > 0$, for any two points $\mathbf{x}_1$, $\mathbf{x}_2 \in \mathbb{R}^N$ so that $B_t(\mathbf{x}_1)$ and $B_t(\mathbf{x}_2)$ are contained in $\mathbb{R}^N \backslash \Sigma$, and for any s, $0 < s < \delta(t)$, then we can find a smooth (for instance, C^1) curve γ connecting $\mathbf{x}_1$ to $\mathbf{x}_2$ so that $B_s(\gamma)$ is contained in $\mathbb{R}^N \backslash \Sigma$ as well.

Let us notice that such an assumption is closed under convergence in the Hausdorff distance and that $\delta(t) \leq t$ for any $t > 0$.

We wish to define suitable classes of admissible scatterers. We begin with some definitions.

Let K be a compact subset of $\mathbb{R}^N$. We say that K is a *mildly Lipschitz hypersurface* , with or without boundary, with positive constants r and L if the following holds.

For any $\mathbf{x} \in K$, there exists a bi-Lipschitz function $\Phi_{\mathbf{x}} : B_r(\mathbf{x}) \to \mathbb{R}^N$ such that

(a) for any $\mathbf{z}_1, \mathbf{z}_2 \in B_r(\mathbf{x})$ we have

$$L^{-1}\|\mathbf{z}_1 - \mathbf{z}_2\| \leq \|\Phi_{\mathbf{x}}(\mathbf{z}_1) - \Phi_{\mathbf{x}}(\mathbf{z}_2)\| \leq L\|\mathbf{z}_1 - \mathbf{z}_2\|;$$

(b) $\Phi_{\mathbf{x}}(\mathbf{x}) = \mathbf{0}$ and $\Phi_{\mathbf{x}}(K \cap B_r(\mathbf{x})) \subset \Pi = \{\mathbf{y} \in \mathbb{R}^N : y_N = 0\}$.

We say that $\mathbf{x} \in K$ belongs to the interior of K if there exists δ, $0 < \delta \leq r$, such that $B_\delta(\mathbf{0}) \cap \Pi \subset \Phi_{\mathbf{x}}(K \cap B_r(\mathbf{x}))$. Otherwise, we say that $\mathbf{x}$ belongs to the boundary of K. We remark that the boundary of K might be empty. Further, we assume that

(c) for any $\mathbf{x}$ belonging to the boundary of K, we have that

$$\Phi_{\mathbf{x}}(K \cap B_r(\mathbf{x})) = \Phi_{\mathbf{x}}(B_r(\mathbf{x})) \cap \Pi^+,$$

where $\Pi^+ = \{\mathbf{y} \in \mathbb{R}^N : y_N = 0,\ y_{N-1} \geq 0\}$.

Let us notice that, by compactness, such an assumption is enough to guarantee that $\mathscr{H}^{N-1}(K)$ is bounded, hence $|K| = 0$. In particular, $\mathscr{H}^{N-1}(K)$ is bounded by a constant depending on the diameter of K, r, and L only. Furthermore, the boundary of K has $\mathscr{H}^{N-2}$ measure bounded by a constant again depending on the diameter of K, r, and L only.

Moreover, K has a finite number of connected components, again bounded by a constant depending on the diameter of K, r, and L only, and the distance between two different connected components of K is bounded from below by a positive constant depending on r and L only.

Let us fix a bounded open set $\Omega \subset \mathbb{R}^N$, $N \geq 2$. We shall call $\mathscr{B}(r, L, \Omega)$ the set of $K \subset \overline{\Omega}$ such that K is a mildly Lipschitz hypersurface with constants r and L. We notice that such a set is compact with respect to the Hausdorff distance; see, for instance, Lemma 3.6 in [18]. We finally remark that such a class is strictly related to a similar one introduced in [8].

Let K be a compact subset of $\mathbb{R}^N$. We say that K is a *Lipschitz hypersurface* , with or without boundary, with positive constants r and L if the following holds.

For any $\mathbf{x} \in K$, there exists a function $\varphi : \mathbb{R}^{N-1} \to \mathbb{R}$, such that $\varphi(\mathbf{0}) = 0$ and which is Lipschitz with Lipschitz constant bounded by L, such that, up to a rigid change of coordinates, we have $\mathbf{x} = \mathbf{0}$ and

$$B_r(\mathbf{x}) \cap K \subset \{\mathbf{y} \in B_r(\mathbf{x}) : y_N = \varphi(\mathbf{y}')\}. \tag{6.2.1}$$

We say that $\mathbf{x} \in K$ belongs to the interior of K if there exists δ, $0 < \delta \leq r$, such that $B_\delta(\mathbf{x}) \cap K = \{\mathbf{y} \in B_\delta(\mathbf{x}) : y_N = \varphi(\mathbf{y}')\}$. Otherwise, we say that $\mathbf{x}$ belongs to the boundary of K. We remark that the boundary of K might be empty. For any $\mathbf{x}$ belonging to the boundary of K, we assume that there exists another function $\varphi_1 : \mathbb{R}^{N-2} \to \mathbb{R}$, such that $\varphi_1(\mathbf{0}) = 0$ and which is Lipschitz with Lipschitz constant bounded by L, such that, up to the previous rigid change of coordinates, we have $\mathbf{x} = \mathbf{0}$ and

$$B_r(\mathbf{x}) \cap K = \{\mathbf{y} \in B_r(\mathbf{x}) : y_N = \varphi(\mathbf{y}'),\ y_{N-1} \leq \varphi_1(\mathbf{y}'')\}. \tag{6.2.2}$$

We call (6.2.1) and (6.2.2) the *L-Lipschitz representation* of K in $B_r(\mathbf{x})$, where (6.2.2) is reserved for points belonging to the boundary of K.

We notice that a Lipschitz hypersurface with constants r and L is also a mildly Lipschitz hypersurface with positive constants $\tilde{r}$ and $\tilde{L}$ depending on r and L only. Furthermore, we call $\mathscr{C} = \mathscr{C}(r, L, \Omega)$ the class of Lipschitz hypersurfaces with constants r and L contained in $\overline{\Omega}$. We notice that $\mathscr{C}$ is compact with respect to the Hausdorff distance, too, and that $\mathscr{C}(r, L, \Omega) \subset \mathscr{B}(\tilde{r}, \tilde{L}, \Omega)$.

We need the following notation. For any direction $\mathbf{v} \in \mathbb{S}^{N-1}$, we denote by $\hat{\mathbf{v}}$ the couple $\hat{\mathbf{v}} = \{\mathbf{v}, -\mathbf{v}\}$. We also define the following distance

$$d(\hat{\mathbf{v}}_1,\hat{\mathbf{v}}_2)=\min\{\|\mathbf{v}_1-\mathbf{v}_2\|,\|\mathbf{v}_1+\mathbf{v}_2\|\}\quad \text{for any } \mathbf{v}_1,\mathbf{v}_2\in\mathbb{S}^{N-1}.$$

Let K be a compact subset of $\mathbb{R}^N$. We say that K is a *strongly Lipschitz hypersurface* , with or without boundary, with positive constants r and L if the following holds.

First, we assume that K is a Lipschitz hypersurface with constants r and L. Then we assume the following further property. For any $\mathbf{x}\in K$, let $\mathbf{e}_1(\mathbf{x}),\dots,\mathbf{e}_N(\mathbf{x})$ be the unit vectors representing the orthonormal base of the coordinate system for which the L-Lipschitz representation of K in $B_r(\mathbf{x})$, (6.2.1) and (6.2.2), holds. Then $\hat{\mathbf{e}}_N(\mathbf{x})$ is a Lipschitz function of $\mathbf{x}\in K$, with Lipschitz constant bounded by L, and $\mathbf{e}_{N-1}(\mathbf{x})$ is a Lipschitz function of $\mathbf{x}$, as $\mathbf{x}$ varies in the boundary of K, with Lipschitz constant bounded by L.

The usefulness of introducing the idea of strongly Lipschitz hypersurfaces is shown in the following proposition.

Proposition 6.2.1. *Let Σ be a scatterer such that $K=\partial\Sigma$ is a strongly Lipschitz hypersurface with positive constants r and L.*

Then there exists a nondecreasing left-continuous function $\delta:(0,+\infty)\to(0,+\infty)$, depending on r and L only, such that Σ satisfies the uniform exterior connectedness with function δ.

Proof. Under these assumptions, the conclusions of Proposition 4.2 in [18] hold, that is, we can find constants $0<a\leq 1\leq b$ and $h_0>0$, depending on r and L only, and a Lipschitz function $\tilde{d}:\mathbb{R}^N\to[0,+\infty)$ such that

$$a\,\mathrm{dist}(\mathbf{x},\Sigma)\leq\tilde{d}(\mathbf{x})\leq b\,\mathrm{dist}(\mathbf{x},\Sigma)\quad \text{for any } \mathbf{x}\in\mathbb{R}^N,$$

and, for any h, $0<h\leq h_0$, $\mathbb{R}^N\backslash\Sigma_h$ is connected, where $\Sigma_h=\{\mathbf{x}\in\mathbb{R}^N:\ \tilde{d}(\mathbf{x})\leq h\}$. Let us notice that the assumption used in [18, Proposition 4.2] that K should be oriented is not really necessary.

Therefore, fixed $t>0$, let $\mathbf{x}_1$, $\mathbf{x}_2\in\mathbb{R}^N$ be any two points so that $B_t(\mathbf{x}_1)$ and $B_t(\mathbf{x}_2)$ are contained in $\mathbb{R}^N\backslash\Sigma$. Then $\tilde{d}(\mathbf{x}_i)\geq at$ for any $i=1,2$. Provided $h=at/2\leq h_0$, then $\mathbf{x}_i\in\mathbb{R}^N\backslash\Sigma_h$ for any $i=1,2$. Then we can find a smooth (for instance, C^1) curve γ connecting $\mathbf{x}_1$ to $\mathbf{x}_2$ so that γ is contained in $\mathbb{R}^N\backslash\Sigma_h$. This means that any point $\mathbf{x}$ of γ is such that $\mathrm{dist}(\mathbf{x},\Sigma)\geq\tilde{d}(\mathbf{x})/b>at/(2b)$. That is, we can choose

$$\delta(t)=\begin{cases} at/(2b) & t\in(0,2h_0/a]\\ h_0/b & t\in[2h_0/a,+\infty),\end{cases}\tag{6.2.3}$$

and the proof is concluded. □

Our next aim is to provide sufficient conditions for a Lipschitz hypersurface to be a strongly Lipschitz hypersurface . We begin with the following lemma.

Lemma 6.2.1. *Let us fix positive constants r and L. Let $K\subset\tilde{K}$ be compact subsets of $\mathbb{R}^N$ such that for any $\mathbf{x}\in K$ there exists a function $\varphi:\mathbb{R}^{N-1}\to\mathbb{R}$, such that $\varphi(\mathbf{0})=0$ and which is Lipschitz with Lipschitz constant bounded by L, such that, up to a rigid change of coordinates, we have $\mathbf{x}=\mathbf{0}$ and*

$$B_r(\mathbf{x})\cap\tilde{K}=\{\mathbf{y}\in B_r(\mathbf{x}):\ y_N=\varphi(\mathbf{y}')\}.\tag{6.2.4}$$

Then there exist positive constants $\tilde{r}$ and $\tilde{L}$, depending on r and L only, such that for any $\mathbf{x}\in K$ there exists a function $\tilde{\varphi}:\mathbb{R}^{N-1}\to\mathbb{R}$, such that $\tilde{\varphi}(\mathbf{0})=0$ and which is Lipschitz with Lipschitz constant bounded by $\tilde{L}$, such that, up to a rigid change of coordinates, we have $\mathbf{x}=\mathbf{0}$ and

$$B_{\tilde{r}}(\mathbf{x})\cap\tilde{K}=\{\mathbf{y}\in B_{\tilde{r}}(\mathbf{x}):\ y_N=\tilde{\varphi}(\mathbf{y}')\},\tag{6.2.5}$$

and the following further property holds. For any $\mathbf{x}\in K$, let $\mathbf{e}_1(\mathbf{x}),\dots,\mathbf{e}_N(\mathbf{x})$ be the unit vectors representing the orthonormal base of the coordinate system for which the $\tilde{L}$-Lipschitz representation of $\tilde{K}$ in $B_{\tilde{r}}(\mathbf{x})$, (6.2.5) holds. Then $\hat{\mathbf{e}}_N(\mathbf{x})$ is a Lipschitz function of $\mathbf{x}\in K$, with Lipschitz constant bounded by $\tilde{L}$.

Proof. Let us fix $\mathbf{x} \in K$. Locally, we can give an orientation to $\tilde{K}$ near $\mathbf{x}$; therefore, without loss of generality we can assume that, locally near $\mathbf{x}$, $\tilde{K}$ is the boundary of a Lipschitz open set. More precisely, we can assume there exists an open set Ω such that $\tilde{K} \cap B_r(\mathbf{x}) \subset \partial\Omega$ and, for any $\mathbf{y} \in \tilde{K}$ whose distance from $\mathbf{x}$ is less than $r/2$, we have $\tilde{K} \cap B_{r/4}(\mathbf{y}) = \partial\Omega \cap B_{r/4}(\mathbf{y})$ and

$$\Omega \cap B_{r/4}(\mathbf{y}) = \{\mathbf{z} \in B_{r/4}(\mathbf{y}) : z_N < \varphi(\mathbf{z}')\},$$

where φ and the orientation depend on $\mathbf{y}$.

Let now $\mathbf{y}_1$ and $\mathbf{y}_2$ be two points belonging to $B_{r/16}(\mathbf{x})$. Let $\mathbf{e}_N^1$ and $\mathbf{e}_N^2$ be the corresponding vectors for which the previous Lipschitz representation holds. Then for any $\mathbf{y} \in B_{r/8}(\mathbf{x})$, we can find two open cones $\mathscr{C}_1$ and $\mathscr{C}_2$, with vertex in $\mathbf{y}$, amplitude given by an angle α_0, $0 < \alpha_0 < \pi/2$ depending on L only, radius $r_0 = r/16$, and bisecting vector given by $\mathbf{e}_N^1$ and $\mathbf{e}_N^2$, respectively, such that $\mathscr{C}_i$ does not intersect $\overline{\Omega}$, whereas the opposite cone is contained in Ω for any $i = 1, 2$. First, we notice that the angle between $\mathbf{e}_N^1$ and $\mathbf{e}_N^2$ is bounded by $\pi - 2\alpha_0$. Then we take any unit vector v on the shorter arc of the great circle on the unit sphere passing through $\mathbf{e}_N^1$ and $\mathbf{e}_N^2$.

Then there exists an absolute constant $\hat{\alpha}_0$, $0 < \hat{\alpha}_0 < \pi/2$, such that, provided $\alpha_0 \leq \hat{\alpha}_0$, we have that the open cone $\mathscr{C}$ with vertex in $\mathbf{y}$, amplitude given by the angle $\alpha_1 = \alpha_0/2$, radius $r_1 = (\alpha_0/3)r_0$, and bisecting vector v does not intersect $\overline{\Omega}$, whereas the opposite cone is contained in Ω. We call this property an interior and exterior cone condition for Ω at $\mathbf{y} \in \partial\Omega$, with amplitude α_1, radius r_1, and bisecting vector v. The proof follows from an elementary, although lengthy, geometric construction and we omit its details.

If one performs such a construction iteratively N times, one obtains

$$\alpha_N = \frac{\alpha_0}{2^N} \quad \text{and} \quad r_N = \frac{\alpha_0^N}{3^N 2^{N(N-1)/2}} r_0.$$

Then we subdivide the whole $\mathbb{R}^N$ into (closed) cubes with sides of length $\tilde{r}_1$ such that their diameter is less than or equal to $r/64$. We then consider only cubes whose intersection with K is not empty. Let us fix one of these and let us call it Q. For any vertex $\mathbf{x}_i$, $i = 1, \ldots, 2^N$, of the cube Q we consider a point $\tilde{\mathbf{x}}_i \in K \cap Q$ such that $\operatorname{dist}(\mathbf{x}_i, K \cap Q) = \|\mathbf{x}_i - \tilde{\mathbf{x}}_i\|$. Then we consider $\mathbf{e}_N^i$ as the vector corresponding to the Lipschitz representation at the point $\tilde{\mathbf{x}}_i$. To illustrate our construction, let us assume for simplicity that $Q = [0, \tilde{r}_1]^N$. We take the points $\mathbf{x}_1 = (0, 0, \ldots, 0)$ and $\mathbf{x}_2 = (\tilde{r}_1, 0, \ldots, 0)$ and we construct a Lipschitz function $\mathbf{e}_N$ on the segment connecting $\mathbf{x}_1$ and $\mathbf{x}_2$ such that $\mathbf{e}_N(\mathbf{x}_i) = \mathbf{e}_N^i$, $i = 1, 2$, and, for any $\mathbf{x}$ in such a segment, $\mathbf{e}_N(\mathbf{x})$ belongs to the shorter arc of the great circle on the unit sphere passing through $\mathbf{e}_N^1$ and $\mathbf{e}_N^2$. Clearly, the Lipschitz constant of such a function $\mathbf{e}_N$ may be bounded by a constant depending on $\tilde{r}_1$ only. Then we perform the same construction on the segment connecting $(0, \tilde{r}_1, 0 \ldots, 0)$ and $(1, \tilde{r}_1, 0 \ldots, 0)$ and, then, on the segments connecting $(t, 0, 0, \ldots, 0)$ to $(t, \tilde{r}_1, 0 \ldots, 0)$, for any t, $0 \leq t \leq \tilde{r}_1$. We iterate such a construction until we find a Lipschitz function $\mathbf{e}_N : Q \to \mathbb{S}^{N-1}$ with Lipschitz constant bounded by a constant depending on $\tilde{r}_1$ only, with the following property. For any $\mathbf{y} \in Q \cap K$, we have that Ω satisfies an interior and exterior cone condition at any $\mathbf{z} \in \tilde{K} \cap B_{r/16}(\mathbf{y})$, with amplitude α_N, radius r_N, and bisecting vector $\mathbf{e}_N(\mathbf{y})$; therefore, we have a Lipschitz representation of $\tilde{K}$ at $\mathbf{y}$ with constants $\tilde{r}$ and $\tilde{L}$ depending on α_N and r_N only, thus on r and L only. Performing the same construction on any cube, the proof can be concluded. □

Let us notice that if K is oriented, then we can choose $\mathbf{e}_N(\mathbf{x})$ itself as a Lipschitz function of $\mathbf{x} \in K$. We also observe that if K is without boundary, then it is oriented, by the Jordan-Brouwer separation theorem, and we can choose $\tilde{K} = K$. Clearly, these remarks apply to any connected component of K. If we limit ourselves to Lipschitz hypersurfaces without boundary, then we have the following corollary.

Corollary 6.2.1. *Let us fix positive constants r and L. Let K be a Lipschitz hypersurface with constants r and L without boundary. Then there exist positive constants $\tilde{r}$ and $\tilde{L}$, depending on r and L only, such that K is a strongly Lipschitz hypersurface with constants $\tilde{r}$ and $\tilde{L}$.*

We conclude this discussion on sufficient conditions for a Lipschitz hypersurface to be a strongly Lipschitz hypersurfaces by proving the following proposition.

Proposition 6.2.2. *Let us fix positive constants r and L. Let K be a Lipschitz hypersurface with constants r and L. For any $\mathbf{x} \in K$, let $\mathbf{e}_1(\mathbf{x}), \ldots, \mathbf{e}_N(\mathbf{x})$ be the unit vectors representing the orthonormal base of the coordinate system for which the L-Lipschitz representation of K in $B_r(\mathbf{x})$,* (6.2.1) *and* (6.2.2)*, holds. Let us assume that $\hat{\mathbf{e}}_N(\mathbf{x})$ is a Lipschitz function of $\mathbf{x} \in K$, with Lipschitz constant bounded by L. Then there exist positive constants $\tilde{r}$ and $\tilde{L}$, depending on r and L only, such that K is a strongly Lipschitz hypersurface with constants $\tilde{r}$ and $\tilde{L}$.*

Proof. Take two couples of orthogonal vectors $\mathbf{e}^1_N$, $\mathbf{e}^1_{N-1}$, and $\mathbf{e}^2_N$, $\mathbf{e}^2_{N-1}$ for which the L-Lipschitz representation holds for the same point $\mathbf{x}$ on the boundary of K in a given ball of radius r. We notice that $\mathbf{e}^2_N = T(\mathbf{e}^1_N)$, where T is a rotation. Then, provided the angle between $\mathbf{e}^1_N$ and $\mathbf{e}^2_N$ is small enough, we have that the same Lipschitz representation holds for $\mathbf{e}^2_N$, $\mathbf{e}^2_{N-1}$, and $\mathbf{e}^2_N$, $T(\mathbf{e}^1_{N-1})$. We then apply the arguments of Lemma 6.2.1 in $\mathbb{R}^{N-1}$ and the proof may be concluded. □

Let us observe that a sufficient condition for the assumptions of Proposition 6.2.2 to hold has been given in Lemma 6.2.1.

We say that an open set $D \subset \mathbb{R}^N$ is *Lipschitz* with constant r and L if the following assumption holds. For any $\mathbf{x} \in \partial D$, there exists a function $\varphi : \mathbb{R}^{N-1} \to \mathbb{R}$, such that $\varphi(\mathbf{0}) = 0$ and which is Lipschitz with Lipschitz constant bounded by L, such that, up to a rigid change of coordinates, we have $\mathbf{x} = \mathbf{0}$ and

$$B_r(\mathbf{x}) \cap D = \{\mathbf{y} \in B_r(\mathbf{x}) : y_N < \varphi(\mathbf{y}')\},$$

and, consequently,

$$B_r(\mathbf{x}) \cap \partial D = \{\mathbf{y} \in B_r(\mathbf{x}) : y_N = \varphi(\mathbf{y}')\}.$$

Clearly, ∂D is a Lipschitz hypersurface, without boundary, with the same constants r and L. Moreover, we notice that D and $\mathbb{R}^N \backslash \overline{D}$ satisfy a uniform cone condition, with a cone depending on r and L only. We recall that, given $\mathscr{C}$ a fixed cone in $\mathbb{R}^N$, we say that an open set $D \subset \mathbb{R}^N$ satisfies the *cone condition with cone* $\mathscr{C}$ if for every $\mathbf{x} \in D$ there exists a cone $\mathscr{C}(\mathbf{x})$ with vertex in $\mathbf{x}$ and congruent to $\mathscr{C}$ such that $\mathscr{C}(\mathbf{x}) \subset D$.

We call $\mathscr{D} = \mathscr{D}(r, L, \Omega)$ the class of sets ∂D where $D \subset \Omega$ is an open set which is Lipschitz with constants r and L. We have that $\mathscr{D}(r, L, \Omega) \subset \mathscr{C}(r, L, \Omega) \subset \mathscr{B}(\tilde{r}, \tilde{L}, \Omega)$, for some constants $\tilde{r}$, $\tilde{L}$ depending on r and L only. Moreover, also $\mathscr{D}(r, L, \Omega)$ is compact with respect to the Hausdorff distance.

We further call $\hat{\mathscr{D}} = \hat{\mathscr{D}}(r, L, \Omega)$ the class of compact sets $\Sigma \subset \overline{\Omega}$ such that $\partial\Sigma \in \mathscr{D}(r, L, \Omega)$. Also this class is compact with respect to the Hausdorff distance.

In the following classes, introduced in [18], we combine different (mildly) Lipschitz hypersurfaces to obtain more general and complex structures.

Definition 6.2.1. Let us fix positive constants r, L, and a bounded open set Ω. Let us also fix $\omega : (0, +\infty) \to (0, +\infty)$ a nondecreasing left-continuous function.

We say that a compact set $K \subset \overline{\Omega}$ belongs to the class $\tilde{\mathscr{B}} = \tilde{\mathscr{B}}(r, L, \Omega, \omega)$, respectively $\tilde{\mathscr{C}} = \tilde{\mathscr{C}}(r, L, \Omega, \omega)$, if it satisfies the following conditions.

(1) $K = \bigcup_{i=1}^M K^i$ where $K^i \in \mathscr{B}(r, L, \Omega)$, respectively, $\mathscr{C}(r, L, \Omega)$, for any $i = 1, \ldots, M$.
(2) For any $i \in \{1, \ldots, M\}$ and any $\mathbf{x} \in K^i$, if its distance from the boundary of K^i is $t > 0$, then the distance of $\mathbf{x}$ from the union of K^j, with $j \neq i$, is greater than or equal to $\omega(t)$.

We say that a compact set $\Sigma \subset \overline{\Omega}$ belongs to the class $\tilde{\mathscr{B}}_1 = \tilde{\mathscr{B}}_1(r, L, \Omega, \omega)$, respectively, $\tilde{\mathscr{C}}_1 = \tilde{\mathscr{C}}_1(r, L, \Omega, \omega)$, if $\partial\Sigma \in \tilde{\mathscr{B}}(r, L, \Omega, \omega)$, respectively, $\partial\Sigma \in \tilde{\mathscr{C}}(r, L, \Omega, \omega)$.

We observe that, for some constants $\tilde{r}$ and $\tilde{L}$ depending on r and L only, we have $\tilde{\mathscr{C}}(r,L,\Omega,\omega) \subset \tilde{\mathscr{B}}(\tilde{r},\tilde{L},\Omega,\omega)$ and $\tilde{\mathscr{C}}_1(r,L,\Omega,\omega) \subset \tilde{\mathscr{B}}_1(\tilde{r},\tilde{L},\Omega,\omega)$.

Let us notice that in the previous definition the number M may depend on K. However, there exists an integer M_0, depending on r, L, the diameter of Ω, and ω only, such that $M \leq M_0$ for any $K \in \tilde{\mathscr{B}}$, respectively, $\tilde{\mathscr{C}}$. As before, we obtain that $\mathscr{H}^{N-1}(K)$ is bounded; hence, $|K| = 0$. In particular $\mathscr{H}^{N-1}(K)$ is bounded by a constant depending on r, L, the diameter of Ω, and M_0 only. Furthermore, if we set as the boundary of K the union of the boundaries of K^i, $i = 1,\ldots,M$, then the boundary of K has $\mathscr{H}^{N-2}$ measure bounded by a constant again depending on r, L, the diameter of Ω, and M_0 only. Finally, the number of connected components of $\mathbb{R}^N \backslash K$ is bounded by a constant M_1 depending on r, L, the diameter of Ω, and ω only.

Without loss of generality, we shall always assume that $\omega(t) \leq t$ for any $t > 0$, and that $\lim_{t \to +\infty} \omega(t)$ is equal to a finite real number which we call $\omega(+\infty)$.

We also remark that, by Condition 2), we have that $K^i \cap K^j$ is contained in the intersection of the boundaries of K^i and K^j, for any $i \neq j$. By [18, Lemma 3.8], we have that the classes $\tilde{\mathscr{B}}$ and $\tilde{\mathscr{C}}$ are closed, and actually compact, under convergence in the Hausdorff distance. In the next lemma we show that this is true for the classes $\tilde{\mathscr{B}}_1$ and $\tilde{\mathscr{C}}_1$ as well.

Lemma 6.2.2. *The classes $\tilde{\mathscr{B}}_1$ and $\tilde{\mathscr{C}}_1$ introduced in Definition* 6.2.1 *are compact under convergence in the Hausdorff distance.*

Moreover, let Σ belong to $\tilde{\mathscr{B}}_1$, or to $\tilde{\mathscr{C}}_1$, and $\mathbf{x} \in \partial\Sigma$. We call $G = \mathbb{R}^N \backslash \Sigma$. For any $r_1 > 0$, the number of connected components U of $B_{r_1}(\mathbf{x}) \cap G$ such that $\mathbf{x} \in \partial U$ is bounded by a constant M_2 depending on r_1, r, L, and ω only. Finally, the number of connected components of $B_{r_1}(\mathbf{x}) \cap G$ intersecting $B_{r_1/2}(\mathbf{x})$ is bounded by a constant M_3 depending on r_1, r, L, and ω only.

Proof. We begin by proving the second part of the lemma. It is clearly enough to consider the case in which $\Sigma \in \tilde{\mathscr{B}}_1$. Let U be a connected component of $B_{r_1}(\mathbf{x}) \cap G$ such that $\mathbf{x} \in \partial U$. We wish to prove that there exists $s_1 > 0$, depending on r_1, r, L, and ω only, and $\mathbf{y}$ such that $B_{s_1}(\mathbf{y}) \subset U$.

Without loss of generality we can assume that $r_1 \leq \tilde{r}_1$ for some $\tilde{r}_1$ depending on r and L only. Let $\mathbf{y}_0 \in U$ be such that $\|\mathbf{y}_0 - \mathbf{x}\| \leq r_1/8$ and let $\mathbf{y}_1 \in \partial\Sigma \cap \partial U$ be such that $\|\mathbf{x} - \mathbf{y}_1\| \leq r_1/4$ and such that $\mathbf{y}_1$ belongs to the interior of K^i for some $i \in \{1,\ldots,M\}$, where $\partial\Sigma = K = \bigcup_{i=1}^{M} K^i$ as in Condition 1). If the distance of $\mathbf{y}_1$ from the boundary of K^i is greater than $r_1/8$, then the conclusion is immediate. Otherwise, let $\mathbf{y}_2$ be a point in the boundary of K^i whose distance from $\mathbf{y}_1$ is less than $r_1/8$. By the local description of K^i near $\mathbf{y}_2$, we can find a point $\mathbf{y}_3 \in K^i \cap \partial U$ such that $\|\mathbf{x} - \mathbf{y}_3\| \leq r_1/2$ and such that its distance from the boundary of K^i is at least r_1/C for some constant $C \geq 8$ depending on L only. Then again we can conclude.

This property immediately implies that the number of connected components U of $B_{r_1}(\mathbf{x}) \cap G$ such that $\mathbf{x} \in \partial U$ is bounded by a constant M_2 depending on r_1, r, L, and ω only. Moreover, it will be crucial to prove the compactness in the Hausdorff distance. We conclude the proof of the second part with the following argument. For any $\mathbf{y} \in B_{r_1/2}(\mathbf{x}) \cap G$, we call $U(\mathbf{y})$ the connected component of $B_{r_1}(\mathbf{x}) \cap G$ containing $\mathbf{y}$. There exists $\mathbf{x}(\mathbf{y}) \in \partial U(\mathbf{y}) \cap \partial\Sigma$, such that $\|\mathbf{y} - \mathbf{x}(\mathbf{y})\| < r_1/2$ and $\|\mathbf{x}(\mathbf{y}) - \mathbf{x}\| < r_1/2$. Therefore, $V(\mathbf{y})$, the connected component of $B_{r_1/2}(\mathbf{x}(\mathbf{y})) \cap G$ containing $\mathbf{y}$, is such that $V(\mathbf{y}) \subset U(\mathbf{y})$ and $\mathbf{x}(\mathbf{y}) \in \partial V(\mathbf{y})$.

We assume that there exist points $\mathbf{y}_n \in B_{r_1/2}(\mathbf{x}) \cap G$, for $n = 1,\ldots,n_0$, such that $U_n = U(\mathbf{y}_n)$ are pairwise disjoint. Therefore, also $V_n = V(\mathbf{y}_n)$ are pairwise disjoint, for $n = 1,\ldots,n_0$. By the previously proved property, we have that n_0 is bounded by a constant M_3 depending on r_1, r, L, and ω only.

About compactness in the Hausdorff distance, it is enough to prove that the class $\tilde{\mathscr{B}}_1$ is closed. We recall that if Σ_n converges to Σ in the Hausdorff distance as $n \to \infty$, and we assume that Σ_n, $n \in \mathbb{N}$, and Σ are compact sets which are uniformly bounded, then

$$\Sigma = \{\mathbf{x} \in \mathbb{R}^N : \text{ there exists } \mathbf{x}_n \in \Sigma_n \text{ such that } \lim_n \mathbf{x}_n = \mathbf{x}\}.$$

We assume that $\Sigma_n \in \tilde{\mathscr{B}}_1$ converges as $n \to \infty$ to Σ. We already know that, up to subsequences that we do not relabel, $\partial\Sigma_n \to \tilde{\Sigma} \in \tilde{\mathscr{B}}$.

It is a general fact that $\partial\Sigma \subset \tilde{\Sigma} \subset \Sigma$. Hence, we just need to show that $\tilde{\Sigma} = \partial\Sigma$. By contradiction, we assume that there exists $\mathbf{x} \in \tilde{\Sigma}\backslash\partial\Sigma$. Clearly, $\mathbf{x}$ belongs to the interior of Σ, that is, for some $d > 0$ we have $B_d(\mathbf{x}) \subset \Sigma$. We can find $\mathbf{x}_n \in \partial\Sigma_n$, $n \in \mathbb{N}$, such that $\lim_n \mathbf{x}_n = \mathbf{x}$. We pick $r_1 = d/4$ and we assume that, for n large enough, $\|\mathbf{x} - \mathbf{x}_n\| < d/4$. For any n large enough, there exists $\mathbf{y}_n$ such that $B_{s_1}(\mathbf{y}_n) \cap \Sigma_n = \emptyset$ and $B_{s_1}(\mathbf{y}_n) \subset B_{d/4}(\mathbf{x}_n) \subset B_{d/2}(\mathbf{x})$. Up to a subsequence, which we do not relabel, $\lim_n \mathbf{y}_n = \mathbf{y} \in \overline{B_{d/2}(\mathbf{x})}$. But $\mathbf{y}$ should belong to Σ; hence, there exists $\tilde{\mathbf{y}}_n \in \Sigma_n$, $n \in \mathbb{N}$, such that $\lim_n \tilde{\mathbf{y}}_n = \mathbf{y}$; therefore, for any n large enough we have that $\|\tilde{\mathbf{y}}_n - \mathbf{y}_n\| < s_1$ and this is a contradiction.

The argument for the class $\tilde{\mathscr{C}}_1$ is completely analogous, and the proof is concluded. □

Finally, we consider the following definition. We recall that $T : D \to D'$, D, and D' being open subsets of $\mathbb{R}^N$ is said to be a bi-$W^{1,\infty}$ mapping with constant L if T is bijective and both $\|JT\|_{L^\infty(D)}$ and $\|J(T^{-1})\|_{L^\infty(D')}$ are bounded by L. Here T^{-1} is the inverse of T and JT denotes the Jacobian matrix of T.

Definition 6.2.2. Let us fix a bounded open set Ω and positive constants r, L, $0 < r_1 < r$, and $\tilde{C} > 0$. Let us also fix $\omega : (0,+\infty) \to (0,+\infty)$, nondecreasing left-continuous functions.

We call $\hat{\mathscr{B}} = \hat{\mathscr{B}}(r,L,\Omega,r_1,\tilde{C},\omega)$ and $\hat{\mathscr{C}} = \hat{\mathscr{C}}(r,L,\Omega,r_1,\tilde{C},\omega)$ the classes of sets satisfying the following assumptions:

(i) any $\Sigma \in \hat{\mathscr{B}}$, respectively, $\hat{\mathscr{C}}$, is a compact set contained in $\overline{\Omega} \subset \mathbb{R}^N$ such that Σ belongs to $\tilde{\mathscr{B}}_1(r,L,\Omega,\omega)$, respectively, $\tilde{\mathscr{C}}_1(r,L,\Omega,\omega)$. We call $G = \mathbb{R}^N\backslash\Sigma$.

(ii) for any $\mathbf{x} \in \partial\Sigma$ and any U connected component of $G \cap B_{r_1}(\mathbf{x})$, with $\mathbf{x} \in \partial U$, we can find an open set U' such that

$$U \subset U' \subset G, \tag{6.2.6}$$

and a bi-$W^{1,\infty}$ mapping $T : (-1,1)^{N-1} \times (0,1) \to U'$, with constant $\tilde{C}$, such that the following properties hold. By the regularity of $Q = (-1,1)^{N-1} \times (0,1)$, the bi-$W^{1,\infty}$ mapping T can be actually extended up to the boundary and we have that $T : \overline{Q} \to \mathbb{R}^N$ is a Lipschitz map with Lipschitz constant bounded by $\tilde{C}$. Furthermore, if we set $\Gamma = [-1,1]^{N-1} \times \{0\}$, we require that

$$T(\mathbf{0}) = \mathbf{x} \quad \text{and} \quad \partial U \cap B_{r_1}(\mathbf{x}) \subset T(\Gamma) \subset \partial G, \tag{6.2.7}$$

and that, for any $0 < s < r_1$ and any $\mathbf{y} \in U \cap B_{r_1 - s}(\mathbf{x})$, we have

$$\operatorname{dist}(T^{-1}(\mathbf{y}), \partial Q\backslash\Gamma) \geq \omega(s). \tag{6.2.8}$$

Remark 6.2.1. We notice that clearly, $T(\partial Q) = \partial U'$ and $\partial U \cap B_{r_1}(\mathbf{x}) \subset \partial G$.

It is pointed out that, up to suitably changing the constants r_1 and $\tilde{C}$ involved, Condition ii) is satisfied provided it holds only for points $\mathbf{x}$ belonging to the boundaries of K_i, $i = 1,\ldots,M$, where $\partial\Sigma = \bigcup_{i=1}^{M} K_i$ by Condition i).

Again, we have $\hat{\mathscr{C}}(r,L,\Omega,r_1,\tilde{C},\omega) \subset \hat{\mathscr{B}}(\tilde{r},\tilde{L},\Omega,r_1,\tilde{C},\omega)$, for some constants $\tilde{r}$ and $\tilde{L}$ depending on r and L only. We also remark that, for some constants and functions depending on r, L, and the diameter of Ω only, we have $\hat{\mathscr{D}}(r,L,\Omega) \subset \hat{\mathscr{C}}(\hat{r},\hat{L},\Omega,r_1,\tilde{C},\omega)$.

It is emphasized that Condition ii) is an extremely weak regularity condition and that it is satisfied by rather complex structures; see, for instance, the discussion on sets in $\mathbb{R}^3$ satisfying this assumption in Sect. 4 of [20], where several examples are shown.

The following compactness result holds true.

Lemma 6.2.3. *The classes $\hat{\mathscr{B}}$ and $\hat{\mathscr{C}}$ introduced in Definition* 6.2.2 *are compact under convergence in the Hausdorff distance.*

Proof. The argument is the same for both classes $\hat{\mathscr{B}}$ and $\hat{\mathscr{C}}$, so we limit ourselves to the first one. It is enough to prove that the class is closed. By Lemma 6.2.2, we just need to prove that also Condition ii) is preserved in the limit. Let $\Sigma_n \in \hat{\mathscr{B}}$, $n \in \mathbb{N}$, be such that $\Sigma_n \to \Sigma \in \tilde{\mathscr{B}}_1$ in the Hausdorff distance as $n \to \infty$.

Let $\mathbf{x} \in \partial\Sigma$ and let U be a connected component of $G \cap B_{r_1}(\mathbf{x})$ with $\mathbf{x} \in \partial U$. Let $\mathbf{y}$ and $s > 0$ be such that $B_s(\mathbf{y}) \subset U$ and $\|\mathbf{y}-\mathbf{x}\| < r_1/2$. We also consider a continuous curve $\gamma\colon [0,1] \to \mathbb{R}^N$ such that $\gamma(0) = \mathbf{y}$, $\gamma(1) = \mathbf{x}$, and $\gamma(t) \in U$ for any $t \in [0,1)$. Let $\tilde{U}_n$, $n \in \mathbb{N}$, be the connected component of G_n containing $\mathbf{y}$, at least for n large enough.

Let $\{t_m\}_{m\in\mathbb{N}} \subset [0,1)$ be an increasing sequence such that $\|\gamma(t_m) - \mathbf{x}\| < 1/m$. Then there exists an increasing sequence $\{n_m\}_{m\in\mathbb{N}}$ of integers such that for any $n \geq n_m$ we have $\gamma([0,t_m]) \subset \tilde{U}_n$. Since there exists $\tilde{\mathbf{x}}_n \in \Sigma_n$ converging to $\mathbf{x}$ as $n \to \infty$, we can conclude that there exists $\mathbf{x}_{n_m} \in \partial\tilde{U}_{n_m} \cap \partial\Sigma_{n_m}$ such that $\lim_m \mathbf{x}_{n_m} = \mathbf{x}$. It is also not difficult to show that, for any m large enough, we can find U_{n_m}, a connected component of $B_{r_1}(\mathbf{x}_{n_m}) \cap G_{n_m}$, such that $\mathbf{x}_{n_m} \in \partial U_{n_m}$, $\mathbf{y} \in U_{n_m}$, $\|\mathbf{y} - \mathbf{x}_{n_m}\| < r_1/2$, and $U_{n_m} \subset \tilde{U}_{n_m}$.

We call $T_m : Q \to U'_{n_m}$ with $U_{n_m} \subset U'_{n_m} \subset G_{n_m}$ as in Condition ii). Clearly, up to a subsequence that we do not relabel, T_m converges uniformly on $\overline{Q}$ to $T : \overline{Q} \to \mathbb{R}^N$, T being a Lipschitz function with constant $\tilde{C}$. Obviously, $T(\mathbf{0}) = \mathbf{x}$ and a straightforward computation shows that $T|_Q$ is actually bi-$W^{1,\infty}$, with constant $\tilde{C}$, between Q and U'. We have that U' is connected and we need to show that $U' \cap \Sigma = \emptyset$, that is $U' \subset G$.

We assume, by contradiction, that there exists $\mathbf{w} \in Q$ such that $T(\mathbf{w}) \in \Sigma$. By the bi-$W^{1,\infty}$ property, we have that $B_s(T_m(\mathbf{w})) \subset T_m(Q)$, for some $s > 0$ independent of m. There exists $\mathbf{y}_n \in \Sigma_n$, $n \in \mathbb{N}$, such that $\lim_n \mathbf{y}_n = T(\mathbf{w})$. On the other hand, $\lim_m T_m(\mathbf{w}) = T(\mathbf{w})$ as well; hence, for m large enough we have $\|\mathbf{y}_{n_m} - T_m(\mathbf{w})\| < s$ and this is a contradiction to the fact that $T_m(Q) \cap \Sigma_{n_m} = \emptyset$.

Next, we prove the first inclusion of (6.2.6). Let $\mathbf{x}_1 \in U$ be fixed. There exists a continuous curve γ_1 in U connecting $\mathbf{x}_1$ with $\mathbf{y}$. We have that, for some $d > 0$, $B_d(\gamma_1) \subset U$; therefore, for any m large enough, $\Sigma_{n_m} \cap B_{d/2}(\gamma_1) = \emptyset$ and $B_{d/2}(\gamma_1) \subset B_{r_1}(\mathbf{x}_{n_m})$. Therefore, $B_{d/2}(\gamma_1) \subset U_{n_m}$ and in particular $B_{d/2}(\mathbf{x}_1) \subset U_{n_m} \subset U'_{n_m}$. By a reasoning completely analogous to the one used to prove that $U' \subset G$, we conclude that $\mathbf{x}_1 \in U'$.

For what concerns (6.2.7) and (6.2.8), these can be proved with straightforward modifications of the above arguments and the proof is complete. □

Now we are ready to define the following classes of admissible scatterers.

Definition 6.2.3. Let us fix positive constants r, L, and R, $0 < r_1 < r$, and $\tilde{C} > 0$. Let us also fix $\omega : (0,+\infty) \to (0,+\infty)$ and $\delta : (0,+\infty) \to (0,+\infty)$ two nondecreasing left-continuous functions.

We call $\hat{\mathscr{B}}_{scat} = \hat{\mathscr{B}}_{scat}(r,L,R,r_1,\tilde{C},\omega,\delta)$ the class of compact sets Σ such that Σ belongs to $\hat{\mathscr{B}}(r,L,B_R,r_1,\tilde{C},\omega)$ and satisfies the uniform exterior connectedness with function δ.

We also define $\tilde{\mathscr{B}}_{scat} = \tilde{\mathscr{B}}_{scat}(r,L,R,\omega,\delta)$ the class of compact sets Σ belonging to $\tilde{\mathscr{B}}_1(r,L,B_R,\omega)$ and satisfying the uniform exterior connectedness with function δ.

Completely analogous definitions may be given for $\hat{\mathscr{C}}_{scat}$ and $\tilde{\mathscr{C}}_{scat}$.

We further call $\hat{\mathscr{D}}_{obst} = \hat{\mathscr{D}}_{obst}(r,L,R)$ the class of compact sets Σ belonging to $\hat{\mathscr{D}}(r,L,B_R)$ and such that $G = \mathbb{R}^3 \backslash \Sigma$ is connected.

Obviously, we have $\hat{\mathscr{B}}_{scat}(r,L,R,r_1,\tilde{C},\omega,\delta) \subset \tilde{\mathscr{B}}_{scat}(r,L,R,\omega,\delta)$ and the same relation holds between $\hat{\mathscr{C}}_{scat}$ and $\tilde{\mathscr{C}}_{scat}$. Moreover, the same relations as before hold between the classes $\hat{\mathscr{B}}_{scat}$ and $\tilde{\mathscr{B}}_{scat}$ and the corresponding classes $\hat{\mathscr{C}}_{scat}$ and $\tilde{\mathscr{C}}_{scat}$. We notice that any scatterer $\Sigma \in \hat{\mathscr{D}}_{obst}$ is indeed an obstacle, that is, Σ is the closure of its interior which is a bounded open set with Lipschitz boundary, with constants r and L. By Corollary 6.2.1 and Proposition 6.2.1, for some constants and functions depending on r, L, and R only, we have $\hat{\mathscr{D}}_{obst}(r,L,R) \subset \hat{\mathscr{C}}_{scat}(\tilde{r},\tilde{L},R,r_1,\tilde{C},\omega,\delta)$.

By our earlier discussion, in particular by Lemmas 6.2.2 and 6.2.3, it is easy to note that all these classes $\tilde{\mathscr{B}}_{scat}$, $\hat{\mathscr{B}}_{scat}$, $\tilde{\mathscr{C}}_{scat}$, $\hat{\mathscr{C}}_{scat}$, and $\hat{\mathscr{D}}_{obst}$ are compact with respect to the Hausdorff distance.

Finally, the sets belonging to the class $\hat{\mathscr{B}}$, thus in particular scatterers belonging to the class $\hat{\mathscr{B}}_{scat}$, satisfy the following property.

Proposition 6.2.3. *Let us fix positive constants r, L, and R, $0 < r_1 < r$, and $\tilde{C} > 0$. Let us also fix $\omega : (0,+\infty) \to (0,+\infty)$ a nondecreasing left-continuous functions.*

Let $\hat{\mathscr{B}} = \hat{\mathscr{B}}(r,L,B_R,r_1,\tilde{C},\omega)$. Then there exist constants $p > 2$ and $\tilde{C}_1 > 0$, depending on $\hat{\mathscr{B}}$ only, such that, for any $\Sigma \in \hat{\mathscr{B}}$, we have

$$\|\mathbf{v}\|_{L^p(B_{R+1}\backslash\Sigma)} \leq \tilde{C}_1\|\mathbf{v}\|_{H^1(B_{R+1}\backslash\Sigma)} \quad \text{for any } \mathbf{v} \in H^1(B_{R+1}\backslash\Sigma). \tag{6.2.9}$$

Moreover, the immersion of $H^1(B_{R+1}\backslash\Sigma)$ into $L^2(B_{R+1}\backslash\Sigma)$ is compact, for any $\Sigma \in \hat{\mathscr{B}}$.

Proof. We fix $\Sigma \in \hat{\mathscr{B}}$ and we call $G = \mathbb{R}^3\backslash\Sigma$. Let us take $\mathbf{v}$ belonging to $H^1(B_{R+1}\backslash\Sigma)$. Without loss of generality, by an easy extension argument around ∂B_{R+1}, we can assume that $\mathbf{v}$ actually belongs to $H^1(\mathbb{R}^3\backslash\Sigma)$, it has bounded support, and its H^1 norm is controlled by a constant C, depending on R only, times the corresponding H^1 norm in $B_{R+1}\backslash\Sigma$.

We start with a local construction. We fix $\mathbf{x} \in \partial\Sigma$ and U a connected component of $B_{r_1}(\mathbf{x}) \cap G$ such that $\mathbf{x} \in \partial U$. We consider U' and $T : Q \to U'$ as in Condition ii) of Definition 6.2.2.

Clearly, we have that U' satisfies

$$\|\mathbf{v}\|_{L^{s_1}(U')} \leq C_1\|\mathbf{v}\|_{H^1(U')} \quad \text{for any } \mathbf{v} \in H^1(U'), \tag{6.2.10}$$

for some constants $s_1 > 2$ and $C_1 > 0$. Since $U \cap B_{3r_1/4}(\mathbf{x}) \subset U'$, we conclude that

$$\|\mathbf{v}\|_{L^{s_1}(U\cap B_{3r_1/4}(\mathbf{x}))} \leq C_1\|\mathbf{v}\|_{H^1(U')} \quad \text{for any } \mathbf{v} \in H^1(U'). \tag{6.2.11}$$

We now consider a covering argument as follows. For any $\mathbf{x} \in \partial\Sigma$, let W_n, $n = 1,\ldots,n_0$, be the connected components of $B_{r_1/2}(\mathbf{x}) \cap G$ such that $W_n \cap B_{r_1/4}(\mathbf{x}) \neq \emptyset$. By Lemma 6.2.2, $n_0 \leq M_3$, where M_3 is a constant depending on r_1, r, L, and ω only. Let $\mathbf{y}_n \in W_n \cap B_{r_1/4}(\mathbf{x})$, $n = 1,\ldots,n_0$. As in the proof of Lemma 6.2.2, there exists $\mathbf{x}_n \in \partial W_n \cap \partial\Sigma$, such that $\|\mathbf{y}_n - \mathbf{x}_n\| < r_1/4$ and $\|\mathbf{x}_n - \mathbf{x}\| < r_1/4$. We call U_n the connected component of $B_{r_1}(\mathbf{x}_n) \cap G$ containing $\mathbf{y}_n$ and we observe that $\mathbf{x}_n \in \partial U_n$ and $W_n \subset U_n$. Actually, $W_n \subset U_n \cap B_{3r_1/4}(\mathbf{x}_n)$.

We conclude that for any $\mathbf{x} \in \partial\Sigma$, there exist n_0 points $\mathbf{x}_1,\ldots,\mathbf{x}_{n_0}$, with $n_0 \leq M_3$, with the following property. For any $n = 1,\ldots,n_0$, there exists U_n, a connected component of $B_{r_1}(\mathbf{x}_n) \cap G$, such that $\mathbf{x}_n \in \partial U_n$, and moreover,

$$B_{r_1/4}(\mathbf{x}) \cap G \subset \bigcup_{n=1}^{n_0}(U_n \cap B_{3r_1/4}(\mathbf{x}_n)).$$

We fix $\delta = r_1/16$ and define the compact set $A_1 = \overline{B_\delta(\partial\Sigma) \cap G}$. We notice that

$$A_1 \subset \bigcup_{\mathbf{x}\in\partial\Sigma} B_{r_1/4}(\mathbf{x}).$$

By the compactness of A_1, we can find a finite number of points $\mathbf{z}_i \in \partial\Sigma$, $i = 1,\ldots,m_1$, such that

$$A_1 \subset \bigcup_{i=1}^{m_1} B_{r_1/4}(\mathbf{z}_i).$$

With a rather simple construction, it is possible to choose m_1 depending on r_1 and R only, for instance, by taking points such that $B_{r_1/16}(\mathbf{z}_i) \cap B_{r_1/16}(\mathbf{z}_j)$ is empty for $i \neq j$.

We further find a finite number of points $\mathbf{z}_i \in \partial B_{R+1}$, $i = m_1+1,\ldots,m_1+m_2$, such that

$$A_2 = \overline{B_{1/16}(\partial B_{R+1})} \subset \bigcup_{i=m_1+1}^{m_1+m_2} B_{1/4}(\mathbf{z}_i),$$

with m_2 depending on R only.

Finally, we call $r_3 = \min\{1, r_1\}$ and

$$A_3 = \{\mathbf{x} \in B_{R+1}\backslash\Sigma : \operatorname{dist}(\mathbf{x}, \partial(B_{R+1}\backslash\Sigma)) \geq r_3/16\}.$$

We can find points $\mathbf{z}_i \in A_3$, $i = m_1 + m_2 + 1, \ldots, m_1 + m_2 + m_3$, such that

$$A_3 \subset \bigcup_{i=m_1+m_2+1}^{m_1+m_2+m_3} B_{r_3/32}(\mathbf{z}_i).$$

Again the number m_3 may be bounded by a constant depending on r_1 and R only.

By applying (6.2.11), at most M_3 times for any $\mathbf{z}_i$, $i = 1, \ldots, m_1$, we have

$$\|\mathbf{v}\|_{L^{s_1}(A_1 \cap G)} \leq C_1(M_3 m_1) C \|\mathbf{v}\|_{H^1(B_{R+1}\backslash\Sigma)}. \tag{6.2.12}$$

By a completely analogous argument, we can find $s_2 > 2$ and C_2 such that

$$\|\mathbf{v}\|_{L^{s_2}(A_2 \cap B_{R+1})} \leq C_2 m_2 \|\mathbf{v}\|_{H^1(B_{R+1}\backslash\Sigma)}. \tag{6.2.13}$$

Applying a classical Sobolev inequality to $D = B_{r_3/32}(\mathbf{z}_i)$, for $i = m_1 + m_2 + 1, \ldots, m_1 + m_2 + m_3$, we can finally find $s_3 > 2$ and C_3 such that

$$\|\mathbf{v}\|_{L^{s_3}(A_3)} \leq C_3 m_3 \|\mathbf{v}\|_{H^1(B_{R+1}\backslash\Sigma)}. \tag{6.2.14}$$

Picking $p = \min\{s_1, s_2, s_3\}$, we obtain that

$$\|\mathbf{v}\|_{L^p(B_{R+1}\backslash\Sigma)} \leq \tilde{C}_1 \|\mathbf{v}\|_{H^1(B_{R+1}\backslash\Sigma)}. \tag{6.2.15}$$

It is an easy remark that p and $\tilde{C}_1$ have the dependence required.

The fact that the immersion of $H^1(B_{R+1}\backslash\Sigma)$ into $L^2(B_{R+1}\backslash\Sigma)$ is compact is an immediate consequence of the property described in (6.2.9). □

We conclude this subsection by introducing suitable classes of polyhedral scatterers. We define a *cell* as the closure of an open subset of an $(N-1)$-dimensional hyperplane. We say that a scatterer Σ is *polyhedral* if the boundary of Σ is given by a finite union of cells $\mathscr{C}_j$, $j = 1, \ldots, M_1$.

In fixed positive constants h and L, we say that a scatterer Σ is *polyhedral with constants h and L* if the boundary of Σ is given by a finite union of cells $\mathscr{C}_j$, $j = 1, \ldots, M_1$, where each $\mathscr{C}_j$ is the closure of a Lipschitz domain with constants h and L contained in an $(N-1)$-dimensional hyperplane and the cells are pairwise internally disjoint, that is, two different cells may intersect only at boundary points.

Let $\hat{\mathscr{B}}_{scat} = \hat{\mathscr{B}}_{scat}(r, L, R, r_1, \tilde{C}, \omega, \delta)$ be the class of scatterers defined in Definition 6.2.3. To fix the size parameter $h > 0$, let $\hat{\mathscr{B}}^h_{scat} = \hat{\mathscr{B}}^h_{scat}(r, L, R, r_1, \tilde{C}, \omega, \delta)$ be the set of scatterers $\Sigma \in \hat{\mathscr{B}}_{scat}$ such that Σ is polyhedral with constants h and L.

Analogously, let $\hat{\mathscr{D}}_{obst} = \hat{\mathscr{D}}_{obst}(r, L, R)$ be the class of obstacles defined in Definition 6.2.3. To fix the size parameter $h > 0$, let $\hat{\mathscr{D}}^h_{obst} = \hat{\mathscr{D}}^h_{obst}(r, L, R)$ be the set of obstacles $\Sigma \in \hat{\mathscr{D}}_{obst}$ such that Σ is polyhedral with constants h and L. Notice that in this case any $\Sigma \in \hat{\mathscr{D}}^h_{obst}$ is formed by a finite number of polyhedra.

6.2.2 Preliminaries

In this subsection we fix positive constants r, L, and R, $0 < r_1 < r$, and $\tilde{C} > 0$ and two nondecreasing left-continuous functions $\omega : (0, +\infty) \to (0, +\infty)$ and $\delta : (0, +\infty) \to (0, +\infty)$. The class of admissible scatterers that we consider will be called $\mathscr{A}$. Here we pick $\mathscr{A}=\hat{\mathscr{B}}_{scat}(r,L,R,r_1,\tilde{C},\omega,\delta)$, as in Definition 6.2.3, and we take Σ and Σ' belonging to $\mathscr{A}$.

We set

$$d=\max\left\{\sup_{\mathbf{x}\in\partial\Sigma\backslash\Sigma'}\operatorname{dist}(\mathbf{x},\partial\Sigma'),\ \sup_{\mathbf{x}\in\partial\Sigma'\backslash\Sigma}\operatorname{dist}(\mathbf{x},\partial\Sigma)\right\}, \tag{6.2.16}$$

and

$$\hat{d}=d_H(\partial\Sigma,\partial\Sigma') \quad\text{and}\quad \tilde{d}=d_H(\Sigma,\Sigma'). \tag{6.2.17}$$

We recall that d_H denotes the Hausdorff distance. We notice that the following relationships among d, $\hat{d}$, and $\tilde{d}$ holds. First, d, $\hat{d}$, and $\tilde{d}$ are all bounded by $2R$. We also obviously have $d\leq\hat{d}$. Up to swapping the role of Σ and Σ', let $\mathbf{x}\in\Sigma'$ be such that $\operatorname{dist}(\mathbf{x},\Sigma)=\tilde{d}$. Clearly, $\operatorname{dist}(\mathbf{x},\partial\Sigma)=\tilde{d}$ as well. If $\mathbf{x}\in\partial\Sigma'$, then we immediately conclude that $\tilde{d}\leq d$. If $\mathbf{x}$ does not belong to $\partial\Sigma'$, then, by using the uniform exterior connectedness property of Σ, for any $s<\delta(\tilde{d})$, we can find a point $\mathbf{x}_1\in\partial\Sigma'$ such that $\operatorname{dist}(\mathbf{x}_1,\Sigma)=\operatorname{dist}(\mathbf{x}_1,\partial\Sigma)\geq s$. Therefore,

$$\delta(\tilde{d})\leq d\leq\hat{d}, \tag{6.2.18}$$

or, in other words

$$\tilde{d}\leq\delta^{-1}(d)\leq\delta^{-1}(\hat{d}), \tag{6.2.19}$$

where $\delta^{-1}:(0,+\infty)\to(0,+\infty)$ is a nondecreasing right-continuous function defined as follows:

$$\delta^{-1}(t)=\min\{\sup\{s:\ \delta(s)\leq t\},2R\}\quad\text{for any } t>0. \tag{6.2.20}$$

On the other hand, let $\mathbf{x}\in\partial\Sigma'$ be such that $\operatorname{dist}(\mathbf{x},\partial\Sigma)=\hat{d}$. If $\mathbf{x}$ does not belong to Σ, then $\operatorname{dist}(\mathbf{x},\Sigma)=\hat{d}$, hence $d=\hat{d}\leq\tilde{d}$. If $\mathbf{x}\in\Sigma$, then $B_{\hat{d}}(\mathbf{x})\subset\Sigma$. Hence, by the properties of the boundary of Σ', there exists a positive constant C_1, depending on the class $\mathscr{A}$ only and a point $\mathbf{x}_1$ such that $B_{C_1\hat{d}}(\mathbf{x}_1)\subset B_{\hat{d}}(\mathbf{x})\backslash\Sigma'$. We can conclude that

$$C_1 d\leq C_1\hat{d}\leq\tilde{d}\leq\delta^{-1}(d)\leq\delta^{-1}(\hat{d}). \tag{6.2.21}$$

Let us notice that we also have the following property that will be of use later on. If $C_2=(C_1+1)/C_1$, then

$$\Sigma\triangle\Sigma'\subset\overline{B_{C_2\tilde{d}}(\partial\Sigma)}\cap\overline{B_{C_2\tilde{d}}(\partial\Sigma')}. \tag{6.2.22}$$

In fact, if $\mathbf{x}\in\Sigma'\backslash\Sigma$, then $\operatorname{dist}(\mathbf{x},\partial\Sigma)\leq\tilde{d}$; therefore, $\mathbf{x}\in\overline{B_{\tilde{d}}(\partial\Sigma)}$. That is, $\operatorname{dist}(\mathbf{x},\partial\Sigma')\leq\tilde{d}+\hat{d}$. Finally, there exists a constant C_3, depending on the class $\mathscr{A}$ only, such that for any t, $0<t\leq 1$, we have

$$|\overline{B_t(\partial\Sigma)}|\leq C_3 t. \tag{6.2.23}$$

We consider the following direct scattering problem. To fix $\Sigma\in\mathscr{A}$, for a fixed wavenumber $k>0$ and a fixed direction of propagation $\mathbf{v}\in\mathbb{S}^{N-1}$, let the incident field u^i be the corresponding time-harmonic plane wave, that is, $u^i(\mathbf{x})=\mathrm{e}^{ik\mathbf{x}\cdot\mathbf{v}}$, $\mathbf{x}\in\mathbb{R}^N$. The incident wave is perturbed by the presence of the scatterer Σ through a scattered wave, characterized by its corresponding scattered field u^s. The total field u is the solution to the following exterior boundary value problem:

$$\begin{cases}\Delta u+k^2u=0 & \text{in } \mathbb{R}^N\backslash\Sigma\\ u=u^i+u^s & \text{in } \mathbb{R}^N\backslash\Sigma\\ B.C. & \text{on } \partial\Sigma\\ \lim_{r\to\infty} r^{(N-1)/2}\left(\dfrac{\partial u^s}{\partial r}-iku^s\right)=0 & r=\|\mathbf{x}\|,\end{cases} \tag{6.2.24}$$

where the last limit is the *Sommerfeld radiation condition* and corresponds to the fact that the scattered wave is radiating. The boundary condition on the boundary of Σ depends on the character of the scatterer Σ. For instance, if Σ is *sound-soft*, then u satisfies the following homogeneous Dirichlet condition:

$$u = 0 \quad \text{on } \partial\Sigma, \tag{6.2.25}$$

whereas if Σ is *sound-hard*, we have

$$\nabla u \cdot \nu = 0 \quad \text{on } \partial\Sigma, \tag{6.2.26}$$

which is a homogeneous Neumann condition. Other conditions such as the impedance boundary condition or transmission conditions for penetrable scatterers may occur in the applications.

We recall that the Sommerfeld radiation condition holds uniformly with respect to all directions $\hat{\mathbf{x}} = \mathbf{x}/\|\mathbf{x}\| \in \mathbb{S}^{N-1}$ and it implies that the scattered field has the asymptotic behavior of an outgoing spherical wave, namely,

$$u^s(\mathbf{x}) = \frac{e^{ik\|\mathbf{x}\|}}{\|\mathbf{x}\|^{(N-1)/2}} \left\{ u_\infty(\hat{\mathbf{x}}) + O\left(\frac{1}{\|\mathbf{x}\|}\right) \right\}, \tag{6.2.27}$$

where $\hat{\mathbf{x}} = \mathbf{x}/\|\mathbf{x}\| \in \mathbb{S}^{N-1}$ and u_∞ is the so-called *far-field pattern* of u^s. In particular, the scattered field satisfies the following decay property for some positive constants E and R_1:

$$|u^s(\mathbf{x})| \le E\|\mathbf{x}\|^{-(N-1)/2} \quad \text{for any } \mathbf{x} \in \mathbb{R}^N \text{ so that } \|\mathbf{x}\| \ge R_1. \tag{6.2.28}$$

We refer to [23] for further details, such as existence and uniqueness of the solution, on the direct scattering problem (6.2.24). For an introduction to the corresponding inverse problems, see, for instance, [4, 12].

Let us fix constants $0 < \underline{k} < \overline{k}$ and let us denote, for any $N \ge 2$,

$$I_N = \begin{cases} [\underline{k}, \overline{k}] & \text{if } N = 2, \\ (0, \overline{k}] & \text{if } N \ge 3. \end{cases} \tag{6.2.29}$$

Proposition 6.2.4. *Let us fix constants* $0 < \underline{k} < \overline{k}$ *and let* I_N *be defined as in* (6.2.29). *Let* $\mathscr{A}$ *be as defined at the beginning of the subsection.*

To fix $\Sigma \in \mathscr{A}$, $k \in I_N$, *and* $\mathbf{v} \in \mathbb{S}^{N-1}$, *let* $u^i(\mathbf{x}) = e^{ik\mathbf{x}\cdot\mathbf{v}}$, $\mathbf{x} \in \mathbb{R}^N$, *and* $u_{\Sigma,k,v}$ *be the solution to* (6.2.24), *with boundary condition* (6.2.25) *or* (6.2.26), *and* $\mathbf{u}^s_{\Sigma,k,v}$ *be its corresponding scattered field.*

Then there exists a constant E, depending on $\mathscr{A}$ *and* I_N *only, such that*

$$|u_{\Sigma,k,v}(\mathbf{x})| \le E \quad \textit{for any } \mathbf{x} \in \mathbb{R}^N \backslash \Sigma. \tag{6.2.30}$$

Furthermore, there exists a constant E_1, *depending on the constant E in* (6.2.30), I_N, *R, and N only, such that for any* $\Sigma \in \mathscr{A}$, *any* $k \in I_N$, *and any* $\mathbf{v} \in \mathbb{S}^{N-1}$, *we have*

$$|\, u^s_{\Sigma,k,v}(\mathbf{x})| \le E_1\|\mathbf{x}\|^{-(N-1)/2} \quad \textit{for any } \mathbf{x} \in \mathbb{R}^N \textit{ so that } \|\mathbf{x}\| \ge R+2, \tag{6.2.31}$$

and

$$\|\nabla u^s_{\Sigma,k,v}(\mathbf{x})\| \le E_1\|\mathbf{x}\|^{-(N-1)/2} \quad \textit{for any } \mathbf{x} \in \mathbb{R}^N \textit{ so that } \|\mathbf{x}\| \ge R+2. \tag{6.2.32}$$

Proof. First of all, we show that there exists a constant E_0, depending on $\mathscr{A}$ and I_N only, such that

$$\|u_{\Sigma,k,v}\|_{L^2(B_{R+3}\backslash\Sigma)} \le E \quad \text{for any } \Sigma \in \mathscr{A}, \text{ any } k \in I_N, \text{ and any } \mathbf{v} \in \mathbb{S}^{N-1}. \tag{6.2.33}$$

This is an immediate consequence of Proposition 3.2 and Theorem 3.9 in [18] for the sound-hard case and of Lemma 3.5 in [19] for the sound-soft case. Already from this first bound we can easily infer that (6.2.31) and (6.2.32) hold true.

The main idea of the proof needed to improve the uniform L^2 bound in (6.2.33) to the uniform L^∞ one contained in (6.2.30) is the following.

Let $\mathbf{x} \in \partial\Sigma$ and let us exploit Condition ii) of Definition 6.2.2. By a change of variables, a reflection argument, and standard regularity estimates, we infer that we can bound $|u|$ almost everywhere in $B_{r_2}(\mathbf{x})$ by a constant $\tilde{C}_1$, where r_2 and $\tilde{C}_1$ depend on r, r_1, $\tilde{C}$, and the L^2 norm of u which is bounded by (6.2.33).

This procedure allows to estimate $|u|$ in a neighborhood of $\partial\Sigma$. Away from $\partial\Sigma$ the estimate is completely standard. □

Let us fix Σ and Σ' belonging $\mathscr{A}$, $\mathscr{A}$ as defined at the beginning of the subsection. We also fix $k > 0$ and a direction of propagation $\mathbf{v} \in \mathbb{S}^{N-1}$. Let u be the solution to (6.2.24) with boundary condition (6.2.25) or (6.2.26). We denote by u^s the corresponding scattered field and by u_∞ its far-field pattern . Moreover, u', $(u^s)'$, and u'_∞ denote the same functions when Σ is replaced by Σ'. Finally, we fix positive R_1 and $\tilde{\rho}$ such that $R+1+\tilde{\rho} \leq R_1$.

By Proposition 6.2.4, we have that

$$|u(\mathbf{x})| + |u'(\mathbf{x})| \leq E \quad \text{for any } \mathbf{x} \in \mathbb{R}^3, \tag{6.2.34}$$

where E depends on k and $\mathscr{A}$ only and it may be assumed to be greater than or equal to 1, and u and u' are extended to $\mathbf{0}$ on Σ and Σ', respectively.

Let us fix a point $\mathbf{x}_0 \in \mathbb{R}^N$ such that $R+1+\tilde{\rho} \leq \|\mathbf{x}_0\| \leq R_1$. For a fixed ε, $0 < \varepsilon \leq E$, let

$$\|u - u'\|_{L^\infty(B_{\tilde{\rho}}(\mathbf{x}_0))} \leq \varepsilon. \tag{6.2.35}$$

We call ε the *near-field error with limited aperture* .

Then, let $\varepsilon_1 > 0$ be such that

$$\|u - u'\|_{L^\infty(B_{\|\mathbf{x}_0\|+\tilde{\rho}} \setminus \overline{B_{\|\mathbf{x}_0\|-\tilde{\rho}}})} \leq \varepsilon_1. \tag{6.2.36}$$

We call ε_1 the *near-field error.*

Finally, if

$$\|u_\infty - u'_\infty\|_{L^2(\mathbb{S}^{N-1})} \leq \varepsilon_0, \tag{6.2.37}$$

here ε_0 will be referred to as the *far-field error.*

We investigate the relations among these errors. First of all, let us recall that a three-sphere inequality holds for the Helmholtz equation, provided the larger ball has a radius bounded by a constant $\tilde{\rho}_0$, $\tilde{\rho}_0$ depending on k only; see, for instance, [2] or for a version suited to our case [21, Lemma 3.5] which we state here for the convenience of the reader.

Lemma 6.2.4. *There exist positive constants $\tilde{\rho}_0$, C and c_1, $0 < c_1 < 1$, depending on k only, such that for every $0 < \rho_1 < \rho < \rho_2 \leq \tilde{\rho}_0$ and any function u such that*

$$\Delta u + k^2 u = 0 \quad \text{in } B_{\rho_2},$$

we have, for any s, $\rho < s < \rho_2$,

$$\|u\|_{L^\infty(B_\rho)} \leq C(1-(\rho/s))^{-N/2} \|u\|^{1-\beta}_{L^\infty(B_{\rho_2})} \|u\|^{\beta}_{L^\infty(B_{\rho_1})}, \tag{6.2.38}$$

for some β such that

$$c_1 (\log(\rho_2/s)) / (\log(\rho_2/\rho_1)) \leq \beta \leq 1 - c_1 (\log(s/\rho_1)) / (\log(\rho_2/\rho_1)). \tag{6.2.39}$$

By an iterated application of this three-sphere inequality , we have that there exist positive constants C and α, $0 < \alpha < 1$, depending on E, $\tilde{\rho}$, R_1 and k only, such that

$$\varepsilon \le \varepsilon_1 \le C\varepsilon^{\alpha}. \tag{6.2.40}$$

Moreover, there exist positive constants $\tilde{\varepsilon}_0 < 1/\mathrm{e}$ and C_1, depending on E, R, $\tilde{\rho}$, R_1, and k only, such that if $0 < \varepsilon_0 \le \tilde{\varepsilon}_0$, then

$$\|u - u'\|_{L^{\infty}(B_{\|\mathbf{x}_0\|+\tilde{\rho}} \setminus \overline{B_{\|\mathbf{x}_0\|-\tilde{\rho}}})} \le \eta_1(\varepsilon_0) = \exp\left(-C_1(-\log \varepsilon_0)^{1/2}\right), \tag{6.2.41}$$

that is,

$$\varepsilon \le \varepsilon_1 \le \eta_1(\varepsilon_0) = \exp\left(-C_1(-\log \varepsilon_0)^{1/2}\right). \tag{6.2.42}$$

This estimate follows immediately by the results in [10] for $N = 3$ and with an easy modification for any other $N \ge 2$; see, for instance, Theorem 4.1 in [22].

If we wish to reduce to obstacles only, we use the class of admissible obstacles $\mathscr{A}_{obst}$. In particular, we set $\mathscr{A}_{obst} = \hat{\mathscr{D}}_{obst}(r, L, R)$.

It is important to notice that δ in this case may be chosen to be as in (6.2.3); therefore, δ^{-1} may be chosen to be $C_2 d$ for any d, for some constant C_2 depending on r, L, and R only, that is, (6.2.21) becomes

$$C_1 d \le C_1 \hat{d} \le \tilde{d} \le C_2 d \le C_2 \hat{d}. \tag{6.2.43}$$

Finally, if we wish to use classes of admissible polyhedral scatterers or obstacles and fixed the size parameter $h > 0$, we use $\mathscr{A}^h = \hat{\mathscr{B}}^h_{scat}(r, L, R, r_1, \tilde{C}, \omega, \delta)$ for general scatterers and $\mathscr{A}^h_{obst} = \hat{\mathscr{D}}^h_{obst}(r, L, R)$ for obstacles.

6.3 The Main Stability Results

In this section we present our stability results for the determination of sound-hard polyhedral scatterers. We distinguish them with respect to the number of scattering measurements used.

In this section we fix positive constants r, L, and R, $0 < r_1 < r$, and $\tilde{C} > 0$. Let us also fix $\omega : (0, +\infty) \to (0, +\infty)$ and $\delta : (0, +\infty) \to (0, +\infty)$ two nondecreasing left-continuous functions. We recall that $\omega(t) \le t$, that $\lim_{t \to +\infty} \omega(t)$ is equal to a finite real number which we call $\omega(+\infty)$, and that $\delta(t) \le t$ for any $t > 0$. We fix the wavenumber $k > 0$. Finally, we fix positive R_1 and $\tilde{\rho}$ such that $R + 1 + \tilde{\rho} \le R_1$. We refer to these constants and functions, including N, as the *a priori data* and we let $\mathscr{A} = \hat{\mathscr{B}}_{scat}(r, L, R, r_1, \tilde{C}, \omega, \delta)$ be the class of scatterers defined in Definition 6.2.3. As before, for any fixed $h > 0$, we call $\mathscr{A}^h = \hat{\mathscr{B}}^h_{scat}(r, L, R, r_1, \tilde{C}, \omega, \delta)$ the set of scatterers $\Sigma \in \mathscr{A}$ such that Σ is polyhedral with constants h and L.

We call $\eta : (0, 1/\mathrm{e}) \to (0, +\infty)$ the following function:

$$\eta(s) = \exp(-(\log(-\log s))^{1/2}) \quad \text{for any } s,\ 0 < s < 1/\mathrm{e}. \tag{6.3.1}$$

6.3.1 Polyhedral Scatterers with N Measurements

We fix N linearly independent unit vectors $\mathbf{v}_1, \ldots, \mathbf{v}_N$. We notice that, given N linearly independent unit vectors $\mathbf{v}_1, \ldots, \mathbf{v}_N$, there exists a positive constant a_0, depending on the vectors $\mathbf{v}_1, \ldots, \mathbf{v}_N$, such that

$$\min_{v \in \mathbb{S}^{N-1}} \left\{ \max_{j \in \{1,\ldots,N\}} |\mathbf{v}_j \cdot v| \right\} \ge a_0. \tag{6.3.2}$$

In fact, $\max_{j \in \{1,\ldots,N\}} |\mathbf{v}_j \cdot v|$ is a continuous function of $v \in \mathbb{S}^{N-1}$ which never vanishes.

We also fix a point $\mathbf{x}_0 \in \mathbb{R}^N$ such that $R + 1 + \tilde{\rho} \le \|\mathbf{x}_0\| \le R_1$.

Theorem 6.3.1. *Let $N \geq 2$. Fix $h > 0$. Let Σ, Σ' belong to $\mathscr{A}^h$ and let d be defined as in* (6.2.16). *For any $j = 1, \ldots, N$, let $u^i(\mathbf{x}) = \mathrm{e}^{\mathrm{i}k\mathbf{x}\cdot\mathbf{v}_j}$, $\mathbf{x} \in \mathbb{R}^N$, and let u_j be the solution to* (6.2.24) *with boundary condition* (6.2.26) *and u'_j be the solution to the same problem with Σ replaced by Σ'.*

If

$$\max_{j=1,\ldots,N} \|u_j - u'_j\|_{L^\infty(B_{\tilde{\rho}}(\mathbf{x}_0))} \leq \varepsilon, \tag{6.3.3}$$

for some $\varepsilon \leq 1/(2\mathrm{e})$, then for some positive constant C depending on the a priori data and on a_0 only, and not on h, we have

$$\min\{d, h\} \leq 2\mathrm{e}R(\eta(\varepsilon))^C. \tag{6.3.4}$$

Therefore,

$$d \leq 2\mathrm{e}R(\eta(\varepsilon))^C, \tag{6.3.5}$$

provided $\varepsilon \leq \hat{\varepsilon}(h)$ where

$$\hat{\varepsilon}(h) = \min\left\{1/(2\mathrm{e}), \eta^{-1}\left(\left(\frac{h}{2\mathrm{e}R}\right)^{1/C}\right)\right\}. \tag{6.3.6}$$

With little modification, we obtain exactly the same stability result if Σ and Σ' are sound-soft scatterers instead of sound-hard ones, even if we reduce the number of measurements from N to 1. That is, as a byproduct of this work, we can significantly extend Theorem 4.1 in [21] to a much more general class of scatterers, namely, $\mathscr{A}^h$, and to any dimension $N \geq 2$. We state such result in the following theorem.

Theorem 6.3.2. *Let $N \geq 2$. Fix $h > 0$. Let Σ, Σ' belong to $\mathscr{A}^h$ and let d be defined as in* (6.2.16). *Let us fix $\mathbf{v} \in \mathbb{S}^{N-1}$ and let $u^i(\mathbf{x}) = \mathrm{e}^{\mathrm{i}k\mathbf{x}\cdot\mathbf{v}}$, $\mathbf{x} \in \mathbb{R}^N$. Let u be the solution to* (6.2.24) *with boundary condition* (6.2.25) *and u' be the solution to the same problem with Σ replaced by Σ'.*

If

$$\|u - u'\|_{L^\infty(B_{\tilde{\rho}}(\mathbf{x}_0))} \leq \varepsilon, \tag{6.3.7}$$

for some $\varepsilon \leq 1/(2\mathrm{e})$ and then for some positive constant C depending on the a priori data only, and not on h, we have

$$\min\{d, h\} \leq 2\mathrm{e}R(\eta(\varepsilon))^C. \tag{6.3.8}$$

6.3.2 Polyhedral Obstacles with Fewer Measurements

It is well known that in general $N-1$ scattering measurements may not be enough to uniquely determine a polyhedral sound-hard screen. However, if we limit ourselves to polyhedral obstacles, that is, to polyhedra, then a single measurement is enough; see [6, 7].

Here we restrict ourselves to obstacles and we aim to obtain corresponding stability estimates with a minimal number of scattering measurements.

For technical reasons we limit ourselves to $N = 2$ or $N = 3$ only. Let us then fix $N \in \{2, 3\}$ and positive constants r, L, and R. We fix the wavenumber $k > 0$. Finally, we fix positive R_1 and $\tilde{\rho}$ and a point $\mathbf{x}_0 \in \mathbb{R}^N$ such that $R + 1 + \tilde{\rho} \leq \|\mathbf{x}_0\| \leq R_1$.

We let $\mathscr{A}_{obst} = \hat{\mathscr{D}}_{obst}(r, L, R)$ be the class of scatterers defined in Definition 6.2.3. For any fixed $h > 0$, we call $\mathscr{A}^h_{obst} = \hat{\mathscr{D}}^h_{obst}(r, L, R)$ the set of obstacles $\Sigma \in \mathscr{A}_{obst}$ such that Σ is polyhedral with constants h and L.

We recall that, for some constants and functions depending on r, L, and R only, we have $\hat{\mathscr{D}}_{obst}(r, L, R) \subset \hat{\mathscr{B}}_{scat}(\tilde{r}, \tilde{L}, R, r_1, \tilde{C}, \omega, \delta)$. Therefore, in this case we may set the constants r, L, R, k, R_1, and $\tilde{\rho}$, including N, as the *a priori data*.

We begin by investigating an intermediate case, namely, the one with $N-1$ scattering measurements.

Theorem 6.3.3. *Let $N = 2,3$. Fix $h > 0$. Let Σ, Σ' belong to $\mathscr{A}^h_{obst}$ and let d be defined as in* (6.2.16).

If $N = 2$, let us fix $\mathbf{v} \in \mathbb{S}^1$ and let $u^i(\mathbf{x}) = \mathrm{e}^{ik\mathbf{x}\cdot\mathbf{v}}$, $\mathbf{x} \in \mathbb{R}^2$. Let u be the solution to (6.2.24) *with boundary condition* (6.2.26) *and u' be the solution to the same problem with Σ replaced by Σ'. We let $\varepsilon > 0$ be such that*

$$\|u - u'\|_{L^\infty(B_{\tilde{\rho}}(\mathbf{x}_0))} \leq \varepsilon. \tag{6.3.9}$$

If $N = 3$, let us fix $\mathbf{v}_1$, $\mathbf{v}_2 \in \mathbb{S}^2$, with $|\mathbf{v}_1 \cdot \mathbf{v}_2| = b_0 < 1$. For any $j = 1,2$, let $u^i(\mathbf{x}) = \mathrm{e}^{ik\mathbf{x}\cdot\mathbf{v}_j}$, $\mathbf{x} \in \mathbb{R}^3$, and let u_j be the solution to (6.2.24) *with boundary condition* (6.2.26) *and u'_j be the solution to the same problem with Σ replaced by Σ'. We let $\varepsilon > 0$ be such that*

$$\max_{j=1,2} \|u_j - u'_j\|_{L^\infty(B_{\tilde{\rho}}(\mathbf{x}_0))} \leq \varepsilon. \tag{6.3.10}$$

There exists a constant $\hat{\varepsilon}_1(h)$, $0 < \hat{\varepsilon}_1(h) \leq 1/(2\mathrm{e})$, depending on the a priori data, on b_0 if $N = 3$, and on h only, such that if $\varepsilon \leq \hat{\varepsilon}_1(h)$, then for some positive constants A_1, depending on the a priori data only, and C, depending on the a priori data, on b_0 if $N = 3$, and on h only, we have

$$d \leq A_1(\eta(\varepsilon))^C. \tag{6.3.11}$$

The main difference with respect to the sound-hard case with N measurements or the sound-soft case is that here we do not have an explicit dependence of $\hat{\varepsilon}_1(h)$ from h, which in Theorems 6.3.1 and 6.3.2 is given by (6.3.6), and that the constant C depends, again in a rather implicit way, on h too.

We finally consider the case of a single scattering measurement. We restrict here to $N = 3$, since $N = 2$ is clearly covered by the previous theorem.

Theorem 6.3.4. *Let $N = 3$. Fix $h > 0$. Let Σ, Σ' belong to $\mathscr{A}^h_{obst}$ and let d be defined as in* (6.2.16). *Let us fix $\mathbf{v} \in \mathbb{S}^2$ and let $u^i(\mathbf{x}) = \mathrm{e}^{ik\mathbf{x}\cdot\mathbf{v}}$, $\mathbf{x} \in \mathbb{R}^3$. Let u be the solution to* (6.2.24) *with boundary condition* (6.2.26) *and u' be the solution to the same problem with Σ replaced by Σ'.*

There exists a constant $\hat{\varepsilon}_2(h)$, $0 < \hat{\varepsilon}_2(h) \leq \hat{\varepsilon}_1(h) \leq 1/(2\mathrm{e})$, depending on the a priori data and on h only, such that if

$$\|u - u'\|_{L^\infty(B_{\tilde{\rho}}(\mathbf{x}_0))} \leq \varepsilon, \tag{6.3.12}$$

for some $\varepsilon \leq \hat{\varepsilon}_2(h)$, then for some positive constants $A_2 \geq A_1$, depending on the a priori data only, and $C_1 \leq C$, depending on the a priori data and on h only, we have

$$d \leq A_2(\eta(\varepsilon))^{C_1}. \tag{6.3.13}$$

Remark 6.3.1. We finally notice that, by the arguments developed in the previous section, we can easily infer corresponding estimates of Theorems 6.3.1 and 6.3.2 and of Theorems 6.3.3 and 6.3.4, if we replace d with $\tilde{d} = d_H(\Sigma, \Sigma')$ or $\hat{d} = d_H(\partial\Sigma, \partial\Sigma')$ or the near-field error with limited aperture ε with a near-field error ε_1 or a far-field error ε_0 on the corresponding solutions. In the first case, it is just enough to use (6.2.21), with δ^{-1} defined as in (6.2.20) and $C_1 > 0$ depending on the a priori data only, for the first two theorems, and to use (6.2.43), with $C_1 > 0$ and C_2 depending on the a priori data only, for the second two theorems. For the second case, by (6.2.42), we have exactly the same results if we replace ε with the near-field error ε_1. If we use the far-field error ε_0 instead, then we need to replace ε with $\eta_1(\varepsilon_0)$, η_1 as in (6.2.42), noting that in this case we can choose $\tilde{\rho}$ and R_1 as depending on the other a priori data.

6.4 The General Geometric Construction

In this section we assume that the assumptions of Theorem 6.3.1 are satisfied. The a priori data will be the one defined at the beginning of Sect. 6.3.

Moreover, for the whole section we shall fix $j \in \{1,\dots,N\}$ and we shall consider the solutions with respect to the incident direction of propagation $\mathbf{v} = \mathbf{v}_j$; therefore, the subscript j will be always dropped.

We call H the connected component of $G \cap G'$, where $G' = \mathbb{R}^N \backslash \Sigma'$, such that $\mathbb{R}^N \backslash \overline{B_R}$ is contained in H. We shall also use the following definition.

Definition 6.4.1. A sequence of balls $B_{\rho_i}(\mathbf{z}_i)$, $i = 0,\dots,n$, forms a *regular chain*, with respect to an open set G, with constants $0 < a_1 < a_2 < a_3 < 1 < a_4$ if the following properties are satisfied

(i) For any $i = 0,1,\dots,n$, $B_{a_4\rho_i}(\mathbf{z}_i) \subset G$.
(ii) For any $i = 1,\dots,n$, we have $\rho_i \leq \rho_{i-1}$ and $B_{a_1\rho_i}(\mathbf{z}_i) \subset B_{a_2\rho_{i-1}}(\mathbf{z}_{i-1})$. And, for any $i = 0,\dots,n-1$, we have $B_{a_1\rho_i}(\mathbf{z}_i) \subset B_{a_3\rho_{i+1}}(\mathbf{z}_{i+1})$.

We have the following lemmas with simple proofs that we leave to the reader.

Lemma 6.4.1. *Let U_1 and $\tilde{U}_1$ be two open sets and let $T : U_1 \to \tilde{U}_1$ be a bi-$W^{1,\infty}$ mapping with constant $\tilde{C}$.*

Let $B_{\tilde{\rho}_i}(\tilde{\mathbf{z}}_i)$, $i = 0,\dots,n$, be a regular chain with respect to $\tilde{U}_1$ with constants $0 < \tilde{a}_1 < \tilde{a}_2 < \tilde{a}_3 < 1 < \tilde{a}_4$. Then, if we call $\mathbf{z}_i = T^{-1}(\tilde{\mathbf{z}}_i)$, $\rho_i = \tilde{\rho}_i/\tilde{C}$, $i = 0,\dots,n$, and $a_1 = \tilde{a}_1$, $a_4 = \tilde{a}_4$, we have that $B_{\rho_i}(\mathbf{z}_i)$, $i = 0,\dots,n$ is a regular chain with respect to U_1 with constants $0 < a_1 < a_2 < a_3 < 1 < a_4$ provided

$$a_1 < \tilde{C}^2\tilde{a}_2 \leq a_2 < a_3 < 1 \quad \text{and} \quad a_1 < \tilde{C}^2\tilde{a}_3 \leq a_3 < 1.$$

Lemma 6.4.2. *Let $\mathscr{C}$ be an open cone with amplitude θ, $0 < \theta < \pi/2$, and radius r. For simplicity we assume that its vertex is in the origin and that its bisecting vector is $\mathbf{e}_N$. We set $0 < a_1 < a_2 < a_3 < 1 < a_4$ and we call $0 < c_1 = \sin(\theta)/a_4 < 1$. Given c_2, $0 < c_2 \leq c_1$, we fix $\mathbf{z}_0 = (r/2)\mathbf{e}_N$ and ρ_0, $c_2(r/2) \leq \rho_0 \leq c_1(r/2)$.*

We can construct a regular chain $B_{\rho_i}(\mathbf{z}_i)$, $i = 0,\dots,n$, with respect to $\mathscr{C}$ with constants $0 < a_1 < a_2 < a_3 < 1 < a_4$, in the following way. For any $i = 0,\dots,n$, we can choose $\mathbf{z}_n = b^n(r/2)\mathbf{e}_N$ and $\rho_n = b^n\rho_0$ provided the constant b satisfies

$$0 < \max\left\{\frac{1-a_2c_2}{1-a_1c_2}, \frac{1+a_1c_2}{1+a_3c_2}\right\} \leq b < 1. \tag{6.4.1}$$

We now proceed to describe the geometric construction needed for the proof of Theorem 6.3.1 and of the other stability results as well. We divide the construction into several steps, proving alongside their corresponding estimates. Without loss of generality, up to swapping Σ with Σ', we can find $\mathbf{x}_1 \in \partial\Sigma' \backslash \Sigma$ such that $d = \operatorname{dist}(\mathbf{x}_1, \partial\Sigma) = \operatorname{dist}(\mathbf{x}_1, \Sigma)$.

6.4.1 Step I: From $\mathbf{x}_0$ to $\mathbf{x}_1$

We construct a sequence of balls $B_{\rho_i}(\mathbf{z}_i)$, $i = \dots, -n, -(n-1), \dots, 0, \dots, n_0$ forming a regular chain with respect to G, with constants $0 < a_1 < a_2 < a_3 < 1 < a_4 = 8$ and ρ_0 depending on the a priori data only and such that the following conditions are satisfied.

First, $\mathbf{z}_0 = \mathbf{x}_0$ and $\mathbf{z}_{n_0} = \mathbf{x}_1$. Second, ρ_0 is a positive constant, depending on the a priori data only, such that $16\rho_0 \leq \min\{\tilde{\rho}, \tilde{\rho}_0, r_1/\tilde{C}\}$, where $\tilde{\rho}_0$ is the positive constant depending on k only that bounds the radius of balls where the three-sphere inequality of Lemma 6.2.4 holds. On the other hand, $\rho_{n_0} = s_0 d$, where s_0 is a positive constant depending on the a priori data only. Third, for any $n = 1,2,\dots$, we pick $\mathbf{z}_{-n} = \mathbf{x}_0 + n(\rho_0/4)(\mathbf{x}_0/\|\mathbf{x}_0\|)$ and $\rho_{-n} = \rho_0$. Finally, n_0 is bounded by a constant, depending on the a priori data only, times $\log(2eR/d)$.

The sequence is constructed as follows. Let $\mathbf{y}_1$ be a point of $\partial\Sigma$ such that $|\mathbf{x}_1 - \mathbf{y}_1| = d$. We recall that $B_d(\mathbf{x}_1) \subset G$.

If $d \geq r_1/3$, then we use the exterior connectedness property of Σ and may easily construct such a chain keeping the radius $\rho_n = \rho_0$ for any $n \leq n_0$, which is simply constant and depending on the a priori data only. In this case we easily infer that n_0 is bounded by a constant depending on the a priori data only as well.

If instead $d \leq r_1/3$, we proceed in the following way. Let U be the connected component of $G \cap B_{r_1}(\mathbf{y}_1)$ containing $B_d(\mathbf{x}_1)$. In particular we have $\mathbf{y}_1 \in \partial U$. By Condition ii) of Definition 6.2.2 applied to $\mathbf{y}_1$, we have the transformation $T : Q \to U'$ and we consider the point $\tilde{\mathbf{x}}_1 = T^{-1}(\mathbf{x}_1)$. We call $\tilde{\mathbf{y}}_2$ the point in Q such that $\tilde{\mathbf{y}}_2 = (\tilde{\mathbf{x}}_1', 3/4)$ and $\mathbf{y}_2 = T(\tilde{\mathbf{y}}_2)$. We have that $B_{d/\tilde{C}}(\tilde{\mathbf{x}}_1) \subset Q$ and that $\mathrm{dist}(\tilde{\mathbf{x}}_1, \partial Q \backslash \Gamma) \geq \omega(2r_1/3)$. In particular, $\|\tilde{\mathbf{x}}_1'\| \leq 1 - \omega(2r_1/3)$. We conclude that $\mathrm{dist}(\mathbf{y}_2, \Sigma)$ is greater than or equal to a constant depending on the a priori data only. By the exterior connectedness property of Σ, we construct such a chain first from $\mathbf{x}_0$ to $\mathbf{y}_2$, keeping the radius constant and depending on the a priori data only. We notice that this part requires a number of balls that may be bounded by a constant depending on the a priori data only. In order to proceed from $\mathbf{y}_2$ to $\mathbf{x}_1$, we use Lemma 6.4.2 to construct a regular chain in Q, with suitable constants, connecting $\tilde{\mathbf{y}}_2$ to $\tilde{\mathbf{x}}_1$. Then our chain in G is obtained by using Lemma 6.4.1 and we easily infer that the number of elements of such a chain may be bounded by a constant, depending on the a priori data only, times $\log(2\mathrm{e}R/d)$; therefore, the claim is proved.

Let us finally notice that here the geometric construction is different from that of [21]. It is actually more general and more complicated and allows us to consider a wider class of admissible scatterers.

Starting from $\mathbf{z}_0 = \mathbf{x}_0$, we take $j_1 \in \{1, \ldots, n_0\}$ such that, for any $i = 0, 1, \ldots, j-1$, $B_{\rho_i}(\mathbf{z}_i) \subset H$ and $B_{\rho_{j_1}}(\mathbf{z}_{j_1}) \cap \Sigma' \neq \emptyset$. We apply the three-sphere inequality of Lemma 6.2.4 as follows. For any $i = 0, 1, \ldots, j-1$,

$$\|u - u'\|_{L^\infty(B_{a_1\rho_{i+1}}(\mathbf{z}_{i+1}))} \leq \|u - u'\|_{L^\infty(B_{a_2\rho_i}(\mathbf{z}_i))} \leq C\|u - u'\|^{1-\beta_i}_{L^\infty(B_{\rho_i}(\mathbf{z}_i))}\|u - u'\|^{\beta_i}_{L^\infty(B_{a_1\rho_i}(\mathbf{z}_i))},$$

where any β_i, $i = 0, \ldots, j-1$, satisfies

$$0 < a \leq \beta_i \leq b < 1,$$

with a and b depending on k only.

If β_i, $i = 0, 1, 2, \ldots$, are positive constants, we shall use the following notation for any $j = 0, 1, 2, \ldots$

$$\mathscr{B}_j = \sum_{r=0}^{j}\prod_{i=r}^{j}\beta_i, \quad \Gamma_j = \prod_{i=0}^{j}\beta_i.$$

Recalling that $\|u - u'\|_{L^\infty(\mathbb{R}^3)} \leq E$ and that $\|u - u'\|_{L^\infty(B_{\tilde{\rho}}(\mathbf{x}_0))} \leq \varepsilon$ and by iterating the previous estimate, we obtain

$$\|u - u'\|_{L^\infty(B_{a_1\rho_{j_1}}(\mathbf{z}_{j_1}))} \leq C^{1+\mathscr{B}_{j_1-1}} E^{1-\Gamma_{j_1-1}} \varepsilon^{\Gamma_{j_1-1}}. \tag{6.4.2}$$

6.4.2 Step II: Towards the Cell and Back

We call $\hat{h} = \min\{d, h\}$. Following the notation of the previous step, we set $\mathbf{z} = \mathbf{z}_{j_1}$ and $\rho = \rho_{j_1}$. Then, $B_{a_1\rho}(\mathbf{z}) \subset H$, $B_{a_4\rho}(\mathbf{z}) \subset G$, with $a_4 = 8$, and there exists $\mathbf{w} \in \partial\Sigma'$ such that $\|\mathbf{z} - \mathbf{w}\| < \rho$ and $B_{\|\mathbf{z}-\mathbf{w}\|}(\mathbf{z}) \subset H$. Let U be the connected component of $G' \cap B_{r_1}(\mathbf{w})$ containing $\mathbf{z}$. Clearly, $\mathbf{w} \in \partial U$. Let $\mathscr{C}'$ be one of the cells of $\partial\Sigma'$ such that $\mathbf{w} \in \mathscr{C}'$ and $\mathscr{C}' \cap B_{r_1/2}(\mathbf{w}) \subset \partial U$.

We call Π' the plane containing $\mathscr{C}'$ and, up to a rigid change of coordinates, without loss of generality, we assume $\Pi' = \{\mathbf{y} \in \mathbb{R}^N : y_N = 0\}$. By the properties of $\mathscr{C}'$, $\mathscr{C}'$ satisfies a uniform cone property, namely, there exists $\omega_1 \in \mathbb{S}^{N-1} \cap \Pi$ and constants c_1, $0 < c_1 \leq 1$, and θ, $0 < \theta < \pi/2$, depending on L and R only, such that $\mathscr{C}(\mathbf{w}, \omega_1, c_1\hat{h}, \theta) \cap \Pi' \subset \mathscr{C}' \cap B_{r_1/2}(\mathbf{w}) \subset \partial U$.

By looking at the points on the bisecting line of $\mathscr{C}(\mathbf{w}, \omega_1, c_1\hat{h}, \theta)$, we may find $\mathbf{w}_1$ on this line, that is, $\mathbf{w}_1 = \mathbf{w} + s_1\hat{h}\omega_1$, such that $B_{s_2\hat{h}}(\mathbf{w}_1) \subset B_{r_1/2}(\mathbf{w}) \cap B_{7\rho/(4\tilde{C}^2)}(\mathbf{w})$, and $B_{s_2\hat{h}}(\mathbf{w}_1) \cap \Pi' \subset \mathscr{C}' \cap B_{r_1/2}(\mathbf{w}) \subset \partial U$, for some positive s_1 and s_2 depending on the a priori data only.

We claim that there exists s_3, $0 < s_3 \leq s_2$, depending on the a priori data only, such that, up to changing the orientation of $\mathbf{e}_N$, we have $B^+_{s_3\hat{h}}(\mathbf{w}_1) \subset U$.

The proof of this claim is the following. We apply Condition ii) of Definition 6.2.2 to $\mathbf{w}$, and we consider the corresponding transformation $T : Q \to U'$. For some $\varepsilon > 0$, possibly taking s_2 slightly smaller but still depending on the a priori data only, we have that $(\overline{B_{s_2\hat{h}}(\mathbf{w}_1) \cap \Pi'}) \times (0, \varepsilon] \subset U \cap B_{r_1/2}(\mathbf{w})$. The function T^{-1} restricted to such a set is bi-Lipschitz onto $A \subset Q$. We observe that A has a positive distance, depending on r_1 and ω only, from $\partial Q \backslash \Gamma$.

It is not difficult to show that T^{-1} can be extended to a function $\tilde{T}^{-1}$ in such a way that $\tilde{T}^{-1}$: $(\overline{B_{s_2\hat{h}}(\mathbf{w}_1) \cap \Pi'}) \times [0, \varepsilon] \to \overline{A}$ is still bijective, with inverse T, and thus bi-Lipschitz. We conclude that also $\tilde{T}^{-1} : (\overline{B_{s_2\hat{h}}(\mathbf{w}_1) \cap \Pi'}) \to \overline{A} \cap \Gamma$ is bijective, with inverse T, and thus bi-Lipschitz. Then let us fix a, $0 < a \leq s_2\hat{h}$ such that $B^+_a(\mathbf{w}_1) \subset U \cap B_{r_1/2}(\mathbf{w}) \subset U'$. If a can be chosen to be $s_2\hat{h}$, the claim is proved; otherwise, we assume that there exists $\mathbf{y} \in \partial U'$ such that $\mathbf{y} \in B^+_{s_2\hat{h}}(\mathbf{w}_1)$ and $\|\mathbf{y} - \mathbf{w}_1\| = a$. Actually, $\mathbf{y} \in \partial U \cap B_{r_1/2}(\mathbf{w})$. Then we call $\tilde{\mathbf{w}}_1 = \tilde{T}^{-1}(\mathbf{w}_1)$; thus, $T(\tilde{\mathbf{w}}_1) = \mathbf{w}_1$, and, by a similar reasoning, extending T^{-1} up to the closure of $B^+_a(\mathbf{w}_1)$, we can find $\tilde{\mathbf{y}} \in \Gamma$ such that $T(\tilde{\mathbf{y}}) = \mathbf{y}$ and $\|\tilde{\mathbf{y}} - \tilde{\mathbf{w}}_1\| \leq \tilde{C}a$. Moreover, we also have that $\tilde{\mathbf{y}} \notin \overline{A} \cap \Gamma$. But $B_{s_2\hat{h}/\tilde{C}}(\tilde{\mathbf{w}}_1) \cap \Gamma \subset \overline{A} \cap \Gamma$; therefore, $\tilde{C}a$ must be greater than $s_2\hat{h}/\tilde{C}$, that is, $a > s_2\hat{h}/\tilde{C}^2$ and the claim is proved.

Then we notice that, through an even reflection, we can extend u' on $B_{s_3\hat{h}}(\mathbf{w}_1)$ by setting $u'(\mathbf{y}) = u'(T_{\Pi'}(\mathbf{y}))$ for any $\mathbf{y} \in B^-_{s_3\hat{h}}(\mathbf{w}_1)$. In this way u' solves the Helmholtz equation on the whole $B_{s_3\hat{h}}(\mathbf{w}_1)$. We notice that $B_{s_3\hat{h}}(\mathbf{w}_1) \subset G$; therefore, on $H \cup B_{s_3\hat{h}}(\mathbf{w}_1)$ both u and u' are well defined and solve the Helmholtz equation.

We construct a regular chain, with respect to $H \cup B_{s_3\hat{h}}(\mathbf{w}_1)$ and with constants depending on the a priori data only, satisfying the following properties. The first ball is centered in $\mathbf{z}$ and it has radius less than or equal to ρ, whereas the last ball is $B_{s_5\hat{h}}(\mathbf{w}_1)$ with s_5 depending on the a priori data only. Finally, the number of balls of such a chain is bounded by a constant depending on the a priori data only times $\log(2eR/\hat{h})$.

The argument is the following. By a reasoning similar to the one used before, we can find $\tilde{\mathbf{w}} \in \Gamma$ such that $T(\tilde{\mathbf{w}}) = \mathbf{w}$ and $\|\tilde{\mathbf{w}}_1 - \tilde{\mathbf{w}}\| \leq \tilde{C}s_1\hat{h}$. Since $B_{7\rho}(\mathbf{w}) \cap U \subset H$, we have $T(B_{7\rho/\tilde{C}}(\tilde{\mathbf{w}}) \cap Q) \subset H$. Since $s_1 \geq s_3$, we have $T^{-1}(B^+_{s_3\hat{h}}(\mathbf{w}_1)) \subset B_{2\tilde{C}s_1\hat{h}}(\tilde{\mathbf{w}})$. Without loss of generality, we require that $2\tilde{C}s_1\hat{h} \leq 7\rho/(2\tilde{C})$, that is, $T^{-1}(B^+_{s_3\hat{h}}(\mathbf{w}_1)) \subset B_{7\rho/(2\tilde{C})}(\tilde{\mathbf{w}}) \cap Q$.

Then we perform the following construction. Take $\mathbf{w}_0$ on the segment connecting $\mathbf{z}$ to $\mathbf{w}$ such that $\|\mathbf{w}_0 - \mathbf{w}\| = 7\rho/(4\tilde{C}^2)$, if this number is less than $\|\mathbf{z} - \mathbf{w}\|$. Otherwise, we pick $\mathbf{w}_0 = \mathbf{z}$. We observe that $T^{-1}(\mathbf{w}_0) \in B_{7\rho/(2\tilde{C})}(\tilde{\mathbf{w}}) \cap Q$.

We construct a regular chain of balls, with a number of balls bounded by a constant depending on the a priori data only, contained in $B_{\|\mathbf{z}-\mathbf{w}\|}(\mathbf{z}) \subset H$ and connecting $\mathbf{z}$ to $\mathbf{w}_0$. We consider $\mathbf{w}_2 = \mathbf{w}_1 + s_4\hat{h}\mathbf{e}_N$ so that $B_{s_5\hat{h}}(\mathbf{w}_2) \subset B^+_{s_3\hat{h}}(\mathbf{w}_1)$. Then with a construction similar to one described before during Step I, which exploits the properties of the change of variables T, we can extend our regular chain, which is still contained in H, from $\mathbf{w}_0$ till we connect to $\mathbf{w}_2$. The number of steps required at this stage is of the order of a constant times $\log(2eR/\hat{h})$. Then we move along the segment connecting $\mathbf{w}_2$ to $\mathbf{w}_1$ and, with a finite number of steps depending only on the a priori data, we are able to reach $\mathbf{w}_1$ and thus conclude the construction.

We notice again that, as in the first Step I, the construction developed here is much more general and much more involved than that used in [21]. Overcoming this technical difficulty is the key ingredient to obtain our results for a more general class of admissible scatterers than the one used in [21].

Again by a repeated use of the three-sphere inequality and by recalling (6.4.2), we obtain that

$$\|u-u'\|_{L^\infty(B_{a_1 s_5 \hat{h}}(\mathbf{w}_1))} \leq C^{1+\mathscr{B}_{n-1}} E^{1-\Gamma_{n-1}} \varepsilon^{\Gamma_{n-1}}, \tag{6.4.3}$$

where, for some constants $\tilde{C}_1$ and a, b, with $0<a<b<1$, depending on the a priori data only, we have

$$n \leq \tilde{C}_1 \left(\log\left(\frac{2eR}{\hat{h}}\right) + \log\left(\frac{2eR}{d}\right) \right) \quad \text{and} \quad a \leq \beta_i \leq b \text{ for any } i=0,\dots,n-1. \tag{6.4.4}$$

We then apply a reflection argument. We call $\Pi_1 = \Pi'$ the hyperplane containing the cell $\mathscr{C}'$. Moreover, ν_1 will be the unit normal to Π_1 and $T_1 = T_{\Pi_1}$ is the reflection in Π_1. We define Σ_1 as the reflection of Σ with respect to the plane Π_1, $G_1 = \mathbb{R}^3 \backslash \Sigma_1$, and u_1 as the even reflection of u with respect to the same plane Π_1, namely, for any $\mathbf{x} \in \mathbb{R}^N$, we set $u_1(\mathbf{x}) = u(T_{\Pi_1}(\mathbf{x}))$. Without loss of generality and since $a_4 = 8$, we can assume that $B_{s_5\hat{h}}(\mathbf{w}_1) \subset B_{2\rho}(\mathbf{z}) \subset B_{8\rho}(\mathbf{z}) \subset G$; therefore, $B_\rho(\mathbf{z}) \subset B_{3\rho}(\mathbf{w}_1) \subset B_{4\rho}(\mathbf{w}_1) \subset G \cap G_1$. Both u and u_1 satisfy the Helmholtz equation on $B_{4\rho}(\mathbf{w}_1)$. Notice that $\nabla u' \cdot \nu_1 = 0$ on $\Pi_1 \cap B_{s_5\hat{h}}(\mathbf{w}_1)$; Therefore, $u' = u' \circ T_{\Pi_1}$ and

$$u - u_1 = u - u' + u' - u_1 = (u-u') - (u-u') \circ T_{\Pi_1}.$$

We can conclude, using (6.4.3), that

$$\|u-u_1\|_{L^\infty(B_{a_1 s_5 \hat{h}}(\mathbf{w}_1))} \leq 2\|u-u'\|_{L^\infty(B_{a_1 s_5 \hat{h}}(\mathbf{w}_1))} \leq 2C^{1+\mathscr{B}_{n-1}} E^{1-\Gamma_{n-1}} \varepsilon^{\Gamma_{n-1}}. \tag{6.4.5}$$

Then by using the arguments of Step IV of Sect. 5 in [21], we obtain that

$$\|u-u_1\|_{L^\infty(B_{\rho_{j_1}}(\mathbf{z}_{j_1}))} \leq C^{1+\mathscr{B}_n} (2E)^{1-\Gamma_n} \varepsilon^{\Gamma_n}, \tag{6.4.6}$$

where $C \geq 1$ and $2E \geq 1$ are constants depending on the a priori data only, (6.4.4) is satisfied and β_n satisfies

$$c_1 \frac{\log(8/7)}{\log(c_2\rho_0/\hat{h})} \leq \beta_n \leq 1 - c_1 + c_1 \frac{\log(8/7)}{\log(c_2\rho_j/\hat{h})},$$

with c_1, c_2 depending on the a priori data only.

Finally, we call $\mathbf{w}_1$ the first reflection point and Π_1 the first reflection hyperplane.

6.4.3 Step III: Returning Back Towards $\mathbf{x}_0$ *and Infinity*

Let us begin by fixing a constant $R_2 \geq \max\{2R_1, 4R\}$, depending on the a priori data and on a_0 only, such that

$$E_1 R_2^{-(N-1)/2} \leq k a_0/2,$$

where E_1 is as in (6.2.32).

Let us now consider the regular chain of balls $B_{\rho_i}(\mathbf{z}_i)$, $i = \dots, -n, \dots, -1, 0, 1 \dots, j_1$, we have constructed in Step I. We have that $B_{\rho_{j_1}}(\mathbf{z}_{j_1})$ is contained in G_1. We proceed backwards along the chain, until we find j_2, $j_2 < j_1$, such that, for any $i = j_2+1, \dots, j_1$, we have $B_{\rho_i}(\mathbf{z}_i) \cap G_1 = \emptyset$, whereas $B_{\rho_{j_2}}(\mathbf{z}_{j_2}) \cap G_1 \neq \emptyset$. Then, we apply Step II to u, u_1, Σ, and Σ_1. We find a second reflection point $\mathbf{w}_2$ and a second reflection hyperplane Π_2, with unit normal ν_2. By reflection in such a hyperplane Π_2, from Σ we obtain Σ_2 and from u we obtain u_2. In a completely analogous way as in (6.4.6), we may estimate $\|u-u_2\|_{L^\infty(B_{\rho_{j_2}}(\mathbf{z}_{j_2}))}$.

We repeat this procedure as many times as needed, until we reach $\mathbf{z}_{-n_1}$, where n_1 is an integer bounded by a constant depending on the a priori data and on a_0 only, with $R_3 = \|\mathbf{z}_{-n_1}\| \geq 2R_2+2$. To fix a hyperplane Π, to be decided later, that is passing through a point belonging to $\overline{B_{R_2+1}}$ and a point $\tilde{\mathbf{z}} \in \partial B_{R_3} \cap \Pi$, by a regular chain of balls with constant radius ρ_0, we proceed from $\mathbf{z}_{-n_1}$ along the boundary of ∂B_{R_3} towards the point $\tilde{\mathbf{z}} \in \partial B_{R_3} \cap \Pi$.

Before reaching $\tilde{\mathbf{z}}$, we have done M reflection procedures as in Step II, where M is a positive integer bounded by n_0 plus a constant depending on the a priori data and on a_0 only.

We now distinguish between two cases. In the first, setting $\Sigma_0 = \Sigma'$ and $u_0 = u'$, we assume that there exists a reflection point $\mathbf{w}_n \in \partial\Sigma_{n-1}$, $1 \leq n \leq M$, as above with $\|\mathbf{w}_n\| \geq R_2+1$. Then we have, since $\nabla u_{n-1}(\mathbf{w}_n)\cdot \nu_n = 0$,

$$\nabla u(\mathbf{w}_n)\cdot \nu_n = \nabla(u-u_{n-1})(\mathbf{w}_n)\cdot \nu_n.$$

Otherwise, in the second case, we can assume that the last reflection point $\mathbf{w}_M$ is such that $\|\mathbf{w}_M\| \leq R_2+1$ and, without loss of generality, we may pick $\Pi = \Pi_M$. Then, since $u_M = \mathbf{u}\circ T_{\Pi_M}$, we have $\nabla u_M(\tilde{\mathbf{z}})\cdot \nu_M = -\nabla u(\tilde{\mathbf{z}})\cdot \nu_M$; hence,

$$2\nabla u(\tilde{\mathbf{z}})\cdot \nu_M = \nabla(u-u_M)(\tilde{\mathbf{z}})\cdot \nu_M.$$

In either cases, picking either $\mathbf{z} = \mathbf{w}_n$ or $\mathbf{z} = \tilde{\mathbf{z}}$, we can prove the following lemma; see [21, Sect. 5] for further details on the computations.

Lemma 6.4.3. *We can find a point* $\mathbf{z}$, $\|\mathbf{z}\| \geq R_2+1$ *and a unit vector* ν *such that*

$$h|\nabla u(\mathbf{z})\cdot \nu| \leq C_0\varepsilon_2, \tag{6.4.7}$$

where for some β_i, $i=0,\ldots,n$,

$$\varepsilon_2 \leq C^{1+\mathscr{B}_n}(2E)^{1-\Gamma_n}\varepsilon^{\Gamma_n}, \tag{6.4.8}$$

with $C \geq 1$, $2E \geq 1$, *and*

$$n \leq \hat{C}\log(2eR/d)(\log(2eR/d)+\log(2eR/\hat{h})).$$

Furthermore, there are at most $M \leq \hat{C}_1\log(2eR/d)$ *of these* β *such that*

$$\beta \geq c_1\frac{\log(8/7)}{\log(c_2\rho_0/\hat{h})},$$

and they are never consecutive ones, and all the others satisfy $0 < a \leq \beta \leq b < 1$. *Here* C_0, C, E, a, b, c_1, c_2, $\hat{C}$, *and* $\hat{C}_1$ *depend on the a priori data only.*

This lemma concludes the general geometric construction that is the basic step for proving our stability results.

6.5 Proofs of the Stability Results

In this section, using the geometric construction of the previous Sect. 6.4 as a starting point, we prove our stability results. For the N measurement case, the conclusion is straightforward, whereas if we consider less than N measurements, we need the following arguments.

6.5.1 The N Measurement Case

We conclude the proof of Theorem 6.3.1 and thus also of Theorem 6.3.2.

The Proof of Theorem 6.3.1. By Lemma 6.4.3 and using its notation, we have for any $j = 1, \dots, N$

$$\max_{j=1,\dots,N} |\nabla u_j(\mathbf{z}) \cdot \nu| \leq \frac{C_0}{h} \varepsilon_2.$$

Therefore, using Proposition 6.2.4 and our choice of R_2, for any $j = 1, \dots, N$, we have

$$k|\mathbf{v}_j \cdot \nu| - ka_0/2 \leq |\nabla u_j(\mathbf{z}) \cdot \nu| \leq \frac{C_0}{h} \varepsilon_2.$$

Therefore, choosing one of the available incident waves, we can infer that

$$ka_0/2 \leq \frac{C_0}{h} \varepsilon_2. \tag{6.5.1}$$

Then, by straightforward although lengthy computations (see [21, Sect. 5] for further details), the proof may be easily concluded. □

6.5.2 The $N-1$ Measurement Case

Here we assume that the hypotheses of Theorem 6.3.3 are satisfied. Since we would like to keep our argument as general as possible, let us assume for the time being that $N \geq 2$ and that we have fixed $N-1$ linearly independent directions $\mathbf{v}_1, \dots, \mathbf{v}_{N-1}$.

We need the following lemma.

Lemma 6.5.1. *There exists a constant $\tilde{a}_0 > 0$, depending on the a priori data only, such that for any direction $\mathbf{v}$ and any polyhedral $\Sigma \in \mathscr{A}_{obst}$, we can find a cell $\mathscr{C}$ in $\partial\Sigma$, with unit normal ν, such that $|\nu \cdot \mathbf{v}| \geq \tilde{a}_0$.*

Proof. Let us assume, by contradiction, that such a positive constant $\tilde{a}_0$ does not exist. Then we can find a sequence of polyhedral obstacles $\Sigma_n \in \mathscr{A}_{obst}$ and of directions $\mathbf{v}_n$, $n \in \mathbb{N}$, such that, for $\mathscr{H}^{N-1}$ almost any point $\mathbf{x}$ of $\partial\Sigma_n$, we have $|\nu(\mathbf{x}) \cdot \mathbf{v}_n| \leq 1/n$. Without loss of generality, we can assume that, as $n \to \infty$, Σ_n converges, in the Hausdorff distance, to $\Sigma \in \mathscr{A}_{obst}$ and that $\mathbf{v}_n \to \mathbf{v} \in \mathbb{S}^{N-1}$. We can conclude that for $\mathscr{H}^{N-1}$ almost any point $\mathbf{x}$ of $\partial\Sigma$, we have $|\nu(\mathbf{x}) \cdot \mathbf{v}| = 0$, which is impossible since Σ is an obstacle. □

We begin with the following interesting and not that difficult case. Let us consider the geometric construction of the previous section, in particular Lemma 6.4.3. If the point $\mathbf{z}$ defined there is a reflection point $\mathbf{w}_n$, then a single measurement would be enough to obtain a stability result. In fact the following result holds.

Proposition 6.5.1. *Let $N \geq 2$. Fix $h > 0$. Let Σ, Σ' belong to $\mathscr{A}^h_{obst}$ and let d be defined as in* (6.2.16). *Let us fix $\mathbf{v} \in \mathbb{S}^{N-1}$ and let $u^i(\mathbf{x}) = \mathrm{e}^{ik\mathbf{x}\cdot\mathbf{v}}$, $\mathbf{x} \in \mathbb{R}^N$. Let u be the solution to* (6.2.24) *with boundary condition* (6.2.26) *and u' be the solution to the same problem with Σ replaced by Σ'. Let us assume that*

$$\|u - u'\|_{L^\infty(B_{\tilde{\rho}}(\mathbf{x}_0))} \leq \varepsilon,$$

for some $\varepsilon \leq 1/(2\mathrm{e})$.

Let us further assume that the point $\mathbf{z}$ *defined in Lemma* 6.4.3 *is a reflection point. Then for some positive constant* C *depending on the a priori data only, and not on* h*, we have*

$$\min\{d,h\} \leq 2\mathrm{e}R(\eta(\varepsilon))^C.$$

Proof. The main idea of the proof is the following. Let $\mathbf{z} = \mathbf{w}_n \in \Sigma_{n-1}$ be the reflection point such that $\|\mathbf{z}\| \geq R_2 + 1$. We consider σ the connected component of Σ_{n-1} containing $\mathbf{z}$ and we find, using Lemma 6.5.1, a point $\tilde{\mathbf{z}} \in \partial\sigma$ such that $|\nu(\tilde{\mathbf{z}}) \cdot \mathbf{v}| \geq \tilde{a}_0$.

Clearly, σ is far away from Σ; therefore, we are able to modify our regular chain by moving around $\partial\sigma$ till we get close to the point $\tilde{\mathbf{z}} \in \partial\sigma$ and we can use such a point $\tilde{\mathbf{z}}$ as a reflection point. Therefore, the proof follows as in the proof of Theorem 6.3.1, simply by replacing a_0 by $\tilde{a}_0$. □

The difficult part arises when the assumptions of Proposition 6.5.1 are not met, namely, when in the geometric construction of the previous section we have M reflection points, all of them contained in B_{R_2+1}.

In the sequel, without loss of generality, we assume that in our geometric construction we have M reflection points, all of them contained in B_{R_2+1}. Using the construction of the previous section, we can reach with a regular chain any point $\tilde{\mathbf{z}} \in (\overline{B_{2R_2+3} \backslash B_{2R_2+2}}) \cap \Pi_M$. Therefore, for any $j = 1, \ldots, N-1$, we have

$$A_j = \max_{\tilde{\mathbf{z}} \in (\overline{B_{2R_2+3} \backslash B_{2R_2+2}}) \cap \Pi_M} |\nabla u_j(\tilde{\mathbf{z}}) \cdot \nu| \leq C_0 \varepsilon_2, \tag{6.5.2}$$

where $\nu = \nu_M$ is the unit normal to Π_M and C_0 and ε_2 satisfy the same properties as those in Lemma 6.4.3.

Let us illustrate what is the difficult point. In order to obtain our stability result, we need to match the upper bound in (6.5.2) with a corresponding lower bound. Let us begin with the following remark. Let $\mathbf{v} = \mathbf{v}_j$, $j \in \{1, \ldots, N-1\}$, be one of the incident directions of propagation, and for the time being, let us drop the subscript j from our solutions. Let us call

$$A = \max_{\tilde{\mathbf{z}} \in (\overline{B_{2R_2+3} \backslash B_{2R_2+2}}) \cap \Pi_M} |\nabla \mathbf{u}(\tilde{\mathbf{z}}) \cdot \nu|.$$

Can A be equal to 0? Indeed this can happen, although only in certain circumstances. Namely, we claim that $A = 0$ if and only if $\mathbf{v} \cdot \nu = 0$ and Σ is symmetric with respect to the hyperplane Π_M. One direction is obvious; let us show the more interesting one, that is, $A = 0$ implies that $\mathbf{v} \cdot \nu = 0$ and Σ is symmetric with respect to the hyperplane Π_M.

In fact, if $A = 0$, then $|\nabla u \cdot \nu| \equiv 0$ on $(\overline{B_{2R_2+3} \backslash B_{2R_2+2}}) \cap \Pi_M$ and, by unique continuation, we actually have that $|\nabla u \cdot \nu| \equiv 0$ on $(\mathbb{R}^N \backslash \overline{B_R}) \cap \Pi_M$. By the decay properties at infinity of ∇u^s, this may hold only if $\mathbf{v} \cdot \nu = 0$. Moreover, we can easily infer that u is even symmetric with respect to Π_M. Let us call $\tilde{\Sigma}$ the complement of the unbounded connected component of $\mathbb{R}^N \backslash (\Sigma \cup T_{\Pi_M}(\Sigma))$. We have that $\tilde{\Sigma}$ is a polyhedral obstacle which is symmetric with respect to Π_M and that u solves (6.2.24) with boundary condition (6.2.26) also with Σ replaced by $\tilde{\Sigma}$. By the uniqueness result for sound-hard polyhedral obstacles with a single measurement, [6, 7], we immediately infer that $\Sigma = \tilde{\Sigma}$; thus, Σ itself is symmetric with respect to Π_M.

Therefore, in order to bound A away from 0, we need to guarantee either that Σ is not symmetric with respect to a hyperplane whose normal is orthogonal to $\mathbf{v}$, or, if Σ is actually symmetric with respect to a hyperplane whose normal is orthogonal to $\mathbf{v}$, that Π_M is different from such a hyperplane. As we shall see, if we use $N-1$ measurements, instead, in order to bound $\max_{j=1,\ldots,N-1} A_j$ away from zero, we need to guarantee either that Σ is not symmetric with respect to a hyperplane whose normal is orthogonal to any $\mathbf{v}_j$, $j = 1, \ldots, N-1$, or, if Σ is actually symmetric with respect to such a hyperplane, that Π_M is different from it.

Here lies the main difference between the $N-1$ measurements and 1 measurement case. In fact, in the $N-1$ measurement case, for any obstacle Σ, there is at most one hyperplane whose normal is orthogonal to any $\mathbf{v}_j$, $j=1,\dots,N-1$, with respect to which Σ is symmetric. On the contrary, with only one measurement, for any obstacle Σ, there might be many hyperplanes whose normal is orthogonal to $\mathbf{v}$ with respect to which Σ is symmetric. This is the main reason why the $N-1$ measurement case is relatively simpler and the corresponding result is somewhat stronger.

We then first consider the $N-1$ measurement case, leaving the 1 measurement case to the next subsection. Another difficulty is that we not only need to bound $\max_{j=1,\dots,N-1} A_j$ away from zero but that we require a suitable quantitative estimate of $\max_{j=1,\dots,N-1} A_j$ from below.

Let us begin with the following definitions. Let us call $\tilde{\Pi}=\mathrm{span}\{\mathbf{v}_1,\dots,\mathbf{v}_{N-1}\}$. For any $\Sigma\in\mathscr{A}_{obst}$, we denote with $P(\Sigma)$ its center of mass and $\tilde{\Pi}(\Sigma)=\tilde{\Pi}+P(\Sigma)$. We define $\mathscr{A}_{sym}$ the set of $\Sigma\in\mathscr{A}_{obst}$ such that Σ is symmetric with respect to $\tilde{\Pi}(\Sigma)$.

We then define the metric space

$$X=\{\Pi:\ \Pi \text{ is a hyperplane in } \mathbb{R}^N \text{ passing through } \overline{B_{R_2+1}}\},$$

with the distance

$$d(\Pi_1,\Pi_2)=d_H(\Pi_1\cap\overline{B_{2R_2+1}},\Pi_2\cap\overline{B_{2R_2+1}})\quad\text{for any } \Pi_1,\Pi_2\in X.$$

Finally, we call $\mathscr{X}=\mathscr{A}_{obst}\times X$, with the standard metric of the product of two metric spaces, and $\mathscr{Y}=\{(\Sigma,\tilde{\Pi}(\Sigma)):\ \Sigma\in\mathscr{A}_{sym}\}\subset\mathscr{X}$.

We have the following preliminary properties.

Proposition 6.5.2. *We have that $\Sigma\to P(\Sigma)$ is a Lipschitz continuous function on $\mathscr{A}_{obst}$ endowed with the Hausdorff distance, with a Lipschitz constant depending on the a priori data only. Consequently, $\mathscr{A}_{sym}$ is a closed subset of $\mathscr{A}_{obst}$ and $\mathscr{Y}$ is closed in $\mathscr{X}$.*

Proof. This is a straightforward consequence of (6.2.22), (6.2.23), and (6.2.43). □

Lemma 6.5.2. *Let $\Sigma\in\mathscr{A}_{sym}$. For simplicity, let us assume that $\tilde{\Pi}(\Sigma)=\{y_N=0\}$. Then we call $G^{\pm}=\{\mathbf{y}\in G:\ y_N\gtrless 0\}$ and we have that $G^{\pm}$ are Lipschitz domains with constants depending on the a priori data only.*

Proof. The difficult part of the proof is to consider the points $\mathbf{z}$ of $\partial G^{\pm}$ such that $\mathbf{z}\in\partial\Sigma\cap\tilde{\Pi}(\Sigma)$.

Let us consider a point $\mathbf{z}\in\partial\Sigma\cap\tilde{\Pi}(\Sigma)$. By the Lipschitz properties of Σ, we have that there exists a given cone $\mathscr{C}$, with vertex in $\mathbf{0}$, such that, for any $\mathbf{y}\in\partial\Sigma\cap B_{r/2}(\mathbf{z})$, $\mathbf{y}+\mathscr{C}\subset G$. Since Σ is symmetric with respect to $\tilde{\Pi}(\Sigma)$, we also have that $\mathbf{y}+\tilde{T}(\mathscr{C})\subset G$, $\tilde{T}$ being the reflection in $\tilde{\Pi}(\Sigma)$. Hence, it is not difficult to show that there exists a cone $\mathscr{C}_1$, with vertex in $\mathbf{0}$ and symmetric with respect to $\tilde{\Pi}(\Sigma)$, such that, for any $\mathbf{y}\in\partial\Sigma\cap B_{\tilde{r}}(\mathbf{z})$, $\mathbf{y}+\mathscr{C}_1\subset G$. We notice that $\tilde{r}>0$ and the amplitude of the cone $\mathscr{C}_1$ depends on r and L only.

Therefore, for any point $\mathbf{z}\in\partial\Sigma\cap\tilde{\Pi}(\Sigma)$, locally in $B_{\tilde{r}_1}(\mathbf{z})$, $\partial\Sigma$ is the graph of a Lipschitz function, with Lipschitz constant bounded by $\tilde{L}_1$, with respect to a Cartesian coordinate system such that $\mathbf{e}_N\in\tilde{\Pi}$. Hence, it is not difficult to show that, locally in $B_{\tilde{r}_2}(\mathbf{z})$ and with respect to a different Cartesian coordinate system, ∂G^{+}, and by symmetry G^{-} as well, is the graph of a Lipschitz function, with Lipschitz constant bounded by $\tilde{L}_2$. Here $\tilde{r}_i$ and $\tilde{L}_i$, $i=1,2$, are positive constants depending on r and L only.

The proof now can be easily concluded. □

In order to obtain the required lower bound on $\max_{j=1,\dots,N-1} A_j$, we distinguish between two cases. The good one is when either Σ is not close to $\mathscr{A}_{sym}$ or, if it is, the hyperplane Π_M is not close to $\tilde{\Pi}(\Sigma)$. The bad one is when Σ is close to $\mathscr{A}_{sym}$ and the hyperplane Π_M is close to $\tilde{\Pi}(\Sigma)$.

In the next proposition we deal with the good case; in the sequel of the proof we shall show that, by a suitable modification of our geometric construction, the bad case actually never occurs.

Let us consider the map

$$\mathscr{X} \ni (\Sigma,\Pi) \mapsto f(\Sigma,\Pi) = \max_{j=1,\ldots,N-1} \left(\max_{\tilde{\mathbf{z}} \in (\overline{B_{2R_2+3}} \setminus B_{2R_2+2}) \cap \Pi} |\nabla \mathbf{u}_j(\tilde{\mathbf{z}}) \cdot \nu| \right),$$

where $\nu = \nu_\Pi$ is the normal to Π and, for any $j = 1,\ldots,N-1$, u_j is the solution to the direct scattering problem (6.2.24) with boundary condition (6.2.26) and incident field $u^i(\mathbf{x}) = \mathrm{e}^{\mathrm{i}k\mathbf{x}\cdot\mathbf{v}_j}$, $\mathbf{x} \in \mathbb{R}^N$. Then the following result holds.

Proposition 6.5.3. *Let us fix a positive constant $\tilde{c}$. For any $a > 0$, let us consider the following subset $\mathscr{X}_a$ of $\mathscr{X}$. We say that $(\Sigma,\Pi) \in \mathscr{X}$ belongs to $\mathscr{X}_a$ if there exists $\hat{\Sigma} \in \mathscr{A}_{sym}$ such that $d_H(\Sigma,\hat{\Sigma}) < a$ and $d(\Pi,\tilde{\Pi}(\Sigma)) < \tilde{c}a$.*

Then there exists a positive constant $\hat{a}_0$, depending on the a priori data, on $\tilde{c}$, on a, and on $\{\mathbf{v}_1,\ldots,\mathbf{v}_{N-1}\}$ only, such that

$$\min\{f(\Sigma,\Pi) : (\Sigma,\Pi) \in \mathscr{X} \setminus \mathscr{X}_a\} \geq \hat{a}_0.$$

Remark 6.5.1. In the previous proposition, if $N = 2$, the result does not depend on the direction $\mathbf{v}$. If $N = 3$, the dependence on $\mathbf{v}_1$ and $\mathbf{v}_2$ is only through the constant $b_0 = |\mathbf{v}_1 \cdot \mathbf{v}_2| < 1$. We also notice that $\hat{a}_0$ does not depend on h.

Proof. We observe that, by the stability result for the direct scattering problem with respect to sound-hard scatterers Σ proved in [18], such a map f is continuous on $\mathscr{X}$.

If $f(\Sigma,\Pi) = 0$, then the unit normal to Π is orthogonal to $\mathbf{v}_j$, for any $j = 1,\ldots,N-1$, and Σ is symmetric with respect to Π, that is, $\Pi = \tilde{\Pi}(\Sigma)$ and $(\Sigma,\Pi) \in \mathscr{Y}$.

Then the proof immediately follows by the fact that $\mathscr{X} \setminus \mathscr{X}_a$ is closed and obviously does not contain any point of $\mathscr{Y}$. □

Up to now, we are able to prove a stability result if either the assumptions of Proposition 6.5.1 are satisfied or, otherwise, if $(\Sigma,\Pi_M) \in \mathscr{X} \setminus \mathscr{X}_a$ for a suitable $a > 0$. In both cases we use the same computation as in the N measurements case, with a_0 replaced by $\tilde{a}_0$ and $\hat{a}_0$, respectively. We notice that in this second case $\hat{a}_0$ depends on a.

Therefore, our strategy is now the following. We choose a suitable value of a and we construct a modified regular chain for Σ as in the general geometric construction such that for any possible reflection hyperplane Π_n, $n = 1,\ldots,M$ (including the first one!), we have that $(\Sigma,\Pi_n) \in \mathscr{X} \setminus \mathscr{X}_a$. As we shall see, actually the first reflection hyperplane is the one that presents the greatest difficulties.

We notice that, so far, all our arguments work for any dimension $N \geq 2$. However, the construction of such a modified regular chain presents some technical challenges, in particular for the proof of Lemma 6.5.4 below. Therefore, in the sequel we limit ourselves to the space dimension $N = 3$ and we notice that when the space dimension is $N = 2$, the result may be proved along the same lines.

A crucial remark is that, unfortunately, we are not able to choose a independently of h. This is the main reason why we lose the precise dependence of our stability result on the size parameter h, which we instead have in the sound-soft case or in the sound-hard case with N measurements.

The following two technical lemmas shall be needed.

Lemma 6.5.3. *Let $N = 3$ and $h > 0$.*

There exist positive constants $\tilde{c}_0$, $\tilde{c}_1$, $\tilde{c}_2 \leq 1$, and $\tilde{L} \geq L$, depending on the a priori data only, such that the following holds.

Let $a = \tilde{c}_0 h$ and let $\Sigma \in \mathscr{A}^h_{obst}$ satisfy the following. We assume that there exists $\hat{\Sigma} \in \mathscr{A}_{sym}$ such that $d_H(\Sigma,\hat{\Sigma}) \leq a$.

For simplicity, let us assume that $\tilde{\Pi}(\Sigma) = \{y_N = 0\}$. Then, for any $\tilde{c}$, $0 \leq \tilde{c} \leq \tilde{c}_1$, if we call $G^{\pm}_a = \{\mathbf{y} \in G : y_N \gtrless \pm\tilde{c}a\}$, we have that $G^{\pm}_a$ are Lipschitz domains with constants $\tilde{r} = \tilde{c}_2 h$ and $\tilde{L}$.

Proof. This is an extension of Lemma 6.5.2, which can be proved by exploiting [21, Proposition 6.1]. □

Lemma 6.5.4. *Let $N = 3$, $h > 0$ and $\tilde{c}_0$, and $\tilde{c}_1$ as in Lemma 6.5.3. Let Σ, Σ' belong to $\mathscr{A}^h_{obst}$ and let d be defined as in (6.2.16). Let $\mathbf{x}_1 \in \partial\Sigma' \backslash \Sigma$ be such that $d = \mathrm{dist}(\mathbf{x}_1, \partial\Sigma) = \mathrm{dist}(\mathbf{x}_1, \Sigma)$.*

There exist positive constants $\tilde{c}_3 \leq 1$, $\tilde{c}_4 \leq \tilde{c}_1$, and $K_1 \leq 1$, depending on the a priori data only, such that the following holds.

Let $a = \tilde{c}_0 h$ and $\tilde{c} = \tilde{c}_4$. Let us assume that there exists $\hat{\Sigma} \in \mathscr{A}_{sym}$ such that $d_H(\Sigma, \hat{\Sigma}) \leq a$. Let us call $G^{\pm}_a = \{\mathbf{y} \in G : \ y_N \gtrless \pm\tilde{c}a\}$, assuming that $\tilde{\Pi}(\Sigma) = \{y_N = 0\}$.

If $d \leq \tilde{c}_3 h$, then, up to swapping the role of G^+_a and G^-_a, there exists $\tilde{\mathbf{x}}_1 \in \partial\Sigma' \backslash \Sigma$ such that $\tilde{\mathbf{x}}_1 \in G^+_a$ and

$$\mathrm{dist}(\tilde{\mathbf{x}}_1, \partial G^+_a) \geq K_1 d^3. \tag{6.5.3}$$

Proof. This is a straightforward consequence of [21, Proposition 6.2]. □

We are now in the position of concluding the proof of the $N-1$ measurement case.

The Proof of Theorem 6.3.3. Without loss of generality we can assume that $h \leq \min\{r, 1\}$.

Let us assume, for the time being, that $d \leq \tilde{c}_3 h \leq h$. Let us set $a = \tilde{c}_0 h$, $\tilde{c}_0$ as in Lemma 6.5.3 and $\tilde{c} = \tilde{c}_4$ as in Lemma 6.5.4.

Then we distinguish between two cases. If there does not exist any $\hat{\Sigma} \in \mathscr{A}_{sym}$ such that $d_H(\Sigma, \hat{\Sigma}) \leq a$, then we conclude using the geometric construction of Sect. 6.4 and the arguments used for the proof of the N measurement case. Here we use either Proposition 6.5.1, replacing a_0 with $\tilde{a}_0$, or Proposition 6.5.3, with $\tilde{c} = \tilde{c}_4$ as in Lemma 6.5.4 and replacing a_0 with $\hat{a}_0$. We have to notice that $\hat{a}_0$ here depends on a and thus on h.

Otherwise, let us assume that there does exist $\hat{\Sigma} \in \mathscr{A}_{sym}$ such that $d_H(\Sigma, \hat{\Sigma}) \leq a$. Then we use the geometric construction and estimates of Sect. 6.4 with the following differences. We replace $\mathbf{x}_1$ with $\tilde{\mathbf{x}}_1$ and G with G^+_a, $\tilde{\mathbf{x}}_1$, and G^+_a as in Lemma 6.5.4. Using Lemma 6.5.3, we further replace r and L with $\tilde{c}_2 h$ and $\tilde{L}$, respectively. Finally, using (6.5.3), we replace d with $K_1 d^3 \leq d$. Then we can repeat the previous argument using either Proposition 6.5.1 or Proposition 6.5.3. In fact any possible reflection point belongs to G^+_a; therefore, any reflection plane is far enough from $\tilde{\Pi}(\Sigma)$.

We conclude that, for any ε, $0 < \varepsilon < 1/(2\mathrm{e})$, provided $d \leq \tilde{c}_3 h$, we have

$$K_1 d^3 \leq 2\mathrm{e}R(\eta(\varepsilon))^{3C},$$

for some constant C depending on the a priori data, on b_0, and on h as well. Therefore,

$$d \leq A_1(\eta(\varepsilon))^C, \tag{6.5.4}$$

where A_1 depends on the a priori data only and C depends on the a priori data, on b_0, and on h.

Finally, we need to drop the assumption that $d \leq \tilde{c}_3 h$. We claim that there exists $\hat{\varepsilon}_1(h)$, $0 < \hat{\varepsilon}_1(h) \leq 1/(2\mathrm{e})$, depending on the a priori data and on h only, such that

$$\hat{\varepsilon}_1(h) < \inf\left\{\|u - u'\|_{L^\infty(B_{\tilde{\rho}}(\mathbf{x}_0))} : \ \mathbf{v} \in \mathbb{S}^{N-1}, \ \Sigma, \Sigma' \in \mathscr{A}^h_{obst} \text{ such that } d_H(\Sigma, \Sigma') \geq C_1 \tilde{c}_3 h\right\}, \tag{6.5.5}$$

where C_1 is as in (6.2.43). If this is true, then obviously we obtain that if $\varepsilon \leq \hat{\varepsilon}_1(h)$, then $d_H(\Sigma, \Sigma') < C_1 \tilde{c}_3 h$, that is, $d \leq d_H(\Sigma, \Sigma')/C_1 < \tilde{c}_3 h$ and the proof would be concluded.

Therefore, we just need to prove the claim in (6.5.5). It is not difficult to show that the infimum on the right-hand side is actually a minimum. Again it is enough to use the stability result of the direct scattering problem with respect to the variation of sound-hard scatterers proved in [18]. Finally, if such a minimum were zero, we would contradict the uniqueness result for the determination of a sound-hard obstacle by a single scattering measurement proved in [6, 7]. □

6.5.3 The Single Measurement Case

We restrict here to $N = 3$, since $N = 2$ is clearly covered by the previous section. We consider the assumptions and notation of Theorem 6.3.4 to hold.

The main technical difficulty we have to tackle if we have only one measurement, compared to the two measurements case, is that we may have several planes whose normal is orthogonal to $\mathbf{v}$ with respect to which Σ might be symmetric. As we discussed in the previous subsection, using two measurements with two directions of propagation $\mathbf{v}_1$ and $\mathbf{v}_2$ allows us to consider only one possible symmetry plane for Σ.

We begin with the following definition. Here we call $\mathscr{A}_{sym}$, respectively, $\mathscr{A}^h_{sym}$, the set of Σ belonging to $\mathscr{A}_{obst}$, respectively $\mathscr{A}^h_{obst}$, such that Σ is symmetric with respect to at least one plane whose normal is orthogonal to the incident direction of propagation $\mathbf{v}$. Moreover, for any $\Sigma \in \mathscr{A}^h_{obst}$, we call $n(\Sigma)$ the number of planes whose normal is orthogonal to $\mathbf{v}$ with respect to which Σ is symmetric. Notice that $n(\Sigma)$ is always a nonnegative integer that, obviously, could also be zero. In other words, $\mathscr{A}^h_{sym}$ is the set of $\Sigma \in \mathscr{A}^h_{obst}$ such that $n(\Sigma) > 0$.

We shall use the following notation. For any $\Sigma \in \mathscr{A}^h_{sym}$, we call $\Pi_i(\Sigma)$, $i = 1,\ldots,n(\Sigma)$, the planes whose normal is orthogonal to $\mathbf{v}$ with respect to which Σ is symmetric. Correspondingly, we define $\nu_i(\Sigma)$, $i = 1,\ldots,n(\Sigma)$, their corresponding unit normals, noticing that they all belong to the plane that is orthogonal to $\mathbf{v}$.

We have the following properties whose proof is elementary and will be omitted.

Proposition 6.5.4. *There exists an integer $M = M(h)$, depending on the a priori data and on h only, such that $n(\Sigma) \leq M$ for any $\Sigma \in \mathscr{A}^h_{sym}$. As a consequence, there exists a constant $\alpha = \alpha(h)$, $0 < \alpha < \pi/2$, such that, for any $\Sigma \in \mathscr{A}^h_{sym}$, the angle between $\nu_i(\Sigma)$ and $\nu_j(\Sigma)$, with $i \neq j$, is bounded from below by α.*

We consider $\mathscr{A}_{obst}$ endowed with the Hausdorff distance. Then the map $\mathscr{A}^h_{obst} \ni \Sigma \mapsto n(\Sigma)$ is upper semicontinuous. Consequently, $\mathscr{A}^h_{sym}$ is a compact subset of $\mathscr{A}^h_{obst}$.

Let us define, for any $n = 1,\ldots,M(h)$, $\mathscr{A}^h_{sym,n}$ as the set of $\Sigma \in \mathscr{A}^h_{sym}$ such that $n(\Sigma) = n$.

The crucial difference with respect to the 2 measurement case is that we need to define the set $\mathscr{X}_a$, for a positive constant a, in a rather more involved way. Next, we describe such a construction, for any positive a and for a fixed constant $\tilde{c}$ to be decided later.

Given $M = M(h)$, we begin in the following way. For any $\hat{\Sigma} \in \mathscr{A}^h_{sym,M}$, we find $r(\hat{\Sigma})$, $0 < r(\hat{\Sigma}) \leq a$, such that for any $\Sigma \in \mathscr{A}^h_{sym}$ with $d_H(\Sigma,\hat{\Sigma}) \leq r(\hat{\Sigma})$, the following holds. For any $i = 1,\ldots,n(\Sigma)$, there exists $j \in \{1,\ldots,n(\hat{\Sigma})\}$ such that $d(\Pi_i(\Sigma),\Pi_j(\hat{\Sigma})) \leq \tilde{c}a/4$.

Then, by compactness, we have that

$$\mathscr{A}^h_{sym,M} \subset \bigcup_{j=1}^{m_M} B_{r(\hat{\Sigma}_j)/2}(\hat{\Sigma}_j) = A^h_M,$$

where, for any $j = 1,\ldots,m_M$, $\hat{\Sigma}_j \in \mathscr{A}^h_{sym,M}$ and $r(\hat{\Sigma}_j) \leq r(\hat{\Sigma}_{j-1})$. Here by convention we set $r(\hat{\Sigma}_0) = a$.

Then we consider $\mathscr{A}^h_{sym,M-1} \backslash A^h_M$, which is again a compact set. We consider a similar construction as before. Namely, for any $\hat{\Sigma} \in \mathscr{A}^h_{sym,M-1} \backslash A^h_M$, we find $r(\hat{\Sigma})$, $0 < r(\hat{\Sigma}) \leq r(\hat{\Sigma}_{m_M})$, such that for any $\Sigma \in \mathscr{A}^h_{sym}$ with $d_H(\Sigma,\hat{\Sigma}) \leq r(\hat{\Sigma})$, the following holds. For any $i = 1,\ldots,n(\Sigma)$, there exists $j \in \{1,\ldots,n(\hat{\Sigma})\}$ such that $d(\Pi_i(\Sigma),\Pi_j(\hat{\Sigma})) \leq \tilde{c}a/4$. Then, by compactness, we have that

$$\mathscr{A}^h_{sym,M-1} \backslash A^h_M \subset \bigcup_{j=m_M+1}^{m_{M-1}} B_{r(\hat{\Sigma}_j)/2}(\hat{\Sigma}_j) = A^h_{M-1},$$

where, for any $j = m_M+1,\ldots,m_{M-1}$, $\hat{\Sigma}_j \in \mathscr{A}^h_{sym,M-1} \backslash A^h_M$ and $r(\hat{\Sigma}_j) \leq r(\hat{\Sigma}_{j-1})$.

We proceed in a completely analogous way until we have that

$$\mathscr{A}^h_{sym} \subset \bigcup_{j=1}^{m_1} B_{r(\hat{\Sigma}_j)/2}(\hat{\Sigma}_j) = \bigcup_{l=1}^{M} A^h_l = A^h.$$

For any $l = 2,\dots,M$ and any $j = m_l+1,\dots,m_{l-1}$, $\hat{\Sigma}_j \in \mathscr{A}^h_{sym,l-1}\setminus\left(\bigcup_{i=l}^M A^h_i\right)$ and $r(\hat{\Sigma}_j) \leq r(\hat{\Sigma}_{j-1})$.

We call $\mathscr{X}^h$ the subset of $(\Sigma,\Pi) \in \mathscr{X}$ such that $\Sigma \in \mathscr{A}^h_{obst}$. We also call $\mathscr{Y}^h$ the subset of $\mathscr{X}^h$ defined as follows:

$$\mathscr{Y}^h = \{(\Sigma,\Pi) \in \mathscr{X}^h : \Sigma \in \mathscr{A}^h_{sym} \text{ and } \Pi = \Pi_i(\Sigma) \text{ for some } i \in \{1,\dots,n(\Sigma)\}\}.$$

Then we define $\mathscr{X}^h_a$ the subset of $\mathscr{X}^h$ with the following properties. We say that $(\Sigma,\Pi) \notin \mathscr{X}^h_a$ either if $\Sigma \notin A^h$ or if $\Sigma \in A^h$ and $d(\Pi,\hat{\Pi}) \geq \tilde{c}a/2$ for any plane $\hat{\Pi}$ such that $\hat{\Pi} = \Pi_i(\hat{\Sigma}_j)$ for some $j \in \{1,\dots,m_1\}$ such that $d_H(\Sigma,\hat{\Sigma}_j) < r(\hat{\Sigma}_j)/2$ and for some $i \in \{1,\dots,n(\hat{\Sigma}_j)\}$.

It is an easy remark that, for any $a>0$, $\mathscr{Y}^h \subset \mathscr{X}^h_a$ and $\mathscr{X}^h\setminus\mathscr{X}^h_a$ is closed. Let us consider the map

$$\mathscr{X}^h \ni (\Sigma,\Pi) \mapsto f(\Sigma,\Pi) = \max_{\tilde{\mathbf{z}}\in(\overline{B_{2R_2+3}\setminus B_{2R_2+2}})\cap\Pi} |\nabla u(\tilde{\mathbf{z}})\cdot \nu|,$$

where $\nu = \nu_\Pi$ is the normal to Π and u is the solution to the direct scattering problem (6.2.24) with boundary condition (6.2.26) and incident field $u^i(\mathbf{x}) = \mathrm{e}^{\mathrm{i}k\mathbf{x}\cdot\mathbf{v}}$, $\mathbf{x}\in\mathbb{R}^N$. Hence, arguing as in the proof of Proposition 6.5.3, we can obtain the following result.

Proposition 6.5.5. *Let us fix a positive constant $\tilde{c}$. For any $a>0$, we define the subset $\mathscr{X}^h_a$ of $\mathscr{X}$ as before.*

Then there exists a positive constant $\hat{a}_0$, depending on the a priori data, on $\tilde{c}$, on a, and on h only, such that

$$\min\left\{f(\Sigma,\Pi) : (\Sigma,\Pi) \in \mathscr{X}^h\setminus\mathscr{X}^h_a\right\} \geq \hat{a}_0.$$

We now consider the corresponding results to Lemmas 6.5.3 and 6.5.4. We need the following notation, recalling that positive constants a and $\tilde{c}$ are fixed. For any $\Sigma \in A^h$, we choose $\hat{\Sigma}(\Sigma)$ as the first $\hat{\Sigma}_j$, $j\in\{1,\dots,m_1\}$, such that $\Sigma \in B_{r(\hat{\Sigma}_j)/2}(\hat{\Sigma}_j)$. For any $i=1,\dots,n(\hat{\Sigma}(\Sigma))$, we define the infinite strips

$$S_i = \Pi_i(\hat{\Sigma}(\Sigma)) + \{ca\nu_i(\hat{\Sigma}(\Sigma)) : |c|\leq\tilde{c}\}.$$

We notice that $\mathbb{R}^3\setminus(\bigcup_{i=1}^{n(\hat{\Sigma}(\Sigma))} S_i)$ consists of $2n(\hat{\Sigma}(\Sigma))$ different connected open sectors that we shall call G^j_a, $j=1,\dots,2n(\hat{\Sigma}(\Sigma))$.

Then, with the notation introduced above, the following important results hold.

Lemma 6.5.5. *Let $N=3$ and $h>0$.*

There exist positive constants $\tilde{c}_0$, depending on the a priori data only, and $\tilde{c}_1$, $\tilde{c}_2\leq 1$, and $\tilde{L}\geq L$, depending on the a priori data and on h only, such that the following holds.

Let $a=\tilde{c}_0 h$ and let $\Sigma\in A^h$. Then, for any $\tilde{c}$, $0\leq\tilde{c}\leq\tilde{c}_1$, we have that, for any $j=1,\dots,2n(\hat{\Sigma}(\Sigma))$, $G^j_a\setminus\Sigma$ is a Lipschitz domain with constants $\tilde{r}=\tilde{c}_2 h$ and $\tilde{L}$.

Proof. If $\hat{\Sigma}(\Sigma)\in\mathscr{A}^h_{sym,1}$, then the result is contained in Lemma 6.5.3. Therefore, without loss of generality, we assume that $\hat{\Sigma}(\Sigma)\in\mathscr{A}^h_{sym,n}$ for some $n\geq 2$.

We begin by proving the following claim, which is the corresponding result to Lemma 6.5.2. We fix an arbitrary $\hat{\Sigma}\in\mathscr{A}^h_{sym,n}$ with $n\geq 2$. Then we call G^j, $j=1,\dots,2n$, the connected components of $\mathbb{R}^3\setminus(\bigcup_{i=1}^n \Pi_i(\hat{\Sigma}))$. We claim that, for any $j=1,\dots,2n$, $G_j\setminus\hat{\Sigma}$ is a Lipschitz domain with constants $\tilde{r}_1$ and $\tilde{L}_1$ depending on the a priori data and on h only.

We deal with the points $\mathbf{z}$ belonging to $\partial\hat{\Sigma}$ and $\Pi_i(\hat{\Sigma})$ for some $i=1,\dots,n$. Let $P(\hat{\Sigma})$ be the center of mass of $\hat{\Sigma}$ and let l be the line defined as follows:

$$l = \{\mathbf{x} \in \mathbb{R}^3 : \mathbf{x} = P(\hat{\Sigma}) + r\mathbf{v},\ r \in \mathbb{R}\}.$$

It is obvious that $l = \Pi_i(\hat{\Sigma}) \cap \Pi_j(\hat{\Sigma})$ for any $i \neq j$.

If we have a point $\mathbf{z}$ belonging to $\partial\hat{\Sigma}$ and $\Pi_i(\hat{\Sigma})$ for some $i = 1,\ldots,n$, which is far enough from l, we can treat it exactly as in the proof of Lemma 6.5.2. Therefore, the most delicate case is the one in which $\mathbf{z} \in \partial\hat{\Sigma} \cap l$. However, following the kind of reasonings used in the proof of Lemma 6.5.2, it is not difficult to show that, locally in $B_{\tilde{r}_2}(\mathbf{z})$, $\partial\Sigma$ is the graph of a Lipschitz function, with Lipschitz constant bounded by $\tilde{L}_2$, with respect to a Cartesian coordinate system such that $\mathbf{e}_3$ is parallel to $\mathbf{v}$, with $\tilde{r}_2$ and $\tilde{L}_2$ depending on the a priori data only.

Then the claim easily follows, with the dependence of $\tilde{r}_1$ and $\tilde{L}_1$ on h essentially given by the angle $\alpha(h)$.

The proof of the proposition can be concluded by using the claim, arguments similar to the ones used to prove the claim, and [21, Proposition 6.1]. □

We notice that the difference with respect to Lemma 6.5.3 is that now $\tilde{c}_1$, $\tilde{c}_2$, and $\tilde{L}$ depend on h as well.

Lemma 6.5.6. *Let $N = 3$, $h > 0$, and $\tilde{c}_0$ and $\tilde{c}_1$ as in Lemma 6.5.5. Let Σ, Σ' belong to $\mathscr{A}^h_{obst}$ and let d be defined as in (6.2.16). Let $\mathbf{x}_1 \in \partial\Sigma'\backslash\Sigma$ be such that $d = \mathrm{dist}(\mathbf{x}_1,\partial\Sigma) = \mathrm{dist}(\mathbf{x}_1,\Sigma)$.*

There exist positive constants $\tilde{c}_3 \leq 1$, $\tilde{c}_4 \leq \tilde{c}_1$, depending on the a priori data and on h only, and $K_1 \leq 1$, depending on the a priori data only, such that the following holds.

Let $a = \tilde{c}_0 h$ and $\tilde{c} = \tilde{c}_4$. Let us assume that $\Sigma \in A^h$.

If $d \leq \tilde{c}_3 h$, then there exist $\tilde{\mathbf{x}}_1 \in \partial\Sigma'\backslash\Sigma$ and $j \in \{1,\ldots,2n(\hat{\Sigma}(\Sigma))\}$ such that $\tilde{\mathbf{x}}_1 \in G^j_a\backslash\Sigma$ and

$$\mathrm{dist}(\tilde{\mathbf{x}}_1, \partial(G^j_a\backslash\Sigma)) \geq K_1 d^3. \tag{6.5.6}$$

Proof. The result is a rather straightforward consequence of [21, Proposition 6.2]. □

Again, it is important to remark that the difference with respect to Lemma 6.5.4 is that now $\tilde{c}_3$ and $\tilde{c}_4$ depend on h too.

We are now in the position of proving our stability result with one measurement.

The Proof of Theorem 6.3.4. The proof follows the same arguments of the proof of Theorem 6.3.3, replacing Proposition 6.5.3, Lemmas 6.5.3, and 6.5.4 with Proposition 6.5.5, Lemmas 6.5.5, and 6.5.6, respectively. We point out the modification that we need to adopt in this case.

Without loss of generality we can assume that $h \leq \min\{r,1\}$. Let us assume, for the time being, that $d \leq \tilde{c}_3 h \leq h$. Let us set $a = \tilde{c}_0 h$, $\tilde{c}_0$ as in Lemma 6.5.5, and $\tilde{c} = \tilde{c}_4$ as in Lemma 6.5.6.

We distinguish two cases. If $\Sigma \notin A^h$, then we conclude using Proposition 6.5.1 and Proposition 6.5.5 with $\tilde{c} = \tilde{c}_4$ as in Lemma 6.5.6.

If instead $\Sigma \in A^h$, we replace $\mathbf{x}_1$ with $\tilde{\mathbf{x}}_1$ and G with $G^j_a\backslash\Sigma$, $\tilde{\mathbf{x}}_1$ and $G^j_a\backslash\Sigma$ as in Lemma 6.5.6. Notice that in this case, any possible reflection point belongs to $G^j_a\backslash\Sigma$; therefore, any reflection plane Π is far from any $\Pi_i(\hat{\Sigma}(\Sigma))$, $i = 1,\ldots,n(\hat{\Sigma}(\Sigma))$, at least $\tilde{c}a$. On the other hand, recalling the construction of A^h and how we choose $\hat{\Sigma}(\Sigma)$, for any $j \in \{1,\ldots,m_1\}$, if $d_H(\Sigma,\hat{\Sigma}_j) < r(\hat{\Sigma}_j)/2$, we have that $d_H(\hat{\Sigma}_j,\hat{\Sigma}(\Sigma)) < r(\hat{\Sigma}(\Sigma))$. We conclude that $d(\Pi,\hat{\Pi}) \geq 3\tilde{c}a/4$ for any plane $\hat{\Pi}$ such that $\hat{\Pi} = \Pi_i(\hat{\Sigma}_j)$ for some $j \in \{1,\ldots,m_1\}$ such that $d_H(\Sigma,\hat{\Sigma}_j) < r(\hat{\Sigma}_j)/2$ and for some $i \in \{1,\ldots,n(\hat{\Sigma}_j)\}$. That is, $(\Sigma,\Pi) \in \mathscr{X}^h\backslash\mathscr{X}^h_a$.

The rest of the argument is the same. However, we notice that, since the domains $G^j_a\backslash\Sigma$ used in Lemmas 6.5.5 and 6.5.6 are Lipschitz with constants both depending on h, the dependence of the stability result on h may be worse than the one in the 2 measurements case. □

References

1. G. Alessandrini and L. Rondi, *Determining a sound-soft polyhedral scatterer by a single far-field measurement*, Proc. Amer. Math. Soc., **35** (2005), 1685–1691.
2. R. Brummelhuis, *Three-spheres theorem for second order elliptic equations*, J. Analyse Math. **65** (1995) 179–206.
3. J. Cheng and M. Yamamoto, *Uniqueness in an inverse scattering problem within non-trapping polygonal obstacles with at most two incoming waves*, Inverse Problems, **19** (2003), 1361–1384.
4. D. Colton and R. Kress, *Inverse acoustic and electromagnetic scattering theory*, volume 93 of Applied Mathematical Sciences. Springer, Cham, 4th edition, 2019.
5. D. Colton and B. Sleeman, *Uniqueness theorems for the inverse problem of acoustic scattering*, IMA J. Appl. Math., **31** (1983), 253–259.
6. J. Elschner and M. Yamamoto, *Uniqueness in determining polygonal sound-hard obstacles with a single incoming wave*, Inverse Problems, **22** (2006), 355–364.
7. J. Elschner and M. Yamamoto, *Uniqueness in determining polyhedral sound-hard obstacles with a single incoming wave*, Inverse Problems, **24** (2008), 035004 (7pp).
8. A. Giacomini, *A stability result for Neumann problems in dimension $N \geq 3$*, J. Convex Anal, **11** (2004), 41–58.
9. V. Isakov, *On uniqueness in the inverse transmission scattering problem*, Commun. Part. Diff. Eqns., **15** (1990), 1565–1587.
10. V. Isakov, *Stability estimates for obstacles in inverse scattering*, J. Comp. Appl. Math. **42** (1992) 79–88.
11. V. Isakov, *New stability results for soft obstacles in inverse scattering*, Inverse Problems **9** (1993) 535–543.
12. V. Isakov, *Inverse Problems for Partial Differential Equations*, 2nd edition, Springer-Verlag, New York, (2006).
13. A. Kirsch and R. Kress, *Uniqueness in inverse obstacle scattering*, Inverse Problems, **9** (1993), 285–299.
14. P. Lax and R. Phillips, *Scattering Theory*, Academic Press, New York and London, (1967).
15. H. Liu, M. Petrini, L. Rondi and J. Xiao, Stable determination of sound-hard polyhedral scatterers by a minimal number of scattering measurements, *J. Differential Equations*, **262(3)** (2017), 1631–1670.
16. H. Liu and J. Zou, *Uniqueness in an inverse acoustic obstacle scattering problem for both sound-hard and sound-soft polyhedral scatterers*, Inverse Problems, **22** (2006), 515–524.
17. H. Liu and J. Zou, *On unique determination of partially coated polyhedral scatterers with far field measurements*, Inverse Problems, **23** (2007), 297–308.
18. G. Menegatti and L. Rondi, *Stability for the acoustic scattering problem for sound-hard scatterers*, Inverse Probl. Imaging **7** (2013) 1307–1329.
19. L. Rondi, *Unique determination of non-smooth sound-soft scatterers by finitely many far-field measurements*, Indiana Univ. Math. J., **52** (2003), 1631–1662.
20. L. Rondi, *A variational approach to the reconstruction of cracks by boundary measurements*, J. Math. Pures Appl. (9), **87** (2007), 324–342.
21. L. Rondi, *Stable determination of sound-soft polyhedral scatterers by a single measurement*, Indiana Univ. Math. J., **57** (2008), 1377–1408.
22. L. Rondi and M. Sini, *Stable determination of a scattered wave from its far-field pattern: the high frequency asymptotics*, Arch. Ration. Mech. Anal. **218** (2015), 1–54.
23. C. H. Wilcox, *Scattering Theory for the d'Alembert Equation in Exterior Domains*, Springer-Verlag, Berlin New York, (1975).

Chapter 7
Stability for Inverse Electromagnetic Obstacle Scattering Problems

7.1 Overview

In this chapter, we deal with the stability issue for the inverse obstacle problem associated with the electromagnetic scattering. We follow the treatment in [21] on a quantitative path argument. We are concerned with the electromagnetic scattering problem, in the time-harmonic case, which is governed by the Maxwell system as follows. Given an incident time-harmonic electromagnetic wave, characterised by the incident electric and magnetic fields $(\mathbf{E}^i, \mathbf{H}^i)$, we seek a pair of functions $(\mathbf{E}, \mathbf{H})$ solving the following exterior boundary value problem

$$\begin{cases} \nabla \wedge \mathbf{E} - ik\mu\mathbf{H} = \mathbf{0} & \text{in } G = \mathbb{R}^3 \setminus \Sigma, \\ \nabla \wedge \mathbf{H} + ik\varepsilon\mathbf{E} = \mathbf{0} & \text{in } G = \mathbb{R}^3 \setminus \Sigma, \\ (\mathbf{E}, \mathbf{H}) = (\mathbf{E}^i, \mathbf{H}^i) + (\mathbf{E}^s, \mathbf{H}^s) & \text{in } G = \mathbb{R}^3 \setminus \Sigma, \\ \nu \wedge \mathbf{E} = \mathbf{0} & \text{on } \partial G = \partial\Sigma, \\ \lim_{r \to +\infty} r \left(\frac{\mathbf{x}}{\|\mathbf{x}\|} \wedge \mathbf{H}^s(\mathbf{x}) + \mathbf{E}^s(\mathbf{x}) \right) = \mathbf{0} & r = \|\mathbf{x}\|. \end{cases} \tag{7.1.1}$$

Here $k > 0$ is the wavenumber; Σ is a scatterer, that is, a compact subset with connected complement, present in the medium; and ε and μ are the electric permittivity and magnetic permeability of the medium, respectively. The first two equations satisfied by $(\mathbf{E}, \mathbf{H})$ outside Σ are the time-harmonic Maxwell equations. The total electric and magnetic fields $(\mathbf{E}, \mathbf{H})$ are the juxtaposition of the incident ones and the reflected, or scattered, ones $(\mathbf{E}^s, \mathbf{H}^s)$. The incident fields $\mathbf{E}^i$ and $\mathbf{H}^i$ are an entire solution of the Maxwell equations with $\varepsilon = \mu = I_3$. Here, and also in what follows, I_3 denotes the 3×3 identity matrix. For instance, one can take $(\mathbf{E}^i, \mathbf{H}^i)$ to be a normalised electromagnetic plane wave with given polarisation vector and direction of propagation. The presence of the scatterer Σ, and of an inhomogeneity in (ε, μ) around it, perturbs the incident wave through the creation of the scattered wave, characterised by the fact that it is radiating, that is, its corresponding fields satisfy the so-called Silver-Müller radiation condition, which is the last limit in (7.1.1). The boundary condition on the boundary of Σ depends on the physical properties of the scatterer. In this case, Σ is a perfectly electric conducting scatterer; however, in a completely equivalent way, we can also treat the case of perfectly magnetic conducting scatterers.

The plan of the chapter is as follows. In Sect. 7.2, we consider the direct electromagnetic scattering problem. In Sect. 7.2.1, we set most of the standard notations needed in the sequel. Moreover, in Proposition 7.2.2, we prove that the subspace $L^\infty \cap H(\mathrm{curl})$ is dense in the space $H(\mathrm{curl})$ for an arbitrary open set, a result that, rather surprisingly, we could not find in this generality in the literature. In Sect. 7.2.2, we formulate the problem and we treat existence and uniqueness of its solution in Sect. 7.2.3. In particular, in Sect. 7.2.4, we discuss the Rellich and Maxwell compactness properties, as well as the unique continuation property for the Maxwell equations.

H. Diao, H. Liu, *Spectral Geometry and Inverse Scattering Theory*,
https://doi.org/10.1007/978-3-031-34615-6_7

Section 7.3 is devoted to some preliminary results. In Sect. 7.3.1, properties of the solutions to the Maxwell equations are considered. In particular, changes of variables and reflection principles for solutions to the Maxwell equations are studied in detail. In Sect. 7.3.2, we discuss suitable classes of admissible scatterers. These kinds of classes were introduced in [26] and further developed in [20].

The main technical section of the chapter is Sect. 7.4. We prove the Mosco convergence result for $H(\mathrm{curl})$ spaces, Theorem 7.4.1 and the higher integrability properties for solutions to the Maxwell equations in nonsmooth domains, Proposition 7.4.1

The main section of the chapter is Sect. 7.5. We consider suitable classes of admissible coefficients for the inhomogeneities, Definition 7.5.1, and discuss their properties. Then we state and prove the main results of the chapter; the stability result, Theorem 7.5.2; and the uniform bounds, Theorem 7.5.1, for the solutions to the direct electromagnetic scattering problem.

In Sect. 7.6, we deal with the inverse electromagnetic scattering problem, for polyhedral scatterers. The stability results, Theorem 7.6.1 for general polyhedral scatterers with two measurements and Theorem 7.6.2 for polyhedra with one measurement, are presented in Sect. 7.6.1. A few details and comments about the proofs of these stability results are in the final Sect. 7.6.2.

7.2 The Electromagnetic Scattering Problem

7.2.1 Notations and Conditions

We shall use the following notations. The integer $N \geq 2$ always denotes the space dimension. Apart from a few instances, N will be equal to 3. In what follows, we always omit the dependence of constants on the space dimension N.

For any $\mathbf{x} \in \mathbb{R}^N$, $N \geq 2$, we denote $\mathbf{x} = (\mathbf{x}', x_N) \in \mathbb{R}^{N-1} \times \mathbb{R}$ and $\mathbf{x} = (\mathbf{x}'', x_{N-1}, x_N) \in \mathbb{R}^{N-2} \times \mathbb{R} \times \mathbb{R}$. For any $s > 0$ and any $\mathbf{x} \in \mathbb{R}^N$, $B_s(\mathbf{x})$ denotes the ball contained in $\mathbb{R}^N$ with radius s and centre $\mathbf{x}$, and $B'_s(\mathbf{x}')$ denotes the ball contained in $\mathbb{R}^{N-1}$ with radius s and centre $\mathbf{x}'$. We also use $B_s = B_s(\mathbf{0})$ and $B'_s = B'_s(\mathbf{0})$. For any $E \subset \mathbb{R}^N$, we denote $B_s(E) = \bigcup_{\mathbf{x} \in E} B_s(\mathbf{x})$. For any ball B centred at zero, we denote $B^{\pm} = B \cap \{\mathbf{y} \in \mathbb{R}^N : \ y_N \gtrless 0\}$. Analogously, for any hyperplane Π in $\mathbb{R}^N$, we use the following notations. If, with respect to a suitable Cartesian coordinate system, one has $\Pi = \{\mathbf{y} \in \mathbb{R}^N : \ y_N = 0\}$, then for any $\mathbf{x} \in \Pi$ and any $r > 0$, we denote $B^{\pm}_r(\mathbf{x}) = B_r(\mathbf{x}) \cap \{y_N \gtrless 0\}$. Furthermore, we denote by T_Π the reflection with respect to Π, namely, in this case for any $\mathbf{y} = (y_1, \ldots, y_{N-1}, y_N) \in \mathbb{R}^N$, we have $T_\Pi(\mathbf{y}) = (y_1, \ldots, y_{N-1}, -y_N)$.

The s-dimensional Hausdorff measure in $\mathbb{R}^N$, $0 \leq s \leq N$, will be denoted by $\mathscr{H}^s$. We recall that $\mathscr{H}^N$ coincides with the Lebesgue measure. For any Borel set $E \subset \mathbb{R}^N$, we denote with $|E|$ its Lebesgue measure.

For any $\mathbf{x} \in \mathbb{R}^3$, we call $r = \|\mathbf{x}\|$ and, provided $\mathbf{x} \neq \mathbf{0}$, $\hat{\mathbf{x}} = \mathbf{x}/\|\mathbf{x}\| \in \mathbb{S}^2 = \{\mathbf{x} \in \mathbb{R}^3 : \ \|\mathbf{x}\| = 1\}$. We finally define $\rho = (1 + r^2)^{1/2}$.

We consider the following direct scattering problem for the electromagnetic (EM) case, in the presence of a scatterer. We say that $\Sigma \subset \mathbb{R}^3$ is a *scatterer* if Σ is a compact set (possibly empty) such that $G = \mathbb{R}^3 \backslash \Sigma$ is connected. On the other hand, an open connected set G whose complement is compact will be referred to as an *exterior domain*. Notice that $\mathbb{R}^3$ itself is an exterior domain.

Let us assume that outside Σ, we have a medium characterised by the electric permittivity ε_1 and the magnetic permeability μ_1. For the time being, we do not assume any regularity or isotropy conditions, that is, they are tensors in G satisfying uniform ellipticity conditions. We recall that $M^{3\times 3}_{sym}(\mathbb{R})$ is the space of real-valued 3×3 symmetric matrices and that, for any open set $D \subset \mathbb{R}^3$, we say that a is a *tensor* in D satisfying *uniform ellipticity conditions* if $a \in L^\infty(D, M^{3\times 3}_{sym}(\mathbb{R}))$ and it satisfies

$$a_0 \|\xi\|^2 \leq a(\mathbf{x})\xi \cdot \xi \leq a_1 \|\xi\|^2 \quad \text{for a.e. } \mathbf{x} \in D \text{ and every } \xi \in \mathbb{R}^3, \tag{7.2.1}$$

where $0 < a_0 < a_1$ are the so-called *ellipticity constants*. In short the above condition may be written as

$$a_0 I_3 \leq a(\mathbf{x}) \leq a_1 I_3 \quad \text{for a.e. } \mathbf{x} \in D.$$

We also assume that, outside a bounded set, the space is homogeneous and isotropic, that is, there exists $R_0 > 0$ and two positive constants ε_∞, μ_∞ such that

$$\Sigma = \mathbb{R}^3 \backslash G \subset \overline{B_{R_0}} \tag{7.2.2}$$

and

$$\varepsilon_1(\mathbf{x}) = \varepsilon_\infty I_3 \text{ and } \mu_1(\mathbf{x}) = \mu_\infty I_3 \quad \text{for every } \mathbf{x} \text{ such that } \|\mathbf{x}\| > R_0. \tag{7.2.3}$$

In the sequel, we fix a *frequency* $\omega > 0$, and we call $k = \omega\sqrt{\varepsilon_\infty \mu_\infty} > 0$ the corresponding *wavenumber* and $\varepsilon = \varepsilon_\infty^{-1}\varepsilon_1$ and $\mu = \mu_\infty^{-1}\mu_1$. We shall take $k > 0$ and assume that ε and μ belong to $L^\infty(G, M^{3\times 3}_{sym}(\mathbb{R}))$ and, for some positive constants R_0 and $0 < \lambda_0 < 1 < \lambda_1$, satisfy

$$\lambda_0 I_3 \leq \varepsilon(\mathbf{x}), \mu(\mathbf{x}) \leq \lambda_1 I_3 \quad \text{for a.e. } \mathbf{x} \in G \tag{7.2.4}$$

and

$$\varepsilon(\mathbf{x}) = \mu(\mathbf{x}) = I_3 \quad \text{for a.e. } \mathbf{x} \in G \text{ such that } \|\mathbf{x}\| > R_0. \tag{7.2.5}$$

We say that $D \subset \mathbb{R}^3$ is a domain if it is an open connected set. We say that a domain D is *Lipschitz* if for any $\mathbf{x} \in \partial D$ there exist $r > 0$ and a Lipschitz function $\varphi : \mathbb{R}^2 \to \mathbb{R}$, with Lipschitz constant L, such that, up to a rigid change of coordinates, we have

$$D \cap B_r(\mathbf{x}) = \{\mathbf{y} = (y_1, y_2, y_3) \in B_r(\mathbf{x}) : \ y_3 < \varphi(y_1, y_2)\}.$$

Clearly, r, L, and the change of coordinates depend on the point $\mathbf{x}$. If we can use the same constant r and L for all $\mathbf{x} \in \partial D$, then we say that D is a with constants r and L. We remark that any Lipschitz domain D with compact boundary is Lipschitz with constants r and L for suitable positive constants r and L depending on D.

Let D and D' be two open sets and $T : D \to D'$ be a function. We say that T is locally $W^{1,\infty}$ if $T \in W^{1,\infty}_{loc}(D)$ and we notice that this is equivalent to requiring that T is locally Lipschitz. As usual we say that T is Lipschitz (with Lipschitz constant L) if $\|T(\mathbf{x}) - T(\mathbf{y})\| \leq L\|\mathbf{x} - \mathbf{y}\|$ for any $\mathbf{x}$, $\mathbf{y} \in D$. We say that T is a *bi-Lipschitz mapping* with constant L if T is bijective and T and T^{-1} are Lipschitz with Lipschitz constant L, on D and D', respectively. We say that T is a *bi-*$W^{1,\infty}$ *mapping* with constant L if T is bijective and $\|JT\|_{L^\infty(D)}$ and $\|J(T^{-1})\|_{L^\infty(D')}$ are both bounded by L. Here, and in what follows, JT denotes the Jacobian matrix of T.

The following remark will be of interest and a proof may be found, for instance, in [32, Chap. 6].

Remark 7.2.1. Let D and D' be open sets and $T : D \to D'$ be a bi-$W^{1,\infty}$ mapping bi-$W^{1,\infty}$ mapping. If D is connected, then either $\det JT(\mathbf{x}) > 0$ for almost every $\mathbf{x} \in D$ or $\det JT(\mathbf{x}) < 0$ for almost every $\mathbf{x} \in D$.

We shall also need the following spaces. Let $D \subset \mathbb{R}^3$ be a domain and let $a \in L^\infty(D, M^{3\times 3}_{sym}(\mathbb{R}))$ satisfying (7.2.1) with constants $0 < a_0 < a_1$.

We define, essentially following [31],

$$H(\text{curl}, D) = \{u \in L^2(D, \mathbb{C}^3) : \ \nabla \wedge u \in L^2(D, \mathbb{C}^3)\}$$

and

$$H(\text{div}, D) = \{u \in L^2(D, \mathbb{C}^3) : \ \nabla \cdot u \in L^2(D)\}.$$

We notice that $\nabla \wedge u$ and $\nabla \cdot u$ are always meant in the sense of distributions. Moreover, $L^2(D) = L^2(D, \mathbb{C})$. We notice that these are Hilbert spaces endowed with the usual graph norm. We also need

$$H_{loc}(\text{curl}, D) = \{u \in L^2_{loc}(D, \mathbb{C}^3) : \ \nabla \wedge u \in L^2_{loc}(D, \mathbb{C}^3)\},$$

and the following two spaces

$$H_0(\mathrm{curl},D)=\{u\in H(\mathrm{curl},D):\ \langle\nabla\wedge u,\phi\rangle-\langle u,\nabla\wedge\phi\rangle=0$$
$$\text{for any }\phi\in H(\mathrm{curl},D)\text{ with bounded support}\},$$

and

$$H_0(\mathrm{div},D)=\{u\in H(\mathrm{div},D):\ \langle\nabla\cdot u,\varphi\rangle+\langle u,\nabla\varphi\rangle=0$$
$$\text{for any }\varphi\in H^1(D)\text{ with bounded support}\}.$$

Here $\langle u,v\rangle=\int_D\overline{u}\cdot v$ is the usual L^2 scalar product, either in $L^2(D)$ or in $L^2(D,\mathbb{C}^3)$. We call $H(\mathrm{div}_a,D)$ and $H_0(\mathrm{div}_a,D)$ the set of $u\in L^2(D,\mathbb{C}^3)$ such that $au\in H(\mathrm{div},D)$ or $au\in H_0(\mathrm{div},D)$, respectively.

We notice that $u\in H_0(\mathrm{curl},D)$ satisfies, in a weak sense, the boundary condition

$$\nu\wedge u=0\quad\text{on }\partial D,$$

whereas $u\in H_0(\mathrm{div}_a,D)$ satisfies, again in a weak sense, the boundary condition

$$\nu\cdot(au)=0\quad\text{on }\partial D.$$

As usual, ν is the exterior unit normal to D on ∂D.

Remark 7.2.2. An important remark related to solutions to the Maxwell equations is the following. If $\mathbf{E}\in H_0(\mathrm{curl},D)$ and $\nabla\wedge\mathbf{E}=ca\mathbf{H}$ in D, for some $c\in\mathbb{C}\backslash\{0\}$, then $\mathbf{H}\in H_0(\mathrm{div}_a,D)$. In fact, first of all we notice that $\nabla\cdot(a\mathbf{H})=0$ in D in the sense of distribution, and therefore, $\mathbf{H}\in H(\mathrm{div}_a,D)$. We need to show that for any $\varphi\in H^1(D)$ with bounded support, we have $\langle a\mathbf{H},\nabla\varphi\rangle=0$. Since $\nabla\varphi\in H(\mathrm{curl},D)$, with $\nabla\wedge(\nabla\varphi)=0$, for any $\varphi\in H^1(D)$,

$$\langle a\mathbf{H},\nabla\varphi\rangle=(1/\overline{c})\langle\nabla\wedge\mathbf{E},\nabla\varphi\rangle=(1/\overline{c})\left(\langle\nabla\wedge\mathbf{E},\nabla\varphi\rangle-\langle\mathbf{E},\nabla\wedge(\nabla\varphi)\rangle\right)=0.$$

Let us notice that, under certain smoothness assumptions on D, we can better characterise these spaces. The smoothness assumption is of Lipschitz type and the result is the following; see, for instance, [27, Chap. 3].

Proposition 7.2.1. *Let D be a bounded Lipschitz domain.*

Then $H(\mathrm{div},D)=\overline{C^\infty(\overline{D})}$ and $H_0(\mathrm{div},D)=\overline{C_0^\infty(D)}$ with respect to the $H(\mathrm{div})$ norm.

For any $u\in H(\mathrm{div},D)$, we can define $\gamma_\nu(u)=\nu\cdot u|_{\partial D}\in H^{-1/2}(\partial D)$ such that

$$\langle\nabla\cdot u,\varphi\rangle+\langle u,\nabla\varphi\rangle=\langle\gamma_\nu(u),\varphi\rangle_{\partial D}\quad\text{for any }\varphi\in H^1(D).$$

Here and in the sequel, $\langle\cdot,\cdot\rangle_{\partial D}$ is the $H^{-1/2}$-$H^{1/2}$ duality on ∂D. Therefore, $H_0(\mathrm{div},D)=\{u\in H(\mathrm{div},D):\ \gamma_\nu(u)=0$ in $H^{-1/2}(\partial D)\}$.

Correspondingly, $H(\mathrm{curl},D)=\overline{C^\infty(\overline{D})}$ and $H_0(\mathrm{curl},D)=\overline{C_0^\infty(D)}$ with respect to the $H(\mathrm{curl})$ norm.

For any $u\in H(\mathrm{curl},D)$, we can define $\gamma_\tau(u)=\nu\wedge u|_{\partial D}\in H^{-1/2}(\partial D,\mathbb{C}^3)$ such that

$$\langle\nabla\wedge u,\phi\rangle-\langle u,\nabla\wedge\phi\rangle=\langle\gamma_\tau(u),\phi\rangle_{\partial D}\quad\text{for any }\phi\in H^1(D,\mathbb{C}^3).$$

Therefore, $H_0(\mathrm{curl},D)=\{u\in H(\mathrm{curl},D):\ \gamma_\tau(u)=0$ in $H^{-1/2}(\partial D,\mathbb{C}^3)\}$.

Also the following density result will be of use. This density is trivial in the case of Lipschitz open sets, but for a general open set, we could not find a reference for it.

Proposition 7.2.2. *Let D be any open set contained in $\mathbb{R}^3$. Then $H(\mathrm{curl},D)\cap L^\infty(D,\mathbb{C}^3)$ is dense in $H(\mathrm{curl},D)$ with respect to the $H(\mathrm{curl})$ norm.*

Proof. For any $M > 0$, we define a truncation operation $F_M : \mathbb{R}^3 \to \mathbb{R}^3$ as follows:

$$F_M(\mathbf{x}) = \begin{cases} M\mathbf{x}/\|\mathbf{x}\| & \text{if } \|\mathbf{x}\| \geq M, \\ \mathbf{x} & \text{if } \|\mathbf{x}\| \leq M. \end{cases}$$

By the general chain rule for vector valued functions proved in [3], for any $M > 0$ and any $v \in W^{1,1}(D,\mathbb{R}^3)$, such that $|\{\mathbf{x} \in D : \ \|v(\mathbf{x})\| = M\}| = 0$, we obtain that $F_M \circ v \in W^{1,1}(D,\mathbb{R}^3)$ and, for almost every $\mathbf{x} \in D$, we have

$$\nabla(F_M \circ v)(\mathbf{x}) = \begin{cases} 0 & \text{if } \|v(\mathbf{x})\| > M, \\ \nabla v(\mathbf{x}) & \text{if } \|v(\mathbf{x})\| < M. \end{cases}$$

In particular, for almost every $\mathbf{x} \in D$, we have

$$\nabla \wedge (F_M \circ v)(\mathbf{x}) = \begin{cases} 0 & \text{if } \|v(\mathbf{x})\| > M, \\ (\nabla \wedge v)(\mathbf{x}) & \text{if } \|v(\mathbf{x})\| < M. \end{cases} \tag{A.1}$$

For any $w \in L^2(D,\mathbb{R}^3)$, we define

$$A(w) = \{t \geq 1 : |\{\mathbf{x} \in D : \ \|v(\mathbf{x})\| = t\}| > 0\}$$

and notice that $A(w)$ is a countable subset of $\mathbb{R}$.

Let $D = \bigcup_{i\in\mathbb{N}} B_i$ where B_i is an open ball compactly contained in D for any $i \in \mathbb{N}$.

Let $u \in H(\text{curl}, D)$. We limit ourselves for simplicity, and without loss of generality, to prove the result for u which has values in $\mathbb{R}^3$ instead of $\mathbb{C}^3$.

For any $i \in \mathbb{N}$, there exists $\{u_n^i\}_{n\in\mathbb{N}} \subset C_0^\infty(\mathbb{R}^3)$ such that the following properties are satisfied. First, as $n \to \infty$, $u_n^i \to u$ and $(\nabla \wedge u_n^i) \to (\nabla \wedge u)$ in $L^2(B_i,\mathbb{R}^3)$ and almost everywhere in B_i. Second, there exists $h^i \in L^2(B_i)$ such that $\|u_n^i\| + \|\nabla \wedge u_n^i\| \leq h^i$ almost everywhere in B_i for any $n \in \mathbb{N}$.

We denote

$$A = A(u) \cup \left(\bigcup_{i,n\in\mathbb{N}} A(u_n^i) \right),$$

which is clearly a countable subset of $\mathbb{R}$. We pick a sequence $\{t_m\}_{m\in\mathbb{N}} \subset \mathbb{R}$ such that, for any $m \in \mathbb{N}$, $1 \leq t_m < t_{m+1}$, $t_m \notin A$ and $\lim_{m\to\infty} t_m = +\infty$.

For any $m \in \mathbb{N}$, we call

$$V_m(\mathbf{x}) = \begin{cases} 0 & \text{if } \|u(\mathbf{x})\| > t_m, \\ (\nabla \wedge u)(\mathbf{x}) & \text{if } \|u(\mathbf{x})\| < t_m. \end{cases}$$

We fix $i \in \mathbb{N}$ and $m \in \mathbb{N}$. Then, for almost every $\mathbf{x} \in B_i$, we have

$$\|F_{t_m} \circ u_n^i - F_{t_m} \circ u\|(\mathbf{x}) \leq h^i(\mathbf{x}) + \|u\|(\mathbf{x}) \quad \text{for any } n \in \mathbb{N}.$$

As $n \to \infty$, since $u_n^i(\mathbf{x}) \to u(\mathbf{x})$, we also have $(F_{t_m} \circ u_n^i)(\mathbf{x}) \to (F_{t_m} \circ u)(\mathbf{x})$. By Lebesgue theorem, we obtain that $(F_{t_m} \circ u_n^i)(\mathbf{x}) \to (F_{t_m} \circ u)(\mathbf{x})$ in $L^2(B_i)$ as $n \to \infty$.

Since, as $n \to \infty$, u_n^i converges to u and $\nabla \wedge u_n^i$ converges to $\nabla \wedge u$ almost everywhere in B_i and $|\{\mathbf{x} \in D : \ \|u(\mathbf{x})\| = t_m\}| = 0$, by (7.2.6), we can immediately infer that $\nabla \wedge (F_{t_m} \circ u_n^i)$ converges to V_m almost everywhere in B_i and, again by Lebesgue theorem, also in $L^2(B_i)$.

Therefore, $F_{t_m} \circ u \in H(\text{curl}, B_i)$, with $\nabla \wedge (F_{t_m} \circ u) = V_m|_{B_i}$ in the sense of distributions in B_i, for any $i \in \mathbb{N}$.

Now let $\varphi \in C_0^\infty(D,\mathbb{R}^3)$ and let $K \subset D$ be compact such that the support of φ is contained in K. Up to changing the order, we can suppose that $K \subset \bigcup_{i=1}^{l} B_i$ and we let $\chi_i \in C_0^\infty(B_i)$, $i = 1, \ldots, l$, be a partition of unity on K. Then

$$\langle F_{t_m} \circ u, \nabla \wedge \varphi \rangle_D = \left\langle F_{t_m} \circ u, \nabla \wedge \left(\sum_i^l \chi_i \varphi \right) \right\rangle_D$$

$$= \sum_{i=1}^l \langle F_{t_m} \circ u, \nabla \wedge (\chi_i \varphi) \rangle_{B_i} = \sum_i^l \langle V_m, \chi_i \varphi \rangle_{B_i} = \langle V_m, \varphi \rangle_D.$$

From here it is easy to conclude that $F_{t_m} \circ u \in H(\mathrm{curl}, D)$, with $\nabla \wedge (F_{t_m} \circ u) = V_m$ in the sense of distributions in D. Finally, it is an obvious remark that $F_{t_m} \circ u \in H(\mathrm{curl}, D) \cap L^\infty(D, \mathbb{R}^3)$ and that $F_{t_m} \circ u$ converges to u, as $m \to \infty$, in the $H(\mathrm{curl})$ norm. □

Finally, let us assume that D is an exterior domain, that is, it contains the exterior of a ball. In order to control the behaviour at infinity, we shall use the following notation. For any $t \in \mathbb{R}$ and any $u \in L^2_{loc}(D)$, either complex-valued or $\mathbb{C}^3$-valued, we set

$$\|u\|_{0,t} = \left(\int_D \rho^{2t} \|u\|^2 \right)^{1/2},$$

where we recall that $\rho = (1 + r^2)^{1/2}$, $r = \|\mathbf{x}\|$. We finally define

$$H(\mathrm{curl}, D, t) = \{ u \in H_{loc}(\mathrm{curl}, D) : \ \|u\|_{0,t} + \|\nabla \wedge u\|_{0,t} < +\infty \}$$

and

$$H(\mathrm{div}, D, t) = \{ u \in H_{loc}(\mathrm{div}, D) : \ \|u\|_{0,t} + \|\nabla \cdot u\|_{0,t} < +\infty \}.$$

7.2.2 Mathematical Formulation

The electromagnetic wave is described by the electric field $\mathscr{E}(\mathbf{x}, t)$ and the magnetic field $\mathscr{H}(\mathbf{x}, t)$ for $(\mathbf{x}, t) \in G \times \mathbb{R}_+$. The electromagnetic wave propagation is governed by the Maxwell equations

$$\nabla \wedge \mathscr{E}(\mathbf{x}, t) + \mu_1 \frac{\partial \mathscr{H}}{\partial t}(\mathbf{x}, t) = 0, \quad \nabla \wedge \mathscr{H}(\mathbf{x}, t) - \varepsilon_1 \frac{\partial \mathscr{E}}{\partial t}(\mathbf{x}, t) = 0. \tag{7.2.6}$$

For time-harmonic electromagnetic waves of the form

$$\mathscr{E}(\mathbf{x}, t) = \Re(\varepsilon_\infty^{-1/2} \mathbf{E}(\mathbf{x}) e^{-\mathrm{i}\omega t}), \quad \mathscr{H}(\mathbf{x}, t) = \Re(\mu_\infty^{-1/2} \mathbf{H}(\mathbf{x}) e^{-\mathrm{i}\omega t}),$$

where $(\mathbf{E}, \mathbf{H})(\mathbf{x}) \in \mathbb{C}^3 \times \mathbb{C}^3$ and $\omega > 0$ is the frequency, it is directly verified that one has the reduced Maxwell equations

$$\nabla \wedge \mathbf{E}(\mathbf{x}) - \mathrm{i}k\mu \mathbf{H}(\mathbf{x}) = \mathbf{0}, \quad \nabla \wedge \mathbf{H}(\mathbf{x}) + \mathrm{i}k\varepsilon \mathbf{E}(\mathbf{x}) = \mathbf{0}, \tag{7.2.7}$$

with the wavenumber $k > 0$ and the coefficients ε and μ defined as in Sect. 7.2.1. Notice that, when the medium is homogenous and isotropic, that is, ε_1 and μ_1 are simply positive constants coinciding with ε_∞ and μ_∞ everywhere, then (7.2.7) reduces to

$$\nabla \wedge \mathbf{E}(\mathbf{x}) - \mathrm{i}k \mathbf{H}(\mathbf{x}) = \mathbf{0}, \quad \nabla \wedge \mathbf{H}(\mathbf{x}) + \mathrm{i}k \mathbf{E}(\mathbf{x}) = \mathbf{0}. \tag{7.2.8}$$

In the sequel we shall consider the system (7.2.7) with $k > 0$ and ε and μ satisfying (7.2.4) and (7.2.5)

The equation (7.2.7) is to be understood in the sense of distribution, and hence, it is well formulated, for instance, for $(\mathbf{E}, \mathbf{H}) \in (H_{loc}(\mathrm{curl}, G))^2$.

We say that a solution $(\mathbf{E},\mathbf{H})$ to (7.2.7) is *outgoing* or *radiating* if it satisfies the following *Silver-Müller radiation condition*:

$$\lim_{r\to+\infty} r\left(\frac{\mathbf{x}}{\|\mathbf{x}\|}\wedge\mathbf{H}(\mathbf{x})+\mathbf{E}(\mathbf{x})\right)=\mathbf{0},\quad r=\|\mathbf{x}\|, \tag{7.2.9}$$

which holds uniformly in all directions $\hat{\mathbf{x}} := \mathbf{x}/\|\mathbf{x}\| \in \mathbb{S}^2$. The Silver-Müller radiation condition characterises the radiating nature of solutions to the Maxwell equations. We recall that this is equivalent to requiring that

$$\lim_{r\to+\infty} r\left(\frac{\mathbf{x}}{\|\mathbf{x}\|}\wedge\mathbf{E}(\mathbf{x})-\mathbf{H}(\mathbf{x})\right)=\mathbf{0},\quad r=\|\mathbf{x}\|; \tag{7.2.10}$$

see, for instance, [10].

In the case of a homogeneous and isotropic medium, the following connection between the Maxwell equations and the vectorial Helmholtz equation is known; see again [10].

Lemma 7.2.1. *If* $(\mathbf{E},\mathbf{H})$ *is a solution to the Maxwell equations* (7.2.8)*, that is,*

$$\nabla\wedge\mathbf{E}-\mathrm{i}k\mathbf{H}=\mathbf{0},\quad \nabla\wedge\mathbf{H}+\mathrm{i}k\mathbf{E}=0,$$

then $\mathbf{E}$ *and* $\mathbf{H}$ *satisfy the following vectorial Helmholtz equations:*

$$\Delta\mathbf{E}+k^2\mathbf{E}=\mathbf{0},\quad \nabla\cdot\mathbf{E}=0, \tag{7.2.11}$$

$$\Delta\mathbf{H}+k^2\mathbf{H}=\mathbf{0},\quad \nabla\cdot\mathbf{H}=0. \tag{7.2.12}$$

Conversely, if $\mathbf{E}$ *satisfies* (7.2.11) (*or* $\mathbf{H}$ *satisfies* (7.2.12))*, then* $\mathbf{E}$ *and* $\mathbf{H}=(\nabla\wedge\mathbf{E})/(\mathrm{i}k)$ (*or* $\mathbf{H}$ *and* $\mathbf{E}=-(\nabla\wedge\mathbf{H})/(\mathrm{i}k)$*, respectively) satisfy the Maxwell equations* (7.2.8).

Moreover, for $(\mathbf{E},\mathbf{H})$*, a solution to the Maxwell equations* (7.2.8)*, the Silver-Müller radiation condition* (7.2.9) *is equivalent to the Sommerfeld radiation condition for all components of* $\mathbf{E}$ *and* $\mathbf{H}$*; that is,*

$$\lim_{r\to+\infty} r\left(\frac{\partial u}{\partial r}-\mathrm{i}ku\right)=0,\quad r=\|\mathbf{x}\|, \tag{7.2.13}$$

where the limit holds uniformly in all directions $\hat{\mathbf{x}}=\mathbf{x}/\|\mathbf{x}\|\in\mathbb{S}^2$*, and* u *denotes any of the Cartesian components of* $\mathbf{E}$ *or* $\mathbf{H}$.

Using the asymptotic behaviour of outgoing solutions to the Helmholtz equation, we can deduce the following asymptotic behaviour of outgoing solutions of the Maxwell equations. That is, as $r=\|\mathbf{x}\|\to+\infty$,

$$\begin{aligned}\mathbf{E}(\mathbf{x}) &= \frac{e^{\mathrm{i}k\|\mathbf{x}\|}}{\|\mathbf{x}\|}\mathbf{E}_\infty(\hat{\mathbf{x}})+\mathcal{O}\left(\frac{1}{\|\mathbf{x}\|^2}\right),\\ \mathbf{H}(\mathbf{x}) &= \frac{e^{\mathrm{i}k\|\mathbf{x}\|}}{\|\mathbf{x}\|}\mathbf{H}_\infty(\hat{\mathbf{x}})+\mathcal{O}\left(\frac{1}{\|\mathbf{x}\|^2}\right),\end{aligned} \tag{7.2.14}$$

which hold uniformly in $\hat{\mathbf{x}}=\mathbf{x}/\|\mathbf{x}\|\in\mathbb{S}^2$. Here $\mathbf{E}_\infty$ and $\mathbf{H}_\infty$ are complex-valued functions defined on $\mathbb{S}^2$ and denote the *electric and magnetic far-field patterns* , respectively, and they satisfy for any $\hat{\mathbf{x}}\in\mathbb{S}^2$,

$$\mathbf{H}_\infty(\hat{\mathbf{x}})=\hat{\mathbf{x}}\wedge\mathbf{E}_\infty(\hat{\mathbf{x}})\quad\text{and}\quad \hat{\mathbf{x}}\cdot\mathbf{E}_\infty(\hat{\mathbf{x}})=\hat{\mathbf{x}}\cdot\mathbf{H}_\infty(\hat{\mathbf{x}})=0. \tag{7.2.15}$$

It is also known that $\mathbf{E}_\infty$ and $\mathbf{H}_\infty$ are real analytic, and, hence if they are given on any open patch of $\mathbb{S}^2$, then they are known on the whole unit sphere by analytic continuation.

For the scatterer Σ defined earlier, the EM wave cannot penetrate inside the scatterer and the Maxwell system (7.2.7) is defined only in $G=\mathbb{R}^3\backslash\Sigma$. Depending on the physical property of the scatterer, one would have the following boundary conditions on $\partial\Sigma$:

$$\nu(\mathbf{x}) \wedge \mathbf{E}(\mathbf{x}) = 0, \quad \mathbf{x} \in \partial\Sigma, \tag{7.2.16}$$

corresponding to a *perfectly electric conducting* (PEC) scatterer Σ, or

$$\nu(\mathbf{x}) \wedge \mathbf{H}(\mathbf{x}) = 0, \quad \mathbf{x} \in \partial\Sigma, \tag{7.2.17}$$

corresponding to a *perfectly magnetic conducting* (PMC) scatterer Σ. Here, $\nu \in \mathbb{S}^2$ denotes the exterior unit normal vector to G (or interior unit normal to Σ) on $\partial G = \partial\Sigma$.

Remark 7.2.3. Let us remark here once for all that if $(\mathbf{E},\mathbf{H})$ solves (7.2.7), then $(\mathbf{H},-\mathbf{E})$ solves the same equation provided we swap ε with μ. Moreover, if $(\mathbf{E},\mathbf{H})$ is an outgoing solution to (7.2.7), then $(\mathbf{H},-\mathbf{E})$ is also outgoing. Therefore, by this kind of change of variables, we can always turn a perfectly magnetic conducting scatterer Σ into a perfectly electric conducting scatterer. Hence, in what follows, we shall only consider the perfectly electric conducting case but it is clear that all the results of this chapter hold for the perfectly magnetic conducting case as well.

Summarising our discussion above, the EM scattering problem we are interested in is the following. We fix $k > 0$ and consider a pair of *incident electric and magnetic fields* $(\mathbf{E}^i,\mathbf{H}^i)$ given by an entire solution to (7.2.8). For example, we can choose

$$\mathbf{E}^i(\mathbf{x}) = \frac{\mathrm{i}}{k}\nabla\wedge\left(\nabla\wedge\mathbf{p}e^{ik\mathbf{x}\cdot\mathbf{d}}\right), \quad \mathbf{H}^i(\mathbf{x}) = \nabla\wedge\mathbf{p}e^{ik\mathbf{x}\cdot\mathbf{d}}, \quad \mathbf{x}\in\mathbb{R}^3. \tag{7.2.18}$$

In this case $(\mathbf{E}^i,\mathbf{H}^i) = (\mathbf{E}^i,\mathbf{H}^i)(\mathbf{p},\mathbf{d})$ is known as the *normalised electromagnetic plane wave* with the *polarisation vector* $\mathbf{p}\in\mathbb{R}^3$, $\mathbf{p}\neq\mathbf{0}$, and the *incident direction* $\mathbf{d}\in\mathbb{S}^2$. Given a perfectly electric conducting scatterer Σ and ε and μ satisfying the above hypotheses, we look for a solution $(\mathbf{E},\mathbf{H})$ to the following exterior boundary value problem:

$$\begin{cases} \nabla\wedge\mathbf{E} - \mathrm{i}k\mu\mathbf{H} = \mathbf{0} & \text{in } G = \mathbb{R}^3\backslash\Sigma, \\ \nabla\wedge\mathbf{H} + ik\varepsilon\mathbf{E} = \mathbf{0} & \text{in } G, \\ (\mathbf{E},\mathbf{H}) = (\mathbf{E}^i,\mathbf{H}^i) + (\mathbf{E}^s,\mathbf{H}^s) & \text{in } G, \\ \nu\wedge\mathbf{E} = \mathbf{0} & \text{on } \partial G, \\ \lim_{r\to+\infty} r\left(\frac{\mathbf{x}}{\|\mathbf{x}\|}\wedge\mathbf{H}^s(\mathbf{x}) + \mathbf{E}^s(\mathbf{x})\right) = \mathbf{0}\ r = \|\mathbf{x}\|. \end{cases} \tag{7.2.19}$$

Here $\mathbf{E}^s$ and $\mathbf{H}^s$ in (7.2.19) are called the *scattered electric and magnetic fields*, respectively, and by the last line they satisfy the Silver-Müller radiation condition the . We call $\mathbf{E}$ and $\mathbf{H}$ the *total electric and magnetic fields*, respectively.

The Maxwell system (7.2.19) describes the following electromagnetic wave scattering. In the homogeneous and isotropic space, the EM incident wave $(\mathbf{E}^i,\mathbf{H}^i)$ would propagate indefinitely, since it is an entire solution to the Maxwell system (7.2.8). The presence of the scatterer Σ, and of possible inhomogeneity around it, perturbs the propagation of the incident wave. Such a perturbation is given by the scattered wave field $(\mathbf{E}^s,\mathbf{H}^s)$, which, outside a large enough ball, is an outgoing, that is, radiating, solution to the Maxwell system (7.2.8).

The existence and uniqueness of a solution to (7.2.19) is classical in the case of a homogeneous and isotropic medium and of smooth scatterers; see, for instance, [10, 29]. More effort is needed for the case of Lipschitz scatterers, and this case is also rather well understood; see, for instance, [27].

In order to deal with the most general scenario, that is, with minimal assumptions on the scatterer Σ and the coefficients ε and μ, we shall make use of a quite general result proved in [31]. The uniqueness and existence result will be treated in the next subsection.

7.2.3 Existence and Uniqueness

Given an exterior domain G, we call $G_R = G \cap B_R(\mathbf{0})$ for any $R > 0$. We assume that (7.2.2), (7.2.4) and (7.2.5) hold for some constants $R_0 > 0$ and $0 < \lambda_0 < \lambda_1$. We say that $(\mathbf{E},\mathbf{H})$ is a (weak) solution to the direct scattering problem (7.2.19) for a given incident field $(\mathbf{E}^i,\mathbf{H}^i)$ if $(\mathbf{E},\mathbf{H}) \in H(\mathrm{curl}, G_R)^2$ for any $R > 0$, the Maxwell system is satisfied in G in the sense of distributions, $\mathbf{E} \in H_0(\mathrm{curl}, G)$, and $(\mathbf{E}^s,\mathbf{H}^s) = (\mathbf{E},\mathbf{H}) - (\mathbf{E}^i,\mathbf{H}^i)$ is an outgoing solution to the Maxwell system (7.2.8) in $\mathbb{R}^3 \backslash \overline{B_{R_0}}$.

To fix an auxiliary function $\chi \in C_0^\infty(\mathbb{R}^3)$ such that $\chi \equiv 1$ in a neighbourhood of $\overline{B_{R_0}}$, we have that $(\mathbf{E},\mathbf{H})$ solves (7.2.19) if and only if $(\mathbf{E}_1,\mathbf{H}_1) = (\mathbf{E},\mathbf{H}) - (1-\chi)(\mathbf{E}^i,\mathbf{H}^i)$ solves, in an analogous weak sense,

$$\begin{cases} \nabla \wedge \mathbf{E}_1 - \mathrm{i}k\mu \mathbf{H}_1 = \mathbf{F} & \text{in } G, \\ \nabla \wedge \mathbf{H}_1 + \mathrm{i}k\varepsilon \mathbf{E}_1 = \mathbf{G} & \text{in } G, \\ \nu \wedge \mathbf{E}_1 = \mathbf{0} & \text{on } \partial G, \\ \lim_{r\to+\infty} r\left(\frac{\mathbf{x}}{\|\mathbf{x}\|} \wedge \mathbf{H}_1(\mathbf{x}) + \mathbf{E}_1(\mathbf{x})\right) = \mathbf{0} \; r = \|\mathbf{x}\|, \end{cases} \tag{7.2.20}$$

where

$$\mathbf{F} = \nabla \wedge ((\chi-1)\mathbf{E}^i) - \mathrm{i}k(\chi-1)\mu\mathbf{H}^i, \quad \mathbf{G} = \nabla \wedge ((\chi-1)\mathbf{H}^i) + \mathrm{i}k(\chi-1)\varepsilon\mathbf{E}^i.$$

Notice that $\mathbf{F}$ and $\mathbf{G}$ belong to $L^2(D,\mathbb{C}^3)$ and have bounded support.

In order to connect to the existence and uniqueness result in [31], we need the following.

Lemma 7.2.2. *Let us fix $R > 0$. Let $(\mathbf{E},\mathbf{H}) \in H_{loc}(\mathrm{curl}, \mathbb{R}^3 \backslash \overline{B_R})^2$ solve (7.2.8) in $\mathbb{R}^3 \backslash \overline{B_R}$. Then the following two facts hold true.*

(A) $(\mathbf{E},\mathbf{H})$ satisfies the Silver-Müller radiation condition if and only if

$$(\mathbf{E},\mathbf{H}) \in \bigcap_{t<-1/2} H(\mathrm{curl}, \mathbb{R}^3 \backslash \overline{B_{R+1}}, t)^2 \tag{7.2.21}$$

and

$$\frac{\mathbf{x}}{\|\mathbf{x}\|} \wedge \mathbf{H}(\mathbf{x}) + \mathbf{E}(\mathbf{x}), \; \frac{\mathbf{x}}{\|\mathbf{x}\|} \wedge \mathbf{E}(\mathbf{x}) - \mathbf{H}(\mathbf{x}) \in \bigcup_{t>-1/2} H(\mathrm{curl}, \mathbb{R}^3 \backslash \overline{B_{R+1}}, t). \tag{7.2.22}$$

(B) If $(\mathbf{E},\mathbf{H})$ satisfies the Silver-Müller radiation condition and

$$(\mathbf{E},\mathbf{H}) \in \bigcap_{t\in\mathbb{R}} H(\mathrm{curl}, \mathbb{R}^3 \backslash \overline{B_{R+1}}, t)^2,$$

then $\mathbf{E} = \mathbf{H} = 0$ in $\mathbb{R}^3 \backslash \overline{B_R}$.

Proof. We begin with part A) and we assume that $(\mathbf{E},\mathbf{H})$ satisfies the Silver-Müller radiation condition. Let u be any Cartesian component of $\mathbf{E}$ or $\mathbf{H}$. Then, by the Sommerfeld radiation condition, we have

$$u(\mathbf{x}) = \frac{\mathrm{e}^{\mathrm{i}k\|\mathbf{x}\|}}{\|\mathbf{x}\|}\mathbf{u}_\infty(\hat{\mathbf{x}}) + \frac{a(r\hat{\mathbf{x}})}{\|\mathbf{x}\|^2}, \quad r = \|\mathbf{x}\| \geq R+1.$$

Here, for some constant C_0,

$$|a(r\hat{\mathbf{x}})| \leq C_0 \quad \text{for any } r \geq R+1, \; \hat{\mathbf{x}} \in \mathbb{S}^2.$$

Since $r \leq \rho \leq C(R)r$ for any $r \geq R > 0$, with $C(R)$ depending on R only, we deduce that, for any $t \in \mathbb{R}$,

$$\int_{\mathbb{R}^3 \backslash \overline{B_{R+1}}} \rho^{2t}\|u\|^2 \leq C_1 \int_{R+1}^{+\infty} r^{2t} \left(\int_{\mathbb{S}^2} \|\mathbf{u}_\infty(\hat{\mathbf{x}}) + a(r\hat{\mathbf{x}})/r\|^2 \mathrm{d}\sigma(\hat{\mathbf{x}})\right) \mathrm{d}r \leq C_2 \int_{R+1}^{+\infty} r^{2t}\mathrm{d}r,$$

with C_1 and C_2 depending on R and t. Since the right-hand side is finite for any $t < -1/2$, we conclude that (7.2.21) holds.

We also notice that, whenever $\mathbf{u}_\infty \not\equiv 0$, then for any $t \in \mathbb{R}$, there exist $\tilde{R} \geq R+1$ and $\tilde{C}_0 > 0$ such that

$$\int_{\mathbb{R}^3\backslash\overline{B_{R+1}}} \rho^{2t}\|u\|^2 \geq \tilde{C}_0 \int_{\tilde{R}}^{+\infty} r^{2t}\mathrm{d}r.$$

The right-hand side is infinite for any $t \geq -1/2$. Therefore, if $\|u\|_{0,t}$ on $\mathbb{R}^3\backslash\overline{B_{R+1}}$ is finite for every $t \in \mathbb{R}$, then we obtain that $\mathbf{u}_\infty \equiv 0$ and, in turn by the Rellich lemma , that $u \equiv 0$ in $\mathbb{R}^3\backslash\overline{B_R}$. Therefore, part B) is also proved.

Now we prove that (7.2.22) holds. This follows immediately by the arguments developed in [10, Chap. 6], since the Silver-Müller radiation condition implies that, for any $\|\mathbf{x}\| = r \geq R+1$,

$$\left\| \frac{\mathbf{x}}{\|\mathbf{x}\|} \wedge \mathbf{H}(\mathbf{x}) + \mathbf{E}(\mathbf{x}) \right\|, \left\| \frac{\mathbf{x}}{\|\mathbf{x}\|} \wedge \mathbf{E}(\mathbf{x}) - \mathbf{H}(\mathbf{x}) \right\| \leq C/r^2$$

for some constant C. We readily conclude that (7.2.22) holds.

We conclude the proof by showing that (7.2.21) and (7.2.22) imply the Silver-Müller radiation condition . We call

$$a(r) = \int_{\partial B_r} \left\| \frac{\mathbf{x}}{\|\mathbf{x}\|} \wedge \mathbf{H}(\mathbf{x}) + \mathbf{E}(\mathbf{x}) \right\|^2 \mathrm{d}\sigma(\mathbf{x}).$$

By (7.2.22), we deduce that

$$\int_{R+1}^{+\infty} r^{2t} a(r)\mathrm{d}r < +\infty \quad \text{for some } t > -1/2. \tag{7.2.23}$$

If $\liminf_{r\to+\infty} a(r) = \overline{a} > 0$, then (7.2.23) is violated. Therefore, we can find $\{r_n\}_{n\in\mathbb{N}}$ such that $R+1 \leq r_n < r_{n+1}$ for any $n \in \mathbb{N}$ and satisfying $\lim_{n\to\infty} r_n = +\infty$ and

$$a(r_n) = \int_{\partial B_{r_n}} \left\| \frac{\mathbf{x}}{\|\mathbf{x}\|} \wedge \mathbf{H}(\mathbf{x}) + \mathbf{E}(\mathbf{x}) \right\|^2 \mathrm{d}\sigma(\mathbf{x}) \to 0 \quad \text{as } n \to \infty.$$

Following the proof of Theorem 6.6 in [10], we can deduce that the Stratton-Chu formulas hold. In turn, they imply the Silver-Müller radiation condition; thus, the proof is concluded. □

Concerning the assumptions on G (or equivalently Σ), we need the following definition, again from [31].

Definition 7.2.1. Let D be a bounded domain. We say that D satisfies the *Rellich compactness property* (in short RCP) if the natural immersion of $H^1(D)$ into $L^2(D)$ is compact.

We say that D satisfies the *Maxwell compactness property* (in short MCP) if the natural immersions of $H_0(\mathrm{curl}, D) \cap H(\mathrm{div}, D)$ and of $H(\mathrm{curl}, D) \cap H_0(\mathrm{div}, D)$ into $L^2(D, \mathbb{C}^3)$ are compact.

Sufficient conditions that guarantee the RCP may be found in many standard reference books on Sobolev spaces, for instance, in [1,25]. A detailed description of sufficient conditions for MCP to hold may be found in [31, Theorem 3.6]. In the next subsection we provide another sufficient condition, which is useful for our purposes; see Proposition 7.2.3.

Now we state the following crucial theorem that is just a rephrasing of the main result of [31].

Theorem 7.2.1 (Theorem 2.10 in [31]). *Fix positive constants k, R_0 and $\lambda_0 < \lambda_1$.*

Let G be an exterior domain satisfying (7.2.2) *and such that, for some $R > R_0$, G_R satisfies the RCP and the MCP.*

Let ε, $\mu \in L^\infty(G, M^{3\times3}_{sym}(\mathbb{R}))$ satisfy (7.2.4) *and* (7.2.5). *Let $\mathbf{F}$, $\mathbf{G} \in L^2(G, \mathbb{C}^3)$ with bounded support.*

Then (7.2.20) *admits a unique solution $(\mathbf{E}_1, \mathbf{H}_1)$ if and only if the corresponding problem with $\mathbf{F} = \mathbf{G} = \mathbf{0}$ in G admits only the trivial solution $\mathbf{E}_1 = \mathbf{H}_1 = \mathbf{0}$ in G.*

Moreover, if $(\mathbf{E}_1, \mathbf{H}_1)$ *solves* (7.2.20) *with* $\mathbf{F} = \mathbf{G} = \mathbf{0}$ *in* G*, then* $(\mathbf{E}_1, \mathbf{H}_1) \in \bigcap_{t\in\mathbb{R}} H(\mathrm{curl}, \mathbb{R}^3 \setminus \overline{B_{R+1}}, t)^2$.

Under the assumptions of Theorem 7.2.1, we have existence and uniqueness of a solution to (7.2.20) and, in turn, of a solution to (7.2.19), if the Maxwell system (7.2.7) admits the *unique continuation property* (in short UCP). In fact, by part *B*) of Lemma 7.2.2 and the last part of Theorem 7.2.1, we obtain that if $(\mathbf{E}_1, \mathbf{H}_1)$ solves (7.2.20) with $\mathbf{F} = \mathbf{G} = \mathbf{0}$ in G, then $\mathbf{E}_1 = \mathbf{H}_1 = \mathbf{0}$ outside a sufficiently large ball.

We summarise the existence and uniqueness result in the following theorem.

Theorem 7.2.2. *Fix positive constants* k, R_0 *and* $\lambda_0 < \lambda_1$.

Let G *be an exterior domain satisfying* (7.2.2) *and such that, for some* $R > R_0$, G_R *satisfies the RCP and the MCP.*

Let ε, $\mu \in L^\infty(G, M^{3\times 3}_{sym}(\mathbb{R}))$ *satisfy* (7.2.4) *and* (7.2.5)*. Moreover, let us assume that* (7.2.7) *satisfies the UCP in* G.

Then, for any $(\mathbf{E}^i, \mathbf{H}^i)$ *entire solution to* (7.2.8)*, the problem* (7.2.19) *admits a unique solution* $(\mathbf{E}, \mathbf{H})$.

7.2.4 Sufficient Conditions for RCP, MCP and UCP

A useful sufficient condition for RCP and MCP to hold is contained in the following proposition.

Proposition 7.2.3. *Let* D *be a bounded domain. Let us assume that for any* $\mathbf{x} \in \partial D$*, there exists an open neighbourhood* $U_\mathbf{x}$ *such that* $U_\mathbf{x} \cap D$ *has a finite number of connected components. Moreover, each connected component of* $U_\mathbf{x} \cap D$ *such that* $\mathbf{x}$ *belongs to its boundary may be mapped onto a Lipschitz domain by a bi-*$W^{1,\infty}$ *mapping.*

Then D *satisfies both the RCP and MCP.*

Proof of Proposition 7.2.3. For any $\mathbf{x} \in \partial D$, let $U_\mathbf{x}$ be as in the assumptions. We call $\tilde{U}_\mathbf{x}$ the union of the connected components of $U_\mathbf{x} \cap D$ such that $\mathbf{x}$ belongs to their boundaries. We call $V_\mathbf{x} = (U_\mathbf{x} \cap D) \setminus \tilde{U}_\mathbf{x}$ and, finally, $\hat{U}_\mathbf{x} = U_\mathbf{x} \setminus \overline{V_\mathbf{x}}$. We have that $\hat{U}_\mathbf{x}$ is an open set containing $\mathbf{x}$ and such that $\hat{U}_\mathbf{x} \cap D = \tilde{U}_\mathbf{x}$. Therefore, $\hat{U}_\mathbf{x} \cap D$ has a finite number of connected components, and each of them may be mapped onto a Lipschitz domain by a bi-$W^{1,\infty}$ mapping.

By compactness, we consider $\partial D \subset \bigcup_{i=1}^n \hat{U}_{\mathbf{x}_i}$. Then $\overline{D} \subset \bigcup_{i=0}^n U_i$ with U_0 a smooth open set compactly contained in D and $U_i = \hat{U}_{\mathbf{x}_i}$ for any $i = 1, \ldots, n$. Let $\chi_i \in C_0^\infty(U_i)$, $i = 0, \ldots, n$, be a partition of unity on $\overline{D}$.

By our previous reasoning, for any $i = 1, \ldots, n$, we have $U_i \cap D = \bigcup_{j=1}^{m_i} D_{i,j}$ where each $D_{i,j}$ is a domain and there exists $T_{i,j} : D_{i,j} \to \tilde{D}_{i,j}$ with $T_{i,j}$ a bi-$W^{1,\infty}$ mapping and $\tilde{D}_{i,j}$ a Lipschitz domain.

By Proposition 7.3.1, through Corollary 7.3.1, we infer that $D_0 = U_0$ and $D_{i,j}$, for any $i = 1, \ldots, n$ and $j = 1, \ldots, m_i$, satisfy the MCP (and, with an analogous reasoning, also the RCP). It is easy to prove that D_0 and $D_i = U_i \cap D$, for any $i = 1, \ldots, n$, satisfy the MCP and RCP as well.

Now, let us consider $\{u^m\}_{m\in\mathbb{N}}$ bounded in $H_0(\mathrm{curl}, D) \cap H(\mathrm{div}, D)$ (or bounded in $H(\mathrm{curl}, D) \cap H_0(\mathrm{div}, D)$, respectively). We define, for any $m \in \mathbb{N}$ and any $i = 0, \ldots, n$, $u_i^m = \chi_i u^m$. It is easy to show that, for any $i = 0, \ldots, n$, $\{u_i^m\}_{m\in\mathbb{N}}$ is bounded in $H_0(\mathrm{curl}, D_i) \cap H(\mathrm{div}, D_i)$ (or bounded in $H(\mathrm{curl}, D_i) \cap H_0(\mathrm{div}, D_i)$, respectively). Therefore, the proof may be easily concluded. □

In the literature there are several sufficient conditions on the coefficients ε and μ in (7.2.7) for UCP to hold. We notice that if $\varepsilon = \mu = I_3$ everywhere, that is, we consider (7.2.8), then UCP trivially holds.

The first result on unique continuation that we wish to recall follows immediately from [30, Theorem 1.1].

Theorem 7.2.3. *Let* D *be a domain. We assume that* ε *and* μ *belong to* $L^\infty(D, M^{3\times 3}_{sym}(\mathbb{R}))$ *and satisfy* (7.2.4) *in* D *for some constants* $0 < \lambda_0 < \lambda_1$.

If ε and μ are locally Lipschitz in D, then (7.2.7) satisfies the UCP in D.

Let us notice that Lipschitz continuity is essentially optimal, as shown by an example in [13]. Inspired by the results and constructions in [4] for the piecewise Lipschitz case, we state the following.

Proposition 7.2.4. *Let D be an open set. Assume that ε and μ belong to $L^\infty(D, M^{3\times3}_{sym}(\mathbb{R}))$ and satisfy (7.2.4) in D for some constants $0 < \lambda_0 < \lambda_1$.*

Assume that:

(i) There exists a family $\{D_i\}$ of domains, which are contained in D and which are pairwise disjoint, such that

$$D \subset \bigcup_i \overline{D_i}.$$

(ii) We have $|\sigma| = 0$ where

$$\sigma = D \cap \left(\bigcup_i \partial D_i \right).$$

(iii) We say that $\mathbf{x} \in \sigma$ separates exactly two partitions *if there exist $\delta > 0$ and two different indexes i and j such that*

$$\left| B_\delta(\mathbf{x}) \setminus (D_i \cup D_j) \right| = 0$$

and $B_\delta(\mathbf{x}) \cap D_i$ and $B_\delta(\mathbf{x}) \cap D_j$ are not empty. We call

$$C = \{\mathbf{x} \in \sigma : \mathbf{x} \text{ does not separate exactly two partitions}\}.$$

We assume that $\tilde{D} = D \setminus C$ is connected.

(iv) $(\varepsilon, \mu) = (\varepsilon_i, \mu_i)$ in D_i where $\{\varepsilon_i\}$ and $\{\mu_i\}$ are families of locally Lipschitz $M^{3\times3}_{sym}(\mathbb{R})$-valued functions in D.

Then (7.2.7) satisfies the UCP in D.

Proof. We begin with a few preliminary remarks. First of all, the family $\{D_i\}$ is countable. Second, $\sigma = D \setminus (\bigcup_i D_i)$ is closed in D. Moreover, C is closed in σ, and thus in D as well; therefore, $\tilde{D}$ is open and D is connected.

Let us assume that there exists an open nonempty set $A \subset D$ such that $(\mathbf{E}, \mathbf{H}) = (\mathbf{0}, \mathbf{0})$ everywhere in A. Without loss of generality, we can find $\hat{\mathbf{x}} \in D$, $r > 0$ and an index $\hat{\imath}$ such that $B_r(\hat{\mathbf{x}}) \subset A \cap D_{\hat{\imath}}$. By Theorem 7.2.3, we deduce that $(\mathbf{E}, \mathbf{H}) = (\mathbf{0}, \mathbf{0})$ everywhere in $D_{\hat{\imath}}$.

Let us assume, by contradiction, that there exists an index $\hat{\jmath}$ such that $(\mathbf{E}, \mathbf{H})$ is not identically equal to $(0,0)$ in $D_{\hat{\jmath}}$. Let us pick any $\hat{\mathbf{y}} \in D_{\hat{\jmath}}$ and let $\gamma : [0,1] \to \tilde{D}$ be a smooth curve such that $\gamma(0) = \hat{\mathbf{x}}$ and $\gamma(1) = \hat{\mathbf{y}}$.

Let us define

$$\hat{t} = \inf\{t \in [0,1] : \gamma(t) \in D_i \text{ for some } i \text{ s.t. } (\mathbf{E}, \mathbf{H}) \not\equiv (\mathbf{0}, \mathbf{0}) \text{ in } D_i\}.$$

We notice that $0 < \hat{t} < 1$. Furthermore, by the definition of $\tilde{D}$ and of C, we obtain that $\gamma(\hat{t}) \in \partial D_i \cap \partial D_j$, for two different indexes i and j, and separates exactly the two partitions D_i and D_j. Moreover, for some positive δ_i and δ_j and up to swapping i with j, we have $\gamma(\hat{t} + \delta_i) \in D_i$ and $(\mathbf{E}, \mathbf{H}) \not\equiv (\mathbf{0}, \mathbf{0})$ in D_i, and $\gamma(\hat{t} - \delta_j) \in D_j$. By the definition of $\hat{t}$, we deduce that $(\mathbf{E}, \mathbf{H}) \equiv (\mathbf{0}, \mathbf{0})$ in D_j.

Then the proof can be concluded by the arguments developed in [4] that we briefly recall here. For simplicity, let us assume that $\gamma(\hat{t}) = 0$. For some $\delta > 0$, we have $\left| B_\delta \cap (D_i \cup D_j) \right| = 0$. We have that $(\varepsilon_i, \mu_i) \in W^{1,\infty}(B_\delta)^2$ and $(\mathbf{E}, \mathbf{H}) \equiv 0$ in D_j. Since $(\mathbf{E}, \mathbf{H})$ solves (7.2.7) in B_δ, it is immediate to notice that $(\mathbf{E}, \mathbf{H})$ solves in B_δ

$$\nabla \wedge \mathbf{E} - \mathrm{i}k\mu_i \mathbf{H} = \mathbf{0}, \quad \nabla \wedge \mathbf{H} + \mathrm{i}k\varepsilon_i \mathbf{E} = \mathbf{0}$$

as well. Since $(\mathbf{E},\mathbf{H}) \equiv 0$ in $B_\delta \cap D_j$, which is not empty, we conclude, again by Theorem 7.2.3, that $(\mathbf{E},\mathbf{H}) \equiv 0$ in B_δ and thus in $B_\delta \cap D_i \neq \emptyset$. Using again Theorem 7.2.3, we conclude that $(\mathbf{E},\mathbf{H}) \equiv 0$ in D_i and we obtain a contradiction, which completes the proof. □

We conclude this excursus on the UCP with the following lemma that provides a sufficient condition for assumption *iii)* in Proposition 7.2.4 to hold.

Lemma 7.2.3. *Let D be a connected open set contained in $\mathbb{R}^N$, $N \geq 2$. Let $C \subset D$ be closed in D. If $D \backslash C$ is not connected, then C has Hausdorff dimension greater than or equal to $N-1$.*

Remark 7.2.4. By Lemma 7.2.3, we can replace assumption *iii)* in Proposition 7.2.4 with the following. We assume that D is connected and that for some $s < 2$ we have $\mathscr{H}^s(C) < +\infty$, where $\mathscr{H}^s$ is the s-dimensional Hausdorff measure .

Proof of Lemma 7.2.3. We begin with the following remark. Let $A \subset \mathbb{R}^N$, $N \geq 2$, be a bounded open set. Then the Hausdorff dimension of ∂A is at least $N-1$. This is a classical result that may be proved as follows. We call $\pi : \mathbb{R}^N \to \mathbb{R}^{N-1}$ the projection onto the first $N-1$ coordinates. We notice that $\pi(A)$ is a bounded open set in $\mathbb{R}^{N-1}$ and that $\pi(\partial A) = \overline{\pi(A)}$. Hence, $0 < \mathscr{H}^{N-1}(\pi(A)) \leq \mathscr{H}^{N-1}(\overline{\pi(A)}) = \mathscr{H}^{N-1}(\pi(\partial A)) \leq C_1 \mathscr{H}^{N-1}(\partial A)$ for some positive constant C_1, and thus, the property is proved.

Let us assume, by contradiction, that the Hausdorff dimension of C is less than $N-1$ and that $D \backslash C$ is not connected; that is, $D \backslash C = D_1 \cup D_2$ where D_1, D_2 are open, nonempty sets which are disjoint. There exist $\mathbf{x} \in C$ and $\delta > 0$ such that $\overline{B_\delta(\mathbf{x})} \subset D$ and $B_\delta(\mathbf{x}) \cap D_i \neq \emptyset$ for $i = 1,2$. We also observe that there exist $\mathbf{x}_i \in \partial B_\delta(\mathbf{x}) \cap D_i$ for $i = 1,2$. In fact, otherwise, $\partial B_\delta(\mathbf{x}) \cap D_i \subset C$; thus, by the property proved above, C has at least dimension $N-1$ and we have a contradiction.

By a bi-Lipschitz map T we transform $\overline{B_\delta(\mathbf{x})}$ onto $\mathbb{S}^+_\delta$ where

$$\mathbb{S}^+_\delta = \{\mathbf{x} = (x_1, \ldots, x_N, x_{N+1}) \in \mathbb{R}^{N+1} : \ \|\mathbf{x}\| = \delta \text{ and } x_{N+1} \geq 0\}.$$

Simply by a reflection in the plane $\Pi = \{\mathbf{x} = (x_1, \ldots, x_N, x_{N+1}) \in \mathbb{R}^{N+1} : \ x_{N+1} = 0\}$, we may find two open subsets of $\mathbb{S}_\delta = \{\mathbf{x} \in \mathbb{R}^{N+1} : \ \|\mathbf{x}\| = \delta\}$, Ω_1 and Ω_2 and a closed set $\tilde{C}$ that are symmetric with respect to Π and such that $\Omega_i \cap \mathbb{S}^+_\delta = T(D_i \cap \overline{B_\delta(\mathbf{x})})$ for $i = 2$ and $\tilde{C} \cap \mathbb{S}^+_\delta = T(C \cap \overline{B_\delta(\mathbf{x})})$. Clearly, Ω_1 and Ω_2 are nonempty and disjoint. Fix $\mathbf{y}_1 \in \Omega_1$ and consider a stereographic projection T_1 with pole $\mathbf{y}_1$. We have that $T_1(\Omega_2)$ is a bounded open set, contained in $\mathbb{R}^{N-1}$. Since $\partial T_1(\Omega_2) \subset T_1(\tilde{C})$ and $T_1(\tilde{C})$ is a bi-Lipschitz image of $\tilde{C}$, we can easily conclude that the Hausdorff dimension of C is at least $N-1$.

The proof is complete. □

7.3 Properties of Solutions to the Maxwell System and Classes of Admissible Scatterers

Before passing to the stability results, in this section we collect a few properties of solutions to the Maxwell system that will be needed in the sequel. Then we define and discuss suitable classes of admissible scatterer for our study.

7.3.1 Properties of Solutions to the Maxwell System

We begin with the following higher integrability and compact immersion results that are proved in [14].

Proposition 7.3.1. *Let $D \subset \mathbb{R}^3$ be a bounded Lipschitz domain . Let $a \in L^\infty(D, M^{3\times 3}_{sym}(\mathbb{R}))$ satisfy* (7.2.1) *for some constants $0 < a_0 < a_1$.*

We call

$$W^{2,r}_{a,\nu}(D) = \{u \in H(\mathrm{curl}, D) \cap H(\mathrm{div}_a, D) : \gamma_\nu(au) \in L^r(\partial D)\}$$

and

$$W^{2,r}_{a,\tau}(D) = \{u \in H(\mathrm{curl}, D) \cap H(\mathrm{div}_a, D) : \gamma_\tau(u) \in L^r(\partial D, \mathbb{C}^3)\}.$$

Then there exists q_1, $2 < q_1 < 6$ depending on D, a_0 and a_1 only, such that for any $r > 4/3$ we have that $W^{2,r}_{a,\nu}(D)$ and $W^{2,r}_{a,\tau}(D)$ are immersed continuously into $L^s(D, \mathbb{C}^3)$, with $s = \min\{3r/2, q_1\} > 2$ and are immersed compactly into $L^2(D, \mathbb{C}^3)$. More precisely, there exists a constant C, depending on D, a_0, a_1 and r only, such that

$$\|u\|_{L^s(D,\mathbb{C}^3)} \leq C\big[\|u\|_{L^2(D,\mathbb{C}^3)} + \|\nabla \wedge u\|_{L^2(D,\mathbb{C}^3)} + \|\nabla \cdot (au)\|_{L^2(D)} + \|\gamma_\nu(au)\|_{L^r(\partial D)}\big] \quad \text{for any } u \in W^{2,r}_{a,\nu}(D)$$

and

$$\|u\|_{L^s(D,\mathbb{C}^3)} \leq C\big[\|u\|_{L^2(D,\mathbb{C}^3)} + \|\nabla \wedge u\|_{L^2(D,\mathbb{C}^3)} + \|\nabla \cdot (au)\|_{L^2(D)} + \|\gamma_\tau(u)\|_{L^r(\partial D,\mathbb{C}^3)}\big] \quad \text{for any } u \in W^{2,r}_{a,\tau}(D).$$

Proof. The first part is a consequence of Corollary 1 in [14], whereas the second follows from Corollary 2 in [14]. □

We remark that, when $a \equiv I_3$ and $r = 2$, these results would follow from the immersions into $H^{1/2}(D)$ proved in [11].

We next investigate how the Maxwell equations are transformed by change of variables. The change of variables we consider is of the following type.

Let D and D' be two connected open sets and let $T : D \to D'$ be a bi-$W^{1,\infty}$ mapping, with constant L. We call $S = T^{-1}$ and $J = JT$ the Jacobian matrix of T. Also, J^T denotes the transpose of J and J^{-T} denotes the transpose of J^{-1}. Recalling Remark 7.2.1, we set $\mathrm{or}(T) = 1$ if $\det J(\mathbf{x}) > 0$ for almost every $\mathbf{x} \in D$ and $\mathrm{or}(T) = -1$ if $\det J(\mathbf{x}) < 0$ for almost every $\mathbf{x} \in D$.

We begin by investigating how the suitable spaces we are dealing with are transformed. It is classical that for any $\varphi \in H^1(D')$, we call $\psi = \tilde{T}(\varphi) = \varphi \circ T$. We have that $\psi \in H^1(D)$ and, for almost every $\mathbf{y} \in D'$,

$$\nabla\varphi(\mathbf{y}) = J^{-T}(T^{-1}(\mathbf{y}))\nabla\psi(T^{-1}(\mathbf{y})).$$

Then we have that $\tilde{T} : H^1(D') \to H^1(D)$ is a linear homeomorphism with inverse $\tilde{S}$.

The change of variables that is suited for $H(\mathrm{curl})$ spaces is the following. For any $u \in L^2(D', \mathbb{C}^3)$, we define $v = \hat{T}(u)$ as follows:

$$v(\mathbf{x}) = \hat{T}(u)(\mathbf{x}) = J^T(\mathbf{x})u(T(\mathbf{x})) \quad \text{for a.e. } \mathbf{x} \in D$$

or, equivalently,

$$u(\mathbf{y}) = \hat{S}(v)(\mathbf{y}) = J^{-T}(T^{-1}(\mathbf{y}))v(T^{-1}(\mathbf{y})) \quad \text{for a.e. } \mathbf{y} \in D'.$$

For $H(\mathrm{div})$ spaces, the change of variables is given by the following. Let $a' \in L^\infty(D', M^{3\times 3}_{sym}(\mathbb{R}))$ satisfy (7.2.1) in D' for some constants $0 < a'_0 < a'_1$. For any $u \in H(\mathrm{div}_{a'}, D')$, then $v = \hat{T}(u) \in H(\mathrm{div}_a, D)$ with the following formulas:

$$a(\mathbf{x}) = T_*(a')(\mathbf{x}) = \left(\frac{J^{-1}a'(T)J^{-T}}{|\det J^{-1}|}\right)(\mathbf{x}) \quad \text{and}$$
$$\nabla\cdot(av)(\mathbf{x}) = \left(\frac{(\nabla\cdot(a'u))(T)}{|\det J^{-1}|}\right)(\mathbf{x}) \quad \text{for a.e. } \mathbf{x}\in D. \quad (7.3.1)$$

We notice that $a \in L^\infty(D, M^{3\times 3}_{sym}(\mathbb{R}))$ and satisfies (7.2.1) in D for some constants $0 < a_0 < a_1$ depending on a'_0, a'_1 and L only.

The following proposition can be proved as in [27] with a little more care since here we are using a bi-$W^{1,\infty}$ mapping instead of a C^1-diffeomorphism.

Proposition 7.3.2. *Under the previous notations and assumptions, if $v \in H(\mathrm{curl}, D)$, then $u = \hat{S}(v) \in H(\mathrm{curl}, D')$ and the following formula holds:*

$$(\nabla\wedge u)(\mathbf{y}) = \left(\frac{J}{\det J}\nabla\wedge v\right)(T^{-1}(\mathbf{y})) \quad \textit{for a.e. } \mathbf{y}\in D'. \quad (7.3.2)$$

By this formula, simple computations lead to the following.

Corollary 7.3.1. *We have that $\hat{T} : H(\mathrm{curl}, D') \to H(\mathrm{curl}, D)$ is a linear homeomorphism with inverse $\hat{S}$. Its norm is bounded by a constant depending on L only. Moreover, $\hat{T} : H_0(\mathrm{curl}, D') \to H_0(\mathrm{curl}, D)$ is bijective. Finally, if u, $\varphi \in H(\mathrm{curl}, D')$ and $v = \hat{T}(u)$, $\psi = \hat{T}(\varphi) \in H(\mathrm{curl}, D)$, then*

$$\langle\nabla\wedge u, \varphi\rangle_{D'} - \langle u, \nabla\wedge\varphi\rangle_{D'} = \mathrm{or}(T)\left[\langle\nabla\wedge v, \psi\rangle_D - \langle v, \nabla\wedge\psi\rangle_D\right]. \quad (7.3.3)$$

We also have that $\hat{T} : H(\mathrm{div}_{a'}, D') \to H(\mathrm{div}_a, D)$ is a linear homeomorphism with inverse $\hat{S}$, where the tensor a is given in (7.3.1)*. Its norm is bounded by a constant depending on L only. Moreover, $\hat{T} : H_0(\mathrm{div}_{a'}, D') \to H_0(\mathrm{div}_a, D)$ is bijective. Finally, if $u \in H(\mathrm{div}_{a'}, D')$ and $\varphi \in H^1(D')$, and $v = \hat{T}(u) \in H(\mathrm{div}_a, D)$ and $\psi = \tilde{T}(\varphi) \in H^1(D)$, then*

$$\langle\nabla\cdot(a'u), \varphi\rangle_{D'} + \langle a'u, \nabla\varphi\rangle_{D'} = \langle\nabla\cdot(av), \psi\rangle_D + \langle av, \nabla\psi\rangle_D. \quad (7.3.4)$$

Let us now investigate the transformation of the Maxwell equations under these changes of variables. Let ε', $\mu' \in L^\infty(D', M^{3\times 3}_{sym}(\mathbb{R}))$ satisfy (7.2.4) in D' for some constants $0 < \lambda'_0 < \lambda'_1$. Let $(\mathbf{E}', \mathbf{H}') \in H(\mathrm{curl}, D')^2$ solve the Maxwell equations for some $k > 0$

$$\nabla\wedge\mathbf{E}' - \mathrm{i}k\mu'\mathbf{H}' = \mathbf{0} \quad \text{and} \quad \nabla\wedge\mathbf{H}' + \mathrm{i}k\varepsilon'\mathbf{E}' = \mathbf{0} \quad \text{in } D'.$$

Then straightforward computations show that $\mathbf{E} = \mathrm{or}(T)\hat{T}(\mathbf{E}')$ and $\mathbf{H} = \hat{T}(\mathbf{H}')$ solve

$$\nabla\wedge\mathbf{E} - \mathrm{i}k\mu\mathbf{H} = \mathbf{0} \quad \text{and} \quad \nabla\wedge\mathbf{H} + \mathrm{i}k\varepsilon\mathbf{E} = \mathbf{0} \quad \text{in } D$$

where

$$\varepsilon(\mathbf{x}) = T_*(\varepsilon')(\mathbf{x}) = \left(\frac{J^{-1}\varepsilon'(T)J^{-T}}{|\det J^{-1}|}\right)(\mathbf{x}) \quad \text{and}$$
$$\mu(\mathbf{x}) = T_*(\mu')(\mathbf{x}) = \left(\frac{J^{-1}\mu'(T)J^{-T}}{|\det J^{-1}|}\right)(\mathbf{x}) \quad \text{for a.e. } \mathbf{x}\in D.$$

Again we have that ε, $\mu \in L^\infty(D, M^{3\times 3}_{sym}(\mathbb{R}))$ and satisfy (7.2.4) in D for some constants $0 < \lambda_0 < \lambda_1$ depending on λ'_0, λ'_1 and L only.

With these results at hand, we also investigate the following reflection principles. Let Π be a plane in $\mathbb{R}^3$ and let $T = T_\Pi$ be the reflection in Π. We notice that $T = T^{-1}$ and JT is identically equal to an orthogonal matrix J with $J = J^T = J^{-1}$ and $\det J = -1$; therefore, $\mathrm{or}(T) = -1$. We call ν one of the two unit vectors orthogonal to Π. Let D' be an open connected set such that

$D' \subset \mathbb{R}^3_+ = \{\mathbf{y} \in \mathbb{R}^3 : \mathbf{y} \cdot \nu > 0\}$. We suppose that there exists Γ, a nonempty open subset of Π such that $\Gamma = \partial D' \backslash (\overline{\partial D' \cap \mathbb{R}^3_+})$. We call $D = T(D)$ and $\Omega = D \cup D' \cup \Gamma$. We notice that Ω is a connected set such that $\Omega \cap \Pi = \Gamma$.

Proposition 7.3.3. *Let $u \in H(\mathrm{curl}, D')$. Then the function*

$$w = \begin{cases} u & \text{in } D' \\ \hat{T}(u) & \text{in } D \end{cases}$$

belongs to $H(\mathrm{curl}, \Omega)$ and we have the following formula

$$\nabla \wedge w = \begin{cases} \nabla \wedge u & \text{in } D' \\ \nabla \wedge \hat{T}(u) & \text{in } D. \end{cases}$$

If $\nu \wedge u = 0$ in $H^{-1/2}(\Gamma_1, \mathbb{C}^3)$ for any open Γ_1 compactly contained in Γ, then we also have that

$$w_1 = \begin{cases} u & \text{in } D' \\ -\hat{T}(u) & \text{in } D \end{cases}$$

belongs to $H(\mathrm{curl}, \Omega)$ and

$$\nabla \wedge w_1 = \begin{cases} \nabla \wedge u & \text{in } D' \\ -\nabla \wedge \hat{T}(u) & \text{in } D. \end{cases}$$

Given $a' \in L^\infty(D', M^{3\times 3}_{sym}(\mathbb{R}))$ satisfying (7.2.1) with constant $0 < a'_0 < a'_1$, we notice that if $u \in H(\mathrm{div}_{a'}, D')$, then the function w_1 defined above belongs to $H(\mathrm{div}_a, \Omega)$, where

$$a = \begin{cases} a' & \text{in } D' \\ T_*(a') & \text{in } D, \end{cases}$$

and we have the following formula

$$\nabla \cdot (a w_1) = \begin{cases} \nabla \cdot (a' u) & \text{in } D' \\ -\nabla \cdot (T_*(a')\hat{T}(u)) & \text{in } D. \end{cases}$$

If $\nu \cdot (a'u) = 0$ in $H^{-1/2}(\Gamma_1)$ for any open Γ_1 compactly contained in Γ, then the function w defined above belongs to $H(\mathrm{div}_a, \Omega)$, and we have the following formula:

$$\nabla \cdot (a w) = \begin{cases} \nabla \cdot (a' u) & \text{in } D' \\ \nabla \cdot (T_*(a')\hat{T}(u)) & \text{in } D. \end{cases}$$

Let us finally notice that if $a' = I_3$, and T is a reflection, then $a = T_*(a') = T_*(I_3) = I_3$.

Proof. Let $\varphi \in C^\infty_0(\Omega, \mathbb{C}^3)$. We can find a bounded Lipschitz domain Ω_1 compactly contained in Ω and containing the support of φ satisfying the following properties. We assume that Ω_1 is symmetric with respect to Π and $D'_1 = \Omega_1 \cap \mathbb{R}^3_+$ is a bounded Lipschitz domain as well. We call $D_1 = T(D_1)$ and $\Gamma_1 = \Omega_1 \cap \Pi$, which is compactly contained in Γ. Our aim is to prove that

$$\begin{aligned}\langle w, \nabla \wedge \varphi \rangle_\Omega = \langle w, \nabla \wedge \varphi \rangle_{\Omega_1} &= \langle u, \nabla \wedge \varphi \rangle_{D'_1} + \langle \hat{T}(u), \nabla \wedge \varphi \rangle_{D_1} \\ &= \langle \nabla \wedge u, \varphi \rangle_{D'_1} + \langle \nabla \wedge \hat{T}(u), \varphi \rangle_{D_1} = \langle \nabla \wedge u, \varphi \rangle_{D'} + \langle \nabla \wedge \hat{T}(u), \varphi \rangle_D.\end{aligned}$$

We know that

$$\langle u, \nabla \wedge \varphi \rangle_{D'_1} - \langle \nabla \wedge u, \varphi \rangle_{D'_1} = -\langle (-\nu) \wedge u, \varphi \rangle_{\Gamma_1}$$

and, analogously,

$$\langle \hat{T}(u), \nabla \wedge \varphi \rangle_{D_1} - \langle \nabla \wedge \hat{T}(u), \varphi \rangle_{D_1} = -\langle \nu \wedge \hat{T}(u), \varphi \rangle_{\Gamma_1}.$$

By an approximation argument, it is not difficult to show that $\nu \wedge u = \nu \wedge \hat{T}(u)$ in $H^{-1/2}(\Gamma_1, \mathbb{C}^3)$. Hence, the first result is proved. The others follow by analogous reasonings, just replacing $\nu \wedge u$ with $\nu \cdot (a'u)$ for the $H(\mathrm{div})$ spaces. □

With the same notations as before, we have that if $(\mathbf{E}', \mathbf{H}') \in H(\mathrm{curl}, D')^2$ solve the Maxwell equations for some $k > 0$

$$\nabla \wedge \mathbf{E}' - ik\mu'\mathbf{H}' = \mathbf{0} \quad \text{and} \quad \nabla \wedge \mathbf{H}' + ik\varepsilon'\mathbf{E}' = \mathbf{0} \quad \text{in } D'$$

and $\nu \wedge \mathbf{E} = \mathbf{0}$ in $H^{-1/2}(\Gamma_1, \mathbb{C}^3)$ for any open Γ_1 compactly contained in Γ, we have that the couple

$$(\mathbf{E}, \mathbf{H}) = \begin{cases} (\mathbf{E}', \mathbf{H}') & \text{in } D' \\ (-\hat{T}(\mathbf{E}'), \hat{T}(\mathbf{H}')) & \text{in } D \end{cases} \tag{7.3.5}$$

belongs to $H(\mathrm{curl}, \Omega)^2$ and solves

$$\nabla \wedge \mathbf{E} - ik\mu\mathbf{H} = \mathbf{0} \quad \text{and} \quad \nabla \wedge \mathbf{H} + ik\varepsilon\mathbf{E} = \mathbf{0} \quad \text{in } \Omega \tag{7.3.6}$$

where

$$(\varepsilon, \mu) = \begin{cases} (\varepsilon', \mu') & \text{in } D' \\ (T_*(\varepsilon'), T_*(\mu')) & \text{in } D. \end{cases} \tag{7.3.7}$$

We conclude our discussion on reflection properties by stating, without proofs, the following lemmas.

Lemma 7.3.1. *Under the previous assumptions, let* $(\mathbf{E}, \mathbf{H})$ *be as in* (7.3.5) *and* (ε, μ) *as in* (7.3.7). *Let* $(\tilde{\mathbf{E}}, \tilde{\mathbf{H}})$ *solve* (7.3.6) *in* Ω *and let us call* $(\tilde{\mathbf{E}}_1, \tilde{\mathbf{H}}_1) = (-\hat{T}(\tilde{\mathbf{E}}), \hat{T}(\tilde{\mathbf{H}}))$ *in* Ω. *We have that* $(\tilde{\mathbf{E}}_1, \tilde{\mathbf{H}}_1)$ *solves* (7.3.6) *in* Ω *and*

$$\|\tilde{\mathbf{E}} - \tilde{\mathbf{E}}_1\|_{L^2(\Omega, \mathbb{C}^3)} \leq \|\tilde{\mathbf{E}} - \mathbf{E}\|_{L^2(\Omega, \mathbb{C}^3)} + \|\mathbf{E} - \tilde{\mathbf{E}}_1\|_{L^2(\Omega, \mathbb{C}^3)} = 2\|\tilde{\mathbf{E}} - \mathbf{E}\|_{L^2(\Omega, \mathbb{C}^3)} \tag{7.3.8}$$

and, analogously,

$$\|\tilde{\mathbf{H}} - \tilde{\mathbf{H}}_1\|_{L^2(\Omega, \mathbb{C}^3)} \leq \|\tilde{\mathbf{H}} - \mathbf{H}\|_{L^2(\Omega, \mathbb{C}^3)} + \|\mathbf{H} - \tilde{\mathbf{H}}_1\|_{L^2(\Omega, \mathbb{C}^3)} = 2\|\tilde{\mathbf{H}} - \mathbf{H}\|_{L^2(\Omega, \mathbb{C}^3)}.$$

Lemma 7.3.2. *Under the assumptions and notations of Lemma* 7.3.1, *we have that, for any open* Γ_1 *compactly contained in* Γ,

$$\nu \wedge (\tilde{\mathbf{E}} - \tilde{\mathbf{E}}_1) = 2(\nu \wedge \tilde{\mathbf{E}}) \quad \text{in } H^{-1/2}(\Gamma_1, \mathbb{C}^3).$$

Moreover, if (7.3.6) *satisfies the UCP in* Ω *and* $\nu \wedge \tilde{\mathbf{E}} = \mathbf{0}$ *in* $H^{-1/2}(\Gamma_1, \mathbb{C}^3)$ *for any open* Γ_1 *compactly contained in* Γ, *then we have that* $(\tilde{\mathbf{E}}, \tilde{\mathbf{H}}) = (\tilde{\mathbf{E}}_1, \tilde{\mathbf{H}}_1)$.

7.3.2 Classes of Admissible Scatterers

We wish to define suitable classes of admissible scatterers. We begin with some definitions that are taken from [20], where these classes are deeply discussed and are defined for any $N \geq 2$. We keep such a generality here as well. Let us also notice that similar classes have been developed earlier in [17] and in [26].

We fix a bounded open set $\Omega \subset \mathbb{R}^N$, $N \geq 2$. Let K be a compact subset of $\mathbb{R}^N$. We say that K is a *mildly Lipschitz surface* , with or without boundary, with positive constants r and L if the following holds.

For any $\mathbf{x} \in K$, there exists a bi-Lipschitz function $\Phi_{\mathbf{x}} : B_r(\mathbf{x}) \to \mathbb{R}^N$ such that

(a) For any $\mathbf{z}_1, \mathbf{z}_2 \in B_r(\mathbf{x})$, we have

$$L^{-1}\|\mathbf{z}_1-\mathbf{z}_2\| \leq \|\Phi_{\mathbf{x}}(\mathbf{z}_1)-\Phi_{\mathbf{x}}(\mathbf{z}_2)\| \leq L\|\mathbf{z}_1-\mathbf{z}_2\|;$$

(b) $\Phi_{\mathbf{x}}(\mathbf{x}) = \mathbf{0}$ and $\Phi_{\mathbf{x}}(K \cap B_r(\mathbf{x})) \subset \Pi = \{\mathbf{y} \in \mathbb{R}^N : \ y_N = 0\}$.

We say that $\mathbf{x} \in K$ belongs to the interior of K if there exists δ, $0 < \delta \leq r$, such that $B_\delta(0) \cap \Pi \subset \Phi_{\mathbf{x}}(K \cap B_r(\mathbf{x}))$. Otherwise, we say that $\mathbf{x}$ belongs to the boundary of K. We remark that the boundary of K might be empty. Further, we assume that

(a) For any $\mathbf{x}$ belonging to the boundary of K, we have that

$$\Phi_{\mathbf{x}}(K \cap B_r(\mathbf{x})) = \Phi_{\mathbf{x}}(B_r(\mathbf{x})) \cap \Pi^+$$

where $\Pi^+ = \{\mathbf{y} \in \mathbb{R}^N : \ y_N = 0, \ y_{N-1} \geq 0\}$.

We shall call $\mathscr{B} = \mathscr{B}(r, L, \Omega)$ the set of $K \subset \overline{\Omega}$ such that K is a mildly Lipschitz hypersurface with constants r and L. We recall that $\mathscr{B}(r, L, \Omega)$ is closed, and actually compact, with respect to the Hausdorff distance ; see [26, Lemma 3.6]. For details about convergence in the Hausdorff distance, we refer, for instance, to [12].

We call $\mathscr{D} = \mathscr{D}(r, L, \Omega)$ the class of sets ∂D where $D \subset \Omega$ is an open set which is Lipschitz with constants r and L. We have that $\mathscr{D}(r, L, \Omega) \subset \mathscr{B}(\tilde{r}, \tilde{L}, \Omega)$, for some constants $\tilde{r}$, $\tilde{L}$ depending on r and L only. Moreover, also $\mathscr{D}(r, L, \Omega)$ is compact with respect to the Hausdorff distance .

We further call $\hat{\mathscr{D}} = \hat{\mathscr{D}}(r, L, \Omega)$ the class of compact sets $\Sigma \subset \overline{\Omega}$ such that $\partial\Sigma \in \mathscr{D}(r, L, \Omega)$. Also this class is compact with respect to the Hausdorff distance.

Again following [26], we combine different mildly Lipschitz hypersurfaces to obtain more complex structures.

Definition 7.3.1. Let us fix positive constants r, L and a bounded open set Ω. Let us also fix $\omega : (0, +\infty) \to (0, +\infty)$, a nondecreasing left-continuous function.

We say that a compact set $K \subset \overline{\Omega}$ belongs to the class $\tilde{\mathscr{B}} = \tilde{\mathscr{B}}(r, L, \Omega, \omega)$ if it satisfies the following conditions:

(1) $K = \bigcup_{i=1}^M K^i$ where $K^i \in \mathscr{B}(r, L, \Omega)$ for any $i = 1, \ldots, M$;
(2) For any $i \in \{1, \ldots, M\}$ and any $\mathbf{x} \in K^i$, if its distance from the boundary of K^i is $t > 0$, then the distance of $\mathbf{x}$ from the union of K^j, with $j \neq i$, is greater than or equal to $\omega(t)$.

We say that a compact set $\Sigma \subset \overline{\Omega}$ belongs to the class $\tilde{\mathscr{B}}_1 = \tilde{\mathscr{B}}_1(r, L, \Omega, \omega)$ if $\partial\Sigma \in \tilde{\mathscr{B}}(r, L, \Omega, \omega)$.

Let us notice that in the previous definition the number M may depend on K. However, there exists an integer M_0, depending on r, L, the diameter of Ω and ω only, such that $M \leq M_0$ for any $K \in \tilde{\mathscr{B}}$. We have that $\mathscr{H}^{N-1}(K)$ is bounded by a constant depending on r, L, the diameter of Ω and M_0 only, hence $|K| = 0$. Furthermore, if we set as the boundary of K the union of the boundaries of K^i, $i = 1, \ldots, M$, then the boundary of K has $\mathscr{H}^{N-2}$ measure bounded by a constant again depending on r, L, the diameter of Ω and M_0 only. Finally, the number of connected components of $\mathbb{R}^N \backslash K$ is bounded by a constant M_1 depending on r, L, the diameter of Ω and ω only.

Without loss of generality, in the sequel we shall always assume that $\omega(t) \leq t$ for any $t > 0$ and that $\lim_{t \to +\infty} \omega(t)$ is equal to a finite real number which we call $\omega(+\infty)$.

The following compactness results hold.

Lemma 7.3.3. *The classes $\tilde{\mathscr{B}}$ and $\tilde{\mathscr{B}}_1$ introduced in Definition 7.3.1 are compact under convergence in the Hausdorff distance .*

Moreover, let $\Sigma \in \tilde{\mathscr{B}}_1$ and $\mathbf{x} \in \partial\Sigma$. We call $G = \mathbb{R}^N \backslash \Sigma$. For any $r_1 > 0$, the number of connected components U of $B_{r_1}(\mathbf{x}) \cap G$ such that $\mathbf{x} \in \partial U$ is bounded by a constant M_2 depending on r_1, r, L and ω only. Finally, the number of connected components of $B_{r_1}(\mathbf{x}) \cap G$ intersecting $B_{r_1/2}(\mathbf{x})$ is bounded by a constant M_3 depending on r_1, r, L and ω only.

Proof. The compactness of the class $\tilde{\mathscr{B}}$ is proved in [26, Lemma 3.8], whereas the compactness of the class $\tilde{\mathscr{B}}_1$ and the second part of the lemma are proved in [20, Lemma 2.6]. □

Finally, we consider the following definition.

Definition 7.3.2. Let us fix a bounded open set Ω and positive constants r, L, $0 < r_1 < r$ and $\tilde{C} > 0$. Let us also fix $\omega : (0, +\infty) \to (0, +\infty)$, nondecreasing left-continuous functions.

We call $\hat{\mathscr{B}} = \hat{\mathscr{B}}(r, L, \Omega, r_1, \tilde{C}, \omega)$ the class of sets satisfying the following assumptions:

(i) Any $\Sigma \in \hat{\mathscr{B}}$ is a compact set contained in $\overline{\Omega} \subset \mathbb{R}^N$ such that $\Sigma \in \tilde{\mathscr{B}}_1(r, L, \Omega, \omega)$. We call $G = \mathbb{R}^N \backslash \Sigma$.

(ii) For any $\mathbf{x} \in \partial\Sigma$ and any U connected component of $G \cap B_{r_1}(\mathbf{x})$, with $\mathbf{x} \in \partial U$, we can find an open set U' such that
$$U \subset U' \subset G, \tag{7.3.9}$$
and a bi-$W^{1,\infty}$ mapping $T : (-1,1)^{N-1} \times (0,1) \to U'$, with constant $\tilde{C}$, such that the following properties hold. By the regularity of $Q = (-1,1)^{N-1} \times (0,1)$, T can be actually extended up to the boundary and we have that $T : \overline{Q} \to \mathbb{R}^N$ is a Lipschitz map with Lipschitz constant bounded by $\tilde{C}$. Furthermore, if we set $\Gamma = [-1,1]^{N-1} \times \{0\}$, we require that
$$T(0) = \mathbf{x} \quad \text{and} \quad \partial U \cap B_{r_1}(\mathbf{x}) \subset T(\Gamma) \subset \partial G, \tag{7.3.10}$$
and that, for any $0 < s < r_1$ and any $\mathbf{y} \in U \cap B_{r_1 - s}(\mathbf{x})$, we have
$$\operatorname{dist}(T^{-1}(\mathbf{y}), \partial Q \backslash \Gamma) \geq \omega(s). \tag{7.3.11}$$

Remark 7.3.1. We notice that clearly, $T(\partial Q) = \partial U'$ and $\partial U \cap B_{r_1}(\mathbf{x}) \subset \partial G$.

It is pointed out that, up to suitably changing the constants r_1 and $\tilde{C}$ involved, Condition ii) is satisfied provided it holds only for points $\mathbf{x}$ belonging to the boundaries of K_i, $i = 1, \ldots, M$, where $\partial\Sigma = \bigcup_{i=1}^{M} K_i$ by Condition i).

We also remark that, for some constants and functions depending on r, L and the diameter of Ω only, we have $\hat{\mathscr{D}}(r, L, \Omega) \subset \hat{\mathscr{B}}(\tilde{r}, \tilde{L}, \Omega, r_1, \tilde{C}, \omega)$.

It is emphasised that Condition ii) is an extremely weak regularity condition and that it is satisfied by rather complex structures; see, for instance, the discussion on sets in $\mathbb{R}^3$ satisfying this assumption in Section 4 of [34], where several examples are shown.

The following compactness result holds true; see [20, Lemma 2.9].

Lemma 7.3.4. *The class $\hat{\mathscr{B}}$ introduced in Definition 7.3.2 is compact under convergence in the Hausdorff distance .*

We are in a position to define a quite general class of scatterers $\Sigma \subset \mathbb{R}^N$, $N \geq 2$. We need a quantitative assumption on the connectedness of $G = \mathbb{R}^N \backslash \Sigma$ as follows. Let $\delta : (0, +\infty) \to (0, +\infty)$ be a nondecreasing left-continuous function. Let Σ be a compact set contained in $\mathbb{R}^N$. We say that Σ satisfies the *uniform exterior connectedness* with function δ if for any $t > 0$, for any two points $\mathbf{x}_1$, $\mathbf{x}_2 \in \mathbb{R}^N$ so that $B_t(\mathbf{x}_1)$ and $B_t(\mathbf{x}_2)$ are contained in $\mathbb{R}^N \backslash \Sigma$, and for any s, $0 < s < \delta(t)$, then we can find a smooth (for instance, C^1) curve γ connecting $\mathbf{x}_1$ to $\mathbf{x}_2$ so that $B_s(\gamma)$ is contained in $\mathbb{R}^N \backslash \Sigma$ as well.

Let us notice that such an assumption is closed under convergence in the Hausdorff distance and that $\delta(t) \leq t$ for any $t > 0$. A detailed investigation on sufficient conditions for such an assumption to hold may be found in Sect. 2, in particular in Proposition 2.1, in [20].

Definition 7.3.3. Let us fix positive constants r, L and R, $0 < r_1 < r$ and $\tilde{C} > 0$. Let us also fix $\omega : (0,+\infty) \to (0,+\infty)$ and $\delta : (0,+\infty) \to (0,+\infty)$, two nondecreasing left-continuous functions.

We call $\hat{\mathscr{B}}_{scat} = \hat{\mathscr{B}}_{scat}(r,L,R,r_1,\tilde{C},\omega,\delta)$ the class of compact sets Σ belonging to $\hat{\mathscr{B}}(r,L,B_R,r_1,\tilde{C},\omega)$ and satisfying the uniform exterior connectedness with function δ.

We also define $\tilde{\mathscr{B}}_{scat} = \tilde{\mathscr{B}}_{scat}(r,L,R,\omega,\delta)$ the class of compact sets Σ belonging to $\tilde{\mathscr{B}}_1(r,L,B_R,\omega)$ and satisfying the uniform exterior connectedness with function δ.

We further call $\hat{\mathscr{D}}_{obst} = \hat{\mathscr{D}}_{obst}(r,L,R)$ the class of compact sets Σ belonging to $\hat{\mathscr{D}}(r,L,B_R)$ and such that $G = \mathbb{R}^3 \backslash \Sigma$ is connected.

Obviously, we have $\hat{\mathscr{B}}_{scat}(r,L,R,r_1,\tilde{C},\omega,\delta) \subset \tilde{\mathscr{B}}_{scat}(r,L,R,\omega,\delta)$. We notice that any scatterer $\Sigma \in \hat{\mathscr{D}}_{obst}$ is indeed an obstacle, that is, Σ is the closure of its interior which is a bounded open set with Lipschitz boundary, with constants r and L. By Corollary 2.3 and Proposition 2.1 in [20], for some constants and functions depending on r, L and R only, we have $\hat{\mathscr{D}}_{obst}(r,L,R) \subset \hat{\mathscr{B}}_{scat}(\tilde{r},\tilde{L},R,r_1,\tilde{C},\omega,\delta)$.

By our earlier discussion, in particular by Lemmas 7.3.3 and 7.3.4, it is easy to note that these classes $\tilde{\mathscr{B}}_{scat}$, $\hat{\mathscr{B}}_{scat}$, and $\hat{\mathscr{D}}_{obst}$ are compact with respect to the Hausdorff distance.

7.4 Mosco Convergence for $H(\mathrm{curl})$ Spaces and Higher Integrability for Solutions to Maxwell Equations

In this section we consider in detail the Mosco convergence for $H(\mathrm{curl})$ spaces. Finally, we discuss higher integrability properties related to solutions to the Maxwell equations. These will be the main ingredients needed to study the stability issue for the electromagnetic scattering problem with respect to variations of scatterers.

The Mosco convergence, introduced in [28], has been widely investigated for H^1 spaces since it is essentially equivalent to the stability of solutions of elliptic Neumann problems with respect to variations of the domain. We recall that, in dimension 2, the problem is completely solved, [6, 7]. The breakthrough was the sufficient condition proved in [8], which is still a quite useful one for the applications. In dimension 2 extensive use is made of complex analytic techniques, in particular duality arguments. In dimension 3 and higher, such a problem has been considered first in [17] and then in [26]. In both cases, conditions of Lipschitz type are used.

In the current article, instead of H^1 spaces, we are mainly concerned with the Mosco convergence of $H(\mathrm{curl})$ spaces, which shall be of fundamental importance to study the solutions of Maxwell equations.

The general abstract definition of Mosco convergence is the following. For a sequence $\{A_n\}_{n\in\mathbb{N}}$ of closed subspaces of a reflexive Banach space X, we call

$$A' = \{\mathbf{x} \in X : \mathbf{x} = w\text{-}\lim_{k\to\infty} x_{n_k},\ x_{n_k} \in A_{n_k}\}$$

and

$$A'' = \{\mathbf{x} \in X : \mathbf{x} = s\text{-}\lim_{n\to\infty} x_n,\ x_n \in A_n\}.$$

The sets A' and A'' are subspaces of X and we have that $A'' \subset A'$ and that A'' is closed. We say that A_n converges, as $n \to \infty$, to a closed subspace A in the sense of Mosco if $A = A' = A''$. Equivalently, the following two conditions need to be satisfied:

(i) For any $\mathbf{x} \in X$, if there exists a subsequence A_{n_k} and a sequence x_k, $k \in \mathbb{N}$, such that x_k converges weakly to $\mathbf{x}$ as $k \to \infty$ and $x_k \in A_{n_k}$ for any $k \in \mathbb{N}$, then $\mathbf{x} \in A$;

(ii) For any $\mathbf{x} \in A$, there exists a sequence $x_n \in A_n$, $n \in \mathbb{N}$, converging strongly to $\mathbf{x}$ as $n \to \infty$.

For any bounded open set D contained in $\mathbb{R}^3$, we call $\mathscr{K}$ the set of all compact subsets of $\overline{D}$. It is well known that $\mathscr{K}$ is compact with respect to the Hausdorff distance . Moreover, if $K_n \in \mathscr{K}$, $n \in \mathbb{N}$, converges in the Hausdorff distance, as $n \to \infty$, to $K \in \mathscr{K}$, then we also have $\lim_{n\to\infty} |K_n \backslash K| = 0$.

For any $K \in \mathscr{K}$, we consider the isometric immersion of $H(\text{curl}, D\backslash K)$ into $L^2(D, \mathbb{C}^6)$ as follows. To each $u \in H(\text{curl}, D\backslash K)$, we associate $(u, \nabla \wedge u) \in L^2(D, \mathbb{C}^6)$ with the convention that u and $\nabla \wedge u$ are extended to zero in K. With this convention, $H(\text{curl}, D\backslash K)$ may be considered as a closed subspace of $L^2(D, \mathbb{C}^6)$.

Given a sequence $\{K_n\}_{n\in\mathbb{N}}$ contained in $\mathscr{K}$ and $K \in \mathscr{K}$, we say that $H(\text{curl}, D\backslash K_n)$ converges, as $n \to \infty$, to $H(\text{curl}, D\backslash K)$ *in the sense of Mosco* if this holds by the previous abstract definition considering $H(\text{curl}, D\backslash K_n)$, $n \in \mathbb{N}$, and $H(\text{curl}, D\backslash K)$ as subspaces of $L^2(D, \mathbb{C}^6)$.

We wish to find general sufficient conditions on K_n, $n \in \mathbb{N}$ and K in order to have such a Mosco convergence for the corresponding $H(\text{curl})$ spaces. The result is the following.

Theorem 7.4.1. *Let us fix positive constants r, L and* $\omega : (0, +\infty) \to (0, +\infty)$, *a nondecreasing left-continuous function. Let D be a bounded open set contained in* $\mathbb{R}^3$. *Let* $\tilde{\mathscr{B}} = \tilde{\mathscr{B}}(r, L, D, \omega)$.

Let, for any $n \in \mathbb{N}$, $K_n \subset \overline{D}$ *be a compact set such that* $\partial(D\backslash K_n) \in \tilde{\mathscr{B}}$.

If, as $n \to \infty$, K_n *converges to a compact K in the Hausdorff distance, then* $H(\text{curl}, D\backslash K_n)$ *converges to* $H(\text{curl}, D\backslash K)$ *in the sense of Mosco.*

Proof. We assume that $\overline{D} \subset B_R$, for some $R > 0$. We call $\tilde{K}_n = K_n \cup (\overline{B_{R+1}} \backslash D)$, $n \in \mathbb{N}$ and $\tilde{K} = K \cup (\overline{B_{R+1}} \backslash D)$. We have that $\partial \tilde{K}_n = \partial B_{R+1} \cup \partial(D\backslash K_n)$, $n \in \mathbb{N}$, and $\partial \tilde{K} = \partial B_{R+1} \cup \partial(D\backslash K)$. Clearly, as $n \to \infty$, $\tilde{K}_n$ converges to $\tilde{K}$ in the Hausdorff distance . We set $D_1 = B_{R+1}$ and we rename $D = B_{R+2}$.

Therefore, without loss of generality, we can assume that D is a Lipschitz domain and there exists D_1, a Lipschitz domain compactly contained in D, such that K_n, $n \in \mathbb{N}$ and K are contained in $\overline{D_1}$.

Since it is not restrictive to pass to subsequences, by the arguments used in the proof of Lemma 7.3.3, we can assume that not only $K_n \to K$ but also $\partial K_n \to \partial K$ in the Hausdorff distance , as $n \to \infty$. Moreover, $\partial K \in \tilde{B}$ as well.

We call $A_n = H(\text{curl}, D\backslash K_n)$, $n \in \mathbb{N}$ and $A = H(\text{curl}, D\backslash K)$. For simplicity and without loss of generality we consider only real vector valued functions.

We begin with the following restrictive assumption. It is assumed that K_n has no interior point, that is, $K_n = \partial K_n$, $n \in \mathbb{N}$ and, consequently, $K = \partial K$ as well.

We first prove that if $|K \backslash K_n| \to 0$ as $n \to \infty$, then $A' \subset A$. By our restrictive hypothesis, we have that $|K| = 0$; hence, we conclude that in this case $A' \subset A$.

For proving the claim, let us consider a subsequence $\{A_{n_k}\}_{k\in\mathbb{N}}$ and let $(u_k, \nabla \wedge u_k) \in A_{n_k}$ for any $k \in \mathbb{N}$. We assume that $(u_k, \nabla \wedge u_k)$ converges to (u, V) weakly in $L^2(D, \mathbb{R}^6)$. For any point $\mathbf{x}$ and any $r > 0$ such that $\overline{B_r(\mathbf{x})} \subset D\backslash K$, it is easy to show that $u|_{B_r(\mathbf{x})} \in H(\text{curl}, B_r(\mathbf{x}))$ and $\nabla \wedge u = V$ in $B_r(\mathbf{x})$. Therefore, $u \in H(\text{curl}, D\backslash K)$ and $\nabla \wedge u = V$ in $D\backslash K$. It remains to prove that (u, V) are identically equal to 0 in K. If $|K \backslash K_n| \to 0$ as $n \to \infty$, this property follows by the arguments of Lemma 2.3 in [26].

The next step is to prove that $A \subset A''$, that is, for every $(u, \nabla \wedge u) \in A$, there exists $(u_n, \nabla \wedge u_n) \in A_n$ such that $(u_n, \nabla \wedge u_n)$ converges, as $n \to \infty$, to $(u, \nabla \wedge u)$ in $L^2(\Omega, \mathbb{R}^6)$. Let us recall that it is enough to prove that for any subsequence A_{n_k}, there exists a further subsequence $A_{n_{k_j}}$ with this property. Hence, without loss of generality, we can always pass to subsequences, which we usually do not relabel.

Since A'' is closed, it is enough to prove the result for any $(u, \nabla \wedge u)$ in a dense subset of A. Let us denote with $\hat{K}_n$, $n \in \mathbb{N}$ and $\hat{K}$ the boundary of K_n and K, respectively, as defined for the elements of the class $\tilde{\mathscr{B}}$ after Definition 7.3.1. We recall that, up to a subsequence, $\hat{K}_n$ converges to $\hat{K}$ in the Hausdorff distance as $n \to \infty$; see [26, Lemma 3.6].

We consider the following subset of A

$$\tilde{A} = \{u \in A : \ u \in L^\infty(D, \mathbb{R}^3) \text{ and } u = 0 \text{ in a neighbourhood of } \hat{K}\}.$$

We shall prove that $\tilde{A}$ is dense in A.

By the density result of Proposition 7.2.2, we need to prove that for any $u \in A \cap L^\infty(D, \mathbb{R}^3)$, we can find a sequence $\{u_n\}_{n\in\mathbb{N}} \in \tilde{A}$ converging to u in A. Since $\mathscr{H}^{N-2}(\hat{K})$ is finite, then $\hat{K}$ has zero capacity. Hence, for any U, an open neighbourhood of $\hat{K}$ compactly contained in D and for any $\varepsilon > 0$, there exists a function χ_ε such that $\chi_\varepsilon \in H^1(D)$, $0 \le \chi_\varepsilon \le 1$ almost everywhere in D, $\chi_\varepsilon = 1$ almost everywhere outside U, $\chi_\varepsilon = 0$ almost everywhere in a neighbourhood of $\hat{K}$, and

$$\int_D \|\nabla \chi_\varepsilon\|^2 \le \varepsilon.$$

Take $u \in H(\text{curl}, D\backslash K) \cap L^\infty(D, \mathbb{R}^3)$. Then $\chi_\varepsilon u \in \tilde{A}$ and

$$\|\chi_\varepsilon u - u\|_{L^2(D\backslash K, \mathbb{R}^3)} \le \|u\|_{L^2(U, \mathbb{R}^3)}$$

and

$$\begin{aligned}\|\nabla \wedge (\chi_\varepsilon u) - \nabla \wedge u\|_{L^2(D\backslash K, \mathbb{R}^3)} &\le \|(\chi_\varepsilon - 1)\nabla \wedge u\|_{L^2(D\backslash K, \mathbb{R}^3)} + \|\nabla \chi_\varepsilon \wedge u\|_{L^2(D\backslash K, \mathbb{R}^3)} \\ &\le \|\nabla \wedge u\|_{L^2(U, \mathbb{R}^3)} + \|u\|_{L^\infty(D, \mathbb{R}^3)} \|\|\nabla \chi_\varepsilon\|_{L^2(U, \mathbb{R}^3)}.\end{aligned}$$

Since U and ε are arbitrary, we conclude that $\tilde{A}$ is dense in A.

Take $u \in \tilde{A}$ and let $\tilde{U}$ be an open set compactly contained in D such that $\hat{K}$ is contained in $\tilde{U}$ and u is zero on $\tilde{U}$. We can find an open subset D_0 compactly contained in $D\backslash K$, a finite number of points $\mathbf{x}_j \in \partial D \cup K$ and positive numbers $\delta_j < \delta'_j$, $j = 1, \ldots, m$, such that $B_{\delta'_j}(\mathbf{x}_j) \cap \hat{K} = \emptyset$ for any $j = 1, \ldots, m$ and

$$\overline{D} \subset \tilde{U} \cup D_0 \cup \left(\bigcup_{j=1}^m B_{\delta_j}(\mathbf{x}_j) \right).$$

Moreover, we assume that for any $j = 1, \ldots, m_0$, $\mathbf{x}_j \in K$ and $B_{\delta'_j}(\mathbf{x}_j) \cap \partial D$ is empty, whereas for any $j = m_0 + 1, \ldots, m$, $\mathbf{x}_j \in \partial D$ and $B_{\delta'_j}(\mathbf{x}_j) \cap K$ is empty.

If $K = \bigcup_{i=1}^M K^i$ as in Definition 7.3.1, for any $j = 1, \ldots, m_0$, we have that $\mathbf{x}_j$ belongs to the interior of $K^{i(j)}$ for some $i = i(j) \in \{1, \ldots, M\}$ and we may assume that $2\delta'_j \le r$ and that $B_{\delta'_j}(\mathbf{x}_j) \cap K = B_{\delta'_j}(\mathbf{x}_j) \cap K^{i(j)}$.

By using a partition of unity, it is enough to consider a function $u \in \tilde{A}$ that is compactly supported either in $B_{\delta_j}(\mathbf{x}_j)$, for some $j \in \{1, \ldots, m\}$, or in D_0. In the latter case, we have that $D_0 \subset (D\backslash K_n)$, and hence, $u \in A_n$, for any n large enough, so the convergence is trivially proved.

We limit ourselves to consider $\mathbf{x}_j \in K$, that is, $j \in \{1, \ldots, m_0\}$, the case in which $\mathbf{x}_j \in \partial D$ is completely analogous.

We show the required convergence for $u \in \tilde{A}$ that is compactly supported in $B_\delta(\mathbf{x})$ for some $0 < \delta < \delta' \le r/2$ and $\mathbf{x} \in K^i$ such that $B_{\delta'}(\mathbf{x}) \cap (\partial D \cup \hat{K}) = \emptyset$ and $B_{\delta'}(\mathbf{x}) \cap K = B_{\delta'}(\mathbf{x}) \cap K^i$, for some $i \in \{1, \ldots, M\}$.

We adapt the reasoning developed in the proof of Theorem 4.2 in [17] to the $H(\text{curl})$ case. Possibly passing to a subsequence, let $\mathbf{x}_n \in K_n$ converge to $\mathbf{x}$ and $\Phi^n_{\mathbf{x}_n}$ converge to a function $\Phi_{\mathbf{x}} : B_r(\mathbf{x}) \to \mathbb{R}^3$ satisfying Conditions a) and b) of the definition of a mildly Lipschitz hypersurface . Without loss of generality, we may assume that $B_\delta(\mathbf{x}) \subset \Phi_{\mathbf{x}}^{-1}(B_{r_1})$ for some positive r_1 such that $B_{r_1} \subset \Phi_{\mathbf{x}}(B_{\delta'}(\mathbf{x}))$. Moreover, we may also assume that $B_{r_1} \cap \Pi \subset \Phi_{\mathbf{x}}(B_{\delta'}(\mathbf{x})) \cap \Pi = \Phi_{\mathbf{x}}(B_{\delta'}(\mathbf{x}) \cap K^i)$ and $B_{r_1} \cap \Pi \subset \Phi^n_{\mathbf{x}_n}(B_{\delta'}(\mathbf{x}_n)) \cap \Pi = \Phi^n_{\mathbf{x}_n}(B_{\delta'}(\mathbf{x}_n) \cap K_n)$ for any n large enough.

Let $v = \widehat{\Phi_{\mathbf{x}}^{-1}}(u) = (J\Phi_{\mathbf{x}}^{-1})^T u(\Phi_{\mathbf{x}}^{-1})$. Then $v \in H(\text{curl}, B_{r_1} \backslash \Pi)$, with bounded support in B_{r_1}. We denote by $v^\pm$ the function v restricted to the half-spaces $T^\pm = \{\mathbf{y} \in \mathbb{R}^3 : \ \pm y_3 > 0\}$, respectively. Then, by a reflection as developed in the previous section, we may define two $H_0(\text{curl}, B_{r_1})$ functions, $\tilde{v}^{\pm}$, such that $\tilde{v}^{\pm} = v^{\pm}$ on T^{+}. Let $\tilde{u}^{\pm} = \widehat{\Phi_{\mathbf{x}}}(\tilde{v}^{\pm}) \in H_0(\mathrm{curl}, B_{\delta'}(\mathbf{x}))$. We can assume that $B_r(\mathbf{x})$ is compactly contained in D; therefore, by extending them to zero, we have that $\tilde{u}^{\pm} \in H_0(\mathrm{curl}, D)$ and have compact support contained in $B_{\delta'}(\mathbf{x})$ and, for n large enough, in $B_{\delta'}(\mathbf{x}_n)$ as well. We then define

$$u_n = \begin{cases} \tilde{u}^{+}(\mathbf{x}) \text{ if } \Phi^n_{\mathbf{x}_n}(\mathbf{x}) \in T^{+} \\ \tilde{u}^{-}(\mathbf{x}) \text{ if } \Phi^n_{\mathbf{x}_n}(\mathbf{x}) \in T^{-}. \end{cases}$$

By construction we have that $u_n \in H(\mathrm{curl}, D\backslash K_n)$ and has compact support contained in $B_{\delta'}(\mathbf{x})$. Furthermore, u_n and $\nabla \wedge u_n$ converge almost everywhere in $D\backslash K$ to u and $\nabla \wedge u$, respectively.

Moreover, almost everywhere in $B_{\delta'}(\mathbf{x})\backslash K_n$, we have

$$\|u_n\| + \|\nabla \wedge u_n\| \leq \|\tilde{u}^{+}\| + \|\tilde{u}^{-}\| + \|\nabla \wedge \tilde{u}^{+}\| + \|\nabla \wedge \tilde{u}^{-}\|.$$

We immediately conclude by Lebesgue theorem that $(u_n, \nabla \wedge u_n)$ converges to $(u, \nabla \wedge u)$, as $n \to \infty$, in $L^2(D, \mathbb{R}^6)$.

The general case, that is, when we drop the assumption that $K_n = \partial K_n$ for any $n \in \mathbb{N}$, is a consequence of the next general lemma. □

Lemma 7.4.1. *Let D be a bounded open set and D_1 a bounded open set compactly contained in D. Let $\{K_n\}_{n\in N}$ be a sequence of compact sets and K and $\tilde{K}$ be two compact sets, all of them contained in $\overline{D_1}$.*

Let us assume that $K_n \to K$ and $\partial K_n \to \tilde{K}$ in the Hausdorff distance as $n \to \infty$.

If, as $n \to \infty$, $\tilde{A}_n = H(\mathrm{curl}, D\backslash \partial K_n)$ converges to $\tilde{A} = H(\mathrm{curl}, D\backslash \tilde{K})$ in the sense of Mosco, then one also has that $A_n = H(\mathrm{curl}, D\backslash K_n)$ converges to $A = H(\mathrm{curl}, D\backslash K)$ in the sense of Mosco.

Proof. The proof that $A' \subset A$ follows from a similar argument of the first part of the proof of Proposition 2.2 in [26], observing that, for any $n \in \mathbb{N}$, we trivially have that $A_n \subset \tilde{A}_n$.

We need to show that $A \subset A''$. Let $u \in A$. We recall that $\partial K \subset \tilde{K} \subset K$; hence, $D\backslash \tilde{K} = (D\backslash K) \cup (\overset{\circ}{K} \backslash \tilde{K})$ and $u \in \tilde{A}$ as well. Then there exists a sequence $\{u_n\}_{n\in\mathbb{N}}$ such that $u_n \in \tilde{A}_n$ for any $n \in \mathbb{N}$ and $(u_n, \nabla \wedge u_n)$ converges to $(u, \nabla \wedge u)$ in $L^2(D, \mathbb{R}^6)$ and, up to a subsequence, also almost everywhere in D, as $n \to \infty$. Let us consider $v_n = u_n$ in $D\backslash K_n$ and zero everywhere else. It is easy to show that $v_n \in A_n$. For any $\mathbf{x} \in D\backslash K$, there exists $r > 0$ and $\bar{n} \in \mathbb{N}$ such that $B_r(x) \cap K_n = \emptyset$ for any $n \geq \bar{n}$. Therefore, we have that $(v_n, \nabla \wedge v_n)$ converges to $(u, \nabla \wedge u)$ almost everywhere in $D\backslash K$ and hence by the Lebesgue theorem also in $L^2(D\backslash K, \mathbb{R}^6)$. On the other hand,

$$\|v_n - u\|_{L^2(K,\mathbb{R}^3)} = \|v_n\|_{L^2(K,\mathbb{R}^3)} \leq \|u_n\|_{L^2(K,\mathbb{R}^3)} = \|u_n - u\|_{L^2(K,\mathbb{R}^3)} \to 0 \quad \text{as } n \to \infty.$$

The same reasoning holds for the curl and the proof is concluded. □

We conclude this section by discussing the following higher integrability property for solutions of the Maxwell equations.

Definition 7.4.1. Let Σ be a compact set contained in $\overline{B_{R_0}}$ for some $R_0 > 0$. We say that Σ satisfies the *Maxwell higher integrability property* if for any constants $0 < a_0 < a_1$, there exist a constant $p > 2$, depending on Σ, R_0, a_0 and a_1 and a constant C, depending on Σ, R_0, p, a_0 and a_1, such that for any $a \in L^\infty(B_{R_0+1}, M^{3\times 3}_{sym}(\mathbb{R}))$ satisfying (7.2.1) in B_{R_0+1}, and any u belonging either to $H_0(\mathrm{curl}, B_{R_0+1}\backslash\Sigma) \cap H(\mathrm{div}_a, B_{R_0+1}\backslash\Sigma)$ or to $H(\mathrm{curl}, B_{R_0+1}\backslash\Sigma) \cap H_0(\mathrm{div}_a, B_{R_0+1}\backslash\Sigma)$, we have

$$\|u\|_{L^p(B_{R_0+1}\backslash\Sigma, \mathbb{C}^3)} \leq C\left[\|u\|_{H(\mathrm{curl}, B_{R_0+1}\backslash\Sigma)} + \|\nabla \cdot (au)\|_{L^2(B_{R_0+1}\backslash\Sigma)}\right]. \tag{7.4.1}$$

By using Proposition 7.3.1, we can immediately deduce the following corollary.

Corollary 7.4.1. *Let Σ be a compact set contained in $\overline{B_{R_0}}$, for some $R_0 > 0$, such that Σ satisfies the Maxwell higher integrability property.*

Then for any constants $0 < \overline{k}$ *and* $0 < \lambda_0 < \lambda_1$, *there exists a constant* $s_0 > 2$, *depending on* R_0, λ_0 *and* λ_1 *only, such that the following holds. If we call* $s = \min\{s_0, p\}$, *with* p *as in the definition of the Maxwell higher integrability property with* $a_0 = \lambda_0$ *and* $a_1 = \lambda_1$, *then* $s > 2$ *and we can find a constant* C_1, *depending on the same* p, s_0, R_0, $\overline{k}$, λ_0, λ_1 *and the constant* C *in* (7.4.1) *only, such that for any* ε, $\mu \in L^\infty(B_{R_0+1}, M^{3\times3}_{sym}(\mathbb{R}))$ *satisfying* (7.2.4) *in* B_{R_0+1} *and any* $(\mathbf{E},\mathbf{H})$ *solving*

$$\begin{cases} \nabla\wedge\mathbf{E} - ik\mu\mathbf{H} = \mathbf{0},\ \nabla\wedge\mathbf{H} + ik\varepsilon\mathbf{E} = \mathbf{0} & \text{in } B_{R_0+1}\backslash\Sigma \\ \nu\wedge\mathbf{E} = \mathbf{0} & \text{on } \partial\Sigma \end{cases}$$

for some $0 < k \le \overline{k}$, *we have*

$$\begin{aligned} \|\mathbf{E}\|_{L^s(B_{R_0+1}\backslash\Sigma,\mathbb{C}^3)} + \|\mathbf{H}\|_{L^s(B_{R_0+1}\backslash\Sigma,\mathbb{C}^3)} \le C_1 \big[&\|\mathbf{E}\|_{L^2(B_{R_0+1}\backslash\Sigma,\mathbb{C}^3)} + \|\mathbf{H}\|_{L^2(B_{R_0+1}\backslash\Sigma,\mathbb{C}^3)} \\ &+ \|\nu\wedge\mathbf{E}\|_{L^2(\partial B_{R_0+1},\mathbb{C}^3)} + \|\nu\wedge\mathbf{H}\|_{L^2(\partial B_{R_0+1},\mathbb{C}^3)}\big]. \end{aligned} \tag{7.4.2}$$

In the next proposition we give a sufficient condition for the Maxwell higher integrability property of Definition 7.4.1 to hold.

Proposition 7.4.1. *Let us fix positive constants* r, L *and* R_0, $0 < r_1 < r$ *and* $\tilde{C} > 0$ *and* $\omega : (0,+\infty) \to (0,+\infty)$, *a nondecreasing left-continuous function. Let* $\hat{\mathscr{B}} = \hat{\mathscr{B}}(r, L, B_{R_0}, r_1, \tilde{C}, \omega)$.

Then any $\Sigma \in \hat{\mathscr{B}}$ *satisfies the Maxwell higher integrability property, with constants in* (7.4.1) $p > 2$, *depending on* $\hat{\mathscr{B}}$, a_0 *and* a_1 *only, and* C, *depending on* $\hat{\mathscr{B}}$, p, a_0 *and* a_1 *only.*

Moreover, there exist constants $p_1 > 2$ *and* C_1, *depending on* $\hat{\mathscr{B}}$ *only, such that, for any* $\Sigma \in \hat{\mathscr{B}}$, *we have*

$$\|v\|_{L^{p_1}(B_{R_0+1}\backslash\Sigma)} \le C_1\|v\|_{H^1(B_{R_0+1}\backslash\Sigma)} \quad \text{for any } v \in H^1(B_{R_0+1}\backslash\Sigma). \tag{7.4.3}$$

Finally, $B_{R_0+1}\backslash\Sigma$ *satisfies the Maxwell and Rellich compactness properties.*

Proof. We observe that (7.4.3) and the Rellich compactness property are proved in [20, Proposition 2.11].

We begin with the following interior estimate. Given two bounded domains D' and D, with D' compactly contained in D and D Lipschitz, let us consider a function $u \in H(\mathrm{curl}, D) \cap H(\mathrm{div}_a, D)$. We fix a cut-off function $\chi \in C^\infty_0(D)$ such that $0 \le \chi \le 1$ everywhere and χ is identically equal to 1 in D'. Then $\chi u \in H_0(\mathrm{curl}, D) \cap H(\mathrm{div}_a, D)$ and

$$\begin{aligned} &\|\chi u\|_{L^2(D,\mathbb{C}^3)} + \|\nabla\wedge(\chi u)\|_{L^2(D,\mathbb{C}^3)} + \|\nabla\cdot(a\chi u)\|_{L^2(D)} \\ &\le C_1\left[\|u\|_{L^2(D,\mathbb{C}^3)} + \|\nabla\wedge u\|_{L^2(D,\mathbb{C}^3)} + \|\nabla\cdot(au)\|_{L^2(D)}\right]. \end{aligned} \tag{7.4.4}$$

Here C_1 depends only on $\|\nabla\chi\|_{L^\infty(D,\mathbb{R}^3)}$ and thus on the distance of D' from the boundary of D.

By Proposition 7.3.1, there exist $s > 2$ and a positive constant C_2, depending on D, a_0 and a_1 only, such that

$$\|\chi u\|_{L^s(D,\mathbb{C}^3)} \le C_2\left[\|\chi u\|_{L^2(D,\mathbb{C}^3)} + \|\nabla\wedge(\chi u)\|_{L^2(D,\mathbb{C}^3)} + \|\nabla\cdot(a\chi u)\|_{L^2(D)}\right]. \tag{7.4.5}$$

We conclude that there exist constants $s > 2$, depending on D, a_0 and a_1 only, and C_3, depending on D, a_0, a_1 and the distance of D' from the boundary of D, such that

$$\|u\|_{L^s(D',\mathbb{C}^3)} \le C_3\left[\|u\|_{L^2(D,\mathbb{C}^3)} + \|\nabla\wedge u\|_{L^2(D,\mathbb{C}^3)} + \|\nabla\cdot(au)\|_{L^2(D)}\right]. \tag{7.4.6}$$

Then we consider the following local construction. We fix $\Sigma \in \hat{\mathscr{B}}$ and we call $G = \mathbb{R}^3\backslash\Sigma$. We fix $\mathbf{x} \in \partial\Sigma$ and U a connected component of $B_{r_1}(\mathbf{x}) \cap G$ such that $\mathbf{x} \in \partial U$. We consider U' and $T : Q \to U'$ as in Condition ii) of Definition 7.3.2.

First of all, by (7.3.11), we infer that $T^{-1}(U \cap B_{3r_1/4}(\mathbf{x}))$ has a positive distance from $\partial Q \backslash \Gamma$ which is bounded from below by a positive constant $r_2 \leq 1/8$ depending on r_1 and ω only.

Let us take u belonging to $H_0(\mathrm{curl}, B_{R_0+1} \backslash \Sigma) \cap H(\mathrm{div}_a, B_{R_0+1} \backslash \Sigma)$, or to $H(\mathrm{curl}, B_{R_0+1} \backslash \Sigma) \cap H_0(\mathrm{div}_a, B_{R_0+1} \backslash \Sigma)$. Without loss of generality, by an easy extension argument around ∂B_{R_0+1}, in either cases we can assume that u and a are defined everywhere outside $\overline{B_{R_0+1}}$ and that $u \in H(\mathrm{curl}, \mathbb{R}^3 \backslash \Sigma) \cap H(\mathrm{div}_a, \mathbb{R}^3 \backslash \Sigma)$ with bounded support and with norms controlled by a constant C times the corresponding ones in $B_{R_0+1} \backslash \Sigma$.

Then $v = \hat{T}(u) \in H(\mathrm{curl}, Q) \cap H(\mathrm{div}_{a_1}, Q)$, where $a_1 = T_*(a)$. Furthermore, $\nabla \wedge v = 0$, or $v \cdot (a_1 v) = 0$, respectively, on any compact subset of the interior of Γ, with respect to the induced topology.

Therefore, by a reflection argument around the plane Π containing Γ, in both cases we can extend v to a function $\tilde{w} \in H(\mathrm{curl}, Q_1) \cap H(\mathrm{div}_{a_2}, Q_1)$ where $Q_1 = [-1,1]^3$ and a_2 satisfies, calling T_Π the reflection in Π,

$$a_2 = \begin{cases} a_1 & \text{in } Q \\ (T_\Pi)_*(a_1) & \text{in } T_\Pi(Q). \end{cases}$$

By using the previous interior estimates applied to Q and $Q' = \{\mathbf{x} \in Q : \mathrm{dist}(\mathbf{x}, \partial Q) > r_2/2\}$, we deduce that there exist constants $s > 2$ and C, depending on a_0, a_1, $\tilde{C}$ and r_2 only, such that

$$\|u\|_{L^s(U \cap B_{3r_1/4}(\mathbf{x}), \mathbb{C}^3)} \leq C \left[\|u\|_{L^2(U', \mathbb{C}^3)} + \|\nabla \wedge u\|_{L^2(U', \mathbb{C}^3)} + \|\nabla \cdot (au)\|_{L^2(U')} \right]. \tag{7.4.7}$$

Then we proceed with the following covering argument, which was developed in the proof of [20, Proposition 2.11]. For any $\mathbf{x} \in \partial \Sigma$, let W_n, $n = 1, \ldots, n_0$, be the connected components of $B_{r_1/2}(\mathbf{x}) \cap G$ such that $W_n \cap B_{r_1/4}(\mathbf{x}) \neq \emptyset$. By Lemma 7.3.3, $n_0 \leq M_3$, where M_3 is a constant depending on r_1, r, L and ω only. As in the proof of Proposition 2.11 in [20], one can show that for any $\mathbf{x} \in \partial \Sigma$, there exist n_0 points $\mathbf{x}_1, \ldots, \mathbf{x}_{n_0}$, with $n_0 \leq M_3$, with the following property. For any $n = 1, \ldots, n_0$, there exists U_n, a connected component of $B_{r_1}(\mathbf{x}_n) \cap G$, such that $\mathbf{x}_n \in \partial U_n$, and moreover,

$$B_{r_1/4}(\mathbf{x}) \cap G \subset \bigcup_{n=1}^{n_0} (U_n \cap B_{3r_1/4}(\mathbf{x}_n)).$$

We fix $\delta = r_1/16$ and define the compact set $A_1 = \overline{B_\delta(\partial \Sigma) \cap G}$. We can find a finite number of points $\mathbf{z}_i \in \partial \Sigma$, $i = 1, \ldots, m_1$, such that

$$A_1 \subset \bigcup_{i=1}^{m_1} B_{r_1/4}(\mathbf{z}_i).$$

With a rather simple construction, it is possible to choose m_1 depending on r_1 and R_0 only, for instance, by taking points such that $B_{r_1/16}(\mathbf{z}_i) \cap B_{r_1/16}(\mathbf{z}_j)$ is empty for $i \neq j$.

We further find a finite number of points $\mathbf{z}_i \in \partial B_{R_0+1}$, $i = m_1 + 1, \ldots, m_1 + m_2$, such that

$$A_2 = \overline{B_{1/16}(\partial B_{R_0+1})} \subset \bigcup_{i=m_1+1}^{m_1+m_2} B_{1/4}(\mathbf{z}_i),$$

with m_2 depending on R_0 only. We call $r_3 = \min\{1, r_1\}$ and $A_3 = \{\mathbf{x} \in B_{R_0+1} \backslash \Sigma : \mathrm{dist}(\mathbf{x}, \partial(B_{R_0+1} \backslash \Sigma)) \geq r_3/16\}$. We can find points $\mathbf{z}_i \in A_3$, $i = m_1 + m_2 + 1, \ldots, m_1 + m_2 + m_3$, such that

$$A_3 \subset \bigcup_{i=m_1+m_2+1}^{m_1+m_2+m_3} B_{r_3/32}(\mathbf{z}_i).$$

Again m_3 may be bounded by a constant depending on r_1 and R_0 only.

We apply the local argument developed at the beginning of the proof, at most M_3 times for any $\mathbf{z}_i$, $i=1,\dots,m_1$, and we can find $s_1>2$ and C_1 such that

$$\|u\|_{L^{s_1}(A_1\cap G)} \leq C_1(M_3m_1)C\big[\|u\|_{L^2(B_{R_0+1}\backslash K,\mathbb{C}^3)} + \|\nabla\wedge u\|_{L^2(B_{R_0+1}\backslash K,\mathbb{C}^3)} + \|\nabla\cdot(au)\|_{L^2(B_{R_0+1}\backslash K)}\big].$$

Analogously, we can find $s_2>2$ and C_2 such that

$$\|u\|_{L^{s_2}(A_2\cap B_{R_0+1})} \leq C_2m_2C\big[\|u\|_{L^2(B_{R_0+1}\backslash K,\mathbb{C}^3)} + \|\nabla\wedge u\|_{L^2(B_{R_0+1}\backslash K,\mathbb{C}^3)} + \|\nabla\cdot(au)\|_{L^2(B_{R_0+1}\backslash K)}\big].$$

Finally, we apply the interior estimate with $D=B_{r_3/16}(\mathbf{z}_i)$ and $D'=B_{r_3/32}(\mathbf{z}_i)$, for $i=m_1+m_2+1,\dots,m_1+m_2+m_3$, and we can find $s_3>2$ and C_3 such that

$$\|u\|_{L^{s_3}(A_3)} \leq C_3m_3\big[\|u\|_{L^2(B_{R_0+1}\backslash K,\mathbb{C}^3)} + \|\nabla\wedge u\|_{L^2(B_{R_0+1}\backslash K,\mathbb{C}^3)} + \|\nabla\cdot(au)\|_{L^2(B_{R_0+1}\backslash K)}\big].$$

Picking $p=\min\{s_1,s_2,s_3\}$, we obtain that

$$\|u\|_{L^p(B_{R_0+1}\backslash K)} \leq C\big[\|u\|_{L^2(B_{R_0+1}\backslash K,\mathbb{C}^3)} + \|\nabla\wedge u\|_{L^2(B_{R_0+1}\backslash K,\mathbb{C}^3)} + \|\nabla\cdot(au)\|_{L^2(B_{R_0+1}\backslash K)}\big].$$

Our arguments clearly show that p and C have the dependence required.

Finally, similar reasonings easily show that $B_{R_0+1}\backslash\Sigma$ satisfies the MCP; therefore, the proof is complete. □

7.5 Stability of Solutions of Electromagnetic Scattering Problems

In this section we investigate the stability of solutions of Maxwell equations, in particular of solutions to electromagnetic scattering problems, with respect to changes both in the exterior domain and in the coefficients.

In this section we shall keep fixed positive constants r, L and R_0, $0<r_1<r$, $\tilde{C}>0$, $0<\lambda_0<1<\lambda_1$ and $0<\underline{k}<\overline{k}$ and two nondecreasing left-continuous functions $\omega:(0,+\infty)\to(0,+\infty)$ and $\delta:(0,+\infty)\to(0,+\infty)$.

We begin by defining the following class of admissible coefficients.

Definition 7.5.1. We say that (ε,μ) is a couple of coefficients belonging to the admissible class $\mathscr{N}=\mathscr{N}(r,L,R_0,\omega,\lambda_0,\lambda_1)$ if the following holds.

We assume that $(\varepsilon,\mu)\in L^\infty(\mathbb{R}^3,M^{3\times 3}_{sym}(\mathbb{R})^2)$ and satisfies (7.2.4) with constants $\lambda_0<\lambda_1$ in $\mathbb{R}^3$.

Then we assume that there exists $K\in\mathscr{B}(r,L,B_{R_0},\omega)$, depending on (ε,μ), with the following properties. We call D_0 the unbounded connected component of $\mathbb{R}^3\backslash K$ and D_i, $i=1,\dots,\tilde{M}$, the bounded connected components of $\mathbb{R}^3\backslash K$. We finally assume that $(\varepsilon,\mu)=(I_3,I_3)$ in D_0 and that, for any $i=1,\dots,\tilde{M}$, $(\varepsilon,\mu)=(\varepsilon_i,\mu_i)$ where (ε_i,μ_i) is a couple of Lipschitz functions from $\overline{B_{R_0+1}}$ to $M^{3\times 3}_{sym}(\mathbb{R})$ with Lipschitz constant bounded by L.

We recall that there exists M_1, depending on r, L, R_0 and ω only, such that $\tilde{M}$ in the previous definition, which depend on K and thus on (ε,μ), satisfies $\tilde{M}\leq M_1$.

The following lemmas justify the previous definition.

Lemma 7.5.1. *Let D be any connected open set contained in $\mathbb{R}^3$. Let $k>0$ and $(\varepsilon,\mu)\in\mathscr{N}=\mathscr{N}(r,L,R_0,\omega,\lambda_0,\lambda_1)$.*

Then the Maxwell system

$$\nabla \wedge \mathbf{E} - \mathrm{i}k\mu\mathbf{H} = \mathbf{0}, \quad \nabla \wedge \mathbf{H} + \mathrm{i}k\varepsilon\mathbf{E} = \mathbf{0} \quad \text{in } D \tag{7.5.1}$$

satisfies the UCP in D.

Proof. This is an easy consequence of Proposition 7.2.4 and Lemma 7.2.3; see Remark 7.2.4. □

Lemma 7.5.2. *The class $\mathscr{N} = \mathscr{N}(r, L, R_0, \omega, \lambda_0, \lambda_1)$ is compact with respect to convergence almost everywhere in $\mathbb{R}^3$, as well with respect to the $L^p(\mathbb{R}^3, M^{3\times 3}_{sym}(\mathbb{R})^2)$ convergence, for any p, $1 \leq p < +\infty$.*

Proof. Let us consider $\{(\varepsilon^n, \mu^n)\}_{n\in\mathbb{N}} \subset \mathscr{N}$. We call, for any $n \in \mathbb{N}$, K_n the corresponding set belonging to $\tilde{\mathscr{B}} = \tilde{\mathscr{B}}(r, L, B_{R_0}, \omega)$, M_n the number of bounded connected components of $\mathbb{R}^3 \backslash K_n$ and D_0^n the unbounded connected component of $\mathbb{R}^3 \backslash K_n$. Without loss of generality, up to passing to a subsequence that we do not relabel, we may assume that $M_n = M \leq M_1$ for any $n \in \mathbb{N}$ and we call D_i^n, $i = 1, \ldots, M$, the bounded connected components of $\mathbb{R}^3 \backslash K_n$. Moreover, again up to subsequences, we can assume that, as $n \to \infty$, $K_n \to K \in \tilde{\mathscr{B}}$ in the Hausdorff distance , and, up to reordering, that $\overline{D_i^n} \to \overline{D_i}$, $i = 0, 1, \ldots, M$, in the Hausdorff distance. Here the sets D_i, $i = 0, \ldots, M$, are open subsets of $\mathbb{R}^3 \backslash K$, which are pairwise disjoint. Moreover, D_0 is the only unbounded one. Namely, each of the D_i is the union of a finite number of connected components of $\mathbb{R}^3 \backslash K$.

For any $i = 1, \ldots, M$, $(\varepsilon^n, \mu^n)|_{D_i^n} = (\varepsilon_i^n, \mu_i^n)$ and we can also suppose that, as $n \to \infty$, $(\varepsilon_i^n, \mu_i^n)$ converges to (ε_i, μ_i) uniformly in $\overline{B_{R_0+1}}$. Clearly, for any $i = 1, \ldots, M$, ε_i and μ_i belong to $L^\infty(\overline{B_{R_0+1}}, M^{3\times 3}_{sym}(\mathbb{R}))$ and are Lipschitz with Lipschitz constant bounded by L.

We then define $(\varepsilon, \mu) \in L^\infty(\mathbb{R}^3, M^{3\times 3}_{sym}(\mathbb{R})^2)$ such that $(\varepsilon, \mu) = (I_3, I_3)$ in D_0 and $(\varepsilon, \mu) = (\varepsilon_i, \mu_i)$ in D_i for any $i = 1, \ldots, M$.

We recall that $|K| = 0$. Now, let $\mathbf{x} \in \mathbb{R}^3 \backslash K$. We have that $\mathbf{x} \in D_i$ for some $i \in \{0, 1, \ldots, M\}$. It is an easy remark that, for any n large enough, $\mathbf{x} \in D_i^n$; therefore, $(\varepsilon^n, \mu^n)(\mathbf{x}) = (\varepsilon_i^n, \mu_i^n)(\mathbf{x})$ converges, as $n \to \infty$, to $(\varepsilon_i, \mu_i)(\mathbf{x}) = (\varepsilon, \mu)(\mathbf{x})$. We conclude that (ε^n, μ^n) converges, as $n \to \infty$, to (ε, μ) almost everywhere in $\mathbb{R}^3$. Then it is not difficult to observe that $(\varepsilon, \mu) \in \mathscr{N}$; thus, the compactness is proved. By the uniform L^∞ bound and since any coefficient coincides with the identity matrix outside a given ball, we can immediately conclude the proof also for the convergence in L^p, $1 \leq p < +\infty$. □

We shall also need the following strong convergence result for solutions to Maxwell systems.

Lemma 7.5.3. *Let D be any open set contained in $\mathbb{R}^3$. Let, for any $n \in \mathbb{N}$, ε_n, $\mu_n \in L^\infty(\mathbb{R}^3, M^{3\times 3}_{sym}(\mathbb{R}))$ satisfy* (7.2.4) *with constants λ_0 and λ_1.*

We assume that, for any $n \in \mathbb{N}$, $(\mathbf{E}_n, \mathbf{H}_n) \in H_{loc}(\mathrm{curl}, D)$ solve

$$\nabla \wedge \mathbf{E}_n - \mathrm{i}k_n\mu_n\mathbf{H}_n = \mathbf{0}, \quad \nabla \wedge \mathbf{H}_n + \mathrm{i}k_n\varepsilon_n\mathbf{E}_n = \mathbf{0} \quad \text{in } D$$

for some $0 < k_n \leq \overline{k}$.

Let us assume that, for some constant C, we have

$$\|\mathbf{E}_n\|_{L^2(D,\mathbb{C}^3)} + \|\mathbf{H}_n\|_{L^2(D,\mathbb{C}^3)} \leq C \quad \text{for any } n \in \mathbb{N}$$

and that $(\varepsilon_n, \mu_n) \to (\varepsilon, \mu)$ almost everywhere in D, as $n \to \infty$.

Then, up to a subsequence that we do not relabel, we have that, as $n \to \infty$,

$$(\mathbf{E}_n, \mathbf{H}_n) \to (\mathbf{E}, \mathbf{H}) \quad \text{strongly in } H_{loc}(\mathrm{curl}, D)$$

where $(\mathbf{E}, \mathbf{H})$ solves, for some $0 \leq k \leq \overline{k}$,

$$\nabla \wedge \mathbf{E} - \mathrm{i}k\mu\mathbf{H} = \mathbf{0}, \quad \nabla \wedge \mathbf{H} + \mathrm{i}k\varepsilon\mathbf{E} = \mathbf{0} \quad \text{in } D. \tag{7.5.2}$$

Proof. Obviously, ε, $\mu \in L^\infty(\mathbb{R}^3, M^{3\times 3}_{sym}(\mathbb{R}))$ and satisfy (7.2.4) with constants λ_0 and λ_1.
By the Maxwell equations, for some constant C_1, we have that

$$\|\mathbf{E}_n\|_{H(\mathrm{curl},D)} + \|\mathbf{H}_n\|_{L^2(\mathrm{curl},D)} \leq C_1 \quad \text{for any } n \in \mathbb{N}.$$

Therefore, we can assume, by passing to a subsequence that we do not relabel, that there exists $(\mathbf{E},\mathbf{H}) \in H(\mathrm{curl},D)^2$ and k, $0 \leq k \leq \overline{k}$, such that, as $n \to \infty$, $k_n \to k$ and, in D, we have $(\mathbf{E}_n,\mathbf{H}_n) \rightharpoonup (\mathbf{E},\mathbf{H})$ weakly in L^2 and $(\nabla\wedge\mathbf{E}_n, \nabla\wedge\mathbf{H}_n) \rightharpoonup (\nabla\wedge\mathbf{E}, \nabla\wedge\mathbf{H})$ weakly in L^2. It easily follows that $(\mathbf{E},\mathbf{H})$ solves (7.5.2).

The difficult part is to prove that, actually, the convergence is strong in L^2, at least locally. Let us fix $D_1 \subset \overline{D_1} \subset D_2 \subset \overline{D_2} \subset D$, with D_1 and D_2 open sets and D_2 with Lipschitz boundary. We consider an auxiliary function $\chi \in C^\infty_0(D_2)$ such that $0 \leq \chi \leq 1$ everywhere and $\chi \equiv 1$ in a neighbourhood of $\overline{D_1}$.

We obviously have

$$\mathbf{E}_n \in H(\mathrm{curl},D) \cap H(\mathrm{div}_{\varepsilon_n},D) \quad \text{and} \quad \mathbf{H}_n \in H(\mathrm{curl},D) \cap H(\mathrm{div}_{\mu_n},D)$$

with a corresponding norm bounded by C, λ_0, λ_1 and $\overline{k}$ only. It is not difficult to show that

$$\chi\mathbf{E}_n \in H_0(\mathrm{curl},D_2) \cap H(\mathrm{div}_{\varepsilon_n},D_2) \quad \text{and} \; \chi\mathbf{H}_n \in H_0(\mathrm{curl},D_2) \cap H(\mathrm{div}_{\mu_n},D_2)$$

with a corresponding norm bounded by C, λ_0, λ_1, $\overline{k}$ and the two sets D_1 and D_2 only. We consider only the case of the electric fields, the one for the magnetic fields being completely analogous. We call $\psi_n = \chi\mathbf{E}_n$. We need to investigate the properties of $\nabla\cdot(\varepsilon\psi_n)$. First of all, we notice that, by Proposition 7.3.1, there exists $q > 2$ such that ψ_n, $n \in \mathbb{N}$, is uniformly bounded in $L^q(D_2,\mathbb{C}^3)$. We have that, for any $\varphi \in C^\infty_0(D_2)$

$$\langle\nabla\cdot(\varepsilon\psi_n),\varphi\rangle_{D_2} = -\langle\varepsilon\psi_n,\nabla\varphi\rangle_{D_2} = \langle(\varepsilon_n-\varepsilon)\psi_n,\nabla\varphi\rangle_{D_2} - \langle\varepsilon_n\psi_n,\nabla\varphi\rangle_{D_2}.$$

We know that $\{\nabla\cdot(\varepsilon_n\psi_n)\}_{n\in\mathbb{N}}$ is bounded in $L^2(D_2)$; therefore, it is compact in $(H^1(D_2))^*$. We also have that, in D_2, $(\varepsilon_n-\varepsilon) \to 0$ strongly in L^p for any p, $1 \leq p < +\infty$, and ψ_n is uniformly bounded in $L^q(D_2,\mathbb{C}^3)$, for some $q > 2$. Hence, we have that $(\varepsilon_n-\varepsilon)\psi_n \to 0$ strongly in $L^2(D_2,\mathbb{C}^3)$. Therefore, $\nabla\cdot((\varepsilon_n-\varepsilon)\psi_n)$ converges to 0, as $n \to \infty$, in $(H^1(D_2))^*$. We conclude that $\{\nabla\cdot(\varepsilon\psi_n)\}_{n\in\mathbb{N}}$ is compact in $(H^1(D_2))^*$. By [14, Lemma 2.11] we can conclude that $\{\psi_n\}_{n\in\mathbb{N}}$ is compact in $L^2(D_2,\mathbb{C}^3)$.

We obtain that $\{\mathbf{E}_n\}_{n\in\mathbb{N}}$ and $\{\mathbf{H}_n\}_{n\in\mathbb{N}}$ are bounded in $L^q(D_1,\mathbb{C}^3)$, for some $q > 2$, and compact in $L^2(D_1,\mathbb{C}^3)$. Since in D_1 they converge weakly in L^2 to $\mathbf{E}$ and $\mathbf{H}$, respectively, it is not difficult to conclude that they actually converge strongly in L^2. Using the Maxwell equations, we also obtain that $\{\nabla\wedge\mathbf{E}_n\}_{n\in\mathbb{N}}$ and $\{\nabla\wedge\mathbf{H}_n\}_{n\in\mathbb{N}}$ are bounded in $L^q(D_1,\mathbb{C}^3)$, for some $q > 2$, and compact in $L^2(D_1,\mathbb{C}^3)$.

The proof is complete. □

We are in the position to prove the following general stability result and uniform bounds for the direct electromagnetic scattering problem. These are the main results of the chapter. We begin with the uniform bounds.

Theorem 7.5.1. *Let $\hat{\mathscr{B}}_{scat} = \hat{\mathscr{B}}_{scat}(r,L,R_0,r_1,\tilde{C},\omega,\delta)$ as in Definition 7.3.3. Let $\mathscr{N} = \mathscr{N}(r,L,R_0,\omega,\lambda_0,\lambda_1)$ as in Definition 7.5.1.*

For any $\Sigma \in \hat{\mathscr{B}}_{scat}$, for any $(\varepsilon,\mu) \in \mathscr{N}$ and for any $(\mathbf{E}^i,\mathbf{H}^i)$ as in (7.2.18) with $\underline{k} \leq k \leq \overline{k}$, $\|\mathbf{p}\| \leq 1$, and $\mathbf{d} \in \mathbb{S}^2$, let $(\mathbf{E},\mathbf{H})$ be the solution to (7.2.19) and $(\mathbf{E}^s,\mathbf{H}^s)$ be the corresponding scattering fields.

Then there exists a constant $C > 0$, depending on r, L, R_0, r_1, $\tilde{C}$, ω, δ, λ_0, λ_1, $\underline{k}$ and $\overline{k}$ only, such that

$$\|\mathbf{E}\|_{L^2(B_{R_0+1}\setminus\Sigma,\mathbb{C}^3)} + \|\mathbf{H}\|_{L^2(R_0+1\setminus\Sigma,\mathbb{C}^3)} \leq C. \tag{7.5.3}$$

Then for any $R \geq R_0+1$, there exists a constant E, E depending on the constant C in (7.5.3), R_0, R and $\overline{k}$ only, such that

$$\|\mathbf{E}\|_{L^2(B_R\setminus\Sigma,\mathbb{C}^3)} + \|\mathbf{H}\|_{L^2(B_R\setminus\Sigma,\mathbb{C}^3)} \leq E. \tag{7.5.4}$$

Furthermore, there exists a constant E_1, depending on the constant C in (7.5.3), $\overline{k}$ and R_0 only, such that for any $\mathbf{x} \in \mathbb{R}^3$, we have

$$\|\mathbf{E}^s(\mathbf{x})\| + \|\mathbf{H}^s(\mathbf{x})\| \leq E_1\|\mathbf{x}\|^{-1} \quad \textit{if } \|\mathbf{x}\| \geq R_0+1/2. \tag{7.5.5}$$

Proof. We argue by contradiction. Let us assume that there exist, for any $n \in \mathbb{N}$, $\Sigma_n \in \hat{\mathscr{B}}_{scat}$, $(\varepsilon_n, \mu_n) \in \mathscr{N}$ and $(\mathbf{E}^i_n, \mathbf{H}^i_n)$ as in (7.2.18) with $\underline{k} \leq k_n \leq \overline{k}$, $\|\mathbf{p}_n\| \leq 1$ and $\mathbf{d}_n \in \mathbb{S}^2$, such that $(\mathbf{E}_n, \mathbf{H}_n)$, the solution to (7.2.19) with these data, satisfies

$$\|\mathbf{E}_n\|_{L^2(B_{R_0+1}\setminus\Sigma_n,\mathbb{C}^3)} + \|\mathbf{H}_n\|_{L^2(R_0+1\setminus\Sigma_n,\mathbb{C}^3)} = b_n \geq n.$$

We call $(\tilde{\mathbf{E}}_n, \tilde{\mathbf{H}}_n) = (\mathbf{E}_n/b_n, \mathbf{H}_\mathbf{n}/b_n)$, $n \in \mathbb{N}$, and, by extending them to 0 in Σ_n, we have that

$$\|\tilde{\mathbf{E}}_n\|_{L^2(B_{R_0+1},\mathbb{C}^3)} + \|\tilde{\mathbf{H}}_n\|_{L^2(B_{R_0+1},\mathbb{C}^3)} = 1 \quad \text{for any } n \in \mathbb{N}. \tag{7.5.6}$$

Up to a subsequence, we have that Σ_n converges, in the Hausdorff distance , to a scatterer $\Sigma \in \hat{\mathscr{B}}_{scat}$ and such that $B_{R_0+1}\setminus\Sigma$ satisfies the RCP and MCP; see Lemma 7.3.4 and Proposition 7.4.1. Moreover, by Lemma 7.5.2, we have that (ε_n, μ_n) converges in the whole $\mathbb{R}^3$ to $(\varepsilon, \mu) \in \mathscr{N}$ in L^p for any $1 \leq p < +\infty$. We also assume that $k_n \to k$, with $\underline{k} \leq k \leq \overline{k}$. In particular we notice that $k > 0$.

We also have, again up to subsequences, that $(\tilde{\mathbf{E}}_n, \tilde{\mathbf{H}}_n)$ converges weakly in $L^2(B_{R_0+1}, \mathbb{C}^6)$ to $(\tilde{\mathbf{E}}, \tilde{\mathbf{H}})$. By the first property of Mosco convergence and Theorem 7.4.1, we easily conclude that $(\tilde{\mathbf{E}}, \tilde{\mathbf{H}}) \in H(\mathrm{curl}, B_{R_0+1}\setminus\Sigma)^2$, where again these functions are extended to 0 in Σ.

It is not difficult to show that $(\tilde{\mathbf{E}}, \tilde{\mathbf{H}})$ solve (7.2.7) in $B_{R_0+1}\setminus\Sigma$. Moreover, by standard regularity estimates in $B_{R_0+1}\setminus\overline{B_{R_0}}$, the Stratton-Chu formulas and the fact that $(\mathbf{E}^i_n, \mathbf{H}^i_n)$ are uniformly bounded, we can also deduce that $(\tilde{\mathbf{E}}_n, \tilde{\mathbf{H}}_n)$ converges in $L^2_{loc}(\mathbb{R}^3\setminus\overline{B_{R_0}}, \mathbb{C}^6)$ to $(\tilde{\mathbf{E}}, \tilde{\mathbf{H}})$, with $(\tilde{\mathbf{E}}, \tilde{\mathbf{H}})$ being an outgoing solution to the Maxwell equations (7.2.8) in $\mathbb{R}^3\setminus\overline{B_{R_0}}$.

We know that $\tilde{\mathbf{E}}_n \in H_0(\mathrm{curl}, \mathbb{R}^3\setminus\Sigma_n)$. Then, let $\phi \in H(\mathrm{curl}, \mathbb{R}^3\setminus\Sigma)$, with bounded support. By the second property of the Mosco convergence and Theorem 7.4.1, we can find $\phi_n \in H(\mathrm{curl}, \mathbb{R}^3\setminus\Sigma_n)$, with bounded support, such that, as $n \to \infty$, $(\phi_n, \nabla\wedge\phi_n) \to (\phi, \nabla\wedge\phi)$ in $L^2(\mathbb{R}^3, \mathbb{C}^6)$, with the usual convention of extending the functions to 0 in Σ_n and Σ, respectively. For any $n \in \mathbb{N}$, we have

$$\langle\nabla\wedge\tilde{\mathbf{E}}_n, \phi_n\rangle_{\mathbb{R}^3} - \langle\nabla\wedge\phi_n, \tilde{\mathbf{E}}_n\rangle_{\mathbb{R}^3} = 0.$$

On the other hand, since

$$\langle\nabla\wedge\tilde{\mathbf{E}}, \phi\rangle_{\mathbb{R}^3} - \langle\nabla\wedge\phi, \tilde{\mathbf{E}}\rangle_{\mathbb{R}^3} = \lim_n \left(\langle\nabla\wedge\tilde{\mathbf{E}}_n, \phi_n\rangle_{\mathbb{R}^3} - \langle\nabla\wedge\phi_n, \tilde{\mathbf{E}}_n\rangle_{\mathbb{R}^3}\right),$$

we can conclude that $\tilde{\mathbf{E}} \in H_0(\mathrm{curl}, \mathbb{R}^3\setminus\Sigma)$.

Then we notice that $(\tilde{\mathbf{E}}, \tilde{\mathbf{H}})$ solves the direct scattering problem (7.2.19) with scatterer Σ, coefficients ε, μ and incident fields $(\tilde{\mathbf{E}}^i, \tilde{\mathbf{H}}^i) \equiv (\mathbf{0}, \mathbf{0})$.

By the properties of Σ and Lemma 7.5.1, as well as Theorem 7.2.2, we conclude that $(\tilde{\mathbf{E}}, \tilde{\mathbf{H}}) \equiv (\mathbf{0}, \mathbf{0})$ in $\mathbb{R}^3$.

We wish to prove that $(\tilde{\mathbf{E}}_n, \tilde{\mathbf{H}}_n)$ converges to $(\tilde{\mathbf{E}}, \tilde{\mathbf{H}})$ not only weakly in $L^2(B_{R_0+1}, \mathbb{C}^6)$ but also strongly in $L^2(B_{R_0+1}, \mathbb{C}^6)$. This would allow us to conclude the proof. In fact, since $(\tilde{\mathbf{E}}, \tilde{\mathbf{H}}) \equiv 0$, it would follow that

$$\|\tilde{\mathbf{E}}_n\|_{L^2(B_{R_0+1},\mathbb{C}^3)} + \|\tilde{\mathbf{H}}_n\|_{L^2(B_{R_0+1},\mathbb{C}^3)} \to 0 \quad \text{as } n \to \infty$$

and this contradicts (7.5.6).

To prove the strong convergence in L^2, we begin with the following argument. For any $\delta > 0$, there exists $\overline{n} \in \mathbb{N}$ large enough such that for any $n \geq \overline{n}$, we have $\Sigma_n \subset B_\delta(\Sigma)$. Hence, by Lemma 7.5.3, we deduce that $(\tilde{\mathbf{E}}_n, \tilde{\mathbf{H}}_n)$ converges to $(\tilde{\mathbf{E}}, \tilde{\mathbf{H}})$ strongly in $L^2(K, \mathbb{C}^6)$ for any compact $K \subset \mathbb{R}^3 \backslash \Sigma$. We also notice that, by Proposition 7.4.1 and Corollary 7.4.1, we can find constants $p > 2$ and $C_1 > 0$ such that

$$\|\tilde{\mathbf{E}}_n\|_{L^p(B_{R_0+1}, \mathbb{C}^3)} + \|\tilde{\mathbf{H}}_n\|_{L^p(B_{R_0+1}, \mathbb{C}^3)} \leq C_1 \quad \text{for any } n \in \mathbb{N}. \tag{7.5.7}$$

Therefore, for any $\delta > 0$,

$$\begin{aligned}\|\tilde{\mathbf{E}}_n - \tilde{\mathbf{E}}\|_{L^2(B_{R_0+1}, \mathbb{C}^3)} &\leq \|\tilde{\mathbf{E}}_n - \tilde{\mathbf{E}}\|_{L^2(B_\delta(\Sigma), \mathbb{C}^3)} + \|\tilde{\mathbf{E}}_n - \tilde{\mathbf{E}}\|_{L^2(B_{R_0+1} \backslash B_\delta(\Sigma), \mathbb{C}^3)} \\ &\leq \|\tilde{\mathbf{E}}_n\|_{L^2(B_\delta(\Sigma) \backslash \Sigma_n, \mathbb{C}^3)} + \|\tilde{\mathbf{E}}\|_{L^2(B_\delta(\Sigma) \backslash \Sigma, \mathbb{C}^3)} + \|\tilde{\mathbf{E}}_n - \tilde{\mathbf{E}}\|_{L^2(B_{R_0+1} \backslash B_\delta(\Sigma), \mathbb{C}^3)}.\end{aligned}$$

We also have that, for any $\delta > 0$, there exists $\tilde{n} \geq \overline{n}$, such that for any $n \geq \tilde{n}$, we have $B_\delta(\Sigma) \backslash \Sigma_n \subset B_\delta(\partial \Sigma)$. Obviously, $B_\delta(\Sigma) \backslash \Sigma \subset B_\delta(\partial \Sigma)$. By Hölder inequality, we conclude that

$$\|\tilde{\mathbf{E}}_n - \tilde{\mathbf{E}}\|_{L^2(B_{R_0+1}, \mathbb{C}^3)} \leq C_1 |B_\delta(\partial \Sigma)|^{(p-2)/p} + \|\tilde{\mathbf{E}}\|_{L^2(B_\delta(\partial \Sigma), \mathbb{C}^3)} + \|\tilde{\mathbf{E}}_n - \tilde{\mathbf{E}}\|_{L^2(B_{R_0+1} \backslash B_\delta(\Sigma), \mathbb{C}^3)},$$

with p and C_1 as in (7.5.7) Since $|\partial \Sigma| = 0$, for any $\varepsilon > 0$, there exists $\delta > 0$ such that

$$C_1 |B_\delta(\partial \Sigma)|^{(p-2)/p} + \|\tilde{\mathbf{E}}\|_{L^2(B_\delta(\partial \Sigma), \mathbb{C}^3)} < \varepsilon/2.$$

To fix such a δ, we can find $\hat{n} \geq \tilde{n}$ such that, for any $n \geq \hat{n}$,

$$\|\tilde{\mathbf{E}}_n - \tilde{\mathbf{E}}\|_{L^2(B_{R_0+1} \backslash B_\delta(\Sigma), \mathbb{C}^3)} \leq \varepsilon/2.$$

Therefore, we obtain that $\tilde{\mathbf{E}}_n \to \tilde{\mathbf{E}}$ in $L^2(B_{R_0+1}, \mathbb{C}^3)$ as $n \to \infty$. The same reasoning applies to $\tilde{\mathbf{H}}_n$; thus, the proof of (7.5.3) is concluded.

From (7.5.3), the estimate (7.5.4) and the uniform decay estimate (7.5.5) easily follow. □

We conclude this section with the general stability result.

Theorem 7.5.2. *Under the assumptions of Theorem 7.5.1, let us consider, for any $n \in \mathbb{N}$, $\Sigma_n \in \hat{\mathscr{B}}_{scat}$, $(\varepsilon_n, \mu_n) \in \mathscr{N}$ and $(\mathbf{E}^i_n, \mathbf{H}^i_n)$ as in (7.2.18) with $\underline{k} \leq k_n \leq \overline{k}$, $\|\mathbf{p}_n\| \leq 1$ and $\mathbf{d}_n \in \mathbb{S}^2$. We call $(\mathbf{E}_n, \mathbf{H}_n)$ the solution to (7.2.19) with these data and we extend them, as well as their curls, to $\mathbf{0}$ in Σ_n.*

We assume that, as $n \to \infty$, Σ_n converges, in the Hausdorff distance , to a scatterer $\Sigma \in \hat{\mathscr{B}}_{scat}$, and (ε_n, μ_n) converges in the whole $\mathbb{R}^3$ to $(\varepsilon, \mu) \in \mathscr{N}$ in L^p for any $1 \leq p < +\infty$, and $k_n \to k$, with $\underline{k} \leq k \leq \overline{k}$ and $\mathbf{p}_n \to \mathbf{p}$, with $\|\mathbf{p}\| \leq 1$, and $\mathbf{d}_n \to \mathbf{d} \in \mathbb{S}^2$. We notice that, under our hypotheses, this is always true up to passing to subsequences. We set $(\mathbf{E}^i, \mathbf{H}^i)$ as in (7.2.18) with k, $\mathbf{p}$ and $\mathbf{d}$, and we call $(\mathbf{E}, \mathbf{H})$ the solution to (7.2.19) and we extend them, as well as their curls, to 0 in Σ.

Then we obtain that, as $n \to \infty$, $(\mathbf{E}_n, \mathbf{H}_n)$ converges to $(\mathbf{E}, \mathbf{H})$, and also $(\nabla \wedge \mathbf{E}_n, \nabla \wedge \mathbf{H}_n)$ converges to $(\nabla \wedge \mathbf{E}, \nabla \wedge \mathbf{H})$, in $L^2(B_R, \mathbb{C}^6)$ for any positive R; thus, in particular $(\mathbf{E}_n, \mathbf{H}_n)$ converges to $(\mathbf{E}, \mathbf{H})$ in $H(\mathrm{curl}, D)$ for any bounded open set D compactly contained in $G = \mathbb{R}^3 \backslash \Sigma$.

Proof. First of all we notice that, as $n \to \infty$, $(\mathbf{E}^i_n, \mathbf{H}^i_n)$ converges to $(\mathbf{E}^i, \mathbf{H}^i)$ in $H(\mathrm{curl}, B_R)$ for any $R > 0$. Using the uniform bound (7.5.3) of Theorem 7.5.1, we can obtain this continuity result by easily adapting the arguments developed in the proof of Theorem 7.5.1 to study the convergence properties of the sequence $\{(\tilde{\mathbf{E}}_n, \tilde{\mathbf{H}}_n)\}_{n \in \mathbb{N}}$. □

7.6 Application to the Inverse Scattering Problem for Polyhedral Scatterers

In this section we fix positive constants r, L and R_0, $0 < r_1 < r$ and $\tilde{C} > 0$. Let us also fix $\omega : (0,+\infty) \to (0,+\infty)$ and $\delta : (0,+\infty) \to (0,+\infty)$, two nondecreasing left-continuous functions. We recall that $\omega(t) \leq t$, that $\lim_{t\to+\infty} \omega(t)$ is equal to a finite real number which we call $\omega(+\infty)$ and that $\delta(t) \leq t$ for any $t > 0$. We fix a wavenumber $k > 0$. Finally, we fix positive R_1 and $\tilde{\rho}$ such that $R_0 + 1 + \tilde{\rho} \leq R_1$. We refer to these constants and functions as the *a priori data*.

We introduce suitable classes of *polyhedral scatterers* in $\mathbb{R}^3$. We define a *cell* as the closure of an open subset of a plane. A scatterer Σ is *polyhedral* if the boundary of Σ is given by a finite union of cells $\mathscr{C}_j$, $j = 1,\ldots,M_1$.

To fix positive constants h and L, we say that a scatterer Σ is *polyhedral with constants h and L* if the boundary of Σ is given by a finite union of cells $\mathscr{C}_j$, $j = 1,\ldots,M_1$, where each $\mathscr{C}_j$ is the closure of a Lipschitz domain with constants h and L contained in a plane and the cells are pairwise internally disjoint, that is, two different cells may intersect only at boundary points.

Let $\hat{\mathscr{B}}_{scat} = \hat{\mathscr{B}}_{scat}(r,L,R_0,r_1,\tilde{C},\omega,\delta)$ be the class of scatterers defined in Definition 7.3.3. We call $\hat{\mathscr{B}}^h_{scat} = \hat{\mathscr{B}}^h_{scat}(r,L,R_0,r_1,\tilde{C},\omega,\delta)$, for a given size parameter $h > 0$, the set of scatterers $\Sigma \in \hat{\mathscr{B}}_{scat}$ such that Σ is polyhedral with constants h and L.

Analogously, let $\hat{\mathscr{D}}_{obst} = \hat{\mathscr{D}}_{obst}(r,L,R_0)$ be the class of obstacles also defined in Definition 7.3.3. To fix the size parameter $h > 0$, we call $\hat{\mathscr{D}}^h_{obst} = \hat{\mathscr{D}}^h_{obst}(r,L,R_0)$ the set of obstacles $\Sigma \in \hat{\mathscr{D}}_{obst}$ such that Σ is polyhedral with constants h and L. Notice that in this case any $\Sigma \in \hat{\mathscr{D}}^h_{obst}$ is formed by a finite number of polyhedra. In this case, we can drop from the set of a priori data r_1, $\tilde{C}$, ω and δ.

We call $\eta : (0,1/\mathrm{e}) \to (0,+\infty)$ the following function

$$\eta(s) = \exp(-(\log(-\log s))^{1/2}) \quad \text{for any } s,\ 0 < s < 1/\mathrm{e}. \tag{7.6.1}$$

We consider two different incident fields $(\mathbf{E}^i_j, \mathbf{H}^i_j)$ $j = 1,2$, given by normalised electromagnetic plane waves with incident directions $\mathbf{d}_j \in \mathbb{S}^2$ and polarisation vectors $\mathbf{p}_j \in \mathbb{R}^3$, $j = 1,2$, respectively. We assume that, for any $j = 1,2$, $\|\mathbf{p}_j\| \leq 1$ and that the two vectors $(\mathbf{d}_j \wedge \mathbf{p}_j) \wedge \mathbf{d}_j$ are linearly independent. For example, this is true if $\mathbf{d}_1 = \mathbf{d}_2$ and the three vectors $\mathbf{d}_1$, $\mathbf{p}_1$ and $\mathbf{p}_2$ are linearly independent. In order to have a quantitative version of these properties, we call

$$b_j = \|(\mathbf{d}_j \wedge \mathbf{p}_j) \wedge \mathbf{d}_j\| > 0, \quad j = 1,2 \tag{7.6.2}$$

and

$$b_0 = \min_{\nu\in\mathbb{S}^2} \left\{ \max_{j\in\{1,2\}} \|\nu \wedge [(\mathbf{d}_j \wedge \mathbf{p}_j) \wedge \mathbf{d}_j]\| \right\} > 0. \tag{7.6.3}$$

We notice that $b_0 > 0$ since $\max_{j\in\{1,2\}} \|\nu \wedge [(\mathbf{d}_j \wedge \mathbf{p}_j) \wedge \mathbf{d}_j]\|$ is a continuous function of $\nu \in \mathbb{S}^2$ which never vanishes.

We assume that the medium is homogeneous and isotropic outside the scatterer, that is, we assume $\varepsilon = \mu = I_3$ everywhere. We also fix a point $\mathbf{x}_0 \in \mathbb{R}^3$ such that $R_0 + 1 + \tilde{\rho} \leq \|\mathbf{x}_0\| \leq R_1$.

We recall that for any two scatterers Σ and Σ' belonging to $\hat{\mathscr{B}}_{scat}$ or to $\hat{\mathscr{D}}_{obst}$, we measure their distance by one of the following quantities:

$$d = \max\left\{ \sup_{x\in\partial\Sigma\setminus\Sigma'} \operatorname{dist}(x,\partial\Sigma'),\ \sup_{x\in\partial\Sigma'\setminus\Sigma} \operatorname{dist}(x,\partial\Sigma) \right\} \tag{7.6.4}$$

and

$$\hat{d} = d_H(\partial\Sigma,\partial\Sigma') \quad \text{and} \quad \tilde{d} = d_H(\Sigma,\Sigma'). \tag{7.6.5}$$

Here d_H denotes the Hausdorff distance . In [20, Sect. 2] there is a detailed analysis of the relationships between these quantities. In particular, for a positive constant C_1 depending on $\hat{\mathscr{B}}_{scat}$ only, we have

$$C_1 d \le C_1 \hat{d} \le \tilde{d} \le \delta^{-1}(d) \le \delta^{-1}(\hat{d}), \tag{7.6.6}$$

where $\delta^{-1} : (0,+\infty) \to (0,+\infty)$ is a nondecreasing right-continuous function defined in the following way

$$\delta^{-1}(t) = \min\{\sup\{s : \delta(s) \le t\}, 2R_0\} \quad \text{for any } t > 0. \tag{7.6.7}$$

Let us also notice that in the case of the class $\hat{\mathscr{D}}_{obst}$, by [20, Corollary 2.4], we can replace (7.6.6) by

$$C_1 d \le C_1 \hat{d} \le \tilde{d} \le C_2 d \le C_2 \hat{d}, \tag{7.6.8}$$

with C_1 and C_2 depending on the class $\hat{\mathscr{D}}_{obst}$ only.

The estimates obtained in Theorem 7.5.1, in particular the uniform bound (7.5.4) for $R = R_0 + 3$ and and the uniform decay (7.5.5), are the crucial preliminary results that are required to extend the stability results obtained in the acoustic case, in [35] for sound-soft scatterers and in [20] for sound-hard scatterers, to the electromagnetic case.

Let Σ, $\Sigma' \in \hat{\mathscr{B}}_{scat}$ be two scatterers and $k > 0$. We recall that ε and μ are identically equal to I_3. Given an incident field $(\mathbf{E}^i, \mathbf{H}^i)$, a normalised electromagnetic plane wave with incident direction $\mathbf{d} \in \mathbb{S}^2$ and polarisation $\mathbf{p}$, with $0 < \|\mathbf{p}\| \le 1$, we call $(\mathbf{E}, \mathbf{H})$ the solution to the direct scattering problem (7.2.19), $(\mathbf{E}^s, \mathbf{H}^s)$ the corresponding scattering fields and $(\mathbf{E}_\infty, \mathbf{H}_\infty)$ their far-field patterns . We call $(\mathbf{E}', \mathbf{H}')$ the solution to (7.2.19) with Σ replaced by Σ', and analogously, we denote $((\mathbf{E}^s)', (\mathbf{H}^s)')$ the corresponding scattering fields, and $(\mathbf{E}'_\infty, \mathbf{H}'_\infty)$ their far-field patterns.

About the measurements to be performed for our inverse problem, there are several possibilities. In our stability results, we use what we refer to as the *near-field error with limited aperture* given by

$$\|\mathbf{E} - \mathbf{E}'\|_{L^2(B_{\tilde{\rho}}(\mathbf{x}_0), \mathbb{C}^3)} \le \varepsilon. \tag{7.6.9}$$

It is also possible to consider the so-called *far-field error* which is the one usually employed in scattering applications and that is defined as

$$\|\mathbf{E}_\infty - \mathbf{E}'_\infty\|_{L^2(\mathbb{S}^{N-1}, \mathbb{C}^3)} \le \varepsilon_0. \tag{7.6.10}$$

We recall that there exist positive constants $\tilde{\varepsilon}_0 < 1/e$ and C_1, depending on E as in (7.5.4) for $R = R_0 + 3$, R_0, $\tilde{\rho}$, R_1 and k only, such that if $0 < \varepsilon_0 \le \tilde{\varepsilon}_0$, then

$$\varepsilon \le \eta_1(\varepsilon_0) = \exp\left(-C_1(-\log \varepsilon_0)^{1/2}\right). \tag{7.6.11}$$

Remark 7.6.1. We wish to notice here that, without any loss of generality, we may also consider errors on the magnetic fields, that is, define

$$\|\mathbf{H} - \mathbf{H}'\|_{L^2(B_{\tilde{\rho}}(\mathbf{x}_0), \mathbb{C}^3)} \le \varepsilon, \tag{7.6.12}$$

and

$$\|\mathbf{H}_\infty - \mathbf{H}'_\infty\|_{L^2(\mathbb{S}^{N-1}, \mathbb{C}^3)} \le \varepsilon_0. \tag{7.6.13}$$

Clearly, (7.6.11) holds true in this case as well. More importantly, all the stability results stated in this section still hold if we replace the errors related to the electric fields with the ones related to the magnetic fields.

7.6.1 Statement of the Stability Results

The following stability result holds for the determination of polyhedral scatterers by two suitable electromagnetic measurements.

Theorem 7.6.1. *Fix $h>0$. Let Σ, Σ' belong to $\hat{\mathscr{B}}^h_{scat}$ and let d be defined as in (7.6.4). For any $j=1,2$, let $(\mathbf{E}^i_j,\mathbf{H}^i_j)$ be normalised electromagnetic plane waves with incident directions $\mathbf{d}_j\in\mathbb{S}^2$ and polarisation vectors $\mathbf{p}_j\in\mathbb{R}^3$, with $0<\|\mathbf{p}_j\|\leq 1$ and such that the two vectors $(\mathbf{d}_j\wedge\mathbf{p}_j)\wedge\mathbf{d}_j$, $j=1,2$, are linearly independent. Let b_0 be defined as in (7.6.3).*

For any $j=1,2$, let $(\mathbf{E}_j,\mathbf{H}_j)$ be the solution to

$$\begin{cases}\nabla\wedge\mathbf{E}_j-\mathrm{i}k\,\mathbf{H}_j=\mathbf{0} & \text{in } G=\mathbb{R}^3\backslash\Sigma\\ \nabla\wedge\mathbf{H}_j+\mathrm{i}k\mathbf{E}_j=\mathbf{0} & \text{in } G\\ (\mathbf{E}_j,\mathbf{H}_j)=(\mathbf{E}^i_j,\mathbf{H}^i_j)+(\mathbf{E}^s_j,\mathbf{H}^s_j) & \text{in } G\\ \nu\wedge\mathbf{E}_j=\mathbf{0} & \text{on } \partial G\\ \lim_{r\to+\infty} r\left(\frac{\mathbf{x}}{\|\mathbf{x}\|}\wedge\mathbf{H}^s_j(\mathbf{x})+\mathbf{E}^s_j(\mathbf{x})\right)=\mathbf{0} & r=\|\mathbf{x}\|\end{cases}\tag{7.6.14}$$

and $(\mathbf{E}'_j,\mathbf{H}'_j)$ be the solution to the same problem with Σ replaced by Σ'.

If

$$\max_{j=1,2}\|\mathbf{E}_j-\mathbf{E}'_j\|_{L^2(B_{\tilde{\rho}}(\mathbf{x}_0),\mathbb{C}^3)}\leq\varepsilon\tag{7.6.15}$$

for some $\varepsilon\leq 1/(2\mathrm{e})$, then for some positive constant C depending on the a priori data and on b_0 only, and not on h, we have

$$\min\{d,h\}\leq 2\mathrm{e}R_0(\eta(\varepsilon))^C.\tag{7.6.16}$$

Therefore,

$$d\leq 2\mathrm{e}R_0(\eta(\varepsilon))^C,\tag{7.6.17}$$

provided $\varepsilon\leq\hat{\varepsilon}(h)$ where

$$\hat{\varepsilon}(h)=\min\left\{1/(2\mathrm{e}),\eta^{-1}\left(\left(\frac{h}{2\mathrm{e}R_0}\right)^{1/C}\right)\right\}.\tag{7.6.18}$$

If we limit ourselves to polyhedral obstacles, that is, to polyhedra, we can reduce the number of electromagnetic measurements to one and have the following stability result.

Theorem 7.6.2. *Fix $h>0$. Let Σ, Σ' belong to $\hat{\mathscr{D}}^h_{obst}$ and let d be defined as in (7.6.4). Let $(\mathbf{E}^i_1,\mathbf{H}^i_1)$ be the normalised electromagnetic plane wave with incident direction $\mathbf{d}_1\in\mathbb{S}^2$ and polarisation vector $\mathbf{p}_1\in\mathbb{R}^3$, with $0<\|\mathbf{p}_1\|\leq 1$ and such that (7.6.2) holds for some positive constant b_1.*

Let $(\mathbf{E}_1,\mathbf{H}_1)$ be the solution to (7.6.14) with $j=1$ and $(\mathbf{E}'_1,\mathbf{H}'_1)$ be the solution to the same problem with Σ replaced by Σ'.

There exists a constant $\hat{\varepsilon}_1(h)$, $0<\hat{\varepsilon}_1(h)\leq 1/(2\mathrm{e})$, depending on the a priori data, on b_1 and on h only, such that if

$$\|\mathbf{E}_1-\mathbf{E}'_1\|_{L^2(B_{\tilde{\rho}}(\mathbf{x}_0),\mathbb{C}^3)}\leq\varepsilon\tag{7.6.19}$$

for some $\varepsilon\leq\hat{\varepsilon}_1(h)$, then for some positive constants A_1, depending on the a priori data only, and C_1, depending on the a priori data, on b_1 and on h only, we have

$$d\leq A_1(\eta(\varepsilon))^{C_1}.\tag{7.6.20}$$

Remark 7.6.2. By the arguments recalled at the beginning of this section, it is an easy task to rephrase the stability estimates of Theorems 7.6.1 and 7.6.2 if we measure the distance between Σ and Σ' with $\tilde{d}=d_H(\Sigma,\Sigma')$ or $\hat{d}=d_H(\partial\Sigma,\partial\Sigma')$ instead of d or if we replace the near-field error with limited aperture ε with the far-field error ε_0 on the corresponding solutions. In the latter case, it is enough to replace ε with $\eta_1(\varepsilon_0)$, η_1 as in (7.6.11), and observe that we may choose $\tilde{\rho}$ and R_1 as depending on the other a priori data.

7.6.2 Remarks and Comments on the Proofs

The following auxiliary propositions are needed. Since they hold in a general case, we state them for any $N \geq 2$. We recall that we always drop the dependence of constants on the dimension N.

The following three-sphere inequality holds true and is a consequence of results proved in [5].

Proposition 7.6.1. *There exist positive constants $\tilde{\rho}_0$, C and c_1, $0 < c_1 < 1$, depending on $k > 0$ only, such that for every $0 < \rho_1 < \rho < \rho_2 \leq \tilde{\rho}_0$ and any function u such that*

$$\Delta u + k^2 u = 0 \quad \text{in } B_{\rho_2} \subset \mathbb{R}^N,$$

we have

$$\|u\|_{L^2(B_\rho)} \leq C \|u\|^{1-\beta}_{L^2(B_{\rho_2})} \|u\|^{\beta}_{L^2(B_{\rho_1})}, \tag{7.6.21}$$

for some β such that

$$c_1 \frac{\log(\rho_2/\rho)}{\log(\rho_2/\rho_1)} \leq \beta \leq 1 - c_1 \frac{\log(\rho/\rho_1)}{\log(\rho_2/\rho_1)}. \tag{7.6.22}$$

The following L^∞-L^2 estimate is a consequence of classical regularity estimates and a simple dilation argument.

Lemma 7.6.1. *Let us fix a positive constant ρ_1. Let us consider ρ, $0 < \rho \leq \rho_1$ and a function u such that*

$$\Delta u + k^2 u = 0 \quad \text{in } B_\rho \subset \mathbb{R}^N.$$

Then, for any constant s, $0 < s < 1$, there exists a constant C, depending on k, ρ_1 and s only, such that

$$\rho^{N/2} \|u\|_{L^\infty(B_{s\rho})} \leq C \|u\|_{L^2(B_\rho)}. \tag{7.6.23}$$

The proofs of Theorems 7.6.1 and 7.6.2 follow the arguments developed for the proofs of Theorem 3.1 and Theorem 3.4 in [20], respectively. We just point out the few differences and leave all the other details to the reader. One difference is that we have an L^2 a priori bound, a consequence of Theorem 7.5.1, instead of an L^∞ a priori bound. However, we can exploit the L^2 three-sphere inequality recalled in Proposition 7.6.1. For arguments concerning reflections in a plane, we can use Lemma 7.3.2 and the estimate given in Lemma 7.3.1. Finally, L^2 estimates provide L^∞ estimates by using Lemma 7.6.1.

Then we can conclude the proofs of our stability theorems in the following way.

Proof of Theorem 7.6.1. We fix a constant $R_2 \geq \max\{2R_1, 4R_0\}$, depending on the a priori data and on b_0 only, such that

$$E_1 R_2^{-1} \leq k b_0/2$$

where E_1 is as in (7.5.5).

By the arguments used to prove [20, Theorem 3.1] and in particular Lemma 4.4 in the same paper, with the slight modification pointed out above, we can find a point $\mathbf{z}$, with $\|\mathbf{z}\| \geq R_2 + 1$ and a unit vector ν such that

$$h^{3/2} \|\nu \wedge \mathbf{E}_j(\mathbf{z})\| \leq C_0 \varepsilon_2, \quad j = 1, 2, \tag{7.6.24}$$

with C_0 and ε_2 as in [20, Lemma 4.4]. Here the term $h^{3/2}$ comes from the application of Lemma 7.6.1.

We recall that for any $\mathbf{x} \in \mathbb{R}^3$ and any $j = 1, 2$, we have

$$\nu \wedge \mathbf{E}^i_j(\mathbf{x}) = \mathrm{i} k e^{\mathrm{i} k \mathbf{x} \cdot \mathbf{d}_j} (\nu \wedge [(\mathbf{d}_j \wedge \mathbf{p}_j) \wedge \mathbf{d}_j]).$$

Therefore, there exists $j \in \{1, 2\}$ such that for any $\mathbf{x} \in \mathbb{R}^3$

$$\|\nu \wedge \mathbf{E}^i_j(\mathbf{x})\| \geq k b_0. \tag{7.6.25}$$

Since for any $j = 1,2$ we have

$$\|\nu \wedge \mathbf{E}^s_j(\mathbf{z})\| \leq kb_0/2,$$

by our definition of R_2, we conclude that there exists $j \in \{1,2\}$ such that

$$kb_0/2 \leq \|\nu \wedge \mathbf{E}_j(\mathbf{z})\| \leq C_0 h^{-3/2}\varepsilon_2. \tag{7.6.26}$$

We can then conclude as in the proof of [20, Theorem 3.1]. □

We conclude by showing the final argument for the proof of the stability result for the determination of polyhedra by a single electromagnetic measurement. As in the sound-hard acoustic case, the single measurement for polyhedra requires a completely nontrivial analysis. We also point out that, contrary to the 2 measurement case, the dependence on the size constant h is no longer explicit.

Proof of Theorem 7.6.2. First of all, we notice that there exists a constant $\tilde{b}_0 > 0$, depending on the a priori data only, such that for any $\omega \in \mathbb{S}^2$ and any polyhedral obstacle $\Sigma \in \hat{\mathscr{D}}_{obst}$, we can find a cell $\mathscr{C}$ in $\partial\Sigma$, with unit normal ν, such that $\|\nu \wedge \omega\| \geq \tilde{b}_0$.

To prove such a claim we argue by contradiction. Assume that there exist polyhedral obstacles $\Sigma_n \in \hat{\mathscr{D}}_{obst}$ and $\omega_n \in \mathbb{S}^2$, $n \in \mathbb{N}$, such that, for almost any $\mathbf{x} \in \partial\Sigma_n$, we have $\|\nu_n(\mathbf{x}) \wedge \omega_n\| < 1/n$, where ν_n denotes the unit normal of $\partial\Sigma_n$. Notice that, up to a subsequence, ω_n converges to $\omega \in \mathbb{S}^2$ and Σ_n converges to $\Sigma \in \hat{\mathscr{D}}_{obst}$ in the Hausdorff distance . Then one obtains that $\|\nu(\mathbf{x}) \wedge \omega\| = 0$ for almost any $\mathbf{x} \in \partial\Sigma$, ν being the unit normal of $\partial\Sigma$. This contradicts the fact that Σ is a solid obstacle.

Then Theorem 7.6.2 can be proved using the argument of the proof of [20, Theorem 3.4] with the same kind of modification needed in the proof of Theorem 7.6.1, which is detailed above, and the following remark.

As in the acoustic case, the issue is the following. Let Π be a plane and ν be its normal. Then we consider

$$\max_{\mathbf{z} \in (\overline{B}_{2R_2+3} \setminus B_{2R_2+2}) \cap \Pi} \|\nu \wedge \mathbf{E}_1(\mathbf{z})\|. \tag{7.6.27}$$

If for one of these $\mathbf{z}$ we have an estimate like in (7.6.24), in order to conclude we need a corresponding lower bound as in (7.6.25). However, it might happen that the quantity in (7.6.27) is actually 0. On the other hand, this happens only on special symmetric cases. Namely, we need that $\nu \wedge [(\mathbf{d}_1 \wedge \mathbf{p}_1) \wedge \mathbf{d}_1] = 0$ and that Σ is symmetric with respect to Π. This can be proved by a simple reflection argument and by using the unique determination of polyhedra with a single scattering electromagnetic measurement proved in [19]. Finally, also a continuous dependence of the quantity in (7.6.27) from Σ and Π would be needed, but this follows quite easily by Theorem 7.5.2. □

References

1. L. Alessandra and V. Felli, *Unique continuation from the edge of a crack*, arXiv:2004.11177
2. G. Alessandrini and L. Rondi, *Determining a sound-soft polyhedral scatterer by a single far-field measurement*, Proc. Amer. Math. Soc., **35** (2005), 1685–1691.
3. L. Ambrosio and G. Dal Maso, *A general chain rule for distributional derivatives*, Proc. Amer. Math. Soc., **108** (1990), 691–702.
4. J. Ball, Y. Capdeboscq and B. Tsering-Xiao, *On uniqueness for time harmonic anisotropic Maxwell's equations with piecewise regular coefficients*, Math. Models Methods Appl. Sci., **22** (2012), 1250036 (11pp).
5. R. Brummelhuis, *Three-spheres theorem for second order elliptic equations*, J. Analyse Math. **65** (1995) 179–206.
6. D. Bucur and N. Varchon, *Stability of the Neumann problem for variations of boundary*, C. R. Acad. Sci. Paris Sér. I Math., **331** (2000), 371–374.
7. D. Bucur and N. Varchon, *Boundary variation for a Neumann problem*, Ann. Scuola Norm. Sup. Pisa Cl. Sci. (4), **29** (2000), 807–821.

8. A. Chambolle and F. Doveri, *Continuity of Neumann linear elliptic problems on varying two dimensional bounded open sets*, Comm. Partial Differential Equations, **22** (1997), 811–840.
9. J. Cheng and M. Yamamoto, *Uniqueness in an inverse scattering problem within non-trapping polygonal obstacles with at most two incoming waves*, Inverse Problems, **19** (2003), 1361–1384.
10. D. Colton and R. Kress, *Inverse acoustic and electromagnetic scattering theory*, volume 93 of Applied Mathematical Sciences. Springer, Cham, 4th edition, 2019.
11. M. Costabel, *A remark on the regularity of solutions of Maxwell's equations on Lipschitz domains*, Math. Methods Appl. Sci., **12** (1990), 365–368.
12. G. Dal Maso, *An Introduction to Γ-Convergence*, Birkhäuser, Boston Basel Berlin, (1993).
13. M. N. Demchenko, *On the nonuniqueness of the continuation of the solution of the* , Zap. Nauchn. Sem. S.-Peterburg. Otdel. Mat. Inst. Steklov. (POMI), **393(41)** (2011), Matematicheskie Voprosy Teorii Rasprostraneniya, 80–100, 261 (Russian). Translation in J. Math. Sci. (N. Y.), **185** (2012), 554–566.
14. P.-É. Druet, *Higher L^p regularity for vector fields that satisfy divergence and rotation constraints in dual Sobolev spaces, and application to some low-frequency Maxwell equations*, Discrete Contin. Dyn. Syst. Ser. S, **8** (2015), 475–496.
15. J. Elschner and M. Yamamoto, *Uniqueness in determining polygonal sound-hard obstacles with a single incoming wave*, Inverse Problems, **22** (2006), 355–364.
16. J. Elschner and M. Yamamoto, *Uniqueness in determining polyhedral sound-hard obstacles with a single incoming wave*, Inverse Problems, **24** (2008), 035004 (7pp).
17. A. Giacomini, *A stability result for Neumann problems in dimension $N \geq 3$*, J. Convex Anal, **11** (2004), 41–58.
18. V. Isakov, *Inverse Problems for Partial Differential Equations*, 2nd edition, Springer-Verlag, New York, (2006).
19. H. Liu, *A global uniqueness for formally determined inverse electromagnetic obstacle scattering*, Inverse Problems, **24** (2008), 035018 (13pp).
20. H. Liu, M. Petrini, L. Rondi and J. Xiao, Stable determination of sound-hard polyhedral scatterers by a minimal number of scattering measurements, *J. Differential Equations*, **262(3)** (2017), 1631–1670.
21. H. Liu, L. Rondi and J. Xiao, Mosco convergence for $H(curl)$ spaces, higher integrability for Maxwell's equations, and stability in direct and inverse EM scattering problems, *J. Eur. Math. Soc.*, *J. Eur. Math. Soc.*, **21** (2019), no. 10, 2945–2993.
22. H. Liu, M. Yamamoto and J. Zou, *Reflection principle for the Maxwell equations and its application to inverse electromagnetic scattering*, Inverse Problems, **23** (2007), 2357–2366.
23. H. Liu and J. Zou, *Uniqueness in an inverse acoustic obstacle scattering problem for both sound-hard and sound-soft polyhedral scatterers*, Inverse Problems, **22** (2006), 515–524.
24. H. Liu and J. Zou, *On unique determination of partially coated polyhedral scatterers with far field measurements*, Inverse Problems, **23** (2007), 297–308.
25. V. Maz'ya, *Sobolev Spaces*, 2nd edition, Springer-Verlag, Berlin, (2011).
26. G. Menegatti and L. Rondi, *Stability for the acoustic scattering problem for sound-hard scatterers*, Inverse Probl. Imaging **7** (2013) 1307–1329.
27. P. Monk, *Finite Element Methods for Maxwell's Equations*, Oxford University Press, Oxford, (2003).
28. U. Mosco, *Convergence of convex sets and of solutions of variational inequalities*, Advances in Math., **3** (1969), 510–585.
29. J.-C. Nédélec, *Acoustic and Electromagnetic Equations*, Springer-Verlag, New York, (2001).
30. T. Nguyen and J.-N. Wang, *Quantitative uniqueness estimate for the Maxwell system with Lipschitz anisotropic media*, Proc. Amer. Math. Soc., **140** (2012), 595–605.
31. R. Picard, N. Weck and K.-J. Witsch, *Time-harmonic Maxwell equations in the exterior of perfectly conducting, irregular obstacles*, Analysis (Munich), **21** (2001), 231–263.
32. Yu. G. Reshetnyak, *Space Mappings with Bounded Distortion*, American Mathematical Society, Providence RI, (1989).
33. L. Rondi, *Unique determination of non-smooth sound-soft scatterers by finitely many far-field measurements*, Indiana Univ. Math. J., **52** (2003), 1631–1662.
34. L. Rondi, *A variational approach to the reconstruction of cracks by boundary measurements*, J. Math. Pures Appl. (9), **87** (2007), 324–342.
35. L. Rondi, *Stable determination of sound-soft polyhedral scatterers by a single measurement*, Indiana Univ. Math. J., **57** (2008), 1377–1408.

Chapter 8
Geometric Structures of Helmholtz's Transmission Eigenfunctions with General Transmission Conditions and Applications

8.1 Overview

Let Ω be a bounded Lipschitz domain in $\mathbb{R}^n$, $n = 2,3$, and $V \in L^\infty(\Omega)$ and $\eta \in L^\infty(\partial\Omega)$ be possibly complex-valued functions. Consider the following interior transmission eigenvalue problem with a conductive boundary condition for v, $w \in H^1(\Omega)$:

$$\begin{cases} \Delta w + k^2(1+V)w = 0 & \text{in } \Omega, \\ \Delta v + k^2 v = 0 & \text{in } \Omega, \\ w = v,\ \partial_\nu v + \eta v = \partial_\nu w & \text{on } \partial\Omega, \end{cases} \tag{8.1.1}$$

where $\nu \in \mathbb{S}^{n-1}$ signifies the exterior unit normal vector to $\partial\Omega$. Clearly, $v = w \equiv 0$ are trivial solutions to (8.1.1). If, for a certain $k \in \mathbb{R}_+$, there exists a pair of nontrivial solutions $(v,w) \in H^1(\Omega) \times H^1(\Omega)$ to (8.1.1), then k is called a conductive transmission eigenvalue, and (v,w) is referred to as the corresponding pair of conductive transmission eigenfunctions. For a special case with $\eta \equiv 0$, (8.1.1) is known to be the interior transmission eigenvalue problem. For terminological convenience, we refer to the nontrivial solutions $(v, w) \in H^1(\Omega)$ to (8.1.1) as the conductive transmission eigenfunctions, and the corresponding k is named as the conductive transmission eigenvalue. The study on the transmission eigenvalue problems arises in the wave scattering theory and has a long and colorful history; see [10,20,23,26,34,35,50,51,53] for the spectral study of the interior transmission eigenvalue problem, and [13,14,32] for the related study of the conductive transmission eigenvalue problem, and a recent survey [15] and the references therein for comprehensive discussions on the state-of-the-art developments. The problem is a type of non-elliptic and non-self-adjoint eigenvalue problem, so its study is mathematically interesting and challenging. The existing results in the literature mainly focus on the spectral properties of the transmission eigenvalues, namely, their existence, discreteness, infiniteness, and Weyl's laws. Roughly speaking, the theorems for the transmission eigenvalues follow in a similar flavor to the results in the spectral theory of the Laplacian on a bounded domain. However, the transmission eigenfunctions reveal certain distinct and intriguing features. In [11,49], it is proved that the interior transmission eigenfunctions cannot be analytically extended across the boundary $\partial\Omega$ if it contains a corner with an interior angle less than π. In [6], geometric structures of interior transmission eigenfunctions were discovered for the first time. It is shown that under certain regularity conditions on the interior transmission eigenfunctions, the eigenfunctions must be locally vanishing near a corner of the domain with an interior angle less than π. With the help of numerics, it is further shown in [5] that if the regularity conditions are not fulfilled, the transmission eigenfunctions may produce a certain localizing phenomenon around the corner. Recently, more geometric

H. Diao, H. Liu, *Spectral Geometry and Inverse Scattering Theory*,
https://doi.org/10.1007/978-3-031-34615-6_8

properties of the interior transmission eigenfunctions were discovered in [8], which are linked with the curvature of a specific boundary point. It is noted that a corner point considered in [5, 6] can be regarded as having an infinite extrinsic curvature since the derivative of the normal vector has a jump singularity there. Recent progress on local and global geometrical structures of interior transmission eigenfunctions can be found in [37].

In addition to the angle of the corner, we would like to emphasize the critical role played by the regularity of the transmission eigenfunctions in the existing studies of the geometric structures in the aforementioned literatures. In [6], the regularity requirements are characterized in two ways. The first one is H^2 smoothness, and the other one is H^1 regularity with a certain Herglotz approximation property. The H^2 regularity requirement can be weakened a bit to be Hölder continuity with any Hölder index $\alpha \in (0, 1)$.

In this chapter, we consider the vanishing property of the conductive transmission eigenfunctions associated with (8.1.1) at a corner as long as its interior angle is not π when the conductive transmission eigenfunctions satisfy certain Herglotz functions' approximation properties. That means, as long as the corner singularity is not degenerate, the vanishing property holds if the underlying conductive transmission eigenfunctions can be approximated by a sequence of Herglotz functions under mild approximation rates. In fact, in the three-dimensional case, the corner singularity is a more general edge singularity. To give the vanishing property, we need to impose certain regularity conditions on the conductive transmission eigenfunctions which basically follow a similar manner to those considered in [6]. That is, the first regularity condition is the Hölder continuity with any Hölder index $\alpha \in (0, 1)$, and the second regularity condition is characterized by the Herglotz approximation. Nevertheless, for the latter case, the regularity requirement is much more relaxed in the present study compared to that in [6]. Finally, we would like to emphasize that in principle, the geometric properties for the conductive transmission eigenfunctions include the results in [6] as a special case by taking the parameter η to be zero. Hence, in the sense described above, the results obtained in this work significantly extend and generalize the ones in [6]. In this chapter, we follow the treatment in [28].

The mathematical argument in [6] is indirect which connects the vanishing property of the interior transmission eigenfunctions with the stability of a certain wave scattering problem with respect to variation of the wave field at the corner point. In [4, 8], direct mathematical arguments based on certain micro-local analysis techniques are developed for dealing with the vanishing properties of the interior transmission eigenfunctions. However, the Hölder continuity on the interior transmission eigenfunctions is an essential assumption in [4, 8]. In this chapter, in order to provide the vanishing property of the conductive transmission eigenfunctions under more general regularity conditions, we basically follow the direct approach.

As an interesting and practical application, we apply the obtained geometric results for the conductive transmission eigenfunctions to an inverse problem associated with the transverse electromagnetic scattering. In a certain scenario, we discuss the unique recovery result by a single far-field measurement in simultaneously determining a polygonal conductive obstacle and its surface conductivity. This contributes to the well-known Schiffer's problem in the inverse scattering theory which is concerned with recovering the shape of an unknown scatterer by a single far-field pattern ; see [2, 7, 9, 16, 17, 19, 21, 29, 33, 36–46, 52] and the references therein for background introduction and the state-of-the-art developments on Schiffer's problem.

The rest of the chapter is organized as follows. In Sects. 8.2 and 8.3, we, respectively, derive the vanishing results of the conductive transmission eigenfunctions near a corner in the two-dimensional and three-dimensional cases. Section 8.4 is devoted to the uniqueness study in determining a polygonal conductive obstacle as well as its surface conductivity by a single far-field pattern.

8.2 Vanishing Near Corners of Conductive Transmission Eigenfunctions: Two-Dimensional Case

In this section, we consider the vanishing near corners of conductive transmission eigenfunctions in the two-dimensional case. First, let us introduce some notations for the subsequent use. Let (r,θ) be the polar coordinates in $\mathbb{R}^2$; that is, $\mathbf{x} = (x_1,x_2) = (r\cos\theta, r\sin\theta) \in \mathbb{R}^2$. For $\mathbf{x} \in \mathbb{R}^2$, $B_h(\mathbf{x})$ denotes an open ball of radius $h \in \mathbb{R}_+$ and centered at x. $B_h := B_h(\mathbf{0})$. Consider an open sector in $\mathbb{R}^2$ with the boundary $\Gamma^\pm$ as follows:

$$W = \left\{\mathbf{x} \in \mathbb{R}^2 \mid \mathbf{x} \neq \mathbf{0}, \quad \theta_m < \arg(x_1 + \mathrm{i}x_2) < \theta_M \right\}, \tag{8.2.1}$$

where $-\pi < \theta_m < \theta_M < \pi$, $\mathrm{i} := \sqrt{-1}$ and Γ^+ and Γ^-, respectively, correspond to (r,θ_M) and (r,θ_m) with $r > 0$. Henceforth, set

$$S_h = W \cap B_h,\ \Gamma_h^\pm = \Gamma^\pm \cap B_h,\ \overline{S}_h = \overline{W} \cap B_h,\ \Lambda_h = W \cap \partial B_h, \text{ and } \Sigma_{\Lambda_h} = S_h \backslash S_{h/2}. \tag{8.2.2}$$

In Fig. 8.1, we give a schematic illustration of the geometry considered here.

For $g \in L^2(\mathbb{S}^{n-1})$, the Herglotz wave function with kernel g is defined by

$$v(\mathbf{x}) = \int_{\mathbb{S}^{n-1}} e^{\mathrm{i}k\xi\cdot\mathbf{x}} g(\xi)\mathrm{d}\sigma(\xi), \quad \xi \in \mathbb{S}^{n-1}, \quad \mathbf{x} \in \mathbb{R}^n. \tag{8.2.3}$$

It can be easily seen that v is an entire solution to the Helmholtz equation $\Delta v + k^2 v = 0$. By Theorem 2 and Remark 2 in [56], we have the following Herglotz approximation result.

Lemma 8.2.1. *Let $\Omega \Subset \mathbb{R}^n$ be a bounded Lipschitz domain and $\mathbf{H}_k$ be the space of all Herglotz wave functions of the form* (8.2.3). *Define*

$$\mathbf{S}_k(\Omega) = \{u \in C^\infty(\Omega) \mid \Delta u + k^2 u = 0\}$$

and

$$\mathbf{H}_k(\Omega) = \{u|_\Omega \mid u \in \mathbf{H}_k\}.$$

Then $\mathbf{H}_k(\Omega)$ is dense in $\mathbf{S}_k(\Omega) \cap L^2(\Omega)$ with respect to the topology induced by the $H^1(\Omega)$ norm.

Remark 8.2.1. From Lemma 8.2.1, for any $v \in H^1(\Omega)$ being a solution to the Helmholtz equation in Ω, we can conclude that there exists a sequence of Herglotz functions which can approximate v to an arbitrary accuracy.

We shall also need the following lemma, which gives a particular type of planar complex geometric optics (CGO) solution whose logarithm is a branch of the square root (cf. [4]).

Lemma 8.2.2 ([4, Lemma 2.2]). *For $x \in \mathbb{R}^2$, denote $r = |\mathbf{x}|$, $\theta = \arg(x_1 + \mathrm{i}x_2)$. Let $s \in \mathbb{R}_+$ and*

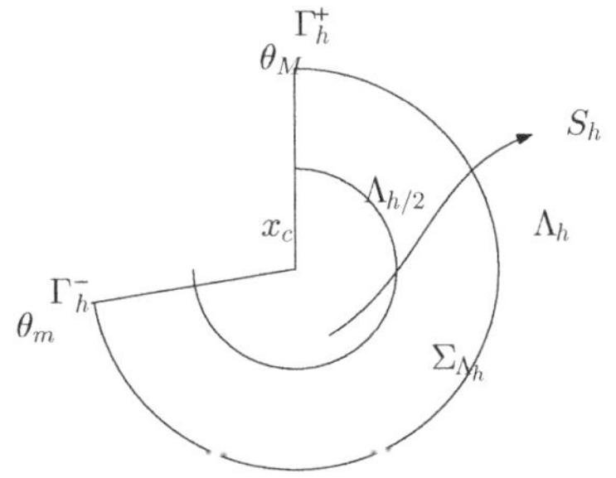

Fig. 8.1 Schematic illustration of the corner in 2D

$$u_0(s\mathbf{x}) := \exp\left(\sqrt{sr}\left(\cos\left(\frac{\theta}{2}+\pi\right)+\mathrm{i}\sin\left(\frac{\theta}{2}+\pi\right)\right)\right). \tag{8.2.4}$$

Then $\Delta u_0 = 0$ *in* $\mathbb{R}^2\backslash\mathbb{R}^2_{0,-}$, *where* $\mathbb{R}^2_{0,-} := \{\mathbf{x}\in\mathbb{R}^2|\mathbf{x}=(x_1,x_2); x_1\leq 0, x_2=0\}$, *and* $s\mapsto u_0(s\mathbf{x})$ *decays exponentially in* $\mathbb{R}_+$. *Let* $\alpha, s>0$. *Then*

$$\int_W |u_0(s\mathbf{x})||\mathbf{x}|^\alpha \mathrm{d}x \leq \frac{2(\theta_M-\theta_m)\Gamma(2\alpha+4)}{\delta_W^{2\alpha+4}}s^{-\alpha-2}, \tag{8.2.5}$$

where $\delta_W = -\max_{\theta_m<\theta<\theta_M}\cos(\theta/2+\pi)>0$. *Moreover,*

$$\int_W u_0(s\mathbf{x})\mathrm{d}x = 6\mathrm{i}(e^{-2\theta_M\mathrm{i}}-e^{-2\theta_m\mathrm{i}})s^{-2}, \tag{8.2.6}$$

and for $h>0$,

$$\int_{W\backslash B_h} |u_0(s\mathbf{x})|\mathrm{d}\mathbf{x} \leq \frac{6(\theta_M-\theta_m)}{\delta_W^4}s^{-2}e^{-\delta_W\sqrt{hs}/2}. \tag{8.2.7}$$

The following lemma states the regularity of the CGO solution $u_0(s\mathbf{x})$ defined in (8.2.4).

Lemma 8.2.3. *Let* S_h *be defined in* (8.2.2) *and* $u_0(s\mathbf{x})$ *be given by* (8.2.4). *Then* $u_0(s\mathbf{x})\in H^1(S_h)$ *and* $\Delta u_0(s\mathbf{x})=0$ *in* S_h. *Furthermore, it holds that*

$$\|u_0(s\mathbf{x})\|^2_{L^2(S_h)} \leq \frac{(\theta_M-\theta_m)e^{-2\sqrt{s\Theta}\delta_W}h^2}{2} \tag{8.2.8}$$

and

$$\||\mathbf{x}|^\alpha u_0(s\mathbf{x})\|^2_{L^2(S_h)} \leq s^{-(2\alpha+2)}\frac{2(\theta_M-\theta_m)}{(4\delta_W^2)^{2\alpha+2}}\Gamma(4\alpha+4), \tag{8.2.9}$$

where $\Theta\in[0,h]$ *and* δ_W *is defined in* (8.2.5).

Proof. Recalling the expression of u_0 given in (8.2.4), using change of variables and the integral mean value theorem, we can deduce that

$$\begin{aligned}\|u_0(s\mathbf{x})\|^2_{L^2(S_h)} &= \int_0^h r\mathrm{d}r\int_{\theta_m}^{\theta_M} e^{2\sqrt{sr}\cos(\theta/2+\pi)}\mathrm{d}\theta \leq \int_0^h r\mathrm{d}r\int_{\theta_m}^{\theta_M} e^{-2\sqrt{sr}\delta_W}\mathrm{d}\theta\\ &= \frac{(\theta_M-\theta_m)e^{-2\sqrt{s\Theta}\delta_W}h^2}{2},\end{aligned}$$

where $\Theta\in[0,h]$ and δ_W is defined in (8.2.5). Furthermore, it can directly verified that

$$\begin{aligned}\frac{\partial u_0(s\mathbf{x})}{\partial r} &= -\frac{s^{1/2}}{2r^{1/2}}e^{-\sqrt{sr}(\cos(\theta/2)+\mathrm{i}\sin(\theta/2))+\mathrm{i}\theta/2},\\ \frac{\partial u_0(s\mathbf{x})}{\partial\theta} &= -\frac{\mathrm{i}\sqrt{\mathrm{sr}}}{2}e^{-\sqrt{sr}(\cos(\theta/2)+\mathrm{i}\sin(\theta/2))+\mathrm{i}\theta/2},\end{aligned}$$

which can be used to obtain that

$$\begin{aligned}\frac{\partial u_0(s\mathbf{x})}{\partial x_1} &= -\frac{s^{1/2}}{2r^{1/2}}e^{-\sqrt{sr}(\cos(\theta/2)+\mathrm{i}\sin(\theta/2))-\mathrm{i}\theta/2},\\ \frac{\partial u_0(s\mathbf{x})}{\partial x_2} &= -\frac{\mathrm{i}s^{1/2}}{2r^{1/2}}e^{-\sqrt{sr}(\cos(\theta/2)+\mathrm{i}\sin(\theta/2))-\mathrm{i}\theta/2}.\end{aligned}$$

Therefore, it yields that

$$\|\nabla u_0(s\mathbf{x})\|^2_{L^2(S_h)} \leq \frac{(\theta_M - \theta_m)sh}{2} e^{-2\sqrt{s\vartheta}\delta_W},$$

by the integral mean value theorem, where $\vartheta \in [0,h]$ and δ_W is defined in (8.2.5). Hence, we know that $u_0(s\mathbf{x}) \in H^1(S_h)$ and $\Delta u_0(s\mathbf{x}) = 0$ in the weak sense.

Using polar coordinate transformation, we can deduce that

$$\||\mathbf{x}|^\alpha u_0(s\mathbf{x})\|^2_{L^2(S_h)} = \int_0^h r\mathrm{d}r \int_{\theta_m}^{\theta_M} r^{2\alpha} e^{2\sqrt{sr}\cos(\theta/2+\pi)} \mathrm{d}\theta$$

$$\leq \int_0^h r\mathrm{d}r \int_{\theta_m}^{\theta_M} r^{2\alpha} e^{-2\sqrt{sr}\delta_W} \mathrm{d}\theta = (\theta_M - \theta_m) \int_0^h r^{2\alpha+1} e^{-2\delta_W\sqrt{sr}} \mathrm{d}r \quad (t = 2\delta_W\sqrt{sr})$$

$$= s^{-(2\alpha+2)} \frac{2(\theta_M - \theta_m)}{(4\delta_W^2)^{2\alpha+2}} \int_0^{2\delta_W\sqrt{sh}} t^{4\alpha+3} e^{-t} \mathrm{d}r \leq s^{-(2\alpha+2)} \frac{2(\theta_M - \theta_m)}{(4\delta_W^2)^{2\alpha+2}} \Gamma(4\alpha+4),$$

where δ_W is defined in (8.2.5). This completes the proof of the lemma. □

Lemma 8.2.4. *For any $\zeta > 0$, if $\omega(\theta) > 0$, then*

$$\int_0^h r^\zeta e^{-\sqrt{sr}\omega(\theta)} \mathrm{d}r = \mathscr{O}(s^{-\zeta-1}), \tag{8.2.10}$$

as $s \to +\infty$.

Proof. Using variable substitution $t = \sqrt{sr}$, it is easy to derive (8.2.10). □

Next, we recall a special type of Green formula for H^1 functions, which shall be needed in providing a key integral identity for deriving the vanishing property of the conductive transmission eigenfunction.

Lemma 8.2.5. *Let $\Omega \Subset \mathbb{R}^n$ be a bounded Lipschitz domain. For any $f, g \in H^1_\Delta := \{f \in H^1(\Omega) | \Delta f \in L^2(\Omega)\}$, there holds the following second Green identity:*

$$\int_\Omega (g\Delta f - f\Delta g)\, \mathrm{d}\mathbf{x} = \int_{\partial\Omega} (g\partial_\nu f - f\partial_\nu g)\, \mathrm{d}\sigma. \tag{8.2.11}$$

Lemma 8.2.5 is a special case of more general results in [24, Lemma 3.4] and [47, Theorem 4.4]. It is pointed out that in Lemma 8.2.5, one need not to require that H^2 regularity as the usual Green formula. In particular, for the transmission eigenfunctions $(v,w) \in H^1(\Omega) \times H^1(\Omega)$ to (8.1.1), we obviously have $v, w \in H^1_\Delta$, and hence, the Green identity (8.2.11) holds for v, w. This fact shall be frequently used in our subsequent analysis.

We proceed to derive several auxiliary lemmas that shall play a key role in providing our first main result in Theorem 8.2.1 in what follows.

Lemma 8.2.6. *Let S_h and $\Gamma_h^\pm$ be defined in (8.2.2). Suppose that $v \in H^1(S_h)$ and $w \in H^1(S_h)$ satisfy the following PDE system:*

$$\begin{cases} \Delta w + k^2 q w = 0 & \text{in } S_h, \\ \Delta v + k^2 v = 0 & \text{in } S_h \\ w = v,\ \partial_\nu v + \eta v = \partial_\nu w & \text{on } \Gamma_h^\pm, \end{cases} \tag{8.2.12}$$

with $\nu \in \mathbb{S}^1$ signifying the exterior unit normal vector to $\Gamma_h^\pm$, $k \in \mathbb{R}_+$, $q \in L^\infty(S_h)$ and $\eta(\mathbf{x}) \in C^\alpha(\overline{\Gamma_h^\pm})$, where

$$\eta(\mathbf{x}) = \eta(\mathbf{0}) + \delta\eta(\mathbf{x}), \quad |\delta\eta(\mathbf{x})| \leq \|\eta\|_{C^\alpha} |\mathbf{x}|^\alpha. \tag{8.2.13}$$

Recall that the CGO solution $u_0(s\mathbf{x})$ *is defined in* (8.2.4) *with the parameter* $s \in \mathbb{R}_+$. *Then the following integral equality holds:*

$$\int_{S_h} u_0(s\mathbf{x})(f_1 - f_2)\mathrm{d}\mathbf{x} = \int_{\Lambda_h} (u_0(s\mathbf{x})\partial_\nu(v-w) - (v-w)\partial_\nu u_0(s\mathbf{x}))\mathrm{d}\sigma - \int_{\Gamma_h^\pm} \eta u_0(s\mathbf{x})v\mathrm{d}\sigma, \tag{8.2.14}$$

where $f_1 = -k^2 v$ *and* $f_2 = -k^2 qw$.

Denote

$$\widetilde{f}_{1j}(\mathbf{x}) = -k^2 v_j(\mathbf{x}), \tag{8.2.15}$$

where

$$v_j(\mathbf{x}) = \int_{\mathbb{S}^1} e^{ik\xi\cdot\mathbf{x}} g_j(\xi)\mathrm{d}\sigma(\xi), \quad \xi \in \mathbb{S}^1, \quad \mathbf{x} \in \mathbb{R}^2, \quad g_j \in L^2(\mathbb{S}^1) \tag{8.2.16}$$

is the Herglotz wave function. Denote

$$\varphi = \angle(\xi, \mathbf{x}). \tag{8.2.17}$$

Then $\widetilde{f}_{1j}(\mathbf{x}) \in C^\alpha(\overline{S}_h)$, *and it has the expansion*

$$\widetilde{f}_{1j}(\mathbf{x}) = -k^2 v_j(\mathbf{x}) = \widetilde{f}_{1j}(\mathbf{0}) + \delta\widetilde{f}_{1j}(\mathbf{x}), \quad |\delta\widetilde{f}_{1j}(\mathbf{x})| \le \|\widetilde{f}_{1j}\|_{C^\alpha}|\mathbf{x}|^\alpha. \tag{8.2.18}$$

Assuming that $f_2 = -k^2 qw \in C^\alpha(\overline{S}_h)$ $(0 < \alpha < 1)$ *satisfying*

$$f_2(\mathbf{x}) = f_2(\mathbf{0}) + \delta f_2(\mathbf{x}), \quad |\delta f_2(\mathbf{x})| \le \|f_2\|_{C^\alpha}|\mathbf{x}|^\alpha, \tag{8.2.19}$$

it holds that

$$(\widetilde{f}_{1j}(\mathbf{0}) - f_2(\mathbf{0}))\int_{S_h} u_0(s\mathbf{x})\mathrm{d}\mathbf{x} + \delta_j(s) = I_3 - I_2^\pm - \int_{S_h} \delta\widetilde{f}_{1j}(\mathbf{x})u_0(s\mathbf{x})\mathrm{d}\mathbf{x} + \int_{S_h} \delta f_2(\mathbf{x})u_0(s\mathbf{x})\mathrm{d}\mathbf{x} - \xi_j^\pm(s), \tag{8.2.20}$$

where

$$\begin{aligned} I_2^\pm &= \int_{\Gamma_h^\pm} \eta(\mathbf{x})u_0(s\mathbf{x})v_j(\mathbf{x})\mathrm{d}\sigma, \quad I_3 = \int_{\Lambda_h} (u_0(s\mathbf{x})\partial_\nu(v-w) - (v-w)\partial_\nu u_0(s\mathbf{x}))\mathrm{d}\sigma, \\ \delta_j(s) &= -k^2\int_{S_h} (v(\mathbf{x}) - v_j(\mathbf{x}))u_0(s\mathbf{x})\mathrm{d}x, \quad \xi_j^\pm(s) = \int_{\Gamma_h^\pm} \eta(\mathbf{x})u_0(s\mathbf{x})(v(\mathbf{x}) - v_j(\mathbf{x}))\mathrm{d}\sigma. \end{aligned} \tag{8.2.21}$$

Furthermore, it yields that

$$\left|\int_{S_h} \delta\widetilde{f}_{1j}(\mathbf{x})u_0(s\mathbf{x})\mathrm{d}x\right| \le \frac{2\sqrt{2\pi}(\theta_M - \theta_m)\Gamma(2\alpha+4)}{\delta_W^{2\alpha+4}} k^2 \mathrm{diam}(S_h)^{1-\alpha} \times (1+k)\|g_j\|_{L^2(\mathbb{S}^1)} s^{-\alpha-2} \tag{8.2.22}$$

and

$$\left|\int_{S_h} \delta f_2(\mathbf{x})u_0(s\mathbf{x})\mathrm{d}\mathbf{x}\right| \le \frac{2(\theta_M - \theta_m)\Gamma(2\alpha+4)}{\delta_W^{2\alpha+4}} \|f_2\|_{C^\alpha} s^{-\alpha-2}, \tag{8.2.23}$$

as $s \to +\infty$. *If* $v - w \in H^2(\Sigma_{\Lambda_h})$, *then one has as* $s \to +\infty$ *that*

$$|I_3| \le Ce^{-c'\sqrt{s}}, \tag{8.2.24}$$

where C and c' are two positive constants.

Proof. From (8.2.12), we have

$$\Delta v = -k^2 v := f_1, \quad \Delta w = -k^2 q w := f_2. \tag{8.2.25}$$

Subtracting the two equations of (8.2.25) together with the use of the boundary conditions of (8.2.12), we deduce that

$$\Delta(v-w) = f_1 - f_2 \text{ in } S_h, \quad v-w=0, \partial_\nu(v-w) = -\eta v \text{ on } \Gamma_h^{\pm}. \tag{8.2.26}$$

Recall that $u_0(s\mathbf{x}) \in H^1(S_h)$ from Lemma 8.2.3. Since $v, w \in H^1(S_h)$ and $q \in L^\infty(S_h)$, it yields that $f_1, f_2 \in L^2(S_h)$. Since S_h is obviously a bounded Lipschitz domain, by virtue of Lemma 8.2.5, we have the following integral identity:

$$\int_{S_h} u_0(s\mathbf{x})\Delta(v-w)\mathrm{d}\mathbf{x} = \int_{\partial S_h} u_0(s\mathbf{x})\partial_\nu(v-w) - (v-w)\partial_\nu u_0(s\mathbf{x})\mathrm{d}\sigma, \tag{8.2.27}$$

by using the fact that $\Delta u_0(s\mathbf{x}) = 0$ in S_h. Substituting (8.2.26) into (8.2.27), it yields (8.2.14).

Recall that v can be approximated by the Herglotz wave function v_j given by (8.2.16) in the topology induced by the H^1 norm. It is clear that

$$\int_{S_h} f_1(\mathbf{x})u_0(s\mathbf{x})\mathrm{d}x = \int_{S_h} \widetilde{f}_{1j}(\mathbf{x})u_0(s\mathbf{x})\mathrm{d}x + \delta_j(s), \tag{8.2.28}$$

where $\widetilde{f}_{1j}(\mathbf{x})$ and $\delta_j(s)$ are defined in (8.2.15) and (8.2.21), respectively. Furthermore, it can be derived that

$$\int_{\Gamma_h^{\pm}} \eta(\mathbf{x})u_0(s\mathbf{x})v(\mathbf{x})\mathrm{d}\sigma = \int_{\Gamma_h^{\pm}} \eta(\mathbf{x})u_0(s\mathbf{x})v_j(\mathbf{x})\mathrm{d}\sigma + \xi_j^{\pm}(s), \tag{8.2.29}$$

where $\xi_j^{\pm}(s)$ is defined in (8.2.21).

Combining (8.2.14) and (8.2.28) with (8.2.29), we have the following integral identity:

$$I_1 + \delta_j(s) = I_3 - I_2^{\pm} - \xi_j^{\pm}(s), \tag{8.2.30}$$

where

$$I_1 = \int_{S_h} u_0(s\mathbf{x})(\widetilde{f}_{1j}(\mathbf{x}) - f_2(\mathbf{x}))\mathrm{d}\mathbf{x}, \tag{8.2.31}$$

$I_2^{\pm}$, I_3, $\delta_j(s)$ and $\xi_j^{\pm}(s)$ are defined in (8.2.21).

It is easy to verify that $v_j \in C^\alpha(\overline{S}_h)$. Therefore, $\widetilde{f}_{1j} \in C^\alpha(\overline{S}_h)$, and for $\mathbf{x} \in S_h$, we have the splitting (8.2.18). Since $f_2 \in C^\alpha(\overline{S}_h)$, substituting (8.2.18) and (8.2.19) into I_1 defined in (8.2.31), we have

$$I_1 = (\widetilde{f}_{1j}(\mathbf{0}) - f_2(\mathbf{0}))\int_{S_h} u_0(s\mathbf{x})\mathrm{d}\mathbf{x} + \int_{S_h} \delta\widetilde{f}_{1j}(\mathbf{x})u_0(s\mathbf{x})\mathrm{d}\mathbf{x} - \int_{S_h} \delta f_2(\mathbf{x})u_0(s\mathbf{x})\mathrm{d}\mathbf{x}. \tag{8.2.32}$$

Substituting (8.2.32) into (8.2.30), we can further deduce the integral equality (8.2.20).

In the following, we shall prove (8.2.22), (8.2.23), and (8.2.24), separately. From (8.2.5) and (8.2.18), it can be derived that

$$\begin{aligned}\left|\int_{S_h} \delta\widetilde{f}_{1j}(\mathbf{x})u_0(s\mathbf{x})\mathrm{d}\mathbf{x}\right| &\leq \int_{S_h} \left|\delta\widetilde{f}_{1j}(\mathbf{x})\right| |u_0(s\mathbf{x})|\mathrm{d}\mathbf{x} \leq \|\widetilde{f}_{1j}\|_{C^\alpha} \int_W |u_0(s\mathbf{x})||\mathbf{x}|^\alpha \mathrm{d}\mathbf{x} \\ &\leq \frac{2(\theta_M - \theta_m)\Gamma(2\alpha+4)}{\delta_W^{2\alpha+4}} \|\widetilde{f}_{1j}\|_{C^\alpha} s^{-\alpha-2}. \end{aligned} \tag{8.2.33}$$

Recall that $\widetilde{f}_{1j} = -k^2 v_j(\mathbf{x})$ and v_j is the Herglotz wave function given by (8.2.16). Using the property of compact embedding of Hölder spaces, we can derive that

$$\|\widetilde{f}_{1j}\|_{C^\alpha} \leq k^2 \mathrm{diam}(S_h)^{1-\alpha} \|v_j\|_{C^1},$$

where $\mathrm{diam}(S_h)$ is the diameter of S_h. By direct computation, we have

$$\|v_j\|_{C^1} \leq \sqrt{2\pi}(1+k)\|g_j\|_{L^2(\mathbb{S}^1)},$$

and therefore, we can deduce (8.2.22). Similarly, using (8.2.5) and (8.2.19), we can derive (8.2.23).

It is easy to see that on Λ_h, one has

$$|u_0(s\mathbf{x})| = e^{\sqrt{sr}\cos(\theta/2+\pi)} \leq e^{-\delta_W\sqrt{sh}},$$
$$|\partial_\nu u_0(s\mathbf{x})| = \left|\frac{\sqrt{s}e^{\mathrm{i}\cos(\theta/2+\pi)}}{2\sqrt{h}} e^{\sqrt{sh}\exp(\mathrm{i}(\theta/2+\pi))}\right| \leq \frac{1}{2}\sqrt{\frac{s}{h}} e^{-\delta_W\sqrt{sh}},$$

both of which decay exponentially as $s \to \infty$. Hence, we know that

$$\|u_0(s\mathbf{x})\|_{L^2(\Lambda_h)} \leq e^{-\delta_W\sqrt{sh}}\sqrt{(\theta_M-\theta_m)h}, \quad \|\partial_\nu u_0(s\mathbf{x})\|_{L^2(\Lambda_h)} \leq \frac{1}{2}e^{-\delta_W\sqrt{sh}}\sqrt{s(\theta_M-\theta_m)}.$$

Under the assumption that $v - w \in H^2(\Sigma_{\Lambda_h})$, using the Cauchy-Schwarz inequality and the trace theorem, we can prove as $s \to +\infty$ that

$$\begin{aligned}|I_3| &\leq \|u_0(s\mathbf{x})\|_{L^2(\Lambda_h)}\|\partial_\nu(v-w)\|_{L^2(\Lambda_h)} + \|\partial_\nu u_0(s\mathbf{x})\|_{L^2(\Lambda_h)}\|v-w\|_{L^2(\Lambda_h)} \\ &\leq \left(\|u_0(s\mathbf{x})\|_{L^2(\Lambda_h)} + \|\partial_\nu u_0(s\mathbf{x})\|_{L^2(\Lambda_h)}\right)\|v-w\|_{H^2(\Sigma_{\Lambda_h})} \leq Ce^{-c'\sqrt{s}},\end{aligned}$$

where C, c' are positive constants.

The proof is complete. □

Lemma 8.2.7. *Under the same setup in Lemma 8.2.6, we assume that a sequence of Herglotz wave functions* $\{v_j\}_{j=1}^{+\infty}$ *possesses the form* (8.2.16)*, which can approximate* v *in* $H^1(S_h)$ *satisfying*

$$\|v - v_j\|_{H^1(S_h)} \leq j^{-1-\Upsilon}, \quad \|g_j\|_{L^2(\mathbb{S}^1)} \leq Cj^\rho, \tag{8.2.34}$$

for some constants $C > 0$, $\Upsilon > 0$, *and* $0 < \rho < 1$. *Then we have the following estimates:*

$$|\delta_j(s)| \leq \frac{\sqrt{\theta_M-\theta_m}k^2 e^{-\sqrt{s\Theta}\delta_W}h}{\sqrt{2}} j^{-1-\Upsilon}, \tag{8.2.35}$$

and

$$|\xi_j^\pm(s)| \leq C\left(|\eta(\mathbf{0})|\frac{\sqrt{\theta_M-\theta_m}e^{-\sqrt{s\Theta}\delta_W}h}{\sqrt{2}} + \|\eta\|_{C^\alpha}s^{-(\alpha+1)}\frac{\sqrt{2(\theta_M-\theta_m)\Gamma(4\alpha+4)}}{(2\delta_W)^{2\alpha+2}}\right) j^{-1-\Upsilon}, \tag{8.2.36}$$

where $\Theta \in [0, h]$ *and* δ_W *is defined in* (8.2.5).

Proof. In the following, we shall prove (8.2.35) and (8.2.36), separately. Recall that $u_0 \in H^1(S_h)$ from Lemma 8.2.3. Clearly, $\widetilde{f}_{1j}(x) \in H^2(S_h)$, which can be embedded into $C^\alpha(\overline{S}_h)$ for $\alpha \in (0,1)$. Moreover, by using the Cauchy-Schwarz inequality, we know that

$$|\delta_j(s)| \leq k^2\|v - v_j\|_{L^2(S_h)}\|u_0(s\mathbf{x})\|_{L^2(S_h)}. \tag{8.2.37}$$

Substituting (8.2.8) into (8.2.37) and using (8.2.34), one readily has (8.2.35).

Since $\eta \in C^\alpha\left(\overline{\Gamma_h^\pm}\right)$, we have the expansion of $\eta(\mathbf{x})$ at the origin as (8.2.13). Therefore, using the Cauchy-Schwarz inequality and the trace theorem, we have

$$
\begin{aligned}
|\xi_j^\pm(s)| &\le |\eta(\mathbf{0})| \int_{\Gamma_h^\pm} |u_0(s\mathbf{x})||v(\mathbf{x})-v_j(\mathbf{x})|\mathrm{d}\sigma + \|\eta\|_{C^\alpha} \int_{\Gamma_h^\pm} |\mathbf{x}|^\alpha |u_0(s\mathbf{x})||v(\mathbf{x})-v_j(\mathbf{x})|\mathrm{d}\sigma \\
&\le |\eta(\mathbf{0})| \|v-v_j\|_{H^{1/2}(\Gamma_h^\pm)} \|u_0(s\mathbf{x})\|_{H^{-1/2}(\Gamma_h^\pm)} \\
&\quad + \|\eta\|_{C^\alpha} \|v-v_j\|_{H^{1/2}(\Gamma_h^\pm)} \||\mathbf{x}|^\alpha u_0(s\mathbf{x})\|_{H^{-1/2}(\Gamma_h^\pm)} \\
&\le |\eta(\mathbf{0})| \|v-v_j\|_{H^1(S_h)} \|u_0(s\mathbf{x})\|_{L^2(S_h)} + \|\eta\|_{C^\alpha} \|v-v_j\|_{H^1(S_h)} \||\mathbf{x}|^\alpha u_0(s\mathbf{x})\|_{L^2(S_h)}.
\end{aligned}
$$

Using (8.2.34), (8.2.8), and (8.2.9), we readily derive (8.2.36).

The proof is complete. □

Lemma 8.2.8. *Recall that $\Gamma_h^\pm$ and $u_0(s\mathbf{x})$ are defined in* (8.2.2) *and* (8.2.4)*, respectively. We have*

$$
\begin{aligned}
\int_{\Gamma_h^+} u_0(s\mathbf{x})\mathrm{d}\sigma &= 2s^{-1}\Big(\mu(\theta_M)^{-2} - \mu(\theta_M)^{-2} e^{-\sqrt{sh}\mu(\theta_M)} \\
&\qquad - \mu(\theta_M)^{-1}\sqrt{sh} e^{-\sqrt{sh}\mu(\theta_M)}\Big), \\
\int_{\Gamma_h^-} u_0(s\mathbf{x})\mathrm{d}\sigma &= 2s^{-1}\Big(\mu(\theta_m)^{-2} - \mu(\theta_m)^{-2} e^{-\sqrt{sh}\mu(\theta_m)} \\
&\qquad - \mu(\theta_m)^{-1}\sqrt{sh} e^{-\sqrt{sh}\mu(\theta_m)}\Big),
\end{aligned}
\tag{8.2.38}
$$

where $\mu(\theta) = -\cos(\theta/2+\pi) - \mathrm{i}\sin(\theta/2+\pi)$.

Proof. Using variable substitution $t = \sqrt{sr}$ and by direct calculations, we can derive (8.2.38). □

Using the Jacobi-Anger expansion (cf. [22, page 75]), for the Herglotz wave function v_j given in (8.2.16), we have

$$
v_j(\mathbf{x}) = v_j(\mathbf{0}) J_0(k|\mathbf{x}|) + 2\sum_{p=1}^\infty \gamma_{pj} i^p J_p(k|\mathbf{x}|), \quad \mathbf{x} \in \mathbb{R}^2, \tag{8.2.39}
$$

where

$$
v_j(\mathbf{0}) = \int_{\mathbb{S}^1} g_j(\theta)\mathrm{d}\sigma(\theta), \quad \gamma_{pj} = \int_{\mathbb{S}^1} g_j(\xi)\cos(p\varphi)\mathrm{d}\sigma(\xi), \quad p, j \in \mathbb{N}, \tag{8.2.40}
$$

$J_p(t)$ is the p-th Bessel function of the first kind [1], g_j is the kernel of v_j defined in (8.2.16), and φ is given by (8.2.17).

Lemma 8.2.9. *Let $\Gamma_h^\pm$ be defined in* (8.2.2)*. Recall that the CGO solution $u_0(s\mathbf{x})$ is defined in* (8.2.4) *with the parameter $s \in \mathbb{R}_+$ and $I_2^\pm$ is defined by* (8.2.21)*. Denote*

$$
\omega(\theta) = -\cos(\theta/2+\pi), \quad \mu(\theta) = -\cos(\theta/2+\pi) - \mathrm{i}\sin(\theta/2+\pi). \tag{8.2.41}
$$

Recall that the Herglotz wave function v_j is given in the form (8.2.16)*. Suppose that $\eta(\mathbf{x}) \in C^\alpha(\overline{\Gamma_h^\pm})$ $(0 < \alpha < 1)$ satisfying* (8.2.13)*, and let*

$$
\begin{aligned}
\mathscr{I}_1^- &= \int_0^h \delta\eta\; J_0(kr)e^{-\sqrt{sr}\mu(\theta_m)}\mathrm{d}r, \quad \mathscr{I}_2^- = 2\sum_{p=1}^{\infty} \mathrm{i}^p \int_0^h \delta\eta\; \gamma_{pj}J_p(kr)e^{-\sqrt{sr}\mu(\theta_m)}\mathrm{d}r,\\
\mathscr{I}_1^+ &= \int_0^h \delta\eta\; J_0(kr)e^{-\sqrt{sr}\mu(\theta_M)}\mathrm{d}r, \quad \mathscr{I}_2^+ = 2\sum_{p=1}^{\infty} \mathrm{i}^p \int_0^h \delta\eta\; \gamma_{pj}J_p(kr)e^{-\sqrt{sr}\mu(\theta_M)}\mathrm{d}r, \qquad (8.2.42)\\
I_\eta^- &= v_j(\mathbf{0})\mathscr{I}_1^- + \mathscr{I}_2^-, \quad I_\eta^+ = v_j(\mathbf{0})\mathscr{I}_1^+ + \mathscr{I}_2^+.
\end{aligned}
$$

Assume that for a fixed $k \in \mathbb{R}_+$, $kh < 1$, where h is the length of $\Gamma_h^{\pm}$, and $-\pi < \theta_m < \theta_M < \pi$, where θ_m and θ_M are defined in (8.2.1). *Then*

$$
\begin{aligned}
I_2^- = & 2\eta(\mathbf{0})v_j(\mathbf{0})s^{-1}\left(\mu(\theta_m)^{-2} - \mu(\theta_m)^{-2}e^{-\sqrt{sh}\mu(\theta_m)} - \mu(\theta_m)^{-1}\sqrt{sh}e^{-\sqrt{sh}\mu(\theta_m)}\right)\\
& + v_j(\mathbf{0})\eta(\mathbf{0})\mathscr{I}_{312}^- + \eta(\mathbf{0})\mathscr{I}_{32}^- + I_\eta^-, \qquad (8.2.43)
\end{aligned}
$$

where

$$
\begin{aligned}
&\mathscr{I}_{312}^- = \sum_{p=1}^{\infty} \frac{(-1)^p k^{2p}}{4^p(p!)^2}\int_0^h r^{2p}e^{-\sqrt{sr}\mu(\theta_m)}\mathrm{d}r, \quad \mathscr{I}_{32}^- = 2\sum_{p=1}^{\infty}\int_0^h \gamma_{pj}\mathrm{i}^p J_p(kr)e^{-\sqrt{sr}\mu(\theta_m)}\mathrm{d}r,\\
&|\mathscr{I}_{312}^-| \le \mathscr{O}(s^{-3}), \quad |\mathscr{I}_{32}^-| \le \mathscr{O}(\|g_j\|_{L^2(\mathbb{S}^1)}s^{-2}),\\
&|I_\eta^-| \le |v_j(0)||\mathscr{I}_1^-| + |\mathscr{I}_2^-|, \quad |\mathscr{I}_1^-| \le \mathscr{O}(\|\eta\|_{C^\alpha}s^{-1-\alpha}), \quad |\mathscr{I}_2^-| \le \mathscr{O}(\|\eta\|_{C^\alpha}\|g_j\|_{L^2(\mathbb{S}^1)}s^{-2-\alpha}),
\end{aligned}
$$

as $s \to +\infty$. Similarly, we have

$$
\begin{aligned}
I_2^+ = & 2\eta(\mathbf{0})v_j(\mathbf{0})s^{-1}\left(\mu(\theta_M)^{-2} - \mu(\theta_M)^{-2}e^{-\sqrt{sh}\mu(\theta_M)} - \mu(\theta_M)^{-1}\sqrt{sh}e^{-\sqrt{sh}\mu(\theta_M)}\right)\\
& + v_j(\mathbf{0})\eta(\mathbf{0})\mathscr{I}_{312}^+ + \eta(\mathbf{0})\mathscr{I}_{32}^+ + I_\eta^+, \qquad (8.2.44)
\end{aligned}
$$

where

$$
\begin{aligned}
&\mathscr{I}_{312}^+ = \sum_{p=1}^{\infty} \frac{(-1)^p k^{2p}}{4^p(p!)^2}\int_0^h r^{2p}e^{-\sqrt{sr}\mu(\theta_M)}\mathrm{d}r, \quad \mathscr{I}_{32}^+ = 2\sum_{p=1}^{\infty}\int_0^h \gamma_{pj}\mathrm{i}^p J_p(kr)e^{-\sqrt{sr}\mu(\theta_M)}\mathrm{d}r,\\
&|\mathscr{I}_{312}^+| \le \mathscr{O}(s^{-3}), \quad |\mathscr{I}_{32}^+| \le \mathscr{O}(\|g_j\|_{L^2(\mathbb{S}^1)}s^{-2}),\\
&|I_\eta^+| \le |v_j(\mathbf{0})||\mathscr{I}_1^+| + |\mathscr{I}_2^+|, \ |\mathscr{I}_1^+| \le \mathscr{O}(\|\eta\|_{C^\alpha}s^{-1-\alpha}), \ |\mathscr{I}_2^+| \le \mathscr{O}(\|\eta\|_{C^\alpha}\|g_j\|_{L^2(\mathbb{S}^1)}s^{-2-\alpha}),
\end{aligned}
$$

as $s \to +\infty$.

Proof. We first investigate the boundary integral I_2^- defined in (8.2.21). Recalling that $\Gamma_h^{\pm}$ is defined in (8.2.2), for $\mathbf{x} \in \Gamma_h^{\pm}$, the polar coordinates $\mathbf{x} = (r\cos\theta, r\sin\theta)$ satisfy $r \in (0,h)$ and $\theta = \theta_m$ or $\theta = \theta_M$ when $\mathbf{x} \in \Gamma_h^-$ or $\mathbf{x} \in \Gamma_h^+$, respectively. Since $\eta \in C^\alpha\left(\overline{\Gamma_h^{\pm}}\right)$, we know that η has the expansion (8.2.13). Denote

$$
I_{21}^- = \int_{\Gamma_h^-} u_0(s\mathbf{x})v_j(\mathbf{x})\mathrm{d}\sigma, \quad I_\eta^- = \int_{\Gamma_h^-} \delta\eta(\mathbf{x})u_0(s\mathbf{x})v_j(\mathbf{x})\mathrm{d}\sigma. \qquad (8.2.45)
$$

Substituting (8.2.13) into the expression of I_2^-, we have

$$
I_2^- = \eta(\mathbf{0})I_{21}^- + I_\eta^-. \qquad (8.2.46)
$$

Recall that γ_{pj} is defined by (8.2.40). Using the Cauchy-Schwarz inequality, it is clear that

$$
|\gamma_{pj}| \le \sqrt{2\pi}\|g\|_{L^2(\mathbb{S}^1)}. \qquad (8.2.47)
$$

For the Bessel function $J_p(t)$, we have from [1] the following series expression:

$$J_p(t) = \frac{t^p}{2^p p!} + \frac{t^p}{2^p} \sum_{\ell=1}^{\infty} \frac{(-1)^\ell t^{2\ell}}{4^\ell (\ell!)^2}, \quad \text{for } p = 1, 2, \ldots, \tag{8.2.48}$$

which is uniformly and absolutely convergent with respect to $t \in [0, +\infty)$.

For $\mathscr{I}_1^-$ and $\mathscr{I}_2^-$ defined in (8.2.42), by substituting (8.2.39) into I_η^- given by (8.2.45), it is directly verified that $I_\eta^- = v_j(0)\mathscr{I}_1^- + \mathscr{I}_2^-$.

For $\omega(\theta)$ defined in (8.2.41), it is easy to see that $\omega(\theta) > 0$ for $-\pi < \theta_m \leq \theta \leq \theta_M < \pi$. By virtue of (8.2.13), we have

$$|\mathscr{I}_1^-| \leq \|\eta\|_{C^\alpha} \int_0^h r^\alpha |J_0(kr)| e^{-\sqrt{sr}\omega(\theta_m)} \mathrm{d}r := \|\eta\|_{C^\alpha} \left(\mathscr{I}_{11}^- + \mathscr{I}_{12}^-\right),$$

where

$$\mathscr{I}_{11}^- = \int_0^h r^\alpha e^{-\sqrt{sr}\omega(\theta_m)} \mathrm{d}r, \quad \mathscr{I}_{12}^- = \sum_{p=1}^{\infty} \frac{k^{2p}}{4^p (p!)^2} \int_0^h r^{\alpha+2p} e^{-\sqrt{sr}\omega(\theta_m)} \mathrm{d}r.$$

From (8.2.10) in Lemma 8.2.4 and noting that $\omega(\theta_m) > 0$, we have

$$\mathscr{I}_{11}^- = \mathscr{O}(s^{-1-\alpha}),$$

as $s \to +\infty$. For $\mathscr{I}_{12}^-$, we have the estimate

$$|\mathscr{I}_{12}^-| \leq \sum_{p=1}^{\infty} \frac{h^{2p-2} k^{2p}}{4^p (p!)^2} \int_0^h r^{\alpha+2} e^{-\sqrt{sr}\omega(\theta_m)} \mathrm{d}r = \mathscr{O}(s^{-3-\alpha}),$$

as $s \to +\infty$, where we suppose that $kh < 1$ for sufficiently small h. Therefore, we conclude that

$$|\mathscr{I}_1^-| \leq \mathscr{O}(\|\eta\|_{C^\alpha} s^{-1-\alpha}) \text{ as } s \to +\infty. \tag{8.2.49}$$

For sufficiently small $h > 0$ fulfilling that $kh < 1$, using (8.2.13), (8.2.47), and (8.2.48), we have

$$\begin{aligned}
|\mathscr{I}_2^-| &\leq 2\sqrt{2\pi} \|\eta\|_{C^\alpha} \|g_j\|_{L^2(\mathbb{S}^1)} \sum_{p=1}^{\infty} \left[\frac{k^p}{2^p p!} \int_0^h r^{p+\alpha} e^{-\sqrt{sr}\omega(\theta_m)} \mathrm{d}r \right. \\
&\quad \left. + \frac{k^p}{2^p} \sum_{\ell=1}^{\infty} \frac{k^{2\ell} h^{2(\ell-1)}}{4^\ell (\ell!)^2} \left(\int_0^h r^{p+\alpha+2} e^{-\sqrt{sr}\omega(\theta_m)} \mathrm{d}r \right) \right] \\
&\leq 2\sqrt{2\pi} \|\eta\|_{C^\alpha} \|g_j\|_{L^2(\mathbb{S}^1)} \sum_{p=1}^{\infty} \left[\frac{k^p}{2^p p!} \int_0^h r^{p+\alpha} e^{-\sqrt{sr}\omega(\theta_m)} \mathrm{d}r \right. \\
&\quad \left. + \frac{(kh)^p}{2^p} \sum_{\ell=1}^{\infty} \frac{k^{2\ell} h^{2(\ell-1)}}{4^\ell (\ell!)^2} \left(\int_0^h r^{\alpha+2} e^{-\sqrt{sr}\omega(\theta_m)} \mathrm{d}r \right) \right] \\
&\leq 2\sqrt{2\pi} \|\eta\|_{C^\alpha} \|g_j\|_{L^2(\mathbb{S}^1)} \sum_{p=1}^{\infty} \left[\frac{k^p h^{p-1}}{2^p p!} \int_0^h r^{\alpha+1} e^{-\sqrt{sr}\omega(\theta_m)} \mathrm{d}r + \mathscr{O}\left(s^{-\alpha-3}\right) \right].
\end{aligned}$$

Using Lemma 8.2.4, we know that

$$\int_0^h r^{\alpha+1} e^{-\sqrt{sr}\omega(\theta_m)} \mathrm{d}r = \mathscr{O}(s^{-\alpha-2}) \text{ as } s \to +\infty.$$

Therefore, we can derive that

$$|\mathscr{I}_2^-| \le \mathscr{O}(\|\eta\|_{C^\alpha}\|g_j\|_{L^2(\mathbb{S}^1)}s^{-\alpha-2}) \text{ as } s \to +\infty. \tag{8.2.50}$$

We proceed to investigate I_{21}^- given by (8.2.46). Denote

$$\mathscr{I}_{31}^- = \int_0^h J_0(kr)e^{-\sqrt{sr}\mu(\theta_m)}\mathrm{d}r, \quad \mathscr{I}_{32}^- = 2\sum_{p=1}^{\infty}\int_0^h \gamma_{pj}\mathrm{i}^p J_p(kr)e^{-\sqrt{sr}\mu(\theta_m)}\mathrm{d}r. \tag{8.2.51}$$

Substituting the expansion (8.2.39) into I_{21}^-, it is easy to see that

$$I_{21}^- = v_j(\mathbf{0})\mathscr{I}_{31}^- + \mathscr{I}_{32}^-.$$

From (8.2.48), we know that

$$J_0(t) = \sum_{p=0}^{\infty}(-1)^p\frac{t^{2p}}{4^p(p!)^2}. \tag{8.2.52}$$

Let

$$\mathscr{I}_{311}^- = \int_0^h e^{-\sqrt{sr}\mu(\theta_m)}\mathrm{d}r, \quad \mathscr{I}_{312}^- = \sum_{p=1}^{\infty}\frac{(-1)^p k^{2p}}{4^p(p!)^2}\int_0^h r^{2p}e^{-\sqrt{sr}\mu(\theta_m)}\mathrm{d}r.$$

Substituting the expansion (8.2.52) of J_0 into $\mathscr{I}_{31}^-$, we have

$$\mathscr{I}_{31}^- = \mathscr{I}_{311}^- + \mathscr{I}_{312}^-,$$

where $\mathscr{I}_{311}^-$ can be derived directly from Lemma 8.2.8.

For $\mathscr{I}_{312}^-$, we have

$$|\mathscr{I}_{312}^-| \le \sum_{p=1}^{\infty}\frac{k^{2p}h^{2p-2}}{4^p(p!)^2}\int_0^h r^2 e^{-\sqrt{sr}\omega(\theta_m)}\mathrm{d}r = \mathscr{O}(s^{-3}) \text{ as } s \to +\infty. \tag{8.2.53}$$

Substituting the expansion (8.2.48) of J_p into $\mathscr{I}_{32}^-$, using (8.2.47), we can deduce that

$$\begin{aligned}
|\mathscr{I}_{32}^-| &\le 2\sqrt{2\pi}\|g_j\|_{L^2(\mathbb{S}^1)}\sum_{p=1}^{\infty}\left[\frac{k^p}{2^p p!}\int_0^h r^p e^{-\sqrt{sr}\omega(\theta_m)}\mathrm{d}r\right.\\
&\qquad\left.+\frac{(kh)^p}{2^p}\sum_{\ell=1}^{\infty}\frac{k^{2\ell}h^{2(\ell-1)}}{4^\ell(\ell!)^2}\int_0^h r^2 e^{-\sqrt{sr}\omega(\theta_m)}\mathrm{d}r\right]\\
&\le 2\sqrt{2\pi}\|g_j\|_{L^2(\mathbb{S}^1)}\sum_{p=1}^{\infty}\left[\frac{k^p h^{p-1}}{2^p p!}\int_0^h re^{-\sqrt{sr}\omega(\theta_m)}\mathrm{d}r + \mathscr{O}\left(s^{-3}\right)\right]\\
&= \mathscr{O}(\|g_j\|_{L^2(\mathbb{S}^1)}s^{-2}),
\end{aligned} \tag{8.2.54}$$

where we suppose that $kh < 1$ for sufficiently small h.

Finally, substituting (8.2.49), (8.2.50), (8.2.38), (8.2.53), and (8.2.54) into (8.2.46), we can obtain the integral equality (8.2.43).

Following a completely similar argument in deriving the integral equality (8.2.43) of I_2^-, we can derive the integral equality (8.2.44) for I_2^+ as well.

The proof is complete. □

Lemma 8.2.10. *Let $\mu(\theta)$ be defined in (8.2.41). Assuming that*

$$-\pi < \theta_m < \theta_M < \pi, \tag{8.2.55}$$

then

$$\mu(\theta_m)^{-2}+\mu(\theta_M)^{-2}\neq 0. \tag{8.2.56}$$

Proof. It can be calculated that

$$\mu(\theta_m)^{-2}+\mu(\theta_M)^{-2}=\frac{(\cos\theta_m+\cos\theta_M)+\mathrm{i}(\sin\theta_m+\sin\theta_M)}{(\cos\theta_m+\mathrm{i}\sin\theta_m)(\cos\theta_M+\mathrm{i}\sin\theta_M)}.$$

Then under the assumption (8.2.55), it is straightforward to verify that

$$\cos\theta_m+\cos\theta_M \text{ and } \sin\theta_m+\sin\theta_M$$

cannot be zero simultaneously, which immediately implies (8.2.56). □

We are in a position to present one of the main theorems in this section.

Theorem 8.2.1. *Let $v\in H^1(\Omega)$ and $w\in H^1(\Omega)$ be a pair of eigenfunctions to* (8.1.1) *associated with $k\in\mathbb{R}_+$. Assume that the Lipschitz domain $\Omega\subset\mathbb{R}^2$ contains a corner $\Omega\cap B_h=\Omega\cap W$ with the vertex being $0\in\partial\Omega$, where W is the sector defined in* (8.2.1) *and $h\in\mathbb{R}_+$. Moreover, there exists a sufficiently small neighborhood $S_h=\Omega\cap B_h=\Omega\cap W$ of 0, such that $qw\in C^\alpha(\overline{S_h})$ with $q:=1+V$ and $\eta\in C^\alpha\left(\overline{\Gamma_h^\pm}\right)$ for $0<\alpha<1$ and $v-w\in H^2(\Sigma_{\Lambda_h})$, where S_h, $\Gamma_h^\pm$, and Σ_{Λ_h} are defined in* (8.2.2). *If the following conditions are fulfilled:*

(a) the transmission eigenfunction v can be approximated in $H^1(S_h)$ by the Herglotz functions v_j, $j=1,2,\ldots$, with kernels g_j satisfying the approximation property (8.2.34)*;*
(b) the function $\eta(\mathbf{x})$ does not vanish at the vertex $\mathbf{0}$, where $\mathbf{0}$ is the vertex of S_h, i.e.,

$$\eta(\mathbf{0})\neq 0; \tag{8.2.57}$$

(c) the open angle of S_h satisfies

$$-\pi<\theta_m<\theta_M<\pi \text{ and } \theta_M-\theta_m\neq\pi; \tag{8.2.58}$$

then one has

$$\lim_{\rho\to+0}\frac{1}{m(B(\mathbf{0},\rho)\cap\Omega)}\int_{B(\mathbf{0},\rho)\cap\Omega}|v(\mathbf{x})|\mathrm{d}\mathbf{x}=0, \tag{8.2.59}$$

where $m(B(\mathbf{0},\rho)\cap\Omega)$ is the area of $B(\mathbf{0},\rho)\cap\Omega$.

Remark 8.2.2. In Theorem 8.2.1, we consider the case that v,w are a pair of conductive transmission eigenfunctions to (8.1.1) and show the vanishing property near a corner. We would like to emphasize that the result can be localized in the sense that as long as v,w satisfy all the conditions stated in Theorem 8.2.1 in $\Omega\cap S_h$, then one has the vanishing property (8.2.59) near the corner. That is, it is not necessary to require that v, w are conductive transmission eigenfunctions, and it is sufficient to require that v, w satisfy the equation in (8.1.1) in $S_h\cap\Omega$ and the conductive transmission conditions on $\overline{S_h}\cap\partial\Omega$, then one has the same vanishing property as stated in Theorem 8.2.1. Indeed, the subsequent proof of Theorem 8.2.1 is for the aforementioned localized problem.

Remark 8.2.3. The condition (8.2.34) signifies a certain regularity condition of the transmission eigenfunction $v\in H^1(\Omega)$. In [6], the following regularity condition was introduced:

$$\|v-v_j\|_{L^2(\Omega)}\leq e^{-j},\quad \|g_j\|_{L^2(\mathbb{S}^{n-1})}\leq C(\ln j)^\beta, \tag{8.2.60}$$

where the constants $C>0$ and $0<\beta<1/(2n+8),(n=2,3)$. Here, we allow the polynomial growth of the kernel functions. Moreover, we would like to remark that $qw\in C^\alpha(\overline{S}_h)$ is technically required in our mathematical argument of proving Theorem 8.2.1. This technical condition can be fulfilled in our study for the unique recovery result of the inverse scattering problem. Indeed, when

q is a constant, it is shown in Lemma 8.4.1 that $qw \in C^{\alpha}(\overline{S}_h)$. The interior regularity requirement $v - w \in H^2(\Sigma_{\Lambda_h})$ can be fulfilled in certain practical scenarios; see Theorem 8.4.1 in what follows on the study of an inverse scattering problem. The introduction of this interior regularity condition shall play a critical role in the proof of Theorem 8.4.1.

Proof of Theorem 8.2.1. It is clear that the transmission eigenfunctions $v \in H^1(\Omega)$ and $w \in H^1(\Omega)$ to (8.1.1) fulfill (8.2.12). From Lemma 8.2.6, we know that (8.2.20) holds. Using the fact that

$$\int_{S_h} u_0(s\mathbf{x})\mathrm{d}\mathbf{x} = \int_W u_0(s\mathbf{x})\mathrm{d}\mathbf{x} - \int_{W\backslash S_h} u_0(s\mathbf{x})\mathrm{d}\mathbf{x}, \tag{8.2.61}$$

where $u_0(s\mathbf{x})$ is defined in (8.2.4), substituting (8.2.61) into (8.2.20), we obtain the following integral equation:

$$\begin{aligned}(\widetilde{f}_{1j}(\mathbf{0}) - f_2(\mathbf{0})) \int_W u_0(s\mathbf{x})\mathrm{d}\mathbf{x} + \delta_j(s) &= I_3 - I_2^{\pm} - \int_{S_h} \delta\widetilde{f}_{1j}(\mathbf{x})u_0(s\mathbf{x})\mathrm{d}\mathbf{x} \\ &+ \int_{S_h} \delta f_2(\mathbf{x})u_0(s\mathbf{x})\mathrm{d}\mathbf{x} + (\widetilde{f}_{1j}(\mathbf{0}) - f_2(\mathbf{0})) \int_{W\backslash S_h} u_0(s\mathbf{x})\mathrm{d}\mathbf{x} - \xi_j^{\pm}(s),\end{aligned}$$

where $\delta\widetilde{f}_{1j}(\mathbf{x})$ and $\delta f_2(\mathbf{x})$ are defined in (8.2.18) and (8.2.19), respectively, and $\delta_j(s)$ and $\xi_j^{\pm}(s)$ are given by (8.2.21).

From (8.2.6), we know that

$$(\widetilde{f}_{1j}(\mathbf{0}) - f_2(\mathbf{0})) \int_W u_0(s\mathbf{x})\mathrm{d}\mathbf{x} = 6i(\widetilde{f}_{1j}(\mathbf{0}) - f_2(\mathbf{0}))(e^{-2\theta_M \mathrm{i}} - e^{-2\theta_m \mathrm{i}})s^{-2}. \tag{8.2.62}$$

From Lemma 8.2.9, it yields (8.2.43) and (8.2.44). Substituting (8.2.43) and (8.2.44) into (8.2.20), multiplying s on both sides of (8.2.20), and rearranging terms, we deduce that

$$\begin{aligned}&2v_j(\mathbf{0})\eta(\mathbf{0})\Bigg[\left(\mu(\theta_M)^{-2} - \mu(\theta_M)^{-2}e^{-\sqrt{sh}\mu(\theta_M)} - \mu(\theta_M)^{-1}\sqrt{sh}e^{-\sqrt{sh}\mu(\theta_M)}\right) \\ &\qquad + \left(\mu(\theta_m)^{-2} - \mu(\theta_m)^{-2}e^{-\sqrt{sh}\mu(\theta_m)} - \mu(\theta_m)^{-1}\sqrt{sh}e^{-\sqrt{sh}\mu(\theta_m)}\right)\Bigg] \\ &= s\Bigg[I_3 - (\widetilde{f}_{1j}(\mathbf{0}) - f_2(\mathbf{0})) \int_{S_h} u_0(s\mathbf{x})\mathrm{d}\mathbf{x} - \delta_j(s) - v_j(\mathbf{0})\eta(\mathbf{0})\left(\mathscr{I}_{312}^- + \mathscr{I}_{312}^+\right) \\ &\quad - \eta(\mathbf{0})(\mathscr{I}_{32}^+ + \mathscr{I}_{32}^-) - I_\eta^+ - I_\eta^- - \int_{S_h} \delta\widetilde{f}_{1j}(\mathbf{x})u_0(s\mathbf{x})\mathrm{d}\mathbf{x} + \int_{S_h} \delta f_2(\mathbf{x})u_0(s\mathbf{x})\mathrm{d}\mathbf{x} - \xi_j^{\pm}(s)\Bigg].\end{aligned} \tag{8.2.63}$$

Taking $s = j$, under the assumption (8.2.34), using (8.2.43) and (8.2.44) in Lemma 8.2.9, we know that

$$\begin{aligned}&j|\mathscr{I}_{32}^-| \leq \mathscr{O}(j^{-1}\|g_j\|_{L^2(\mathbb{S}^1)}) \leq \mathscr{O}(j^{-1+\rho}), \quad j|\mathscr{I}_{32}^+| \leq \mathscr{O}(j^{-1}\|g_j\|_{L^2(\mathbb{S}^1)}) \leq \mathscr{O}(j^{-1+\rho}), \\ &j|I_\eta^-| \leq \|\eta\|_{C^\alpha}\left(|v_j(\mathbf{0})|\mathscr{O}(j^{-\alpha}) + \mathscr{O}(\|g_j\|_{L^2(\mathbb{S}^1)}j^{-1-\alpha})\right), \\ &\quad \leq \|\eta\|_{C^\alpha}\left(|v_j(\mathbf{0})|\mathscr{O}(j^{-\alpha}) + \mathscr{O}(j^{-1-\alpha+\rho})\right), \\ &j|I_\eta^+| \leq \|\eta\|_{C^\alpha}\left(|v_j(\mathbf{0})|\mathscr{O}(j^{-\alpha}) + \mathscr{O}(\|g_j\|_{L^2(\mathbb{S}^1)}j^{-1-\alpha})\right), \\ &\quad \leq \|\eta\|_{C^\alpha}\left(|v_j(\mathbf{0})|\mathscr{O}(j^{-\alpha}) + \mathscr{O}(j^{-1-\alpha+\rho})\right), \\ &j\mathscr{I}_{312}^- \leq \mathscr{O}(j^{-2}), \quad j\mathscr{I}_{312}^+ \leq \mathscr{O}(j^{-2}).\end{aligned} \tag{8.2.64}$$

Clearly, when $s=j$, under the assumption (8.2.34), by virtue of (8.2.22), (8.2.23), and (8.2.24) in Lemma 8.2.6, we can obtain

$$\begin{aligned} j\left|\int_{S_h}\delta\widetilde{f}_{1j}(\mathbf{x})u_0(j\mathbf{x})\mathrm{d}\mathbf{x}\right| &\leq \frac{2\sqrt{2\pi}(\theta_M-\theta_m)\Gamma(2\alpha+4)}{\delta_W^{2\alpha+4}}k^2\mathrm{diam}(S_h)^{1-\alpha} \\ &\quad\times(1+k)\|g_j\|_{L^2(\mathbb{S}^1)}j^{-\alpha-1}\leq\mathscr{O}(j^{-1-\alpha+\rho}), \\ j\left|\int_{S_h}\delta f_2(\mathbf{x})u_0(j\mathbf{x})\mathrm{d}\mathbf{x}\right| &\leq \frac{2(\theta_M-\theta_m)\Gamma(2\alpha+4))}{\delta_W^{2\alpha+4}}\|f_2\|_{C^\alpha}j^{-\alpha-1},\ j|I_3|\leq Cje^{-c'\sqrt{j}}, \end{aligned} \tag{8.2.65}$$

where $C,c'>0$ and δ_W is defined in (8.2.5). Similarly, under the assumption (8.2.34), when $s=j$, in view of (8.2.35) and (8.2.36) in Lemma 8.2.7, it can be derived that

$$\begin{aligned} j|\xi_j^{\pm}(j)| &\leq C\left(|\eta(\mathbf{0})|\frac{\sqrt{\theta_M-\theta_m}e^{-\sqrt{j\Theta}\delta_W h}}{\sqrt{2}}j+\|\eta\|_{C^\alpha}j^{-\alpha}\frac{\sqrt{2(\theta_M-\theta_m)\Gamma(4\alpha+4)}}{(2\delta_W)^{2\alpha+2}}\right)j^{-1-\Upsilon}, \\ j|\delta_j(j)| &\leq \frac{\sqrt{\theta_M-\theta_m}k^2e^{-\sqrt{j\Theta}\delta_W h}}{\sqrt{2}}j^{-\Upsilon},\quad \Theta\in[0,h]. \end{aligned} \tag{8.2.66}$$

Furthermore, taking $s=j$ and using (8.2.61), from (8.2.6) and (8.2.7), we can deduce that

$$j\left|\int_{S_h}u_0(j\mathbf{x})\mathrm{d}\mathbf{x}\right|\leq 6|e^{-2\theta_M\mathrm{i}}-e^{-2\theta_m\mathrm{i}}|j^{-1}+\frac{6(\theta_M-\theta_m)}{\delta_W^4}j^{-1}e^{-\delta_W\sqrt{hj}/2}. \tag{8.2.67}$$

The coefficient of $v_j(\mathbf{0})$ of (8.2.63) with respect to the zeroth order of s is

$$2\eta(\mathbf{0})\left(\mu(\theta_m)^{-2}+\mu(\theta_M)^{-2}\right).$$

Under the assumption (8.2.58), from Lemma 8.2.10, we have

$$\mu(\theta_m)^{-2}+\mu(\theta_M)^{-2}\neq 0. \tag{8.2.68}$$

We take $s=j$ in (8.2.63). By letting $j\to\infty$ in (8.2.63), from (8.2.64) and (8.2.67), we can prove that

$$\eta(\mathbf{0})\left(\mu(\theta_m)^{-2}+\mu(\theta_M)^{-2}\right)\lim_{j\to\infty}v_j(\mathbf{0})=0.$$

Since $\eta(\mathbf{0})\neq 0$ and using (8.2.68), it is easy to see that

$$\lim_{j\to\infty}v_j(\mathbf{0})=0.$$

Using the fact that

$$\begin{aligned} \lim_{\rho\to+0}\frac{1}{m(B(\mathbf{0},\rho)\cap\Omega)}\int_{B(\mathbf{0},\rho)\cap\Omega}|v(\mathbf{x})|\mathrm{d}\mathbf{x} &\leq \lim_{j\to\infty}\Big(\lim_{\rho\to+0}\frac{1}{m(B(\mathbf{0},\rho)\cap\Omega)} \\ &\quad\times\int_{B(\mathbf{0},\rho)\cap\Omega}|v(\mathbf{x})-v_j(\mathbf{x})|\mathrm{d}\mathbf{x}+\lim_{\rho\to+0}\frac{1}{m(B(\mathbf{0},\rho)\cap\Omega)}\int_{B(\mathbf{0},\rho)\cap\Omega}|v_j(\mathbf{x})|\mathrm{d}\mathbf{x}\Big), \end{aligned} \tag{8.2.69}$$

we readily finish the proof of this theorem. □

We next consider the degenerate case of Theorem 8.2.1 with $\eta\equiv 0$. The conductive transmission eigenvalue problem (8.1.1) is reduced to the following interior transmission eigenvalue problem:

$$\begin{cases} \Delta w + k^2(1+V)w = 0 & \text{in } \Omega, \\ \Delta v + k^2 v = 0 & \text{in } \Omega, \\ w = v, \quad \partial_\nu v = \partial_\nu w & \text{on } \partial\Omega. \end{cases} \tag{8.2.70}$$

By slightly modifying our proof of Theorem 8.2.1, we can show the following result.

Corollary 8.2.1. *Let $\Omega \Subset \mathbb{R}^2$ be a bounded Lipschitz domain containing a corner $\Omega \cap B_h = \Omega \cap W$ with the vertex being $0 \in \partial\Omega$, where W is the sector defined in (8.2.1) and $h \in \mathbb{R}_+$. Suppose $v \in H^1(\Omega)$ and $w \in H^1(\Omega)$ are a pair of interior transmission eigenfunctions to (8.2.70). Let W and S_h be the same as described in Theorem 8.2.1. Assume that $v - w \in H^2(\Sigma_{\Lambda_h})$ and $qw \in C^\alpha(\overline{S}_h)$ for $0 < \alpha < 1$. Under the conditions (8.2.58) and that the transmission eigenfunction v can be approximated in $H^1(S_h)$ by the Herglotz functions v_j, $j = 1, 2, \ldots$, with kernels g_j satisfying*

$$\|v - v_j\|_{H^1(S_h)} \le j^{-2-\Upsilon}, \quad \|g_j\|_{L^2(\mathbb{S}^1)} \le C j^\rho, \tag{8.2.71}$$

for some constants $C > 0$, $\Upsilon > 0$, and $0 < \rho < \alpha$, one has

$$\lim_{\rho \to +0} \frac{1}{m(B(\mathbf{0}, \rho) \cap \Omega)} \int_{B(\mathbf{0}, \rho) \cap \Omega} V(\mathbf{x}) w(\mathbf{x}) \mathrm{d}\mathbf{x} = 0.$$

Remark 8.2.4. As discussed in the introduction, the vanishing near a corner of the interior transmission eigenfunctions was considered in [6]. Compared to the main result in [6], Corollary 8.2.1 is more general in two aspects. First, the corner in [6] must be a convex one, whereas in Corollary 8.2.1, the corner could be an arbitrary one as long as the corner is not degenerate, namely, (8.2.58) is fulfilled. Second, the regularity requirement on the eigenfunction v is relaxed from (8.2.60) to (8.2.71). Moreover, technical condition $qw \in C^\alpha(\overline{S}_h)$ in Corollary 8.2.1 can be readily fulfilled when we consider the unique recovery of the inverse scattering problem under the condition that q is a constant. Please refer to Lemma 8.4.1.

Proof of Corollary 8.2.1. The proof follows from the one for Theorem 8.2.1 with some necessary modifications, and we only outline it in the following. It is clear that the transmission eigenfunctions $v \in H^1(\Omega)$ and $w \in H^1(\Omega)$ to (8.2.70) fulfill (8.2.12) for $\eta \equiv 0$. Since $\eta(\mathbf{x}) \equiv 0$ near the corner, similar to (8.2.20) in Lemma 8.2.6, we have the following integral identity:

$$(\widetilde{f}_{1j}(\mathbf{0}) - f_2(\mathbf{0})) \int_{S_h} u_0(s\mathbf{x}) \mathrm{d}\mathbf{x} + \delta_j(s) = I_3 - \int_{S_h} \delta \widetilde{f}_{1j}(\mathbf{x}) u_0(s\mathbf{x}) \mathrm{d}\mathbf{x} + \int_{S_h} \delta f_2(\mathbf{x}) u_0(s\mathbf{x}) \mathrm{d}\mathbf{x}, \tag{8.2.72}$$

where $f_2(\mathbf{x})$, $\widetilde{f}_{1j}(\mathbf{x})$, $\delta_j(s)$, I_3, $\delta \widetilde{f}_{1j}(\mathbf{x})$, and $\delta f_2(\mathbf{x})$ are defined in (8.2.25), (8.2.15), (8.2.21), and (8.2.19), respectively.

From (8.2.6), it follows that

$$\begin{aligned} (\widetilde{f}_{1j}(\mathbf{0}) - f_2(\mathbf{0})) \int_{S_h} u_0(s\mathbf{x}) \mathrm{d}\mathbf{x} &= (\widetilde{f}_{1j}(\mathbf{0}) - f_2(\mathbf{0})) \int_W u_0(s\mathbf{x}) \mathrm{d}\mathbf{x} \\ &\quad - (\widetilde{f}_{1j}(\mathbf{0}) - f_2(\mathbf{0})) \int_{W \setminus S_h} u_0(s\mathbf{x}) \mathrm{d}\mathbf{x} \\ &= 6i(\widetilde{f}_{1j}(\mathbf{0}) - f_2(\mathbf{0}))(e^{-2\theta_M \mathrm{i}} - e^{-2\theta_m \mathrm{i}}) s^{-2} \\ &\quad - (\widetilde{f}_{1j}(\mathbf{0}) - f_2(\mathbf{0})) \int_{W \setminus S_h} u_0(s\mathbf{x}) \mathrm{d}\mathbf{x}. \end{aligned} \tag{8.2.73}$$

Combining (8.2.71) with (8.2.8) in Lemma 8.2.3, one can see that

$$j^2 |\delta_j(s)| \le \frac{\sqrt{\theta_M - \theta_m} k^2 e^{-\sqrt{s\Theta}\delta_W} h}{\sqrt{2}} j^{-\Upsilon}, \tag{8.2.74}$$

where $\Theta \in [0,h]$ and δ_W is defined in (8.2.5). By (8.2.22) in Lemma 8.2.6, we can also deduce that

$$j^2\left|\int_{S_h}\delta\widetilde{f}_{1j}(\mathbf{x})u_0(j\mathbf{x})\mathrm{d}\mathbf{x}\right| \leq \frac{2\sqrt{2\pi}(\theta_M-\theta_m)\Gamma(2\alpha+4)}{\delta_W^{2\alpha+4}}k^2\mathrm{diam}(S_h)^{1-\alpha}$$
$$\times(1+k)\|g_j\|_{L^2(\mathbb{S}^1)}j^{-\alpha}\leq\mathscr{O}(j^{-(\alpha-\rho)}), \tag{8.2.75}$$

for $0<\rho<\alpha$. After substituting (8.2.73) into (8.2.72), we take $s=j$. Since (8.2.73), multiply j^2 on both sides of (8.2.72). Using the assumptions (8.2.71) and (8.2.58), by letting $j\to\infty$, from (8.2.7) in Lemma 8.2.2; (8.2.22), (8.2.23), and (8.2.24) in Lemma 8.2.6; and (8.2.74), we prove that

$$\lim_{j\to\infty}v_j(\mathbf{0})=\frac{f_2(\mathbf{0})}{-k^2}.$$

Since

$$\begin{aligned}\lim_{j\to\infty}v_j(\mathbf{0})&=\lim_{j\to\infty}\lim_{\rho\to+0}\frac{1}{m(B(\mathbf{0},\rho)\cap\Omega)}\int_{B(\mathbf{0},\rho)\cap\Omega}v_j(\mathbf{x})\mathrm{d}\mathbf{x}\\&=\lim_{\rho\to+0}\frac{1}{m(B(\mathbf{0},\rho)\cap\Omega)}\int_{B(\mathbf{0},\rho)\cap\Omega}v(\mathbf{x})\mathrm{d}\mathbf{x},\\\frac{f_2(\mathbf{0})}{-k^2}&=\lim_{\rho\to+0}\frac{1}{m(B(\mathbf{0},\rho)\cap\Omega)}\int_{B(0,\rho)\cap\Omega}qw(\mathbf{x})\mathrm{d}\mathbf{x},\end{aligned}$$

together with

$$\lim_{\rho\to+0}\frac{1}{m(B(\mathbf{0},\rho)\cap\Omega)}\int_{B(0,\rho)\cap\Omega}v(\mathbf{x})\mathrm{d}\mathbf{x}=\lim_{\rho\to+0}\frac{1}{m(B(\mathbf{0},\rho)\cap\Omega)}\int_{B(\mathbf{0},\rho)\cap\Omega}w(\mathbf{x})\mathrm{d}\mathbf{x},$$

we finish the proof of this corollary. □

Remark 8.2.5. If $V(\mathbf{x})$ is continuous near the corner $\mathbf{0}$ and $V(\mathbf{0})\neq 0$, from the fact that

$$\begin{aligned}&\lim_{\rho\to+0}\frac{1}{m(B(\mathbf{0},\rho)\cap\Omega)}\int_{B(\mathbf{0},\rho)\cap\Omega}V(\mathbf{x})w(\mathbf{x})\mathrm{d}\mathbf{x}\\&=V(\mathbf{x}_c)\lim_{\rho\to+0}\frac{1}{m(B(\mathbf{0},\rho)\cap\Omega)}\int_{B(\mathbf{0},\rho)\cap\Omega}w(\mathbf{x})\mathrm{d}\mathbf{x},\end{aligned}$$

we can prove the vanishing property near the corner 0 of the interior transmission eigenfunctions $v\in H^1(\Omega)$ and $w\in H^1(\Omega)$ under the assumptions (8.2.58) and (8.2.71).

If stronger regularity conditions are satisfied by the conductive transmission eigenfunctions v and w to (8.1.1), we can show that more apparent vanishing properties hold at the corner. The rest of this section is devoted to this case. In fact, we have the following theorem.

Theorem 8.2.2. *Let $v\in H^2(\Omega)$ and $w\in H^1(\Omega)$ be eigenfunctions to* (8.1.1). *Assume that $\Omega\subset\mathbb{R}^2$ contains a corner $\Omega\cap B_h=\Omega\cap W$ with the vertex being $\mathbf{0}\in\partial\Omega$, where W is the sector defined in* (8.2.1) *and $h\in\mathbb{R}_+$. Moreover, there exists a sufficiently small neighborhood S_h (i.e., $h>0$ is sufficiently small) of $\mathbf{0}$ in Ω, such that $qw\in C^\alpha(\overline{S}_h)$ and $\eta\in C^\alpha\left(\overline{\Gamma_h^\pm}\right)$ for $0<\alpha<1$ and $v-w\in H^2(\Sigma_{\Lambda_h})$. Under the following assumptions:*

(a) the function $\eta(\mathbf{x})$ does not vanish at the vertex $\mathbf{0}$, i.e.,

$$\eta(\mathbf{0})\neq 0, \tag{8.2.76}$$

(b) the open angle of S_h containing the corner satisfies

$$-\pi<\theta_m<\theta_M<\pi \text{ and } \theta_M-\theta_m\neq\pi, \tag{8.2.77}$$

then we have $v(\mathbf{0})=w(\mathbf{0})=0$.

Proof. It is clear that the transmission eigenfunctions $v \in H^2(\Omega)$ and $w \in H^1(\Omega)$ to (8.1.1) fulfill (8.2.12). Recall that f_1 and f_2 are defined by (8.2.25). From Lemma 8.2.6, we know that (8.2.14) is satisfied, which can be further formulated as

$$\int_{S_h}(f_1 - f_2)u_0(s\mathbf{x})\mathrm{d}\mathbf{x} = I_3 - \int_{\Gamma_h^\pm} \eta(\mathbf{x})u_0(s\mathbf{x})v(\mathbf{x})\mathrm{d}\sigma, \tag{8.2.78}$$

where I_3 is defined in (8.2.21). Since $f_2 \in C^\alpha(\overline{S}_h)$ and $\eta \in C^\alpha\left(\overline{\Gamma_h^\pm}\right)$, we know that η and f_2 have the expansions (8.2.13) and (8.2.19) around the origin, respectively. Furthermore, due to the fact that $v \in H^2(S_h)$, which can be embedded into $C^\alpha(\overline{S}_h)$, we have the following expansions:

$$\begin{aligned} f_1(\mathbf{x}) &= f_1(\mathbf{0}) + \delta f_1(\mathbf{x}), \quad |\delta f_1(\mathbf{x})| \le \|f_1\|_{C^\alpha}|\mathbf{x}|^\alpha, \\ v(\mathbf{x}) &= v(\mathbf{0}) + \delta v(\mathbf{x}), \quad |\delta v(\mathbf{x})| \le \|v\|_{C^\alpha}|\mathbf{x}|^\alpha. \end{aligned} \tag{8.2.79}$$

Substituting (8.2.13), (8.2.19), and (8.2.79) into (8.2.78), we can derive that

$$\begin{aligned} &(f_1(\mathbf{0}) - f_2(\mathbf{0}))\int_{S_h} u_0(s\mathbf{x})\mathrm{d}\mathbf{x} + \int_{S_h}(\delta f_1 - \delta f_2)u_0(s\mathbf{x})\mathrm{d}\mathbf{x} \\ &= I_3 - \eta(\mathbf{0})v(\mathbf{0})\int_{\Gamma_h^\pm} u_0(s\mathbf{x})\mathrm{d}\sigma - \eta(\mathbf{0})\int_{\Gamma_h^\pm}\delta v(\mathbf{x})u_0(s\mathbf{x})\mathrm{d}\sigma - v(\mathbf{0})\int_{\Gamma_h^\pm}\delta\eta(\mathbf{x})u_0(s\mathbf{x})\mathrm{d}\sigma \\ &\quad - \int_{\Gamma_h^\pm}\delta\eta(\mathbf{x})\delta v(\mathbf{x})u_0(s\mathbf{x})\mathrm{d}\sigma. \end{aligned} \tag{8.2.80}$$

Using (8.2.38) in Lemma 8.2.8, it is easy to see that

$$\begin{aligned} \eta(\mathbf{0})v(\mathbf{0})\int_{\Gamma_h^+} u_0(s\mathbf{x})\mathrm{d}\sigma &= 2s^{-1}v(\mathbf{0})\eta(\mathbf{0})\Big(\mu(\theta_M)^{-2} - \mu(\theta_M)^{-2}e^{-\sqrt{sh}\mu(\theta_M)} \\ &\quad -\mu(\theta_M)^{-1}\sqrt{sh}e^{-\sqrt{sh}\mu(\theta_M)}\Big), \\ \eta(\mathbf{0})v(\mathbf{0})\int_{\Gamma_h^-} u_0(s\mathbf{x})\mathrm{d}\sigma &= 2s^{-1}v(\mathbf{0})\eta(\mathbf{0})\Big(\mu(\theta_m)^{-2} - \mu(\theta_m)^{-2}e^{-\sqrt{sh}\mu(\theta_m)} \\ &\quad -\mu(\theta_m)^{-1}\sqrt{sh}e^{-\sqrt{sh}\mu(\theta_m)}\Big), \end{aligned} \tag{8.2.81}$$

where $\mu(\theta)$ is defined in (8.2.41). Besides, from (8.2.79), using (8.2.10), we can estimate

$$\begin{aligned} s\left|\int_{\Gamma_h^-}\delta v(\mathbf{x})u_0(s\mathbf{x})\mathrm{d}\sigma\right| &\le s\|v\|_{C^\alpha}\int_0^h r^\alpha e^{-\sqrt{sr}\omega(\theta_m)}\mathrm{d}r = \mathcal{O}(s^{-\alpha}), \\ s\left|\int_{\Gamma_h^-}\delta\eta(\mathbf{x})u_0(s\mathbf{x})\mathrm{d}\sigma\right| &\le s\|\eta\|_{C^\alpha}\int_0^h r^\alpha e^{-\sqrt{sr}\omega(\theta_m)}\mathrm{d}r = \mathcal{O}(s^{-\alpha}), \\ s\left|\int_{\Gamma_h^-}\delta v(\mathbf{x})\delta\eta(\mathbf{x})u_0(s\mathbf{x})\mathrm{d}\sigma\right| &\le s\|v\|_{C^\alpha}\|\eta\|_{C^\alpha}\int_0^h r^{2\alpha} e^{-\sqrt{sr}\omega(\theta_m)}\mathrm{d}r = \mathcal{O}(s^{-2\alpha}), \\ s\left|\int_{S_h}\delta f_1 u_0(s\mathbf{x})\mathrm{d}\mathbf{x}\right| &\le s\cdot\|f_1\|_{C^\alpha}\int_W |u_0(s\mathbf{x})||\mathbf{x}|^\alpha\mathrm{d}\mathbf{x} \\ &\le \frac{2\|f_1\|_{C^\alpha}(\theta_M - \theta_m)\Gamma(2\alpha+4)}{\delta_W^{2\alpha+4}}s^{-\alpha-1} \\ s\left|\int_{S_h}\delta f_2 u_0(s\mathbf{x})\mathrm{d}\mathbf{x}\right| &\le s\cdot\|f_2\|_{C^\alpha}\int_W |u_0(s\mathbf{x})||\mathbf{x}|^\alpha\mathrm{d}\mathbf{x} \\ &\le \frac{2\|f_2\|_{C^\alpha}(\theta_M - \theta_m)\Gamma(2\alpha+4)}{\delta_W^{2\alpha+4}}s^{-\alpha-1}. \end{aligned} \tag{8.2.82}$$

Substituting (8.2.81) into (8.2.80) and multiplying s on both sides of (8.2.80), after arranging terms, we obtain that

$$\begin{aligned}&2v(\mathbf{0})\eta(\mathbf{0})\left(\mu(\theta_M)^{-2}+\mu(\theta_m)^{-2}\right)=2v(\mathbf{0})\eta(\mathbf{0})\Big(\mu(\theta_M)^{-2}e^{-\sqrt{sh}\mu(\theta_M)}\\&+\mu(\theta_M)^{-1}\sqrt{sh}e^{-\sqrt{sh}\mu(\theta_M)}+\mu(\theta_m)^{-2}e^{-\sqrt{sh}\mu(\theta_m)}+\mu(\theta_m)^{-1}\sqrt{sh}e^{-\sqrt{sh}\mu(\theta_m)}\Big)\\&+s\Big[I_3-(f_1(\mathbf{0})-f_2(\mathbf{0}))\int_{S_h}u_0(s\mathbf{x})\mathrm{d}\mathbf{x}-\int_{S_h}(\delta f_1-\delta f_2)u_0(s\mathbf{x})\mathrm{d}\mathbf{x}\\&-\eta(\mathbf{0})\int_{\Gamma_h^\pm}\delta v(\mathbf{x})u_0(s\mathbf{x})\mathrm{d}\sigma-v(\mathbf{0})\int_{\Gamma_h^\pm}\delta\eta(\mathbf{x})u_0(s\mathbf{x})\mathrm{d}\sigma-\int_{\Gamma_h^\pm}\delta\eta(\mathbf{x})\delta v(\mathbf{x})u_0(s\mathbf{x})\mathrm{d}\sigma\Big].\end{aligned} \tag{8.2.83}$$

Since $v-w\in H^2(\Sigma_{\Lambda_h})$, (8.2.24) still holds. In (8.2.83), letting $s\to\infty$, from (8.2.7), (8.2.24) in Lemma 8.2.6, (8.2.73), and (8.2.82), we can show that

$$\eta(\mathbf{0})\left(\mu(\theta_M)^{-2}+\mu(\theta_m)^{-2}\right)v(\mathbf{0})=0.$$

Under the assumption (8.2.77), from Lemma 8.2.10, we have $\mu(\theta_M)^{-2}+\mu(\theta_m)^{-2}\neq 0$. Since $\eta(\mathbf{0})\neq 0$ from (8.2.76), we finish the proof of this theorem. □

Remark 8.2.6. Under the H^2 regularity, the interior transmission eigenfunctions to (8.2.70) have been shown that they always vanish at a corner point if the interior angle of the corner is not π; see [4, Theorem 4.2] for more details.

8.3 Vanishing Near Corners of Conductive Transmission Eigenfunctions: Three-Dimensional Case

In this section, we study the vanishing property of the conductive transmission eigenfunctions for the 3D case. In principle, we could also consider a generic corner in the usual sense as the one for the 2D case. However, in what follows, we introduce a more general corner geometry that is described by $S_h\times(-M,M)$, where S_h is defined in (8.2.2) and $M\in\mathbb{R}_+$. It is readily seen that $S_h\times(-M,M)$ actually describes an edge singularity and we call it a 3D corner for notational unification. Suppose that the Lipschitz domain $\Omega\subset\mathbb{R}^3$ with $\mathbf{0}\in\partial\Omega$ possesses a 3D corner. Let $\mathbf{0}\in\mathbb{R}^2$ be the vertex of S_h and $x_3^c\in(-M,M)$. Then $(\mathbf{0},x_3^c)$ is defined as the edge point of $S_h\times(-M,M)$. In Fig. 8.2, we give a schematic illustration of the geometry considered in 3D. In this section, under some appropriate assumptions, we show that the conductive transmission eigenfunctions v and w vanish at $(\mathbf{0},x_3^c)$. Since the CGO solution constructed in Lemma 8.2.2 is only two-dimensional, in order to make use of the similar arguments of Theorem 8.2.1, we introduce the following dimension reduction operator. The dimension reduction operator technique is also introduced in [4, Lemma 3.4] for studying the vanishing property of non-radiating sources and the transmission eigenfunctions at edges in three dimensions. Similar to Theorem 8.2.1, we first assume that v is only H^1 smooth but can be approximated by the Herglotz wave functions with some mild assumptions, whereas in Theorem 8.3.1, the interior angle of S_h cannot be π. Besides, if v has H^2 regularity near the edge point, in Theorem 8.3.2, we also prove the vanishing property of v and w near the edge point.

Definition 8.3.1. Let $S_h\subset\mathbb{R}^2$ be defined in (8.2.2), $M>0$. Assume that g is a function defined in the domain $S_h*(-M,M)$ with the domain $S_h\times(-M,M)$. Pick up any point $x_3^c\in(-M,M)$. Suppose that $\psi\in C_0^\infty((x_3^c-L,x_3^c+L))$ is a nonnegative function and $\psi\not\equiv 0$, where L is sufficiently

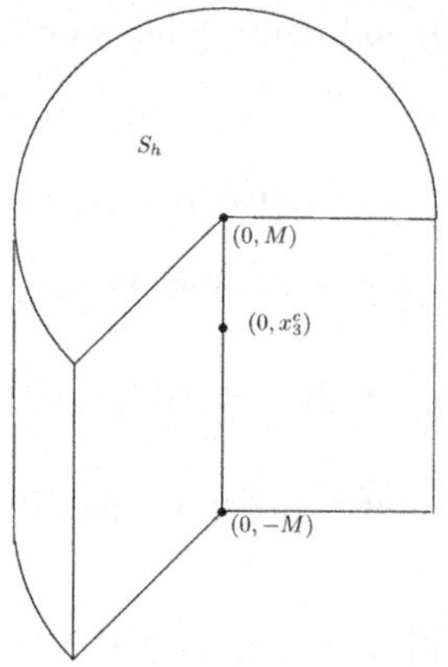

Fig. 8.2 Schematic illustration of the corner in 3D

small such that $(x_3^c - L, x_3^c + L) \subset (-M, M)$, and write $\mathbf{x} = (\mathbf{x}', x_3) \in \mathbb{R}^3$, $\mathbf{x}' \in \mathbb{R}^2$. The dimension reduction operator $\mathscr{R}$ is defined by

$$\mathscr{R}(g)(\mathbf{x}') = \int_{x_3^c - L}^{x_3^c + L} \psi(x_3) g(\mathbf{x}', x_3)\, \mathrm{d}x_3, \tag{8.3.1}$$

where $\mathbf{x}' \in S_h$.

Remark 8.3.1. The assumption on the nonnegativity of ψ plays an important role in our proof of Theorem 8.3.1 in what follows, where we use the integral mean value theorem to carefully investigate the asymptotic property of the parameter s appearing in the CGO solution $u_0(s\mathbf{x}')$ given in Lemma 8.2.2 as $s \to \infty$. In order to use the two-dimensional CGO solution $u_0(s\mathbf{x}')$ to prove the vanishing property of the conductive transmission eigenfunctions in $\mathbb{R}^3$, we need the dimension reduction operator defined in Definition 8.3.1 in our proof of Theorem 8.3.1.

Before presenting the main results of this section, we first analyze the regularity of the functions after applying the dimension reduction operator. Using a similar argument of [4, Lemma 3.4], we can prove the following lemma, whose detailed proof is omitted.

Lemma 8.3.1. *Let $g \in H^2(S_h \times (-M, M)) \cap C^\alpha(\overline{S}_h \times [-M, M])$, where $0 < \alpha < 1$. Then*

$$\mathscr{R}(g)(\mathbf{x}') \in H^2(S_h) \cap C^\alpha(\overline{S}_h).$$

Similarly, if $g \in H^1(S_h \times (-M, M))$, we have $\mathscr{R}(g)(\mathbf{x}') \in H^1(S_h)$.

In Theorem 8.3.1, we shall prove the vanishing property of conductive transmission eigenfunctions at an edge corner in 3D. Let us first introduce the mathematical setup.

Let $S_h \subset \mathbb{R}^2$ be defined in (8.2.2), $M > 0$, $0 < \alpha < 1$. For any fixed $x_3^c \in (-M, M)$ and $L > 0$ defined in Definition 8.3.1, we suppose that L is sufficiently small such that $(x_3^c - L, x_3^c + L) \subset (-M, M)$. Write $\mathbf{x} = (\mathbf{x}', x_3) \in \mathbb{R}^3$, $\mathbf{x}' \in \mathbb{R}^2$. Let $v, w \in H^1(S_h \times (-M, M))$ fulfill that

$$\begin{cases} \Delta v + k^2 v = 0, & \mathbf{x}' \in S_h, -M < x_3 < M, \\ \Delta w + k^2 q w = 0, & \mathbf{x}' \in S_h, -M < x_3 < M, \\ w = v \quad \partial_\nu v + \eta v = \partial_\nu w, & \mathbf{x}' \in \Gamma_h^\pm, -M < x_3 < M, \end{cases} \tag{8.3.2}$$

where $\Gamma_h^\pm$ are defined in (8.2.2), ν is the outward normal vector to $\Gamma_h^\pm \times (-M, M)$, $q \in L^\infty(S_h \times (-M, M))$, and $\eta \in L^\infty(\Gamma_h^\pm \times (-M, M))$ is independent of x_3.

Lemmas 8.3.2 and 8.3.7 will be used to prove Theorem 8.3.1 in what follows.

Lemma 8.3.2. *Suppose that* $v,w \in H^1(S_h \times (-M,M))$ *fulfill* (8.3.2). *Denote*

$$\begin{aligned} G(\mathbf{x}') = & \int_{-L}^{L} \psi''(x_3)v(\mathbf{x}',x_3)\mathrm{d}x_3 - k^2\mathscr{R}(v)(\mathbf{x}'), \\ \widetilde{G}(\mathbf{x}') = & \int_{-L}^{L} \psi''(x_3)w(\mathbf{x}',x_3)\mathrm{d}x_3 - k^2\mathscr{R}(qw)(\mathbf{x}'), \end{aligned} \tag{8.3.3}$$

where $\mathscr{R}$ *is the dimension reduction operator associated with* ψ *defined in* (8.3.1). *Then there hold that*

$$\begin{cases} \Delta_{\mathbf{x}'}\mathscr{R}(v)(\mathbf{x}') = G(\mathbf{x}') & \text{in } S_h, \\ \Delta_{\mathbf{x}'}\mathscr{R}(w)(\mathbf{x}') = \widetilde{G}(\mathbf{x}') & \text{in } S_h, \\ \mathscr{R}(w)(\mathbf{x}') = \mathscr{R}(v)(\mathbf{x}'),\ \partial_\nu\mathscr{R}(v)(\mathbf{x}') + \eta(\mathbf{x}')\mathscr{R}(v)(\mathbf{x}') = \partial_\nu\mathscr{R}(w)(\mathbf{x}') & \text{on } \Gamma_h^{\pm}, \end{cases} \tag{8.3.4}$$

in the distributional sense, where ν *signifies the exterior unit normal vector to* $\Gamma_h^{\pm}$. *Let*

$$G(\mathbf{x}') - \widetilde{G}(\mathbf{x}') = F_1(\mathbf{x}') + F_2(\mathbf{x}') + F_3(\mathbf{x}'), \tag{8.3.5}$$

where

$$\begin{aligned} & F_1(\mathbf{x}') = \int_{-L}^{L} \psi''(x_3)(v(\mathbf{x}',x_3) - w(\mathbf{x}',x_3))\mathrm{d}x_3,\ F_2(\mathbf{x}') = k^2\mathscr{R}(qw)(\mathbf{x}'), \\ & F_3(\mathbf{x}') = -k^2\mathscr{R}(v)(\mathbf{x}'). \end{aligned}$$

Recall that the CGO solution $u_0(s\mathbf{x}')$ *is defined in* (8.2.4) *with the parameter* $s \in \mathbb{R}_+$. *There holds the following integral identity:*

$$\int_{S_h} (F_1(\mathbf{x}') + F_2(\mathbf{x}') + F_3(\mathbf{x}'))u_0(s\mathbf{x}')\mathrm{d}\mathbf{x}' = I_3 - \int_{\Gamma_h^{\pm}} \eta(\mathbf{x}')\mathscr{R}(v)(\mathbf{x}')u_0(s\mathbf{x}')\mathrm{d}\sigma, \tag{8.3.6}$$

where $I_3 = \int_{\Lambda_h}(u_0(s\mathbf{x}')\partial_\nu\mathscr{R}(v-w)(\mathbf{x}') - \mathscr{R}(v-w)(\mathbf{x}')\partial_\nu u_0(s\mathbf{x}'))\mathrm{d}\sigma$. *If* $qw \in C^\alpha(\overline{S}_h \times [-M,M])$ *for* $0 < \alpha < 1$ *and* $v - w \in H^2(S_h \times (-M,M))$, *then we have* $F_1(\mathbf{x}') \in C^\alpha(\overline{S}_h)$ *for* $\alpha \in (0,1)$ *and* $F_2(\mathbf{x}') \in C^\alpha(\overline{S}_h)$.

Proof. For the edge point $(0,x_3^c) \in S_h \times (-M,M)$, where $x_3^c \in (-M,M)$, without loss of generality, in the subsequent analysis, we assume that $x_3^c = 0$. Since $\Delta_{\mathbf{x}'}v = -k^2v - \partial_{x_3}^2 v$ and $\Delta_{\mathbf{x}'}w = -k^2qw - \partial_{x_3}^2 w$, by the dominated convergence theorem, integration by parts gives

$$\Delta_{\mathbf{x}'}\mathscr{R}(v)(\mathbf{x}') = \int_{-L}^{L} \psi''(x_3)v(\mathbf{x}',x_3)\mathrm{d}x_3 - k^2\mathscr{R}(v)(\mathbf{x}') = G(\mathbf{x}'), \tag{8.3.7}$$

$$\Delta_{\mathbf{x}'}\mathscr{R}(w)(\mathbf{x}') = \int_{-L}^{L} \psi''(x_3)w(\mathbf{x}',x_3)\mathrm{d}x_3 - k^2\mathscr{R}(qw)(\mathbf{x}') = \widetilde{G}(\mathbf{x}'). \tag{8.3.8}$$

Moreover, we have

$$\mathscr{R}(w)(\mathbf{x}') = \mathscr{R}(v)(\mathbf{x}') \text{ on } \Gamma_h^{\pm}, \tag{8.3.9}$$

in the sense of distribution, since $w(\mathbf{x}',x_3) = v(\mathbf{x}',x_3)$ when $\mathbf{x}' \in \Gamma_h^{\pm}$ and $-L < x_3 < L$. Similarly, using the fact that η is independent of x_3, we can easily show that

$$\partial_\nu\mathscr{R}(v)(\mathbf{x}') + \eta(\mathbf{x}')\mathscr{R}(v)(\mathbf{x}') = \partial_\nu\mathscr{R}(w)(\mathbf{x}') \text{ on } \Gamma_h^{\pm}, \tag{8.3.10}$$

in the sense of distribution.

Subtracting (8.3.8) from (8.3.7), combining with the boundary condition (8.3.9) and (8.3.10), we deduce that

$$\begin{aligned}&\Delta_{\mathbf{x}'}(\mathscr{R}(v)(\mathbf{x}')-\mathscr{R}(w)(\mathbf{x}'))=F_1(\mathbf{x}')+F_2(\mathbf{x}')+F_3(\mathbf{x}') \text{ in } S_h,\\&\mathscr{R}(v)(\mathbf{x}')-\mathscr{R}(w)(\mathbf{x}')=0,\ \partial_\nu\mathscr{R}(v)(\mathbf{x}')-\partial_\nu\mathscr{R}(w)(\mathbf{x}')=-\eta(\mathbf{x}')\mathscr{R}(v)(\mathbf{x}') \text{ on } \Gamma_h^{\pm}.\end{aligned} \tag{8.3.11}$$

Recall that $u_0(s\mathbf{x}')\in H^1(S_h)$ from Lemma 8.2.3. Since $v, w\in H^1(S_h\times(-M,M))$ and $q\in L^\infty(S_h\times(-M,M))$, by virtue of Lemma 8.3.1, it yields that F_1, F_2, $F_3\in L^2(S_h)$. By Lemma 8.2.5 and using the fact that $\Delta_{\mathbf{x}'}u_0(s\mathbf{x}')=0$ in S_h, we have the following Green identity:

$$\int_{S_h}u_0(s\mathbf{x}')\Delta_{\mathbf{x}'}\mathscr{R}(v-w)(\mathbf{x}')\mathrm{d}\mathbf{x}'=\int_{\partial S_h}(u_0(s\mathbf{x}')\partial_\nu\mathscr{R}(v-w)(\mathbf{x}')-\mathscr{R}(v-w)(\mathbf{x}')\partial_\nu u_0(s\mathbf{x}'))\mathrm{d}\sigma. \tag{8.3.12}$$

Substituting (8.3.11) into (8.3.12), it yields (8.3.6).

Recall that F_1 and F_2 are defined in (8.3.5). Since $v-w\in H^2(S_h\times(-M,M))$, from Lemma 8.3.1, we know that $F_1(\mathbf{x}')\in H^2(S_h)$, which can be embedded into $C^\alpha(\overline{S}_h)$ for $\alpha\in(0,1)$. Moreover, from Lemma 8.3.1, we have $F_2(\mathbf{x}')\in C^\alpha(\overline{S}_h)$, since $qw\in C^\alpha(\overline{S}_h\times[-M,M])$ and $0<\alpha<1$. □

Lemma 8.3.3. *Let S_h and $\Gamma_h^\pm$ be defined in (8.2.2). Suppose that $v,w\in H^1(S_h\times(-M,M))$ fulfill (8.3.2). Recall that the CGO solution $u_0(s\mathbf{x})$ is defined in (8.2.4) with the parameter $s\in\mathbb{R}_+$. Let*

$$F_{3j}(\mathbf{x}')=-k^2\mathscr{R}(v_j)(\mathbf{x}'),$$

where v_j is the Herglotz wave function given by

$$v_j(\mathbf{x})=\int_{\mathbb{S}^2}e^{ik\mathbf{d}\cdot\mathbf{x}}g_j(\mathbf{d})\mathrm{d}\sigma(\mathbf{d}),\quad \mathbf{d}\in\mathbb{S}^2. \tag{8.3.13}$$

Then $F_{3j}(\mathbf{x}')\in C^\alpha(\overline{S}_h)$ and it has the expansion

$$F_{3j}(\mathbf{x}')=F_{3j}(\mathbf{0})+\delta F_{3j}(\mathbf{x}'),\quad |\delta F_{3j}(\mathbf{x}')|\leq\|F_{3j}\|_{C^\alpha}|\mathbf{x}'|^\alpha. \tag{8.3.14}$$

Recall that $F_1(\mathbf{x}')$ and $F_2(\mathbf{x}')$ are defined in (8.3.5). Assuming that $F_1(\mathbf{x}')\in C^\alpha(\overline{S}_h)$ and $F_2(\mathbf{x}')\in C^\alpha(\overline{S}_h)$ $(0<\alpha<1)$ satisfying

$$\begin{aligned}F_1(\mathbf{x}')&=F_1(\mathbf{0})+\delta F_1(\mathbf{x}'),\quad |\delta F_1(\mathbf{x}')|\leq\|F_1\|_{C^\alpha}|\mathbf{x}'|^\alpha,\\F_2(\mathbf{x}')&=F_2(\mathbf{0})+\delta F_2(\mathbf{x}'),\quad |\delta F_2(\mathbf{x}')|\leq\|F_2\|_{C^\alpha}|\mathbf{x}'|^\alpha,\end{aligned} \tag{8.3.15}$$

then there holds that

$$\begin{aligned}(F_1(\mathbf{0})+F_2(\mathbf{0})+F_{3j}(\mathbf{0}))&\int_{S_h}u_0(s\mathbf{x}')\mathrm{d}\mathbf{x}'+\delta_j(s)=I_3-I_2^\pm-\varepsilon_j^\pm(s)\\&-\int_{S_h}\delta F_1(\mathbf{x}')u_0(s\mathbf{x}')\mathrm{d}\mathbf{x}'-\int_{S_h}\delta F_2(\mathbf{x}')u_0(s\mathbf{x}')\mathrm{d}\mathbf{x}'-\int_{S_h}\delta F_{3j}(\mathbf{x}')u_0(s\mathbf{x}')\mathrm{d}\mathbf{x}',\end{aligned} \tag{8.3.16}$$

where

$$\begin{aligned}I_2^\pm&=\int_{\Gamma_h^\pm}\eta(\mathbf{x}')u_0(s\mathbf{x}')\mathscr{R}(v_j)(\mathbf{x}')\mathrm{d}\sigma,\\\delta_j(s)&=-k^2\int_{S_h}(\mathscr{R}(v)(\mathbf{x}')-\mathscr{R}(v_j)(\mathbf{x}'))u_0(s\mathbf{x}')\mathrm{d}\mathbf{x}',\\\varepsilon_j^\pm(s)&=\int_{\Gamma_h^\pm}\eta(\mathbf{x}')u_0(s\mathbf{x}')\mathscr{R}(v(\mathbf{x}',x_3)-v_j(\mathbf{x}',x_3))\mathrm{d}\sigma,\end{aligned} \tag{8.3.17}$$

and I_3 is defined in (8.3.6). *Furthermore, assuming that the transmission eigenfunction v can be approximated in $H^1(S_h \times (-M,M))$ by the Herglotz wave functions v_j defined in* (8.3.13), *$j = 1,2,\ldots$, with kernels g_j satisfying*

$$\|v - v_j\|_{H^1(S_h\times(-M,M))} \le j^{-1-\Upsilon}, \quad \|g_j\|_{L^2(\mathbb{S}^2)} \le Cj^{1+\rho}, \tag{8.3.18}$$

for some positive constants C, Υ and ρ, then there hold that

$$\left|\int_{S_h} \delta F_{3j}(\mathbf{x})u_0(s\mathbf{x}')\mathrm{d}\mathbf{x}'\right| \le \frac{8L\sqrt{\pi}\|\psi\|_{L^\infty}(\theta_M-\theta_m)\Gamma(2\alpha+4)}{\delta_W^{2\alpha+4}}k^2\mathrm{diam}(S_h)^{1-\alpha} \times (1+k)\|g_j\|_{L^2(\mathbb{S}^2)}s^{-\alpha-2}, \tag{8.3.19a}$$

$$\left|\int_{S_h} \delta F_1(\mathbf{x})u_0(s\mathbf{x}')\mathrm{d}\mathbf{x}'\right| \le \frac{2\|F_1\|_{C^\alpha}(\theta_M-\theta_m)\Gamma(2\alpha+4)}{\delta_W^{2\alpha+4}}s^{-\alpha-2}, \tag{8.3.19b}$$

$$\left|\int_{S_h} \delta F_2(\mathbf{x})u_0(s\mathbf{x}')\mathrm{d}\mathbf{x}'\right| \le \frac{2\|F_2\|_{C^\alpha}(\theta_M-\theta_m)\Gamma(2\alpha+4)}{\delta_W^{2\alpha+4}}s^{-\alpha-2}, \tag{8.3.19c}$$

where $\Theta \in [0,h]$ and δ_W is defined in (8.2.5), *as $s \to +\infty$. If $v - w \in H^2(S_h \times (-M,M))$, then one has*

$$|I_3| \le Ce^{-c'\sqrt{s}} \quad as \ s \to +\infty, \tag{8.3.20}$$

where $C > 0$ and $c' > 0$ are constants.

Proof. It is clear that the Herglotz wave functions $v_j \in H^2(S_h \times (-M,M))$. From Lemma 8.3.1, we have $\mathscr{R}(v_j)(\mathbf{x}') \in H^2(S_h)$, which can be embedded into $C^\alpha(\overline{S}_h)$ satisfying (8.3.14).

Since $v \in H^1(S_h \times (-M,M))$ is a solution to the Helmholtz equation in $S_h \times (-M,M)$, from Lemma 8.2.1, v can be approximated by the Herglotz wave functions $v_j(x)$ given in (8.3.13) in the H^1 topology. Therefore, we deduce that

$$\int_{S_h} -k^2\mathscr{R}(v)(\mathbf{x}')u_0(s\mathbf{x}')\mathrm{d}\mathbf{x}' = \int_{S_h} -k^2\mathscr{R}(v_j)(\mathbf{x}')u_0(s\mathbf{x}')\mathrm{d}\mathbf{x}' + \delta_j(s), \tag{8.3.21}$$

where $\delta_j(s)$ is defined in (8.3.17).

Let

$$I_1 = \int_{S_h} u_0(s\mathbf{x}')(F_1(\mathbf{x}') + F_2(\mathbf{x}') + F_{3j}(\mathbf{x}'))\mathrm{d}\mathbf{x}'.$$

Substituting (8.3.14) and (8.3.15) into I_1, it yields that

$$I_1 = (F_1(\mathbf{0}) + F_2(\mathbf{0}) + F_{3j}(\mathbf{0}))\int_{S_h} u_0(s\mathbf{x}')\mathrm{d}\mathbf{x}' + \int_{S_h} \delta F_1(\mathbf{x}')u_0(s\mathbf{x}')\mathrm{d}\mathbf{x}' + \int_{S_h} \delta F_2(\mathbf{x}')u_0(s\mathbf{x}')\mathrm{d}\mathbf{x}' + \int_{S_h} \delta F_{3j}(\mathbf{x}')u_0(s\mathbf{x}')\mathrm{d}\mathbf{x}'.$$

Let $F_3(\mathbf{x}')$ be defined in (8.3.5). By virtue of (8.3.21), we have

$$\begin{aligned}\int_{S_h} (F_1(\mathbf{x}') + F_2(\mathbf{x}') + F_3(\mathbf{x}'))u_0(s\mathbf{x}')\mathrm{d}\mathbf{x}' &= (F_1(\mathbf{0}) + F_2(\mathbf{0}) + F_{3j}(\mathbf{0}))\int_{S_h} u_0(s\mathbf{x}')\mathrm{d}\mathbf{x}' \\ &+ \int_{S_h} \delta F_1(\mathbf{x}')u_0(s\mathbf{x}')\mathrm{d}\mathbf{x}' + \int_{S_h} \delta F_2(\mathbf{x}')u_0(s\mathbf{x}')\mathrm{d}\mathbf{x}' \\ &+ \int_{S_h} \delta F_{3j}(\mathbf{x}')u_0(s\mathbf{x}')\mathrm{d}\mathbf{x}' + \delta_j(s).\end{aligned} \tag{8.3.22}$$

Recall that v can be approximated by the Herglotz wave functions v_j given in (8.3.13) in the H^1 norm. Then

$$\int_{\Gamma_h^\pm} \eta(\mathbf{x}')u_0(s\mathbf{x}')\mathscr{R}(v)(\mathbf{x}')\mathrm{d}\sigma = \int_{\Gamma_h^\pm} \eta(\mathbf{x}')u_0(s\mathbf{x}')\mathscr{R}(v_j)(\mathbf{x}')\mathrm{d}\sigma + \varepsilon_j^\pm(s), \tag{8.3.23}$$

$$\varepsilon_j^\pm(s) = \int_{\Gamma_h^\pm} \eta(\mathbf{x}')u_0(s\mathbf{x}')\mathscr{R}(v(\mathbf{x}',x_3) - v_j(\mathbf{x}',x_3))\mathrm{d}\sigma.$$

Plugging (8.3.22) and (8.3.23) into (8.3.6) in Lemma 8.3.2, we can obtain (8.3.16).

Recall that $F_{3j}(\mathbf{x}') = -k^2\mathscr{R}(v_j)(\mathbf{x}')$. Using the property of compact embedding of Hölder spaces, we can derive that for $0 < \alpha < 1$,

$$\|F_{3j}\|_{C^\alpha} \le k^2 \mathrm{diam}(S_h)^{1-\alpha}\|\mathscr{R}(v_j)\|_{C^1},$$

where $\mathrm{diam}(S_h)$ is the diameter of S_h. By the definition of the dimension reduction operator (8.3.1), it is easy to see that

$$|\mathscr{R}(v_j)(\mathbf{x}')| \le 4L\sqrt{\pi}\|\psi\|_{L^\infty}\|g_j\|_{L^2(\mathbb{S}^2)}, \quad |\partial_{\mathbf{x}'}\mathscr{R}(v_j)(\mathbf{x}')| \le 4kL\sqrt{\pi}\|\psi\|_{L^\infty}\|g_j\|_{L^2(\mathbb{S}^2)}.$$

Thus, we have

$$\|\mathscr{R}(v_j)\|_{C^1} \le 4L\sqrt{\pi}\|\psi\|_{L^\infty}(1+k)\|g_j\|_{L^2(\mathbb{S}^2)}.$$

Therefore, from (8.2.5) in Lemma 8.2.2, we have (8.3.19a). Using similar arguments, we can deduce (8.3.19b) and (8.3.19c).

Since $v - w \in H^2(S_h \times (-M,M))$, which implies that $\mathscr{R}(v-w) \in H^2(S_h)$ by using Lemma 8.3.1, from (8.2.24) in Lemma 8.2.6, we can prove (8.3.20).

The proof is complete. □

Lemma 8.3.4. *Under the same setup in Lemma 8.3.3, we assume that the transmission eigenfunction v to (8.3.2) can be approximated by a sequence of the Herglotz wave functions $\{v_j\}_{j=1}^{+\infty}$ with the form (8.3.13) in $H^1(S_h \times (-M,M))$ satisfying*

$$\|v - v_j\|_{H^1(S_h\times(-M,M))} \le j^{-1-\Upsilon}, \quad \|g_j\|_{L^2(\mathbb{S}^2)} \le Cj^{1+\rho}, \tag{8.3.24}$$

for some positive constant C, $\Upsilon > 0$ and $\rho > 0$. Let $\delta_j(s)$ and $\varepsilon_j^\pm(s)$ be defined in (8.3.17). Then we have the following estimate:

$$|\delta_j(s)| \le \frac{k^2\|\psi\|_{L^\infty}\sqrt{C(L,h)(\theta_M - \theta_m)}e^{-\sqrt{s\Theta}\delta_W h}}{\sqrt{2}} j^{-1-\Upsilon}. \tag{8.3.25}$$

Furthermore, assuming that the boundary parameter η in (8.3.2) fulfills $\eta(\mathbf{x}) \in C^\alpha(\overline{\Gamma_h^\pm} \times [-M,M])$, which indicates that $\eta(\mathbf{x}') \in C^\alpha(\overline{\Gamma_h^\pm})$ and has the expansion

$$\eta(\mathbf{x}') = \eta(\mathbf{0}) + \delta\eta(\mathbf{x}'), \quad |\delta\eta(\mathbf{x}')| \le \|\eta\|_{C^\alpha}|\mathbf{x}'|^\alpha, \tag{8.3.26}$$

then under the assumption (8.3.24), we have

$$|\varepsilon_j^\pm(s)| \le C\|\psi\|_{L^\infty}\left(|\eta(\mathbf{0})|\frac{\sqrt{\theta_M - \theta_m}e^{-\sqrt{s\Theta}\delta_W h}}{\sqrt{2}} + \|\eta\|_{C^\alpha}s^{-(\alpha+1)}\frac{\sqrt{2(\theta_M - \theta_m)\Gamma(4\alpha+4)}}{(2\delta_W)^{2\alpha+2}}\right) j^{-1-\Upsilon}. \tag{8.3.27}$$

Proof. We first prove (8.3.25). Indeed, by using the Cauchy-Schwarz inequality, we have

$$\begin{aligned}\|\mathscr{R}(v)-\mathscr{R}(v_j)\|^2_{L^2(S_h)} &= \int_{S_h}\left|\int_{-L}^{L}\psi(x_3)(v(\mathbf{x}',x_3)-v_j(\mathbf{x}',x_3))\mathrm{d}x_3\right|^2 \mathrm{d}\mathbf{x}'\\ &\le C(L,h)\|\psi\|^2_{L^\infty}\|v-v_j\|^2_{L^2(S_h\times(-L,L))},\end{aligned} \tag{8.3.28}$$

where $C(L,h)$ is a positive constant depending on L and h. Since the L^2 norm of u_0 in S_h can be estimated by (8.2.8) in Lemma 8.2.3, recalling that $\delta_j(s)$ is defined in (8.3.17), and using the Cauchy-Schwarz inequality again, by virtue of (8.3.24), we can deduce (8.3.25).

Note that $\varepsilon_j^{\pm}(s)$ is defined in (8.3.17) and η has the expansion (8.3.26). Using the Cauchy-Schwarz inequality and the trace theorem, we have

$$\begin{aligned}|\varepsilon_j^{\pm}(s)| &\le |\eta(\mathbf{0})|\int_{\Gamma_h^{\pm}}|u_0(s\mathbf{x}')||\mathscr{R}(v(\mathbf{x}',x_3)-v_j(\mathbf{x}',x_3))|\mathrm{d}\sigma\\ &\quad+\|\eta\|_{C^\alpha}\int_{\Gamma_h^{\pm}}|\mathbf{x}'|^\alpha|u_0(s\mathbf{x}')||\mathscr{R}(v(\mathbf{x}',x_3)-v_j(\mathbf{x}',x_3))|\mathrm{d}\sigma\\ &\le |\eta(\mathbf{0})|\|\mathscr{R}(v-v_j)\|_{H^{1/2}(\Gamma_h^{\pm})}\|u_0(s\mathbf{x}')\|_{H^{-1/2}(\Gamma_h^{\pm})}\\ &\quad+\|\eta\|_{C^\alpha}\|\mathscr{R}(v-v_j)\|_{H^{1/2}(\Gamma_h^{\pm})}\||\mathbf{x}'|^\alpha u_0(s\mathbf{x}')\|_{H^{-1/2}(\Gamma_h^{\pm})}\\ &\le |\eta(\mathbf{0})|\|\mathscr{R}(v-v_j)\|_{H^1(S_h)}\|u_0(s\mathbf{x}')\|_{L^2(S_h)}\\ &\quad+\|\eta\|_{C^\alpha}\|\mathscr{R}(v-v_j)\|_{H^1(S_h)}\||\mathbf{x}'|^\alpha u_0(s\mathbf{x}')\|_{L^2(S_h)}\\ &\le C\|\psi\|_{L^\infty}\|v-v_j\|_{H^1(S_h\times(-L,L))}(|\eta(\mathbf{0})|\|u_0(s\mathbf{x}')\|_{L^2(S_h)}+\|\eta\|_{C^\alpha}\||\mathbf{x}'|^\alpha u_0(s\mathbf{x}')\|_{L^2(S_h)}),\end{aligned} \tag{8.3.29}$$

where C is a positive constant and the last inequality comes from Lemma 8.3.1. Substituting (8.2.8), (8.2.9), and (8.3.24) into (8.3.29), we obtain (8.3.27). □

Lemma 8.3.5. *Let $j_\ell(t)$ be the ℓ-th spherical Bessel function with the form*

$$j_\ell(t)=\frac{t^\ell}{(2\ell+1)!!}\left(1-\sum_{l=1}^{\infty}\frac{(-1)^l t^{2l}}{2^l l! N_{\ell,l}}\right), \tag{8.3.30}$$

where $N_{\ell,l}=(2\ell+3)\cdots(2\ell+2l+1)$, and $\mathscr{R}$ be the dimension reduction operator defined in (8.3.1). Then

$$\mathscr{R}(j_0)(\mathbf{x}')=C(\psi)\left[1-\sum_{l=1}^{\infty}\frac{(-1)^l k^{2l}}{2^l l!(2l+1)!!}\left(|\mathbf{x}'|^2+a_{0,l}^2\right)^l\right], \tag{8.3.31a}$$

$$\mathscr{R}(j_\ell)(\mathbf{x}')=\frac{k^\ell(|\mathbf{x}'|^2+a_\ell^2)^{(\ell-1)/2}}{(2\ell+1)!!}\left[1-\sum_{l=1}^{\infty}\frac{(-1)^l k^{2l}(|\mathbf{x}'|^2+a_{\ell,l}^2)^l}{2^l l! N_{\ell,l}}\right]C_1(\psi)|\mathbf{x}'|^2, \tag{8.3.31b}$$

where $j_0=j_0(k|\mathbf{x}|)$, $j_\ell=j_\ell(k|\mathbf{x}|)$, $\ell\in\mathbb{N}$, $a_{0,l}\in[-L,L]$, a_ℓ, $a_{\ell,l}\in[-L,L]$, and

$$C(\psi)=\int_{-L}^{L}\psi(x_3)\mathrm{d}x_3,\quad C_1(\psi)=\int_{-\arctan L/|\mathbf{x}'|}^{\arctan L/|\mathbf{x}'|}\psi(|\mathbf{x}'|\tan\varpi)\sec^3\varpi\mathrm{d}\varpi. \tag{8.3.32}$$

Furthermore, there holds that

$$0<C_1(\psi)<2^{5/2}\|\psi\|_{L^\infty}\arctan L. \tag{8.3.33}$$

Proof. From the definition of the dimension reduction operator (8.3.1) and the integral mean value theorem, we know that

$$\begin{aligned}\mathscr{R}(j_0)(\mathbf{x}') &= \int_{-L}^{L} \psi(x_3) j_0(k|\mathbf{x}|)\mathrm{d}x_3 \\ &= \int_{-L}^{L} \psi(x_3)\mathrm{d}x_3 - \sum_{l=1}^{\infty} \frac{(-1)^l k^{2l}}{2^l l!(2l+1)!!} \int_{-L}^{L} \psi(x_3)\left(|\mathbf{x}'|^2 + x_3^2\right)^l \mathrm{d}x_3,\end{aligned}$$

from which we obtain (8.3.31a).

For $\mathscr{R}(j_\ell)(\mathbf{x}') = \int_{-L}^{L} \psi(x_3) j_\ell(k|\mathbf{x}|)\mathrm{d}x_3$, using the integral mean value theorem, we can deduce that for $\ell \in \mathbb{N}$,

$$\begin{aligned}\int_{-L}^{L} \psi(x_3)(|\mathbf{x}'|^2 + x_3^2)^{\ell/2}\mathrm{d}x_3 &= (|\mathbf{x}'|^2 + a_\ell^2)^{(\ell-1)/2} \int_{-L}^{L} \psi(x_3)(|\mathbf{x}'|^2 + x_3^2)^{1/2}\mathrm{d}x_3 \\ &= C_1(\psi)|\mathbf{x}'|^2(|\mathbf{x}'|^2 + a_\ell^2)^{(\ell-1)/2},\end{aligned} \tag{8.3.34}$$

where $a_\ell \in [-L, L]$. Thus, for $\ell \in \mathbb{N}$, from (8.3.30), we have

$$\begin{aligned}\mathscr{R}(j_\ell)(\mathbf{x}') &= \int_{-L}^{L} \psi(x_3) j_\ell(k|\mathbf{x}|)\mathrm{d}x_3 \\ &= \frac{k^\ell}{(2\ell+1)!!} \int_{-L}^{L} \psi(x_3)(|\mathbf{x}'|^2 + x_3^2)^{\ell/2} \left(1 - \sum_{l=1}^{\infty} \frac{(-1)^l k^{2l}(|\mathbf{x}'|^2 + x_3^2)^l}{2^l l! N_{\ell,l}}\right)\mathrm{d}x_3.\end{aligned} \tag{8.3.35}$$

Substituting (8.3.34) into (8.3.35), we can obtain (8.3.31b).

Clearly, if $L < |\mathbf{x}'|$, we know that $0 < \sec\varpi < \sqrt{\frac{L^2}{|\mathbf{x}'|^2} + 1}$, where

$$\varpi \in [-\arctan L/|\mathbf{x}'|, \arctan L/|\mathbf{x}'|].$$

Therefore, we can deduce (8.3.33). □

Using the Jacobi-Anger expansion (cf. [22, Page 75]), for the Herglotz wave function v_j given in (8.3.13), we have

$$v_j(\mathbf{x}) = v_j(\mathbf{0}) j_0(k|\mathbf{x}|) + \sum_{\ell=1}^{\infty} \gamma_{\ell j} i^\ell (2\ell+1) j_\ell(k|\mathbf{x}|), \quad \mathbf{x} \in \mathbb{R}^3, \tag{8.3.36}$$

where

$$v_j(\mathbf{0}) = \int_{\mathbb{S}^2} g_j(d)\mathrm{d}\sigma(d), \quad \gamma_{\ell j} = \int_{\mathbb{S}^2} g_j(\mathbf{d}) P_\ell(\cos(\varphi))\mathrm{d}\sigma(\mathbf{d}), \quad \mathbf{d} \in \mathbb{S}^2,$$

and $j_\ell(t)$ is the ℓ-th spherical Bessel function [1] and φ is the angle between $\mathbf{x}$ and $\mathbf{d}$.

In the next lemma, we characterize the integrals $I_2^\pm$ defined by (8.3.17), which shall play a critical role in the proof of our main Theorem 8.3.1 in what follows.

Lemma 8.3.6. *Let $\Gamma_h^\pm$ be defined in* (8.2.2) *and $u_0(s\mathbf{x})$ be the CGO solution defined in* (8.2.4) *with the parameter $s \in \mathbb{R}_+$ and $I_2^\pm$ be defined by* (8.3.17)*. Recall that the Herglotz wave function v_j is given in the form* (8.3.13)*. Suppose that $\eta(\mathbf{x}) \in C^\alpha(\overline{\Gamma_h^\pm} \times [-M, M])$ $(0 < \alpha < 1)$ satisfying* (8.3.26)*, and let*

$$
\begin{aligned}
I_{\eta,1}^{-} &= \int_{\Gamma_h^-} \delta\eta(\mathbf{x}')u_0(s\mathbf{x}')\mathscr{R}(j_0)(\mathbf{x}')\mathrm{d}\sigma, \quad I_{\eta,1}^{+} = \int_{\Gamma_h^+} \delta\eta(\mathbf{x}')u_0(s\mathbf{x}')\mathscr{R}(j_0)(\mathbf{x}')\mathrm{d}\sigma, \\
I_{\eta,2}^{-} &= \sum_{\ell=1}^{\infty} \gamma_{\ell j} i^{\ell}(2\ell+1) \int_{\Gamma_h^-} \delta\eta(\mathbf{x}')u_0(s\mathbf{x}')\mathscr{R}(j_\ell)(\mathbf{x}')\mathrm{d}\sigma, \\
I_{\eta,2}^{+} &= \sum_{\ell=1}^{\infty} \gamma_{\ell j} i^{\ell}(2\ell+1) \int_{\Gamma_h^+} \delta\eta(\mathbf{x}')u_0(s\mathbf{x}')\mathscr{R}(j_\ell)(\mathbf{x}')\mathrm{d}\sigma, \\
I_{\eta}^{-} &= v_j(\mathbf{0})I_{\eta,1}^{-} + I_{\eta,2}^{-}, \quad I_{\eta}^{+} = v_j(\mathbf{0})I_{\eta,1}^{+} + I_{\eta,2}^{+}.
\end{aligned}
\tag{8.3.37}
$$

Assume that for a fixed $k \in \mathbb{R}_+$, h is sufficiently small such that $k^2(h^2+L^2) < 1$ and

$$kL < 1, \tag{8.3.38}$$

where $2L$ is the length of the interval of the dimension reduction operator $\mathscr{R}$ in (8.3.1), *and $-\pi < \theta_m < \theta_M < \pi$, where θ_m and θ_M are defined in* (8.2.1). *Then*

$$
\begin{aligned}
I_2^{-} &= 2\eta(\mathbf{0})v_j(\mathbf{0})s^{-1}\left(\mu(\theta_m)^{-2} - \mu(\theta_m)^{-2}e^{-\sqrt{sh}\mu(\theta_m)} - \mu(\theta_m)^{-1}\sqrt{sh}e^{-\sqrt{sh}\mu(\theta_m)}\right)C(I_{311}^{-}) \\
&\quad + v_j(\mathbf{0})\eta(\mathbf{0})I_{312}^{-} + \eta(\mathbf{0})I_{32}^{-} + I_{\eta}^{-},
\end{aligned}
\tag{8.3.39}
$$

where $\mu(\theta_m)$ is defined in (8.2.41),

$$
\begin{aligned}
C(I_{311}^{-}) &= C(\psi)\left[1 - \sum_{l=1}^{\infty} \frac{(-1)^l k^{2l}}{(2l+1)!!} a_{0,l}^{2l}\right], \\
I_{32}^{-} &= \sum_{\ell=1}^{\infty} \gamma_{\ell j} i^{\ell}(2\ell+1) \int_{\Gamma_h^-} u_0(s\mathbf{x}')\mathscr{R}(j_\ell)(\mathbf{x}')\mathrm{d}\sigma, \\
I_{312}^{-} &= -C(\psi)\sum_{l=1}^{\infty} \frac{(-1)^l k^{2l}}{(2l+1)!!}\left(\sum_{i_1=1}^{l} C(l,i_1)a_{0,l}^{2(l-i_1)} \int_0^h r^{2i_1} e^{-\sqrt{sr}\mu(\theta_m)}\mathrm{d}r\right),
\end{aligned}
\tag{8.3.40}
$$

with $C(l,i_1) = \frac{l!}{i_1!(l-i_1)!}$ being the combinatorial number of the order l. Here, $a_{0,l}$, $\mathscr{R}(j_\ell)$, and $C(\psi)$ are defined in (8.3.31a), (8.3.31b), *and* (8.3.32), *respectively. It holds as $s \to +\infty$ that*

$$I_{312}^{-} \leq \mathscr{O}(s^{-3}), \tag{8.3.41a}$$

$$I_{32}^{-} \leq \mathscr{O}(\|g_j\|_{L^2(\mathbb{S}^2)} s^{-3}), \tag{8.3.41b}$$

$$\left|I_{\eta,1}^{-}\right| \leq \mathscr{O}(s^{-1-\alpha}), \tag{8.3.41c}$$

$$\left|I_{\eta,2}^{-}\right| \leq \mathscr{O}(\|g_j\|_{L^2(\mathbb{S}^2)} s^{-3-\alpha}), \tag{8.3.41d}$$

$$|I_{\eta}^{-}| \leq \|\eta\|_{C^\alpha}\left(|v_j(\mathbf{0})||I_{\eta,1}^{-}| + |I_{\eta,2}^{-}|\right). \tag{8.3.41e}$$

Similarly, we have

$$
\begin{aligned}
I_2^{+} &= 2\eta(\mathbf{0})v_j(\mathbf{0})s^{-1}\left(\mu(\theta_M)^{-2} - \mu(\theta_M)^{-2}e^{-\sqrt{sh}\mu(\theta_M)} - \mu(\theta_M)^{-1}\sqrt{sh}e^{-\sqrt{sh}\mu(\theta_M)}\right)C(I_{311}^{+}) \\
&\quad + v_j(\mathbf{0})\eta(\mathbf{0})I_{312}^{+} + \eta(\mathbf{0})I_{32}^{+} + I_{\eta}^{+},
\end{aligned}
\tag{8.3.42}
$$

where

$$C(I_{311}^{+}) = C(\psi)\left[1-\sum_{l=1}^{\infty}\frac{(-1)^l k^{2l}}{(2l+1)!!}a_{0,l,+}^{2l}\right], \quad a_{0,l,+}\in[-L,L],$$

$$I_{312}^{+} = -C(\psi)\sum_{l=1}^{\infty}\frac{(-1)^l k^{2l}}{(2l+1)!!}\sum_{i_1=1}^{l}C(l,i_1)a_{0,l,+}^{2(l-i_1)}\int_0^h r^{2i_1}e^{-\sqrt{sr}\mu(\theta_M)}\mathrm{d}r, \tag{8.3.43}$$

$$I_{32}^{+} = \sum_{\ell=1}^{\infty}\gamma_{\ell j}i^{\ell}(2\ell+1)\int_{\Gamma_h^+}u_0(s\mathbf{x}')\mathscr{R}(j_\ell)(\mathbf{x}')\mathrm{d}\sigma.$$

There hold as $s\to+\infty$ *that*

$$\begin{aligned}
&|I_{312}^{+}|\le\mathscr{O}(s^{-3}),\quad |I_{32}^{+}|\le\mathscr{O}(\|g_j\|_{L^2(\mathbb{S}^2)}s^{-3}),\\
&\left|I_{\eta,1}^{+}\right|\le\mathscr{O}(s^{-1-\alpha}),\quad \left|I_{\eta,2}^{+}\right|\le\mathscr{O}(\|g_j\|_{L^2(\mathbb{S}^2)}s^{-3-\alpha}),\\
&|I_{\eta}^{+}|\le\|\eta\|_{C^\alpha}\left(|v_j(0)||I_{\eta,1}^{+}|+|I_{\eta,2}^{+}|\right).
\end{aligned}\tag{8.3.44}$$

Proof. We first investigate the boundary integral I_2^- which is given by (8.3.17). In this situation, the polar coordinates $\mathbf{x}=(r\cos\theta, r\sin\theta)$ satisfy $r\in(0,h)$ and $\theta=\theta_m$ or $\theta=\theta_M$ when $\mathbf{x}\in\Gamma_h^-$ or $\mathbf{x}\in\Gamma_h^+$, respectively. Since $\eta\in C^\alpha(\overline{\Gamma_h^{\pm}}\times[-M,M])$, we have the expansion (8.3.26). Substituting (8.3.26) into the expression of I_2^-, we have

$$I_2^- = \eta(\mathbf{0})I_{21}^- + I_\eta^-, \tag{8.3.45}$$

where

$$I_{21}^- = \int_{\Gamma_h^-}u_0(s\mathbf{x}')\mathscr{R}(v_j)(\mathbf{x}')\mathrm{d}\sigma,\quad I_\eta^- = \int_{\Gamma_h^-}\delta\eta(\mathbf{x}')u_0(s\mathbf{x}')\mathscr{R}(v_j)(\mathbf{x}')\mathrm{d}\sigma.$$

By virtue of (8.3.36), it can be verified that $I_\eta^- = v_j(\mathbf{0})I_{\eta,1}^- + I_{\eta,2}^-$.

Let

$$I_{31}^- = \int_{\Gamma_h^-}u_0(s\mathbf{x}')\mathscr{R}(j_0)(\mathbf{x}')\mathrm{d}\sigma.$$

Substituting (8.3.31b) and (8.3.36) into the expression of I_{21}^- defined in (8.3.45), we can deduce that

$$I_{21}^- = v_j(0)I_{31}^- + I_{32}^-, \tag{8.3.46}$$

where I_{32}^- is defined in (8.3.40). Substituting the expansion (8.3.31a) into I_{31}^- and recalling that $\mu(\theta) = -\cos(\theta/2+\pi) - \mathrm{i}\sin(\theta/2+\pi)$, we have

$$\begin{aligned}
I_{31}^- &= C(\psi)\int_0^h\left[1-\sum_{l=1}^{\infty}\frac{(-1)^l k^{2l}}{(2l+1)!!}(r^2+a_{0,l}^2)^l\right]e^{-\sqrt{sr}\mu(\theta_m)}\mathrm{d}r := I_{311}^- + I_{312}^-,\\
I_{311}^- &= C(\psi)\left[1-\sum_{l=1}^{\infty}\frac{(-1)^l k^{2l}}{(2l+1)!!}a_{0,l}^{2l}\right]\int_0^h e^{-\sqrt{sr}\mu(\theta_m)}\mathrm{d}r,
\end{aligned}\tag{8.3.47}$$

where $a_{0,l}$, $C(\psi)$, and I_{312}^- are defined in (8.3.31a), (8.3.32), and (8.3.40), respectively. Moreover, using (8.2.38) in Lemma 8.2.8, we obtain that

$$I_{311}^- = 2s^{-1}\left(\mu(\theta_m)^{-2} - \mu(\theta_m)^{-2}e^{-\sqrt{sh}\mu(\theta_m)} - \mu(\theta_m)^{-1}\sqrt{sh}e^{-\sqrt{sh}\mu(\theta_m)}\right)C(I_{311}^-), \tag{8.3.48}$$

where $C(I_{311}^-) = C(\psi)\left[1 - \sum_{l=1}^{\infty} \frac{(-1)^l k^{2l}}{(2l+1)!!} a_{0,l}^{2l}\right]$.

Finally, substituting (8.3.46), (8.3.47), and (8.3.48) into (8.3.45), we have the integral equality (8.3.39).

Following a similar argument for deriving the integral equality (8.3.39) of I_2^-, one can derive the integral equality (8.3.42) for I_2^+.

In the following, we derive the estimate for I_η^- in (8.3.41e) by investigating (8.3.41c) and (8.3.41d). Substituting (8.3.31a) into $I_{\eta,1}^-$, we can derive that

$$\begin{aligned}|I_{\eta,1}^-| &\leq |C(\psi)| \|\eta\|_{C^\alpha} \int_0^h r^\alpha \left| 1 - \sum_{l=1}^{\infty} \frac{(-1)^l k^{2l}}{(2l+1)!!} \left(r^2 + a_{0,l}^2\right)^l \right| e^{-\sqrt{sr}\omega(\theta_m)} \mathrm{d}r \\ &= 2L\|\psi\|_{L^\infty} \|\eta\|_{C^\alpha} \left| 1 - \sum_{l=1}^{\infty} \frac{(-1)^l k^{2l}}{(2l+1)!!} \left(\beta_{0,l}^2 + a_{0,l}^2\right)^l \right| \int_0^h r^\alpha e^{-\sqrt{sr}\omega(\theta_m)} \mathrm{d}r,\end{aligned}$$

where $\beta_{0,l} \in [0,h]$ such that $k^2(\beta_{0,l}^2 + a_{0,l}^2) \leq k^2(h^2+L^2) < 1$ for sufficiently small h and L. From (8.2.10) in Lemma 8.2.2, we obtain (8.3.41c). Substituting (8.3.31b) into $I_{\eta,2}^-$, and using (8.3.33), we can deduce that

$$\begin{aligned}I_{\eta,2}^- &\leq C_1(\psi)\|\eta\|_{C^\alpha} \\ &\quad \times \sum_{\ell=1}^{\infty} |\gamma_{\ell j}| \int_0^h r^\alpha \frac{k^\ell (r^2 + a_\ell^2)^{(\ell-1)/2}}{(2\ell-1)!!} \left| 1 - \sum_{l=1}^{\infty} \frac{k^{2l}(r^2 + a_{\ell,l}^2)^l}{2^l l! N_{\ell,l}} \right| r^2 e^{-\sqrt{sr}\omega(\theta_m)} \mathrm{d}r \\ &\leq C_1(\psi)\|\eta\|_{C^\alpha} \|g_j\|_{L^2(\mathbb{S}^2)} \int_0^h r^{2+\alpha} e^{-\sqrt{sr}\omega(\theta_m)} \mathrm{d}r \\ &\quad \times \sum_{\ell=1}^{\infty} \frac{k^\ell (\beta_\ell^2 + a_\ell^2)^{(\ell-1)/2}}{(2\ell-1)!!} \left| 1 - \sum_{l=1}^{\infty} \frac{k^{2l}(\beta_{\ell,l}^2 + a_{\ell,l}^2)^l}{2^l l! N_{\ell,l}} \right| \\ &= \mathcal{O}(\|g_j\|_{L^2(\mathbb{S}^2)} s^{-\alpha-3}),\end{aligned}$$

where $\beta_\ell, \beta_{\ell,l} \in [0,h]$ such that $k^2(\beta_\ell^2 + a_\ell^2) \leq k^2(h^2+L^2) < 1$ and $k^2(\beta_{\ell,l}^2 + a_{\ell,l}^2) \leq k^2(h^2+L^2) < 1$ for sufficiently small h and L, by utilizing the claim that

$$|\gamma_{\ell j}| \leq \|g_j\|_{L^2(\mathbb{S}^2)},$$

where we use the fact that $|P_\ell(t)| \leq 1$ when $|t| \leq 1$. Consequently, (8.3.41e) can be derived.

For I_{312}^-, we can deduce that

$$\begin{aligned}|I_{312}^-| &\leq |C(\psi)| \sum_{l=1}^{\infty} \frac{k^{2l}}{(2l+1)!!} \sum_{i_1=1}^{l} C(l,i_1) h^{2(i_1-1)} L^{2(l-i_1)} \int_0^h r^2 e^{-\sqrt{sr}\omega(\theta_m)} \mathrm{d}r \\ &= |C(\psi)| \sum_{l=1}^{\infty} \frac{k^{2l}}{(2l+1)!! h^2} \sum_{i_1=1}^{l} C(l,i_1) h^{2i_1} L^{2(l-i_1)} \int_0^h r^2 e^{-\sqrt{sr}\omega(\theta_m)} \mathrm{d}r \\ &= |C(\psi)| \sum_{l=1}^{\infty} \frac{k^{2l}}{(2l+1)!! h^2} ((h^2+L^2)^l - L^{2l}) \int_0^h r^2 e^{-\sqrt{sr}\omega(\theta_m)} \mathrm{d}r \\ &\leq 2L\|\psi\|_{L^\infty} \sum_{l=1}^{\infty} \frac{k^{2l}}{(2l+1)!! h^2} ((h^2+L^2)^l - L^{2l}) \cdot \mathcal{O}(s^{-3}) \\ &- \mathcal{O}(s^{-3}),\end{aligned}$$

where we choose h and L such that $k^2(h^2+L^2) < 1$ and $kL < 1$.

Substituting the expansion (8.3.31b) of j_ℓ into I_{32}^-, we have

$$
\begin{aligned}
|I_{32}^-| &\le C_1(\psi)\|g_j\|_{L^2(\mathbb{S}^2)} \\
&\quad \times \sum_{\ell=1}^{\infty}\int_0^h r^2 e^{-\sqrt{s}r\omega(\theta_m)}\frac{k^\ell(|r|^2+a_\ell^2)^{(\ell-1)/2}}{(2\ell-1)!!}\left|1-\sum_{l=1}^{\infty}\frac{(-1)^l k^{2l}(|r|^2+a_{\ell,l}^2)^l}{2^l l! N_{\ell,l}}\right|\mathrm{d}r \\
&= C_1(\psi)\|g_j\|_{L^2(\mathbb{S}^2)}\sum_{\ell=1}^{\infty}\frac{k^\ell(|\beta_\ell|^2+a_\ell^2)^{(\ell-1)/2}}{(2\ell-1)!!}\left|1-\sum_{l=1}^{\infty}\frac{(-1)^l k^{2l}(|\beta_{\ell,l}|^2+a_{\ell,l}^2)^l}{2^l l! N_{\ell,l}}\right| \\
&\quad \times \int_0^h r^2 e^{-\sqrt{s}r\omega(\theta_m)}\mathrm{d}r \\
&= \mathcal{O}(\|g_j\|_{L^2(\mathbb{S}^2)}s^{-3}),
\end{aligned}
$$

where $\beta_\ell, \beta_{\ell,l}\in[0,h]$ such that $k^2(\beta_\ell^2+a_\ell^2)\le k^2(h^2+L^2)<1$ and $k^2(|\beta_{\ell,l}|^2+a_{\ell,l}^2)\le k^2(h^2+L^2)<1$ for sufficiently small h and L.

The asymptotic analysis (8.3.44) for I_{312}^+, I_{32}^+, $I_{\eta,1}^+$, and $I_{\eta,2}^+$ can be analyzed in a similar way, which is omitted. □

In the next proposition, we deduce the lower and upper bounds for $C(I_{311}^-)$ and $C(I_{311}^+)$, where $C(I_{311}^-)$ and $C(I_{311}^+)$ are defined in (8.3.40) and (8.3.43), respectively, which shall be used to prove Lemma 8.3.7 in the following.

Proposition 8.3.1. *Let $C(I_{311}^-)$ and $C(I_{311}^+)$ be defined in* (8.3.40) *and* (8.3.43), *respectively. Assuming that the condition* (8.3.38) *is fulfilled for a sufficiently small $L\in\mathbb{R}_+$, then*

$$
0<C(I_{311}^-)\le\frac{C(\psi)}{1-(kL)^2},\quad 0<C(I_{311}^+)\le\frac{C(\psi)}{1-(kL)^2}, \tag{8.3.49}
$$

where $C(\psi)$ is defined in (8.3.51).

Proof. Recall that $|a_{0,l}|\le L$ is defined in (8.3.31a). By (8.3.38), we know that

$$
\left|\sum_{l=1}^{\infty}\frac{(-1)^l k^{2l}}{(2l+1)!!}a_{0,l}^{2l}\right|\le\sum_{l=1}^{\infty}(kL)^{2l}=\frac{(kL)^2}{1-(kL)^2}. \tag{8.3.50}
$$

From (8.3.50), we know that

$$
0<\frac{C(\psi)(1-2(kL)^2)}{1-(kL)^2}\le C(I_{311}^-)\le\frac{C(\psi)}{1-(kL)^2}, \tag{8.3.51}
$$

where $C(\psi)=\int_{-L}^{L}\psi(x_3)\mathrm{d}x_3>0$, since $\psi\not\equiv 0$ is a nonnegative function.

For sufficiently small L, similar to (8.3.51), we know that

$$
0<\frac{C(\psi)(1-2(kL)^2)}{1-(kL)^2}\le C(I_{311}^+)\le\frac{C(\psi)}{1-(kL)^2}, \tag{8.3.52}
$$

where $C(\psi)$ is defined in (8.3.51). □

Lemma 8.3.7. *Let θ_m and θ_M be defined in* (8.2.2). *Assuming that θ_m and θ_M fulfil the condition* (8.2.55), *and moreover* (8.3.38) *is satisfied for a sufficiently small $L\in\mathbb{R}_+$, then*

$$
C(I_{311}^-)\mu(\theta_m)^{-2}+C(I_{311}^+)\mu(\theta_M)^{-2}\neq 0, \tag{8.3.53}
$$

where $C(I_{311}^-)$ and $C(I_{311}^+)$ are defined in (8.3.40) *and* (8.3.43), *respectively.*

Proof. It can be calculated that

$$\begin{aligned}
&C(I_{311}^-)\mu(\theta_m)^{-2}+C(I_{311}^+)\mu(\theta_M)^{-2}\\
&=\frac{(C(I_{311}^+)\cos\theta_m+C(I_{311}^-)\cos\theta_M)+\mathrm{i}(C(I_{311}^+)\sin\theta_m+C(I_{311}^-)\sin\theta_M)}{(\cos\theta_m+\mathrm{i}\sin\theta_m)(\cos\theta_M+\mathrm{i}\sin\theta_M)}.
\end{aligned}$$

Therefore, under the assumption (8.2.55), we know that

$$\cos\theta_m+\cos\theta_M \text{ and } \sin\theta_m+\sin\theta_M$$

cannot be zero simultaneously. Without loss of generality, we assume that $\cos\theta_m+\cos\theta_M\neq 0$. Then we consider the following two cases:

(i) Case A: $\cos\theta_m+\cos\theta_M>0$,
(ii) Case B: $\cos\theta_m+\cos\theta_M<0$.

For Case A, we first consider that $\cos\theta_m$ and $\cos\theta_M$ have the same sign. From (8.3.49) in Proposition 8.3.1, it is not difficult to see that the real part of the numerator of $C(I_{311}^-)\mu(\theta_m)^{-2}+C(I_{311}^+)\mu(\theta_M)^{-2}$ cannot be zero. Therefore, (8.3.53) is proved when $\cos\theta_m$ and $\cos\theta_M$ have the same sign.

In the following, we assume that $\cos\theta_m$ and $\cos\theta_M$ have different signs. Then it implies that $\cos\theta_m\leq 0$ and $\cos\theta_M>0$. From (8.3.51) and (8.3.52), we can deduce that

$$\begin{aligned}
\frac{C(\psi)}{1-(kL)^2}(\cos\theta_m+(1-2(kL)^2)\cos\theta_M)&\leq C(I_{311}^+)\cos\theta_m+C(I_{311}^-)\cos\theta_M\\
&\leq\frac{C(\psi)}{1-(kL)^2}((1-2(kL)^2)\cos\theta_m+\cos\theta_M).
\end{aligned}$$

Since L is flexible, for a given $0<\varepsilon<1$, we can choose L such that $0<kL<\sqrt{\varepsilon/2}$, from which we can derive the bounds as follows:

$$\begin{aligned}
\frac{C(\psi)}{1-(kL)^2}(\cos\theta_m+(1-\varepsilon)\cos\theta_M)&\leq C(I_{311}^+)\cos\theta_m+C(I_{311}^-)\cos\theta_M\\
&\leq\frac{C(\psi)}{1-(kL)^2}((1-\varepsilon)\cos\theta_m+\cos\theta_M).
\end{aligned}\tag{8.3.54}$$

Since $\cos\theta_m+\cos\theta_M>0$, we can consider the lower bound in (8.3.54). Denote $\varepsilon_0=\min\{\frac{\cos\theta_m+\cos\theta_M}{2\cos\theta_M},1\}$, and choose $\varepsilon\in(0,\varepsilon_0)$. It can be verified that

$$C(I_{311}^+)\cos\theta_m+C(I_{311}^-)\cos\theta_M\geq\frac{C(\psi)}{1-(kL)^2}(\cos\theta_m+(1-\varepsilon)\cos\theta_M)>0,$$

which means that (8.3.53) still holds.

For Case B, if $\cos\theta_m=0$ or $\cos\theta_M=0$ is satisfied, from the upper bound of (8.3.54), we can easily show that

$$C(I_{311}^+)\cos\theta_m+C(I_{311}^-)\cos\theta_M<0.\tag{8.3.55}$$

Otherwise, if $|\cos\theta_m|\leq|\cos\theta_M|$, from the fact that $(1-\varepsilon)|\cos\theta_m|\leq|\cos\theta_M|$, we know that (8.3.55) still holds from the upper bound of (8.3.54). If $|\cos\theta_m|>|\cos\theta_M|$, we can choose ε such that $\varepsilon>1-|\cos\theta_M|/|\cos\theta_m|>0$ to make (8.3.55) also be fulfilled from the upper bound of (8.3.54). Therefore, for Case B, we know that (8.3.53) is fulfilled.

The proof is complete. □

We are in a position to present another main result of this chapter on the vanishing of conductive transmission eigenfunctions at an edge corner in 3D.

Theorem 8.3.1. *Let* $\Omega \Subset \mathbb{R}^3$ *be a bounded Lipschitz domain with* $\mathbf{0} \in \partial\Omega$ *and* $S_h \subset \mathbb{R}^2$ *be defined in* (8.2.2), $M > 0$, $0 < \alpha < 1$. *For any fixed* $x_3^c \in (-M, M)$ *and* $L > 0$ *defined in Definition 8.3.1, we suppose that L is sufficiently small such that* $(x_3^c - L, x_3^c + L) \subset (-M, M)$ *and*

$$(B_h \times (-M,M)) \cap \Omega = S_h \times (-M,M),$$

where $B_h \Subset \mathbb{R}^2$ *is the central ball of radius* $h \in \mathbb{R}_+$. *Assume that* $v, w \in H^1(\Omega)$ *are the transmission eigenfunctions to* (8.1.1) *and there exists a sufficiently small neighborhood* $S_h \times (-M,M)$ *(i.e.,* $h > 0$ *is sufficiently small) of* $(\mathbf{0}, x_3^c)$ *with* $x_3^c \in (-M,M)$ *such that* $qw \in C^\alpha(\overline{S}_h \times [-M,M])$ *and* $\eta \in C^\alpha(\overline{\Gamma}_h^\pm \times [-M,M])$ *for* $0 < \alpha < 1$ *and* $v - w \in H^2(S_h \times (-M,M))$, *where* $\mathbf{0}$ *is the vertex of* S_h *defined in* (8.2.2). *Write* $\mathbf{x} = (\mathbf{x}', x_3) \in \mathbb{R}^3$, $\mathbf{x}' \in \mathbb{R}^2$. *If the following conditions are fulfilled:*

(a) the transmission eigenfunction v can be approximated in $H^1(S_h \times (-M,M))$ *by the Herglotz functions* v_j, $j = 1, 2, \ldots$, *with kernels* g_j *satisfying the approximation property* (8.3.24), *where the parameter* ρ *in* (8.3.24) *fulfills that* $0 < \rho < \alpha$,

(b) the function $\eta = \eta(\mathbf{x}')$ *is independent of* x_3 *and*

$$\eta(\mathbf{0}) \neq 0, \tag{8.3.56}$$

(c) the angles θ_m *and* θ_M *of* S_h *satisfy*

$$-\pi < \theta_m < \theta_M < \pi \text{ and } \theta_M - \theta_m \neq \pi, \tag{8.3.57}$$

then for every edge point $(\mathbf{0}, x_3^c) \in \mathbb{R}^3$ *of* $S_h \times (-M,M)$ *where* $x_3^c \in (-M,M)$, *one has*

$$\lim_{\rho \to +0} \frac{1}{B((\mathbf{0}, x_3^c), \rho) \cap \Omega} \int_{m(B((\mathbf{0}, x_3^c), \rho) \cap \Omega)} |v(\mathbf{x})| \mathrm{d}\mathbf{x} = 0,$$

where $m(B((\mathbf{0}, x_3^c), \rho) \cap \Omega)$ *is the volume of* $B((\mathbf{0}, x_3^c), \rho) \cap \Omega$.

Proof. One can easily see that the transmission eigenfunctions $v, w \in H^1(\Omega)$ to (8.1.1) fulfill the PDE system (8.3.2). Using the dimension reduction operator $\mathscr{R}$ given by (8.3.1), we can show that $\mathscr{R}(v)$ and $\mathscr{R}(w)$ satisfy (8.3.4). By virtue of Lemma 8.3.2 and Lemma 8.3.3, we know that the integral equality (8.3.16) holds under the assumption that $qw \in C^\alpha(\overline{S}_h \times [-M,M])$ for $0 < \alpha < 1$ and $v - w \in H^2(S_h \times (-M,M))$. Since $\eta \in C^\alpha(\overline{\Gamma}_h^\pm \times [-M,M])$ for $0 < \alpha < 1$, from Lemma 8.3.6, we can obtain (8.3.39) and (8.3.42). Recall that $\mu(\theta_m)$ and $\mu(\theta_M)$ are defined in (8.2.41). Therefore, substituting (8.3.39) and (8.3.42) into (8.3.16), after rearranging terms and multiplying s on both sides of (8.3.16), we deduce that

$$\begin{aligned}
&2v_j(\mathbf{0})\eta(\mathbf{0})\Big[\Big(\mu(\theta_M)^{-2} - \mu(\theta_M)^{-2}e^{-\sqrt{sh}\mu(\theta_M)} - \mu(\theta_M)^{-1}\sqrt{sh}e^{-\sqrt{sh}\mu(\theta_M)}\Big)C(I_{311}^+) \\
&\quad + \Big(\mu(\theta_m)^{-2} - \mu(\theta_m)^{-2}e^{-\sqrt{sh}\mu(\theta_m)} - \mu(\theta_m)^{-1}\sqrt{sh}e^{-\sqrt{sh}\mu(\theta_m)}\Big)C(I_{311}^-)\Big] \\
&= s\Big[I_3 - (F_1(\mathbf{0}) + F_2(\mathbf{0}) + F_{3j}(\mathbf{0}))\int_{S_h} u_0(s\mathbf{x}')\mathrm{d}\mathbf{x}' - \delta_j(s) \\
&\quad - \eta(\mathbf{0})(I_{32}^+ + I_{32}^-) - I_\eta^+ - I_\eta^- - \int_{S_h} \delta F_1(\mathbf{x}')u_0(s\mathbf{x}')\mathrm{d}\mathbf{x}' - \int_{S_h} \delta F_2(\mathbf{x}')u_0(s\mathbf{x}')\mathrm{d}\mathbf{x}' \\
&\quad - \int_{S_h} \delta F_{3j}(\mathbf{x}')u_0(s\mathbf{x}')\mathrm{d}\mathbf{x}' - v_j(\mathbf{0})\eta(0)\left(I_{312}^- + I_{312}^+\right) - \varepsilon_j^\pm(s)\Big],
\end{aligned} \tag{8.3.58}$$

where $C(I_{311}^-)$ and $C(I_{311}^+)$ are defined in (8.3.40) and (8.3.43), respectively. Here, δF_{3j}, δF_1, and δF_2 are given by (8.3.14) and (8.3.15).

When $s = j$, under the assumption (8.3.24), using (8.3.41a), (8.3.41b), (8.3.41e), and (8.3.44) in Lemma 8.3.6, we know that

$$
\begin{aligned}
&j|I_{32}^-| \le \mathscr{O}(j^{-2}\|g_j\|_{L^2(\mathbb{S}^2)}) \le \mathscr{O}(j^{-1+\rho}), \quad j|I_{32}^+| \le \mathscr{O}(j^{-2}\|g_j\|_{L^2(\mathbb{S}^2)}) \le \mathscr{O}(j^{-1+\rho}),\\
&jI_{312}^- \le \mathscr{O}(j^{-2}), \quad jI_{312}^+ \le \mathscr{O}(j^{-2}),
\end{aligned} \tag{8.3.59}
$$

and

$$
\begin{aligned}
j|I_\eta^-| &\le \|\eta\|_{C^\alpha}\left(|v_j(\mathbf{0})|\mathscr{O}(j^{-\alpha}) + \mathscr{O}(\|g_j\|_{L^2(\mathbb{S}^2)}j^{-2-\alpha})\right)\\
&\le \|\eta\|_{C^\alpha}\left(|v_j(\mathbf{0})|\mathscr{O}(j^{-\alpha}) + \mathscr{O}(j^{-1-(\alpha-\rho)})\right),\\
j|I_\eta^+| &\le \|\eta\|_{C^\alpha}\left(|v_j(\mathbf{0})|\mathscr{O}(j^{-\alpha}) + \mathscr{O}(\|g_j\|_{L^2(\mathbb{S}^2)}j^{-2-\alpha})\right),\\
&\le \|\eta\|_{C^\alpha}\left(|v_j(\mathbf{0})|\mathscr{O}(j^{-\alpha}) + \mathscr{O}(j^{-1-(\alpha-\rho)})\right),
\end{aligned} \tag{8.3.60}
$$

as $j \to +\infty$. Clearly, when $s = j$, under the assumption (8.3.24), from (8.2.6), (8.2.7), (8.3.25), and (8.3.27) in Lemma 8.3.4 and (8.3.20), (8.3.19a), and (8.3.19c) in Lemma 8.3.3, we can derive that

$$
\begin{aligned}
&j|I_3| \le Cje^{-c'\sqrt{j}}, \; j\left|\int_{S_h} u_0(j\mathbf{x})\mathrm{d}\mathbf{x}\right| \le \frac{6|e^{-2\theta_M \mathrm{i}} - e^{-2\theta_m \mathrm{i}}|}{j} + \frac{6(\theta_M-\theta_m)e^{-\delta_W\sqrt{hj}/2}}{j\delta_W^4},\\
&j\left|\int_{S_h}\delta F_{3j}(\mathbf{x})u_0(s\mathbf{x}')\mathrm{d}\mathbf{x}'\right| \le \frac{8L\sqrt{\pi}\|\psi\|_{L^\infty}(\theta_M-\theta_m)\Gamma(2\alpha+4)}{\delta_W^{2\alpha+4}}k^2\mathrm{diam}(S_h)^{1-\alpha}\\
&\qquad\qquad \times(1+k)\|g_j\|_{L^2(\mathbb{S}^2)}j^{-\alpha-1} \le \mathscr{O}\left(j^{-(\alpha-\rho)}\right),\\
&j\left|\int_{S_h}\delta F_1(\mathbf{x})u_0(s\mathbf{x}')\mathrm{d}\mathbf{x}'\right| \le \frac{2\|F_1\|_{C^\alpha}(\theta_M-\theta_m)\Gamma(2\alpha+4)}{\delta_W^{2\alpha+4}}j^{-\alpha-1},\\
&j\left|\int_{S_h}\delta F_2(\mathbf{x})u_0(s\mathbf{x}')\mathrm{d}\mathbf{x}'\right| \le \frac{2\|F_2\|_{C^\alpha}(\theta_M-\theta_m)\Gamma(2\alpha+4)}{\delta_W^{2\alpha+4}}j^{-\alpha-1},
\end{aligned} \tag{8.3.61}
$$

and

$$
\begin{aligned}
j|\varepsilon_j^\pm(j)| &\le C\|\psi\|_{L^\infty}\Bigg(|\eta(\mathbf{0})|\frac{\sqrt{\theta_M-\theta_m}e^{-\sqrt{j\Theta}\delta_W h}}{\sqrt{2}}j\\
&\qquad + \|\eta\|_{C^\alpha}j^{-\alpha}\frac{\sqrt{2(\theta_M-\theta_m)\Gamma(4\alpha+4)}}{(2\delta_W)^{2\alpha+2}}\Bigg)j^{-1-\Upsilon},\\
j|\delta_j(j)| &\le \frac{k^2\|\psi\|_{L^\infty}\sqrt{C(L,h)(\theta_M-\theta_m)}e^{-\sqrt{s\Theta}\delta_W h}}{\sqrt{2}}j^{-\Upsilon}, \quad \Theta\in[0,h],
\end{aligned} \tag{8.3.62}
$$

as $j \to +\infty$, where $c' > 0$ and δ_W are defined in (8.3.20) and (8.2.5), respectively.

The coefficient of $v_j(0)$ of (8.3.58) with respect to the zeroth order of s is

$$
2\eta(\mathbf{0})\left(C(I_{311}^-)\mu(\theta_m)^{-2} + C(I_{311}^+)\mu(\theta_M)^{-2}\right).
$$

In (8.3.58), we take $s = j$ and let $j \to \infty$, combining with (8.3.59), (8.3.60), (8.3.61), and (8.3.62), we can prove that

$$
\lim_{j\to\infty}\eta(\mathbf{0})\left(C(I_{311}^-)\mu(\theta_m)^{-2} + C(I_{311}^+)\mu(\theta_M)^{-2}\right)v_j(\mathbf{0}) = 0. \tag{8.3.63}
$$

Under the assumption (8.3.57), from Lemma 8.3.7, we have $C(I_{311}^-)\mu(\theta_m)^{-2} + C(I_{311}^+)\mu(\theta_M)^{-2} \neq 0$. Therefore, from (8.3.56) and (8.3.63), we prove that

$$
\lim_{j\to\infty} v_j(\mathbf{0}) = 0.
$$

Using the a similar argument of (8.2.69), we finish the proof of this theorem. □

Remark 8.3.2. Similar to Remark 8.2.2, Theorem 8.3.1 can be localized. Moreover, we would like to mention that in contrast to the regularity assumption on $v-w$ near the corner in 2D of Theorem 8.2.1, we impose that $v-w \in H^2(S_h \times (-M,M))$ in Theorem 8.3.1, where we need to use the C^α continuity of $\mathscr{R}(v-w)$ to investigate the asymptotical order of s with respect to $s \to \infty$ for the volume integral of $F_1(\mathbf{x}')$ over S_h in (8.3.16).

Similar to Corollary 8.2.1, we consider the vanishing property of the interior transmission eigenfunctions $v \in H^1(W \times (-M,M))$ and $w \in H^1(W \times (-M,M))$ to (8.2.70) on the edge point under the assumptions (8.3.57) and (8.3.64).

Corollary 8.3.1. *Let $\Omega \Subset \mathbb{R}^3$ be a bounded Lipschitz domain with $\mathbf{0} \in \partial\Omega$ and $S_h \subset \mathbb{R}^2$ be defined in* (8.2.2), *$M > 0$, $0 < \alpha < 1$. For any fixed $x_3^c \in (-M,M)$ and $L > 0$ defined in Definition 8.3.1, we suppose that L is sufficiently small such that $(x_3^c - L, x_3^c + L) \subset (-M,M)$ and*

$$(B_h \times (-M,M)) \cap \Omega = S_h \times (-M,M),$$

where $B_h \Subset \mathbb{R}^2$ is the central ball of radius $h \in \mathbb{R}_+$. Suppose $v \in H^1(\Omega)$ and $w \in H^1(\Omega)$ are the interior transmission eigenfunctions to (8.2.70) *in $\mathbb{R}^3$. Suppose that there exists a sufficiently small neighborhood $S_h \times (-M,M)$ (i.e., $h > 0$ is sufficiently small) of $(\mathbf{0}, x_3^c)$ with $x_3^c \in (-M,M)$ such that $qw \in C^\alpha(\overline{S_h} \times [-M,M])$ for $0 < \alpha < 1$ and $v - w \in H^2(S_h \times (-M,M))$, where $\mathbf{0}$ is the vertex of S_h defined in* (8.2.2). *If the following conditions are fulfilled:*

(a) the transmission eigenfunction v can be approximated in $H^1(S_h \times (-M,M))$ by the Herglotz waves v_j, $j = 1,2,\ldots$, with kernels g_j satisfying

$$\|v - v_j\|_{H^1(S_h \times (-M,M))} \leq j^{-2-\Upsilon}, \quad \|g_j\|_{L^2(\mathbb{S}^2)} \leq C j^{\rho}, \tag{8.3.64}$$

for some positive constants C, $\Upsilon > 0$ and $0 < \rho < \alpha$,
(b) the angles θ_m and θ_M of S_h satisfy

$$-\pi < \theta_m < \theta_M < \pi \text{ and } \theta_M - \theta_m \neq \pi, \tag{8.3.65}$$

then we have

$$\lim_{\rho \to +0} \frac{1}{m(B((\mathbf{0},x_3^c),\rho) \cap P(\Omega))} \int_{B((\mathbf{0},x_3^c),\rho) \cap P(\Omega)} \mathscr{R}(Vw)(\mathbf{x}')\mathrm{d}\mathbf{x}' = 0,$$

where $P(\Omega)$ is the projection set of Ω on $\mathbb{R}^2$ and $q(\mathbf{x}', x_3) = 1 + V(\mathbf{x}', x_3)$.

Proof. It is clear that the transmission eigenfunctions $v \in H^1(\Omega)$ and $w \in H^1(\Omega)$ to (8.2.70) fulfill (8.3.2) for $\eta \equiv 0$. Using the dimension reduction operator $\mathscr{R}$ given by (8.3.1), we can show that $\mathscr{R}(v)$ and $\mathscr{R}(w)$ satisfy (8.3.4) for $\eta \equiv 0$. By virtue of Lemmas 8.3.2 and 8.3.3, due to $\eta(\mathbf{x}) \equiv 0$, from (8.3.16), we have the following integral equality:

$$\begin{aligned}&(F_1(\mathbf{0}) + F_2(\mathbf{0}) + F_{3j}(\mathbf{0})) \int_{S_h} u_0(s\mathbf{x}')\mathrm{d}\mathbf{x}' + \delta_j(s) \\ &= I_3 - \int_{S_h} \delta F_1(\mathbf{x}') u_0(s\mathbf{x}')\mathrm{d}\mathbf{x}' - \int_{S_h} \delta F_2(\mathbf{x}') u_0(s\mathbf{x}')\mathrm{d}\mathbf{x}' - \int_{S_h} \delta F_{3j}(\mathbf{x}') u_0(s\mathbf{x}')\mathrm{d}\mathbf{x}',\end{aligned} \tag{8.3.66}$$

where $\delta_j(s)$ is defined in (8.3.21); $\delta F_1(\mathbf{x}')$, $\delta F_2(\mathbf{x}')$, and $\delta F_{3j}(\mathbf{x}')$ are defined in (8.3.15); and I_3 is given in (8.3.16). Since $v = w$ on $\Gamma_h^\pm \times (-M,M)$, it is easy to see that

$$F_1(\mathbf{0}) = \int_{-L}^{L} \psi''(x_3)(v(\mathbf{0},x_3) - w(\mathbf{0},x_3))\mathrm{d}x_3 = 0.$$

Therefore, using (8.2.6), from (8.3.66), we deduce that

$$
\begin{aligned}
&6\mathrm{i}(F_2(\mathbf{0})+F_{3j}(\mathbf{0}))(e^{-2\theta_M \mathrm{i}}-e^{-2\theta_m \mathrm{i}})s^{-2}-(F_2(\mathbf{0})+F_{3j}(\mathbf{0}))\int_{W\setminus S_h} u_0(s\mathbf{x}')\mathrm{d}\mathbf{x}' \\
&=I_3-\int_{S_h}\delta F_1(\mathbf{x}')u_0(s\mathbf{x}')\mathrm{d}\mathbf{x}'-\int_{S_h}\delta F_2(\mathbf{x}')u_0(s\mathbf{x}')\mathrm{d}\mathbf{x}'-\int_{S_h}\delta F_{3j}(\mathbf{x}')u_0(s\mathbf{x}')\mathrm{d}\mathbf{x}'-\delta_j(s).
\end{aligned}
\tag{8.3.67}
$$

In (8.3.67), we take $s=j$ and multiply j^2 on both sides of (8.3.67), and then we have

$$
\begin{aligned}
&6\mathrm{i}(F_2(\mathbf{0})+F_{3j}(\mathbf{0}))(e^{-2\theta_M \mathrm{i}}-e^{-2\theta_m \mathrm{i}})=j^2\Big[I_3+(F_2(\mathbf{0})+F_{3j}(\mathbf{0}))\int_{W\setminus S_h} u_0(s\mathbf{x}')\mathrm{d}\mathbf{x}' \\
&-\int_{S_h}\delta F_1(\mathbf{x}')u_0(s\mathbf{x}')\mathrm{d}\mathbf{x}'-\int_{S_h}\delta F_2(\mathbf{x}')u_0(s\mathbf{x}')\mathrm{d}\mathbf{x}'-\int_{S_h}\delta F_{3j}(\mathbf{x}')u_0(s\mathbf{x}')\mathrm{d}\mathbf{x}'-\delta_j(s)\Big].
\end{aligned}
\tag{8.3.68}
$$

Using (8.3.25) in Lemma 8.3.4 and under the assumption (8.3.64), it is easy to see that

$$
j^2|\delta_j(s)|\leq \frac{|k|^2\|\psi\|_{L^\infty}\sqrt{C(L,h)(\theta_M-\theta_m)}e^{-\sqrt{s\Theta}\delta_W h}}{\sqrt{2}}j^{-\Upsilon},
\tag{8.3.69}
$$

where $C(L,h)$ is a positive number defined in (8.3.28), $\Theta\in[0,h]$, and δ_W is defined in (8.2.5).

Under the assumption (8.3.64), by virtue of (8.3.19a), (8.3.19b), (8.3.19c), and (8.3.20) in Lemma 8.3.3, we can obtain the following estimates:

$$
\begin{aligned}
j^2\left|\int_{S_h}\delta F_{3j}(\mathbf{x})u_0(s\mathbf{x}')\mathrm{d}\mathbf{x}'\right| &\leq \frac{8L\sqrt{\pi}\|\psi\|_{L^\infty}(\theta_M-\theta_m)\Gamma(2\alpha+4)}{\delta_W^{2\alpha+4}}k^2\mathrm{diam}(S_h)^{1-\alpha} \\
&\quad\times(1+k)\|g_j\|_{L^2(\mathbb{S}^2)}j^{-\alpha}\leq\mathscr{O}\left(j^{-(\alpha-\rho)}\right), \\
j^2\left|\int_{S_h}\delta F_1(\mathbf{x})u_0(s\mathbf{x}')\mathrm{d}\mathbf{x}'\right| &\leq \frac{2\|F_1\|_{C^\alpha}(\theta_M-\theta_m)\Gamma(2\alpha+4)}{\delta_W^{2\alpha+4}}j^{-\alpha}, \\
j^2\left|\int_{S_h}\delta F_2(\mathbf{x})u_0(s\mathbf{x}')\mathrm{d}\mathbf{x}'\right| &\leq \frac{2\|F_2\|_{C^\alpha}(\theta_M-\theta_m)\Gamma(2\alpha+4)}{\delta_W^{2\alpha+4}}j^{-\alpha}.
\end{aligned}
\tag{8.3.70}
$$

Under the assumption (8.3.65), it is easy to see that

$$
\left|e^{-2\theta_M\mathrm{i}}-e^{-2\theta_m\mathrm{i}}\right|=\left|1-e^{-2(\theta_M-\theta_m)\mathrm{i}}\right|\neq 0,
$$

since $\theta_M-\theta_m\neq\pi$. In (8.3.68), by letting $j\to\infty$, from (8.2.7), (8.3.69), and (8.3.70), we prove that

$$
\lim_{j\to\infty}F_{3j}(\mathbf{0})=-F_2(\mathbf{0}),
$$

which implies

$$
\lim_{j\to\infty}\mathscr{R}(v_j)(\mathbf{0})=\mathscr{R}(qw)(\mathbf{0})
\tag{8.3.71}
$$

through recalling that F_2 and F_{3j} are given in (8.3.5). From (8.3.9), we have

$$
\begin{aligned}
&\lim_{\rho\to+0}\frac{1}{m(B(\mathbf{0},\rho)\cap P(\Omega))}\int_{B(\mathbf{0},\rho)\cap P(\Omega)}\mathscr{R}(v)(\mathbf{x}')\mathrm{d}\mathbf{x}' \\
&=\lim_{\rho\to+0}\frac{1}{m(B(\mathbf{0},\rho)\cap P(\Omega))}\int_{B(\mathbf{0},\rho)\cap P(\Omega)}\mathscr{R}(w)(\mathbf{x}')\mathrm{d}\mathbf{x}'.
\end{aligned}
$$

Since

$$\lim_{j\to\infty}\mathscr{R}(v_j)(\mathbf{0}) = \lim_{j\to\infty}\lim_{\rho\to+0}\frac{1}{m(B(\mathbf{0},\rho)\cap P(\Omega))}\int_{B(\mathbf{0},\rho)\cap P(\Omega)}\mathscr{R}(v_j)(\mathbf{x}')\mathrm{d}\mathbf{x}'$$
$$= \lim_{\rho\to+0}\frac{1}{m(B(\mathbf{0},\rho)\cap P(\Omega))}\int_{B(\mathbf{0},\rho)\cap P(\Omega)}\mathscr{R}(v)(\mathbf{x}')\mathrm{d}\mathbf{x}',$$
$$\mathscr{R}(qw)(\mathbf{0}) = \lim_{\rho\to+0}\frac{1}{m(B(\mathbf{0},\rho)\cap P(\Omega))}\int_{B(\mathbf{0},\rho)\cap P(\Omega)}\mathscr{R}(qw)(\mathbf{x}')\mathrm{d}\mathbf{x}',$$

and from (8.3.71), we finish the proof of this corollary. □

Remark 8.3.3. Corollary 8.3.1 states that the average value of the function Vw over the cylinder centered at the edge point $(\mathbf{0},x_3^c)$ with the height L vanishes in the distribution sense. In addition, if $V(\mathbf{x}',x_3)$ is continuous near the edge point $(\mathbf{0},x_3^c)$ where $x_3^c\in(-M,M)$ and $V(\mathbf{0},x_3^c)\neq 0$, from the dominated convergence theorem and the definition of the dimension reduction operator $\mathscr{R}$, we can prove that

$$\lim_{\rho\to+0}\frac{1}{m(B(\mathbf{0},\rho)\cap P(\Omega))}\int_{B(\mathbf{0},\rho)\cap P(\Omega)}\int_{x_3^c-L}^{x_3^c+L}\psi(x_3)w(\mathbf{x}',x_3)\mathrm{d}\mathbf{x}'\mathrm{d}x_3 = 0,$$

under the assumptions in Corollary 8.3.1, which also describes the vanishing property of the interior eigenfunctions v and w near the edge point in 3D. Furthermore, if $\psi(x_3^c)\neq 0$, one can prove that

$$\lim_{\rho\to+0}\frac{1}{m(B(\mathbf{0},\rho)\cap P(\Omega))}\int_{B(\mathbf{0},\rho)\cap P(\Omega)}\int_{x_3^c-L}^{x_3^c+L}w(\mathbf{x}',x_3)\mathrm{d}\mathbf{x}'\mathrm{d}x_3 = 0.$$

In the following theorem, we impose a stronger regularity requirement for the conductive transmission eigenfunction v of (8.3.2), i.e., v has H^2 regularity near the considering edge point. Using the dimension reduction operator given in Definition 8.3.1, as well as the Hölder continuity of the considering functions, we can prove the following theorem in a similar way of proving Theorem 8.2.2. The detailed proof of Theorem 8.3.2 is omitted here.

Theorem 8.3.2. *Let $\Omega\Subset\mathbb{R}^3$ be a bounded domain with $0\in\partial\Omega$ and $S_h\subset\mathbb{R}^2$ be defined in* (8.2.2), *$M>0$, $0<\alpha<1$. For any fixed $x_3^c\in(-M,M)$ and $L>0$ defined in Definition 8.3.1, we suppose that L is sufficiently small such that $(x_3^c-L,x_3^c+L)\subset(-M,M)$ and*

$$(B_h\times(-M,M))\cap\Omega = S_h\times(-M,M),$$

where $B_h\Subset\mathbb{R}^2$ is the central ball of radius $h\in\mathbb{R}_+$. Let $v\in H^2(\Omega)$ and $w\in H^1(\Omega)$ be the eigenfunctions to (8.3.2). *Moreover, there exists a sufficiently smaller neighborhood $S_h\times(-M,M)$ (i.e., $h>0$ is sufficiently small) of $(\mathbf{0},x_3^c)$ with $x_3^c\in(-M,M)$, such that $qw\in C^\alpha(\overline{S}_h\times[-M,M])$ and $\eta\in C^\alpha(\overline{\Gamma}_h^\pm\times[-M,M])$ for $0<\alpha<1$ and $v-w\in H^2(S_h\times(-M,M))$. Under the following assumptions:*

(a) the function $\eta=\eta(\mathbf{x}',x_3)$ is independent of x_3 and does not vanish on the edge of $W\times(-M,M)$, i.e.,

$$\eta(\mathbf{0})\neq 0,$$

(b) the angles θ_m and θ_M of S_h satisfy

$$-\pi<\theta_m<\theta_M<\pi \text{ and } \theta_M-\theta_m\neq\pi,$$

then we have v and w vanish at the edge point $(\mathbf{0},x_3^c)\in\mathbb{R}^3$ of $S_h\times(-M,M)$, where $x_3^c\in(-M,M)$.

Remark 8.3.4. When $\eta\equiv 0$ near the edge point, under the H^2 regularity of the interior transmission eigenfunctions v and w, the vanishing property of v and w is investigated in [4].

8.4 Unique Recovery Results for the Inverse Acoustic Medium Scattering Problem

In this section, we apply the vanishing property of the conductive transmission eigenfunctions at a corner in 2D to investigate the unique recovery in the inverse problem associated with the corresponding conductive scattering problem. Before that, we first describe the relevant physical background.

The time-harmonic electromagnetic wave scattering from a conductive medium body arises in the application of practical importance, for example, the modeling of an electromagnetic object coated with a thin layer of a highly conducting material (cf. [3]). In what follows, we let ε, μ, and σ denote the electric permittivity, the magnetic permeability, and the conductivity of a medium, respectively. Let Ω be a bounded Lipschitz domain in $\mathbb{R}^2$ with a connected complement $\mathbb{R}^2\backslash\overline{\Omega}$. Consider a cylinder-like medium body $D := \Omega \times \mathbb{R}$ in $\mathbb{R}^3$ with the cross section being Ω along the x_3-axis for $\mathbf{x} = (x_j)_{j=1}^3 \in D$. In the following discussions, with a bit abuse of notation, we shall also use $\mathbf{x} = (x_1, x_2)$ in the 2D case, which should be clear from the context. Let $\Omega_\delta := \{\mathbf{x} + h\nu(x); \mathbf{x} \in \partial\Omega \text{ and } h \in (0,\delta)\}$ with $\delta \in \mathbb{R}_+$ being sufficiently small, where $\nu \in \mathbb{S}^1$ signifies the exterior unit normal vector to $\partial\Omega$. Set $D_\delta = \Omega_\delta \times \mathbb{R}$ to denote a layer of thickness δ coated on the medium body D. The material configuration associated with the above medium structure is given as follows:

$$\varepsilon, \mu, \sigma = \varepsilon_1, \mu_0, \sigma_1 \text{ in } D;\ \varepsilon_2, \mu_0, \frac{\gamma}{\delta} \text{ in } D_\delta;\ \varepsilon_0, \mu_0, 0 \text{ in } \mathbb{R}^3\backslash\overline{(D \cup D_\delta)}, \tag{8.4.1}$$

where $\varepsilon_j > 0$, $j = 0, 1, 2$, $\mu_0, \gamma > 0$, and $\sigma_1 \geq 0$ are all constants. Consider a time-harmonic incidence:

$$\nabla \times \mathbf{E}^i - \mathrm{i}\omega\mu_0\mathbf{H}^i = \mathbf{0}, \quad \nabla \times \mathbf{H}^i + \mathrm{i}\omega\varepsilon_0\mathbf{E}^i = \mathbf{0} \quad \text{in } \mathbb{R}^3, \tag{8.4.2}$$

where $\mathrm{i} := \sqrt{-1}$, $\mathbf{E}^i$ and $\mathbf{H}^i$ are, respectively, the electric and magnetic fields and $\omega \in \mathbb{R}_+$ is the angular frequency. The electromagnetic scattering is generated by the impingement of the incident field $(\mathbf{E}^i, \mathbf{H}^i)$ on the medium body described in (8.4.1) as follows:

$$\begin{cases} \nabla \times \mathbf{E} - \mathrm{i}\omega\mu\mathbf{H} = \mathbf{0}, \quad \nabla \times \mathbf{H} + \mathrm{i}\omega\varepsilon\mathbf{E} = \sigma\mathbf{E}, & \text{in } \mathbb{R}^3, \\ \mathbf{E} = \mathbf{E}^i + \mathbf{E}^s, \quad \mathbf{H} = \mathbf{H}^i + \mathbf{H}^s, & \text{in } \mathbb{R}^3, \\ \lim_{r\to\infty} r\,(\mathbf{H}^s \wedge \hat{\mathbf{x}} - \mathbf{E}^s) = \mathbf{0}, & r := |\mathbf{x}|, \hat{\mathbf{x}} := \mathbf{x}/|\mathbf{x}|, \end{cases} \tag{8.4.3}$$

where the tangential components of the electric field $\mathbf{E}$ and the magnetic field $\mathbf{H}$ are continuous across the material interfaces ∂D and ∂D_δ. The last limit in (8.4.3) is known as the Silver-Müller radiation condition.

Under the transverse-magnetic (TM) polarization, namely,

$$\mathbf{E}^i = \begin{bmatrix} 0 \\ 0 \\ u^i(x_1, x_2) \end{bmatrix}, \quad \mathbf{H}^i = \begin{bmatrix} H_1(x_1, x_2) \\ H_2(x_1, x_2) \\ 0 \end{bmatrix},$$

and

$$\mathbf{E} = \begin{bmatrix} 0 \\ 0 \\ u(x_1, x_2) \end{bmatrix}, \quad \mathbf{H} = \begin{bmatrix} H_1(x_1, x_2) \\ H_2(x_1, x_2) \\ 0 \end{bmatrix}, \tag{8.4.4}$$

it is rigorously verified in [12] that as $\delta \to +0$, one has

$$
\begin{cases}
\Delta u^- + k^2 q u^- = 0 & \text{in } \Omega, \\
\Delta u^+ + k^2 u^+ = 0 & \text{in } \mathbb{R}^2 \backslash \Omega, \\
u^+ = u^-, \quad \partial_\nu u^+ + \eta u^+ = \partial_\nu u^- & \text{on } \partial\Omega, \\
u^+ = u^i + u^s & \text{in } \mathbb{R}^2 \backslash \Omega, \\
\lim_{r\to\infty} r^{1/2} (\partial_r u^s - \mathrm{i} k u^s) = 0, & r = |\mathbf{x}|,
\end{cases}
\tag{8.4.5}
$$

where

$$
u^- = u|_\Omega, u^+ = u|_{\mathbb{R}^2\backslash\overline{\Omega}} \text{ and } k = \omega\sqrt{\varepsilon_0\mu_0}, \eta = \mathrm{i}\omega\gamma\mu_0. \tag{8.4.6}
$$

The last limit in (8.4.5) is known as the Sommerfeld radiation condition. According to (8.4.1), as $\delta \to +0$, it is clear that the conductivity in the thin layer D_δ goes to infinity or, equivalently, its resistivity goes to zero. This in general would lead to the so-called perfectly electric conducting (PEC) boundary, which prevents the electric field from penetrating inside the medium body and instead generates a certain boundary current. However, it is noted in our case that the thickness of the coating layer D_δ also goes to zero, and this allows the electromagnetic waves to penetrate inside the medium body. Nevertheless, the thin highly conducting layer effectively produces a transmission boundary condition across the material interface ∂D involving a conductive parameter η, which is referred to as the conductive transmission condition. It is known to us that a perfect conductor does not exist in nature and hence the conductive medium body provides a more realistic means to model the electromagnetic scattering from an object coated with a thin layer of a highly conducting material; see [31,54] more relevant discussion about this aspect.

The well-posedness of the direct problem (8.4.5) is known (cf. [14]), and there exists a unique solution $u := u^-\chi_\Omega + u^+\chi_{\mathbb{R}^2\backslash\Omega} \in H^1_{loc}(\mathbb{R}^2)$. Moreover, there holds the following asymptotic expansion:

$$
u^s(\mathbf{x}) = \frac{e^{\mathrm{i}k|\mathbf{x}|}}{|\mathbf{x}|^{1/2}} u^\infty(\hat{\mathbf{x}}) + \mathscr{O}\left(\frac{1}{|\mathbf{x}|}\right), |\mathbf{x}| \to +\infty
$$

uniformly in all directions $\hat{\mathbf{x}} = \mathbf{x}/|\mathbf{x}| \in \mathbb{S}^1$. The real analytic function $u^\infty(\hat{\mathbf{x}})$ is referred to as the *far-field pattern* or the *scattering amplitude* associated with u^i. The inverse scattering problem is concerned with the recovery of the scatterer $(\Omega; q, \eta)$ by knowledge of the far-field pattern $u^\infty(\hat{\mathbf{x}}; u^i)$; that is,

$$
u^\infty(\hat{\mathbf{x}}; u^i) \to (\Omega; q, \eta). \tag{8.4.7}
$$

In (8.4.7), if the far-field pattern is given corresponding to a single incident wave u^i, then it is referred to as a single far-field measurement; otherwise, it is referred to as many far-field measurements. It is known that the inverse problem (8.4.7) is nonlinear and ill-conditioned. For the reconstruction of the shape of the scatterer Ω by using the factorization method for (8.4.7), uniqueness issue has been studied in [14]. The inverse spectral problem of gaining the information about the material properties associated with the conductive transmission eigenvalue problem has been studied in [13]. In [32], the method of uniquely recovering the conductive boundary parameter η from the measured scattering data as well as the convergence of the conductive transmission eigenvalues as the conductivity parameters which tend to zero has also been studied. In all of the aforementioned literatures, the unique determination results are based on the far-field patterns of all incident plane waves at a fixed frequency, which means that infinitely many far-field measurements have been used. In what follows, we show that in a rather general and practical scenario, the polyhedral shape of the scatterer, namely, Ω, can be uniquely recovered by a single far-field measurement without knowing its material contents, namely, q and η. Moreover, if the surface conductive parameter η is constant, then it can be recovered as well. When $\eta \equiv 0$, the corresponding unique determination of q by many measurements can be found in [48,55].

Our main unique recovery results for the inverse scattering problem (8.4.7) are contained in Theorems 8.4.1 and 8.4.2. In Theorem 8.4.1, we consider the unique recovery results by a single far-field measurement in determining a 2D polygonal conductive scatterer without knowing its contents. In Theorem 8.4.2, the surface conductive parameter η of the scatterer can be further recovered if it is a constant. Before presenting the main results, we first show in Proposition 8.4.1 that the conductive parameter η in (8.4.5) has a close relationship with the wavenumber k from the practical point view of the TM-mode (transverse magnetic) for the time-harmonic Maxwell system [3]. This relationship helps us to show that our assumption in Theorem 8.4.1 can be fulfilled when the wavenumber k is sufficiently small.

In view of (8.4.6), we readily have the following observation.

Remark 8.4.1. The conductive boundary parameter η of (8.4.5) satisfies

$$\eta = \mathrm{i}\omega\gamma\mu_0 = \mathscr{O}(k), \tag{8.4.8}$$

where $k := \omega\sqrt{\varepsilon_0\mu_0}$ is the wavenumber in (10.4.1).

Remark 8.4.1 basically indicates that when considering the conductive scattering problem (8.4.5), one may impose the low-frequency dependence behavior (8.4.8) on the surface conductive parameter. As remarked earlier, Remark 8.4.1 only considers the simple model (8.4.3) for illustration of the low-frequency behavior (8.4.8). For more complex Maxwell models, one can derive the conductive scattering system (8.4.5) of a general form.

$$\begin{aligned} &\Gamma_h^{\pm}(\mathbf{x}_c) := \partial W_{\mathbf{x}_c}(\theta_W)\cap B_h(\mathbf{x}_c), \quad S_h(\mathbf{x}_c) := \Omega\cap B_h(\mathbf{x}_c) = \Omega\cap W_{\mathbf{x}_c}(\theta_W), \\ &S_{h/2}(\mathbf{x}_c) := \Omega\cap B_{h/2}(\mathbf{x}_c) = \Omega\cap W_{\mathbf{x}_c}(\theta_W), \quad \Sigma_{\Lambda_h}(\mathbf{x}_c) := S_h(\mathbf{x}_c)\backslash S_{h/2}(\mathbf{x}_c). \end{aligned} \tag{8.4.9}$$

Definition 8.4.1. Let $(\Omega;k,d,q,\eta)$ be a conductive scatterer associated with the incident plane wave $u^i = e^{\mathrm{i}kx\cdot d}$ with $d\in\mathbb{S}^1$ and $k\in\mathbb{R}_+$. Consider the scattering problem (8.4.5) and u is the total wave field therein. The scatterer is said to be admissible if it fulfils the following conditions:

(a) Ω is a bounded simply connected Lipschitz domain in $\mathbb{R}^2$, and $q\in L^\infty(\Omega)$, $\eta\in L^\infty(\partial\Omega)$.
(b) Following the notations in Theorem 8.2.1, if Ω possesses a corner $B_h(\mathbf{x}_c)\cap\Omega = \Omega\cap W_{\mathbf{x}_c}(\theta_W)$ where x_c is the vertex of the sector $W_{\mathbf{x}_c}(\theta_W)$ and the open angle θ_W of $W_{x_c}(\theta_W)$ satisfies $\theta_W\neq\pi$, then $q|_{S_h(\mathbf{x}_c)}$ is a constant, $\eta\in C^\alpha(\overline{\Gamma_h^{\pm}(\mathbf{x}_c)})$, where $S_h(x_c)$ and $\Gamma_h^{\pm}(\mathbf{x}_c)$ are defined in (8.4.9).
(c) The total wave field u is non-vanishing everywhere in the sense that for any $\mathbf{x}\in\mathbb{R}^2$,

$$\lim_{\rho\to+0}\frac{1}{m(B(\mathbf{x},\rho))}\int_{B(\mathbf{x},\rho)}|u(\mathbf{x})|\mathrm{d}\mathbf{x}\neq 0. \tag{8.4.10}$$

We would like to point out that the conditions stated in Definition 8.4.1 can be fulfilled by the conductive scatterer $(\Omega;k,d,q,\eta)$ and the scattering problem (8.4.5) in certain general and practical scenarios. In particular, the condition (8.4.10) in (c) can be fulfilled at least when k is sufficiently small. In fact, it has been shown in Proposition 8.4.1 that if $\eta\neq 0$, then $\eta=\mathscr{O}(k)$. For the scattered field u^s of (8.4.5), from [14, Theorem 2.4], it is proved that

$$\|u^s\|_{H^1(B)} \leq C(\|\eta u^i\|_{H^{-1/2}(\partial\Omega)} + k^2\|qu^i\|_{L^2(\Omega)}) = \mathscr{O}(k)\|u^i\|_{L^2(\Omega)},$$

where C is a positive number and B is a large ball containing Ω. Hence, if the incident field u^i is non-vanishing everywhere, say $u^i = e^{\mathrm{i}kx\cdot d}$ with $d\in\mathbb{S}^1$ being a plane wave, and k is sufficiently small, then (8.4.10) is obviously fulfilled. Nevertheless, by Definition 8.4.1, we may include more general situations into our subsequent study of the inverse problem (8.4.7).

The determination of the geometric shape of a conductive scatterer can be discussed in Theorem 8.4.1 by using Theorem 8.2.2 and a contradiction argument. The technical requirement $qw\in C^\alpha(\overline{S}_h)$ in Theorem 8.2.2 can be easily fulfilled. Indeed, this condition can be derived by using the classical results on the singular behaviors of the solutions to elliptic PDEs in a corner domain [18, 25, 27, 30]. In fact, it is known that the solution can be decomposed into a singular part and a regular part, where the singular part is of a Hölder form that depends on the corner geometry as well as the boundary and the right-hand inputs. For our subsequent use, we first give the following result in a relatively simple scenario.

In (8.4.5), by the standard PDE theory (see, e.g., [47]), we know that the solution u is real analytic away from the conductive interface. In Lemma 8.4.1 as follows, we further give the Hölder regularity of the solution up to the conductive interface, especially to the vertex corner point. Denote

$$S_{2h} = W \cap B_{2h}, \quad \Gamma_{2h}^{\pm} = \Gamma^{\pm} \cap B_{2h},$$

where W is the sector defined in (8.2.1), B_{2h} is an open ball centered at 0 in $\mathbb{R}^2$ with the radius $2h$, and $\Gamma^{\pm}$ are the boundaries of W.

Lemma 8.4.1 (**[17, Lemma 3.4]**). *Suppose that $u \in H^1(B_{2h})$ satisfies*

$$\begin{cases} \Delta u^- + k^2 q_- u^- = 0 & \text{in } S_{2h}, \\ \Delta u^+ + k^2 q_+ u^+ = 0 & \text{in } B_{2h} \backslash \overline{S_{2h}}, \\ u^+ = u^- & \text{on } \Gamma_{2h}^{\pm}, \end{cases} \tag{8.4.11}$$

where $u^+ = u|_{B_{2h} \backslash \overline{S_{2h}}}$, $u^- = u|_{S_{2h}}$, and $q_{\pm}$, k are complex constants. Assume that u^+ and u^- are, respectively, real analytic in $B_{2h} \backslash \overline{S_{2h}}$ and S_{2h}. There exists $\alpha \in (0,1)$ such that $u^- \in C^{\alpha}(\overline{S_h})$, where S_h is defined in (8.2.2).

Theorem 8.4.1. *Consider the conductive scattering problem* (8.4.5) *associated with two conductive scatterers $(\Omega_j; k, d, q_j, \eta_j)$, $j = 1, 2$, in $\mathbb{R}^2$. Let $u_\infty^j(\hat{x}; u^i)$ be the far-field pattern associated with the scatterer $(\Omega_j; k, d, q_j, \eta_j)$ and the incident field u^i. Suppose that $(\Omega_j; k, d, q_j, \eta_j)$, $j = 1, 2$ are admissible and*

$$u_\infty^1(\hat{\mathbf{x}}; u^i) = u_\infty^2(\hat{\mathbf{x}}; u^i), \tag{8.4.12}$$

for all $\hat{\mathbf{x}} \in \mathbb{S}^1$ and a fixed incident wave u^i. Then

$$\Omega_1 \Delta \Omega_2 := (\Omega_1 \backslash \Omega_2) \cup (\Omega_2 \backslash \Omega_1) \tag{8.4.13}$$

cannot possess a corner. Hence, if Ω_1 and Ω_2 are convex polygons in $\mathbb{R}^2$, one must have

$$\Omega_1 = \Omega_2. \tag{8.4.14}$$

Proof. By contradiction, we assume that there is a corner contained in $\Omega_1 \Delta \Omega_2$. Without loss of generality, we may assume that the vertex O of the corner $\Omega_2 \cap W$ is such that $O \in \partial \Omega_2$ and $O \notin \overline{\Omega}_1$. Without loss of generality, we may assume that O is the origin of $\mathbb{R}^2$. Please refer to Fig. 8.3 for a schematic illustration.

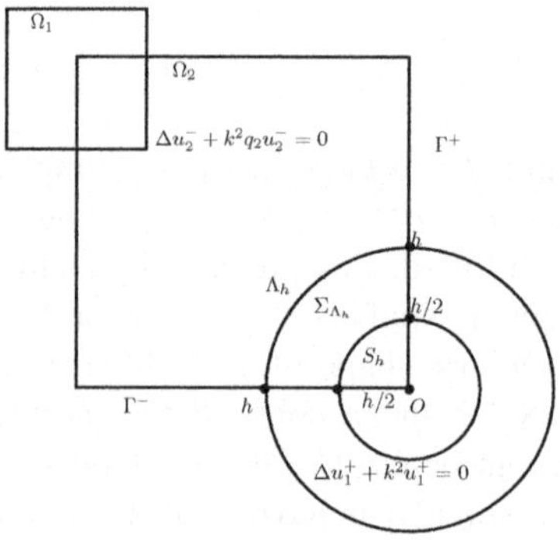

Fig. 8.3 Schematic illustration of the geometry setup in the proof of Theorem 8.4.1

Since $u_\infty^1(\hat{\mathbf{x}};u^i)=u_\infty^2(\hat{\mathbf{x}};u^i)$ for all $\hat{\mathbf{x}}\in\mathbb{S}^1$, applying Rellich's theorem (see [22]), we know that $u_1^s=u_2^s$ in $\mathbb{R}^2\backslash(\overline{\Omega}_1\cup\overline{\Omega}_2)$. Thus,

$$u_1(\mathbf{x})=u_2(\mathbf{x}), \tag{8.4.15}$$

for all $\mathbf{x}\in\mathbb{R}^2\backslash(\overline{\Omega}_1\cup\overline{\Omega}_2)$. Following the notations in (8.2.2), we have from (8.4.15) that

$$u_2^-=u_2^+=u_1^+,\quad \partial u_2^-=\partial u_2^++\eta_2u_2^+=\partial u_1^++\eta_2u_1^+ \text{ on } \Gamma_h^\pm,$$

where the superscripts $(\cdot)^-,(\cdot)^+$ stand for the limits taken from Ω_2 and $\mathbb{R}^2\backslash\overline{\Omega_2}$, respectively. Moreover, suppose the neighborhood B_{2h} is sufficiently small such that

$$\Delta u_1^++k^2u_1^+=0,\quad \Delta u_2^-+k^2q_2u_2^-=0 \text{ in } B_{2h}.$$

It is clear that u_1^+ and u_2^- are, respectively, real analytic in $B_{2h}\backslash\overline{S_{2h}}$ and S_{2h}. Since $q_2\big|_{S_h}$ is a constant, by virtue of Lemma 8.4.1, we know that $u_2^-\in C^\alpha(\overline{S_h})$, which implies that

$$q_2u_2^-\in C^\alpha(\overline{S_h}). \tag{8.4.16}$$

Clearly, $u_1^+\in H^2(S_h)$ and $u_2^-\in H^1(S_h)$. Now we prove that

$$u_1^+-u_2^-\in H^2(\Sigma_{\Lambda_h}),$$

where Σ_{Λ_h} is defined in (8.2.2). We first note that on the boundary $\Gamma_h^\pm$, one has $u_2^-=u_1^+$, where $u_1^+\in H^{3/2}(\Gamma_h^\pm)$ from the trace theorem. Denote

$$D_1^+=B_{h/4}(x_{\Gamma^+})\cap\Sigma_{\Lambda_h},\quad D_1^-=B_{h/4}(x_{\Gamma^-})\cap\Sigma_{\Lambda_h}, \tag{8.4.17}$$

where x_{Γ^+} and x_{Γ^-} are the mid-points of $\Gamma_{(h/2,h)}^+$ and $\Gamma_{(h/2,h)}^-$, respectively.

Since $\Gamma_h^+\in C^{1,1}$, from [47, Theorem 4.18], we have the following regularity estimate for u_2^- up to the boundary $\Gamma_{(h/2,h)}^+$ of Σ_{Λ_h}:

$$\|u_2^-\|_{H^2(D_1^+)}\le C\left(\|u_2^-\|_{H^1(S_h)}+\|u_1^+\|_{H^{3/2}(\Gamma_h^+)}\right),$$

where $C>0$ is a constant and D_1^+ is defined in (8.4.17). Using the similar argument, we can prove that u_2^- has H^2 regularity up to the boundary $\Gamma_{(h/2,h)}^-$ of Σ_{Λ_h}. Therefore, $u_2^-\in H^2(\Sigma_{\Lambda_h})$ by using the interior regularity of the standard elliptic PDE theory, which means that $u_1^+-u_2^-\in H^2(\Sigma_{\Lambda_h})$. Since $(\Omega_j;k,d,q_j,\eta_j)$, $j=1,2$, are admissible, we know that $\eta_j\in C^\alpha(\overline{\Gamma}_h^\pm)$. Noting (8.4.16) and by applying Theorem 8.2.2, if $\eta_2(\mathbf{0})\neq 0$, and Remark 8.2.6 if $\eta_2(\mathbf{0})=0$ on $\Gamma_h^\pm$, and also utilizing the fact that u_1 is continuous at the vertex 0, we have

$$u_1(\mathbf{0})=0,$$

which contradicts to the admissibility condition (c) in Definition 8.4.1.

The proof is complete. □

Based on Definition 8.4.1, if we further assume that the surface conductive parameter η is constant, we can recover η simultaneously once the admissible conductive scatterer Ω is determined. However, in determining the surface conductive parameter, we need to assume that $q_1=q_2:=q$ are known.

Theorem 8.4.2. *Consider the conductive scattering problem* (8.4.5) *associated with the admissible conductive scatterers* $(\Omega_j;k,d,q,\eta_j)$, *where* $\Omega_j=\Omega$ *for* $j=1,2$ *and* $\eta_j\neq 0$, $j=1,2$, *are two* *constants. Let $u_\infty^j(\hat{\mathbf{x}};u^i)$ be the far-field pattern associated with the scatterer $(\Omega;k,d,q,\eta_j)$ and the incident field u^i. Suppose that $(\Omega;k,d,q,\eta_j)$, $j=1,2$, are admissible and*

$$u_\infty^1(\hat{\mathbf{x}};u^i) = u_\infty^2(\hat{\mathbf{x}};u^i), \tag{8.4.18}$$

for all $\hat{\mathbf{x}} \in \mathbb{S}^1$ and a fixed incident wave u^i. Then if k is not an eigenvalue of the partial differential operator $\Delta + k^2 q$ in $H_0^1(\Omega)$, we have $\eta_1 = \eta_2$.

Proof. Since $u_\infty^1(\hat{\mathbf{x}};u^i) = u_\infty^2(\hat{\mathbf{x}};u^i)$ for all $\hat{\mathbf{x}} \in \mathbb{S}^1$, we can derive that $u_1^+ = u_2^+$ for all $\mathbf{x} \in \mathbb{R}^2 \backslash \overline{\Omega}$ and thus $\partial_\nu u_1^+ = \partial_\nu u_2^+$ on $\partial\Omega$. Combining with the transmission condition in the scattering problem (8.4.5), we deduce that

$$u_1^- = u_1^+ = u_2^+ = u_2^- \quad \text{on } \partial\Omega.$$

Thus, we have

$$\partial_\nu(u_1^- - u_2^-) = \partial_\nu(u_1^+ - u_2^+) + \eta_1 u_1^+ - \eta_2 u_2^+ = (\eta_1 - \eta_2)u_1^- \quad \text{on } \partial\Omega.$$

Define $v := u_1^- - u_2^-$. Then v fulfills

$$\begin{cases} (\Delta + k^2 q)v = 0 & \text{in } \Omega, \\ v = 0 & \text{on } \partial\Omega, \\ \partial_\nu v = (\eta_1 - \eta_2)u_1^- & \text{on } \partial\Omega. \end{cases} \tag{8.4.19}$$

Since k is not an eigenvalue of the operator $\Delta + k^2 q$ in $H_0^1(\Omega)$, hence, one must have $v = 0$ to (8.4.19). Substituting this into the Neumann boundary condition of (8.4.19), we know that $(\eta_1 - \eta_2)u_1^- = \partial_\nu v = 0$ on $\partial\Omega$.

Next, we prove the uniqueness of η by contradiction. Assume that $\eta_1 \neq \eta_2$. Since $(\eta_1 - \eta_2)u_1^- = 0$ on $\partial\Omega$ and η_j, $j = 1,2$ are constants, we can deduce that $u_1^- = 0$ on $\partial\Omega$. Then u_1^- satisfies

$$\begin{cases} (\Delta + k^2 q)u_1^- = 0 & \text{in } \Omega, \\ u_1^- = 0 & \text{on } \partial\Omega. \end{cases}$$

Similar to (8.4.19), this Dirichlet problem also only has a trivial solution $u_1^- = 0$ in Ω, since k is not an eigenvalue of $\Delta + k^2 q$. Then, we can derive $u_1^+ = u_1^- = 0$ and

$$\partial_\nu u_1^- = \partial_\nu u_1^+ + \eta_1 u_1^+ = \partial_\nu u_1^+ = 0 \text{ on } \partial\Omega,$$

which implies that $u_1 \equiv 0$ in $\mathbb{R}^2$ and thus $u_1^s = -u^i$. This contradicts to the fact that u_1^s satisfies the Sommerfeld radiation condition.

The proof is complete. □

Remark 8.4.2. In Theorem 8.4.2, it is required that k is not an eigenvalue of $\Delta + k^2 q$ in $H_0^1(\Omega)$. Clearly, if q is negative-valued in Ω or $\Im q \neq 0$ in Ω, this condition is fulfilled. On the other hand, if q is positive-valued in Ω, then this condition can be readily fulfilled when $k \in \mathbb{R}_+$ is sufficiently small.

References

1. M. Abramowitz and I. A. Stegun, *Handbook of mathematical functions: with formulas, graphs, and mathematical tables*, vol. 55, Courier Corporation, (1964).
2. G. Alessandrini and L. Rondi, *Determining a sound-soft polyhedral scatterer by a single far-field measurement*, Proc. Amer. Math. Soc., **35** (2005), 1685–1691.

3. T. S. Angell and A. Kirsch, *The conductive boundary condition for Maxwells equations*, SIAM Journal on Applied Mathematics, **52(6)** (1992), 1597–1610.
4. E. Blåsten, *Nonradiating sources and transmission eigenfunctions vanish at corners and edges*, SIAM J. Math. Anal., **50(6)** (2018), 6255–6270.
5. E. Blåsten, X. Li,H. Liu and Y. Wang, On vanishing and localization near cusps of transmission eigenfunctions: a numerical study, Inverse Problems, **33** (2017), 105001.
6. E. Blåsten and H. Liu, *On vanishing near corners of transmission eigenfunctions*, J. Funct. Anal., **273(11)** (2017), 3616–3632. Addendum, arXiv:1710.08089
7. E. Blåsten and H. Liu, *Recovering piecewise constant refractive indices by a single far-field pattern*,Inverse Problems, **36** (2020), 085005.
8. E. Blåsten and H. Liu, *Scattering by curvatures, radiationless sources, transmission eigenfunctions and inverse scattering problems*, SIAM Journal on Mathematical Analysis, **53** (2021), no. 4, 3801–3837.
9. E. Blåsten and H. Liu, *On corners scattering stably and stable shape determination by a single far-field pattern*, Indiana Univ. Math. J., **70** (2021), no. 3, 907–947.
10. E. Blåsten and L. Päivärinta, *Completeness of generalized transmission eigenfunctions*, Inverse Problems, **29** (2013), 104002.
11. E. Blåsten, L. Päivärinta and J. Sylvester, *Corners always scatter*, Comm. Math. Phys., **331** (2014), 725–753.
12. O. Bondarenko, *The Factorization Method for Conducting Transmission Conditions*, M.Sc. Dissertation, KIT, 2016.
13. O. Bondarenko, I. Harris, and A. Kleefeld, *The interior transmission eigenvalue problem for an inhomogeneous media with a conductive boundary*, Applicable Analysis, **96(1)** (2017), 2–22.
14. O. Bondarenko and X. Liu, *The factorization method for inverse obstacle scattering with conductive boundary condition*, Inverse Problems, **29(9)** (2013), 095021.
15. F. Cakoni and H. Haddar, *Transmission eigenvalues in inverse scattering theory*, in "Inverse Problems and Applications: Inside Out II", Math. Sci. Res. Inst. Publ., Vol. 60, pp. 529–580, Cambridge Univ. Press., Cambridge, (2013).
16. X. Cao, H. Diao, and J. Li, *Some recent progress on inverse scattering problems within general polyhedral geometry*, Electronic Research Archive, **29** (1) (2021), 1753–1782.
17. X. Cao, H. Diao and H. Liu, *Determining a piecewise conductive medium body by a single far-field measurement*, CSIAM Trans. Appl. Math., **1** (2020), 740–765.
18. T. Chaumont-Frelet and S. Nicaise, *High-frequency behaviour of corner singularities in Helmholtz problems*, ESAIM: Math. Model. Numer. Anal., **52(5)**(2018), 1803–1845.
19. J. Cheng and M. Yamamoto, *Uniqueness in an inverse scattering problem within non-trapping polygonal obstacles with at most two incoming waves*, Inverse Problems, **19** (2003), 1361–1384.
20. D. Colton, A. Kirsch and L. Päivärinta, *Far-field patterns for acoustic waves in an inhomogeneous medium*, SIAM J. Math. Anal., **20** (1989),1472–1483.
21. D. Colton and R. Kress, *Looking back on inverse scattering theory*, SIAM Review, **60(40)** (2018), 779–807.
22. D. Colton and R. Kress, *Inverse acoustic and electromagnetic scattering theory*, volume 93 of Applied Mathematical Sciences. Springer, Cham, 4th edition, 2019.
23. D. Colton and P. Monk, *The inverse scattering problem for time-harmonic acoustic waves in an inhomogeneous medium*, Quart. J. Mech. Appl. Math., **41** (1988), 97–125.-
24. M. Costabel, *Boundary integral operators on Lipschitz domains: elementary results*, SIAM J. Math. Anal., **19(3)** (1988), 613–626.
25. M. Costabel, *A remark on the regularity of solutions of Maxwell's equations on Lipschitz domains*, Math. Methods Appl. Sci., **12** (1990), 365–368.
26. A. Cossonniére and H. Haddar, *The electromagnetic interior transmission problem for regions with cavities*, SIAM J. Math. Anal., **43(4)** (2011), 1698–1715.
27. M. Dauge, *Elliptic Boundary Value Problems in Corner Domains-Smoothness and Asymptotics of Solutions*, Lecture Notes in Mathematics, **1341**. Springer-Verlag, Berlin (1988).
28. H. Diao, X. Cao and H. Liu, *On the geometric structures of conductive transmission eigenfunctions and their application*, Comm. Partial Differential Equations, **46(4)** (2021), 630–679.
29. H. Diao, H. Liu and L. Wang, *On generalized Holmgren's principle to the Lamé operator with applications to inverse elastic problems*, Calculus of Variations and Partial Differential Equations, **59** (2020), 179.
30. P. Grisvard, *Boundary Value Problems in Non-Smooth Domains*, Pitman, London (1985).
31. R. F. Harrington and J. R. Mautz, *An impedance sheet approximation for thin dielectric shells*, IEEE Trans. Antennas and Propagation, **AP-23** (1975), 531–534.
32. F. Hettlich and A. Kirsch, *Schiffers theorem in inverse scattering for periodic structures*, Inverse Problems **13** (1997), 351–361.
33. G. Hu, M. Salo and E. Vesalainen, *Shape identification in inverse medium scattering problems with a single far-field pattern*, SIAM J. Math. Anal., **48** (2016),152–165.
34. A. Kirsch, *The denseness of the far field patterns for the transmission problem*, IMA J. Appl. Math., **37** (1986), 213–225.

35. E. Lakshtanov and B. Vainberg. *Applications of elliptic operator theory to the isotropic interior transmission eigenvalue problem*, Inverse Problems, **29** (2013), 104003.
36. H. Liu, *A global uniqueness for formally determined inverse electromagnetic obstacle scattering*, Inverse Problems, **24** (2008), 035018 (13pp).
37. H. Liu, *On local and global structures of transmission eigenfunctions and beyond*, J. Inverse Ill-Posed Probl., **30(2)** (2022), 287–305.
38. H. Liu, M. Petrini, L. Rondi and J. Xiao, Stable determination of sound-hard polyhedral scatterers by a minimal number of scattering measurements, *J. Differential Equations*, **262(3)** (2017), 1631–1670.
39. H. Liu, L. Rondi and J. Xiao, Mosco convergence for $H(curl)$ spaces, higher integrability for Maxwell's equations, and stability in direct and inverse EM scattering problems, *J. Eur. Math. Soc.*, *J. Eur. Math. Soc.*, **21** (2019), no. 10, 2945–2993.
40. H. Liu and J. Xiao, Decoupling elastic waves and its applications, *J. Differential Equations*, **265(8)** (2017), 4442–4480.
41. H. Liu and J. Xiao, *On electromagnetic scattering from a penetrable corner*, SIAM J. Math. Anal., **49(6)** (2017), 5207–5241.
42. H. Liu, M. Yamamoto and J. Zou, *Reflection principle for the Maxwell equations and its application to inverse electromagnetic scattering*, Inverse Problems, **23** (2007), 2357–2366.
43. H. Liu, M. Yamamoto and J. Zou, *New reflection principles for Maxwell's equations and their applications*, Numer. Math. Theor. Meth. Appl., **2(1)** (2009), 1–17.
44. H. Liu and J. Zou, *Uniqueness in an inverse acoustic obstacle scattering problem for both sound-hard and sound-soft polyhedral scatterers*, Inverse Problems, **22** (2006), 515–524.
45. H. Liu and J. Zou, *On uniqueness in inverse acoustic and electromagnetic obstacle scattering problems*, Journal of Physics: Conference Series, Applied Inverse Problems, Vancouver, Canada, (2007).
46. H. Liu and J. Zou, *On unique determination of partially coated polyhedral scatterers with far field measurements*, Inverse Problems, **23** (2007), 297–308.
47. W. Mclean, *Strongly Elliptic Systems and Boundary Integral Equation*, Cambridge University Press, Cambridge, (2000).
48. A. I. Nachman, Reconstructions from boundary measurements, *Ann. of Math. (2)* **128(3)** (1988), 531–576.
49. L. Päivärinta, M. Salo and E. V. Vesalainen, *Strictly convex corners scatter*, Rev. Mat. Iberoam. **33(4)** (2017), 1369–1396.
50. L. Päivärinta and J. Sylvester, *Transmission eigenvalues*, SIAM J. Math. Anal., **40** (2008), 738–753.
51. L. Robbiano, *Spectral analysis of the interior transmission eigenvalue problem*, Inverse Problems, **29** (2013), 104001.
52. L. Rondi, *Stable determination of sound-soft polyhedral scatterers by a single measurement*, Indiana Univ. Math. J., **57** (2008), 1377–1408.
53. B. P. Rynne and B. D. Sleeman, *The interior transmission problem and inverse scattering from inhomogeneous media*, SIAM J. Math. Anal., **22** (1991), 1755–1762.
54. T. B. A. Senior, *A note on the impedance boundary condition*, Canad. J. Phys., **40** (1962), 663–665.
55. J. Sylvester and G. Uhlmann, *A global uniqueness theorem for an inverse boundary value problem*, Ann. of Math. (2) **125(1)** (1987), 153–169.
56. N. Weck, *Approximation by Herglotz wave functions*, Math. Methods Appl. Sci., **27(2)** (2004), 155–162.

Chapter 9
Geometric Structures of Maxwell's Transmission Eigenfunctions and Applications

9.1 Overview

In this chapter, we are mainly concerned with the time-harmonic Maxwell system, where we follow the treatment in [8]. Let $\omega \in \mathbb{R}_+$ denote the frequency, and let $\varepsilon_0, \mu_0 \in \mathbb{R}_+$, respectively, signify the electric permittivity and magnetic permeability of a uniformly homogeneous space. Throughout, we let $\mathbf{E}$ and $\mathbf{H}$ denote, respectively, the electric and magnetic fields, which are $\mathbb{C}^3$-valued functions. We first introduce the notion of the corner and the corresponding Maxwell system locally around the corner.

Let $w_j \in \mathbb{S}^2 := \{\mathbf{x} \in \mathbb{R}^3; |\mathbf{x}| = 1\}$, $j = 1, \ldots, n$, be n number of unit vectors with $n \geq 3$ such that they are triple-wise linearly independent. For a given point $\mathbf{x}_0 \in \mathbb{R}^3$, we define

$$\mathscr{K} = \mathscr{K}_{w_1,\ldots,w_n;\mathbf{x}_0} := \{\mathbf{x} = \mathbf{x}_0 + \sum_{j=1}^{n} c_j w_j;\ c_j > 0, j = 1, \ldots, n\} \subset \mathbb{R}^3. \tag{9.1.1}$$

We assume that $\mathscr{K}$ is strictly convex: it is convex and must fit into a spherical cone of opening angle less than π. We may assume that none of the w_js are redundant. Then $\mathscr{K}$ is called a convex polyhedral cone with n edges in $\mathbb{R}^3$. The point x_0 is the apex of the cone, and w_j, $j = 1, 2, \ldots, n$, are the n directions of the corresponding edges. Given a constant $r_0 \in \mathbb{R}_+$, we define the truncated parallelized $\mathscr{K}^{r_0} = \mathscr{K}_{\mathbf{x}_0}^{r_0}$ as

$$\mathscr{K}^{r_0} = \mathscr{K}_{\mathbf{x}_0}^{r_0} = \mathscr{K}_{w_1,\ldots,w_n;x_0}^{r_0} := \mathscr{K}_{w_1,\ldots,w_n;\mathbf{x}_0} \cap B_{r_0}(\mathbf{x}_0). \tag{9.1.2}$$

Let $\mathbf{F}_1$ and $\mathbf{F}_2$ be two $\mathbb{C}^3$-valued functions such that $\mathbf{F}_j \in C^\alpha(\overline{\mathscr{K}^{r_0}})^3$, $j = 1, 2$, where the Hölder index $\alpha \in (0, 1)$. We start from the following Maxwell system in the corner $\mathscr{K}_{\mathbf{x}_0}^{r_0}$:

$$\begin{cases} \nabla \wedge \mathbf{E} - \mathrm{i}\omega\mu_0 \mathbf{H} = \mathbf{F}_1 & \text{in } \mathscr{K}_{\mathbf{x}_0}^{r_0}, \\ \nabla \wedge \mathbf{H} + \mathrm{i}\omega\varepsilon_0 \mathbf{E} = \mathbf{F}_2 & \text{in } \mathscr{K}_{\mathbf{x}_0}^{r_0}, \\ \nu \wedge \mathbf{E} = \nu \wedge \mathbf{H} = \mathbf{0} & \text{on } \partial\mathscr{K}_{\mathbf{x}_0}^{r_0} \setminus \partial B_{r_0}(\mathbf{x}_0), \end{cases} \tag{9.1.3}$$

where $\mathrm{i} := \sqrt{-1}$ denotes the imaginary unit and $\nu \in \mathbb{S}^2$ is the exterior unit normal vector to $\partial\mathscr{K}_{\mathbf{x}_0}^{r_0} \setminus \partial B_{r_0}(\mathbf{x}_0)$. We emphasize that the homogeneous boundary conditions in (9.1.3) are imposed only on the faces of the corner $\mathscr{K}_{\mathbf{x}_0}^{r_0}$ around the apex $\mathbf{x}_0$.

It can be seen from later discussion that the problem (9.1.3) is the key for several separate topics in time-harmonic electromagnetic scattering, including the geometric characterization of non-radiating sources and the uniqueness in determining the support of radiating sources. Moreover, mathematical models for invisible medium scatters, geometric structures of interior transmission eigenfunctions, as well as the unique determination of the support of inhomogeneous medium scat-

H. Diao, H. Liu, *Spectral Geometry and Inverse Scattering Theory*,
https://doi.org/10.1007/978-3-031-34615-6_9

ters can be also reduced into the local problem (9.1.3). These topics, in particular those associated with medium scattering, have received considerable interest in the literature recently and have been extensively investigated from different perspectives [1–7,9,10,14,20,24,25,30,37,42], mainly for the Helmholtz system governing the acoustic scattering. In this chapter, we aim to extend some of the earlier results in [1,2,4–7] for the acoustic (medium) scattering to the electromagnetic medium as well as source scattering. Recently, visibility, invisibility, and unique recovery of inverse electromagnetic problems with conical and polyhedral singularities can be found in [21,22].

The key result concerning the problem (9.1.3), which will be substantially used later to obtain results for the aforementioned related topics, is stated as follows.

Theorem 9.1.1. *For any given* $\mathbf{F}_j \in C^{\alpha}(\overline{\mathscr{K}^{r_0}})^3$, $j=1,2$, *with* $\alpha \in (0,1)$, *we consider the Maxwell problem* (9.1.3). *Suppose there exists a pair of solutions* $(\mathbf{E},\mathbf{H}) \in H(\mathrm{curl},\mathscr{K}^{r_0}) \times H(\mathrm{curl},\mathscr{K}^{r_0})$ *to* (9.1.3). *Then there must hold that*

$$\mathbf{F}_1(\mathbf{x}_0) = \mathbf{F}_2(\mathbf{x}_0) = \mathbf{0}. \tag{9.1.4}$$

We would like to emphasize that Theorem 9.1.1 and the problem (9.1.3) concern only the Maxwell equations locally in the truncated polyhedral cone $\mathscr{K}_{\mathbf{x}_0}^{r_0}$. This allows for great generality for various applications in source or medium scattering as mentioned before. Roughly speaking, only local conditions of coefficients or fields around the corner need to be assumed, in order for results on scattering or inverse scattering in latter sections to be valid. Moreover, we allow corners with three or more than three number of edges.

The rest of the chapter is organized as follows. In Sect. 9.2, we give the proof of Theorem 9.1.1. In Sect. 9.3, we consider the electromagnetic scattering from active sources. We provide a geometric characterization of non-radiating sources; see Theorem 9.3.1 and its corollaries. The unique recovery results of the supports of radiating sources are included in Theorems 9.3.2 and 9.3.3. In Sect. 9.4, we consider the electromagnetic scattering from medium scatters due to incident waves. We derive a geometric structure of electromagnetic interior transmission eigenfunctions in Theorem 9.4.1. An implication to invisibility cloaking is given in Theorem 9.4.2 and that to the unique determination of the supports of electromagnetic media in Theorem 9.4.3.

9.2 Proof of Theorem 9.1.1

In this section, we give the proof of Theorem 9.1.1. To that end, we first present two auxiliary lemmas.

Lemma 9.2.1. *Let* Ω *be a bounded Lipschitz domain in* $\mathbb{R}^3$ *and* $\mathbf{J}_j \in L^2(\Omega;\mathbb{C}^3)$. *Suppose that* $(\mathbf{E},\mathbf{H}) \in H(\mathrm{curl},\Omega) \times H(\mathrm{curl},\Omega)$ *is a solution to the Maxwell system*

$$\nabla \wedge \mathbf{E} - \mathrm{i}\omega\mu_0\mathbf{H} = \mathbf{J}_1, \quad \nabla \wedge \mathbf{H} + \mathrm{i}\omega\nu_0\mathbf{E} = \mathbf{J}_2 \qquad \text{in } \Omega. \tag{9.2.1}$$

Then one has

$$\int_{\Omega} \mathbf{J}_1 \cdot \mathbf{W} + \int_{\Omega} \mathbf{J}_2 \cdot \mathbf{V} = \int_{\partial\Omega} \mathbf{W} \cdot (\nu \wedge \mathbf{E}) + \int_{\partial\Omega} \mathbf{V} \cdot (\nu \wedge \mathbf{H}) \tag{9.2.2}$$

and

$$\varepsilon_0 \int_{\Omega} \mathbf{J}_1 \cdot \mathbf{V} - \mu_0 \int_{\Omega} \mathbf{J}_2 \cdot \mathbf{W} = \varepsilon_0 \int_{\partial\Omega} \mathbf{V} \cdot (\nu \wedge \mathbf{E}) - \mu_0 \int_{\partial\Omega} \mathbf{W} \cdot (\nu \wedge \mathbf{H}), \tag{9.2.3}$$

for any $(\mathbf{V},\mathbf{W}) \in H(\mathrm{curl},\Omega) \times H(\mathrm{curl},\Omega)$ *satisfying*

$$\nabla \wedge \mathbf{V} - \mathrm{i}\omega\mu_0\mathbf{W} = \mathbf{0}, \quad \nabla \wedge \mathbf{W} + \mathrm{i}\omega\varepsilon_0\mathbf{V} = \mathbf{0} \qquad \text{in } \Omega. \tag{9.2.4}$$

Proof. Applying $\mathbf{V} \in H(\mathrm{curl}, \Omega)$ as a test function of the Maxwell system (9.2.1) and integrating by parts yield

$$\begin{aligned} \int_\Omega \mathbf{J}_1 \cdot \mathbf{V} &= \int_\Omega (\nabla \wedge \mathbf{E} - \mathrm{i}\omega\mu_0 \mathbf{H}) \cdot \mathbf{V} \\ &= \int_{\partial\Omega} \mathbf{V} \cdot (\nu \wedge \mathbf{E}) + \int_\Omega \mathrm{i}\omega (\mu_0 \mathbf{W} \cdot \mathbf{E} - \mu_0 \mathbf{H} \cdot \mathbf{V}) \end{aligned} \tag{9.2.5}$$

and

$$\begin{aligned} \int_\Omega \mathbf{J}_2 \cdot \mathbf{V} &= \int_\Omega (\nabla \wedge \mathbf{H} + \mathrm{i}\omega\varepsilon_0 \mathbf{E}) \cdot \mathbf{V} \\ &= \int_{\partial\Omega} \mathbf{V} \cdot (\nu \wedge \mathbf{H}) + \int_\Omega \mathrm{i}\omega (\mu_0 \mathbf{W} \cdot \mathbf{H} + \varepsilon_0 \mathbf{E} \cdot \mathbf{V}), \end{aligned} \tag{9.2.6}$$

where we have used the property that $(\mathbf{V}, \mathbf{W})$ satisfies the Maxwell system (9.2.4). In a similar way, one can obtain

$$\int_\Omega \mathbf{J}_1 \cdot \mathbf{W} = \int_{\partial\Omega} \mathbf{W} \cdot (\nu \wedge \mathbf{E}) - \int_\Omega \mathrm{i}\omega (\varepsilon_0 \mathbf{V} \cdot \mathbf{E} + \mu_0 \mathbf{H} \cdot \mathbf{W}) \tag{9.2.7}$$

and

$$\int_\Omega \mathbf{J}_2 \cdot \mathbf{W} = \int_{\partial\Omega} \mathbf{W} \cdot (\nu \wedge \mathbf{H}) - \int_\Omega \mathrm{i}\omega (\varepsilon_0 \mathbf{V} \cdot \mathbf{H} - \varepsilon_0 \mathbf{E} \cdot \mathbf{W}). \tag{9.2.8}$$

Now, the identities (9.2.2) and (9.2.3) can be straightforwardly verified by using (9.2.5)–(9.2.8). □

Lemma 9.2.2. *Let $\mathscr{K} = \mathscr{K}_{w_1,\ldots,w_n;\mathbf{0}}$ be a convex polyhedral cone with n number of edges, $n \geq 3$. Given a constant k and a nontrivial constant complex vector $\mathbf{F}_0$, there exist positive constants $c_{\mathscr{K}}$ and $C_{\mathscr{K}}$ and vectors $\mathbf{d}, \mathbf{d}^\perp \in \mathbb{S}^2$ with $\mathbf{d}^\perp \perp \mathbf{d}$, which satisfy that*

$$\mathbf{d} \cdot \theta < -c_{\mathscr{K}} \quad \textit{for any } \theta \in \mathscr{K} \cap \mathbb{S}^2 \tag{9.2.9}$$

and that

$$\left| \int_{\mathscr{K}} e^{\rho \cdot \mathbf{x}} \mathrm{d}\mathbf{x} \right| \geq C_{\mathscr{K}} \tau^{-3}, \quad \textit{for any } \tau \geq k, \tag{9.2.10}$$

where the complex vector ρ is given by

$$\rho := \tau \mathbf{d} + \mathrm{i}\sqrt{\tau^2 + k^2}\,\mathbf{d}^\perp. \tag{9.2.11}$$

Moreover, the vectors d and $d^\perp$ can be chosen in such a way that, denoting

$$\mathbf{p} := \mathbf{d}^\perp - \mathrm{i}\sqrt{1 + k^2/\tau^2}\,\mathbf{d}, \tag{9.2.12}$$

one has

$$\mathbf{p} \cdot \rho = 0, \qquad \rho \wedge \mathbf{p} = -k^2 \left(\mathbf{d} \wedge \mathbf{d}^\perp \right) / \tau, \tag{9.2.13}$$

and the limit below exists and satisfies

$$\lim_{\tau \to \infty} \mathbf{F}_0 \cdot \mathbf{p} \neq 0. \tag{9.2.14}$$

We postpone the proof of Lemma 9.2.2 to the end of this section and first present the proof of Theorem 9.1.1.

Proof of Theorem 9.1.1. We assume without loss of generality that $\mathbf{x}_0 = \mathbf{0}$, which can be achieved by a rigid change of coordinates. Recall from Lemma 9.2.1 and the boundary condition in (9.1.3) that, for any $(\mathbf{V}, \mathbf{W})$ satisfying (9.2.4), there holds

$$\int_{\mathscr{K}^{r_0}} \mathbf{F}_1 \cdot \mathbf{W} + \int_{\mathscr{K}^{r_0}} \mathbf{F}_2 \cdot \mathbf{V} = \int_{\partial \mathscr{K}^{r_0} \cap \partial B_{r_0}} \mathbf{W} \cdot (\nu \wedge \mathbf{E}) + \int_{\partial \mathscr{K}^{r_0} \cap \partial B_{r_0}} \mathbf{V} \cdot (\nu \wedge \mathbf{H}). \tag{9.2.15}$$

We first prove $\mathbf{F}_2(\mathbf{0}) = 0$ by contradiction.

Since $\mathbf{F}_j \in C^\alpha(\overline{\mathscr{K}^{r_0}})^3$, we can write

$$\mathbf{F}_2 = \mathbf{F}_0 + \tilde{\mathbf{F}},$$

with $\mathbf{F}_0$ a constant vector and $\tilde{\mathbf{F}}$ a vector field satisfying

$$|\tilde{\mathbf{F}}(\mathbf{x})| \leq C|\mathbf{x}|^\alpha, \quad \mathbf{x} \in \mathscr{K}^{r_0}.$$

Under this splitting of $\mathbf{F}_2$, the equation (9.2.15) can be written as

$$\begin{aligned} \int_{\mathscr{K}} \mathbf{F}_0 \cdot \mathbf{V} = & \int_{\partial \mathscr{K}^{r_0} \cap \partial B_{r_0}} \mathbf{W} \cdot (\nu \wedge \mathbf{E}) + \int_{\partial \mathscr{K}^{r_0} \cap \partial B_{r_0}} \mathbf{V} \cdot (\nu \wedge \mathbf{H}) \\ & + \int_{\mathscr{K} \setminus \mathscr{K}^{r_0}} \mathbf{F}_0 \cdot \mathbf{V} - \int_{\mathscr{K}^{r_0}} \tilde{\mathbf{F}} \cdot \mathbf{V} - \int_{\mathscr{K}^{r_0}} \mathbf{F}_1 \cdot \mathbf{W}. \end{aligned} \tag{9.2.16}$$

We choose $(\mathbf{V}, \mathbf{W})$, a pair of solutions to the Maxwell system (9.2.4), as

$$\mathbf{V}(\mathbf{x}) = \mathbf{p} e^{\rho \cdot \mathbf{x}} \quad \text{and} \quad \mathbf{W}(\mathbf{x}) = \frac{1}{\mathrm{i}\omega\mu_0} \rho \wedge \mathbf{p} e^{\rho \cdot \mathbf{x}},$$

with the complex vectors ρ and $\mathbf{p}$ given in Lemma 9.2.2 for $k^2 = \omega^2 \varepsilon_0 \mu_0$.

We assume $\mathbf{F}_2(\mathbf{0}) \neq 0$, namely, $\mathbf{F}_0 \neq 0$. Concerning the LHS of (9.2.16), we obtain from Lemma 9.2.2, in particular from (9.2.10) and (9.2.14), that

$$\left| \int_{\mathscr{K}} \mathbf{F}_0 \cdot \mathbf{V} \right| = |\mathbf{F}_0 \cdot p| \left| \int_{\mathscr{K}} e^{\rho \cdot \mathbf{x}} \mathrm{d}\mathbf{x} \right| \geq \mathbf{F}_0 \cdot \mathbf{p} C_{\mathscr{K}} \tau^{-3} \geq C_0 \tau^{-3} > 0 \tag{9.2.17}$$

holds for τ sufficiently large, with a constant C_0 (strictly) positive and independent of τ. We shall show in the rest of the proof that the RHS of (9.2.16) is bounded by $C\tau^{-(3+\alpha)}$ and hence leads to a contradiction.

We first deal with the terms in (9.2.16) concerning $\mathbf{V}$. For the integral over $\mathscr{K}^{r_0}$, we have

$$\left| \int_{\mathscr{K}^{r_0}} \tilde{\mathbf{F}} \cdot \mathbf{V} \right| \leq \|\mathbf{F}_2\|_{C^\alpha} |\mathbf{p}| \int_{\mathscr{K}^{r_0}} |\mathbf{x}|^\alpha e^{\tau \mathbf{d} \cdot \mathbf{x}} \mathrm{d}\mathbf{x} \leq 3 \|\mathbf{F}_2\|_{C^\alpha} \tau^{-(3+\alpha)} \int_{\mathscr{K}} |\mathbf{y}|^\alpha e^{\mathbf{d} \cdot \mathbf{y}} \mathrm{d}\mathbf{y}.$$

Recalling (9.2.9) from Lemma 9.2.2, one obtains

$$\int_{\mathscr{K}} |\mathbf{y}|^\alpha e^{\mathbf{d} \cdot \mathbf{y}} \mathrm{d}\mathbf{y} \leq \int_{\mathscr{K}} |\mathbf{y}|^\alpha e^{-c_{\mathscr{K}} |\mathbf{y}|} \mathrm{d}\mathbf{y} \leq 2\pi \int_0^\infty r^{2+\alpha} e^{-c_{\mathscr{K}} r} \mathrm{d}r = C_{\mathscr{K},\alpha} < \infty.$$

As a consequence, we have

$$\left| \int_{\mathscr{K}^{r_0}} \tilde{\mathbf{F}} \cdot \mathbf{V} \right| \leq C_{\mathscr{K},\alpha} \|\mathbf{F}_2\|_{C^\alpha} \tau^{-(3+\alpha)}. \tag{9.2.18}$$

For the integral over $\mathscr{K} \setminus \mathscr{K}^{r_0}$, we can derive in a similar way that

$$
\begin{aligned}
\left|\int_{\mathscr{K}\setminus\mathscr{K}^{r_0}} \mathbf{F}_0\cdot\mathbf{V}\right| \leq & |\mathbf{p}\cdot\mathbf{F}_0|\int_{\mathscr{K}\setminus\mathscr{K}^{r_0}} e^{\tau\mathbf{d}\cdot\mathbf{x}}\mathrm{d}\mathbf{x} \leq |\mathbf{p}\cdot\mathbf{F}_0|\int_{\mathscr{K}\setminus\mathscr{K}^{r_0}} e^{-c_{\mathscr{K}}\tau|\mathbf{x}|}\mathrm{d}\mathbf{x} \\
\leq & 2\pi\,|\mathbf{p}\cdot\mathbf{F}_0|\int_{r_0}^{\infty} r^2 e^{-c_{\mathscr{K}}\tau r}\mathrm{d}r \\
= & 2\pi\,|\mathbf{p}\cdot\mathbf{F}_0|\,e^{-c_{\mathscr{K}}\tau r_0}\int_0^{\infty}(r+r_0)^2 e^{-c_{\mathscr{K}}\tau r}\mathrm{d}r \\
\leq & C_{\mathscr{K},r_0}\,|\mathbf{p}\cdot\mathbf{F}_0|\,e^{-c_{\mathscr{K}}\tau r_0} < \infty,
\end{aligned} \tag{9.2.19}
$$

when τ is sufficiently large. As for the boundary integral in (9.2.16), we have the estimate

$$
\begin{aligned}
\left|\int_{\partial\mathscr{K}^{r_0}\cap\partial B_{r_0}} \mathbf{V}\cdot(\nu\wedge\mathbf{H})\right| \leq & 3\int_{\partial\mathscr{K}^{r_0}\cap\partial B_{r_0}} |\nu\wedge\mathbf{H}|\,e^{-c_{\mathscr{K}}\tau|\mathbf{x}|}\mathrm{d}\mathbf{x} \\
\leq & 3C_{\mathscr{K},r_0}\tau^{1/2}e^{-c_{\mathscr{K}}r_0\tau}\|\mathbf{H}\|_{H(\mathrm{curl},\mathscr{K}^{r_0})}.
\end{aligned} \tag{9.2.20}
$$

Recall the identity (9.2.13) from Lemma 9.2.2 that the modulus of $\mathbf{W}$ has the order of τ^{-1}, with respect to τ. Hence, similar to (9.2.20), we have

$$
\left|\int_{\partial\mathscr{K}^{r_0}\cap\partial B_{r_0}} \mathbf{W}\cdot(\nu\wedge\mathbf{E})\right| \leq 3C_{\mathscr{K},r_0}k^2\tau^{-1/2}e^{-c_{\mathscr{K}}r_0\tau}\|\mathbf{E}\|_{H(\mathrm{curl},\mathscr{K}^{r_0})}. \tag{9.2.21}
$$

Lastly, by using a similar argument as for deriving (9.2.18), one can obtain

$$
\left|\int_{\mathscr{K}^{r_0}} \mathbf{F}_1\cdot\mathbf{W}\right| \leq \omega\varepsilon_0\tau^{-1}\|\mathbf{F}_1\|_{C^0}\int_{\mathscr{K}^{r_0}} e^{\tau\mathbf{d}\cdot\mathbf{x}}\mathrm{d}\mathbf{x} \leq C_{\mathscr{K}}\,\omega\varepsilon_0\|\mathbf{F}_1\|_{C^0}\tau^{-4}. \tag{9.2.22}
$$

In summary, the assumption $\mathbf{F}_2(\mathbf{0}) \neq 0$ implies

$$
C_0\tau^{-3} \leq \tilde{C}_{\mathbf{F}_1,\mathbf{F}_2,\mathscr{K},r_0,\alpha}\tau^{-(3+\alpha)}, \tag{9.2.23}
$$

with $C_0 > 0$, and this holds for any τ sufficiently large, which is impossible. Therefore, we have shown by contradiction that $\mathbf{F}_2(\mathbf{0}) = \mathbf{0}$. Finally, one can verify $\mathbf{F}_1(\mathbf{0}) = \mathbf{0}$ in the same way and by taking

$$
\mathbf{V}(\mathbf{x}) = -\frac{1}{\mathrm{i}\omega\varepsilon_0}\rho\wedge\mathbf{p}e^{\rho\cdot\mathbf{x}} \quad \text{and} \quad \mathbf{W}(\mathbf{x}) = \mathbf{p}e^{\rho\cdot\mathbf{x}}.
$$

□

9.2.1 Proof of Lemma 9.2.2

Proof. We assume, up to some exchanges of notations, that the plane $span\{w_2, w_3\}$ separates w_1 to the other side of space from w_j, $j = 4, \ldots, n$. Notice that when the polyhedral cone contains only $n = 3$ number of edges, this assumption holds automatically. Since the convex polyhedral cone $\mathscr{K} = \mathscr{K}_{w_1,\ldots,w_n;0}$ fits into a half-space, we can find a positive constant κ and a vector $z \in \mathbb{S}^2$ satisfying

$$
z\cdot w_1 = 0, \quad \text{and} \quad z\cdot w_j < -\kappa,\ j = 2,\ldots,n. \tag{9.2.24}
$$

For any constant $s > 0$, we define

$$
\mathbf{d} = \mathbf{d}_s := \frac{z - sw_1}{|z - sw_1|} = \frac{z - sw_1}{\sqrt{1+s^2}} \tag{9.2.25}
$$

and $\mathbf{d}^\perp := w_1 \wedge z$. It is noticed that $\mathbf{d}^\perp$ is perpendicular to $\mathbf{d}_s$, independent of the choice of s. We also set

$$\rho = \rho_s := \tau \mathbf{d}_s + \mathrm{i}\sqrt{\tau^2 + k^2}\mathbf{d}^\perp. \tag{9.2.26}$$

Denote

$$\mathscr{K}_0 := \{c_1 w_1 + c_2 w_2 + c_3 w_3;\ c_1, c_2, c_3 > 0\}.$$

It is observed by straightforward calculation (see also [7, Proof of Lemma 3.4]) that

$$\int_{\mathscr{K}_0} e^{\rho\cdot\mathbf{x}} \mathrm{d}\mathbf{x} = \frac{|w_1 \wedge w_2 \wedge w_3|}{(-\rho\cdot w_1)(-\rho\cdot w_2)(-\rho\cdot w_3)} \tag{9.2.27}$$

and that

$$\left|\frac{\tau}{\rho\cdot w_1}\right| = \left|\frac{1}{d_s\cdot w_1}\right| = \frac{\sqrt{1+s^2}}{s} > \frac{1}{s}. \tag{9.2.28}$$

Moreover, we have for $\tau \geq k$ that

$$\begin{aligned}
\left|\frac{\tau}{\rho\cdot w_j}\right| &= \frac{\sqrt{1+s^2}}{\left|z\cdot w_j - s w_1\cdot w_j + \mathrm{i}\sqrt{1+k^2/\tau^2}\sqrt{1+s^2} d^\perp\cdot w_j\right|} \\
&= \frac{\sqrt{1+s^2}}{\sqrt{\left|z\cdot w_j - s w_1\cdot w_j\right|^2 + (1+k^2/\tau^2)(1+s^2)\left|d^\perp\cdot w_j\right|^2}} \\
&\geq \frac{\sqrt{1+s^2}}{\sqrt{2(1+s^2)+(1+k^2/\tau^2)(1+s^2)}} \geq \frac{1}{2}, \qquad j = 2,3.
\end{aligned} \tag{9.2.29}$$

Combining (9.2.27)–(9.2.29) yields

$$\left|\tau^3 \int_{\mathscr{K}_0} e^{\rho\cdot\mathbf{x}} \mathrm{d}\mathbf{x}\right| > \frac{|w_1\wedge w_2 \wedge w_3|}{4s}. \tag{9.2.30}$$

Let us prove an s-independent upper bound for the integral of $\exp(\rho\cdot\mathbf{x})$ over $\mathscr{K}\setminus\mathscr{K}_0$ next. This cone is generated by the vectors $w_2, \ldots, w_n$. Hence, for any unit vector $\theta \in \mathscr{K}\setminus\mathscr{K}_0$, there are $\alpha_2, \ldots, \alpha_n \in \mathbb{R}_+ \cup \{0\}$ such that

$$\theta = \alpha_2 w_2 + \ldots + \alpha_n w_n.$$

Then

$$\mathbf{d}_s\cdot\theta = \sum_{j=2}^n \alpha_j \frac{z\cdot w_j - s w_1\cdot w_j}{\sqrt{1+s^2}}. \tag{9.2.31}$$

Recall that $z\cdot w_j < -\kappa$ when $j = 2, \ldots, n$. If

$$0 < s \leq \kappa/3 < 1/3, \tag{9.2.32}$$

then we have

$$z\cdot w_j - s w_1\cdot w_j \leq -2\kappa/3, \quad j = 2, \ldots, n. \tag{9.2.33}$$

As a consequence, we obtain

$$\mathbf{d}_s\cdot\theta \leq -\sum_{j=2}^n \alpha_j \frac{2\kappa}{3\sqrt{1+1/3^2}} \leq -\frac{1}{2}\kappa, \tag{9.2.34}$$

where we have used the fact that

$$1 = |\theta| \leq \sum_{j=2}^{n} \alpha_j |w_j| = \sum_{j=2}^{n} \alpha_j.$$

Finally, this gives

$$\begin{aligned}\left|\tau^3 \int_{\mathscr{K}\setminus\mathscr{K}_0} e^{\rho\cdot\mathbf{x}}\mathrm{d}\mathbf{x}\right| &\leq \tau^3 \int_{(\mathscr{K}\setminus\mathscr{K}_0)\cap\mathbb{S}^2} \int_0^\infty e^{r\tau\mathbf{d}_s\cdot\hat{\mathbf{x}}} r^2 \mathrm{d}r\mathrm{d}\hat{\mathbf{x}} \\ &\leq \left|(\mathscr{K}\setminus\mathscr{K}_0)\cap\mathbb{S}^2\right| \tau^3 \int_0^\infty e^{-\kappa\tau r/2} r^2 \mathrm{d}r \leq 16\pi\kappa^{-3}.\end{aligned} \tag{9.2.35}$$

Let us have s so small that the right-hand side of (9.2.30) is larger than the one in (9.2.35), for example, if

$$s < s_0 := \frac{|w_1 \wedge w_2 \wedge w_3|}{128\pi}\kappa^3, \tag{9.2.36}$$

and $s_0 \leq \kappa/3$ to satisfy (9.2.32). In this case, for any $s \in (0, s_0)$, we obtain

$$\begin{aligned}\left|\int_{\mathscr{K}} e^{\rho\cdot\mathbf{x}}\mathrm{d}\mathbf{x}\right| &\geq \left|\int_{\mathscr{K}_0} e^{\rho\cdot\mathbf{x}}\mathrm{d}\mathbf{x}\right| - \left|\int_{\mathscr{K}\setminus\mathscr{K}_0} e^{\rho\cdot\mathbf{x}}\mathrm{d}\mathbf{x}\right| \\ &> \frac{|w_1 \wedge w_2 \wedge w_3|}{4s}\tau^{-3} - 16\pi\kappa^{-3}\tau^{-3} > 16\pi\kappa^{-3}\tau^{-3}.\end{aligned} \tag{9.2.37}$$

We are left to verify the property (9.2.14). Let

$$\mathbf{p}_s := \mathbf{d}^\perp - \mathrm{i}\sqrt{1 + k^2/\tau^2}\mathbf{d}_s.$$

We shall show that $\lim_{\tau\to\infty} \mathbf{F}_0 \cdot \mathbf{p}_s$ exists and cannot vanish for all $s \in (0, s_0)$. We write $\mathbf{F}_0$ as

$$\mathbf{F}_0 = b_1 w_1 + b_2 z + b_3 d^\perp.$$

Then

$$\mathbf{F}_0 \cdot p_s = b_3 + \mathrm{i}\sqrt{1 + k^2/\tau^2}\frac{sb_1 - b_2}{\sqrt{1+s^2}} \longrightarrow b_3 + \mathrm{i}\frac{sb_1 - b_2}{\sqrt{1+s^2}},$$

as $\tau \to \infty$. The right-hand side is a real analytic function of s. If it would vanish in the open interval $(0, s_0)$, then it must be zero everywhere. Considering its values at $s = 0$, $s = 1$, and $s \to \infty$ yields $b_1 = b_2 = b_3 = 0$ and hence leads to the contradiction that $\mathbf{F}_0 = \mathbf{0}$.

In conclusion, one can find $s \in (0, s_0)$, $d = d_s$ as in (9.2.25), $\mathbf{d}^\perp \in \mathbb{S}^2$ with $\mathbf{d}^\perp \cdot \mathbf{d} = 0$ satisfying (9.2.9), (9.2.10), and (9.2.14). Lastly, the equation (9.2.13) can be verified by straightforward computations. □

9.3 Non-radiating Sources and Inverse Source Scattering Problems

In this section, we are concerned with the electromagnetic scattering induced by an active source. The source is characterized by two vectorial functions $\mathbf{J}_1 \in L^2(\Omega; \mathbb{C}^3)$ and $\mathbf{J}_2 \in L^2(\Omega; \mathbb{C}^3)$, which are, respectively, referred to as the electric and the magnetic current densities. Here, in order to appeal for a general mathematical study, we consider the possible presence of both electric and magnetic sources, though only the electric source might be the physically meaningful one. The source radiates electromagnetic waves and satisfies the following Maxwell system:

$$
\begin{aligned}
&\nabla \wedge \mathbf{E}(\mathbf{x}) - \mathrm{i}\omega\mu_0 \mathbf{H}(x) = \mathbf{J}_1(\mathbf{x}), \quad \mathbf{x} \in \mathbb{R}^3, \\
&\nabla \wedge \mathbf{H}(\mathbf{x}) + \mathrm{i}\omega\varepsilon_0 \mathbf{E}(\mathbf{x}) = \mathbf{J}_2(\mathbf{x}), \quad \mathbf{x} \in \mathbb{R}^3, \\
&\lim_{|\mathbf{x}|\to\infty} |\mathbf{x}| \left(\sqrt{\mu_0}\mathbf{H} \times \frac{\mathbf{x}}{|\mathbf{x}|} - \sqrt{\varepsilon_0}\mathbf{E} \right) = \mathbf{0},
\end{aligned}
\tag{9.3.1}
$$

where $\omega \in \mathbb{R}_+$ signifies the frequency of the wave. The last limit in (9.3.1) is known as the Silver-Müller radiation condition which holds uniformly in all directions $\hat{\mathbf{x}} := \mathbf{x}/|\mathbf{x}| \in \mathbb{S}^2$, $\mathbf{x} \in \mathbb{R}^3 \setminus \{0\}$, and characterizes the outgoing nature of the electromagnetic waves. As a consequence, one has the following asymptotic as $|\mathbf{x}| \to +\infty$ (cf. [16]):

$$
(\mathbf{E}, \mathbf{H})(\mathbf{x}) = \frac{e^{\mathrm{i}k|\mathbf{x}|}}{|\mathbf{x}|} (\mathbf{E}_\infty, \mathbf{H}_\infty)(\hat{\mathbf{x}}) + \mathscr{O}\left(\frac{1}{|\mathbf{x}|^2} \right), \tag{9.3.2}
$$

where $k := \omega\sqrt{\varepsilon_0\mu_0}$ is known as the wavenumber. The two fields $\mathbf{E}_\infty(\hat{\mathbf{x}})$ and $\mathbf{H}_\infty(\hat{\mathbf{x}})$ are known as, respectively, the electric and the magnetic far-field patterns. By the Rellich theorem (cf. [16]), they encode all the information of the scattered wave fields $\mathbf{E}$ and $\mathbf{H}$ in the exterior of any Lipschitz domain that encloses the support of $\mathbf{J}_1$ and $\mathbf{J}_2$. Moreover, $\mathbf{E}_\infty$ and $\mathbf{H}_\infty$ are analytic functions on the unit sphere $\mathbb{S}^2$ and satisfy the following one-to-one correspondence:

$$
\mathbf{H}_\infty(\hat{\mathbf{x}}) = \hat{\mathbf{x}} \wedge \mathbf{E}_\infty(\hat{\mathbf{x}}) \quad \text{and} \quad \mathbf{E}_\infty(\hat{\mathbf{x}}) = -\hat{\mathbf{x}} \wedge \mathbf{H}_\infty(\hat{\mathbf{x}}), \qquad \forall \hat{\mathbf{x}} \in \mathbb{S}^2.
$$

The Maxwell system (9.3.1) is well understood. We refer to [38, 41] for the existence of a unique pair of solutions $(\mathbf{E}, \mathbf{H}) \in H_{\mathrm{loc}}(\mathrm{curl}, \mathbb{R}^3) \times H_{\mathrm{loc}}(\mathrm{curl}, \mathbb{R}^3)$.

An important inverse scattering problem that arises in practical applications is to recover the unknown/inaccessible source from its associated far-field measurement. That is,

$$
\mathbf{E}_\infty(\hat{\mathbf{x}}), \hat{\mathbf{x}} \in \mathbb{S}^2 \mapsto (\Omega; \mathbf{J}_1, \mathbf{J}_2). \tag{9.3.3}
$$

Since $\mathbf{E}$ is real analytic on the unit sphere, we see that $\mathbb{S}^2$ can actually be replaced by any open subset of the unit sphere by virtue of analytic continuation. In the generic case, the dimensions of the measurement data $\mathbf{E}(\hat{\mathbf{x}})$ (associated with a fixed frequency $\omega \in \mathbb{R}_+$) and the unknown source $(\mathbf{J}_1, \mathbf{J}_2)$ in (9.3.3) are, respectively, two and three. Here, by dimension, we mean the number of free variables of the underlying quantity. Hence, it is impractical to ask for the unique recovery of the inverse problem (9.3.3), and a more practical inverse scattering problem could be posed as follows:

$$
\mathbf{E}_\infty(\hat{\mathbf{x}}),\ \hat{\mathbf{x}} \in \mathbb{S}^2 \mapsto \Omega. \tag{9.3.4}
$$

That is, instead of seeking to completely recover the unknown source functions, one intends to recover the location and the shape of the support of the source. In determining Ω, it suffices to recover $\partial\Omega$, and hence, one can easily verify that for any fixed frequency $\omega \in \mathbb{R}_+$, the inverse scattering problem in (9.3.4) is formally posed. Nevertheless, we would like to point out that the inverse problem (9.3.3) is linear, whereas the inverse problem (9.3.4) is nonlinear.

Associated with the inverse scattering problem (9.3.4), we are mainly concerned with the following two fundamental issues:

1. For what kind of source there is no radiation, namely, $\mathbf{E}_\infty \equiv \mathbf{0}$? In such a case, the source is invisible to exterior measurements since if $\mathbf{E}_\infty \equiv \mathbf{0}$, one actually has by the Rellich theorem that $\mathbf{E} = \mathbf{H} \equiv \mathbf{0}$ in $\mathbb{R}^3 \setminus \overline{\Omega}$. This kind of source is referred to as non-radiating or radiationless in the literature.
2. If the source is not invisible, namely, that it is detectable, can one really identify it by using the corresponding far-field observation? This is the identifiable and unique recovery issue. Mathematically, it can be stated as follows. Suppose that $(\Omega; \mathbf{J}_1, \mathbf{J}_2)$ and $(\Omega'; \mathbf{J}_1', \mathbf{J}_2')$ are two electromagnetic source configurations and $\mathbf{E}_\infty$ and $\mathbf{E}_\infty'$ are the associated far-field patterns, respectively. Can one conclude that

$$\mathbf{E}_\infty(\hat{\mathbf{x}}) = \mathbf{E}'_\infty(\hat{\mathbf{x}}), \hat{\mathbf{x}} \in \mathbb{S}^2 \quad \text{only if} \quad \Omega = \Omega'? \tag{9.3.5}$$

We mention in passing some related uniqueness results in [1, 32, 35, 36] for a similar inverse problem (9.3.4) posed for the acoustic scattering. We would also like to mention that for the linear inverse problem (9.3.3), however with the measurement data given by $\mathbf{E}_\infty(\hat{\mathbf{x}}, \omega)$ for all $\hat{\mathbf{x}} \in \mathbb{S}^2$ and $\omega \in \mathbb{R}_+$, there is a vast amount of literature devoted to it, both theoretically and computationally. It is of different nature from the focus of the current study which is mainly concerned with a single far-field pattern. So it would be impossible for us to give a comprehensive review of that interesting topic.

Let us first consider the geometric characterization of radiationless sources. The study of radiationless sources has a long and colorful history, which dates back to Sommerfeld's theory of extended rigid electron in 1904 [45,46]. Many physicists had theoretically predicated the existence of non-radiating sources, and it was even postulated that non-radiating charge distributions might be used as models for elementary particles and might lead to a "theory of nature" [11,12,19,23,26–29, 31, 33, 40]. As an easy example, for any Ψ_1 and Ψ_2 being $\mathbb{C}^3$-valued smooth functions with compact supports in $\mathbb{R}^3$, if one sets

$$\mathbf{J}_1 := \nabla \wedge \Psi_1 - \mathrm{i}\omega\mu_0\Psi_2, \quad \mathbf{J}_2 := \nabla \wedge \Psi_2 + \mathrm{i}\omega\varepsilon_0\Psi_1, \tag{9.3.6}$$

then $(\mathbf{J}_1, \mathbf{J}_2)$ is radiationless. Using Theorem 9.1.1 in the previous section, we can derive a geometric characterization of radiationless sources. To that end, the following definition of admissible sources shall be needed for our subsequent study.

Definition 9.3.1. Given a source function $\mathbf{J}$, it is said to belong to the class $\mathscr{A}$ if the following conditions are fulfilled:

1. There exists a bounded Lipschitz domain $\Omega \subset \mathbb{R}^3$ with a connected complement and a function $\Phi \in L^2_{loc}(\mathbb{R}^3; \mathbb{C}^3)$ satisfying that $\mathbf{J} = \chi_{\overline{\Omega}}\Phi$.
2. There exists a polyhedral cone $\mathscr{K} = \mathscr{K}_{\mathbf{x}_0}$ with the apex $x_0 \in \partial\Omega$ such that

$$\Omega \cap B_{2r_0}(\mathbf{x}_0) = \mathscr{K}_{\mathbf{x}_0} \cap B_{2r_0}(\mathbf{x}_0) \text{ for some } r_0 \in \mathbb{R}_+$$

and

$$\Phi \in C^\alpha(B_{2r_0}(\mathbf{x}_0)) \text{ for some } \alpha \in (0,1).$$

3. There is a path in $\mathbb{R}^3 \setminus \overline{\Omega}$ joining x_0 to infinity.

In this case, we also say that $(\mathbf{x}_0; \mathscr{K}^{r_0}_{\mathbf{x}_0})$ is a (generalized) corner of $\mathbf{J}$ and $\mathbf{J}$ is C^α regular at the corner.

Remark 9.3.1. We use the expression "generalized" because a corner $(\mathbf{x}_0; \mathscr{K}^{r_0}_{\mathbf{x}_0})$ of a source function $\mathbf{J}$ might be degenerated in the sense that the case $supp\mathbf{J} \cap B_{\varepsilon_0}(\mathbf{x}_0) = \emptyset$ with a constant $\varepsilon_0 > 0$ is admitted in Definition 9.3.1. As a simplest example, the trivial source function $\mathbf{J} \equiv \mathbf{0}$ belongs to the class $\mathscr{A}$, and any point $\mathbf{x}_0 \in \mathbb{R}^3$ is a (generalized) corner of $\mathbf{J}$.

Then we have a geometric characterization and classification of radiationless electromagnetic sources as follows.

Theorem 9.3.1. *Consider an electric source $\mathbf{J}_1$ and a magnetic source $\mathbf{J}_2$ that are both supported in Ω. Suppose that both $\mathbf{J}_1$ and $\mathbf{J}_2$ belong to the class $\mathscr{A}$, and let $(\mathbf{x}_0, \mathscr{K}^{r_0}_{\mathbf{x}_0})$ be a (generalized) corner of $\mathbf{J}_1$ and $\mathbf{J}_2$. If $(\Omega; \mathbf{J}_1, \mathbf{J}_2)$ is radiationless, namely, the far-field pattern $(\mathbf{E}_\infty, \mathbf{H}_\infty)$ of the Maxwell system* (9.3.1) *associated with the source $(\Omega; \mathbf{J}_1, \mathbf{J}_2)$ is identically zero, then one must have that*

$$\mathbf{J}_1(\mathbf{x}_0) = \mathbf{J}_2(\mathbf{x}_0) = \mathbf{0}. \tag{9.3.7}$$

The following is a direct consequence of Theorem 9.3.1 on electromagnetic sources which always radiate.

Corollary 9.3.1. *Consider an electromagnetic source $(\Omega;\mathbf{J}_1,\mathbf{J}_2)$ as described in Theorem 9.3.1. For each fixed $j\in\{1,2\}$, if $(\mathbf{x}_0;\mathscr{K}_{\mathbf{x}_0}^{r_0})$ is a corner of $\mathbf{J}_j$ in the sense of Definition 9.3.1, then the source must radiate a nonzero far-field pattern whenever $\mathbf{J}_j(\mathbf{x}_0)\neq\mathbf{0}$.*

Proof of Theorem 9.3.1. Suppose that $(\Omega;\mathbf{J}_1,\mathbf{J}_2)$ is radiationless, namely, $(\mathbf{E}_\infty,\mathbf{H}_\infty)\equiv\mathbf{0}$. Then by the Rellich theorem (cf. [16]), we know that $\mathbf{E}=\mathbf{H}=\mathbf{0}$ in the component of $\mathbb{R}^3\setminus\overline{\Omega}$ that is unbounded. Hence, we see from (9.3.1) and the definition of $\mathscr{A}$ (Definition 9.3.1) that

$$\begin{cases} \nabla\wedge\mathbf{E}(\mathbf{x})-\mathrm{i}\omega\mu_0\mathbf{H}(\mathbf{x})=\mathbf{J}_1(\mathbf{x}), & \mathbf{x}\in\mathscr{K}_{\mathbf{x}_0}^{r_0},\\ \nabla\wedge\mathbf{H}(\mathbf{x})+\mathrm{i}\omega\varepsilon_0\mathbf{E}(\mathbf{x})=\mathbf{J}_2(\mathbf{x}), & \mathbf{x}\in\mathscr{K}_{\mathbf{x}_0}^{r_0},\\ \nu(\mathbf{x})\wedge\mathbf{E}(\mathbf{x})=\nu(\mathbf{x})\wedge\mathbf{H}(\mathbf{x})=\mathbf{0}, & \mathbf{x}\in\partial\mathscr{K}_{\mathbf{x}_0}^{r_0}\setminus\partial B_{r_0}(\mathbf{x}_0), \end{cases} \tag{9.3.8}$$

for any (external) corner point $\mathbf{x}_0$ of Ω. Therefore, we readily have from Theorem 9.1.1 that

$$\mathbf{J}_1(\mathbf{x}_0)=\mathbf{J}_2(\mathbf{x}_0)=\mathbf{0}.$$

□

We note here that Theorem 9.3.1 is still valid when one of the sources, $\mathbf{J}_1$ or $\mathbf{J}_2$, is trivially zero. Hence, we can obtain similar geometric characterization of pure electric or magnetic sources which are radiationless. This also provides the following geometric characterization of a certain function space.

Corollary 9.3.2. *Let $L_c^2(\mathbb{R}^3;\mathbb{C}^3)$ denote the space of $L^2(\mathbb{R}^3;\mathbb{C}^3)$ functions which are compactly supported. We introduce the following function space:*

$$\mathscr{F}:=\{\mathbf{F}\in L_c^2(\mathbb{R}^3;\mathbb{C}^3);\ \mathbf{F}=\nabla\wedge\nabla\wedge\mathbf{M}+c_0\mathbf{M},\ \ \mathbf{M}\in H_{loc}(\mathrm{curl},\mathbb{R}^3),c_0\in\mathbb{R}\}, \tag{9.3.9}$$

where the equality holds in the weak sense. Then any $\mathbf{F}\in\mathscr{F}\cap\mathscr{A}$ must vanish at its corner points.

Remark 9.3.2. Notice that $\mathscr{F}$ in fact represents the space of compactly supported pure electric or magnetic sources with constant electric permittivity and magnetic permeability. In particular, if there is mere electric source, which means that $\mathbf{J}_2\equiv 0$ in (9.3.1), then the Maxwell equations can be reduced to

$$\nabla\wedge\nabla\wedge\mathbf{H}-\omega^2\mu_0\varepsilon_0\mathbf{H}=-\mathrm{i}\omega\varepsilon_0\mathbf{J}_1.$$

Hence, $\mathbf{J}_1\in\mathscr{F}$. The same is true when there is only magnetic source $\mathbf{J}_2$ while the electric source $\mathbf{J}_1\equiv 0$, in which case we have

$$\nabla\wedge\nabla\wedge\mathbf{E}-\omega^2\mu_0\varepsilon_0\mathbf{E}=\mathrm{i}\omega\mu_0\mathbf{J}_2.$$

Proof of Corollary 9.3.2. We first consider the case that $c_0\in\mathbb{R}_+$. Suppose that $\mathbf{F}\in\mathscr{F}\cap\mathscr{A}$. Set $\omega=\sqrt{c_0}$ and $\varepsilon_0=\mu_0=1$, and

$$\mathbf{J}_1=-\frac{1}{\mathrm{i}\omega\varepsilon_0}\mathbf{F},\quad \mathbf{J}_2\equiv\mathbf{0}.$$

By straightforward calculations, one can verify that $\mathbf{J}_1$ is a radiationless magnetic source and hence by Theorem 9.3.1, it must be vanishing at its corner points. Similarly, for the case $c_0\in\mathbb{R}_-$, by setting

$$\mathbf{J}_1\equiv\mathbf{0},\quad \mathbf{J}_2=\frac{1}{\mathrm{i}\omega\mu_0}\mathbf{F},$$

one can directly verify that $\mathbf{J}_2$ is a radiationless electric source and hence it must be vanishing at its corner points. □

We proceed with the unique recovery issue of the nonlinear inverse scattering problem (9.3.4). We first present a local unique recovery result as follows.

Theorem 9.3.2. *Let $(\Omega;\mathbf{J}_1,\mathbf{J}_2)$ and $(\Omega';\mathbf{J}_1',\mathbf{J}_2')$ be two electromagnetic source configurations where $\mathbf{J}_1,\mathbf{J}_2,\mathbf{J}_1',\mathbf{J}_2'$ belong to the class $\mathscr{A}$ in terms of Definition 9.3.1 such that $\mathbf{J}_1,\mathbf{J}_2$ are supported on Ω and $\mathbf{J}_1',\mathbf{J}_2'$ on Ω'. Let $\mathbf{E}_\infty$ and $\mathbf{E}'_\infty$ be the electric far-field patterns associated with $(\Omega;\mathbf{J}_1,\mathbf{J}_2)$ and $(\Omega';\mathbf{J}_1',\mathbf{J}_2')$, respectively. If $\mathbf{E}_\infty(\hat{\mathbf{x}})=\mathbf{E}'_\infty(\hat{\mathbf{x}})$ for all $\hat{x}\in\mathbb{S}^2$, then the set difference*

$$\Omega\Delta\Omega' := (\Omega\setminus\overline{\Omega'})\cup(\Omega'\setminus\overline{\Omega}) \tag{9.3.10}$$

cannot contain a corner whose apex, say x_0, connects to infinity in the unbounded component of $\mathbb{R}^3\setminus\overline{\Omega\cup\Omega'}$ and satisfies

$$\begin{aligned}&(i).\ \mathbf{J}_1(\mathbf{x}_0)\neq\mathbf{0} \text{ or } \mathbf{J}_2(\mathbf{x}_0)\neq\mathbf{0} \text{ if } \mathbf{x}_0 \text{ is a corner of } \Omega,\\ &(ii).\ \mathbf{J}_1'(\mathbf{x}_0)\neq\mathbf{0} \text{ or } \mathbf{J}_2'(\mathbf{x}_0)\neq\mathbf{0} \text{ if } \mathbf{x}_0 \text{ is a corner of } \Omega'.\end{aligned} \tag{9.3.11}$$

Proof. We prove the theorem by a reductio ad absurdum. Let $(\mathbf{E},\mathbf{H})$ and $(\mathbf{E}',\mathbf{H}')$ be the electromagnetic fields of the Maxwell system (9.3.1) associated with $(\Omega;\mathbf{J}_1,\mathbf{J}_2)$ and $(\Omega';\mathbf{J}_1',\mathbf{J}_2')$, respectively. Set $\mathbf{G}$ to denote the unbounded connected component of $\mathbb{R}^3\setminus\overline{\Omega\cup\Omega'}$. By the Rellich theorem and the fact that $\mathbf{E}_\infty=\mathbf{E}'_\infty$, we readily have that

$$(\mathbf{E},\mathbf{H})=(\mathbf{E}',\mathbf{H}') \quad \text{in } \mathbf{G}. \tag{9.3.12}$$

Without loss of generality, we assume that $\mathbf{x}_0\in\overline{\Omega}\setminus\Omega'$, and we let $\mathscr{K}_{\mathbf{x}_0}^{r_0}$ with $r_0\in\mathbb{R}_+$ sufficiently small, be a corner of Ω such that $\mathscr{K}_{\mathbf{x}_0}^{r_0}\subset\overline{\Omega}\setminus\Omega'$. Clearly, there hold

$$\begin{cases}\nabla\wedge\mathbf{E}(\mathbf{x})-\mathrm{i}\omega\mu_0\mathbf{H}(\mathbf{x})=\mathbf{J}_1(\mathbf{x}), & \mathbf{x}\in\mathscr{K}_{\mathbf{x}_0}^{r_0},\\ \nabla\wedge\mathbf{H}(\mathbf{x})+\mathrm{i}\omega\varepsilon_0\mathbf{E}(\mathbf{x})=\mathbf{J}_2(\mathbf{x}), & \mathbf{x}\in\mathscr{K}_{\mathbf{x}_0}^{r_0},\end{cases} \tag{9.3.13}$$

and

$$\begin{cases}\nabla\wedge\mathbf{E}'(\mathbf{x})-\mathrm{i}\omega\mu_0\mathbf{H}'(\mathbf{x})=\mathbf{0}, & \mathbf{x}\in\mathscr{K}_{\mathbf{x}_0}^{r_0},\\ \nabla\wedge\mathbf{H}'(\mathbf{x})+\mathrm{i}\omega\varepsilon_0\mathbf{E}'(\mathbf{x})=\mathbf{0}, & \mathbf{x}\in\mathscr{K}_{\mathbf{x}_0}^{r_0},\end{cases} \tag{9.3.14}$$

Set

$$\widetilde{\mathbf{E}}=\mathbf{E}-\mathbf{E}', \quad \widetilde{\mathbf{E}}=\mathbf{H}-\mathbf{H}'.$$

By (9.3.12)–(9.3.14), one sees that

$$\begin{cases}\nabla\wedge\widetilde{\mathbf{E}}(\mathbf{x})-\mathrm{i}\omega\mu_0\widetilde{\mathbf{H}}(\mathbf{x})=\mathbf{J}_1(\mathbf{x}), & \mathbf{x}\in\mathscr{K}_{\mathbf{x}_0}^{r_0},\\ \nabla\wedge\widetilde{\mathbf{H}}(\mathbf{x})+\mathrm{i}\omega\varepsilon_0\widetilde{\mathbf{E}}(\mathbf{x})=\mathbf{J}_2(\mathbf{x}), & \mathbf{x}\in\mathscr{K}_{\mathbf{x}_0}^{r_0},\\ \nu(\mathbf{x})\wedge\widetilde{\mathbf{E}}(\mathbf{x})=\nu(\mathbf{x})\wedge\widetilde{\mathbf{H}}(\mathbf{x})=\mathbf{0}, & \mathbf{x}\in\partial\mathscr{K}_{\mathbf{x}_0}^{r_0}\cap\partial\mathbf{G}.\end{cases} \tag{9.3.15}$$

Hence, by Theorem 9.1.1, we readily have that

$$\mathbf{J}_1(\mathbf{x}_0)=\mathbf{J}_2(\mathbf{x}_0)=\mathbf{0},$$

which is a contradiction to (9.3.11). □

As an immediate consequence of Theorem 9.3.2, we have the following:

Theorem 9.3.3. *Let $(\Omega;\mathbf{J}_1,\mathbf{J}_2)$ be an electromagnetic source from the class $\mathscr{A}$ in terms of Definition 9.3.1. Suppose that Ω is a convex polyhedron and at each of its apexes, either $\mathbf{J}_1$ or $\mathbf{J}_2$ is non-vanishing. Then $\mathbf{E}_\infty$ uniquely determines Ω.*

Proof. Suppose that there exists another electromagnetic source $(\Omega';\mathbf{J}_1',\mathbf{J}_2')$ with $\mathbf{E}_\infty=\mathbf{E}'_\infty$. If $\Omega\neq\Omega'$, then there is a corner $\mathbf{x}_0$ of let's say Ω such that $\mathbf{x}_0\in\overline{\Omega}\setminus\Omega'$, and it can be connected to infinity outside of $\Omega\cup\Omega'$. Since $\mathbf{J}_1(\mathbf{x}_0)\neq 0$ or $\mathbf{J}_2(\mathbf{x}_0)\neq 0$, Theorem 9.3.2 implies that $\mathbf{x}_0$ cannot be a corner of $\Omega\Delta\Omega'$. In other words, $\mathbf{x}_0\in\overline{\Omega'}$ too. The contradiction gives $\Omega=\Omega'$. □

It is remarked that one can show the same unique recovery result as that in Theorem 9.3.3 for a bit more general case where the source support Ω consists of finitely many disjoint convex polyhedron.

9.4 Inverse Medium Scattering and Interior Transmission Eigenvalue Problem

In this section, we consider another scenario of practical importance where the electromagnetic scattering is induced by an inhomogeneous medium and an incident wave field. Suppose an inhomogeneous medium is embedded in a uniformly homogeneous space with electric permittivity ε_0 and magnetic permeability μ_0. The inhomogeneous medium is supported in Ω and is characterized by its material parameters including the electric permittivity $\varepsilon \in L^\infty(\Omega;\mathbb{R}_+)$, magnetic permeability $\mu \in L^\infty(\Omega;\mathbb{R}_+)$, and electric conductivity $\sigma \in L^\infty(\Omega;\mathbb{R}_+^0)$. Throughout the rest of the chapter, it is assumed that Ω is a bounded Lipschitz domain in $\mathbb{R}^3$ with a connected complement $\mathbb{R}^3\setminus\overline{\Omega}$. In what follows, for notational convenience, we extend ε,μ, and σ to the whole space $\mathbb{R}^3$ by setting $\varepsilon(\mathbf{x})=\varepsilon_0$, $\mu(\mathbf{x})=\mu_0$, and $\sigma(\mathbf{x})=0$ for $\mathbf{x}\in\mathbb{R}^3\setminus\overline{\Omega}$. Associated with the scattering medium $(\Omega;\varepsilon,\mu,\sigma)$ described above, the electromagnetic scattering is then induced by sending a wave field $(\mathbf{E}^i,\mathbf{H}^i)$ impinging on Ω. It is a pair of entire solutions to the following homogeneous Maxwell system:

$$\operatorname{curl}\mathbf{E}^i-\mathrm{i}\omega\mu_0\mathbf{H}^i=\mathbf{0},\quad \operatorname{curl}\mathbf{H}^i+\mathrm{i}\omega\varepsilon_0\mathbf{E}^i=\mathbf{0}\quad\text{in }\mathbb{R}^3. \tag{9.4.1}$$

The interaction of the incident field $(\mathbf{E}^i,\mathbf{H}^i)$ and the scattering body $(\Omega;\varepsilon,\mu,\sigma)$ generates electromagnetic wave scattering. We let $(\mathbf{E},\mathbf{H})$ and $(\mathbf{E}^t,\mathbf{H}^t)$ denote, respectively, the scattered and the total electromagnetic fields. There hold

$$(\mathbf{E}^t,\mathbf{H}^t)=(\mathbf{E}^i,\mathbf{H}^i)+(\mathbf{E},\mathbf{H})\quad\text{in }\mathbb{R}^3$$

and the following Maxwell system

$$\begin{aligned}
&\nabla\wedge\mathbf{E}^t(x)-\mathrm{i}\omega\mu(\mathbf{x})\mathbf{H}^t(\mathbf{x})=\mathbf{0}, && \mathbf{x}\in\mathbb{R}^3,\\
&\nabla\wedge\mathbf{H}^t(\mathbf{x})+(\mathrm{i}\omega\varepsilon(\mathbf{x})-\sigma(\mathbf{x}))\mathbf{E}^t(\mathbf{x})=\mathbf{0}, && \mathbf{x}\in\mathbb{R}^3,\\
&\lim_{|\mathbf{x}|\to\infty}|\mathbf{x}|\left(\sqrt{\mu_0}\mathbf{H}\times\frac{\mathbf{x}}{|\mathbf{x}|}-\sqrt{\varepsilon_0}\mathbf{E}\right)=\mathbf{0}.
\end{aligned} \tag{9.4.2}$$

We refer to [16, 38, 41] for the existence of a unique pair of solutions $(\mathbf{E},\mathbf{H})\in H_{\mathrm{loc}}(\mathrm{curl},\mathbb{R}^3)\times H_{\mathrm{loc}}(\mathrm{curl},\mathbb{R}^3)$ and the following far-field expansion:

$$(\mathbf{E},\mathbf{H})(\mathbf{x})=\frac{e^{\mathrm{i}k|\mathbf{x}|}}{|\mathbf{x}|}(\mathbf{E}_\infty,\mathbf{H}_\infty(\hat{\mathbf{x}})+\mathscr{O}\left(\frac{1}{|\mathbf{x}|^2}\right), \tag{9.4.3}$$

where $k=\omega\sqrt{\varepsilon_0\mu_0}$.

Similar to the inverse source scattering problem (9.3.3), the inverse medium scattering problem can be stated as follows:

$$\mathbf{E}_\infty(\hat{\mathbf{x}}),\ \hat{\mathbf{x}}\in\mathbb{S}^2\mapsto(\Omega;\varepsilon,\mu,\sigma). \tag{9.4.4}$$

It can be verified directly that the inverse medium scattering problem (9.4.4) is nonlinear and under-determined in the generic case. In what follows, similar to the inverse source scattering case, we first consider the invisibility issue for the inverse medium scattering problem (9.4.4), namely, $\mathbf{E}_\infty\equiv\mathbf{0}$.

If $\mathbf{E}_\infty \equiv \mathbf{0}$, by the Rellich theorem, one has $(\mathbf{E},\mathbf{H}) = \mathbf{0}$ in $\mathbb{R}^3 \setminus \overline{\Omega}$. Hence, it is straightforward to show that in this case, there holds

$$\begin{cases} \nabla \wedge \mathbf{E}^t - \mathrm{i}\omega\mu\mathbf{H}^t = 0, & \nabla \wedge \mathbf{H}^t + \mathrm{i}\omega\gamma\mathbf{E}^t = 0 \quad \text{in } \Omega, \\ \nabla \wedge \mathbf{E}^0 - \mathrm{i}\omega\mu_0\mathbf{H}^0 = 0, & \nabla \wedge \mathbf{H}^0 + \mathrm{i}\omega\varepsilon_0\mathbf{E}^0 = 0 \quad \text{in } \Omega, \\ \nu \wedge \mathbf{E}^t = \nu \wedge \mathbf{E}^0, & \nu \wedge \mathbf{H}^t = \nu \wedge \mathbf{H}^0 \quad \text{on } \partial\Omega, \end{cases} \tag{9.4.5}$$

where $\gamma := \varepsilon + \mathrm{i}\sigma/\omega$ with $(\Omega;\varepsilon,\mu,\sigma)$ introduced earlier. Equation (9.4.5) is known as the interior transmission eigenvalue problem in the literature. If, for a certain $\omega \in \mathbb{R}_+$, there exist nontrivial $(\mathbf{E}^t,\mathbf{H}^t) \in H_{\mathrm{loc}}(\mathrm{curl},\mathbb{R}^3) \times H_{\mathrm{loc}}(\mathrm{curl},\mathbb{R}^3)$ and $(\mathbf{E}^0,\mathbf{H}^0) \in H_{\mathrm{loc}}(\mathrm{curl},\mathbb{R}^3) \times H_{\mathrm{loc}}(\mathrm{curl},\mathbb{R}^3)$ satisfying (9.4.5), then ω is called an interior transmission eigenvalue, and $(\mathbf{E}^t,\mathbf{H}^t), (\mathbf{E}^0,\mathbf{H}^0)$ are called the corresponding eigenfunctions. From our discussion above, it is seen that if invisibility occurs, then ω is an interior transmission eigenvalue and the restrictions of the total wave field $(\mathbf{E}^t,\mathbf{H}^t)$ and incident wave field $(\mathbf{E}^0,\mathbf{H}^0)$ form the corresponding eigenfunctions. In order to gain more insights about the invisibility, we first provide a geometric characterization of the interior transmission eigenfunctions.

Theorem 9.4.1. *Consider the interior transmission eigenvalue problem* (9.4.5), *and suppose that* $(\mathbf{E}^t,\mathbf{H}^t)$ *and* $(\mathbf{E}^0,\mathbf{H}^0)$ *are a pair of eigenfunctions associated with the eigenvalue* $\omega \in \mathbb{R}_+$. *Assume that* Ω *possesses a corner* $\mathscr{K}_{\mathbf{x}_0}^{r_0}$ *and moreover,*

$$(\mu - \mu_0)\mathbf{H}^t,\ (\gamma - \varepsilon_0)\mathbf{E}^t \in C^\alpha(\overline{\mathscr{K}_{\mathbf{x}_0}^{r_0}})^3, \tag{9.4.6}$$

for some $\alpha \in (0,1)$. *Then there holds*

$$(\mu - \mu_0)(\mathbf{x}_0)\mathbf{H}^t(\mathbf{x}_0) = (\gamma - \varepsilon_0)(\mathbf{x}_0)\mathbf{E}^t(\mathbf{x}_0) = 0. \tag{9.4.7}$$

Proof. By straightforward calculations, one can show by virtue of (9.4.6) that $(\widehat{\mathbf{E}},\widehat{\mathbf{H}}) := (\mathbf{E}^t,\mathbf{H}^t) - (\mathbf{E}^0,\mathbf{H}^0)$ satisfies

$$\begin{cases} \nabla \wedge \widehat{\mathbf{E}} - \mathrm{i}\omega\mu_0\widehat{\mathbf{H}} = \mathbf{F}_1 & \text{in } \mathscr{K}_{\mathbf{x}_0}^{r_0}, \\ \nabla \wedge \widehat{\mathbf{H}} + \mathrm{i}\omega\varepsilon_0\widehat{\mathbf{E}} = \mathbf{F}_2 & \text{in } \mathscr{K}_{\mathbf{x}_0}^{r_0}, \\ \nu \wedge \widehat{\mathbf{E}} = \nu \wedge \widehat{\mathbf{H}} = 0 & \text{on } \partial\mathscr{K}_{\mathbf{x}_0}^{r_0} \cap \partial\Omega, \end{cases} \tag{9.4.8}$$

where

$$\mathbf{F}_1 = (\mu - \mu_0)\mathbf{H}^t \quad \text{and} \quad \mathbf{F}_2 = (\gamma - \varepsilon_0)\mathbf{E}^t. \tag{9.4.9}$$

Hence, by Theorem 9.1.1 and (9.4.6), one readily has (9.4.7). □

In Theorem 9.4.1, we reveal the intrinsic geometric structure of the interior transmission eigenfunctions associated with the Maxwell system. By assuming that $(\mu - \mu_0)(\mathbf{x}_0) \neq 0$ and $(\gamma - \varepsilon_0)(\mathbf{x}_0) \neq 0$, one readily has from (9.4.7) that

$$\mathbf{E}^t(\mathbf{x}_0) = \mathbf{H}^t(\mathbf{x}_0) = 0. \tag{9.4.10}$$

This vanishing property at the corner point is consistent with most of the existing results for the interior transmission eigenfunctions associated with the acoustic scattering.

Now, we are in a position to consider the practical implication of the geometric property in Theorem 9.4.1 to invisibility in wave scattering. We have the following:

Theorem 9.4.2. *Consider the electromagnetic scattering problem* (9.4.2). *If the medium* $(\Omega;\varepsilon,\mu,\sigma)$ *possesses a corner* $\mathscr{K}_{x_0}^{r_0}$ *in its support and moreover,*

$$(\mu - \mu_0)\mathbf{H}^t,\ (\gamma - \varepsilon_0)\mathbf{E}^t \in C^\alpha(\overline{\mathscr{K}_{\mathbf{x}_0}^{r_0}})^3, \tag{9.4.11}$$

for some $\alpha \in (0,1)$ *and*

$$(\mu-\mu_0)(\mathbf{x}_0)\mathbf{H}_0(\mathbf{x}_0)\neq 0, \quad or \quad (\gamma-\varepsilon_0)(\mathbf{x}_0)\mathbf{E}_0(\mathbf{x}_0)\neq 0, \tag{9.4.12}$$

then the corresponding far-field pattern cannot be identically vanishing; that is, invisibility does not occur.

Proof. This is a direct consequence of the fact that if the far-field pattern is identically vanishing, then one has both a) scattered wave fields $\mathbf{H},\mathbf{E}$ vanishing at the boundary and b) the interior transmission eigenvalue problem (9.4.5). Hence, one arrives at a contradiction by using the vanishing property in Theorem 9.4.1 and the non-vanishing condition in (9.4.12). □

Theorem 9.4.2 shows that if a medium scatterer possesses a corner, then it is detectable. Next, we show that the detectability also implies the identifiability, namely, the unique recovery of the inverse scattering problem (9.4.4). In fact, we have the following:

Theorem 9.4.3. *Let $(\Omega;\gamma,\mu)$ and $(\Omega';\gamma',\mu)$ be two medium scatterers and $(\mathbf{E}_t,\mathbf{H}_t,\mathbf{E}_\infty)$ and $(\mathbf{E}_t',\mathbf{H}_t',\mathbf{E}_\infty')$ be the associated total and far fields. If $\mathbf{E}_\infty(\hat{\mathbf{x}})=\mathbf{E}'_\infty(\hat{\mathbf{x}})$ for all $\hat{\mathbf{x}}\in\mathbb{S}^2$, then the set difference $\Omega\Delta\Omega'$ as defined in* (9.3.10) *cannot contain a corner whose apex, say $\mathbf{x}_0$, connects to infinity in the unbounded component of $\mathbb{R}^3\setminus\overline{\Omega\cup\Omega'}$, where* (9.3.11) *and $\mathbf{J}_j,\mathbf{J}'_j\in C^\alpha(\mathscr{K}_{\mathbf{x}_0}^{r_0})^3$, $j=1,2$, are satisfied with*

$$\mathbf{J}_1:=(\mu-\mu_0)\mathbf{H}_t,\ \mathbf{J}_2:=(\gamma-\varepsilon_0)\mathbf{E}_t;\ \mathbf{J}'_1:=(\mu'-\mu_0)\mathbf{H}'_t,\ \mathbf{J}'_2:=(\gamma'-\varepsilon_0)\mathbf{E}'_t. \tag{9.4.13}$$

Proof. The proof follows from a similar argument as that of Theorem 9.3.2 along with the use of the same reduction strategy in Theorem 9.4.1 in transforming the medium scattering problem to a source scattering problem. □

References

1. E. Blåsten, *Nonradiating sources and transmission eigenfunctions vanish at corners and edges*, SIAM J. Math. Anal., **50(6)** (2018), 6255–6270.
2. E. Blåsten, X. Li,H. Liu and Y. Wang, On vanishing and localization near cusps of transmission eigenfunctions: a numerical study, Inverse Problems, **33** (2017), 105001.
3. E. Blåsten and Y.-H. Lin, *Radiating and non-radiating sources in elasticity*, Inverse Problems, 35, 1 (2019) 015005.
4. E. Blåsten and H. Liu, *On vanishing near corners of transmission eigenfunctions*, J. Funct. Anal., **273(11)** (2017), 3616–3632. Addendum, arXiv:1710.08089
5. E. Blåsten and H. Liu, *On corners scattering stably and stable shape determination by a single far-field pattern*, Indiana Univ. Math. J., **70** (2021), no. 3, 907–947.
6. E. Blåsten and H. Liu, *Scattering by curvatures, radiationless sources, transmission eigenfunctions and inverse scattering problems*, SIAM Journal on Mathematical Analysis, **53** (2021), no. 4, 3801–3837.
7. E. Blåsten and H. Liu, *Recovering piecewise constant refractive indices by a single far-field pattern*,Inverse Problems, **36** (2020), 085005.
8. E. Blåsten, H. Liu and J. Xiao, *On an electromagnetic problem in a corner and its applications*, Anal. PDE, **14(7)** (2021), 2207–2224.
9. E. Blåsten, L. Päivärinta and J. Sylvester, *Corners always scatter*, Comm. Math. Phys., **331** (2014), 725–753.
10. E. Blåsten and E. Vesalainen, *Non-scattering energies and transmission eigenvalues in $\mathbb{H}^n$*, Annales Academiae Scientiarum Fennicae. Mathematica, **45(1)** (2020), 547–576. Suomalainen Tiedeakatemia.
11. N. Bleistein and J.K. Cohen, *Nonuniqueness in the inverse source problem in acoustics and electromagnetics*, J. Math. Phys., **18** (1977), 194–201.
12. D. Bohm and M. Weinstein, *The self-oscillations of a charged particle*, Phys. Rev., 74:1789–1798, 1948.
13. F. Cakoni and H. Haddar, *Transmission eigenvalues in inverse scattering theory*, in "Inverse Problems and Applications: Inside Out II", Math. Sci. Res. Inst. Publ., Vol. 60, pp. 529–580, Cambridge Univ. Press., Cambridge, (2013).
14. F. Cakoni and J. Xiao, *On corner scattering for operators of divergence form and applications to inverse scattering*, Comm. Partial Differential Equations, in press, (2020).
15. D. Colton, A. Kirsch and L. Päivärinta, *Far-field patterns for acoustic waves in an inhomogeneous medium*, SIAM J. Math. Anal., **20** (1989),1472–1483.

16. D. Colton and R. Kress, *Inverse acoustic and electromagnetic scattering theory*, volume 93 of Applied Mathematical Sciences. Springer, Cham, 4th edition, 2019.
17. D. Colton and P. Monk, *The inverse scattering problem for time-harmonic acoustic waves in an inhomogeneous medium*, Quart. J. Mech. Appl. Math., **41** (1988), 97–125.
18. A. Cossonniére and H. Haddar, *The electromagnetic interior transmission problem for regions with cavities*, SIAM J. Math. Anal., **43(4)** (2011), 1698–1715.
19. A.J. Devaney and E. Wolf, *Non-radiating stochastic scalar sources*, In L. Mandel and E. Wolf, editors, Coherence and Quantum Optics V, 417–421, New York, (1984). Plenum Press.
20. H. Diao, X. Cao and H. Liu, *On the geometric structures of conductive transmission eigenfunctions and their application*, Comm. Partial Differential Equations, **46(4)** (2021), 630–679.
21. H. Diao, X. Fei, H. Liu and K.Yang, *Visibility, invisibility and unique recovery of inverse electromagnetic problems with conical singularities*, arxiv 2204.02835, (2022).
22. H. Diao, H. Liu, X. Wang and K. Yang, *On vanishing and localizing around corners of electromagnetic transmission resonance*, Partial Differ. Equ., **2**(2021), 78.
23. P. Ehrenfest, *Ungleichförmige Elektrizitätsbewegungen ohne Magnet- und Strahlungsfeld*, Physik. Zeit., 11:708–709, (1910).
24. J. Elschner and G. Hu, *Corners and edges always scatter*, Inverse Problems **31(1)** (2015), 015003.
25. J. Elschner and G. Hu, *Acoustic scattering from corners, edges and circular cones*, Arch. Ration. Mech. Anal., **228(2)** (2018), 653–690.
26. F.G. Friedlander, *An inverse problem for radiation fields*, Proc. London Math. Soc., **2** (1973), 551–576.
27. A. Gamliel, K. Kim, A.I. Nachman, and E. Wolf, *A new method for specifying nonradiating monochromatic sources and their fields*, J. Opt. Soc. Am. A, **6** (1989), 1388–1393.
28. G. Gbur, *Nonradiating Sources and the Inverse Source Problem*, PhD Thesis, Univ. Rochester, (2001).
29. G.H. Goedecke, *Classically radiationless motions and possible implications for quantum theory*, Phys. Rev., 135:B281–B288, (1964).
30. G. Hu, M. Salo and E. Vesalainen, *Shape identification in inverse medium scattering problems with a single far-field pattern*, SIAM J. Math. Anal., **48** (2016),152–165.
31. B.J. Hoenders and H.P. Baltes, *The scalar theory of nonradiating partially coherent sources*, Lettere al Nuovo Cimento, **23** (1979), 206–208.
32. M. Ikehata, *Reconstruction of a source domain from the Cauchy data*, Inverse Problems, **15** (1999), 637–645.
33. K. Kim and E. Wolf, *Non-radiating monochromatic sources and their fields*, Opt. Commun., **59** (1986), 1–6.
34. A. Kirsch, *The denseness of the far field patterns for the transmission problem*, IMA J. Appl. Math., **37** (1986), 213–225.
35. S. Kusiak and J. Sylvester, *The scattering support*, Comm. Pure Appl. Math., **56** (2003), 1525–1548.
36. S. Kusiak and J. Sylvester, *The convex scattering support in a background medium*, SIAM J. Math. Anal., **36** (2005), 1142–1158.
37. J. Li, X. Li and H. Liu, *Reconstruction via the intrinsic geometric structures of interior transmission eigenfunctions*, arXiv:1706.04418, (2017).
38. H. Liu, L. Rondi and J. Xiao, Mosco convergence for $H(curl)$ spaces, higher integrability for Maxwell's equations, and stability in direct and inverse EM scattering problems, *J. Eur. Math. Soc.*, *J. Eur. Math. Soc.*, **21** (2019), no. 10, 2945–2993.
39. H. Liu and J. Xiao, *On electromagnetic scattering from a penetrable corner*, SIAM J. Math. Anal., **49(6)** (2017), 5207–5241.
40. E.A. Marengo and R.W. Ziolkowski, *On the radiating and nonradiating components of scalar, electromagnetic, and weak gravitational sources*, Phys. Rev. Lett., **83** (1999), 3345–3349.
41. J.-C. Nédélec, *Acoustic and Electromagnetic Equations*, Springer-Verlag, New York, (2001).
42. L. Päivärinta, M. Salo and E. V. Vesalainen, *Strictly convex corners scatter*, Rev. Mat. Iberoam. **33(4)** (2017), 1369–1396.
43. L. Päivärinta and J. Sylvester, *Transmission eigenvalues*, SIAM J. Math. Anal., **40** (2008), 738–753.
44. B. P. Rynne and B. D. Sleeman, *The interior transmission problem and inverse scattering from inhomogeneous media*, SIAM J. Math. Anal., **22** (1991), 1755–1762.
45. A. Sommerfeld, *Zur elektronentheorie. I. Allgemeine untersuchung des feldes eines beliebig bewegten elektrons*, Akad. der Wiss. (Gött.), Math. Phys. Klasse, Nach., 99–130, (1904).
46. A. Sommerfeld, *Zur elektronentheorie. II. Grundlagen für eine allgemeine dynamik des elektrons*, Akad. der Wiss. (Gött.), Math. Phys. Klasse, Nach., 363–439, (1904).

Chapter 10
Geometric Structures of Lamé's Transmission Eigenfunctions with General Transmission Conditions and Applications

10.1 Overview

10.1.1 Background and Motivation

We first introduce the elastic scattering due to an embedded inhomogeneous medium and an incident wave field, which is the physical origin of our study. Let λ, μ be real constants satisfying the following strong convexity condition:

$$\mu > 0, \quad n\lambda + 2\mu > 0 \quad \text{for} \quad n = 2, 3.$$

Let $\Omega \subset \mathbb{R}^n$, $n = 2, 3$, be a bounded Lipschitz domain with a connected complement $\mathbb{R}^n \backslash \overline{\Omega}$. Suppose that $V \in L^\infty(\mathbb{R}^n)$ is a real-valued function with $\text{supp}(V) \subset \Omega$. The parameters λ, μ, and V characterize the elastic medium configuration of the space $\mathbb{R}^n$, with λ, μ, and $1+V$, respectively, denoting the bulk modulus and density. Throughout, we assume that V is nontrivial, which is also referred to as a scattering potential. $(\Omega; \lambda, \mu, V)$ signifies an inhomogeneous scatter embedded in the uniformly homogeneous space $\mathbb{R}^n$. Let $\mathbf{u}^i$ be an incident field which is a $\mathbb{C}^n$-valued entire solution to the following Lamé system:

$$\mathscr{L}\mathbf{u}^i + \omega^2 \mathbf{u}^i = 0 \quad \text{in } \mathbb{R}^n, \quad \mathscr{L}\mathbf{u}^i := \lambda \Delta \mathbf{u}^i + (\lambda + \mu) \nabla \nabla \cdot \mathbf{u}^i, \tag{10.1.1}$$

where $\omega \in \mathbb{R}_+$ signifies the angular frequency of the time-harmonic wave propagation. The interaction between the incident field $\mathbf{u}^i$ and the scatter $(\Omega; \lambda, \mu, V)$ generates the elastic scattering, which is governed by the following Lamé system:

$$\mathscr{L}\mathbf{u} + \omega^2 (1+V)\mathbf{u} = 0 \quad \text{in } \mathbb{R}^n; \quad \mathbf{u} = \mathbf{u}^i + \mathbf{u}^{\text{sc}}; \quad \mathbf{u}^{\text{sc}} \text{ is radiating.} \tag{10.1.2}$$

Here by radiating, we mean that $\mathbf{u}^{\text{sc}}$ satisfies the following Kupradze radiation condition:

$$\lim_{r \to \infty} r^{\frac{n-1}{2}} \left(\frac{\partial \mathbf{u}_\beta^{\text{sc}}}{\partial r} - \mathrm{i} k_\beta \mathbf{u}_\beta^{\text{sc}} \right) = \mathbf{0}, \quad r := |\mathbf{x}|, \quad \beta = p, s, \tag{10.1.3}$$

where

$$\mathbf{u}^{\text{sc}} = \mathbf{u}_p^{\text{sc}} + \mathbf{u}_s^{\text{sc}}, \quad \mathbf{u}_p^{\text{sc}} := -\frac{1}{k_p^2} \nabla (\nabla \cdot \mathbf{u}^{\text{sc}}), \quad \mathbf{u}_s^{\text{sc}} := \begin{cases} \frac{1}{k_s^2} \nabla \times \nabla \times \mathbf{u}^{\text{sc}} & \text{(3D)} \\ \frac{1}{k_s^2} \mathbf{curl}\, \text{curl}\, \mathbf{u}^{\text{sc}} & \text{(2D)} \end{cases}, \tag{10.1.4}$$

and

H. Diao, H. Liu, *Spectral Geometry and Inverse Scattering Theory*,
https://doi.org/10.1007/978-3-031-34615-6_10

$$k_p := \frac{\omega}{\sqrt{2\mu+\lambda}}, \quad k_s := \frac{\omega}{\sqrt{\mu}}. \tag{10.1.5}$$

In (10.1.4), the two-dimensional operators **curl** and curl are defined, respectively, by

$$\operatorname{curl}\mathbf{u} = \partial_1 u_2 - \partial_2 u_1, \quad \mathbf{curl}\, u = (\partial_2 u, -\partial_1 u)^\top,$$

with $\mathbf{u} = (u_1, u_2)$ and u being vector-valued and scalar functions, respectively. It is noted that (10.1.4) is the Helmholtz decomposition of the vector field $\mathbf{u}^{sc}$, where $\mathbf{u}_p^{sc}$ and $\mathbf{u}_s^{sc}$ are, respectively, referred to as the compression and shear parts of $\mathbf{u}^{sc}$. This decomposition shall also play an important role in our subsequent analysis. The Kupradze radiation condition characterizes the outward propagating nature of the scattered field $\mathbf{u}^{sc}$. The well-possession of the scattering system (10.1.2) is known [29], and in particular, there exists a unique solution $\mathbf{u} \in H^1_{loc}(\mathbb{R}^n)$ which admits the following asymptotic expansions:

$$\begin{aligned}
\mathbf{u}_p^{sc}(\mathbf{x}) &= \frac{e^{ik_p r}}{r^{\frac{n-1}{2}}}\left\{u_p^\infty(\hat{\mathbf{x}})\hat{\mathbf{x}} + O\left(\frac{1}{r}\right)\right\}, \quad \hat{\mathbf{x}} := \mathbf{x}/|\mathbf{x}|,\\
\mathbf{u}_s^{sc}(\mathbf{x}) &= \frac{e^{ik_p r}}{r^{\frac{n-1}{2}}}\left\{U_s^\infty(\hat{\mathbf{x}}) + O\left(\frac{1}{r}\right)\right\}, \quad \hat{\mathbf{x}}\cdot U_s^\infty(\hat{\mathbf{x}}) = 0,\\
\mathbf{u}^{sc}(\mathbf{x}) &= \frac{e^{ik_p r}}{r^{\frac{n-1}{2}}} u_p^\infty(\hat{\mathbf{x}})\hat{\mathbf{x}} + \frac{e^{ik_s r}}{r^{\frac{n-1}{2}}} U_s^\infty(\hat{\mathbf{x}}) + O\left(\frac{1}{r^{(n+1)/2}}\right),
\end{aligned} \tag{10.1.6}$$

as $r = |\mathbf{x}| \to \infty$, where u_p^∞ and u_s^∞ are both scalar functions defined on the unit sphere $\mathbb{S}^{n-1} := \{\hat{\mathbf{x}} \in \mathbb{R}^n \,|\, |\hat{\mathbf{x}}| = 1\}$. Define the far-field pattern $\mathbf{u}_t^\infty$ of $\mathbf{u}^{sc}$ as $\mathbf{u}_t^\infty(\hat{\mathbf{x}}) := u_p^\infty(\hat{\mathbf{x}})\hat{\mathbf{x}} + U_s^\infty(\hat{\mathbf{x}})$. One clearly has the unique correspondences: $\mathbf{u}_p^\infty(\hat{\mathbf{x}}) = \mathbf{u}_t^\infty(\hat{\mathbf{x}}) \cdot \hat{\mathbf{x}}$ and $U_s^\infty(\hat{\mathbf{x}}) = \mathbf{u}_t^\infty(\hat{\mathbf{x}}) \cdot \hat{\mathbf{x}}^\perp$. Moreover, due to the Rellich theorem [19], the correspondence between $\mathbf{u}_t^\infty$ and $\mathbf{u}^{sc}$ is also one-to-one.

Define the scattering operator $\mathscr{S}$ as

$$\mathscr{S}\big((\Omega;\lambda,\mu,V),\mathbf{u}^i\big) = \mathbf{u}_t^\infty(\hat{\mathbf{x}}), \quad \hat{\mathbf{x}} \in \mathbb{S}^{n-1}, \tag{10.1.7}$$

which is implicitly defined by the scattering system (10.1.2). It can be directly verified that $\mathscr{S}$ is nonlinear with respect to the argument $(\Omega;\lambda,\mu,V)$, whereas it is linear with respect to the argument $\mathbf{u}^i$. An inverse problem of industrial importance in geophysical exploration and medical imaging is to recover $(\Omega;\lambda,\mu,V)$ by knowledge of $\mathbf{u}_t^\infty$. In this chapter, we are mainly concerned with the kernel space of $\mathscr{S}$, namely,

$$\mathscr{S}\big((\Omega;\lambda,\mu,V),\mathbf{u}^i\big) = \mathbf{0}, \tag{10.1.8}$$

which corresponds to the physical scenario that no scattering, a.k.a. invisibility , occurs. In particular, we consider the geometric structures of the kernel space $\mathrm{Ker}(\mathscr{S})$, namely, the quantitative relationships between the configurations in $\mathrm{Ker}(\mathscr{S})$ and the intrinsic geometries of Ω.

Consider the following PDE system for $\mathbf{v} = (v_\ell)_{\ell=1}^n \in H^1(\Omega)^n$ and $\mathbf{w} = (w_\ell)_{\ell=1}^n \in H^1(\Omega)^n$:

$$\begin{cases}
\lambda\Delta\mathbf{v} + (\lambda+\mu)\nabla\nabla\cdot\mathbf{v} + \omega^2\mathbf{v} = \mathbf{0} & \text{in } \Omega,\\
\lambda\Delta\mathbf{w} + (\lambda+\mu)\nabla\nabla\cdot\mathbf{w} + \omega^2(1+V)\mathbf{w} = \mathbf{0} & \text{in } \Omega,\\
\mathbf{w} = \mathbf{v}, \quad T_\nu\mathbf{v} = T_\nu\mathbf{w} & \text{on } \partial\Omega,
\end{cases} \tag{10.1.9}$$

with ν signifying the outward unit normal to $\partial\Omega$, and the *boundary traction* operator $T_\nu\mathbf{v}$ defined as

$$T_\nu\mathbf{v} = \begin{cases}
2\mu\partial_\nu\mathbf{v} + \lambda\nu(\nabla\cdot\mathbf{v}) + \mu(\partial_2 v_1 - \partial_1 v_2)\nu^\perp, & \text{for } n = 2,\\
2\mu\partial_\nu\mathbf{v} + \lambda\nu(\nabla\cdot\mathbf{v}) + \mu\nu\times(\nabla\times\mathbf{v}), & \text{for } n = 3,
\end{cases} \tag{10.1.10}$$

where for $n = 2$, $\nu^{\perp} \in \mathbb{R}^2$ denotes the unit vector obtained via rotating ν anticlockwise by $\pi/2$. It is clear that $\mathbf{v} = \mathbf{w} \equiv \mathbf{0}$ are a pair of trivial solutions to (10.1.9). If, for a certain $\omega \in \mathbb{R}_+$, there exists a pair of nontrivial solutions $(\mathbf{v}, \mathbf{w}) \in H^1(\Omega)^n \times H^1(\Omega)^n$ to (10.1.9), then ω is called an elastic transmission eigenvalue, and $(\mathbf{v}, \mathbf{w})$ is referred to as the corresponding pair of elastic transmission eigenfunctions. The connection between Ker($\mathscr{S}$) and the transmission eigenvalue problem (10.1.9) can be described as follows. If $\mathbf{u}_t^{\infty} \equiv 0$, one clearly has $\mathbf{u}^{sc} = 0$ in $\mathbb{R}^n \backslash \overline{\Omega}$. This in turn yields that $\mathbf{u} = \mathbf{u}^i$ in $\mathbb{R}^n \backslash \overline{\Omega}$ for the scattering system (10.1.2). In such a case, one can show that $\mathbf{v} = \mathbf{u}^i|_{\Omega}$ and $\mathbf{w} = \mathbf{u}|_{\Omega}$ are a pair of transmission eigenfunctions. However, if $(\mathbf{v}, \mathbf{w})$ are a pair of transmission eigenfunctions, one cannot directly have the non-scattering/invisibility unless $\mathbf{v}$ can be (analytically) extended to an entire solution to (10.1.1) to generate an incident field. Nevertheless, $(\mathbf{v}, \mathbf{w})$ is located in any small neighborhood of Ker($\mathscr{S}$) in the sense that $\mathbf{v}$ and $\mathbf{w}$ can, respectively, approximate $\mathbf{u}^i$ and $\mathbf{u}$ in Ω within ε-accuracy for any $\varepsilon > 0$ such that the corresponding scattering amplitude is also of order ε. This viewpoint has been verified in [9,11,32] for the acoustic scattering governed by the Helmholtz equation, and one should be able to show similar results for the elastic scattering by following a similar spirit. However, this is beyond the scope of the current study.

In [8], the authors proved that the elastic transmission eigenfunctions locally vanish around a corner under the condition that the eigenfunctions are Hölder continuous in the domain. The result was applied to deriving a uniqueness result in determining an unknown elastic source from its far-field pattern. In this chapter, we shall provide a much more comprehensive study of this important geometric property for the elastic transmission eigenfunctions, where we follow the treatment in [24]. First, we shall consider a more general formulation of the elastic transmission eigenvalue problem which includes (10.1.9) as a special case. Second, we establish the local vanishing property under both the Hölder continuity and the Fourier extension property of the generalized transmission eigenfunctions. Third, we apply the results to deriving two unique identifiable results for the inverse elastic problem in determining the polygonal/polyhedral support of an inhomogeneous medium independent of its content. Here, we would like to mention in passing some related practical background on the inverse determinations in elasticity [25,26]. Finally, we would like to briefly discuss the mathematical strategies that we develop to derive the results. In order to establish the geometric property, one needs to track the singularity of the transmission eigenfunctions (with respect to their smoothness) induced by the geometric singularity of the domain, namely, the corner. To achieve that end, we develop a micro-local argument that can be localized around the corner. An integral identity involving the difference of the two transmission eigenfunctions as well as a special type of CGO (complex geometric optics) solutions is a critical ingredient in our study. Compared to most of the existing studies mentioned earlier, there is a boundary integral term due to the generalized transmission conditions in our study, which, together with the more complicated nature of the Lamé system, makes the related analysis and estimates highly technical and subtle. In tracking the order of the asymptotic parameter in the phase of the CGO solutions, we manage to derive the desired results.

10.1.2 Mathematical Setup and Summary of the Main Results

Let $\eta \in L^{\infty}(\partial\Omega)$ be a real-valued function. Instead of (10.1.9), we consider the following generalized elastic transmission eigenvalue problem for $\mathbf{v} = (v_\ell)_{\ell=1}^n, \mathbf{w} = (w_\ell)_{\ell=1}^n \in H^1(\Omega)^n$:

$$\begin{cases} \lambda\Delta\mathbf{v} + (\lambda+\mu)\nabla\nabla\cdot\mathbf{v} + \omega^2\mathbf{v} = \mathbf{0} & \text{in } \Omega, \\ \lambda\Delta\mathbf{w} + (\lambda+\mu)\nabla\nabla\cdot\mathbf{w} + \omega^2(1+V)\mathbf{w} = \mathbf{0} & \text{in } \Omega, \\ \mathbf{w} = \mathbf{v}, \quad T_\nu\mathbf{v} + \eta\mathbf{v} = T_\nu\mathbf{w} & \text{on } \Gamma. \end{cases} \tag{10.1.11}$$

where $\Gamma \subset \partial\Omega$ is an open subset. We note that if $\eta \equiv 0$ and $\Gamma = \partial\Omega$, the transmission eigenvalue problem (10.1.11) is reduced to (10.1.9). Hence, we refer to (10.1.11) as the generalized transmission eigenvalue problem. It is particular to note that the transmission condition $T_\nu \mathbf{v} + \eta\mathbf{v} = T_\nu \mathbf{w}$ not only brings mathematical generalization but also is physically meaningful. In fact, it is referred to as the conductive transmission condition in the context of electromagnetic scattering, which arises in effectively describing a thin layer of highly conducting coating [2, 17]. In (10.1.11), the generalized transmission condition can also be used to effectively describe a thin layer of highly loss elastic coating. However, we shall not explore more about this point since it is not the focus of the current article. If $\eta \equiv 0$, the existence and properties of the transmission eigenvalues to (10.1.11) have been studied in [4,5]. In this chapter, we shall study the intrinsic geometric properties of the transmission eigenfunctions assuming their existence in the general case.

The major geometric result can be sketched as follows. Let $\Gamma = \Gamma^- \cup \Gamma^+$, where $\Gamma^\pm$ are two non-collinear/non-coplanar line segments or planes in 2D and 3D, respectively. That is, Γ^- and Γ^+ form a (non-degenerate) corner on $\partial\Omega$. Under mild conditions on V and η as well as the necessary regularity requirements on $\mathbf{v}$ and $\mathbf{w}$ as discussed above, it is shown that $\mathbf{v}$ and $\mathbf{w}$ are vanishing around the corner. The 2D results are contained in Theorems 10.2.1 and 10.2.2 and Corollary 10.2.1 for $\eta \neq 0$ and $\eta \equiv 0$, respectively, whereas the corresponding 3D results are contained in Theorem 10.3.1 and Corollaries 10.3.1 and 10.3.2. According to our earlier discussion, the geometric results imply that when non-scattering or nearly non-scattering occurs, the incident and the total wave fields propagate in a peculiar manner that avoids the corner places of the inhomogeneous medium. Moreover, the results can be used to obtain unique identifiable results the geometrical inverse elastic problem, which are contained in Theorems 10.4.1 and 10.4.2.

The rest of the chapter is organized as follows. In Sects. 10.2 and 10.3, we present the studies in two and three dimensions, respectively. Section 10.4 is devoted to the study of the inverse elastic problem.

10.2 Vanishing Near Corners of Generalized Elastic Transmission Eigenfunctions: Two-Dimensional Case

In this section, we consider the geometric property of the generalized elastic transmission eigenfunctions to (10.1.11) in two dimensions. First, we introduce the geometric setup of our study. For $\mathbf{x} = (x_1, x_2)^\top \in \mathbb{R}^2$, the polar coordinate of $\mathbf{x}$ is given by $\mathbf{x} = (r\cos\theta, r\sin\theta)^\top$. Denote an open sector $W \Subset \mathbb{R}^2$ and its boundary $\Gamma^\pm$ as follows:

$$\begin{aligned} W &= \{\mathbf{x} \in \mathbb{R}^2 | \mathbf{x} \neq \mathbf{0},\ \ \theta_m < \arg(x_1 + \mathrm{i}x_2) < \theta_M\}, \\ \Gamma^- &= \{\mathbf{x} \in \mathbb{R}^2 | \mathbf{x} \neq \mathbf{0},\ \ \arg(x_1 + \mathrm{i}x_2) = \theta_m\}, \\ \Gamma^+ &= \{\mathbf{x} \in \mathbb{R}^2 | \mathbf{x} \neq \mathbf{0},\ \ \arg(x_1 + \mathrm{i}x_2) = \theta_M\}, \end{aligned} \tag{10.2.1}$$

where $-\pi < \theta_m < \theta_M < \pi$. Let B_h and B_ε denote open disks centered at $\mathbf{0}$ of radii $h \in \mathbb{R}_+$ and $\varepsilon \in \mathbb{R}_+$ with $\varepsilon < h$, respectively. In the sequel, we set

$$S_h = W \cap B_h,\ \Gamma_h^\pm = \Gamma^\pm \cap B_h,\ \overline{S}_h = \overline{W} \cap \overline{B_h},\ \text{ and } \Lambda_h = \partial S_h \backslash (\Gamma_h^+ \cup \Gamma_h^-). \tag{10.2.2}$$

The elastic Herglotz wave function $\mathbf{v_g}$ in $\mathbb{R}^2$ is defined by

$$\mathbf{v_g} = e^{-\frac{\mathrm{i}\pi}{4}} \int_{\mathbb{S}^1} \left\{ \sqrt{\frac{k_p}{\omega}} e^{\mathrm{i}k_p \mathbf{d}\cdot\mathbf{x}} g_p(\mathbf{d})\mathbf{d} + \sqrt{\frac{k_s}{\omega}} e^{\mathrm{i}k_s \mathbf{d}\cdot\mathbf{x}} g_s(\mathbf{d})\mathbf{d}^\perp \right\} \mathrm{d}\sigma(\mathbf{d}), \tag{10.2.3}$$

where the kernel $\mathbf{g} = (g_p, g_s)^\top$ satisfies $g_p,\ g_s \in L^2(\mathbb{S}^1)$, $\mathbf{d}, \mathbf{d}^\perp \in \mathbb{S}^1$, and $\mathbf{d} \perp \mathbf{d}^\perp$.

Lemma 10.2.1 ([3, Theorem 3.4]). *Let $D \Subset \mathbb{R}^2$ be a bounded Lipschitz domain with a connected complement. Then the set of elastic Herglotz wave functions is dense with respect to the $H^1(D)^2$ norm in the set of solutions to the Lamé system*

$$\mathscr{L}\mathbf{v} + \omega^2\mathbf{v} = \mathbf{0}, \quad \mathscr{L} := \lambda\Delta + (\lambda+\mu)\nabla(\nabla\cdot) \quad \textit{in } D. \tag{10.2.4}$$

By virtue of Lemma 10.2.1, for any generalized elastic transmission eigenfunctions $(\mathbf{v},\mathbf{w}) \in H^1(\Omega)^2 \times H^1(\Omega)^2$ to (10.1.11), there exists a sequence of elastic Herglotz wave functions $\mathbf{v}_j$ given by

$$\mathbf{v}_j(\mathbf{x}) = e^{-\frac{\mathrm{i}\pi}{4}} \int_{\mathbb{S}^1} \left\{ \sqrt{\frac{k_p}{\omega}} e^{\mathrm{i}k_p \mathbf{d}\cdot\mathbf{x}} g_{jp}(\mathbf{d})\mathbf{d} + \sqrt{\frac{k_s}{\omega}} e^{\mathrm{i}k_s \mathbf{d}\cdot\mathbf{x}} g_{js}(\mathbf{d})\mathbf{d}^\perp \right\} \mathrm{d}\sigma(\mathbf{d}), \tag{10.2.5}$$

which can approximate $\mathbf{v}$ to an arbitrary accuracy in $H^1(\Omega)^2$. It is clear that $\mathbf{v}_j$ can be regarded as the Fourier extension of $\mathbf{v}$. Let

$$\mathbf{v}_j(\mathbf{x}) = \mathbf{v}_j^{\mathsf{R}}(\mathbf{x}) + \mathrm{i}\mathbf{v}_j^{\mathsf{I}}(\mathbf{x}), \tag{10.2.6}$$

where $\mathbf{v}_j^{\mathsf{R}}(\mathbf{x})$ and $\mathbf{v}_j^{\mathsf{I}}(\mathbf{x})$ are real-valued functions.

In what follows, we shall split the real and imaginary parts of the elastic transmission eigenfunctions $(\mathbf{v},\mathbf{w})$ to (10.1.11) as

$$\mathbf{v} = \mathbf{v}_{\mathsf{R}} + \mathrm{i}\mathbf{v}_{\mathsf{I}}, \quad \mathbf{w} = \mathbf{w}_{\mathsf{R}} + \mathrm{i}\mathbf{w}_{\mathsf{I}}. \tag{10.2.7}$$

It is straightforward to verify that both $(\mathbf{v}_{\mathsf{R}},\mathbf{w}_{\mathsf{R}})$ and $(\mathbf{v}_{\mathsf{R}},\mathbf{w}_{\mathsf{R}})$ satisfy (10.1.11). In a similar manner, we let the real and imaginary parts of the kernel functions $g_{j\beta}(\mathbf{d})$ $(\beta = p,s)$ of (10.2.5) be given by

$$g_{j\beta}(\mathbf{d}) = g_{j\beta}^{\mathsf{R}}(\mathbf{d}) + \mathrm{i}g_{j\beta}^{\mathsf{I}}(\mathbf{d}). \tag{10.2.8}$$

Proposition 10.2.1. *Let the elastic Herglotz wave function $\mathbf{v}_j$ be defined by* (10.2.5). *Denote*

$$\begin{aligned}
\mathbf{v}_{jp}^{\mathsf{R}}(\mathbf{0}) &= \frac{1}{\sqrt{2}}\int_{\mathbb{S}^1}\sqrt{\frac{k_p}{\omega}} g_{jp}^{\mathsf{R}}(\mathbf{d})\mathbf{d}\mathrm{d}\sigma(\mathbf{d}), \quad \mathbf{v}_{js}^{\mathsf{R}}(\mathbf{0}) = \frac{1}{\sqrt{2}}\int_{\mathbb{S}^1}\sqrt{\frac{k_s}{\omega}} g_{js}^{\mathsf{R}}(\mathbf{d})\mathbf{d}^\perp\mathrm{d}\sigma(\mathbf{d}),\\
\mathbf{v}_{jp}^{\mathsf{I}}(\mathbf{0}) &= \frac{1}{\sqrt{2}}\int_{\mathbb{S}^1}\sqrt{\frac{k_p}{\omega}} g_{jp}^{\mathsf{I}}(\mathbf{d})\mathbf{d}\mathrm{d}\sigma(\mathbf{d}), \quad \mathbf{v}_{js}^{\mathsf{I}}(\mathbf{0}) = \frac{1}{\sqrt{2}}\int_{\mathbb{S}^1}\sqrt{\frac{k_s}{\omega}} g_{js}^{\mathsf{I}}(\mathbf{d})\mathbf{d}^\perp\mathrm{d}\sigma(\mathbf{d}).
\end{aligned} \tag{10.2.9}$$

Then

$$\begin{aligned}
\mathbf{v}_j(\mathbf{0}) = \mathbf{v}_j^{\mathsf{R}}(\mathbf{0}) + \mathrm{i}\mathbf{v}_j^{\mathsf{I}}(\mathbf{0}) &:= (\mathbf{v}_{jp}^{\mathsf{R}}(\mathbf{0}) + \mathbf{v}_{js}^{\mathsf{R}}(\mathbf{0}) + \mathbf{v}_{jp}^{\mathsf{I}}(\mathbf{0}) + \mathbf{v}_{js}^{\mathsf{I}}(\mathbf{0}))\\
&\quad + \mathrm{i}(\mathbf{v}_{jp}^{\mathsf{I}}(\mathbf{0}) + \mathbf{v}_{js}^{\mathsf{I}}(\mathbf{0}) - \mathbf{v}_{jp}^{\mathsf{R}}(\mathbf{0}) - \mathbf{v}_{js}^{\mathsf{R}}(\mathbf{0})),
\end{aligned} \tag{10.2.10}$$

where $g_{j\beta}^{\mathsf{R}}(\mathbf{d})$ and $g_{j\beta}^{\mathsf{I}}(\mathbf{d})$ $(\beta = p,s)$ are defined in (10.2.8). *Let $J_\ell(t)$ be the ℓ-th Bessel function of the first kind for $\ell \in \mathbb{N}\cup\{0\}$. Denote $J_{\ell,\beta} = J_\ell(k_\beta|\mathbf{x}|)$, $\beta = p$, s. Furthermore, we have*

$$\begin{aligned}
\mathbf{v}_j^{\mathsf{R}}(\mathbf{x}) =& \mathbf{v}_{jp}^{\mathsf{R}}(\mathbf{0})J_{0,p} + \mathbf{v}_{jp}^{\mathsf{I}}(\mathbf{0})J_{0,p} + \mathbf{v}_{js}^{\mathsf{R}}(\mathbf{0})J_{0,s} + \mathbf{v}_{js}^{\mathsf{I}}(\mathbf{0})J_{0,s}\\
&+ \sqrt{2}\sum_{\ell=1}^{+\infty}(-1)^\ell\left(J_{2\ell,p}\mathbf{A}_{jp,1}^{(\ell)} + J_{2\ell,s}\mathbf{A}_{js,1}^{(\ell)} + J_{2\ell-1,p}\mathbf{A}_{jp,2}^{(\ell)} + J_{2\ell-1,s}\mathbf{A}_{js,2}^{(\ell)}\right),
\end{aligned} \tag{10.2.11}$$

and

$$\begin{aligned}
\mathbf{v}_j^{\mathsf{I}}(\mathbf{x}) =& \mathbf{v}_{jp}^{\mathsf{I}}(\mathbf{0})J_{0,p} + \mathbf{v}_{jp}^{\mathsf{I}}(\mathbf{0})J_{0,p} - \mathbf{v}_{js}^{\mathsf{R}}(\mathbf{0})J_{0,s} - \mathbf{v}_{js}^{\mathsf{R}}(\mathbf{0})J_{0,s}\\
&+ \sqrt{2}\sum_{\ell=1}^{+\infty}[(-1)^\ell\left(J_{2\ell,p}\mathbf{A}_{jp,3}^{(\ell)} + J_{2\ell,s}\mathbf{A}_{js,3}^{(\ell)} + J_{2\ell-1,p}\mathbf{A}_{jp,2}^{(\ell)} + J_{2\ell-1,s}\mathbf{A}_{js,2}^{(\ell)}\right),
\end{aligned} \tag{10.2.12}$$

where $\mathbf{p}_p = \mathbf{d}$, $\mathbf{p}_s = \mathbf{d}^\perp$, and

$$\mathbf{A}_{j\beta,1}^{(\ell)} = \int_{\mathbb{S}^1} \sqrt{\frac{k_\beta}{\omega}} g_{j\beta}^{\mathsf{R}}(\mathbf{d}) \cos(2\ell\theta)\mathbf{p}_\beta \mathrm{d}\sigma(\mathbf{d}) + \int_{\mathbb{S}^1} \sqrt{\frac{k_\beta}{\omega}} g_{j\beta}^{\mathsf{I}}(\mathbf{d}) \cos(2\ell\theta))\mathbf{p}_\beta \mathrm{d}\sigma(\mathbf{d}),$$

$$\mathbf{A}_{j\beta,2}^{(\ell)} = \int_{\mathbb{S}^1} \sqrt{\frac{k_\beta}{\omega}} g_{j\beta}^{\mathsf{I}}(\mathbf{d}) \cos((2\ell-1)\theta)\mathbf{p}_\beta \mathrm{d}\sigma(\mathbf{d}) - \int_{\mathbb{S}^1} \sqrt{\frac{k_\beta}{\omega}} g_{j\beta}^{\mathsf{R}}(\mathbf{d}) \cos((2\ell-1)\theta)\mathbf{p}_\beta \mathrm{d}\sigma(\mathbf{d}),$$

$$\mathbf{A}_{j\beta,3}^{(\ell)} = \int_{\mathbb{S}^1} \sqrt{\frac{k_\beta}{\omega}} g_{j\beta}^{\mathsf{R}}(\mathbf{d}) \cos(2\ell\theta)\mathbf{p}_\beta \mathrm{d}\sigma(\mathbf{d}) - \int_{\mathbb{S}^1} \sqrt{\frac{k_\beta}{\omega}} g_{j\beta}^{\mathsf{I}}(\mathbf{d}) \cos(2\ell\theta))\mathbf{p}_\beta \mathrm{d}\sigma(\mathbf{d}), \beta = p \text{ or } s,$$

and θ is the angle between $\mathbf{x}$ and $\mathbf{d}$ in (10.2.5). *It holds that*

$$|\mathbf{A}_{jp,i}^{(\ell)}| \leq 2\sqrt{\frac{k_p\pi}{\omega}} \|g_{jp}\|_{L^2(\mathbb{S}^1)}, \quad |\mathbf{A}_{js,i}^{(\ell)}| \leq 2\sqrt{\frac{k_s\pi}{\omega}} \|g_{js}\|_{L^2(\mathbb{S}^1)}, \quad i = 1,\ldots,3. \tag{10.2.13}$$

Proof. In view of (10.2.5), we can directly derive (10.2.10). Using the Jacobi-Anger expansion (cf. [19]), we have

$$\begin{aligned} e^{\mathrm{i}k_\beta \mathbf{d}\cdot\mathbf{x}} &= J_0(k_\beta|\mathbf{x}|) + 2\sum_{n=1}^{+\infty} \mathrm{i}^n J_n(k_\beta|\mathbf{x}|)\cos n\theta, \\ &= J_0(k_\beta|\mathbf{x}|) + 2\sum_{\ell=1}^{+\infty} \Big[(-1)^\ell J_{2\ell}(k_\beta|\mathbf{x}|)\cos(2\ell\theta) \\ &\quad - \mathrm{i}(-1)^\ell J_{2\ell-1}(k_\beta|\mathbf{x}|)\cos((2\ell-1)\theta)\Big], \end{aligned} \tag{10.2.14}$$

where $\theta = \angle(\mathbf{x},\mathbf{d})$, $\beta = p,s$. Substituting (10.2.10) and (10.2.14) into (10.2.5), we can obtain (10.2.11) and (10.2.12) by direct calculations. (10.2.13) can be obtained by using the Cauchy-Schwartz inequality. □

We shall make use of the complex geometric optics (CGO) solution $\mathbf{u}(\mathbf{x})$ introduced in [8]. We next review some quantitative properties of $\mathbf{u}(\mathbf{x})$, which shall be used in our subsequent analysis.

Lemma 10.2.2 ([8, Proposition 3.1]). *Let $\Omega \subset \mathbb{R}^2$ such that $\Omega \cap (\mathbb{R}_- \cup \{\mathbf{0}\}) = \emptyset$. Denote*

$$\mathbf{u}(\mathbf{x}) = \begin{pmatrix} \exp(-s\sqrt{z}) \\ \mathrm{i}\exp(-s\sqrt{z}) \end{pmatrix} := \begin{pmatrix} u_1(\mathbf{x}) \\ u_2(\mathbf{x}) \end{pmatrix}, \quad \mathbf{x} = (x_1,x_2)^\top, \tag{10.2.15}$$

where $z = x_1 + \mathrm{i}x_2$ and $s \in \mathbb{R}_+$. The complex square root is defined as

$$\sqrt{z} = \sqrt{|z|}\left(\cos\frac{\theta}{2} + \mathrm{i}\sin\frac{\theta}{2}\right),$$

where $-\pi < \theta \leq \pi$ is the argument of z. Then $\mathbf{u}$ satisfies

$$\mathscr{L}\mathbf{u} = \mathbf{0} \quad in \quad \Omega.$$

Let the open sector W be defined in (10.2.1). *Then*

$$\int_W u_1(\mathbf{x})\mathrm{d}\mathbf{x} = 6\mathrm{i}(e^{-2\theta_M \mathrm{i}} - e^{-2\theta_m \mathrm{i}})s^{-4}. \tag{10.2.16}$$

In addition to $\alpha, h > 0$ and $j \in \{1,2\}$, we have the upper bounds

$$\int_W |u_j(\mathbf{x})||\mathbf{x}|^\alpha \mathrm{d}\mathbf{x} \leq \frac{2(\theta_M - \theta_m)\Gamma(2\alpha+4)}{\delta_W^{2\alpha+4}} s^{-2\alpha-4} \tag{10.2.17}$$

and

$$\int_{W\setminus S_h} |u_j(\mathbf{x})|\mathrm{d}\mathbf{x} \le \frac{6(\theta_M-\theta_m)}{\delta_W^4} s^{-4} e^{-\delta_W s\sqrt{h}/2}, \tag{10.2.18}$$

where $\delta_W = \min_{\theta_m<\theta<\theta_M} \cos\left(\frac{\theta}{2}\right)$ *is a positive constant.*

The following lemma states the regularity of the CGO solution $\mathbf{u}(\mathbf{x})$ defined in (10.2.15), where the proof follows from a similar argument to that of [23, Lemma 2.3] and we skip the details.

Lemma 10.2.3. *Let* S_h *be defined in* (10.2.2) *and* $\mathbf{u}(\mathbf{x})$ *be given in* (10.2.15). *Then* $\mathbf{u}(\mathbf{x}) \in H^1(S_h)^2$ *and* $\mathscr{L}\mathbf{u}(\mathbf{x}) = \mathbf{0}$ *in* S_h. *Furthermore, it holds that*

$$\|\mathbf{u}(\mathbf{x})\|_{L^2(S_h)^2} \le \sqrt{\theta_M-\theta_m} e^{-s\sqrt{\Theta}\delta_W} h \tag{10.2.19}$$

and

$$\||\mathbf{x}|^\alpha \mathbf{u}(\mathbf{x})\|_{L^2(S_h)^2} \le s^{-2(\alpha+1)} \frac{2\sqrt{(\theta_M-\theta_m)\Gamma(4\alpha+4)}}{(2\delta_W)^{2\alpha+2}}, \tag{10.2.20}$$

where $\Theta \in [0,h]$ *and* δ_W *is defined in* (10.2.18).

Lemma 10.2.4. *Suppose that* Λ_h *and* $\mathbf{u}(\mathbf{x})$ *are defined by* (10.2.2) *and* (10.2.15), *respectively. Recall that* $\delta_W > 0$ *is given in* (10.2.18). *We have*

$$\|\mathbf{u}(\mathbf{x})\|_{H^1(\Lambda_h)^2} \le \sqrt{h+\frac{s^2}{2}}\sqrt{\theta_M-\theta_m} e^{-s\sqrt{h}\delta_W}, \tag{10.2.21a}$$

$$\|T_\nu(\mathbf{u})\|_{L^2(\Lambda_h)^2} \le \frac{s\mu}{\sqrt{2}}\sqrt{\theta_M-\theta_m} e^{-s\sqrt{h}\delta_W}, \tag{10.2.21b}$$

both of which decay exponentially as $s \to +\infty$.

Proof. By (10.2.15), one has

$$\|u_1(\mathbf{x})\|_{L^2(\Lambda_h)^2} \le \sqrt{h} e^{-s\sqrt{h}\delta_W} \sqrt{\theta_M-\theta_m}. \tag{10.2.22}$$

Using the polar coordinate $\mathbf{x} = (x_1, x_2) = r(\cos\theta, \sin\theta)$, by virtue of (10.2.15), it is directly verified that

$$\frac{\partial u_1(\mathbf{x})}{\partial r} = -\frac{s}{2r^{1/2}} e^{-s\sqrt{r}(\cos(\theta/2)+\mathrm{i}\sin(\theta/2))+\mathrm{i}\theta/2},$$
$$\frac{\partial u_1(\mathbf{x})}{\partial \theta} = -\frac{\mathrm{i}s\sqrt{r}}{2} e^{-s\sqrt{r}(\cos(\theta/2)+\mathrm{i}\sin(\theta/2))+\mathrm{i}\theta/2},$$

which can be used to obtain that

$$\begin{aligned} \frac{\partial u_1(\mathbf{x})}{\partial x_1} &= -\frac{s}{2r^{1/2}} e^{-s\sqrt{r}(\cos(\theta/2)+\mathrm{i}\sin(\theta/2))-\mathrm{i}\theta/2}, \\ \frac{\partial u_1(\mathbf{x})}{\partial x_2} &= -\frac{is}{2r^{1/2}} e^{-s\sqrt{r}(\cos(\theta/2)+\mathrm{i}\sin(\theta/2))-\mathrm{i}\theta/2}. \end{aligned} \tag{10.2.23}$$

Therefore,

$$\|\nabla u_1(\mathbf{x})\|_{L^2(\Lambda_h)^2} \le \frac{s}{2} e^{-s\sqrt{h}\delta_W} \sqrt{\theta_M-\theta_m}. \tag{10.2.24}$$

Combining (10.2.22) and (10.2.24), as well as noting $u_2(s\mathbf{x}) = \mathrm{i}u_2(s\mathbf{x})$, we can prove (10.2.21a).

Using (10.1.10), it is directly calculated that $T_\nu\mathbf{u} = \mu\nabla\mathbf{u}\ \nu$. Therefore, by virtue of (10.2.23), (10.2.24), and the Cauchy-Schwartz inequality, we have

$$\|T_\nu(\mathbf{u})\|^2_{L^2(\Lambda_h)^2} - 2\mu^2\|\nabla u_1(\mathbf{x})\|^2_{L^2(\Lambda_h)^2} \le \frac{s\mu^2}{2} e^{-2s\sqrt{h}\delta_W}(\theta_M-\theta_m).$$

The proof is complete. □

We proceed to derive several key lemmas in order to establish the main geometric result of this section. It is first recalled the following Green formula for the Lamé operator; see [20, Lemm 3.4] and [37, Theorem 4.4].

Lemma 10.2.5. *Suppose that $\Omega \Subset \mathbb{R}^n (n=2,3)$ is a bounded Lipschitz domain. Let $\mathbf{u_1} \in H^1(\Omega)^2$ and $\mathbf{v}_1 \in H^1(\Omega)^2$ satisfying $\mathscr{L}\mathbf{u_1} \in L^2(\Omega)^n$ and $\mathscr{L}\mathbf{v_1} \in L^2(\Omega)^n$. The following Green identity holds:*

$$\int_{\Omega} (\mathscr{L}\mathbf{u}_1 \cdot \mathbf{v}_1 - \mathscr{L}\mathbf{v}_1 \cdot \mathbf{u}_1) \mathrm{d}\mathbf{x} = \int_{\partial\Omega} (T_\nu \mathbf{u}_1 \cdot \mathbf{v}_1 - T_\nu \mathbf{v}_1 \cdot \mathbf{u}_1) \mathrm{d}\sigma. \tag{10.2.25}$$

Recall the splitting (10.2.7). In what follows, we shall mainly focus on establishing the relevant results for $(\mathbf{v}_\mathrm{R}, \mathbf{w}_\mathrm{R})$. Due to the symmetric role of $(\mathbf{v}_\mathrm{R}, \mathbf{w}_\mathrm{R})$ and $(\mathbf{v}_\mathrm{I}, \mathbf{w}_\mathrm{I})$, those results hold equally for $(\mathbf{v}_\mathrm{I}, \mathbf{w}_\mathrm{I})$ and hence $(\mathbf{v}, \mathbf{w})$.

Lemma 10.2.6. *Let $\mathbf{v}_\mathrm{R} \in H^1(\Omega)^2$ and $\mathbf{w}_\mathrm{R} \in H^1(\Omega)^2$ be a pair of generalized elastic transmission eigenfunctions to* (10.1.11)*. Let the CGO solution $\mathbf{u}$ and the elastic Herglotz wave function $\mathbf{v}_j^\mathrm{R}$ be defined in* (10.2.15) *and* (10.2.11)*, respectively. Assume that the Lipschitz domain $\Omega \subset \mathbb{R}^2$ contains a corner $S_h \Subset \Omega \cap W$, where S_h is defined in* (10.2.2) *and W is a sector defined in* (10.2.1)*. Denote $q = 1 + V$, where V is defined in* (10.1.11)*. Then the following integral equality holds:*

$$I_1 + I_2 = I_{\Lambda_h} - I_\pm - I_\pm^\Delta, \tag{10.2.26}$$

where

$$I_1 = \omega^2 \int_{S_h} (q\mathbf{w}_\mathrm{R} - \mathbf{v}_j^\mathrm{R}) \cdot \mathbf{u}(\mathbf{x}) \mathrm{d}\mathbf{x}, \quad I_2 = -\omega^2 \int_{S_h} (\mathbf{v}_\mathrm{R} - \mathbf{v}_j^\mathrm{R}) \cdot \mathbf{u}(\mathbf{x}) \mathrm{d}\mathbf{x},$$
$$I_{\Lambda_h} = \int_{\Lambda_h} (T_\nu(\mathbf{v}_\mathrm{R} - \mathbf{w}_\mathrm{R})) \cdot \mathbf{u} - (T_\nu(\mathbf{u})) \cdot (\mathbf{v}_\mathrm{R} - \mathbf{w}_\mathrm{R}) \mathrm{d}\sigma,$$
$$I_\pm = \int_{\Gamma_h^\pm} \eta \mathbf{u} \cdot \mathbf{v}_j^\mathrm{R} \mathrm{d}\sigma, \quad I_\pm^\Delta = \int_{\Gamma_h^\pm} \eta \mathbf{u} \cdot (\mathbf{v}_\mathrm{R} - \mathbf{v}_j^\mathrm{R}) \mathrm{d}\sigma.$$

Here, Λ_h and $\Gamma_h^\pm$ are defined in (10.2.2)*.*

Proof. Recall that the differential operator $\mathscr{L}$ is defined in (10.2.4). In view of the first and second equations in (10.1.11), we have

$$\mathscr{L}\mathbf{v}_\mathrm{R} = -\omega^2 \mathbf{v}_\mathrm{R}, \quad \mathscr{L}\mathbf{u}_\mathrm{R} = -\omega^2 q \mathbf{w}_\mathrm{R} \quad \text{in } S_h. \tag{10.2.27}$$

Using the boundary condition in (10.1.11), it yields that

$$\mathbf{v}_\mathrm{R} - \mathbf{w}_\mathrm{R} = 0, \quad T_\nu(\mathbf{v}_\mathrm{R} - \mathbf{w}_\mathrm{R}) = -\eta \mathbf{v}_\mathrm{R} \quad \text{on } \Gamma_h^\pm. \tag{10.2.28}$$

Using Green's formula (10.2.25) on the domain S_h together with $\mathscr{L}\mathbf{u} = \mathbf{0}$ in S_h, we have

$$\int_{S_h} (\mathscr{L}(\mathbf{v}_\mathrm{R} - \mathbf{w}_\mathrm{R})) \cdot \mathbf{u}(\mathbf{x}) \mathrm{d}\mathbf{x} = \int_{\Gamma_h^\pm \cup \Lambda_h} (T_\nu(\mathbf{v}_\mathrm{R} - \mathbf{w}_\mathrm{R})) \cdot \mathbf{u} - (T_\nu(\mathbf{u})) \cdot (\mathbf{v}_\mathrm{R} - \mathbf{w}_\mathrm{R}) \mathrm{d}\sigma. \tag{10.2.29}$$

By virtue of (10.2.28), we have

$$\begin{aligned} &\int_{\Gamma_h^\pm} (T_\nu(\mathbf{v}_\mathrm{R} - \mathbf{w}_\mathrm{R})) \cdot \mathbf{u} - (T_\nu(\mathbf{u})) \cdot (\mathbf{v}_\mathrm{R} - \mathbf{w}_\mathrm{R}) \mathrm{d}\sigma \\ &= \int_{\Gamma_h^\pm} -\eta \mathbf{u} \cdot \mathbf{v}_\mathrm{R} \mathrm{d}\sigma = \int_{\Gamma_h^\pm} -\eta \mathbf{u} \cdot \mathbf{v}_j^\mathrm{R} \mathrm{d}\sigma + \int_{\Gamma_h^\pm} -\eta \mathbf{u} \cdot (\mathbf{v}_\mathrm{R} - \mathbf{v}_j^\mathrm{R}) \mathrm{d}\sigma. \end{aligned} \tag{10.2.30}$$

From (10.2.27), we have

$$\begin{aligned}\int_{S_h}(\mathscr{L}(\mathbf{v}_{\mathrm{R}}-\mathbf{w}_{\mathrm{R}}))\cdot\mathbf{u}(\mathbf{x})\mathrm{d}\mathbf{x}&=\int_{S_h}(-\omega^2\mathbf{v}_{\mathrm{R}}+\omega^2q\mathbf{w}_{\mathrm{R}})\cdot\mathbf{u}(\mathbf{x})\mathrm{d}\mathbf{x}\\&=\int_{S_h}(-\omega^2\mathbf{v}_j^{\mathrm{R}}+\omega^2q\mathbf{w}_{\mathrm{R}})\cdot\mathbf{u}(\mathbf{x})\mathrm{d}\mathbf{x}+\int_{S_h}-\omega^2(\mathbf{v}_{\mathrm{R}}-\mathbf{v}_j^{\mathrm{R}})\cdot\mathbf{u}(\mathbf{x})\mathrm{d}\mathbf{x}.\end{aligned}\tag{10.2.31}$$

By (10.2.29), (10.2.30), and (10.2.31), we can derive (10.2.26).
The proof is complete. □

By using the Hölder inequality, Lemma 10.2.4, and the trace theorem, one has the following:

Lemma 10.2.7. *Let I_{Λ_h} be defined in* (10.2.26)*. Under the same setup as that in Lemma 10.2.6, we have the following estimate:*

$$|I_{\Lambda_h}|\leq C\frac{\sqrt{2h+s^2}+\mu s}{\sqrt{2}}\sqrt{\theta_M-\theta_m}e^{-s\sqrt{h}\delta_W}\|\mathbf{v}_{\mathrm{R}}-\mathbf{w}_{\mathrm{R}}\|_{H^1(S_h)^2},\tag{10.2.32}$$

where C is a positive constant coming from the trace theorem and S_h and $\delta_W>0$ are defined in (10.2.2) *and* (10.2.18)*, respectively.*

The proof of Lemma 10.2.8 is based on the Cauchy-Schwarz inequality and the trace theorem together with Lemma 8.2.3.

Lemma 10.2.8. *Under the same setup as that in Lemma 10.2.6, we further suppose that the boundary parameter η of* (10.1.11) *satisfies $\eta\in C^\alpha(\overline{\Gamma_h^\pm})$ for $0<\alpha<1$. For any given constants β_1, β_2, and γ satisfying $\gamma>\max\{\beta_1,\beta_2\}>0$, assume that there exists a sequence of $\{\mathbf{v}_j^{\mathrm{R}}\}_{j=1}^{+\infty}$ defined by* (10.2.6) *with kernels g_{jp} and g_{js} which can approximate $\mathbf{v}_{\mathrm{R}r}$ in $H^1(S_h)$ fulfilling*

$$\|\mathbf{v}_{\mathrm{R}}-\mathbf{v}_j^{\mathrm{R}}\|_{H^1(S_h)^2}\leq j^{-\gamma},\ \|g_{jp}\|_{L^2(\mathbb{S}^1)}\leq j^{\beta_1},\ \|g_{js}\|_{L^2(\mathbb{S}^1)}\leq j^{\beta_2}.\tag{10.2.33}$$

Recall that I_2 and $I_\pm^\Delta$ are defined in (10.2.26)*. Then the following integral estimates hold:*

$$|I_2|\leq\omega^2h\sqrt{\theta_M-\theta_m}e^{-s\sqrt{\Theta}\delta_W}j^{-\gamma}\tag{10.2.34}$$

and

$$\begin{aligned}\left|I_\pm^\Delta\right|\leq&\Big(|\eta(\mathbf{0})|h\sqrt{\theta_M-\theta_m}e^{-s\sqrt{\Theta}\delta_W}\\&+\|\eta\|_{C^\alpha}\frac{2\sqrt{(\theta_M-\theta_m)\Gamma(4\alpha+4)}}{(2\delta_W)^{2\alpha+2}}s^{-2(\alpha+1)}\Big)j^{-\gamma},\end{aligned}\tag{10.2.35}$$

where $\Theta\in[0,h]$, δ_W is defined in (10.2.18)*, and θ_m and θ_M are defined in* (10.1.11)*.*

Proof. By using the Cauchy-Schwarz inequality, we have

$$|I_2|\leq\omega^2\|\mathbf{v}_{\mathrm{R}}-\mathbf{v}_j^{\mathrm{R}}\|_{L^2(S_h)^2}\|\mathbf{u}(s\mathbf{x})\|_{L^2(S_h)^2}\leq\omega^2\|\mathbf{u}(s\mathbf{x})\|_{L^2(S_h)^2}j^{-\gamma}.\tag{10.2.36}$$

In view of (10.2.19), we can immediately obtain (10.2.34).
Since $\eta\in C^\alpha(\overline{\Gamma_h^\pm})$, we have the following expansion of $\eta(\mathbf{x})$ at the origin as

$$\eta(\mathbf{x})=\eta(\mathbf{0})+\delta\eta(\mathbf{x}),|\ \delta\eta(\mathbf{x})\ |\leq\|\eta\|_{C^\alpha}|\mathbf{x}|^\alpha.\tag{10.2.37}$$

By using the Cauchy-Schwartz inequality and the trace theorem, we have

$$\begin{aligned}\left|I_\pm^\Delta\right|\leq&|\eta(\mathbf{0})|\int_{\Gamma_h^\pm}|\mathbf{u}||(\mathbf{v}_{\mathrm{R}}-\mathbf{v}_j^{\mathrm{R}})|\mathrm{d}\sigma+\|\eta\|_{C^\alpha}\int_{\Gamma_h^\pm}|\mathbf{x}|^\alpha|\mathbf{u}||(\mathbf{v}_{\mathrm{R}}-\mathbf{v}_j^{\mathrm{R}})|\mathrm{d}\sigma\\\leq&|\eta(\mathbf{0})|\|\mathbf{v}_{\mathrm{R}}-\mathbf{v}_j^{\mathrm{R}}\|_{H^{\frac{1}{2}}(\Gamma_h^\pm)^2}\|\mathbf{u}\|_{H^{-\frac{1}{2}}(\Gamma_h^\pm)^2}+\|\eta\|_{C^\alpha}\|\mathbf{v}_{\mathrm{R}}-\mathbf{v}_j^{\mathrm{R}}\|_{H^{\frac{1}{2}}(\Gamma_h^\pm)^2}\||\mathbf{x}|^\alpha\mathbf{u}\|_{H^{-\frac{1}{2}}(\Gamma_h^\pm)^2}\end{aligned}$$

$$
\begin{aligned}
&\leq |\eta(\mathbf{0})| \|\mathbf{v} - \mathbf{v}_j^{\mathsf{R}}\|_{H^1(S_h)^2} \|\mathbf{u}\|_{L^2(S_h)^2} + \|\eta\|_{C^\alpha} \|\mathbf{v}_{\mathsf{R}} - \mathbf{v}_j^{\mathsf{R}}\|_{H^1(S_h)^2} \||\mathbf{x}|^\alpha \mathbf{u}\|_{L^2(S_h)^2} \\
&\leq \left(|\eta(\mathbf{0})| \|\mathbf{u}\|_{L^2(S_h)^2} + \|\eta\|_{C^\alpha} \||\mathbf{x}|^\alpha \mathbf{u}\|_{L^2(S_h)^2} \right) j^{-\gamma}.
\end{aligned} \tag{10.2.38}
$$

By (10.2.19), (10.2.20), and (10.2.38), it readily yields (10.2.35). □

Under the Hölder-continuous assumption on $q\mathbf{w}_{\mathsf{R}}(\mathbf{x})$ near the corner S_h, by Lemma 10.2.2, we can obtain the following lemma.

Lemma 10.2.9. *Under the same setup as that in Lemma 10.2.8, we further suppose that $q\mathbf{w}_{\mathsf{R}} \in C^\alpha(S_h)^2$ $(0 < \alpha < 1)$ and hence*

$$
\mathbf{f}_{\mathsf{R}}(\mathbf{x}) := q\mathbf{w}_{\mathsf{R}}(\mathbf{x}) = \mathbf{f}_{\mathsf{R}}(\mathbf{0}) + \delta \mathbf{f}_{\mathsf{R}}(\mathbf{x}), \quad |\delta \mathbf{f}_{\mathsf{R}}| \leq \|\mathbf{f}_{\mathsf{R}}(\mathbf{x})\|_{C^\alpha(S_h)^2} |\mathbf{x}|^\alpha. \tag{10.2.39}
$$

Then the following integral estimate holds:

$$
\begin{aligned}
|I_1| \leq & \omega^2 \Bigg(4 \left(\sqrt{\frac{\pi k_p}{\omega}} (1 + k_p) \|g_{jp}\|_{L^2(S_h)} + \sqrt{\frac{\pi k_s}{\omega}} (1 + k_s) \|g_{js}\|_{L^2(S_h)} \right) \\
& \times \frac{(\theta_M - \theta_m)\Gamma(2\alpha + 4)}{\delta_W^{2\alpha+4}} s^{-2\alpha-4} \\
& + \|\mathbf{f}_{\mathsf{R}}(\mathbf{x})\|_{C^\alpha(\Omega)^2} \frac{2\sqrt{2}(\theta_M - \theta_m)\Gamma(4\alpha + 4)}{\delta_W^{2\alpha+4}} s^{-2\alpha-4} + 6\sqrt{2}(|\mathbf{f}_{\mathsf{R}}(\mathbf{0})| + |\mathbf{v}_j^{\mathsf{R}}(\mathbf{0}|) \\
& \times |e^{-2\theta_M \mathrm{i}} - e^{-2\theta_m \mathrm{i}}| s^{-4} \Bigg),
\end{aligned} \tag{10.2.40}
$$

where δ_W is defined in (10.2.18) *and θ_m and θ_M are defined in* (10.1.11).

Proof. Since $\mathbf{v}_j^{\mathsf{R}} \in C^\alpha(S_h)^2$, $\alpha \in (0, 1)$, we have the following splitting:

$$
\mathbf{v}_j^{\mathsf{R}}(\mathbf{x}) = \mathbf{v}_j^{\mathsf{R}}(\mathbf{0}) + \delta \mathbf{v}_j^{\mathsf{R}}(\mathbf{x}), \quad |\delta \mathbf{v}_j^{\mathsf{R}}| \leq \|\mathbf{v}_j^{\mathsf{R}}\|_{C^\alpha(\Omega)^2} |\mathbf{x}|^\alpha. \tag{10.2.41}
$$

Therefore, it holds that

$$
\begin{aligned}
\int_{S_h} (q\mathbf{w}_{\mathsf{R}} - \mathbf{v}_j^{\mathsf{R}}) \cdot \mathbf{u}(s\mathbf{x}) \mathrm{d}\mathbf{x} = & \int_{S_h} (\mathbf{f}_{\mathsf{R}}(\mathbf{0}) - \mathbf{v}_j^{\mathsf{R}}(\mathbf{0})) \cdot \mathbf{u}(s\mathbf{x}) \mathrm{d}\mathbf{x} \\
& - \int_{S_h} \delta \mathbf{v}_j^{\mathsf{R}}(\mathbf{x}) \cdot \mathbf{u}(s\mathbf{x}) \mathrm{d}\mathbf{x} + \int_{S_h} \delta \mathbf{f}_{\mathsf{R}}(\mathbf{x}) \cdot \mathbf{u}(s\mathbf{x}) \mathrm{d}\mathbf{x}.
\end{aligned} \tag{10.2.42}
$$

Using the compact embedding of Hölder spaces, one can obtain that

$$
\|\mathbf{v}_j^{\mathsf{R}}\|_{C^\alpha(S_h)} \leq \mathrm{diam}(S_h)^{1-\alpha} \|\mathbf{v}_j^{\mathsf{R}}\|_{C^1(S_h)},
$$

where $\mathrm{diam}(S_h)$ is the diameter of S_h. By direct computations, we have

$$
\begin{aligned}
\|\mathbf{v}_j^{\mathsf{R}}\|_{C^1(S_h)} \leq & \sqrt{\frac{\pi k_p}{\omega}} (1 + k_p) (\|g_{jp}^{\mathsf{R}}\|_{L^2(S_h)} + \|g_{jp}^{\mathsf{I}}\|_{L^2(S_h)}) \\
& + \sqrt{\frac{\pi k_s}{\omega}} (1 + k_s) (\|g_{js}^{\mathsf{R}}\|_{L^2(S_h)} + \|g_{js}^{\mathsf{I}}\|_{L^2(S_h)}) \\
\leq & \sqrt{2} \sqrt{\frac{\pi k_p}{\omega}} (1 + k_p) \|g_{jp}\|_{L^2(S_h)} + \sqrt{2} \sqrt{\frac{\pi k_s}{\omega}} (1 + k_s) \|g_{js}\|_{L^2(S_h)}.
\end{aligned} \tag{10.2.43}
$$

Due to (10.2.17), (10.2.39), and (10.2.43), one can verify that

$$
\begin{aligned}
\left|\int_{S_h} \delta \mathbf{v}_j^{\mathrm{R}}(\mathbf{x}) \cdot \mathbf{u}(s\mathbf{x}) \mathrm{d}\mathbf{x}\right| &\le \|\mathbf{v}_j^{\mathrm{R}}\|_{C^\alpha(\Omega)^2} \int_{S_h} |\mathbf{u}(s\mathbf{x})||\mathbf{x}|^\alpha \mathrm{d}\mathbf{x} \\
&\le 4\sqrt{\frac{\pi}{\omega}} \left(k_p^{1/2}(1+k_p)\|g_{jp}\|_{L^2(S_h)} + k_s^{1/2}(1+k_s)\|g_{js}\|_{L^2(S_h)} \right) \\
&\quad \times \frac{(\theta_M-\theta_m)\Gamma(2\alpha+4)}{\delta_W^{2\alpha+4}} s^{-2\alpha-4}
\end{aligned} \tag{10.2.44}
$$

and

$$
\begin{aligned}
\left|\int_{S_h} \delta \mathbf{f}_{\mathrm{R}}(\mathbf{x}) \cdot \mathbf{u}(s\mathbf{x}) \mathrm{d}\mathbf{x}\right| &\le \|\mathbf{f}_{\mathrm{R}}(\mathbf{x})\|_{C^\alpha(\Omega)^2} \int_{W} |\mathbf{u}(s\mathbf{x})||\mathbf{x}|^\alpha \mathrm{d}\mathbf{x} \\
&\le \|\mathbf{f}_{\mathrm{R}}(\mathbf{x})\|_{C^\alpha(\Omega)^2} \frac{2\sqrt{2}(\theta_M-\theta_m)\Gamma(2\alpha+4)}{\delta_W^{2\alpha+4}} s^{-2\alpha-4}.
\end{aligned} \tag{10.2.45}
$$

Finally, by (10.2.16), (10.2.44), and (10.2.45), one can arrive at (10.2.40).

The proof is complete. □

Lemma 10.2.10 ([23, Lemma 2.4]). *For any $\zeta > 0$, if $\omega(\theta) > 0$, then*

$$
\int_0^h r^\zeta e^{-s\sqrt{r}\omega(\theta)} \mathrm{d}r = O(s^{-2\zeta-2}) \quad \text{as } s \to +\infty. \tag{10.2.46}
$$

Lemma 10.2.11 ([23, Lemma 2.8]). *Recall that $\Gamma_h^\pm$ and $u_1(\mathbf{x})$ are defined in* (10.2.2) *and* (10.2.15)*, respectively. We have*

$$
\begin{aligned}
\int_{\Gamma_h^+} u_1(\mathbf{x}) \mathrm{d}\sigma &= 2s^{-2}\left(\mu(\theta_M)^{-2} - \mu(\theta_M)^{-2} e^{-s\sqrt{h}\mu(\theta_M)} - \mu(\theta_M)^{-1} s\sqrt{h} e^{-s\sqrt{h}\mu(\theta_M)}\right), \\
\int_{\Gamma_h^-} u_1(\mathbf{x}) \mathrm{d}\sigma &= 2s^{-2}\left(\mu(\theta_m)^{-2} - \mu(\theta_m)^{-2} e^{-s\sqrt{h}\mu(\theta_m)} - \mu(\theta_m)^{-1} s\sqrt{h} e^{-s\sqrt{h}\mu(\theta_m)}\right),
\end{aligned} \tag{10.2.47}
$$

where $\mu(\theta) = \cos(\theta/2) + \mathrm{i}\sin(\theta/2)$.

The following series expression for the Bessel function $J_p(t)$ can be found in [1] as

$$
J_p(t) = \frac{t^p}{2^p p!} + \frac{t^p}{2^p} \sum_{\ell=1}^{\infty} \frac{(-1)^\ell t^{2\ell}}{4^\ell (\ell!)^2}, \quad \text{for } p = 1, 2, \ldots, \tag{10.2.48}
$$

which is uniformly and absolutely convergent with respect to $t \in [0, +\infty)$.

By virtue of (10.2.11) and the Cauchy-Schwartz inequality, it can be direct to obtain the following two lemmas after some tedious calculations.

Lemma 10.2.12. *Consider the same setup as that in Lemma 10.2.6, and suppose that $\eta \in C^\alpha(\overline{\Gamma_h^\pm})$ and has expansion at the origin as*

$$
\eta(\mathbf{x}) = \eta(\mathbf{0}) + \delta\eta(\mathbf{x}), |\delta\eta(\mathbf{x})| \le \|\eta\|_{C^\alpha} |\mathbf{x}|^\alpha. \tag{10.2.49}
$$

Recall that $I_\pm$ is defined (10.2.26)*, and denote*

$$
I_\pm = \mathscr{I}_1^\pm + \eta(\mathbf{0}) \mathscr{I}_2^\pm, \tag{10.2.50}
$$

where $\mathscr{I}_1^\pm = \int_{\Gamma_h^\pm} \delta\eta \mathbf{u} \cdot \mathbf{v}_j^{\mathrm{R}} \mathrm{d}\sigma$ and $\mathscr{I}_2^\pm = \int_{\Gamma_h^\pm} \mathbf{u} \cdot \mathbf{v}_j^{\mathrm{R}} \mathrm{d}\sigma$. Then the following estimate holds:

$$
|\mathscr{I}_1^\pm| \le O(s^{-2\alpha-2}) + \left(\|g_{jp}\|_{L^2(\mathbb{S}^1)} + \|g_{js}\|_{L^2(\mathbb{S}^1)}\right) \times O(s^{-2\alpha-4}) \quad \text{as } s \to +\infty. \tag{10.2.51}
$$

Lemma 10.2.13. *Consider the same setup as that in Lemma 10.2.6, and recall that* $\mathscr{I}_2^\pm$ *is defined* (10.2.50). *When* $s \to +\infty$, *the following results hold:*

$$\mathscr{I}_{211}^{\pm(p)} + \mathscr{I}_{211}^{\pm(s)} + \mathscr{I}_{212}^{\pm(p)} + \mathscr{I}_{212}^{\pm(s)} = \eta(\mathbf{0})\left(\mathbf{v}_{jp}^{\mathsf{R}}(\mathbf{0}) + \mathbf{v}_{js}^{\mathsf{R}}(\mathbf{0}) + \mathbf{v}_{jp}^{\mathsf{I}}(\mathbf{0}) + \mathbf{v}_{js}^{\mathsf{I}}(\mathbf{0})\right)\cdot\begin{pmatrix}1\\ \mathrm{i}\end{pmatrix}$$
$$\times 2s^{-2}\Big(\mu(\theta_M)^{-2} - \mu(\theta_M)^{-2}e^{-s\sqrt{h}\mu(\theta_M)} - \mu(\theta_M)^{-1}s\sqrt{h}e^{-s\sqrt{h}\mu(\theta_M)}$$
$$+\mu(\theta_m)^{-2} - \mu(\theta_m)^{-2}e^{-s\sqrt{h}\mu(\theta_m)} - \mu(\theta_m)^{-1}s\sqrt{h}e^{-s\sqrt{h}\mu(\theta_m)}\Big),$$
$$\left|\mathscr{I}_{212}^{\pm(p)}\right| \le O(s^{-6}),\ \left|\mathscr{I}_{212}^{\pm(s)}\right| \le O(s^{-6}), \tag{10.2.52}$$

where

$$\mathscr{I}_{211}^{\pm(\beta)} = \int_{\Gamma_h^\pm}\mathbf{u}\cdot\mathbf{v}_{j\beta}^{\mathsf{R}}(\mathbf{0})\mathrm{d}\sigma,\ \mathscr{I}_{212}^{\pm(\beta)} = \int_{\Gamma_h^\pm}\mathbf{u}\cdot\mathbf{v}_{j\beta}^{\mathsf{R}}(\mathbf{0})S_\beta\mathrm{d}\sigma,\ S_\beta = \sum_{n=1}^{\infty}\frac{(-1)^n k_\beta^{2n}|\mathbf{x}|^{2n}}{4^n(n!)^2}.$$

Denote

$$\mathscr{I}_{213}^{\pm(\beta)} = \int_{\Gamma_h^\pm}\mathbf{u}\cdot\mathbf{v}_{j\beta}^{\mathsf{I}}(\mathbf{0})\mathrm{d}\sigma,\ \mathscr{I}_{214}^{\pm(\beta)} = \int_{\Gamma_h^\pm}\mathbf{u}\cdot\mathbf{v}_{j\beta}^{\mathsf{I}}(\mathbf{0})S_\beta\mathrm{d}\sigma,$$
$$\mathscr{I}_{22}^{\pm(\beta)} = \sqrt{2}\sum_{\ell=1}^{+\infty}(-1)^\ell\int_{\Gamma_h^\pm}\mathbf{u}\cdot\mathbf{A}_{j\beta,1}^{(\ell)}J_{2\ell}(k_\beta|\mathbf{x}|)\mathrm{d}\sigma,$$
$$\mathscr{I}_{23}^{\pm(\beta)} = \sqrt{2}\sum_{\ell=1}^{+\infty}(-1)^\ell\int_{\Gamma_h^\pm}\mathbf{u}\cdot\mathbf{A}_{j\beta,1}^{(\ell)}J_{2\ell-1}(k_\beta|\mathbf{x}|)\mathrm{d}\sigma.$$

It holds that

$$\left|\mathscr{I}_{214}^{\pm(p)}\right| \le O(s^{-6}),\ \left|\mathscr{I}_{214}^{\pm(s)}\right| \le O(s^{-6}),$$
$$\left|\mathscr{I}_{22}^{\pm(p)}\right| \le \|g_{jp}\|_{L^2(\mathbb{S}^1)}\times O(s^{-4}),\ \left|\mathscr{I}_{22}^{\pm(s)}\right| \le \|g_{js}\|_{L^2(\mathbb{S}^1)}\times O(s^{-4}),$$
$$\left|\mathscr{I}_{23}^{\pm(p)}\right| \le \|g_{jp}\|_{L^2(\mathbb{S}^1)}\times O(s^{-4}),\ \left|\mathscr{I}_{23}^{\pm(s)}\right| \le \|g_{js}\|_{L^2(\mathbb{S}^1)}\times O(s^{-4})$$

as $s \to +\infty$.

Theorem 10.2.1. *Let* $(\mathbf{v},\mathbf{w}) \in H^1(\Omega)^2 \times H^1(\Omega)^2$ *be a pair of eigenfunctions to* (10.1.11) *associated with* $\omega \in \mathbb{R}_+$. *Assume that the domain* $\Omega \subset \mathbb{R}^2$ *contains a corner* $\Omega \cap B_h \Subset \Omega \cap W$ *with* $h \ll 1$. *By rigid motions if necessary, we can assume that the vertex of the corner is* $\mathbf{0} \in \partial\Omega$. *Let* W *be the sector defined in* (10.2.1) *and* $S_h = \Omega \cap B_h \Subset \Omega \cap W$ *in* Ω. *Suppose that* $q\mathbf{w} \in C^\alpha(\overline{S_h})^2$ *with* $q := 1+V$ *and* $\eta \in C^\alpha(\overline{\Gamma_h^\pm})$ *for* $0 < \alpha < 1$. *If the following conditions are fulfilled:*

(a) *For any given constants* $\gamma > \max\{\beta_1, \beta_2\} > 0$, *assume that there exists a sequence of Herglotz functions* $\{\mathbf{v}_j\}_{j=1}^\infty$, *where* $\mathbf{v}_j$ *is defined in* (10.2.5), *which can approximate* $\mathbf{v}$ *in* $H^1(S_h)^2$ *fulfilling*

$$\|\mathbf{v}-\mathbf{v}_j\|_{H^1(S_h)^2} \le j^{-\gamma},\ \|g_{jp}\|_{L^2(\mathbb{S}^1)} \le j^{\beta_1},\ \|g_{js}\|_{L^2(\mathbb{S}^1)} \le j^{\beta_2}. \tag{10.2.53}$$

(b) *The function* $\eta(\mathbf{x})$ *does not vanish at the corner point, i.e.,*

$$\eta(\mathbf{0}) \neq 0. \tag{10.2.54}$$

(c) *the corner is non-degenerate, namely, the angles* θ_m *and* θ_M *of the sector* W *satisfy*

$$-\pi < \theta_m < \theta_M < \pi \text{ and } \theta_M - \theta_m \neq \pi. \tag{10.2.55}$$

then we have

$$\lim_{\rho\to+0}\frac{1}{m(B(\mathbf{0},\rho)\cap\Omega)}\int_{B(\mathbf{0},\rho)\cap\Omega}|\mathbf{v}(\mathbf{x})|\mathrm{d}\mathbf{x}=0, \tag{10.2.56}$$

where $m(B(\mathbf{0},\rho)\cap\Omega)$ *is the measure of* $B(\mathbf{0},\rho)\cap\Omega$.

Proof. As remarked earlier, we shall make use of the splitting (10.2.7), and it is sufficient for us to show that $\mathbf{v}_{\mathsf{R}}$ satisfies the geometric property (10.2.56). First, it is easy to see that $q\mathbf{w}_{\mathsf{R}}\in C^{\alpha}(\overline{S}_h)^2$ and $\mathbf{v}_{\mathsf{R}}\in H^1(S_h)^2$ can be approximated by $\{\mathbf{v}_j^{\mathsf{R}}\}_{j=1}^{+\infty}$ defined in (10.2.6) satisfying (10.2.33). Therefore, the assumptions in Lemmas 10.2.6–10.2.13 are fulfilled.

Substituting (10.2.50) and (10.2.52) into (10.2.26) and rearranging the terms, we have

$$\begin{aligned}
&\eta(\mathbf{0})\left(\mathbf{v}_{jp}^{\mathsf{R}}(\mathbf{0})+\mathbf{v}_{js}^{\mathsf{R}}(\mathbf{0})+\mathbf{v}_{jp}^{\mathsf{I}}(\mathbf{0})+\mathbf{v}_{js}^{\mathsf{I}}(\mathbf{0})\right)\cdot\begin{pmatrix}1\\ \mathrm{i}\end{pmatrix}\\
&\times 2s^{-2}\Big(\mu(\theta_M)^{-2}-\mu(\theta_M)^{-2}e^{-s\sqrt{h}\mu(\theta_M)}-\mu(\theta_M)^{-1}s\sqrt{h}e^{-s\sqrt{h}\mu(\theta_M)}\\
&+\mu(\theta_m)^{-2}-\mu(\theta_m)^{-2}e^{-s\sqrt{h}\mu(\theta_m)}-\mu(\theta_m)^{-1}s\sqrt{h}e^{-s\sqrt{h}\mu(\theta_m)}\Big)\\
&=-\eta(\mathbf{0})(\mathscr{I}_{22}^{\pm(p)}+\mathscr{I}_{22}^{\pm(s)}+\mathscr{I}_{23}^{\pm(p)}+\mathscr{I}_{23}^{\pm(s)}+\mathscr{I}_{211}^{\pm(p)}+\mathscr{I}_{212}^{\pm(s)}+\mathscr{I}_{214}^{\pm(p)}+\mathscr{I}_{214}^{\pm(s)})\\
&-\mathscr{I}_1^{\pm}+I_{\Lambda_h}-I_1-I_2-I_{\pm}^{\Delta}.
\end{aligned} \tag{10.2.57}$$

Multiplying s^2 on both sides of (10.2.57), by virtue of (10.2.32), (10.2.34), (10.2.35), (10.2.40), (10.2.50), (10.2.51), (10.2.52), and (10.2.33), and letting $s=j^{\rho/2}$ ($\max\{\beta_1,\beta_2\}<\rho<\gamma$) with $j\to+\infty$, we have

$$\lim_{j\to\infty}\eta(\mathbf{0})\left(\mathbf{v}_{jp}^{\mathsf{R}}r(\mathbf{0})+\mathbf{v}_{js}^{\mathsf{R}}(\mathbf{0})+\mathbf{v}_{jp}^{\mathsf{I}}(\mathbf{0})+\mathbf{v}_{js}^{\mathsf{I}}(\mathbf{0})\right)\cdot\begin{pmatrix}1\\ \mathrm{i}\end{pmatrix}\left(\mu^{-2}(\theta_m)+\mu^{-2}(\theta_M)\right)=0, \tag{10.2.58}$$

which further implies that

$$\lim_{j\to\infty}\eta(\mathbf{0})\mathbf{v}_j^{\mathsf{R}}\mathbf{0})\left(\mu^{-2}(\theta_m)+\mu^{-2}(\theta_M)\right)=0. \tag{10.2.59}$$

Here, we use the fact that

$$\mathbf{v}_j^{\mathsf{R}}(\mathbf{0})=\mathbf{v}_{jp}^{\mathsf{R}}(\mathbf{0})+\mathbf{v}_{js}^{\mathsf{R}}(\mathbf{0})+\mathbf{v}_{jp}^{\mathsf{I}}(\mathbf{0})+\mathbf{v}_{js}^{\mathsf{I}}(\mathbf{0}),$$

according to Proposition 10.2.1.

Under the condition (10.2.55), from [23, Lemma 2.10], we know that

$$\mu^{-2}(\theta_m)+\mu^{-2}(\theta_M)\neq 0. \tag{10.2.60}$$

Since η is a real-valued function, by virtue of (10.2.54) and (10.2.60), from (10.2.59), one has

$$\lim_{j\to\infty}\mathbf{v}_j^{\mathsf{R}}(\mathbf{0})\cdot\begin{pmatrix}1\\ \mathrm{i}\end{pmatrix}=0, \tag{10.2.61}$$

which implies that

$$\lim_{j\to\infty}\mathbf{v}_j^{\mathsf{R}}(\mathbf{0})=\mathbf{0}. \tag{10.2.62}$$

Finally, it can be directly deduced that

$$\begin{aligned}
&\lim_{\rho\to 0}\frac{1}{m(B(\mathbf{0},\rho)\cap\Omega)}\int_{B(\mathbf{0},\rho)\cap\Omega}|\mathbf{v}_{\mathrm{R}}(\mathbf{x})|\mathrm{d}\mathbf{x}\\
&\leq \lim_{j\to\infty}\lim_{\rho\to 0}\frac{1}{m(B(\mathbf{0},\rho)\cap\Omega}\int_{B(\mathbf{0},\rho)\cap\Omega}\left|\mathbf{v}_{\mathrm{R}}(\mathbf{x})-\mathbf{v}_j^{\mathrm{R}}(\mathbf{x})\right|\mathrm{d}\mathbf{x} \qquad (10.2.63)\\
&+\lim_{\rho\to 0}\frac{1}{m(B(\mathbf{0},\rho)\cap\Omega)}\int_{B(\mathbf{0},\rho)\cap\Omega}\left|\mathbf{v}_j^{\mathrm{R}}(\mathbf{x})\right|\mathrm{d}\mathbf{x}.
\end{aligned}$$

Combining (10.2.33) and (10.2.62) with (10.2.63), one can prove (10.2.56).

The proof is complete. □

Remark 10.2.1. We would like to point out that the Fourier extension property (10.2.53) can be generalized as follows:

$$\|\mathbf{v}-\mathbf{v}_j\|_{H^1(S_h)^2}\leq\phi_1(j),\ \|g_{jp}\|_{L^2(\mathbb{S}^1)}\leq\frac{1}{\phi_2(j)},\ \|g_{js}\|_{L^2(\mathbb{S}^1)}\leq\frac{1}{\phi_3(j)}, \qquad (10.2.64)$$

where $\phi_\ell(j)\in\mathbb{R}_+$ are strictly decreasing functions with respect to j and $\lim_{j\to+\infty}\phi_\ell(j)=0$, $\ell=1,2,3$, satisfying

$$(\phi_1(j))^\gamma\leq\phi_4(j)=\min\{\phi_2(j),\phi_3(j)\},\quad 0<\gamma<1,\quad \forall j\in\mathbb{N}.$$

In fact, by letting

$$s=\frac{1}{(\phi_1(j))^{\rho/2}},\quad \gamma<\rho<1, \qquad (10.2.65)$$

one can show that

$$s^2\|\mathbf{v}-\mathbf{v}_j\|_{H^1(S_h)^2}\leq(\phi_1(j))^{1-\rho},\ s^{-2}\|g_{jp}\|_{L^2(\mathbb{S}^1)}\leq(\phi_1(j))^{\rho-\gamma},\ s^{-2}\|g_{js}\|_{L^2(\mathbb{S}^1)}\leq(\phi_1(j))^{\rho-\gamma} \qquad (10.2.66)$$

By virtue of (10.2.64), under the same setup of Theorem 10.2.1, and using a similar argument as that in Lemma 10.2.8, one can prove that

$$\begin{aligned}
|I_2|&\leq\omega^2h\sqrt{\Theta_M-\Theta_m}e^{-s\sqrt{\Theta}\delta_W}\phi_1(j),\\
\left|I_\pm^\Delta\right|&\leq\Big(|\eta(\mathbf{0})|h\sqrt{\Theta_M-\Theta_m}e^{-s\sqrt{\Theta}\delta_W} \qquad (10.2.67)\\
&+\|\eta\|_{C^\alpha}\frac{2\sqrt{\Theta_M-\Theta_m}\Gamma(4\alpha+4)}{(2\delta_W)^{2\alpha+2}}s^{-2(\alpha+1)}\Big)\phi_1(j).
\end{aligned}$$

In the integral identity (10.2.57), we can multiply s^2 on both sides of (10.2.57) and choose s by (10.2.65). Then by virtue of (10.2.66), (10.2.67), (10.2.32), (10.2.40), (10.2.50), and (10.2.51), we can obtain

$$\lim_{j\to\infty}\mathbf{v}_j(\mathbf{0})=\mathbf{0},$$

by letting $j\to+\infty$. Therefore, we can prove (10.2.56) under the same setup of Theorem 10.2.1, where the condition (10.2.33) is replaced by (10.2.64) and $q\mathbf{w}\in C^\alpha(\overline{S_h})^2$.

We next consider the degenerate case of (10.1.11) with $\eta\equiv 0$ in (10.1.11). The proof of Corollary 10.2.1 follows from the one of Theorem 10.2.1 with some necessary modifications. One can establish the integral identity by noting $\eta\equiv 0$ in (10.2.26). Hence, the asymptotic analysis with respect to the parameter s in the CGO for each term in the integral equality can be derived.

Corollary 10.2.1. *Under the same setup as that in Theorem 10.2.1 with $\eta\equiv 0$, we have*

$$\lim_{\rho\to 0}\frac{1}{m(B(\mathbf{0},\rho)\cap\Omega)}\int_{B(\mathbf{0},\rho)\cap\Omega}|V(\mathbf{x})\mathbf{w}(\mathbf{x})|\mathrm{d}\mathbf{x}=0. \qquad (10.2.68)$$

Remark 10.2.2. The difference between Theorem 10.2.1 and Corollary 10.2.1 lies in their conclusions (10.2.56) and (10.2.68). If one further assumes in Corollary 10.2.1 that $V(\mathbf{x}) \in C(\overline{S_h})$ and $V(\mathbf{0}) \neq 0$, it is easy to show that

$$\lim_{\rho\to+0} \frac{1}{m(B(\mathbf{0},\rho)\cap\Omega)} \int_{B(\mathbf{0},\rho)\cap\Omega} |\mathbf{w}(\mathbf{x})| \mathrm{d}\mathbf{x} = 0,$$

which together with the transmission condition $\mathbf{v} = \mathbf{w}$ on Γ further implies that (10.2.56) holds as well.

Proof of Corollary 10.2.1. The proof follows from the one of Theorem 10.2.1 with some necessary modifications. Similar to (10.2.26), we have the integral identity as follows:

$$I_1 + I_2 = I_{\Lambda_h}, \tag{10.2.69}$$

where I_1, I_2, and I_{Λ_h} are defined in (10.2.26). Substituting (10.2.39) and (10.2.41) into (10.2.69), one can see that

$$\begin{aligned}(-\omega^2\mathbf{v}_j^{\mathsf{R}}(\mathbf{0}) + \omega^2\mathbf{f}_{\mathsf{R}}(\mathbf{0})) \int_{S_h} \mathbf{u}(\mathbf{x})\mathrm{d}\mathbf{x} + I_2 =& I_{\Lambda_h} + \int_{S_h} \omega^2 \delta\mathbf{f}_{\mathsf{R}}(\mathbf{x}) \cdot \mathbf{u}(\mathbf{x})\mathrm{d}\mathbf{x} \\ &- \int_{S_h} \omega^2 \delta\mathbf{v}_j^{\mathsf{R}}(\mathbf{x}) \cdot \mathbf{u}(\mathbf{x})\mathrm{d}\mathbf{x},\end{aligned} \tag{10.2.70}$$

where $\mathbf{v}_j^{\mathsf{R}}(\mathbf{0}), \mathbf{u}(\mathbf{x}), and \mathbf{f}_{\mathsf{R}}(\mathbf{0})$ are defined in (10.2.11), (10.2.15), and (10.2.39).

From Lemma 10.2.2, it is easy to obtain that

$$\begin{aligned}&(-\omega^2\mathbf{v}_j^{\mathsf{R}}(\mathbf{0}) + \omega^2\mathbf{f}_{\mathsf{R}}(\mathbf{0})) \int_{S_h} \mathbf{u}(\mathbf{x})\mathrm{d}\mathbf{x} \\ =&\omega^2(\mathbf{f}_{\mathsf{R}}(\mathbf{0}) - \mathbf{v}_j^{\mathsf{R}}(\mathbf{0})) \cdot \begin{bmatrix}1\\ \mathrm{i}\end{bmatrix} \left(\int_W u_1(\mathbf{x})\mathrm{d}\mathbf{x} - \int_{W\backslash S_h} u_1(\mathbf{x})\mathrm{d}\mathbf{x}\right) \\ =&\omega^2(\mathbf{f}_{\mathsf{R}}(\mathbf{0}) - \mathbf{v}_j^{\mathsf{R}}(\mathbf{0})) \cdot \begin{bmatrix}1\\ \mathrm{i}\end{bmatrix} \left(6\mathrm{i}(e^{-2\theta_M \mathrm{i}} - e^{-2\theta_m \mathrm{i}})s^{-4} - \int_{W\backslash S_h} u_1(\mathbf{x})\mathrm{d}\mathbf{x}\right).\end{aligned} \tag{10.2.71}$$

Substituting (10.2.71) into (10.2.70) and rearranging the terms, we have

$$\begin{aligned}&\omega^2(\mathbf{f}_{\mathsf{R}}(\mathbf{0}) - \mathbf{v}_j^{\mathsf{R}}(\mathbf{0})) \cdot \begin{bmatrix}1\\ \mathrm{i}\end{bmatrix} 6\mathrm{i}(e^{-2\theta_M \mathrm{i}} - e^{-2\theta_m \mathrm{i}})s^{-4} \\ =& - I_2 + I_{\Lambda_h} + \int_{S_h} \omega^2 \delta\mathbf{f}_{\mathsf{R}}(\mathbf{x}) \cdot \mathbf{u}(\mathbf{x})\mathrm{d}\mathbf{x} - \int_{S_h} \omega^2 \delta\mathbf{v}_j^{\mathsf{R}}(\mathbf{x}) \cdot \mathbf{u}(\mathbf{x})\mathrm{d}\mathbf{x} \\ &+ \omega^2(\mathbf{f}_{\mathsf{R}}(\mathbf{0}) - \mathbf{v}_j^{\mathsf{R}}(\mathbf{0})) \cdot \begin{bmatrix}1\\ \mathrm{i}\end{bmatrix} \int_{W\backslash S_h} u_1(\mathbf{x})\mathrm{d}\mathbf{x}.\end{aligned} \tag{10.2.72}$$

Note that (10.2.53) holds. Multiplying s^4 on both sides of (10.2.72); using (10.2.18), (10.2.34), (10.2.44), and (10.2.45); and letting $s = j^{\rho/4}$ $(\max\{\beta_1, \beta_2\} < \rho < \gamma)$ with $j \to +\infty$, we have

$$\lim_{j\to\infty} \mathbf{v}_j^{\mathsf{R}}(\mathbf{0}) = \mathbf{f}_{\mathsf{R}}(\mathbf{0}). \tag{10.2.73}$$

Since

$$\begin{aligned}\lim_{j\to\infty} \mathbf{v}_j^{\mathsf{R}}(\mathbf{0}) &= \lim_{j\to\infty} \lim_{\rho\to+0} \frac{1}{m(B(\mathbf{0},\rho)\cap\Omega)} \int_{B(\mathbf{0},\rho)\cap\Omega} \mathbf{v}_j^{\mathsf{R}}(\mathbf{x})\mathrm{d}\mathbf{x} \\ &= \lim_{\rho\to+0} \frac{1}{m(B(\mathbf{0},\rho)\cap\Omega)} \int_{B(\mathbf{0},\rho)\cap\Omega} \mathbf{v}_{\mathsf{R}}(\mathbf{x})\mathrm{d}\mathbf{x},\end{aligned}$$

and

$$\mathbf{f}_{\mathrm{R}}(\mathbf{0}) = \lim_{\rho \to +0} \frac{1}{m(B(\mathbf{0},\rho)\cap\Omega)} \int_{B(\mathbf{0},\rho)\cap\Omega} q\mathbf{w}_{\mathrm{R}}(\mathbf{x})\mathrm{d}\mathbf{x},$$

and

$$\lim_{\rho \to +0} \frac{1}{m(B(\mathbf{0},\rho)\cap\Omega)} \int_{B(\mathbf{0},\rho)\cap\Omega} \mathbf{v}_{\mathrm{R}}(\mathbf{x})\mathrm{d}\mathbf{x} = \lim_{\rho \to +0} \frac{1}{m(B(\mathbf{0},\rho)\cap\Omega)} \int_{B(\mathbf{0},\rho)\cap\Omega} \mathbf{w}_{\mathrm{R}}(\mathbf{x})\mathrm{d}\mathbf{x},$$

we can finish the proof of this corollary by using (10.2.73). □

Finally, we provide the vanishing property under the Hölder regularity of the transmission eigenfunctions, which shall be useful for our study of the inverse elastic problem in Sect. 10.4. Substituting the Hölder expansion of $\mathbf{v}$ in (10.2.26), following a similar idea in the proof of Theorem 10.2.1 to characterize the asymptotic analysis with respect to the parameter s in the CGO for each term in the integral identity, we are able to show that $\mathbf{v}$ must vanish near the planner corner.

Theorem 10.2.2. *Let $(\mathbf{v},\mathbf{w}) \in H^1(\Omega)^2 \times H^1(\Omega)^2$ be a pair of eigenfunctions to* (10.1.11) *associated with $\omega \in \mathbb{R}_+$. Let W, S_h and q be those described in Theorem 10.2.1. Suppose that $\mathbf{v} \in C^\alpha(\overline{S_h})^2$, $q\mathbf{w} \in C^\alpha(\overline{S_h})^2$, $\eta \in C^\alpha(\overline{\Gamma_h^\pm})$ with $\eta(\mathbf{0}) \neq 0$, for $\alpha \in (0,1)$ and the corner W is non-degenerate. Then we have*

$$\mathbf{v}(\mathbf{0}) = \mathbf{0}. \tag{10.2.74}$$

Proof. It is sufficient for us to show that (10.2.74) holds for $\mathbf{v}_{\mathrm{R}}$. Similar to (10.2.26), one can give the following integral identity:

$$-\omega^2 \int_{S_h} (\mathbf{v}_{\mathrm{R}}(\mathbf{x}) - \mathbf{f}_{\mathrm{R}}(\mathbf{x}))\cdot\mathbf{u}(\mathbf{x})\mathrm{d}\mathbf{x} = I_{\Lambda_h} - \int_{\Gamma_h^\pm} \eta\mathbf{u}\cdot\mathbf{v}_{\mathrm{R}}\mathrm{d}\sigma, \tag{10.2.75}$$

where I_{Λ_h} is defined in (10.2.26) and $\mathbf{f}_{\mathrm{R}}(\mathbf{x})$ is given in (10.2.39).

Since $\mathbf{f}_{\mathrm{R}}(\mathbf{0}) \in C^\alpha(\overline{S}_h)$ and $\eta \in C^\alpha\left(\overline{\Gamma_h^\pm}\right)$, we know that η and $\mathbf{f}_{\mathrm{R}}(\mathbf{x})$ have the expansions (10.2.49) and (10.2.39) around the origin, respectively. Furthermore, due to the fact that $\mathbf{v} \in C^\alpha(\overline{S}_h)^2$, we have the following expansions:

$$\mathbf{v}_{\mathrm{R}}(\mathbf{x}) = \mathbf{v}_{\mathrm{R}}(\mathbf{0}) + \delta\mathbf{v}_{\mathrm{R}}(\mathbf{x}), \quad |\delta\mathbf{v}_{\mathrm{R}}| \le \|\mathbf{v}_{\mathrm{R}}(\mathbf{x})\|_{C^\alpha(\Omega)^2}|\mathbf{x}|^\alpha. \tag{10.2.76}$$

Substituting (10.2.49), (10.2.39), and (10.2.76) into (10.2.75), we can derive that

$$\begin{aligned}
&-\omega^2(\mathbf{v}_{\mathrm{R}}(\mathbf{0}) - \mathbf{f}_{\mathrm{R}}(\mathbf{0}))\int_{S_h}\mathbf{u}(\mathbf{x})\mathrm{d}\mathbf{x} - \omega^2\int_{S_h}(\delta\mathbf{v}_{\mathrm{R}}(\mathbf{x}) - \delta\mathbf{f}_{\mathrm{R}}(\mathbf{x}))\cdot\mathbf{u}(\mathbf{x})\mathrm{d}\mathbf{x} \\
&= I_{\Lambda_h} - \eta(\mathbf{0})\mathbf{v}_{\mathrm{R}}(\mathbf{0})\int_{\Gamma_h^\pm}\mathbf{u}(\mathbf{x})\mathrm{d}\sigma - \eta(\mathbf{0})\int_{\Gamma_h^\pm}\mathbf{u}(\mathbf{x})\cdot\delta\mathbf{v}_{\mathrm{R}}(\mathbf{x})\mathrm{d}\sigma \\
&\quad - \mathbf{v}_{\mathrm{R}}(\mathbf{0})\int_{\Gamma_h^\pm}\delta\eta(\mathbf{x})\mathbf{u}(\mathbf{x})\mathrm{d}\sigma - \int_{\Gamma_h^\pm}\delta\eta(\mathbf{x})\mathbf{u}(\mathbf{x})\cdot\delta\mathbf{v}_{\mathrm{R}}(\mathbf{x})\mathrm{d}\sigma.
\end{aligned} \tag{10.2.77}$$

From (10.2.77), after rearranging terms, we have

$$\begin{aligned}
&\eta(\mathbf{0})\mathbf{v}_{\mathrm{R}}(\mathbf{0})\int_{\Gamma_h^\pm}\mathbf{u}\mathrm{d}\sigma \\
&= -\omega^2(\mathbf{v}_{\mathrm{R}}(\mathbf{0}) - \mathbf{f}_{\mathrm{R}}(\mathbf{0}))\int_{S_h}\mathbf{u}(\mathbf{x})\mathrm{d}\mathbf{x} - \int_{S_h}\omega^2(\delta\mathbf{v}_{\mathrm{R}}(\mathbf{x}) - \delta\mathbf{f}_{\mathrm{R}}(\mathbf{x}))\cdot\mathbf{u}(\mathbf{x})\mathrm{d}\mathbf{x} \\
&\quad + I_{\Lambda_h} - \mathbf{v}_{\mathrm{R}}(\mathbf{0})\int_{\Gamma_h^\pm}\delta\eta(\mathbf{x})\mathbf{u}(\mathbf{x})\mathrm{d}\sigma - \int_{\Gamma_h^\pm}\delta\eta(\mathbf{x})\mathbf{u}(\mathbf{x})\cdot\delta\mathbf{v}_{\mathrm{R}}(\mathbf{x})\mathrm{d}\sigma \\
&\quad - \eta(\mathbf{0})\int_{\Gamma_h^\pm}\mathbf{u}(\mathbf{x})\cdot\delta\mathbf{v}_{\mathrm{R}}(\mathbf{x})\mathrm{d}\sigma.
\end{aligned} \tag{10.2.78}$$

Using (10.2.47) in Lemma 10.2.11, we have

$$\begin{aligned}\int_{\Gamma_h^-} \mathbf{u} \mathrm{d}\sigma &= \begin{pmatrix}1\\ \mathrm{i}\end{pmatrix} \int_0^h e^{-s\sqrt{r}\mu(\theta_m)} \mathrm{d}r \\ &= \begin{pmatrix}1\\ \mathrm{i}\end{pmatrix} 2s^{-2}\left(\mu(\theta_m)^{-2} - \mu(\theta_m)^{-2} e^{-s\sqrt{h}\mu(\theta_m)} - \mu(\theta_m)^{-1} s\sqrt{h} e^{-s\sqrt{h}\mu(\theta_m)}\right)\end{aligned} \tag{10.2.79}$$

and

$$\begin{aligned}\int_{\Gamma_h^+} \mathbf{u} \mathrm{d}\sigma &= \begin{pmatrix}1\\ \mathrm{i}\end{pmatrix} \int_0^h e^{-s\sqrt{r}\mu(\theta_M)} \mathrm{d}r \\ &= \begin{pmatrix}1\\ \mathrm{i}\end{pmatrix} 2s^{-2}\left(\mu(\theta_M)^{-2} - \mu(\theta_M)^{-2} e^{-s\sqrt{h}\mu(\theta_M)} - \mu(\theta_M)^{-1} s\sqrt{h} e^{-s\sqrt{h}\mu(\theta_M)}\right),\end{aligned} \tag{10.2.80}$$

where $\mu(\theta)$ is defined in (10.2.47). By virtue of (10.2.46), using (10.2.49) and (10.2.76), we have the estimates

$$\begin{aligned}\left|\int_{\Gamma_h^-} \mathbf{u}(\mathbf{x}) \cdot \delta \mathbf{v}_{\mathrm{R}}(\mathbf{x}) \mathrm{d}\sigma\right| &\leq \sqrt{2}\|\mathbf{v}^{\mathrm{R}}\|_{C^\alpha} \int_0^h r^\alpha e^{-s\sqrt{r}\cos\frac{\theta_m}{2}} \mathrm{d}r = O(s^{-2\alpha-2}), \\ \left|\int_{\Gamma_h^-} \delta\eta(\mathbf{x}) \mathbf{u}(\mathbf{x}) \mathrm{d}\sigma\right| &\leq \sqrt{2}\|\eta\|_{C^\alpha} \int_0^h r^\alpha e^{-s\sqrt{r}\cos\frac{\theta_m}{2}} \mathrm{d}r = O(s^{-2\alpha-2}), \\ \left|\int_{\Gamma_h^-} \delta\eta(\mathbf{x}) \mathbf{u}(\mathbf{x}) \cdot \delta \mathbf{v}_{\mathrm{R}}(\mathbf{x}) \mathrm{d}\sigma\right| &\leq \sqrt{2}\|\mathbf{v}_{\mathrm{R}}\|_{C^\alpha}\|\eta\|_{C^\alpha} \int_0^h r^{2\alpha} e^{-s\sqrt{r}\cos\frac{\theta_m}{2}} \mathrm{d}r = O(s^{-4\alpha-2}),\end{aligned} \tag{10.2.81}$$

and

$$\begin{aligned}\left|\int_{\Gamma_h^+} \mathbf{u}(\mathbf{x}) \cdot \delta \mathbf{v}_{\mathrm{R}}(\mathbf{x}) \mathrm{d}\sigma\right| &\leq \sqrt{2}\|\mathbf{v}^{\mathrm{R}}\|_{C^\alpha} \int_0^h r^\alpha e^{-s\sqrt{r}\cos\frac{\theta_M}{2}} \mathrm{d}r = O(s^{-2\alpha-2}), \\ \left|\int_{\Gamma_h^+} \delta\eta(\mathbf{x}) \mathbf{u}(\mathbf{x}) \mathrm{d}\sigma\right| &\leq \sqrt{2}\|\eta\|_{C^\alpha} \int_0^h r^\alpha e^{-s\sqrt{r}\cos\frac{\theta_M}{2}} \mathrm{d}r = O(s^{-2\alpha-2}), \\ \left|\int_{\Gamma_h^+} \delta\eta(\mathbf{x}) \mathbf{u}(\mathbf{x}) \cdot \delta \mathbf{v}_{\mathrm{R}}(\mathbf{x}) \mathrm{d}\sigma\right| &\leq \sqrt{2}\|\mathbf{v}_{\mathrm{R}}\|_{C^\alpha}\|\eta\|_{C^\alpha} \int_0^h r^{2\alpha} e^{-s\sqrt{r}\cos\frac{\theta_M}{2}} \mathrm{d}r = O(s^{-4\alpha-2}),\end{aligned} \tag{10.2.82}$$

as $s \to +\infty$.

Similarly, using (10.2.17), (10.2.39), and (10.2.76), one has

$$\begin{aligned}\left|\int_{S_h} \delta \mathbf{v}_{\mathrm{R}}(\mathbf{x}) \cdot \mathbf{u}(\mathbf{x}) \mathrm{d}\mathbf{x}\right| &\leq \sqrt{2}\|\mathbf{v}_{\mathrm{R}}\|_{C^\alpha} \int_W |u_1(\mathbf{x})||\mathbf{x}|^\alpha \mathrm{d}\mathbf{x} \\ &\leq 2\sqrt{2}\|\mathbf{v}_{\mathrm{R}}\|_{C^\alpha} \frac{(\theta_M - \theta_m)\Gamma(2\alpha+4)}{\delta_W^{2\alpha+4}} s^{-2\alpha-4}, \\ \left|\int_{S_h} \delta \mathbf{f}_{\mathrm{R}}(\mathbf{x}) \cdot \mathbf{u}(\mathbf{x}) \mathrm{d}\mathbf{x}\right| &\leq \sqrt{2}\|\mathbf{f}_{\mathrm{R}}\|_{C^\alpha} \int_W |u_1(\mathbf{x})||\mathbf{x}|^\alpha \mathrm{d}\mathbf{x} \\ &\leq 2\sqrt{2}\|\mathbf{f}_{\mathrm{R}}\|_{C^\alpha} \frac{(\theta_M - \theta_m)\Gamma(2\alpha+4)}{\delta_W^{2\alpha+4}} s^{-2\alpha-4},\end{aligned} \tag{10.2.83}$$

as $s \to +\infty$.

Multiplying s^2 on both sides of (10.2.78), by virtue of (10.2.32), (10.2.79), (10.2.80), (10.2.81), (10.2.82), and (10.2.83), and letting $s \to +\infty$, we have

$$\eta(\mathbf{0}) \mathbf{v}_{\mathrm{R}}(\mathbf{0}) \cdot \begin{pmatrix}1\\ \mathrm{i}\end{pmatrix} \left(\mu(\theta_M)^{-2} + \mu(\theta_m)^{-2}\right) = 0. \tag{10.2.84}$$

Since the corner W is non-degenerate, namely, $\theta_M - \theta_m \neq \pi$, from [23, Lemma 2.10], we know that

$$\mu^{-2}(\theta_M) + \mu^{-2}(\theta_m) \neq 0.$$

Finally, by noting $\eta(\mathbf{0}) \neq \mathbf{0}$, it is easy to infer from (10.2.84) that $\mathbf{v}_{\mathrm{R}}(\mathbf{0}) = \mathbf{0}$.

The proof is complete. □

Remark 10.2.3. Similar to Corollary 10.2.1, under the same setup of Theorem 10.2.2, for the degenerate case of (10.1.11) with $\eta \equiv 0$ in (10.1.11), if V is Hölder-regular near the corner and $V(\mathbf{0}) \neq 0$, one can prove that $\mathbf{v}(\mathbf{0}) = \mathbf{0}$, where $\mathbf{v} \in H^1(\Omega)^2 \cap C^\alpha(\overline{S_h})$ and $\mathbf{w} \in H^1(\Omega)^2$ are a pair of the generalized elastic transmission eigenfunctions to (10.1.11) associated with $\omega \in \mathbb{R}_+$ such that $q\mathbf{w} \in C^\alpha(\overline{S_h})$. We choose not to discuss the details in this chapter. In [8, Theorem 1.5], the corresponding vanishing property was considered when $\mathbf{v} - \mathbf{w} \in H^2(\Omega)^2$, V is Hölder continuous around the corner such that$V(\mathbf{0}) \neq 0$, and either $\mathbf{v}$ or $\mathbf{w}$ is Hölder continuous around the corner. Compared with [8, Theorem 1.5], the assumption $\mathbf{v} - \mathbf{w} \in H^2(\Omega)^2$ can be removed in our setting.

10.3 Vanishing Near Corners of Generalized Elastic Transmission Eigenfunctions: Three-Dimensional Case

In this section, we discuss the vanishing property of the generalized elastic transmission eigenfunctions for the 3D case. In principle, we could also consider a generic corner in the usual sense as the one for the 2D case. However, we consider a 3D edge corner described by $S_h \times (-M, M)$, where S_h is defined in (10.2.2) and $M \in \mathbb{R}_+$. It is readily seen that $S_h \times (-M, M)$ actually describes an edge singularity. In what follows, we suppose that the Lipschitz domain $\Omega \subset \mathbb{R}^3$ with $\mathbf{0} \in \partial\Omega$ possesses a 3D edge corner. Let $\mathbf{0} \in \mathbb{R}^2$ be the vertex of S_h and $x_3 \in (-M, M)$. Then $(\mathbf{0}, x_3)$ is defined as an edge point of $S_h \times (-M, M)$.

In order to make use of the CGO solution $\mathbf{u}(\mathbf{x})$ introduced in Lemma 10.2.2 to study the vanishing property of $(\mathbf{v}, \mathbf{w})$ to (10.1.11) at a 3D edge corner, we define the following dimension reduction operator.

Definition 10.3.1. Let $S_h \subset \mathbb{R}^2$ be defined in (10.2.2), $M > 0$. Assume that g is a function defined in the domain $S_h \times (-M, M)$. Pick up any point $x_3 \in (-M, M)$. Suppose that $\phi \in C_0^\infty((x_3 - L, x_3 + L))$ is a non-negative function and $\phi \not\equiv 0$, where L is sufficiently small such that $(x_3 - L, x_3 + L) \subset (-M, M)$, and write $\mathbf{x} = (\mathbf{x}', x_3) \in \mathbb{R}^3$, $\mathbf{x}' \in \mathbb{R}^2$. The dimension reduction operator $\mathscr{R}$ is defined by

$$\mathscr{R}(\mathbf{g})(\mathbf{x}') = \int_{x_3 - L}^{x_3 + L} \phi(x_3) \mathbf{g}(\mathbf{x}', x_3) \mathrm{d}x_3, \tag{10.3.1}$$

where $\mathbf{x}' \in S_h$.

Before presenting the main results of this section, we first analyze the regularity of the functions after applying the dimension reduction operator. Using a similar argument of [6, Lemma 3.4], we can prove the following lemma, whose detailed proof is omitted.

Lemma 10.3.1. *Let* $\mathbf{g} \in H^m(S_h \times (-M, M))^3$, $m = 1, 2$. *Then*

$$\mathscr{R}(\mathbf{g})(\mathbf{x}') \in H^m(S_h)^3.$$

Similarly, if $\mathbf{g} \in C^\alpha(\overline{S_h} \times [-M, M])^3$, $0 < \alpha < 1$, *then*

$$\mathscr{R}(\mathbf{g})(\mathbf{x}') \in C^\alpha(\overline{S_h})^3.$$

The elastic Herglotz wave function $\mathbf{v_g}$ in $\mathbb{R}^3$ is defined by

$$\mathbf{v_g}(\mathbf{x}) = \int_{\mathbb{S}^2} \{e^{\mathrm{i}k_p\mathbf{x}\cdot\mathbf{d}}\mathbf{g}_p(\mathbf{d}) + e^{\mathrm{i}k_s\mathbf{x}\cdot\mathbf{d}}\mathbf{g}_s(\mathbf{d})\}\mathrm{d}\sigma(\mathbf{d}), \tag{10.3.2}$$

where the kernel $\mathbf{g} = \mathbf{g}_p + \mathbf{g}_s$ with $\mathbf{g}_p \in L^2(\mathbb{S}^2)^3$ and $\mathbf{g}_s \in L^2(\mathbb{S}^2)^3$ and $\mathbf{d} \in \mathbb{S}^2$. Lemma 10.2.1 holds equally in the three dimensions. In view of this lemma, for any pair of the generalized elastic transmission eigenfunctions $(\mathbf{v}, \mathbf{w})$ to (10.1.11), there exists a sequence of Herglotz wave function $\{\mathbf{v}_j\}_{j=1}^{+\infty}$ defined by

$$\mathbf{v}_j(\mathbf{x}) = \int_{\mathbb{S}^2} \{e^{ik_p\mathbf{x}\cdot\mathbf{d}}\mathbf{g}_{jp}(\mathbf{d}) + e^{ik_s\mathbf{x}\cdot\mathbf{d}}\mathbf{g}_{js}(\mathbf{d})\}\mathrm{d}\sigma(\mathbf{d}), \tag{10.3.3}$$

where $\mathbf{g}_j = \mathbf{g}_{jp} + \mathbf{g}_{js}$ with $\mathbf{g}_{jp} \in L^2(\mathbb{S}^2)^3$ and $\mathbf{g}_{js} \in L^2(\mathbb{S}^2)^3$ which can approximate $\mathbf{v}$ to an arbitrary accuracy in $H^1(\Omega)^3$. Henceforth, we let the real and imaginary parts of the kernel functions $\mathbf{g}_{j\beta}(\mathbf{d})$ ($\beta = p, s$) in (10.3.3) be defined by

$$\mathbf{g}_{j\beta}(\mathbf{d}) = \mathbf{g}_{\mathsf{R},j\beta}(\mathbf{d}) + \mathrm{i}\mathbf{g}_{\mathsf{I},j\beta}(\mathbf{d}), \tag{10.3.4}$$

where

$$\begin{aligned} \mathbf{g}_{\mathsf{R},jp}(\mathbf{d}) &= \begin{pmatrix} \mathbf{g}_{\mathsf{R},jp}^{(1,2)}(\mathbf{d}) \\ g_{\mathsf{R},jp}^{(3)}(\mathbf{d}) \end{pmatrix} \in \mathbb{R}^3, \ \mathbf{g}_{\mathsf{R},js}(\mathbf{d}) = \begin{pmatrix} \mathbf{g}_{\mathsf{R},js}^{(1,2)}(\mathbf{d}) \\ g_{\mathsf{R},js}^{(3)}(\mathbf{d}) \end{pmatrix} \in \mathbb{R}^3, \\ \mathbf{g}_{\mathsf{I},jp}(\mathbf{d}) &= \begin{pmatrix} \mathbf{g}_{\mathsf{I},jp}^{(1,2)}(\mathbf{d}) \\ g_{\mathsf{I},jp}^{(3)}(\mathbf{d}) \end{pmatrix} \in \mathbb{R}^3, \ \mathbf{g}_{\mathsf{I},js}(\mathbf{d}) = \begin{pmatrix} \mathbf{g}_{\mathsf{I},js}^{(1,2)}(\mathbf{d}) \\ g_{\mathsf{I},js}^{(3)}(\mathbf{d}) \end{pmatrix} \in \mathbb{R}^3, \end{aligned} \tag{10.3.5}$$

with $\mathbf{g}_{\mathsf{R},j\beta}^{(1,2)}(\mathbf{d}) \in \mathbb{R}^2$ and $\mathbf{g}_{\mathsf{I},j\beta}^{(1,2)}(\mathbf{d}) \in \mathbb{R}^2$.

Similar to Proposition 10.2.1, using the Jacobi-Anger expansion , we have the expansion of the real and imaginary parts of $\mathbf{v}_j(\mathbf{x})$ defined in (10.3.3) as follows.

Proposition 10.3.1. *Let the elastic Herglotz wave function* $\mathbf{v}_j$ *be defined by* (10.3.3)*. Denote*

$$\begin{aligned} \mathbf{v}_{jp}^{\mathsf{R}}(\mathbf{0}) &= \int_{\mathbb{S}^2} g_{jp}^{\mathsf{R}}(\mathbf{d})\mathrm{d}\sigma(\mathbf{d}), \ \mathbf{v}_{js}^{\mathsf{R}}(\mathbf{0}) = \int_{\mathbb{S}^2} g_{js}^{\mathsf{R}}(\mathbf{d})\mathrm{d}\sigma(\mathbf{d}), \\ \mathbf{v}_{jp}^{\mathsf{I}}i(\mathbf{0}) &= \int_{\mathbb{S}^2} g_{jp}^{\mathsf{I}}(\mathbf{d})\mathrm{d}\sigma(\mathbf{d}), \ \mathbf{v}_{js}^{\mathsf{I}}(\mathbf{0}) = \int_{\mathbb{S}^2} g_{js}^{\mathsf{I}}(\mathbf{d})\mathrm{d}\sigma(\mathbf{d}). \end{aligned}$$

Then

$$\mathbf{v}_j(\mathbf{0}) = \mathbf{v}_j^{\mathsf{R}}(\mathbf{0}) + \mathrm{i}\mathbf{v}_j^{\mathsf{I}}(\mathbf{0}) = (\mathbf{v}_{jp}^{\mathsf{R}}(\mathbf{0}) + \mathbf{v}_{js}^{\mathsf{R}}(\mathbf{0})) + \mathrm{i}(\mathbf{v}_{jp}^{\mathsf{R}}(\mathbf{0}) + \mathbf{v}_{js}^{\mathsf{R}}(\mathbf{0})), \tag{10.3.6}$$

where $g_{j\beta}^{\mathsf{R}}(\mathbf{d})$ *and* $g_{j\beta}^{\mathsf{R}}(\mathbf{d})$ ($\beta = p, s$) *are defined in* (10.3.4)*. Let* $j_\ell(t)$ *be the* ℓ*-th Bessel function for* $\ell \in \mathbb{N} \cup \{0\}$*.*

In view of the explicit expression $\mathbf{v}_j(x)$ defined in (10.3.3), using the Euler formula, one can handily compute the following result.

Proposition 10.3.2. *Let the elastic Herglotz wave function* $\mathbf{v}_j$ *be defined in* (10.3.3)*, where* $\mathbf{v}_j^{\mathsf{R}}(\mathbf{x})$ *and* $\mathbf{v}_j^{\mathsf{I}}(\mathbf{x})$ *are the real and imaginary parts of* $\mathbf{v}_j$*, respectively. Then it holds that*

$$\begin{aligned} \mathbf{v}_j^{\mathsf{R}}(\mathbf{x}) = \int_{\mathbb{S}^2} \Big(&\cos(k_p\mathbf{d}\cdot\mathbf{x})\mathbf{g}_{jp}^{\mathsf{R}}(\mathbf{d}) - \sin(k_p\mathbf{d}\cdot\mathbf{x})\mathbf{g}_{jp}^{\mathsf{I}}(\mathbf{d}) + \cos(k_s\mathbf{d}\cdot\mathbf{x})\mathbf{g}_{js}^{\mathsf{R}}(\mathbf{d}) \\ &- \sin(k_s\mathbf{d}\cdot\mathbf{x})\mathbf{g}_{js}^{\mathsf{I}}(\mathbf{d}) \Big)\mathrm{d}\sigma(\mathbf{d}) \end{aligned} \tag{10.3.7}$$

and

$$\begin{aligned} \mathbf{v}_j^{\mathsf{I}}(\mathbf{x}) = \int_{\mathbb{S}^2} \Big(&\cos(k_p\mathbf{d}\cdot\mathbf{x})\mathbf{g}_{jp}^{\mathsf{I}}(\mathbf{d}) + \sin(k_p\mathbf{d}\cdot\mathbf{x})\mathbf{g}_{jp}^{\mathsf{R}}(\mathbf{d}) + \cos(k_s\mathbf{d}\cdot\mathbf{x})\mathbf{g}_{js}^{\mathsf{I}}(\mathbf{d}) \\ &+ \sin(k_s\mathbf{d}\cdot\mathbf{x})\mathbf{g}_{js}^{\mathsf{R}}(\mathbf{d}) \Big)\mathrm{d}\sigma(\mathbf{d}). \end{aligned}$$

Let $S_h \subset \mathbb{R}^2$ be defined in (10.2.2) and $M > 0$. For any fixed $x_3 \in (-M, M)$ and $L > 0$ defined in Definition 10.3.1, we suppose that L is sufficiently small such that $(x_3 - L, x_3 + L) \subset (-M, M)$. Write $\mathbf{x} = (\mathbf{x}', x_3) \in \mathbb{R}^3$, $\mathbf{x}' \in \mathbb{R}^2$. In what follows, we consider the transmission eigenvalue problem for $\mathbf{v}, \mathbf{w} \in H^1(S_h \times (-M, M))^3$:

$$\begin{cases} \lambda\Delta\mathbf{v} + (\lambda+\mu)\nabla\nabla\cdot\mathbf{v} + \omega^2\mathbf{v} = \mathbf{0} & \mathbf{x}' \in S_h,\ -M < x_3 < M, \\ \lambda\Delta\mathbf{w} + (\lambda+\mu)\nabla\nabla\cdot\mathbf{w} + \omega^2(1+V)\mathbf{w} = \mathbf{0} & \mathbf{x}' \in S_h,\ -M < x_3 < M, \\ \mathbf{w} = \mathbf{v}, \quad T_\nu\mathbf{v} + \eta\mathbf{v} = T_\nu\mathbf{w} & \mathbf{x}' \in \Gamma_h^\pm,\ -M < x_3 < M, \end{cases} \tag{10.3.8}$$

where $\Gamma_h^\pm$ are defined in (10.2.2), T_ν is the boundary traction operator to $\Gamma_h^\pm \times (-M, M)$, $q \in L^\infty(S_h \times (-M, M))$ defined in (10.1.10), and $\eta \in L^\infty(\Gamma_h^\pm \times (-M, M))$ is independent of x_3. Similar to the 2D case, we let $(\mathbf{v}_\mathrm{R}, \mathbf{w}_\mathrm{R})$ and $(\mathbf{v}_\mathrm{I}, \mathbf{w}_\mathrm{I})$, respectively, signify the real and imaginary parts of $(\mathbf{v}, \mathbf{w})$, and both of them satisfy the Lamé system (10.3.8). We shall mainly focus on dealing with $(\mathbf{v}_\mathrm{R}, \mathbf{w}_\mathrm{R})$, and all the results hold equally for $(\mathbf{v}_\mathrm{I}, \mathbf{w}_\mathrm{I})$ and hence $(\mathbf{v}, \mathbf{w})$.

Noting that η in (10.3.8) is independent of the x_3 variable, and by applying the reduction operator $\mathscr{R}$ defined in Definition 10.3.1, one can show by direct verifications the following lemma.

Lemma 10.3.2. *Denote*

$$\begin{aligned} \mathbf{G}_1(\mathbf{x}') =& -\omega^2\mathscr{R}(\mathbf{v}_\mathrm{R})(\mathbf{x}') - \int_{-L}^{L} \phi''(x_3) \begin{bmatrix} \lambda v_1 \\ \lambda v_2 \\ (2\lambda+\mu)v_3 \end{bmatrix} (\mathbf{x}', x_3)\mathrm{d}x_3 \\ &+ (\lambda+\mu)\int_{-L}^{L} \phi'(x_3) \begin{bmatrix} \partial_1 v_3 \\ \partial_2 v_3 \\ \partial_1 v_1 + \partial_2 v_2 \end{bmatrix} (\mathbf{x}', x_3)\mathrm{d}x_3 = \begin{bmatrix} \mathbf{G}_1^{(1,2)}(\mathbf{x}') \\ G_1^{(3)}(\mathbf{x}') \end{bmatrix}, \\ \mathbf{G}_2(\mathbf{x}') =& -\omega^2\mathscr{R}(\mathbf{f}_\mathrm{R})(\mathbf{x}') - \int_{-L}^{L} \phi''(x_3) \begin{pmatrix} \lambda w_1 \\ \lambda w_2 \\ (2\lambda+\mu)w_3 \end{pmatrix} (\mathbf{x}', x_3)\mathrm{d}x_3 \\ &+ (\lambda+\mu)\int_{-L}^{L} \phi'(x_3) \begin{pmatrix} \partial_1 w_3 \\ \partial_2 w_3 \\ \partial_1 w_1 + \partial_2 w_2 \end{pmatrix} (\mathbf{x}', x_3)\mathrm{d}x_3 = \begin{bmatrix} \mathbf{G}_2^{(1,2)}(\mathbf{x}') \\ G_2^{(3)}(\mathbf{x}') \end{bmatrix}, \end{aligned} \tag{10.3.9}$$

where $\mathbf{G}_\ell^{(1,2)}(\mathbf{x}') \in \mathbb{R}^2$ *(*$\ell = 1, 2$*),* $\mathbf{f}_\mathrm{R} := (1+V)\mathbf{w}_\mathrm{R} = q\mathbf{w}_\mathrm{R}$*, and* $\mathscr{R}$ *is the dimension reduction operator associated with* ϕ *defined in Definition 10.3.1. Denote*

$$\mathbf{v}_\mathrm{R} = \begin{pmatrix} v_1 \\ v_2 \\ v_3 \end{pmatrix} = \begin{pmatrix} \mathbf{v}^{(1,2)} \\ v_3 \end{pmatrix} \in \mathbb{R}^3, \quad \mathbf{w}_\mathrm{R} = \begin{pmatrix} w_1 \\ w_2 \\ w_3 \end{pmatrix} = \begin{pmatrix} \mathbf{w}^{(1,2)} \\ w_3 \end{pmatrix} \in \mathbb{R}^3, \tag{10.3.10}$$

where $\mathbf{v}^{(1,2)} \in \mathbb{R}^2$ *and* $\mathbf{w}^{(1,2)} \in \mathbb{R}^2$*. Assume that* $\eta \in L^\infty(\Gamma_h^\pm \times (-M, M))$ *in* (10.3.8) *is independent of* x_3*. Then it holds that*

$$\begin{cases} \widetilde{\mathscr{L}}\mathscr{R}(\mathbf{v}_\mathrm{R})(\mathbf{x}') = \mathbf{G}_1(\mathbf{x}') & \text{in } S_h, \\ \widetilde{\mathscr{L}}\mathscr{R}(\mathbf{w}_\mathrm{R})(\mathbf{x}') = \mathbf{G}_2(\mathbf{x}') & \text{in } S_h, \\ \mathscr{R}(\mathbf{w}_\mathrm{R})(\mathbf{x}') = \mathscr{R}(\mathbf{v}_\mathrm{R})(\mathbf{x}'), & \text{on } \Gamma_h^\pm, \\ \begin{bmatrix} T_\nu\mathscr{R}(\mathbf{v}^{(1,2)}) + \lambda\mathscr{R}(\partial_3 v_3)\nu \\ \mu\partial_\nu\mathscr{R}(v_3) + \mu\begin{bmatrix} \mathscr{R}(\partial_3 v_1) \\ \mathscr{R}(\partial_3 v_2) \end{bmatrix}\nu \end{bmatrix} + \eta\mathscr{R}(\mathbf{v}_\mathrm{R}) = \begin{bmatrix} T_\nu\mathscr{R}(\mathbf{w}^{(1,2)}) + \lambda\mathscr{R}(\partial_3 w_3)\nu \\ \mu\partial_\nu\mathscr{R}(w_3) + \mu\begin{bmatrix} \mathscr{R}(\partial_3 w_1) \\ \mathscr{R}(\partial_3 w_2) \end{bmatrix}\nu \end{bmatrix} & \text{on } \Gamma_h^\pm, \end{cases} \tag{10.3.11}$$

in the distributional sense, where ν *signifies the exterior unit normal vector to* $\Gamma_h^{\pm}$, T_ν *is the two-dimensional boundary traction operator defined in* (10.1.10), *and*

$$\widetilde{\mathscr{L}} = \begin{pmatrix} \lambda\Delta' + (\lambda+\mu)\partial_1^2 & (\lambda+\mu)\partial_1\partial_2 & 0 \\ (\lambda+\mu)\partial_1\partial_2 & \lambda\Delta' + (\lambda+\mu)\partial_2^2 & 0 \\ 0 & 0 & \lambda\Delta' \end{pmatrix} := \begin{pmatrix} \mathscr{L} & 0 \\ 0 & \lambda\Delta' \end{pmatrix}, \tag{10.3.12}$$

with $\Delta' := \partial_1^2 + \partial_2^2$ *being the Laplace operator with respect to the* $\mathbf{x}'$ *variables. Here,* $\mathscr{L}$ *is the two-dimensional Lamé operator with respect to the* $\mathbf{x}'$ *variable.*

Lemma 10.3.3 is a direct result of Lemma 10.3.2.

Lemma 10.3.3. *Under the same setup in Lemma 10.3.2, the PDE system* (10.3.11) *is equivalent to*

$$\begin{cases} \mathscr{L}\mathscr{R}(\mathbf{v}^{(1,2)})(\mathbf{x}') = \mathbf{G}_1^{(1,2)}(\mathbf{x}') & \text{in } S_h, \\ \mathscr{L}\mathscr{R}(\mathbf{w}^{(1,2)})(\mathbf{x}') = \mathbf{G}_2^{(1,2)}(\mathbf{x}') & \text{in } S_h, \\ \mathscr{R}(\mathbf{w}^{(1,2)})(\mathbf{x}') = \mathscr{R}(\mathbf{v}^{(1,2)})(\mathbf{x}'), & \text{on } \Gamma_h^{\pm}, \\ T_\nu\mathscr{R}(\mathbf{v}^{(1,2)})(\mathbf{x}') + \eta\mathscr{R}(\mathbf{v}^{(1,2)})(\mathbf{x}') = T_\nu\mathscr{R}(\mathbf{w}^{(1,2)})(\mathbf{x}') & \text{on } \Gamma_h^{\pm}, \end{cases} \tag{10.3.13}$$

and

$$\begin{cases} \lambda\Delta'\mathscr{R}(v_3)(\mathbf{x}') = \mathbf{G}_1^{(3)}(\mathbf{x}') & \text{in } S_h, \\ \lambda\Delta'\mathscr{R}(w_3)(\mathbf{x}') = \mathbf{G}_2^{(3)}(\mathbf{x}') & \text{in } S_h, \\ \mathscr{R}(w_3)(\mathbf{x}') = \mathscr{R}(v_3)(\mathbf{x}'), \partial_\nu\mathscr{R}(v_3)(\mathbf{x}') + \mu^{-1}\eta\mathscr{R}(v_3)(\mathbf{x}') = \partial_\nu\mathscr{R}(w_3)(\mathbf{x}') & \text{on } \Gamma_h^{\pm}, \end{cases} \tag{10.3.14}$$

where $\mathbf{G}_\ell^{(1,2)}$ *and* $\mathbf{G}_\ell^{(3)}$ *are defined in* (10.3.9), $\ell = 1,2$.

Next we mainly study the system (10.3.13). Using Green's formula (10.2.25) on the domain S_h and the boundary condition in (10.3.13), one can obtain a similar integral equality as in Lemma 10.2.6.

Lemma 10.3.4. *Let* S_h, Λ_h, *and* $\Gamma_h^{\pm}$ *be defined in* (10.2.2). *Suppose that* $\mathbf{v}^{(1,2)}, \mathbf{w}^{(1,2)} \in H^1(S_h \times (-M,M))^2$ *fulfill* (10.3.13). *Recall that the CGO solution* $\mathbf{u}(s\mathbf{x})$ *is defined in* (10.2.15) *and* $P_n(t)$ *is the Legendre polynomial. Let* $\beta = p$ *or* s, $j_{\ell,\beta} = j_\ell(k_\beta|\mathbf{x}|)$, *and*

$$\mathbf{v}_{jp}^{(1,2)}(\mathbf{0}) = \int_{\mathbb{S}^2} \mathbf{g}_{\mathrm{R},jp}^{(1,2)}(\mathbf{d})\mathrm{d}\sigma(\mathbf{d}),\ \mathbf{v}_{js}^{(1,2)}(\mathbf{0}) = \int_{\mathbb{S}^2} \mathbf{g}_{\mathrm{R},js}^{(1,2)}(\mathbf{d})\mathrm{d}\sigma(\mathbf{d}),$$
$$\mathbf{B}_{j\beta,1}^{(\ell)} = \int_{\mathbb{S}^2} \mathbf{g}_{\mathrm{R},j\beta}^{(1,2)}(\mathbf{d})P_{2\ell}(\cos\varphi)\mathrm{d}\sigma(\mathbf{d}), \quad \mathbf{B}_{j\beta,2}^{(\ell)} = \int_{\mathbb{S}^2} \mathbf{g}_{\mathrm{I},j\beta}^{(1,2)}(\mathbf{d})P_{2\ell-1}(\cos\varphi)\mathrm{d}\sigma(\mathbf{d}).$$

Denote

$$\mathbf{v}_j^{\mathrm{R}}(\mathbf{x}) = \begin{pmatrix} \mathbf{v}_j^{(1,2)}(\mathbf{x}) \\ v_j^{(3)}(\mathbf{x}) \end{pmatrix} \in \mathbb{R}^3, \quad \mathbf{v}_j^{(1,2)}(\mathbf{x}) \in \mathbb{R}^2,$$

where

$$\begin{aligned} \mathbf{v}_j^{(1,2)}(\mathbf{x}) =& \mathbf{v}_{jp}^{(1,2)}(\mathbf{0})j_0(k_p|\mathbf{x}|) + \mathbf{v}_{js}^{(1,2)}(\mathbf{0})j_0(k_s|\mathbf{x}|) + \sum_{\beta=p,s}\sum_{\ell=1}^{+\infty}(-1)^\ell(4\ell+1)j_{2\ell,\beta}\mathbf{B}_{j\beta,1}^{(\ell)} \\ &+ 2\sum_{\beta=p,s}\sum_{\ell=1}^{+\infty}(-1)^\ell(4\ell-1)j_{2\ell-1,\beta}\mathbf{B}_{j\beta,2}^{(\ell)}, \end{aligned} \tag{10.3.15}$$

and

$$
\begin{aligned}
v_j^{(3)}(\mathbf{x}) =& j_0(k_p|\mathbf{x}|)\int_{\mathbb{S}^2} g_{\mathrm{R},jp}^{(3)}(\mathbf{d})\mathrm{d}\sigma(\mathbf{d}) + j_0(k_s|\mathbf{x}|)\int_{\mathbb{S}^2} g_{\mathrm{R},js}^{(3)}(\mathbf{d})\mathrm{d}\sigma(\mathbf{d})\\
&+\sum_{\beta=p,s}\sum_{\ell=1}^{+\infty}(-1)^{\ell}(4\ell+1)j_{2\ell}(k_\beta|\mathbf{x}|)\int_{\mathbb{S}^2} g_{\mathrm{R},j\beta}^{(3)}(\mathbf{d})P_{2\ell}(\cos\varphi)\mathrm{d}\sigma(\mathbf{d})\\
&+\sum_{\beta=p,s}\sum_{\ell=1}^{+\infty}(-1)^{\ell}(4\ell-1)j_{2\ell-1}(k_\beta|\mathbf{x}|)\int_{\mathbb{S}^2} g_{\mathrm{I},j\beta}^{(3)}(\mathbf{d})P_{2\ell-1}(\cos\varphi)\mathrm{d}\sigma(\mathbf{d}),
\end{aligned}
$$

with $\mathbf{g}_{\mathrm{R},j\beta}^{(1,2)}$, $\mathbf{g}_{\mathrm{I},j\beta}^{(1,2)}$, $g_{\mathrm{R},j\beta}^{(3)}$, *and* $g_{\mathrm{I},j\beta}^{(3)}$ *defined in* (10.3.5). *Then the following integral equality holds:*

$$
\widetilde{I}_1+\widetilde{I}_2=\widetilde{I}_{\Lambda_h}-\widetilde{I}_{\pm}-\widetilde{I}_{\pm}^{\Delta}, \tag{10.3.16}
$$

where

$$
\begin{aligned}
&\widetilde{I}_1=\int_{S_h}(\mathbf{f}_{1j}+\mathbf{f}_2+\mathbf{f}_3+\mathbf{f}_4)\cdot\mathbf{u}(s\mathbf{x}')\mathrm{d}\mathbf{x}',\ \widetilde{I}_2=\int_{S_h}(\mathbf{f}_1-\mathbf{f}_{1j})\cdot\mathbf{u}(s\mathbf{x}')\mathrm{d}\mathbf{x}', \qquad (10.3.17)\\
&\widetilde{I}_{\Lambda_h}=\int_{\Lambda_h}(T_\nu\mathscr{R}(\mathbf{v}^{(1,2)}-\mathbf{w}^{(1,2)}))\cdot\mathbf{u}-(T_\nu(\mathbf{u}))\cdot\mathscr{R}(\mathbf{v}^{(1,2)}-\mathbf{w}^{(1,2)})\mathrm{d}\sigma,\\
&\widetilde{I}_{\pm}=\int_{\Gamma_h^{\pm}}\eta(\mathbf{x}')\mathscr{R}(\mathbf{v}_j^{(1,2)})\cdot\mathbf{u}(s\mathbf{x}')\mathrm{d}\sigma,\ \widetilde{I}_{\pm}^{\Delta}=\int_{\Gamma_h^{\pm}}\eta(\mathbf{x}')\mathscr{R}(\mathbf{v}^{(1,2)}-\mathbf{v}_j^{(1,2)})\cdot\mathbf{u}(s\mathbf{x}')\mathrm{d}\sigma,\\
&\mathbf{f}_1=-\omega^2\mathscr{R}(\mathbf{v}^{(1,2)}),\ \mathbf{f}_2=\omega^2\mathscr{R}(q\mathbf{w}^{(1,2)}),\ \mathbf{f}_3=-\int_{-L}^{L}\phi''(x_3)\begin{pmatrix}\lambda(v_1-w_1)\\ \lambda(v_2-w_2)\end{pmatrix}(\mathbf{x}',x_3)\mathrm{d}x_3,\\
&\mathbf{f}_{1j}=-\omega^2\mathscr{R}(\mathbf{v}_j^{(1,2)}),\ \mathbf{f}_4=(\lambda+\mu)\int_{-L}^{L}\phi'(x_3)\begin{pmatrix}\partial_1(v_3-w_3)\\ \partial_2(v_3-w_3)\end{pmatrix}(\mathbf{x}',x_3)\mathrm{d}x_3.
\end{aligned}
$$

Proof. Recall that $\mathscr{L}$ is defined in (10.3.12). Since $\mu>0$, we see that $3\lambda+2\mu>0$ implies $3(\lambda+\mu)>0$ and hence also $2\lambda+2\mu>0$, so $\mathscr{L}$ is strongly elliptic. Using Green's formula (10.2.25) on the domain S_h together with $\mathscr{L}\mathbf{u}=\mathbf{0}$ in S_h, we have

$$
\begin{aligned}
&\int_{S_h}\mathscr{L}\mathscr{R}(\mathbf{v}^{(1,2)}-\mathbf{w}^{(1,2)})\cdot\mathbf{u}(\mathbf{x}')\mathrm{d}\mathbf{x}'\\
&=\int_{\Gamma_h^{\pm}\cup\Lambda_h}(T_\nu\mathscr{R}(\mathbf{v}^{(1,2)}-\mathbf{w}^{(1,2)}))\cdot\mathbf{u}-(T_\nu(\mathbf{u}))\cdot\mathscr{R}(\mathbf{v}^{(1,2)}-\mathbf{w}^{(1,2)})\mathrm{d}\sigma.
\end{aligned}\tag{10.3.18}
$$

Using the boundary condition in (10.3.13), it yields that

$$
\begin{aligned}
&\int_{\Gamma_h^{\pm}}(T_\nu\mathscr{R}(\mathbf{v}^{(1,2)}-\mathbf{w}^{(1,2)}))\cdot\mathbf{u}-(T_\nu(\mathbf{u}))\cdot\mathscr{R}(\mathbf{v}^{(1,2)}-\mathbf{w}^{(1,2)})\mathrm{d}\sigma\\
&=-\int_{\Gamma_h^{\pm}}\eta(\mathbf{x}')\mathscr{R}(\mathbf{v}^{(1,2)})\cdot\mathbf{u}(\mathbf{x}')\mathrm{d}\sigma\\
&=-\int_{\Gamma_h^{\pm}}\eta(\mathbf{x}')\mathscr{R}(\mathbf{v}_j^{(1,2)})\cdot\mathbf{u}(\mathbf{x}')\mathrm{d}\sigma-\int_{\Gamma_h^{\pm}}\eta(\mathbf{x}')\mathscr{R}(\mathbf{v}^{(1,2)}-\mathbf{v}_j^{(1,2)})\cdot\mathbf{u}(\mathbf{x}')\mathrm{d}\sigma.
\end{aligned}\tag{10.3.19}
$$

By virtue of (10.3.13), we have

$$
\begin{aligned}
&\int_{S_h}\mathscr{L}\mathscr{R}(\mathbf{v}^{(1,2)}-\mathbf{w}^{(1,2)})\cdot\mathbf{u}(\mathbf{x}')\mathrm{d}\mathbf{x}'\\
&=\int_{S_h}(\mathbf{G}_1^{(1,2)}-\mathbf{G}_2^{(1,2)})\cdot\mathbf{u}(\mathbf{x}')\mathrm{d}\mathbf{x}'=\int_{S_h}(\mathbf{f}_1+\mathbf{f}_2+\mathbf{f}_3+\mathbf{f}_4)\cdot\mathbf{u}(\mathbf{x}')\mathrm{d}\mathbf{x}'\\
&=\int_{S_h}(\mathbf{f}_{1j}+\mathbf{f}_2+\mathbf{f}_3+\mathbf{f}_4)\cdot\mathbf{u}(\mathbf{x}')\mathrm{d}\mathbf{x}'+\int_{S_h}(\mathbf{f}_1-\mathbf{f}_{1j})\cdot\mathbf{u}(\mathbf{x}')\mathrm{d}\mathbf{x}'.
\end{aligned}\tag{10.3.20}
$$

From (10.3.18), (10.3.19), and (10.3.20), we can derive (10.3.16). □

Similar to Lemma 10.2.7, for the integral $\widetilde{I}_{\Lambda_h}$ defined in (10.3.16), one has the following:

Lemma 10.3.5. *Recall that $\widetilde{I}_{\Lambda_h}$ is defined in* (10.3.17). *Under the same setup in Lemma 10.3.4, the following integral estimate holds:*

$$|\widetilde{I}_{\Lambda_h}| \leq C\frac{\sqrt{2h+s^2}+\mu s}{\sqrt{2}}\sqrt{\theta_M-\theta_m}e^{-s\sqrt{h}\delta_W}\|\mathbf{v}^{(1,2)}-\mathbf{w}^{(1,2)}\|_{H^1(S_h)^2}, \tag{10.3.21}$$

where C is a positive constant coming from the trace theorem and S_h and $\delta_W > 0$ are defined in (10.2.2) *and* (10.2.18), *respectively.*

Lemma 10.3.6. *Under the same setup in Lemma 10.3.4, for any given positive constants γ, β_1, and β_2, we assume that there exists a sequence of Herglotz wave functions $\{\mathbf{v}_j\}_{j=1}^{+\infty}$, where $\mathbf{v}_j$ is defined by* (10.3.3), *which can approximate $\mathbf{v}$ in $H^1(S_h\times(-M,M))^3$ satisfying*

$$\|\mathbf{v}-\mathbf{v}_j\|_{H^1(S_h\times(-M,M))^3} \leq j^{-\gamma},\ \|\mathbf{g}_{jp}\|_{L^2(\mathbb{S}^2)^3} \leq j^{\beta_1},\ \|\mathbf{g}_{js}\|_{L^2(\mathbb{S}^2)^3} \leq j^{\beta_2}. \tag{10.3.22}$$

Furthermore, suppose that $\eta \in C^{\alpha}(\overline{\Gamma_h^{\pm}}\times[-M,M])$, $0<\alpha<1$, and hence it holds that

$$\eta(\mathbf{x}') = \eta(\mathbf{0}) + \delta\eta(\mathbf{x}'),\ |\delta\eta(\mathbf{x}')| \leq \|\eta(\mathbf{x}')\|_{C^{\alpha}}|\mathbf{x}'|^{\alpha}, \tag{10.3.23}$$

for $0<\alpha<1$. Recall that $\widetilde{I}_2$ and $\widetilde{I}_{\pm}^{\Delta}$ are defined in (10.3.17). *Then we have the following estimates:*

$$\left|\widetilde{I}_2\right| \leq h\sqrt{2L(\theta_M-\theta_m)}e^{-s\sqrt{\Theta}\delta_W}\|\phi\|_{L^\infty}j^{-\gamma}, \tag{10.3.24}$$
$$\left|\widetilde{I}_{\pm}^{\Delta}\right| \leq C\|\phi\|_{\infty}(|\eta(\mathbf{0})|h\sqrt{\theta_M-\theta_m}e^{-s\sqrt{\Theta}\delta_W} + \|\eta\|_{C^\alpha}\frac{2\sqrt{\theta_M-\theta_m}\Gamma(4\alpha+4)}{(2\delta_W)^{2\alpha+2}}s^{-2\alpha-2})j^{-\gamma},$$

where C is a positive constant depending on L and h.

Proof. Clearly, $\mathbf{v}_{\mathrm{R}}$ can be approximated by $\{\mathbf{v}_j^{\mathrm{R}}\}_{j=1}^{+\infty}$ in the sense of (10.3.22). Therefore, by the definition of $\mathbf{v}_{\mathrm{R}}^{(1,2)}$ given by (10.3.10) and $\mathbf{v}_j^{(1,2)}$ given by (10.3.15), it can be directly seen that

$$\|\mathbf{v}_{\mathrm{R}}^{(1,2)}-\mathbf{v}_j^{(1,2)}\|_{H^1(S_h\times(-L,L))^2} \leq \|\mathbf{v}_{\mathrm{R}}-\mathbf{v}_j^{\mathrm{R}}\|_{H^1(S_h\times(-M,M))^3} \leq j^{-\gamma}. \tag{10.3.25}$$

By (10.3.25) and using the Cauchy-Schwartz inequality, we have

$$\begin{aligned}
&\|\mathscr{R}(\mathbf{v}^{(1,2)})-\mathscr{R}(\mathbf{v}_j^{(1,2)})\|_{L^2(S_h)^2}^2\\
&=\int_{S_h}\left|\int_{-L}^{L}\phi(x_3)(\mathbf{v}^{(1,2)}(\mathbf{x}',x_3)-\mathbf{v}_j^{(1,2)}(\mathbf{x}',x_3))\mathrm{d}x_3\right|^2\mathrm{d}\mathbf{x}'\\
&\leq 2L\|\phi\|_{L^\infty}^2\|\mathbf{v}-\mathbf{v}_j\|_{L^2(S_h\times(-L,L))^3}^2.
\end{aligned}\tag{10.3.26}$$

By virtue of (10.2.19) and (10.3.26), we have

$$\begin{aligned}
\left|\widetilde{I}_2\right| &= \left|\int_{S_h}(\omega^2\mathscr{R}(\mathbf{v}^{(1,2)})-\omega^2\mathscr{R}(\mathbf{v}_j^{(1,2)}))\cdot\mathbf{u}(s\mathbf{x}')\mathrm{d}\mathbf{x}'\right|\\
&\leq\omega^2\|\mathscr{R}(\mathbf{v}^{(1,2)})-\mathscr{R}(\mathbf{v}_j^{(1,2)})\|_{L^2(S_h)^2}\|\mathbf{u}(s\mathbf{x}')\|_{L^2(S_h)^2}\\
&\lesssim Ch\sqrt{\theta_M-\theta_m}e^{-s\sqrt{\Theta}\delta_W}\|\phi\|_{L^\infty}j^{-\gamma}.
\end{aligned}$$

Using the Cauchy-Schwartz inequality and the trace theorem, according to (10.3.23), we have

$$
\begin{aligned}
\left|\widetilde{I}_{\pm}^{\Delta}\right| \leq & |\eta(\mathbf{0})| \int_{\Gamma_h^{\pm}} |\mathbf{u}||\mathscr{R}(\mathbf{v}^{(1,2)}-\mathbf{v}_j^{(1,2)})|\mathrm{d}\sigma + \|\eta\|_{C^\alpha} \int_{\Gamma_h^{\pm}} |\mathbf{x}'|^\alpha |\mathbf{u}||\mathscr{R}(\mathbf{v}^{(1,2)}-\mathbf{v}_j^{(1,2)})|\mathrm{d}\sigma \\
\leq & |\eta(\mathbf{0})| \|\mathscr{R}(\mathbf{v}^{(1,2)}-\mathbf{v}_j^{(1,2)})\|_{H^{\frac{1}{2}}(\Gamma_h^{\pm})^2} \|\mathbf{u}\|_{H^{-\frac{1}{2}}(\Gamma_h^{\pm})^2} \\
& + \|\eta\|_{C^\alpha} \|\mathscr{R}(\mathbf{v}^{(1,2)}-\mathbf{v}_j^{(1,2)})\|_{H^{\frac{1}{2}}(\Gamma_h^{\pm})^2} \| |\mathbf{x}'|^\alpha \mathbf{u}\|_{H^{-\frac{1}{2}}(\Gamma_h^{\pm})^2} \\
\leq & |\eta(\mathbf{0})| \|\mathscr{R}(\mathbf{v}^{(1,2)}-\mathbf{v}_j^{(1,2)})\|_{H^1(S_h)^2} \|\mathbf{u}\|_{L^2(S_h)^2} \\
& + \|\eta\|_{C^\alpha} \|\mathscr{R}(\mathbf{v}^{(1,2)}-\mathbf{v}_j^{(1,2)})\|_{H^1(S_h)^2} \| |\mathbf{x}'|^\alpha \mathbf{u}\|_{L^2(S_h)^2} \\
\leq & C\|\phi\|_{L^\infty} \|\mathbf{v}^{(1,2)}-\mathbf{v}_j^{(1,2)}\|_{H^1(S_h)^2} (|\eta(\mathbf{0})| \|\mathbf{u}\|_{L^2(S_h)^2} + \|\eta\|_{C^\alpha} \||\mathbf{x}'|^\alpha \mathbf{u}\|_{L^2(S_h)^2}) \\
\leq & C\|\phi\|_{L^\infty} (|\eta(\mathbf{0})| h\sqrt{\theta_M-\theta_m} e^{-s\sqrt{\Theta}\delta_W} + \|\eta\|_{C^\alpha} \frac{2\sqrt{\theta_M-\theta_m}\Gamma(4\alpha+4)}{(2\delta_W)^{2\alpha+2}} s^{-2\alpha-2}) j^{-\gamma},
\end{aligned}
$$

where C is a positive constant and the last second inequality comes from Lemma 3.4.1. □

By virtue of Lemma 10.3.1, the regularity results on $\mathbf{f}_j$ defined in (10.3.17) can be easily obtained in the following lemma, whose proof is omitted.

Lemma 10.3.7. *Let* $\mathbf{f}_2(\mathbf{x}')$, $\mathbf{f}_3(\mathbf{x}')$, *and* $\mathbf{f}_4(\mathbf{x}')$ *be defined in* (10.3.17). *If*

$$
q\mathbf{w}_{\mathrm{R}} \in C^\alpha(\overline{S}_h \times [-M,M]), \tag{10.3.27}
$$

then $\mathbf{f}_2(\mathbf{x}') \in C^\alpha(S_h)$. *Furthermore, we assume that*

$$
\mathbf{v}_{\mathrm{R}} - \mathbf{w}_{\mathrm{R}} \in C^{1,\alpha}(\overline{S}_h \times [-M,M])^3, \tag{10.3.28}
$$

where $\alpha \in (0,1)$. *Then one has* $\mathbf{f}_3(\mathbf{x}') \in C^\alpha(\overline{S}_h)$ *and* $\mathbf{f}_4(\mathbf{x}') \in C^\alpha(\overline{S}_h)$.

Proof. According to (10.3.28), one handily computes that

$$
\begin{aligned}
\left|\mathbf{f}_3(\mathbf{x}') - \mathbf{f}_3(\mathbf{y}')\right| &= |\lambda| \left| \int_{-L}^{L} \phi''(x_3) \left(\begin{pmatrix} (v_1-w_1) \\ (v_2-w_2) \end{pmatrix} (\mathbf{y}',x_3) - \begin{pmatrix} (v_1-w_1) \\ (v_2-w_2) \end{pmatrix} (\mathbf{x}',x_3) \right) \mathrm{d}x_3 \right| \\
&\leq 2C|\lambda| L \|\phi\|_{C^2} \|\mathbf{v}_{\mathrm{R}} - \mathbf{w}_{\mathrm{R}}\|_{C^{1,\alpha}((S_h\times(-M,M))^3)} \|\mathbf{x}'-\mathbf{y}'\|^\alpha, \\
\left|\mathbf{f}_4(\mathbf{x}') - \mathbf{f}_4(\mathbf{y}')\right| &= |\lambda+\mu| \left| \int_{-L}^{L} \phi'(x_3) \left(\begin{pmatrix} \partial_1(v_3-w_3) \\ \partial_2(v_3-w_3) \end{pmatrix} (\mathbf{x}',x_3) \right.\right. \\
& \qquad \left.\left. - \begin{pmatrix} \partial_1(v_3-w_3) \\ \partial_2(v_3-w_3) \end{pmatrix} (\mathbf{y}',x_3) \right) \mathrm{d}x_3 \right| \\
&\leq 2|\lambda+\mu| L \|\phi\|_{C^1} \|\mathbf{v}_{\mathrm{R}} - \mathbf{w}_{\mathrm{R}}\|_{C^{1,\alpha}((S_h\times(-M,M))^3)} \|\mathbf{x}'-\mathbf{y}'\|^\alpha,
\end{aligned}
$$

where C is a positive constant. Therefore, we know that $\mathbf{f}_3(\mathbf{x}') \in C^\alpha(\overline{S}_h)$ and $\mathbf{f}_4(\mathbf{x}') \in C^\alpha(\overline{S}_h)$.

Due to (10.3.27), $\mathbf{f}_2(\mathbf{x}') \in C^\alpha(\overline{S}_h)$ can be obtained directly by virtue of Lemma 3.4.1. □

The integral estimate for $\widetilde{I}_1$ defined in (10.3.17) is given in the following lemma. Using the Hölder continuity of $\mathbf{f}_\ell$, the definition of the dimension reduction operator, and (10.2.17), we can analyze the integral asymptotic behavior with the parameter s in the CGO $\mathbf{u}$.

Lemma 10.3.8. *Recall that* $\mathbf{f}_2(\mathbf{x}')$, $\mathbf{f}_3(\mathbf{x}')$, *and* $\mathbf{f}_4(\mathbf{x}')$ *are defined in* (10.3.17). *Suppose that* $\mathbf{f}_\ell(\mathbf{x}') \in C^\alpha(\overline{S}_h)$ *(*$\ell = 2,3,4$*),* $0<\alpha<1$. *Recalling that* $\widetilde{I}_1$ *is defined in* (10.3.17), *then the following integral estimates hold:*

$$\begin{aligned}\left|\widetilde{I}_1\right| \leq & 6(\mathbf{f}_{1j}(\mathbf{0})+\mathbf{f}_2(\mathbf{0})+\mathbf{f}_3(\mathbf{0})+\mathbf{f}_4(\mathbf{0}))|e^{-2\theta_M \mathrm{i}}-e^{-2\theta_m \mathrm{i}}|s^{-4} \\ & +\sum_{\ell=2}^{4}\|\mathbf{f}_\ell\|_{C^\alpha(S_h)^2}\frac{2\sqrt{2}(\theta_M-\theta_m)\Gamma(2\alpha+4)}{\delta_W^{2\alpha+4}}s^{-2\alpha-4} \\ & +4L\omega^2\mathrm{diam}(S_h)^{1-\alpha}\sqrt{\pi}\|\phi\|_{L^\infty}\sum_{\beta=p,s}(1+k_\beta)\left(\|\mathbf{g}_{\mathrm{R},j\beta}^{(1,2)}\|_{L^2(\mathbb{S}^2)}+\|\mathbf{g}_{\mathrm{I},j\beta}^{(1,2)}\|_{L^2(\mathbb{S}^2)}\right),\end{aligned} \tag{10.3.29}$$

where $\mathbf{g}_{\mathrm{R},j\beta}^{(1,2)}$ *and* $\mathbf{g}_{\mathrm{I},j\beta}^{(1,2)}$ *are defined in* (10.3.5) ($\beta = p,\ s$)*,* δ_W *is defined in* (10.2.18)*,* θ_m*, and* θ_M *are defined in* (10.1.11).

Proof. Due to $\mathbf{f}_\ell(\mathbf{x}') \in C^\alpha(\overline{S}_h)$ ($\ell = 2,3,4$), one has

$$\begin{aligned}\mathbf{f}_2(\mathbf{x}') &= \mathbf{f}_2(\mathbf{0})+\delta\mathbf{f}_2(\mathbf{x}'),\ |\delta\mathbf{f}_2(\mathbf{x}')| \leq \|\mathbf{f}_2\|_{C^\alpha(S_h)^2}|\mathbf{x}'|^\alpha, \\ \mathbf{f}_3(\mathbf{x}') &= \mathbf{f}_3(\mathbf{0})+\delta\mathbf{f}_3(\mathbf{x}'),\ |\delta\mathbf{f}_3(\mathbf{x}')| \leq \|\mathbf{f}_3\|_{C^\alpha(S_h)^2}|\mathbf{x}'|^\alpha, \\ \mathbf{f}_4(\mathbf{x}') &= \mathbf{f}_4(\mathbf{0})+\delta\mathbf{f}_4(\mathbf{x}'),\ |\delta\mathbf{f}_4(\mathbf{x}')| \leq \|\mathbf{f}_4\|_{C^\alpha(S_h)^2}|\mathbf{x}'|^\alpha.\end{aligned} \tag{10.3.30}$$

It is easy to see that $\mathbf{f}_{1j}(\mathbf{x}') \in C^\alpha(\overline{S}_h)$. Hence, it yields that

$$\mathbf{f}_{1j}(\mathbf{x}') = \mathbf{f}_{1j}(\mathbf{0})+\delta\mathbf{f}_{1j}(\mathbf{x}'),\ |\delta\mathbf{f}_{1j}(\mathbf{x}')| \leq \|\mathbf{f}_{1j}\|_{C^\alpha(S_h)^2}|\mathbf{x}'|^\alpha. \tag{10.3.31}$$

By virtue of (10.3.30) and (10.3.31), we have

$$\begin{aligned}\widetilde{I}_1 =&(\mathbf{f}_{1j}(\mathbf{0})+\mathbf{f}_2(\mathbf{0})+\mathbf{f}_3(\mathbf{0})+\mathbf{f}_4(\mathbf{0}))\int_{S_h}\mathbf{u}(\mathbf{x}')\mathrm{d}\mathbf{x}'+\int_{S_h}\delta\mathbf{f}_{1j}\cdot\mathbf{u}(\mathbf{x}')\mathrm{d}\mathbf{x}' \\ &+\int_{S_h}\delta\mathbf{f}_2\cdot\mathbf{u}(\mathbf{x}')\mathrm{d}\mathbf{x}'+\int_{S_h}\delta\mathbf{f}_3\cdot\mathbf{u}(\mathbf{x}')\mathrm{d}\mathbf{x}'+\int_{S_h}\delta\mathbf{f}_4\cdot\mathbf{u}(\mathbf{x}')\mathrm{d}\mathbf{x}'.\end{aligned} \tag{10.3.32}$$

Recall that $\mathbf{f}_{1j} = -\omega^2\mathscr{R}(\mathbf{v}_j^{(1,2)})$. Using the property of compact embedding of Hölder spaces, we can derive that for $0<\alpha<1$,

$$\|\mathbf{f}_{1j}\|_{C^\alpha} \leq \omega^2\mathrm{diam}(S_h)^{1-\alpha}\|\mathscr{R}(\mathbf{v}_j^{(1,2)})\|_{C^1},$$

where $\mathrm{diam}(S_h)$ is the diameter of S_h. By the definition of the dimension reduction operator (10.3.1), using (10.3.5) and (10.3.7), it is easy to see that

$$\begin{aligned}|\mathscr{R}(\mathbf{v}_j^{(1,2)})(\mathbf{x}')| \leq & 4L\sqrt{\pi}\|\phi\|_{L^\infty}\Big(\|\mathbf{g}_{\mathrm{R},jp}^{(1,2)}\|_{L^2(\mathbb{S}^2)}+\|\mathbf{g}_{\mathrm{I},jp}^{(1,2)}\|_{L^2(\mathbb{S}^2)} \\ & +\|\mathbf{g}_{\mathrm{R},js}^{(1,2)}\|_{L^2(\mathbb{S}^2)}+\|\mathbf{g}_{\mathrm{I},js}^{(1,2)}\|_{L^2(\mathbb{S}^2)}\Big), \\ |\partial_{x_i}\mathscr{R}(\mathbf{v}_j^{(1,2)})(\mathbf{x}')| \leq & 4L\sqrt{\pi}\|\phi\|_{L^\infty}\Big(k_p\|\mathbf{g}_{\mathrm{R},jp}^{(1,2)}\|_{L^2(\mathbb{S}^2)}+k_p\|\mathbf{g}_{\mathrm{I},jp}^{(1,2)}\|_{L^2(\mathbb{S}^2)} \\ & +k_s\|\mathbf{g}_{\mathrm{R},js}^{(1,2)}\|_{L^2(\mathbb{S}^2)}+k_s\|\mathbf{g}_{\mathrm{I},js}^{(1,2)}\|_{L^2(\mathbb{S}^2)}\Big).\end{aligned}$$

Thus, we have

$$\begin{aligned}\|\mathscr{R}(\mathbf{v}_j^{(1,2)})\|_{C^1} \leq & 4L\sqrt{\pi}\|\phi\|_{L^\infty}\Big((1+k_p)\|\mathbf{g}_{\mathrm{R},jp}^{(1,2)}\|_{L^2(\mathbb{S}^2)}+(1+k_p)\|\mathbf{g}_{\mathrm{I},jp}^{(1,2)}\|_{L^2(\mathbb{S}^2)} \\ & +(1+k_s)\|\mathbf{g}_{\mathrm{R},js}^{(1,2)}\|_{L^2(\mathbb{S}^2)}+(1+k_s)\|\mathbf{g}_{\mathrm{I},js}^{(1,2)}\|_{L^2(\mathbb{S}^2)}\Big).\end{aligned} \tag{10.3.33}$$

From (10.2.17) and (10.3.33), we have

$$\left|\int_{S_h} \delta \mathbf{f}_{1j} \cdot \mathbf{u}(\mathbf{x}')\mathrm{d}\mathbf{x}'\right| \le \|\mathbf{f}_{1j}\|_{C^\alpha(S_h)^2} \left|\int_W \mathbf{u}(\mathbf{x}')|\mathbf{x}'|^\alpha \mathrm{d}\mathbf{x}'\right|$$

$$\le \|\mathbf{f}_{1j}\|_{C^\alpha(S_h)^2} \frac{2\sqrt{2}(\theta_M-\theta_m)\Gamma(2\alpha+4)}{\delta_W^{2\alpha+4}} s^{-2\alpha-4}.$$

$$\le \omega^2 \mathrm{diam}(S_h)^{1-\alpha} 4L\sqrt{\pi}\|\phi\|_{L^\infty}\Big((1+k_p)\|\mathbf{g}_{\mathrm{R},jp}^{(1,2)}\|_{L^2(\mathbb{S}^2)} + (1+k_p)\|\mathbf{g}_{\mathrm{I},jp}^{(1,2)}\|_{L^2(\mathbb{S}^2)}$$

$$+(1+k_s)\|\mathbf{g}_{\mathrm{R},js}^{(1,2)}\|_{L^2(\mathbb{S}^2)} + (1+k_s)\|\mathbf{g}_{\mathrm{I},js}^{(1,2)}\|_{L^2(\mathbb{S}^2)}\Big) s^{-2\alpha-4}. \tag{10.3.34}$$

Similarly, we have

$$\left|\int_{S_h} \delta \mathbf{f}_2 \cdot \mathbf{u}(\mathbf{x}')\mathrm{d}\mathbf{x}'\right| \le \|\mathbf{f}_2\|_{C^\alpha(S_h)^2} \left|\int_W \mathbf{u}(\mathbf{x}')|\mathbf{x}'|^\alpha \mathrm{d}\mathbf{x}'\right|$$
$$\le \|\mathbf{f}_2\|_{C^\alpha(S_h)^2} \frac{2\sqrt{2}(\theta_M-\theta_m)\Gamma(2\alpha+4)}{\delta_W^{2\alpha+4}} s^{-2\alpha-4}, \tag{10.3.35}$$

$$\left|\int_{S_h} \delta \mathbf{f}_3 \cdot \mathbf{u}(\mathbf{x}')\mathrm{d}\mathbf{x}'\right| \le \|\mathbf{f}_3\|_{C^\alpha(S_h)^2} \left|\int_W \mathbf{u}(\mathbf{x}')|\mathbf{x}'|^\alpha \mathrm{d}\mathbf{x}'\right|$$
$$\le \|\mathbf{f}_3\|_{C^\alpha(S_h)^2} \frac{2\sqrt{2}(\theta_M-\theta_m)\Gamma(2\alpha+4)}{\delta_W^{2\alpha+4}} s^{-2\alpha-4}, \tag{10.3.36}$$

$$\left|\int_{S_h} \delta \mathbf{f}_4 \cdot \mathbf{u}(\mathbf{x}')\mathrm{d}\mathbf{x}'\right| \le \|\mathbf{f}_4\|_{C^\alpha(S_h)^2} \left|\int_W \mathbf{u}(\mathbf{x}')|\mathbf{x}'|^\alpha \mathrm{d}\mathbf{x}'\right|$$
$$\le \|\mathbf{f}_4\|_{C^\alpha(S_h)^2} \frac{2\sqrt{2}(\theta_M-\theta_m)\Gamma(2\alpha+4)}{\delta_W^{2\alpha+4}} s^{-2\alpha-4}. \tag{10.3.37}$$

From (10.3.32), (10.3.34), (10.3.35), (10.3.36), and (10.3.37), we can derive (10.3.29). □

Using the integral mean value theorem, we have the following lemma.

Lemma 10.3.9. *Let $j_\ell(t)$ be the ℓ-th spherical Bessel function with the form*

$$j_\ell(t) = \frac{t^\ell}{(2\ell+1)!!}\left(1 - \sum_{l=1}^{\infty} \frac{(-1)^l t^{2l}}{2^l l! N_{\ell,l}}\right),$$

where $N_{\ell,l} = (2\ell+3)\dots(2\ell+2l+1)$, and $\mathscr{R}$ be the dimension reduction operator defined in Definition 10.3.1. Then

$$\mathscr{R}(j_0(k_\beta|\mathbf{x}|))(\mathbf{x}') = C(\phi) + C_1(\phi)\sum_{l=1}^{\infty} \frac{(-1)^l k_\beta^{2l}(|\mathbf{x}'|^2 + a_{0,l}^2)^{l-\frac{1}{2}}}{2^l l!(2l+1)!!}|\mathbf{x}'|^2, \tag{10.3.38a}$$

$$\mathscr{R}(j_{2\ell}(k_\beta|\mathbf{x}|))(\mathbf{x}') = C_1(\phi)|\mathbf{x}'|^2 \frac{k_\beta^{2\ell}(|\mathbf{x}'|^2 + a_{\ell,1}^2)^{\ell-\frac{1}{2}}}{(4\ell+1)!!}\left[1 - \sum_{l=1}^{\infty} \frac{(-1)^l k_\beta^{2l}(|\mathbf{x}'|^2 + a_{\ell,l,1}^2)^l}{2^l l! N_{\ell,l,1}}\right], \tag{10.3.38b}$$

$$\mathscr{R}(j_{2\ell-1}(k_\beta|\mathbf{x}|))(\mathbf{x}') = C_1(\phi)|\mathbf{x}'|^2 \frac{k_\beta^{2\ell-1}(|\mathbf{x}'|^2 + a_{\ell,1}^2)^{\ell-1}}{(4\ell-1)!!}\left[1 - \sum_{l=1}^{\infty} \frac{(-1)^l k_\beta^{2l}(|\mathbf{x}'|^2 + a_{\ell,l,1}^2)^l}{2^l l! N_{\ell,l,2}}\right], \tag{10.3.38c}$$

where $\beta = p, s$, $\ell \in \mathbb{N}$, $a_{0,l}, \xi, a_{\ell,1}, a_{\ell,l,1} \in (-L, L)$, $N_{\ell,l,1} = (4\ell+3)\ldots(4\ell+2l+1)$, $N_{\ell,l,2} = (4\ell+1)\ldots(4\ell+2l-1)$ *and*

$$C(\phi) = \int_{-L}^{L} \phi(x_3)\mathrm{d}x_3 > 0, \quad C_1(\phi) = \int_{-\arctan L/|\mathbf{x}'|}^{\arctan L/|\mathbf{x}'|} \phi(|\mathbf{x}'|\tan\omega)\sec^3\omega \mathrm{d}\omega. \tag{10.3.39}$$

Furthermore, it holds that

$$0 < C_1(\phi) < \sec^3 \varpi C(\phi), \tag{10.3.40}$$

where $\varpi \in (-\arctan L/|\mathbf{x}'|, \arctan L/|\mathbf{x}'|)$.

Proof. By virtue of (10.3.1), we have

$$\begin{aligned}
\mathscr{R}(j_0)(\mathbf{x}') &= \int_{-L}^{L} \phi(x_3)\left(1 - \sum_{l=1}^{\infty} \frac{(-1)^l k_\beta^{2l}(|\mathbf{x}'|^2 + x_3^2)^l}{2^l l!(2l+1)!!}\right)\mathrm{d}x_3 \\
&= \int_{-L}^{L} \phi(x_3)\mathrm{d}x_3 + \int_{-L}^{L} \phi(x_3) \sum_{l=1}^{\infty} \frac{(-1)^l k_\beta^{2l}(|\mathbf{x}'|^2 + x_3^2)^l}{2^l l!(2l+1)!!}\mathrm{d}x_3.
\end{aligned} \tag{10.3.41}$$

For $\int_{-L}^{L} \phi(x_3) \sum_{l=1}^{\infty} \frac{(-1)^l k_\beta^{2l}(|\mathbf{x}'|^2 + x_3^2)^l}{2^l l!(2l+1)!!}\mathrm{d}x_3$, using integral mean value theorem, we have

$$\begin{aligned}
&\int_{-L}^{L} \phi(x_3) \sum_{l=1}^{\infty} \frac{(-1)^l k_\beta^{2l}(|\mathbf{x}'|^2 + x_3^2)^l}{2^l l!(2l+1)!!}\mathrm{d}x_3 \\
&= \sum_{l=1}^{\infty} \frac{(-1)^l k_\beta^{2l}(|\mathbf{x}'|^2 + a_{0,l}^2)^{l-\frac{1}{2}}}{2^l l!(2l+1)!!} \int_{-L}^{L} \phi(x_3)(|\mathbf{x}'|^2 + x_3^2)^{\frac{1}{2}}\mathrm{d}x_3,
\end{aligned} \tag{10.3.42}$$

where $a_{0,l} \in (-L, L)$. By change of variables, where $x_3 = |x'|\tan\omega$, we have

$$\int_{-L}^{L} \phi(x_3)(|\mathbf{x}'|^2 + x_3^2)^{\frac{1}{2}}\mathrm{d}x_3 = \int_{-\arctan\frac{L}{|\mathbf{x}'|}}^{\arctan\frac{L}{|\mathbf{x}'|}} \phi(|\mathbf{x}'|\tan\omega)|\mathbf{x}'|^2\sec^3\omega\mathrm{d}\omega = |\mathbf{x}'|^2 C_1(\phi). \tag{10.3.43}$$

Plugging (10.3.42) and (10.3.43) into (10.3.41), it yields (10.3.38a).

We proceed to prove (10.3.40). Using the integral mean value theorem and variable substitution $\omega = \arctan\frac{x_3}{|\mathbf{x}'|}$, we have

$$C_1(\phi) = \sec^3\varpi \int_{-\arctan L/|\mathbf{x}'|}^{\arctan L/|\mathbf{x}'|} \phi(|\mathbf{x}'|\tan\omega)\mathrm{d}\omega = \sec^3\varpi \int_{-L}^{L} \phi(x_3)\frac{|\mathbf{x}'|^2}{|\mathbf{x}'|^2 + x_3^2}\mathrm{d}x_3,$$

where $\varpi \in (-\arctan L/|\mathbf{x}'|, \arctan L/|\mathbf{x}'|)$. Using the integral mean value theorem again, we have

$$C_1(\phi) = \frac{|\mathbf{x}'|^2}{|\mathbf{x}'|^2 + \xi^2}\sec^3\varpi \int_{-L}^{L} \phi(x_3)\mathrm{d}x_3 \le \sec^3\varpi C(\phi),$$

where $\xi \in (-L, L)$.

The series expansion of $\mathscr{R}(j_{2\ell})(\mathbf{x}')$ and $\mathscr{R}(j_{2\ell-1})(\mathbf{x}')$ in (10.3.38b) and (10.3.38c) can be obtained by a similar argument. □

Remark 10.3.1. We should emphasize that $C_1(\phi)$ is uniformly bounded with respect to $|\mathbf{x}'|$ since $\varpi \in (-\arctan\frac{L}{|\mathbf{x}'|}, \arctan\frac{L}{|\mathbf{x}'|})$ is fixed, which shall be used in the proof of Theorem 10.3.1 in what follows.

By virtue of (10.3.15), the reduction operator $\mathscr{R}$ defined in (10.3.1), the integral mean value theorem, and the Cauchy-Schwartz inequality, one can derive the integral equality (10.3.44) and obtain the asymptotic analysis (10.3.45) with respect to s in the CGO $\mathbf{u}$.

Lemma 10.3.10. *Consider the same setup in Lemma 10.3.4 and recall that* $\widetilde{I}_\pm$ *is defined* (10.3.17). *Supposing that* $\eta \in C^\alpha(\overline{\Gamma_h^\pm} \times [-M,M])$ *is independent of* x_3 *and has the expansion* (10.3.23), *then it holds that*

$$\widetilde{I}_\pm = \widetilde{\mathscr{I}}_1^\pm + \eta(\mathbf{0})\widetilde{\mathscr{I}}_2^\pm, \tag{10.3.44}$$

where $\widetilde{I}_\pm$ *are defined in* (10.3.17) *and*

$$\widetilde{\mathscr{I}}_1^\pm = \int_{\Gamma_h^\pm} \delta\eta(\mathbf{x}')\mathscr{R}(\mathbf{v}_j^{(1,2)}) \cdot \mathbf{u}(s\mathbf{x}')\mathrm{d}\sigma, \quad \widetilde{\mathscr{I}}_2^\pm = \int_{\Gamma_h^\pm} \mathscr{R}(\mathbf{v}_j^{(1,2)}) \cdot \mathbf{u}(s\mathbf{x}')\mathrm{d}\sigma.$$

Furthermore, the following estimate holds as $s \to +\infty$*:*

$$\left|\widetilde{\mathscr{I}}_1^\pm\right| \leq O(s^{-2\alpha-6}) + \sum_{\substack{\alpha=\mathrm{R,I}\\ \beta=p,s}} \left(\|\mathbf{g}_{\alpha,j\beta}^{(1,2)}\|_{L^2(\mathbb{S}^2)^2} + \|\mathbf{g}_{\alpha,j\beta}^{(1,2)}\|_{L^2(\mathbb{S}^2)^2}\right) \times O(s^{-2\alpha-6}), \tag{10.3.45}$$

where $\mathbf{g}_{\mathrm{R},j\beta}^{(1,2)}$ *and* $\mathbf{g}_{\mathrm{I},j\beta}^{(1,2)}$ *are defined in* (10.3.5).

Proof. One can immediately deduce (10.3.44) by (10.3.23). By virtue of (10.3.15) and the reduction operator $\mathscr{R}$ defined in (10.3.1), we deduce that

$$\begin{aligned}\mathscr{R}(\mathbf{v}_j^{(1,2)}(\mathbf{x})) &= \mathbf{v}_{jp}^{(1,2)}(\mathbf{0})\mathscr{R}(j_0(k_p|\mathbf{x}|)) + \mathbf{v}_{js}^{(1,2)}(\mathbf{0})\mathscr{R}(j_0(k_s|\mathbf{x}|)) \\ &+ 2\sum_{\ell=1}^{+\infty}(-1)^\ell(4\ell+1)\mathscr{R}(j_{2\ell}(k_p|\mathbf{x}|))\mathbf{B}_{jp,1}^{(\ell)} + 2\sum_{\ell=1}^{+\infty}(-1)^\ell(4\ell+1) \\ &\times \mathscr{R}(j_{2\ell}(k_s|\mathbf{x}|))\mathbf{B}_{js,1}^{(\ell)} + 2\sum_{\ell=1}^{+\infty}(-1)^\ell(4\ell-1)\mathscr{R}(j_{2\ell-1}(k_p|\mathbf{x}|))\mathbf{B}_{jp,2}^{(\ell)} \\ &+ 2\sum_{\ell=1}^{+\infty}(-1)^\ell(4\ell-1)\mathscr{R}(j_{2\ell-1}(k_s|\mathbf{x}|))\mathbf{B}_{js,2}^{(\ell)}.\end{aligned} \tag{10.3.46}$$

Plugging (10.3.46) into $\widetilde{\mathscr{I}}_1^\pm$, together with the use of (10.2.15), one can directly verify that

$$\widetilde{\mathscr{I}}_1^\pm = \widetilde{\mathscr{I}}_{11}^\pm + \widetilde{\mathscr{I}}_{12}^\pm + \widetilde{\mathscr{I}}_{13}^\pm + \widetilde{\mathscr{I}}_{14}^\pm + \widetilde{\mathscr{I}}_{15}^\pm + \widetilde{\mathscr{I}}_{16}^\pm, \tag{10.3.47}$$

where

$$\begin{aligned}\widetilde{\mathscr{I}}_{11}^\pm &= \int_{\Gamma_h^\pm} \begin{pmatrix}1\\ \mathrm{i}\end{pmatrix} \cdot \mathbf{v}_{jp}^{(1,2)}(\mathbf{0})\delta\eta(\mathbf{x}')\mathscr{R}(j_0(k_p|\mathbf{x}|))u_1\mathrm{d}\sigma, \\ \widetilde{\mathscr{I}}_{12}^\pm &= \int_{\Gamma_h^\pm} \begin{pmatrix}1\\ \mathrm{i}\end{pmatrix} \cdot \mathbf{v}_{js}^{(1,2)}(\mathbf{0})\delta\eta(\mathbf{x}')\mathscr{R}(j_0(k_s|\mathbf{x}|))u_1\mathrm{d}\sigma, \\ \widetilde{\mathscr{I}}_{13}^\pm &= \sum_{\ell=1}^{\infty}\int_{\Gamma_h^\pm} \begin{pmatrix}1\\ \mathrm{i}\end{pmatrix} \cdot \mathbf{B}_{jp,1}^{(\ell)}(-1)^\ell(4\ell+1)\delta\eta(\mathbf{x}')\mathscr{R}(j_{2\ell}(k_p|\mathbf{x}|))u_1\mathrm{d}\sigma, \\ \widetilde{\mathscr{I}}_{14}^\pm &= \sum_{\ell=1}^{\infty}\int_{\Gamma_h^\pm} \begin{pmatrix}1\\ \mathrm{i}\end{pmatrix} \cdot \mathbf{B}_{js,1}^{(\ell)}(-1)^\ell(4\ell+1)\delta\eta(\mathbf{x}')\mathscr{R}(j_{2\ell}(k_s|\mathbf{x}|))u_1\mathrm{d}\sigma, \\ \widetilde{\mathscr{I}}_{15}^\pm &= \sum_{\ell=1}^{\infty}\int_{\Gamma_h^\pm} \begin{pmatrix}1\\ \mathrm{i}\end{pmatrix} \cdot \mathbf{B}_{jp,2}^{(\ell)}(-1)^\ell(4\ell-1)\delta\eta(\mathbf{x}')\mathscr{R}(j_{2\ell-1}(k_p|\mathbf{x}|))u_1\mathrm{d}\sigma, \\ \widetilde{\mathscr{I}}_{16}^\pm &= \sum_{\ell=1}^{\infty}\int_{\Gamma_h^\pm} \begin{pmatrix}1\\ \mathrm{i}\end{pmatrix} \cdot \mathbf{B}_{js,2}^{(\ell)}(-1)^\ell(4\ell-1)\delta\eta(\mathbf{x}')\mathscr{R}(j_{2\ell-1}(k_s|\mathbf{x}|))u_1\mathrm{d}\sigma.\end{aligned}$$

By virtue of (10.2.46), (10.3.23), (10.3.38a), and (10.3.40), using the integral mean value theorem, one has as $s \to +\infty$

$$\begin{aligned}\left|\widetilde{\mathscr{I}}_{11}^{-}\right| &= \left|\int_{\Gamma_h^-}\begin{pmatrix}1\\ \mathrm{i}\end{pmatrix}\cdot\mathbf{v}_{jp}^{(1,2)}(\mathbf{0})\delta\eta(\mathbf{x}')\left[C(\phi)+C_1(\phi)\sum_{l=1}^{\infty}\frac{(-1)^l k_p^{2l}(|\mathbf{x}'|^2+a_{0,l}^2)^{l-\frac{1}{2}}}{2^l l!(2l+1)!!}|\mathbf{x}'|^2\right]u_1\mathrm{d}\sigma\right|\\
&\leq \sqrt{2}|C(\phi)|\|\eta\|_{C^\alpha}\left|\mathbf{v}_{jp}^{(1,2)}(\mathbf{0})\right|\left|\int_0^h r^\alpha\left[1+\sec^3\varpi\sum_{l=1}^{\infty}\frac{(-1)^l k_p^{2l}(|\mathbf{x}'|^2+a_{0,l}^2)^{l-\frac{1}{2}}}{2^l l!(2l+1)!!}|\mathbf{x}'|^2\right]u_1\mathrm{d}r\right|\\
&\leq \sqrt{2}|C(\phi)|\|\eta\|_{C^\alpha}\left|\mathbf{v}_{jp}^{(1,2)}(\mathbf{0})\right|\left|\int_0^h r^\alpha e^{-s\sqrt{r}\cos\frac{\theta}{2}}\mathrm{d}r+\sqrt{2}|C(\phi)|\|\eta\|_{C^\alpha}\right.\\
&\quad\times\left|\mathbf{v}_{jp}^{(1,2)}(\mathbf{0})\right|\left|\sec^3\varpi\sum_{l=1}^{\infty}\frac{(-1)^l k_p^{2l}(\beta_{0l}^2+a_{0,l}^2)^{l-\frac{1}{2}}}{2^l l!(2l+1)!!}\right|\int_0^h r^{\alpha+2}e^{-s\sqrt{r}\cos\frac{\theta}{2}}\mathrm{d}r\\
&= O(s^{-2\alpha-6}),\end{aligned}\tag{10.3.48}$$

where $\beta_{0l}\in(0,h)$ and $a_{0,l}\in(-L,L)$. Similarly, we can derive that

$$\left|\widetilde{\mathscr{I}}_{11}^{+}\right|\leq O(s^{-2\alpha-6}),\quad \left|\widetilde{\mathscr{I}}_{12}^{\pm}\right|\leq O(s^{-2\alpha-6})\quad \text{as } s\to+\infty.\tag{10.3.49}$$

From (10.3.7), using the Cauchy-Schwartz inequality, we have

$$\begin{aligned}&\left|\mathbf{B}_{jp,1}^{(\ell)}\right|\leq 2\sqrt{\pi}\|\mathbf{g}_{\mathrm{R},jp}^{(1,2)}\|_{L^2(\mathbb{S}^2)^2},\quad \left|\mathbf{B}_{js,1}^{(\ell)}\right|\leq 2\sqrt{\pi}\|\mathbf{g}_{\mathrm{R},js}^{(1,2)}\|_{L^2(\mathbb{S}^2)^2},\\
&\left|\mathbf{B}_{jp,2}^{(\ell)}\right|\leq 2\sqrt{\pi}\|\mathbf{g}_{\mathrm{I},jp}^{(1,2)}\|_{L^2(\mathbb{S}^2)^2},\quad \left|\mathbf{B}_{js,2}^{(\ell)}\right|\leq 2\sqrt{\pi}\|\mathbf{g}_{\mathrm{I},js}^{(1,2)}\|_{L^2(\mathbb{S}^2)^2}.\end{aligned}\tag{10.3.50}$$

From (10.2.46), (10.3.38a), and (10.3.50), we have

$$\begin{aligned}\left|\widetilde{\mathscr{I}}_{13}^{-}\right| &\leq |\sec^3\varpi|\,|C(\phi)|\,\|\eta\|_{C^\alpha}\sum_{\ell=1}^{\infty}|4\ell+1|\left|\begin{pmatrix}1\\ \mathrm{i}\end{pmatrix}\cdot\mathbf{B}_{jp,1}\right|\\
&\quad\times\left|\int_0^h r^{\alpha+2}\frac{k_p^{2\ell}(r^2+a_\ell^2)^{\ell-\frac{1}{2}}}{(4\ell+1)!!}\left[1-\sum_{l=1}^{\infty}\frac{k_p^{2l}(r^2+a_{\ell,l}^2)^l}{2^l l!N_{\ell,l,1}}\right]e^{-s\sqrt{r}\mu(\theta_m)}\mathrm{d}r\right|\\
&\leq \sqrt{2}|\sec^3\varpi|\,|C(\phi)|\,\|\eta\|_{C^\alpha}\sum_{\ell=1}^{\infty}|4\ell+1|\left|\mathbf{B}_{jp,1}\right|\\
&\quad\times\frac{k_p^{2\ell}(\beta_\ell^2+\alpha_\ell^2)^{\ell-\frac{1}{2}}}{(2\ell+1)!!}\left|1-\sum_{l=1}^{\infty}\frac{k_p^{2l}(\beta_{\ell,l}^2+\alpha_{\ell,l}^2)^l}{2^l l!N_{\ell,l}}\right|\int_0^h r^{\alpha+2}e^{-s\sqrt{r}\cos\frac{\theta}{2}}\mathrm{d}r\\
&\leq \|\mathbf{g}_{\mathrm{R},jp}^{(1,2)}\|_{L^2(\mathbb{S}^2)^2}\times O(s^{-2\alpha-6}),\end{aligned}\tag{10.3.51}$$

as $s\to+\infty$, where $\beta_{\ell l}$, $\beta_\ell\in(0,h)$ and $a_{\ell,l}\in(-L,L)$. Similarly, we can derive that

$$\left|\widetilde{\mathscr{I}}_{13}^{+}\right|\leq\|\mathbf{g}_{\mathrm{R},jp}^{(1,2)}\|_{L^2(\mathbb{S}^2)^2}\times O(s^{-2\alpha-6}),\quad \left|\widetilde{\mathscr{I}}_{14}^{\pm}\right|\leq\|\mathbf{g}_{\mathrm{R},js}^{(1,2)}\|_{L^2(\mathbb{S}^2)^2}\times O(s^{-2\alpha-6}),\tag{10.3.52}$$

as $s\to+\infty$. By (10.2.46) and (10.3.38a), we have

$$\begin{aligned}\left|\widetilde{\mathscr{I}}_{15}^{-}\right| &\leq |C(\phi)|\|\eta\|_{C^\alpha}\sum_{\ell=1}^{\infty}|4\ell-1|\left|\begin{pmatrix}1\\ \mathrm{i}\end{pmatrix}\cdot\mathbf{B}_{jp,2}^{(\ell)}\right|\\
&\quad\times\left|\int_0^h r^\alpha\frac{k_p^{2\ell-1}(r^2+a_\ell^2)^{\ell-1}}{(4\ell-1)!!}\left[1-\sum_{l=1}^{\infty}\frac{k_p^{2l}(r^2+a_{\ell,l}^2)^l}{2^l l!N_{\ell,l,1}}\right]e^{-s\sqrt{r}\mu(\theta_m)}\mathrm{d}r\right|\end{aligned}$$

$$\leq\sqrt{2}|C(\phi)|\|\eta\|_{C^\alpha}\sum_{\ell=1}^{\infty}|4\ell-1|\left|\mathbf{B}_{jp,2}^{(\ell)}\right|$$
$$\times\frac{k_p^{2\ell-1}(\beta_\ell^2+\alpha_\ell^2)^{\ell-1}}{(2\ell+1)!!}\left|1-\sum_{l=1}^{\infty}\frac{k_p^{2l}(\beta_{\ell,l}^2+a_{\ell,l}^2)^l}{2^l l! N_{\ell,l}}\right|\int_0^h r^{\alpha+2}e^{-s\sqrt{r}\cos\frac{\theta}{2}}\mathrm{d}r$$
$$\leq\|\mathbf{g}_{\mathrm{l},jp}^{(1,2)}\|_{L^2(\mathbb{S}^2)^2}\times O(s^{-2\alpha-6}), \tag{10.3.53}$$

as $s\to+\infty$, where $\beta_{\ell l},\beta_\ell\in(0,h)$ and $a_{\ell,l}\in(-L,L)$. Similarly, we can derive that

$$\left|\widetilde{\mathscr{I}}_{15}^{+}\right|\leq\|\mathbf{g}_{\mathrm{l},jp}^{(1,2)}\|_{L^2(\mathbb{S}^2)^2}\times O(s^{-2\alpha-6}),\quad \left|\widetilde{\mathscr{I}}_{16}^{\pm}\right|\leq\|\mathbf{g}_{\mathrm{l},js}^{(1,2)}\|_{L^2(\mathbb{S}^2)^2}\times O(s^{-2\alpha-6}), \tag{10.3.54}$$

as $s\to+\infty$. From (10.3.47), (10.3.48), (10.3.49), (10.3.51), (10.3.52), (10.3.53), and (10.3.54), we obtain (10.3.45). □

Lemma 10.3.11. *Consider the same setup in Lemma 10.3.4, and let $\widetilde{\mathscr{I}}_2^{\pm}$ be defined in* (10.3.44). *Let*

$$\widetilde{I}_{21,\beta}^{\pm}=\int_{\Gamma_h^{\pm}}\begin{pmatrix}1\\ \mathrm{i}\end{pmatrix}\cdot\mathbf{v}_{j\beta}^{(1,2)}(\mathbf{0})\mathscr{R}(j_0(k_\beta|\mathbf{x}|))u_1\mathrm{d}\sigma,$$
$$\widetilde{I}_{23,\beta}^{\pm}=\sum_{\ell=1}^{\infty}\int_{\Gamma_h^{\pm}}\begin{pmatrix}1\\ \mathrm{i}\end{pmatrix}\cdot\mathbf{B}_{j\beta,1}^{(\ell)}(-1)^\ell(4\ell+1)\mathscr{R}(j_{2\ell}(k_\beta|\mathbf{x}|))u_1\mathrm{d}\sigma,$$
$$\widetilde{I}_{25,\beta}^{\pm}=\sum_{\ell=1}^{\infty}\int_{\Gamma_h^{\pm}}\begin{pmatrix}1\\ \mathrm{i}\end{pmatrix}\cdot\mathbf{B}_{j\beta,2}^{(\ell)}(-1)^\ell(4\ell-1)\mathscr{R}(j_{2\ell-1}(k_\beta|\mathbf{x}|))u_1\mathrm{d}\sigma,$$

where $\mathbf{B}_{j\beta,i}^{(\ell)}$ ($\beta=p,s$, $i=1,2$) are defined in (10.3.15). *It holds that*

$$\widetilde{I}_2^{\pm}=\widetilde{I}_{21,p}^{\pm}+\widetilde{I}_{21,s}^{\pm}+\widetilde{I}_{23,p}^{\pm}+\widetilde{I}_{23,s}^{\pm}+\widetilde{I}_{25,p}^{\pm}+\widetilde{I}_{25,s}^{\pm}. \tag{10.3.55}$$

Proof. Plugging (10.3.46) into $\widetilde{I}_2^{\pm}$, one can readily obtain (10.3.55). □

Using Lemma 10.3.9 and the integral mean value theorem, after some tedious calculations, we obtain the following lemma.

Lemma 10.3.12. *Consider the same setup in Lemma 10.3.4, and recall that $\widetilde{I}_2^{\pm}$ is defined* (10.3.44). *Denote*

$$\widetilde{I}_{211}^{\pm}=C(\phi)\begin{pmatrix}1\\ \mathrm{i}\end{pmatrix}\cdot\mathbf{v}_{jp}^{(1,2)}(\mathbf{0})\int_{\Gamma_h^{\pm}}u_1\mathrm{d}\sigma,\quad \widetilde{I}_{221}^{\pm}=C(\phi)\begin{pmatrix}1\\ \mathrm{i}\end{pmatrix}\cdot\mathbf{v}_{js}^{(1,2)}(\mathbf{0})\int_{\Gamma_h^{\pm}}u_1\mathrm{d}\sigma,$$
$$\widetilde{I}_{212}^{\pm}=\begin{pmatrix}1\\ \mathrm{i}\end{pmatrix}\cdot\mathbf{v}_{jp}^{(1,2)}(\mathbf{0})\int_{\Gamma_h^{\pm}}C_1(\phi)\sum_{l=1}^{\infty}\frac{(-1)^l k_p^{2l}(|\mathbf{x}'|^2+a_{0,l}^2)^{l-\frac{1}{2}}}{2^l l!(2l+1)!!}|\mathbf{x}'|^2u_1\mathrm{d}\sigma,$$
$$\widetilde{I}_{222}^{\pm}=\begin{pmatrix}1\\ \mathrm{i}\end{pmatrix}\cdot\mathbf{v}_{js}^{(1,2)}(\mathbf{0})\int_{\Gamma_h^{\pm}}C_1(\phi)\sum_{l=1}^{\infty}\frac{(-1)^l k_s^{2l}(|\mathbf{x}'|^2+a_{0,l}^2)^{l-\frac{1}{2}}}{2^l l!(2l+1)!!}|\mathbf{x}'|^2u_1\mathrm{d}\sigma,$$

where $C(\phi)$ and $C_1(\phi)$ are defined in (10.3.38a) *and $a_{0,l}\in(-L,L)$. Then it holds that*

$$\widetilde{I}_{211}^{-}+\widetilde{I}_{211}^{+}+\widetilde{I}_{221}^{-}+\widetilde{I}_{221}^{+}=2s^{-2}\left(C(\phi)\begin{pmatrix}1\\ \mathrm{i}\end{pmatrix}\cdot\mathbf{v}_{jp}^{(1,2)}(\mathbf{0})+C(\phi)\begin{pmatrix}1\\ \mathrm{i}\end{pmatrix}\cdot\mathbf{v}_{js}^{(1,2)}(\mathbf{0})\right) \tag{10.3.56}$$
$$\Big(-s\sqrt{h}e^{-s\sqrt{h}\mu(\theta_m)}\mu^{-1}(\theta_m)-e^{-s\sqrt{h}\mu(\theta_m)}\mu^{-2}(\theta_m)+\mu^{-2}(\theta_m)$$

$$- s\sqrt{h}e^{-s\sqrt{h}\mu(\theta_M)}\mu^{-1}(\theta_M) - e^{-s\sqrt{h}\mu(\theta_M)}\mu^{-2}(\theta_M) + \mu^{-2}(\theta_M)\Big),$$

where

$$\mu(\theta) = \cos(\theta/2) + \mathrm{i}\sin(\theta/2). \tag{10.3.57}$$

Furthermore, the following estimates hold:

$$|\widetilde{I}_{212}^{\pm}| \le O(s^{-4}), \quad |\widetilde{I}_{222}^{\pm}| \le O(s^{-4}), \quad \left|\widetilde{I}_{23,\beta}^{\pm}\right| \le \|\mathbf{g}_{\mathrm{R}r,j\beta}^{(1,2)}\|_{L^2(\mathbb{S}^2)^2} \times O(s^{-2\alpha-6}),$$
$$\left|\widetilde{I}_{25,\beta}^{\pm}\right| \le \|\mathbf{g}_{\mathrm{I},j\beta}^{(1,2)}\|_{L^2(\mathbb{S}^2)^2} \times O(s^{-2\alpha-6}) \quad \text{as } s \to +\infty. \tag{10.3.58}$$

Proof. Substituting (10.3.38a) into $\widetilde{I}_{21,p}^{\pm}$ in (10.3.55), one can obtain that

$$\widetilde{I}_{21,p}^{\pm} = \widetilde{I}_{211}^{\pm} + \widetilde{I}_{212}^{\pm}, \tag{10.3.59}$$

where $\widetilde{I}_{211}^{\pm}$ and $\widetilde{I}_{212}^{\pm}$ are defined in (10.3.56).

By (10.3.57), it is directly seen that

$$\begin{aligned}\widetilde{I}_{211}^{-} =& C(\phi)\begin{pmatrix}1\\ \mathrm{i}\end{pmatrix}\cdot \mathbf{v}_{jp}^{(1,2)}(\mathbf{0})\int_{\Gamma_h^-} u_1 \mathrm{d}\sigma \\ =& C(\phi)\begin{pmatrix}1\\ \mathrm{i}\end{pmatrix}\cdot \mathbf{v}_{jp}^{(1,2)}(\mathbf{0})\frac{2}{s^2}\Big(-s\sqrt{h}\frac{e^{-s\sqrt{h}\mu(\theta_m)}}{\mu(\theta_m)} - \frac{e^{-s\sqrt{h}\mu(\theta_m)}}{\mu^2(\theta_m)} + \frac{1}{\mu^2(\theta_m)}\Big).\end{aligned} \tag{10.3.60}$$

Similarly, using (10.3.57), we have

$$\begin{aligned}\widetilde{I}_{211}^{+} &= C(\phi)\begin{pmatrix}1\\ \mathrm{i}\end{pmatrix}\cdot \mathbf{v}_{jp}^{(1,2)}(\mathbf{0})\frac{2}{s^2}\Big(-s\sqrt{h}\frac{e^{-s\sqrt{h}\mu(\theta_M)}}{\mu(\theta_M)} - \frac{e^{-s\sqrt{h}\mu(\theta_M)}}{\mu^2(\theta_M)} + \frac{1}{\mu^2(\theta_M)}\Big), \\ \widetilde{I}_{221}^{-} &= C(\phi)\begin{pmatrix}1\\ \mathrm{i}\end{pmatrix}\cdot \mathbf{v}_{js}^{(1,2)}(\mathbf{0})\frac{2}{s^2}\Big(-s\sqrt{h}\frac{e^{-s\sqrt{h}\mu(\theta_m)}}{\mu(\theta_m)} - \frac{e^{-s\sqrt{h}\mu(\theta_m)}}{\mu^2(\theta_m)} + \frac{1}{\mu^2(\theta_m)}\Big), \\ \widetilde{I}_{221}^{+} &= C(\phi)\begin{pmatrix}1\\ \mathrm{i}\end{pmatrix}\cdot \mathbf{v}_{js}^{(1,2)}(\mathbf{0})\frac{2}{s^2}\Big(-s\sqrt{h}\frac{e^{-s\sqrt{h}\mu(\theta_M)}}{\mu(\theta_M)} - \frac{e^{-s\sqrt{h}\mu(\theta_M)}}{\mu^2(\theta_M)} + \frac{1}{\mu^2(\theta_M)}\Big).\end{aligned} \tag{10.3.61}$$

Combining (10.3.60) with (10.3.61), we obtain (10.3.56).

For $\widetilde{I}_{212}^{-}$, using Lemma 10.2.10 and the integral mean value theorem, we can deduce that

$$\begin{aligned}|\widetilde{I}_{212}^{-}| \le& \left|\begin{pmatrix}1\\ \mathrm{i}\end{pmatrix}\cdot \mathbf{v}_{jp}^{(1,2)}(\mathbf{0})\int_{\Gamma_h^{\pm}} C_1(\phi)\sum_{l=1}^{\infty}\frac{(-1)^l k_p^{2l}(|\mathbf{x}'|^2 + a_{0,l}^2)^{l-\frac{1}{2}}}{2^l l!(2l+1)!!}|\mathbf{x}'|^2 u_1 \mathrm{d}\sigma\right| \\ \le& \sqrt{2}|\sec^3\varpi|\,|C(\phi)|\left|\mathbf{v}_{jp}^{(1,2)}(\mathbf{0})\right|\sum_{l=1}^{\infty}\frac{k_p^{2l}(|\beta_l|^2 + a_{0,l}^2)^{l-\frac{1}{2}}}{2^l l!(2l+1)!!}\int_0^h r^2 e^{-s\sqrt{r}\cos\frac{\theta}{2}}\mathrm{d}r \\ \le& O(s^{-4}) \quad \text{as } s \to +\infty,\end{aligned} \tag{10.3.62}$$

where $\beta_l \in (0,h)$, $a_{0,l} \in (-L,L)$ and $\varpi \in (-\arctan L/|\mathbf{x}'|, \arctan L/|\mathbf{x}'|)$ given by (10.3.40).

Using a similar argument for $\widetilde{I}_{212}^{-}$, we can derive that

$$|\widetilde{I}_{212}^{+}| \le O(s^{-4}), \quad |\widetilde{I}_{222}^{\pm}| \le O(s^{-4}) \quad \text{as } s \to +\infty. \tag{10.3.63}$$

By virtue of (10.2.46), (10.3.38b), and (10.3.50), one has

$$
\begin{aligned}
\left|\widetilde{I_{23,p}^{-}}\right| \leq & \sum_{\ell=1}^{\infty}\left|\begin{pmatrix}1\\ \mathrm{i}\end{pmatrix}\cdot\mathbf{B}_{jp}^{(\ell)}\right||4\ell+1|\int_0^h\frac{k_p^{2\ell}(\beta_\ell^2+a_\ell^2)^{\ell-\frac{1}{2}}}{(4\ell+1)!!}\\
&\times\left[1-\sum_{l=1}^{\infty}\frac{(-1)^l k_p^{2l}(\beta_{\ell,l}^2+a_{\ell,l}^2)^l}{2^l l! N_{\ell,l,1}}\right]|\sec^3\varpi||C(\phi)|\int_0^h r^2 e^{-s\sqrt{r}\cos\frac{\theta_m}{2}}\mathrm{d}r \\
\leq & \|\mathbf{g}_{\mathrm{R},jp}^{(1,2)}\|_{L^2(\mathbb{S}^2)^2}\times O(s^{-2\alpha-6}) \quad \text{as } s\to+\infty,
\end{aligned}
\tag{10.3.64}
$$

where $\beta_\ell, \beta_{\ell,l}\in(0,h)$. Please be noted that according to Remark 10.3.1, $|\sec^3\varpi|$ is uniformly bounded with respect to $|\mathbf{x}'|$. Similarly, using (10.2.46), (10.3.38b), and (10.3.50), we obtain that

$$
\left|\widetilde{I_{23,p}^{+}}\right| \leq \|\mathbf{g}_{\mathrm{R},jp}^{(1,2)}\|_{L^2(\mathbb{S}^2)^2}\times O(s^{-2\alpha-6}),\ \left|\widetilde{I_{23,s}^{\pm}}\right| \leq \|\mathbf{g}_{\mathrm{R},js}^{(1,2)}\|_{L^2(\mathbb{S}^2)^2}\times O(s^{-2\alpha-6}) \text{ as } s\to+\infty.
\tag{10.3.65}
$$

By virtue of (10.2.46), (10.3.38c), and (10.3.50), one has

$$
\begin{aligned}
\left|\widetilde{I_{25,p}^{-}}\right| \leq & \left|\sum_{\ell=1}^{\infty}\int_{\Gamma_h^-}\begin{pmatrix}1\\ \mathrm{i}\end{pmatrix}\cdot\mathbf{B}_{jp,2}^{(\ell)}(-1)^\ell(4\ell-1)\mathscr{R}(j_\ell(k_p|\mathbf{x}|))u_1\mathrm{d}\sigma\right|\\
\leq & \sum_{\ell=1}^{\infty}\left|\begin{pmatrix}1\\ \mathrm{i}\end{pmatrix}\cdot\mathbf{B}_{jp,2}^{(\ell)}\right||4\ell-1|\int_0^h\frac{k_p^{2\ell-1}(\beta_\ell^2+a_\ell^2)^{\ell-\frac{3}{2}}}{(4\ell+1)!!}\\
&\times\left[1-\sum_{l=1}^{\infty}\frac{(-1)^l k_p^{2l}(\beta_{\ell,l}^2+a_{\ell,l}^2)^l}{2^l l! N_{\ell,l,2}}\right]|\sec^3\varpi||C(\phi)|\int_0^h r^2 e^{-s\sqrt{r}\cos\frac{\theta_m}{2}}\mathrm{d}r \\
= & \|\mathbf{g}_{\mathrm{I},jp}^{(1,2)}\|_{L^2(\mathbb{S}^2)^2}\times O(s^{-2\alpha-6}) \quad \text{as } s\to+\infty,
\end{aligned}
\tag{10.3.66}
$$

where $\beta_\ell, \beta_{\ell,l}\in(0,h)$. Again according to (8.2.10), (10.3.38c), and (10.3.50), we can conclude that

$$
\left|\widetilde{I_{25,p}^{+}}\right| \leq \|\mathbf{g}_{\mathrm{I},jp}^{(1,2)}\|_{L^2(\mathbb{S}^2)^2}\times O(s^{-2\alpha-6}),\ \left|\widetilde{I_{25,s}^{+}}\right| \leq \|\mathbf{g}_{\mathrm{I},js}^{(1,2)}\|_{L^2(\mathbb{S}^2)^2}\times O(s^{-2\alpha-6}),
\tag{10.3.67}
$$

as $s\to+\infty$.

In view of (10.3.62)–(10.3.67), we derive (10.3.58). □

By virtue of Lemma 10.3.3, using a similar argument as that for Lemma 10.3.7 and [23, Theorem 3.1], we can prove the following lemma.

Lemma 10.3.13. *Consider the same setup in Lemma 10.3.2, and suppose that* $\mathbf{v}_{\mathrm{R}}$ *can be approximated by a sequence of the functions* $\{\mathbf{v}_j^{\mathrm{R}}\}_{j=1}^{+\infty}$ *defined by* (10.3.3) *in* $H^1(S_h\times(-M,M))^3$ *satisfying* (10.3.22). *If* $\lambda\neq 0$, $\eta(\mathbf{0})\neq 0$, *and* $-\pi<\theta_m<\theta_M<\pi$ *satisfying* $\theta_M-\theta_m\neq\pi$, *then we have*

$$
\lim_{j\to+\infty} v_j^{(3)}(\mathbf{0}) = 0.
\tag{10.3.68}
$$

Proof. From Lemma 10.3.3, we know that $(v_3, w_3)\in H^1(S_h\times(-M,M))$ fulfills (10.3.14). Since (10.3.27) and (10.3.28) are satisfied, using a similar argument as that for Lemma 10.3.7, we can prove that $\mathbf{G}_1^{(3)}(\mathbf{x}')\in C^\alpha(\overline{S}_h)$ and $\mathbf{G}_2^{(3)}(\mathbf{x}')\in C^\alpha(\overline{S}_h)$ $(0<\alpha<1)$, where $\mathbf{G}_1^{(3)}(\mathbf{x}')$ and $\mathbf{G}_2^{(3)}(\mathbf{x}')$ are the RHS terms of (10.3.14). According to (10.3.22), one has

$$
\|v_3 - v_j^{(3)}\|_{H^1(S_h\times(-M,M))} \leq j^{-\gamma}.
\tag{10.3.69}
$$

By virtue of (10.3.69), under the assumption $\theta_M-\theta_m\neq\pi$ and $\eta(\mathbf{0})\neq 0$, adopting a similar argument for [23, Theorem 3.1], we can prove (10.3.68). □

Theorem 10.3.1. *Let Ω, $S_h \times (-M,M)$ be described above and $\alpha \in (0,1)$. For any fixed $x_3^c \in (-M,M)$ and $L > 0$ defined in Definition 10.3.1, we suppose that L is sufficiently small such that $(x_3^c - L, x_3^c + L) \subset (-M,M)$ and*

$$(B_h \times (-M,M)) \cap \Omega = S_h \times (-M,M),$$

where $B_h \Subset \mathbb{R}^2$ is the central ball of radius $h \in \mathbb{R}_+$. Assume that $\mathbf{v},\mathbf{w} \in H^1(\Omega)^3$ are the generalized elastic transmission eigenfunctions to (10.1.11), where the Lamé constant $\lambda \neq 0$. Assume further that $q\mathbf{w} \in C^\alpha(\overline{S}_h \times [-M,M])^3$ and $\mathbf{v}-\mathbf{w} \in C^{1,\alpha}(\overline{S}_h \times [-M,M])^3$, where $\mathbf{0}$ is the vertex of S_h defined in (10.2.1). Write $\mathbf{x} = (\mathbf{x}', x_3) \in \mathbb{R}^3$, $\mathbf{x}' \in \mathbb{R}^2$. If the following conditions are fulfilled:

(a) *$\mathbf{v}$ can be approximated in $H^1(S_h \times (-M,M))^3$ by the Herglotz functions $\mathbf{v}_j, j = 1,2,\ldots,$ with kernels $\mathbf{g}_{jp}$ and $\mathbf{g}_{js}$ satisfying* (10.3.22).

(b) *The function $\eta = \eta(\mathbf{x}')$ is independent of x_3 and*

$$\eta(\mathbf{0}) \neq 0. \tag{10.3.70}$$

(c) *the angles θ_m and θ_M of the sector S_h satisfy*

$$-\pi < \theta_m < \theta_M < \pi \text{ and } \theta_M - \theta_m \neq \pi. \tag{10.3.71}$$

then for every edge point $(\mathbf{0}, x_3^c) \in \mathbb{R}^3$ of $S_h \times (-M,M)$, namely, $x_3^c \in (-M,M)$, one has

$$\lim_{\rho \to +0} \frac{1}{m(B((\mathbf{0},x_3^c),\rho) \cap \Omega)} \int_{m(B((\mathbf{0},x_3^c),\rho)\cap\Omega)} |\mathbf{v}(\mathbf{x})| \mathrm{d}\mathbf{x} = 0, \tag{10.3.72}$$

where $m(B((\mathbf{0},x_3^c),\rho) \cap \Omega)$ is the measure of $B((\mathbf{0},x_3^c),\rho) \cap \Omega$.

Remark 10.3.2. Similar to Remark 10.2.1 in the 2D case, a more general Fourier extension property (cf. (10.2.64)) can be proposed to study the vanishing property in Theorem 10.3.1. However, this will involve rather lengthy and complicate analysis, and we choose not to explore more along that direction in this section.

Proof of Theorem 10.3.1. As remarked earlier, it is sufficient for us to consider $\mathbf{v}_\mathrm{R}$ and $\mathbf{w}_\mathrm{R}$ and prove that (10.3.72) holds for $\mathbf{v}_\mathrm{R}$. Since $q\mathbf{w} \in C^\alpha(\overline{S}_h \times [-M,M])^3$, it is easy to see that $q\mathbf{w}_\mathrm{R} \in C^\alpha(\overline{S}_h \times [-M,M])^3$. Similarly, we know that $\mathbf{v}_\mathrm{R} - \mathbf{w}_\mathrm{R} \in C^{1,\alpha}(\overline{S}_h \times [-M,M])^3$ under the assumption $\mathbf{v}-\mathbf{w} \in C^{1,\alpha}(\overline{S}_h \times [-M,M])^3$.

Since the Herglotz functions $\mathbf{v}_j, j = 1,2,\ldots,$ defined in (10.3.3) with kernels $\mathbf{g}_{jp}$ and $\mathbf{g}_{js}$ can approximate $\mathbf{v}$ under the condition (10.3.22), we see that $\mathbf{v}_\mathrm{R} \in H^1(S_h \times (-M,M))^3$ can be approximated by $\{\mathbf{v}_j^\mathrm{R}\}_{j=1}^{+\infty}$ defined in (10.3.3) satisfying (10.3.22). Therefore, the assumptions in Lemmas 10.3.4–10.3.12 are fulfilled.

We divide the proof into two parts.

Part I. First, we shall prove that

$$\lim_{j \to +\infty} \mathbf{v}_{jp}^{(1,2)}(\mathbf{0}) + \mathbf{v}_{js}^{(1,2)}(\mathbf{0}) = \mathbf{0}, \tag{10.3.73}$$

where $\mathbf{v}_{jp}^{(1,2)}(\mathbf{0})$ and $\mathbf{v}_{js}^{(1,2)}(\mathbf{0})$ are defined in (10.3.15).

By Lemmas 10.3.2 and 10.3.3, we consider the PDE system (10.3.13). Recall that $C(\phi)$ and $\mu(\theta)$ are defined in (10.3.39) and (10.3.57), respectively. Substituting (10.3.44) and (10.3.55) into (10.3.16), after arranging terms, we have

$$
\begin{aligned}
&\eta(\mathbf{0})2s^{-2}\left(C(\phi)\begin{pmatrix}1\\ \mathrm{i}\end{pmatrix}\cdot\mathbf{v}_{jp}^{(1,2)}(\mathbf{0})+C(\phi)\begin{pmatrix}1\\ \mathrm{i}\end{pmatrix}\cdot\mathbf{v}_{js}^{(1,2)}(\mathbf{0})\right)\\
&\Big(-s\sqrt{h}e^{-s\sqrt{h}\mu(\theta_m)}\mu^{-1}(\theta_m)-e^{-s\sqrt{h}\mu(\theta_m)}\mu^{-2}(\theta_m)+\mu^{-2}(\theta_m)\\
&\quad -s\sqrt{h}e^{-s\sqrt{h}\mu(\theta_M)}\mu^{-1}(\theta_M)-e^{-s\sqrt{h}\mu(\theta_M)}\mu^{-2}(\theta_M)+\mu^{-2}(\theta_M)\Big),\\
&=-\eta(\mathbf{0})(\widetilde{\mathscr{I}}_{23,p}^{\pm}+\widetilde{\mathscr{I}}_{23,s}^{\pm}+\mathscr{I}_{25,p}^{\pm}+\mathscr{I}_{25,s}^{\pm}+\widetilde{\mathscr{I}}_{211}^{\pm}+\widetilde{\mathscr{I}}_{212}^{\pm}+\widetilde{\mathscr{I}}_{221}^{\pm}+\widetilde{\mathscr{I}}_{222}^{\pm})\\
&\quad-\widetilde{\mathscr{I}}_{1}^{\pm}+\widetilde{I}_{\Lambda_h}-\widetilde{I}_1-\widetilde{I}_2-\widetilde{I}_{\pm}^{\Delta},
\end{aligned}
\tag{10.3.74}
$$

where $\widetilde{\mathscr{I}}_{2\ell,\beta}^{\pm}$ ($\ell=3,5$, $\beta=p,s$), $\widetilde{\mathscr{I}}_{21\ell}^{\pm}$ ($\ell=1,2$), $\widetilde{\mathscr{I}}_{22\ell}^{\pm}$ ($\ell=1,2$), $\widetilde{\mathscr{I}}_{1}^{\pm}$, $\widetilde{I}_{\Lambda_h}$, $\widetilde{I}_\ell$ ($\ell=1,2$), $\widetilde{I}_{\pm}^{\Delta}$ are defined in (10.3.56), (10.3.55), (10.3.44), and (10.3.17), respectively.

Multiplying s^2 on both sides of (10.3.74), by virtue of (10.3.21), (10.3.24), (10.3.29), (10.3.45), and (10.3.56), and letting $s=j^{\rho/2}$ ($\max\{\beta_1,\beta_2\}<\rho<\gamma$) with $j\to+\infty$, we have

$$
\lim_{j\to\infty} C(\phi)\eta(\mathbf{0})\left(\mathbf{v}_{jp}^{(1,2)}(\mathbf{0})+\mathbf{v}_{js}^{(1,2)}(\mathbf{0})\right)\cdot\begin{pmatrix}1\\ \mathrm{i}\end{pmatrix}\left(\mu^{-2}(\theta_m)+\mu^{-2}(\theta_M)\right)=0.
$$

Recalling the definition of ϕ in Definition 10.3.1, we know that $C(\phi)>0$. Under the condition (10.3.71), from [23, Lemma 2.10], we know that

$$
\mu^{-2}(\theta_m)+\mu^{-2}(\theta_M)\neq 0.
$$

Under the condition (10.3.70), since η is a real-valued function, we have

$$
\lim_{j\to\infty}\left(\mathbf{v}_{jp}^{(1,2)}(\mathbf{0})+\mathbf{v}_{js}^{(1,2)}(\mathbf{0})\right)\cdot\begin{pmatrix}1\\ \mathrm{i}\end{pmatrix}=0,
$$

which readily implies (10.3.73) by noting that $\mathbf{v}_{jp}^{(1,2)}(\mathbf{0})+\mathbf{v}_{js}^{(1,2)}(\mathbf{0})$ is a real vector.

Part II. By Lemma 10.3.13, we have (10.3.68). Combining (10.3.73) with (10.3.68), by using the definition of $\mathbf{v}_j^{\mathrm{R}}(\mathbf{0})$ in (10.3.6), we can prove that

$$
\lim_{j\to\infty}\mathbf{v}_j^{\mathrm{R}}(\mathbf{0})=\mathbf{0}.
$$

Using a similar argument to that for (10.2.63), we can finish the proof of this theorem. □

We next consider the degenerate case of Theorem 10.3.1 with $\eta\equiv 0$. Indeed, the proof follows from the one of Theorem 10.3.1 with some necessary modifications by noting $\eta\equiv 0$ in the integral equality (10.3.16).

Corollary 10.3.1. *Consider the same setup in Theorem 10.3.1 with $\eta\equiv 0$. If the following conditions are fulfilled:*

(a) *For any given constants $\gamma>\max\{\beta_1,\beta_2\}>0$, there exists a sequence of Herglotz functions $\mathbf{v}_j, j=1,2,\ldots$, with kernels $\mathbf{g}_{jp}$ and $\mathbf{g}_{js}$ which can approximate $\mathbf{v}$ in $H^1(S_h\times(-M,M))$ satisfying*

$$
\|\mathbf{v}-\mathbf{v}_j\|_{H^1(S_h\times(-M,M))^3}\le j^{-\gamma},\ \|\mathbf{g}_{jp}\|_{L^2(\mathbb{S}^2)^3}\le j^{\beta_1}\ \text{and}\ \|\mathbf{g}_{js}\|_{L^2(\mathbb{S}^2)^3}\le j^{\beta_2}. \tag{10.3.75}
$$

(b) *The angles θ_m and θ_M of the sector W satisfy*

$$
-\pi<\theta_m<\theta_M<\pi\quad\text{and}\quad\theta_M-\theta_m\neq\pi. \tag{10.3.76}
$$

then for every edge point $(\mathbf{0}, x_3^c) \in \mathbb{R}^3$ of $S_h \times (-M,M)$, namely, $x_3^c \in (-M,M)$, one has

$$\lim_{\rho \to +0} \frac{1}{m(B((\mathbf{0},x_3^c),\rho) \cap P(\Omega))} \int_{B((\mathbf{0},x_3^c),\rho) \cap P(\Omega))} \mathscr{R}(V\mathbf{w})(\mathbf{x}')\mathrm{d}\mathbf{x}' = \mathbf{0}, \tag{10.3.77}$$

where $P(\Omega)$ is the projection set of Ω on $\mathbb{R}^2$.

Proof. The proof follows from the one of Theorem 10.3.1 with some necessary modifications. It is sufficient for us to show that (10.3.77) holds for $\mathbf{w}_\mathrm{R}$. Clearly, we have

$$q\mathbf{w}_\mathrm{R} \in C^\alpha(\overline{S_h})^2, \quad \mathbf{v}_\mathrm{R} - \mathbf{w}_\mathrm{R} \in C^\alpha(\overline{S}_h \times [-M,M])^3, \tag{10.3.78}$$

and the condition (10.3.75) gives

$$\|\mathbf{v}^\mathrm{R} - \mathbf{v}_j^\mathrm{R}\|_{H^1(S_h \times (-M,M))^3} \leq j^{-\gamma}, \|\mathbf{g}_{jp}\|_{L^2(\mathbb{S}^2)^3} \leq j^{\beta_1} \text{ and } \|\mathbf{g}_{js}\|_{L^2(\mathbb{S}^2)^3} \leq j^{\beta_2},$$

for any given constants $\gamma > \max\{\beta_1, \beta_2\} > 0$.

According to the proof of Theorem 10.3.1, we divide the proof into two parts.

Part I. First, let $\mathbf{v}^{(1,2)}$ and $\mathbf{w}^{(1,2)}$ be defined in (10.3.10), which fulfill (10.3.13) with $\eta \equiv 0$. From (10.3.16), it follows that

$$\widetilde{I}_1 + \widetilde{I}_2 = \widetilde{I}_{\Lambda_h}, \tag{10.3.79}$$

where $\widetilde{I}_1$, $\widetilde{I}_2$, and $\widetilde{I}_{\Lambda_h}$ are defined in (10.3.17). From (10.2.16), one can see that

$$\begin{aligned} \int_{S_h} \mathbf{u}(\mathbf{x}')\mathrm{d}\mathbf{x}' &= \int_W \mathbf{u}(\mathbf{x}')\mathrm{d}\mathbf{x}' + \int_{W \setminus S_h} \mathbf{u}(\mathbf{x}')\mathrm{d}\mathbf{x}' \\ &= 6\mathrm{i}(e^{-2\theta_M \mathrm{i}} - e^{-2\theta_m i})s^{-4} \begin{pmatrix} 1 \\ \mathrm{i} \end{pmatrix} + \int_{W \setminus S_h} \mathbf{u}(\mathbf{x}')\mathrm{d}\mathbf{x}'. \end{aligned} \tag{10.3.80}$$

Under the condition (10.3.78), by Lemma 10.3.7, we know that $\mathbf{f}_2(\mathbf{x}')$, $\mathbf{f}_3(\mathbf{x}')$, and $\mathbf{f}_4(\mathbf{x}')$ have the expansion (10.3.30), where $\mathbf{f}_\ell(\mathbf{x}')$ is defined in (10.3.17). Therefore, $\widetilde{I}_1$ can be rewritten as (10.3.32).

Since $\mathbf{v}_\mathrm{R} = \mathbf{w}_\mathrm{R}$ on $\Gamma_h^\pm \times (-M,M)$, recalling the definitions of $\mathbf{f}_3$ and $\mathbf{f}_4$ in (10.3.17), it is easy to see that

$$\begin{aligned} \mathbf{f}_3(\mathbf{0}) &= -\int_{-L}^{L} \phi''(x_3) \begin{pmatrix} \lambda(v_1(\mathbf{0},x_3) - w_1(\mathbf{0},x_3)) \\ \lambda(v_2(\mathbf{0},x_3) - w_2(\mathbf{0},x_3)) \end{pmatrix} (\mathbf{0},x_3)\mathrm{d}x_3 = \mathbf{0}, \\ \mathbf{f}_4(\mathbf{0}) &= (\lambda + \mu) \int_{-L}^{L} \phi'(x_3) \begin{pmatrix} \partial_1(v_3(\mathbf{0},x_3) - w_3(\mathbf{0},x_3)) \\ \partial_2(v_3(\mathbf{0},x_3) - w_3(\mathbf{0},x_3)) \end{pmatrix} (\mathbf{0},x_3)\mathrm{d}x_3 = \mathbf{0}. \end{aligned} \tag{10.3.81}$$

By virtue of (10.3.81), substituting (10.3.32) and (10.3.80) into (10.3.79), we can deduce that

$$\begin{aligned} & 6\mathrm{i}(e^{-2\theta_M \mathrm{i}} - e^{-2\theta_m i})s^{-4} \begin{pmatrix} 1 \\ \mathrm{i} \end{pmatrix} \cdot (\mathbf{f}_{1j}(\mathbf{0}) + \mathbf{f}_2(\mathbf{0})) \\ &= -\widetilde{I}_2 + \widetilde{I}_{\Lambda_h} - (\mathbf{f}_{1j}(\mathbf{0}) + \mathbf{f}_2(\mathbf{0})) \int_{W \setminus S_h} \mathbf{u}(\mathbf{x}')\mathrm{d}\mathbf{x}' - \int_{S_h} \delta \mathbf{f}_{1j} \cdot \mathbf{u}(\mathbf{x}')\mathrm{d}\mathbf{x}' \\ &\quad - \int_{S_h} \delta \mathbf{f}_2 \cdot \mathbf{u}(\mathbf{x}')\mathrm{d}\mathbf{x}' - \int_{S_h} \delta \mathbf{f}_3 \cdot \mathbf{u}(\mathbf{x}')\mathrm{d}\mathbf{x}' - \int_{S_h} \delta \mathbf{f}_4 \cdot \mathbf{u}(\mathbf{x}')\mathrm{d}\mathbf{x}'. \end{aligned} \tag{10.3.82}$$

In (10.3.82), we take

$$s = j^{\rho/4}, \quad \max\{\beta_1, \beta_2\} < \rho < \gamma, \tag{10.3.83}$$

and multiply j^ρ on both sides of (10.3.82), and then it yields that

$$
\begin{aligned}
&6\mathrm{i}(e^{-2\theta_M \mathrm{i}} - e^{-2\theta_m i}) \begin{pmatrix} 1 \\ \mathrm{i} \end{pmatrix} \cdot (\mathbf{f}_{1j}(\mathbf{0}) + \mathbf{f}_2(\mathbf{0})) \\
&= j^{\rho} \Big(-\widetilde{I}_2 + \widetilde{I}_{\Lambda_h} - (\mathbf{f}_{1j}(\mathbf{0}) + \mathbf{f}_2(\mathbf{0})) \int_{W \setminus S_h} \mathbf{u}(\mathbf{x}')\mathrm{d}\mathbf{x}' - \int_{S_h} \delta \mathbf{f}_{1j} \cdot \mathbf{u}(\mathbf{x}')\mathrm{d}\mathbf{x}' \\
&\quad - \int_{S_h} \delta \mathbf{f}_2 \cdot \mathbf{u}(\mathbf{x}')\mathrm{d}\mathbf{x}' - \int_{S_h} \delta \mathbf{f}_3 \cdot \mathbf{u}(\mathbf{x}')\mathrm{d}\mathbf{x}' - \int_{S_h} \delta \mathbf{f}_4 \cdot \mathbf{u}(\mathbf{x}')\mathrm{d}\mathbf{x}' \Big).
\end{aligned} \tag{10.3.84}
$$

By virtue of (10.3.24) in Lemma 10.3.6, we have

$$
j^{\rho} | -\widetilde{I}_2 | \le h\sqrt{2L(\theta_M - \theta_m)} e^{-s\sqrt{\Theta}\delta_W} \|\phi\|_{L^\infty} j^{-\gamma+\rho}. \tag{10.3.85}
$$

Under the assumption (10.3.75), in (10.3.34), (10.3.35), (10.3.36), and (10.3.37), in view of (10.3.83), we can obtain the following estimates:

$$
\begin{aligned}
j^{\rho} \left| \int_{S_h} \delta \mathbf{f}_{1j} \cdot \mathbf{u}(\mathbf{x}')\mathrm{d}\mathbf{x}' \right| &\le \omega^2 \mathrm{diam}(S_h)^{1-\alpha} 4L\sqrt{\pi} \|\phi\|_{L^\infty} \Big(2(1+k_p) j^{\beta_1 - \frac{1}{2}\alpha\rho - \rho} \\
&\quad + 2(1+k_s) j^{\beta_2 - \frac{1}{2}\alpha\rho - \rho} \Big), \\
j^{\rho} \left| \int_{S_h} \delta \mathbf{f}_2 \cdot \mathbf{u}(\mathbf{x}')\mathrm{d}\mathbf{x}' \right| &\le \|\mathbf{f}_2\|_{C^\alpha(S_h)^2} \frac{2\sqrt{2}(\theta_M - \theta_m)\Gamma(2\alpha+4)}{\delta_W^{2\alpha+4}} j^{-\frac{1}{2}\alpha\rho - \rho}, \\
j^{\rho} \left| \int_{S_h} \delta \mathbf{f}_3 \cdot \mathbf{u}(\mathbf{x}')\mathrm{d}\mathbf{x}' \right| &\le \|\mathbf{f}_3\|_{C^\alpha(S_h)^2} \frac{2\sqrt{2}(\theta_M - \theta_m)\Gamma(2\alpha+4)}{\delta_W^{2\alpha+4}} j^{-\frac{1}{2}\alpha\rho - \rho}, \\
j^{\rho} \left| \int_{S_h} \delta \mathbf{f}_4 \cdot \mathbf{u}(\mathbf{x}')\mathrm{d}\mathbf{x}' \right| &\le \|\mathbf{f}_4\|_{C^\alpha(S_h)^2} \frac{2\sqrt{2}(\theta_M - \theta_m)\Gamma(2\alpha+4)}{\delta_W^{2\alpha+4}} j^{-\frac{1}{2}\alpha\rho - \rho}.
\end{aligned} \tag{10.3.86}
$$

Under the assumption (10.3.76), it is easy to see that

$$
\left| e^{-2\theta_M \mathrm{i}} - e^{-2\theta_m \mathrm{i}} \right| = \left| 1 - e^{-2(\theta_M - \theta_m)\mathrm{i}} \right| \neq 0,
$$

since $\theta_M - \theta_m \neq \pi$. Letting $j \to \infty$, by virtue of (10.2.18), (10.3.21), (10.3.85), and (10.3.86), we have

$$
\lim_{j \to \infty} \mathbf{f}_{1j}(\mathbf{0}) = -\mathbf{f}_2(\mathbf{0}).
$$

Since $\mathbf{f}_{1j}(\mathbf{x}') = -\omega^2 \mathscr{R}(\mathbf{v}_j^{(1,2)})(\mathbf{x}')$ and $\mathbf{f}_2(\mathbf{x}') = \omega^2 \mathscr{R}(q\mathbf{w}^{(1,2)})(\mathbf{x}')$, we obtain that

$$
\lim_{j \to \infty} \mathscr{R}(\mathbf{v}_j^{(1,2)})(\mathbf{0}) = \mathscr{R}(q\mathbf{w}^{(1,2)})(\mathbf{0}). \tag{10.3.87}
$$

From the boundary condition in (10.3.13) for $\eta \equiv 0$, we have

$$
\begin{aligned}
&\lim_{\rho \to +0} \frac{1}{m(B(\mathbf{0},\rho) \cap P(\Omega))} \int_{B(\mathbf{0},\rho) \cap P(\Omega)} \mathscr{R}(\mathbf{v}_j^{(1,2)})(\mathbf{x}')\mathrm{d}\mathbf{x}' \\
&= \lim_{\rho \to +0} \frac{1}{m(B(\mathbf{0},\rho) \cap P(\Omega))} \int_{B(\mathbf{0},\rho) \cap P(\Omega)} \mathscr{R}(\mathbf{w}^{(1,2)})(\mathbf{x}')\mathrm{d}\mathbf{x}'.
\end{aligned} \tag{10.3.88}
$$

Furthermore, it yields that

$$\lim_{j\to\infty}\mathscr{R}(\mathbf{v}_j^{(1,2)})(\mathbf{0}) = \lim_{j\to\infty}\lim_{\rho\to+0}\frac{1}{m(B(\mathbf{0},\rho))\cap P(\Omega)}\int_{B(\mathbf{0},\rho)\cap P(\Omega)}\mathscr{R}(\mathbf{v}_j^{(1,2)})(\mathbf{x}')\mathrm{d}\mathbf{x}'$$
$$= \lim_{\rho\to+0}\frac{1}{m(B(\mathbf{0},\rho))\cap P(\Omega)}\int_{B(\mathbf{0},\rho)\cap P(\Omega)}\mathscr{R}(\mathbf{v}^{(1,2)})(\mathbf{x}')\mathrm{d}\mathbf{x}', \tag{10.3.89}$$
$$\mathscr{R}(q\mathbf{w}^{(1,2)})(\mathbf{0}) = \lim_{\rho\to+0}\frac{1}{m(B(\mathbf{0},\rho))\cap P(\Omega)}\int_{B(\mathbf{0},\rho)\cap P(\Omega)}\mathscr{R}(q\mathbf{w}^{(1,2)})(\mathbf{x}')\mathrm{d}\mathbf{x}'.$$

Due to (10.3.87), combining (10.3.88) with (10.3.89), we can prove that

$$\lim_{\rho\to+0}\frac{1}{m(B(\mathbf{0},\rho)\cap P(\Omega))}\int_{B(\mathbf{0},\rho)\cap P(\Omega)}\mathscr{R}(V\mathbf{w}^{(1,2)})(\mathbf{x}')\mathrm{d}\mathbf{x}' = 0. \tag{10.3.90}$$

Part II. Since the Lamé constant $\lambda \neq 0$, similar to Lemma 10.3.13, consider the PDE system (10.3.14) for $\eta \equiv 0$, under the conditions (10.3.75) and (10.3.76), using a similar argument of [23, Corollary 3.1], we can prove that

$$\lim_{\rho\to+0}\frac{1}{m(B(\mathbf{0},\rho)\cap P(\Omega))}\int_{B(\mathbf{0},\rho)\cap P(\Omega)}\mathscr{R}(Vw_3)(\mathbf{x}')\mathrm{d}\mathbf{x}' = 0, \tag{10.3.91}$$

where w_3 is the third component of $\mathbf{w}_\mathrm{R}$. According to (10.3.90) and (10.3.91), we finish the proof of this corollary. □

Remark 10.3.3. According to Corollary 10.3.1, the average value of each component of the function $V\mathbf{w}$ over the cylinder centered at the edge point $(\mathbf{0},x_3^c)$ with the height L vanishes in the distribution sense. Moreover, if we assume that $V(\mathbf{x}',x_3)$ is continuous near the edge point $(\mathbf{0},x_3^c)$ where $x_3^c \in (-M,M)$ and $V(\mathbf{0},x_3^c) \neq 0$, then by the dominant convergence theorem and the definition of $\mathscr{R}$, one can show that

$$\lim_{\rho\to+0}\frac{1}{m(B(\mathbf{0},\rho)\cap P(\Omega))}\int_{B(\mathbf{0},\rho)\cap P(\Omega)}\left(\int_{x_3^c-L}^{x_3^c+L}\phi(x_3)\mathbf{w}(\mathbf{x}',x_3)\mathrm{d}x_3\right)\mathrm{d}\mathbf{x}' = \mathbf{0}. \tag{10.3.92}$$

under the assumptions in Corollary 10.3.1. Since $\mathbf{w}\in H^1(\Omega)^3$, it can be readily seen that

$$\lim_{\rho\to+0}\frac{1}{m(B(\mathbf{0},\rho)\cap P(\Omega))}\int_{B(\mathbf{0},\rho)\cap P(\Omega)}\mathbf{w}(\mathbf{x}',x_3)\mathrm{d}\mathbf{x}' = \mathbf{0},\quad \forall x_3\in(-M,M),$$

which also describes the vanishing property of the interior elastic transmission eigenfunctions $\mathbf{v}$ and $\mathbf{w}$ near the edge point in 3D. We would like to point out that $\mathbf{v}$ and $\mathbf{w}$ must vanish at an edger corner point $(\mathbf{0},x_3^c)$ with $x_3^c\in(-M,M)$ if V and either one of $\mathbf{v}$, $\mathbf{w}$ are C^α smooth near the corner, $\mathbf{v}-\mathbf{w}\in H^2(\Omega)^3$, $V((\mathbf{0},x_3^c))\neq 0$ (cf. [8, Theorem 1.5]). Indeed, the assumption that V and $\mathbf{w}$ are C^α smooth near the corner can imply $q\mathbf{w}$ is C^α smooth near the corner. Comparing Corollary 10.3.1 with [8, Theorem 1.5], we remove the assumption $\mathbf{v}-\mathbf{w}\in H^2(\Omega)^3$ and give (10.3.92).

Similar to Corollary 10.3.1, if we assume that $\mathbf{v}$ has the C^α regularity near a 3D edge corner, we can use a similar argument of Corollary 10.3.1 to prove that $\mathbf{v}$ and $\mathbf{w}$ must vanish at the underlying edge corner point.

Corollary 10.3.2. *Consider the same setup in Corollary 10.3.1, and assume that* $\mathbf{v}\in C^\alpha(\overline{S}_h\times[-M,M])^3$, $q\mathbf{w}\in C^\alpha(\overline{S}_h\times[-M,M])^3$, *and* $\mathbf{v}-\mathbf{w}\in C^{1,\alpha}(\overline{S}_h\times[-M,M])^3$, *for* $0<\alpha<1$. *If* $V(\mathbf{x}',x_3)=q(\mathbf{x}',x_3)-1$ *is continuous near the edge point* $(\mathbf{0},x_3^c)$ *with* $V(\mathbf{0},x_3^c)\neq 0$ *for* $x_3^c\in(-M,M)$, *then for every edge point* $(\mathbf{0},x_3^c)\in\mathbb{R}^3$ *of* $S_h\times(-M,M)$, *one has*

$$\mathbf{v}((\mathbf{0},x_3^c)) - \mathbf{w}((\mathbf{0},x_3^c)) = \mathbf{0} \tag{10.3.93}$$

Proof. According to Corollary 10.3.1, we know that (10.3.90) holds. In what follows, let us consider two separate cases.

Case 1: $\lambda \neq 0$. Combining the argument in **Part II** of the proof of Corollary 10.3.1 with a similar argument for Remark 10.3.3, one can show that (10.3.93) holds by noting that $\mathbf{w}$ is continuous near the edge point.

Case 2: $\lambda = 0$. Since $\mathbf{v}_{\mathrm{R}} = \mathbf{w}_{\mathrm{R}}$ on $\Gamma_h^{\pm} \times (-L,L)$, we have $\partial_\ell(v_j - w_j) = 0$ for $\ell = 1,2$ and $j = 1,2$, where v_j and w_j the $j-$th components of $\mathbf{v}_{\mathrm{R}}$ and $\mathbf{w}_{\mathrm{R}}$, respectively. Therefore, under $\lambda = 0$, subtracting the first equation of (10.3.14) from the second one of (10.3.14), it yields that

$$\mathscr{R}(qw_3) = \mathscr{R}(v_3) \text{ on } \Gamma_h^{\pm},$$

which can be used to further deduce that $\mathscr{R}(w_3(\mathbf{0})) = 0$. Therefore, we have $w_3(\mathbf{0}, x_3^c) = 0$ for all $x_3^c \in (-L,L)$. Due to (10.3.90), we can show that $\mathbf{v}_{\mathrm{R}}((\mathbf{0},x_3^c)) = \mathbf{w}_{\mathrm{R}}((\mathbf{0},x_3^c)) = \mathbf{0}$. Using a similar argument, we can prove (10.3.93) when $\lambda = 0$.

The proof is complete. □

If we further require that $\mathbf{v}$ is Hölder continuous at the edge corner, similar to Theorem 10.3.1, we have the following theorem. Indeed, by exploring the Hölder continuity of $\mathbf{f}_\ell$ ($\ell = 2,3,4$) near the corner, adopting a similar argument in Theorem 10.3.1, we can show that $\mathbf{v}$ must vanish at the edge corner point.

Theorem 10.3.2. *Consider the same setup in Theorem 10.3.1, and assume that* $\mathbf{v} \in H^1(\Omega)^3 \cap C^\alpha(\overline{S}_h \times [-M,M])^3$, $q\mathbf{w} \in C^\alpha(\overline{S}_h \times [-M,M])^3$, *and* $\eta \in C^\alpha(\overline{\Gamma}_h^{\pm} \times [-M,M])$ *for* $0 < \alpha < 1$. *If* η *is independent of* x_3 *and* $\eta(\mathbf{0}) \neq 0$, *and the corner is non-degenerate, then* $\mathbf{v}$ *vanishes at the edge point* $(\mathbf{0}, x_3^c) \in \mathbb{R}^3$ *of* $S_h \times (-M,M)$, *where* $x_3^c \in (-M,M)$.

Proof. It is sufficient to show the vanishing property for $\mathbf{v}_{\mathrm{R}}$. Clearly, one has $q\mathbf{w}_{\mathrm{R}} \in C^\alpha(\overline{S}_h \times [-M,M])^3$ and $\mathbf{v}_{\mathrm{R}} - \mathbf{w}_{\mathrm{R}} \in C^{1,\alpha}(\overline{S_h} \times [-M,M])^3$. Therefore, the assumptions in Lemmas 10.3.4–10.3.12 are fulfilled. We next divide the proof into two parts.

Part I. Let $\mathbf{v}^{(1,2)})$ be defined in (10.3.10). First, we shall prove

$$\mathbf{v}^{(1,2)}(\mathbf{0}) = \mathbf{0}. \tag{10.3.94}$$

From (10.3.18), (10.3.19), and (10.3.20), it follows that

$$\int_{S_h} (\mathbf{f}_1 + \mathbf{f}_2 + \mathbf{f}_3 + \mathbf{f}_4) \cdot \mathbf{u}(\mathbf{x}') \mathrm{d}\mathbf{x}' = \widetilde{I}_{\Lambda_h} - \int_{\Gamma_h^{\pm}} \eta(\mathbf{x}') \mathscr{R}(\mathbf{v}^{(1,2)}) \cdot \mathbf{u}(\mathbf{x}') \mathrm{d}\sigma, \tag{10.3.95}$$

where $\widetilde{I}_{\Lambda_h}$ is defined in (10.3.17).

Using the conditions $q\mathbf{w}_{\mathrm{R}} \in C^\alpha(\overline{S}_h \times [-M,M])^3$ and $\mathbf{v}_{\mathrm{R}} - \mathbf{w}_{\mathrm{R}} \in C^{1,\alpha}(S_h \times (-M,M))^3$, by Lemma 10.3.7, we know that $\mathbf{f}_\ell(\mathbf{x}') \in C^\alpha(\overline{S}_h)$, where $\mathbf{f}_\ell(\mathbf{x}')$ is defined in (10.3.17). Since $\eta \in C^\alpha\left(\overline{\Gamma}_h^{\pm}\right)$, we know that η, $\mathbf{f}_2, \mathbf{f}_3$, and $\mathbf{f}_4$ have the expansions (10.3.23) and (10.3.30) around the origin. Recall the definition of $\mathbf{f}_1$, i.e., $\mathbf{f}_1 = -\omega^2 \mathscr{R}(\mathbf{v}^{(1,2)})$. Due to the fact that $\mathbf{v}^{(1,2)} \in C^\alpha(\overline{S}_h)$, we have the following expansions:

$$\mathscr{R}(\mathbf{v}^{(1,2)})(\mathbf{x}') = \mathscr{R}(\mathbf{v}^{(1,2)})(\mathbf{0}) + \delta\mathscr{R}(\mathbf{v}^{(1,2)})(\mathbf{x}'),\ |\delta\mathscr{R}(\mathbf{v}^{(1,2)})| \le \|\mathscr{R}(\mathbf{v}^{(1,2)})(\mathbf{x}')\|_{C^\alpha(S_h)^2} |\mathbf{x}'|^\alpha. \tag{10.3.96}$$

Substituting (10.3.23), (10.3.30), and (10.3.96) into (10.3.95), we have

$$
\begin{aligned}
&\Big(- \omega^2 \mathscr{R}(\mathbf{v}^{(1,2)})(\mathbf{0}) + \mathbf{f}_2(\mathbf{0}) + \mathbf{f}_3(\mathbf{0}) + \mathbf{f}_4(\mathbf{0}) \Big) \int_{S_h} \mathbf{u}(\mathbf{x}')\mathrm{d}\mathbf{x}' \\
&+ \int_{S_h} \Big(- \omega^2 \delta \mathscr{R}(\mathbf{v}^{(1,2)})(\mathbf{x}') + \delta \mathbf{f}_2(\mathbf{x}') + \delta \mathbf{f}_3(\mathbf{x}') + \delta \mathbf{f}_4(\mathbf{x}') \Big) \cdot \mathbf{u}(\mathbf{x}')\mathrm{d}\mathbf{x}' \\
=& \widetilde{I}_{\Lambda_h} - \eta(\mathbf{0}) \mathscr{R}(\mathbf{v}^{(1,2)})(\mathbf{0}) \int_{\Gamma_h^{\pm}} \mathbf{u}(\mathbf{x}')\mathrm{d}\sigma - \eta(\mathbf{0}) \int_{\Gamma_h^{\pm}} \mathbf{u}(\mathbf{x}') \cdot \delta \mathscr{R}(\mathbf{v}^{(1,2)})(\mathbf{x}')\mathrm{d}\sigma \\
&- \mathscr{R}(\mathbf{v}^{(1,2)})(\mathbf{0}) \int_{\Gamma_h^{\pm}} \delta \eta(\mathbf{x}') \mathbf{u}(\mathbf{x}')\mathrm{d}\sigma - \int_{\Gamma_h^{\pm}} \delta \eta(\mathbf{x}') \mathbf{u}(\mathbf{x}') \cdot \delta \mathscr{R}(\mathbf{v}^{(1,2)})(\mathbf{x}')\mathrm{d}\sigma .
\end{aligned} \tag{10.3.97}
$$

Combining (10.2.17) with (10.2.46), we have the following integral estimates:

$$
\left| \int_{\Gamma_h^{\pm}} \delta \eta(\mathbf{x}') \mathbf{u}(\mathbf{x}')\mathrm{d}\sigma \right| \leq \sqrt{2} \|\eta\|_{C^\alpha} \int_0^h r^\alpha e^{-s\sqrt{r}\cos\frac{\theta_m}{2}} \mathrm{d}r = O(s^{-2\alpha-2}), \tag{10.3.98}
$$

$$
\begin{aligned}
&\left| \int_{\Gamma_h^{\pm}} \mathbf{u}(\mathbf{x}') \cdot \delta \mathscr{R}(\mathbf{v}^{(1,2)})(\mathbf{x}')\mathrm{d}\sigma \right| \leq \sqrt{2} \|\mathscr{R}(\mathbf{v}^{(1,2)})\|_{C^\alpha} \int_0^h r^\alpha e^{-s\sqrt{r}\cos\frac{\theta_m}{2}} \mathrm{d}r \\
&= O(s^{-2\alpha-2}),
\end{aligned} \tag{10.3.99}
$$

$$
\begin{aligned}
&\left| \int_{\Gamma_h^{\pm}} \delta \eta(\mathbf{x}') \mathbf{u}(\mathbf{x}') \cdot \delta \mathscr{R}(\mathbf{v}^{(1,2)})(\mathbf{x}')\mathrm{d}\sigma \right| \leq \sqrt{2} \|\eta\|_{C^\alpha} \|\mathscr{R}(\mathbf{v}^{(1,2)})\|_{C^\alpha} \int_0^h r^{2\alpha} e^{-s\sqrt{r}\cos\frac{\theta_m}{2}} \mathrm{d}r \\
&= O(s^{-4\alpha-2}),
\end{aligned} \tag{10.3.100}
$$

$$
\begin{aligned}
&\left| \int_{S_h} \delta \mathscr{R}(\mathbf{v}^{(1,2)})(\mathbf{x}') \cdot \mathbf{u}(\mathbf{x}')\mathrm{d}\mathbf{x}' \right| \leq \sqrt{2} \|\mathscr{R}(\mathbf{v}^{(1,2)})\|_{C^\alpha} \int_W |u_1(\mathbf{x}')| |\mathbf{x}|^\alpha \mathrm{d}\mathbf{x}' \\
&\leq 2\sqrt{2} \|\mathscr{R}(\mathbf{v}^{(1,2)})\|_{C^\alpha} \frac{(\theta_M - \theta_m)\Gamma(2\alpha+4)}{\delta_W^{2\alpha+4}} s^{-2\alpha-4},
\end{aligned} \tag{10.3.101}
$$

$$
\begin{aligned}
&\left| \int_{S_h} \delta \mathbf{f}_2(\mathbf{x}') \cdot \mathbf{u}(\mathbf{x}')\mathrm{d}\mathbf{x}' \right| \leq \sqrt{2} \|\mathbf{f}_2\|_{C^\alpha} \int_W |u_1(\mathbf{x}')| |\mathbf{x}|^\alpha \mathrm{d}\mathbf{x}' \\
&\leq 2\sqrt{2} \|\mathbf{f}_2\|_{C^\alpha} \frac{(\theta_M - \theta_m)\Gamma(2\alpha+4)}{\delta_W^{2\alpha+4}} s^{-2\alpha-4},
\end{aligned} \tag{10.3.102}
$$

$$
\begin{aligned}
&\left| \int_{S_h} \delta \mathbf{f}_3(\mathbf{x}') \cdot \mathbf{u}(\mathbf{x}')\mathrm{d}\mathbf{x}' \right| \leq \sqrt{2} \|\mathbf{f}_3\|_{C^\alpha} \int_W |u_1(\mathbf{x}')| |\mathbf{x}|^\alpha \mathrm{d}\mathbf{x}' \\
&\leq 2\sqrt{2} \|\mathbf{f}_3\|_{C^\alpha} \frac{(\theta_M - \theta_m)\Gamma(2\alpha+4)}{\delta_W^{2\alpha+4}} s^{-2\alpha-4},
\end{aligned} \tag{10.3.103}
$$

$$
\begin{aligned}
&\left| \int_{S_h} \delta \mathbf{f}_4(\mathbf{x}') \cdot \mathbf{u}(\mathbf{x}')\mathrm{d}\mathbf{x}' \right| \leq \sqrt{2} \|\mathbf{f}_4\|_{C^\alpha} \int_W |u_1(\mathbf{x}')| |\mathbf{x}|^\alpha \mathrm{d}\mathbf{x}' \\
&\leq 2\sqrt{2} \|\mathbf{f}_4\|_{C^\alpha} \frac{(\theta_M - \theta_m)\Gamma(2\alpha+4)}{\delta_W^{2\alpha+4}} s^{-2\alpha-4},
\end{aligned} \tag{10.3.104}
$$

as $s \to +\infty$. From (10.3.97), after rearranging terms, we have

$$
\begin{aligned}
&\eta(\mathbf{0})\mathscr{R}(\mathbf{v}^{(1,2)})(\mathbf{0})\int_{\Gamma_h^{\pm}}\mathbf{u}(\mathbf{x}')\mathrm{d}\sigma\\
=&\widetilde{I}_{\Lambda_h}-\Big(-\omega^2\mathscr{R}(\mathbf{v}^{(1,2)})(\mathbf{0})+\mathbf{f}_2(\mathbf{0})+\mathbf{f}_3(\mathbf{0})+\mathbf{f}_4(\mathbf{0})\Big)\int_{S_h}\mathbf{u}(\mathbf{x}')\mathrm{d}\mathbf{x}'\\
&-\int_{S_h}\Big(-\omega^2\delta\mathscr{R}(\mathbf{v}^{(1,2)})(\mathbf{x}')+\delta\mathbf{f}_2(\mathbf{x}')+\delta\mathbf{f}_3(\mathbf{x}')+\delta\mathbf{f}_4(\mathbf{x}')\Big)\cdot\mathbf{u}(\mathbf{x}')\mathrm{d}\mathbf{x}' \qquad (10.3.105)\\
&-\eta(\mathbf{0})\int_{\Gamma_h^{\pm}}\mathbf{u}(\mathbf{x}')\cdot\delta\mathscr{R}(\mathbf{v}^{(1,2)})(\mathbf{x}')\mathrm{d}\sigma-\mathscr{R}(\mathbf{v}^{(1,2)})(\mathbf{0})\int_{\Gamma_h^{\pm}}\delta\eta(\mathbf{x}')\mathbf{u}(\mathbf{x}')\mathrm{d}\sigma\\
&-\int_{\Gamma_h^{\pm}}\delta\eta(\mathbf{x}')\mathbf{u}(\mathbf{x}')\cdot\delta\mathscr{R}(\mathbf{v}^{(1,2)})(\mathbf{x}')\mathrm{d}\sigma.
\end{aligned}
$$

Multiplying s^2 on both sides of (10.3.105), by virtue of (10.2.32), (10.2.79), (10.2.80), (10.3.98), (10.3.99), (10.3.100), (10.3.101), (10.3.102), (10.3.103), and (10.3.104), and letting $s \to \infty$, we have

$$\eta(\mathbf{0})\mathscr{R}(\mathbf{v}^{(1,2)})(\mathbf{0})\cdot\begin{pmatrix}1\\ \mathrm{i}\end{pmatrix}\Big(\mu(\theta_M)^{-2}+\mu(\theta_m)^{-2}\Big)=0. \qquad (10.3.106)$$

Since the corner is non-degenerate, we know that

$$\mu^{-2}(\theta_m)+\mu^{-2}(\theta_M)\neq 0.$$

Hence, we can obtain (10.3.94), which implies that $\mathbf{v}^{(1,2)}((\mathbf{0},x_3^c))=\mathbf{0}$.

Part II. Consider the PDE system (10.3.14) when $\lambda \neq 0$. According to [23, Theorem 3.2], under the condition $q\mathbf{w}_\mathrm{R} \in C^\alpha(\overline{S_h})^2$ and $\mathbf{v}_\mathrm{R}-\mathbf{w}_\mathrm{R} \in C^\alpha(\overline{S}_h \times [-M,M])^3$, we know that $v_3((\mathbf{0},x_3^c))=0$ for all $x_3^c \in (-L,L)$. For $\lambda=0$, we can use a similar argument in the proof of Corollary 10.3.2 for proving the second case to consider the vanishing property of v_3 at the edge point.

The proof is complete. □

In Corollary 10.3.3, if global Hölder-continuous regularities for $\mathbf{v}$ and $\mathbf{w}$ are fulfilled, similar to Theorem 10.3.2, we can prove that $\mathbf{v}$ and $\mathbf{w}$ must vanish at the edge corner point by removing the assumption that $\mathbf{v}-\mathbf{w}$ is $C^{1,\alpha}$ continuous at the edge corner. The detailed proof of the following corollary is omitted.

Lemma 10.3.14 ([8, Proposition 2.7]). *Let $D \subset \mathbb{R}^3$ be a bounded Lipschitz domain. Let $\mathbf{U} \in H^1(D)^3$ solve*

$$\begin{cases}\mathscr{L}\mathbf{U}=\mathbf{f} \text{ in } D,\\ \mathbf{U}-\mathbf{g}\in H_0^1(D)^3,\end{cases} \qquad (10.3.107)$$

for some $\mathbf{f} \in C^\alpha(D)$, $\mathbf{g} \in C^{1,\alpha}(\overline{D})^3$, and $\alpha \in (0,1)$. Then it holds that $\mathbf{U} \in C^{1,\alpha}(\overline{D})$.

Corollary 10.3.3. *Consider the same setup in Theorem 10.3.2. Assume that $\mathbf{v},\mathbf{w} \in H^1(\Omega)^3 \cap C^\alpha(\overline{\Omega})^3$ $(0<\alpha<1)$ are the generalized elastic transmission eigenfunctions to* (10.1.11). *Suppose further that $\eta \in C^\alpha(\overline{\Gamma}_h^{\pm} \times [-M,M])$ with $\eta(\mathbf{0}) \neq 0$ and $q\mathbf{w} \in C^\alpha(\overline{\Omega})$ for $0<\alpha<1$. If the corner is non-degenerate, then $\mathbf{v}$ and $\mathbf{w}$ vanish at the edge point $(\mathbf{0},x_3^c) \in \mathbb{R}^3$ of $S_h \times (-M,M)$, where $x_3^c \in (-M,M)$.*

Proof. Let $\mathbf{U}=\mathbf{v}-\mathbf{w}$. It can be verified that $\mathbf{U}$ solves (10.3.107) for $\mathbf{f}=\omega^2 q\mathbf{w}-\omega^2\mathbf{v}$, $\mathbf{g}=\mathbf{0}$, and $D=\Omega$. Due to $q\mathbf{w} \in C^\alpha(\overline{\Omega})^3$ and $\mathbf{v} \in C^\alpha(\overline{\Omega})^3$, we have $\mathbf{f} \in C^\alpha(\overline{\Omega})$. Hence, by Lemma 10.3.14, we further have $\mathbf{v}-\mathbf{w} \in C^{1,\alpha}(\overline{\Omega})$. Therefore, the assumptions in Theorem 10.3.2 are fulfilled. We readily finish the proof of this corollary by Theorem 10.3.2. □

10.4 Unique Recovery Results for the Inverse Elastic Medium Problem

In this section, we apply the geometric property of generalized elastic transmission eigenfunctions discussed in the previous sections to the study of the unique recovery for the inverse elastic problem (10.1.8).

We first introduce a more general formulation of the inverse elastic problem. Let Ω, $\mathbf{u}^i$, V, and $q = 1+V$ be those introduced in Sect. 10.1.1. Let $\eta \in L^\infty(\partial\Omega)$ with $\Im\eta \geq 0$. Consider the following elastic scattering system:

$$\begin{cases} \mathscr{L}\mathbf{u}^- + \omega^2 q\mathbf{u}^- = \mathbf{0} & \text{in } \Omega, \\ \mathscr{L}\mathbf{u}^+ + \omega^2\mathbf{u}^+ = \mathbf{0} & \text{in } \mathbb{R}^n\backslash\Omega, \\ \mathbf{u}^+ = \mathbf{u}^-, \quad T_\nu\mathbf{u}^+ + \eta\mathbf{u}^+ = T_\nu\mathbf{u}^- & \text{on } \partial\Omega, \\ \mathbf{u}^+ = \mathbf{u}^i + \mathbf{u}^{sc} & \text{in } \mathbb{R}^n\backslash\Omega, \\ \lim\limits_{r\to\infty} r^{\frac{n-1}{2}}\left(\dfrac{\partial\mathbf{u}_\beta^{sc}}{\partial r} - ik_\beta\mathbf{u}_\beta^{sc}\right) = \mathbf{0}, & \beta = p, s. \end{cases} \tag{10.4.1}$$

If $\eta \equiv 0$, (10.4.1) is reduced to be (10.1.2). The well-posedness of (10.4.1) for the case $\eta \equiv 0$ was investigated in [28, 29]. By following a standard variation argument in [28, 29], one can show the unique existence of a solution $\mathbf{u} = \mathbf{u}^-\chi_\Omega + \mathbf{u}^+\chi_{\mathbb{R}^n\backslash\overline{\Omega}} \in H^1_{loc}(\mathbb{R}^n)^n$ to (10.4.1). However, it is not the focus of this chapter, and in what follows, we always assume the well-posedness of the system (10.4.1). We write $\mathbf{u}_\beta^\infty(\hat{\mathbf{x}};\mathbf{u}^i)$, $\beta = t, p,$ or s to signify the far-field patterns associated with (10.4.1) and consider the following inverse problem :

$$\mathscr{F}(\Omega; q, \eta) = \mathbf{u}_\beta^\infty(\hat{\mathbf{x}};\mathbf{u}^i), \quad \beta = t, p, \text{ or } s, \tag{10.4.2}$$

where $\mathscr{F}$ is implicitly defined by the scattering system (10.4.1). We are particularly interested in the geometrical inverse problem of recovering Ω independent of q and η. On the other hand, it is pointed out that if Ω can be recovered, the boundary parameter η can be recovered as well by a standard argument. If the far-field pattern is given associated with a single incident wave, then it is referred to as a single far-field measurement for the inverse problem (10.4.2); otherwise, it is referred to as multiple measurements. It can be directly verified that with a single far-field measurement, the inverse problem (10.4.2) is formally determined since both $\mathbb{S}^{n-1}$ (on which the far-field pattern is given) and $\partial\Omega$ (it completely determines the shape of Ω) are $(n-1)$-dimensional manifolds. The inverse shape problem associated with a single far-field measurement constitutes a longstanding challenging problem in the inverse scattering theory [18], a fortify the one described above for the elastic scattering. In the rest of the chapter, we shall apply the geometric results derived in the previous sections to give several unique identifiable results to the inverse shape problem (10.4.2) associated with a single far-field measurement within a certain generic (though still specific) scenario. We follow the treatment in [24]. Indeed, unique shape and boundary parameter determination for a convex polygonal or polyhedral elastic medium scatter by a single far-field measurement is established in [24]. Before that, we would like to mention that the unique result in [28, 30] concerning the inverse problem (10.4.2) associated with multiple far-field measurements and $\eta \equiv 0$. Recently, effective medium theory for elastic medium scattering with an embedded obstacle was developed in [38].

First, we introduce the admissible class of elastic scatters in our study. Let $W_{\mathbf{x}_c}(\theta_W)$ be an open sector in $\mathbb{R}^2$ with the vertex $\mathbf{x}_c$ and an open angle θ_W. Denote

$$\Gamma_h^\pm(\mathbf{x}_c) := \partial W_{\mathbf{x}_c}(\theta_W) \cap B_h(\mathbf{x}_c), \quad S_h(\mathbf{x}_c) := B_h(\mathbf{x}_c) \cap W_{\mathbf{x}_c}(\theta_W), \tag{10.4.3}$$

where $B_h(\mathbf{x}_c)$ is an open disk centered at $\mathbf{x}_c$ with the radius $h \in \mathbb{R}_+$.

Definition 10.4.1. Let $(\Omega; q, \eta)$ be an elastic scatter. Consider the scattering problem (10.4.1) and $\mathbf{u}^i$ is the incident wave field therein. The scattering configuration is said to be admissible if the following conditions are fulfilled:

(a) Ω is a bounded simply connected Lipschitz domain in $\mathbb{R}^n$ with a connected complement, and $q \in L^\infty(\Omega)$ and $\eta \in L^\infty(\partial\Omega)$ are real-valued functions.
(b) Following the notations in Theorem 10.2.2, if $\Omega \Subset \mathbb{R}^2$ possesses a planar corner $B_h(\mathbf{x}_c) \cap \Omega = \Omega \cap W_{\mathbf{x}_c}(\theta_W)$ where $\mathbf{x}_c$ is the vertex of the sector $W_{\mathbf{x}_c}(\theta_W)$ and the open angle θ_W of $W_{\mathbf{x}_c}(\theta_W)$ satisfies $\theta_W \in (0, \pi)$, then $q \in C^\alpha(\overline{S}_h(\mathbf{x}_c))$ and $\eta \in C^\alpha(\overline{\Gamma_h^\pm(\mathbf{x}_c)})$ for $\alpha \in (0,1)$ with $q(\mathbf{x}_c) \neq 1$ and $\eta(\mathbf{x}_c) \neq 0$, where $S_h(\mathbf{x}_c)$ and $\Gamma_h^\pm(\mathbf{x}_c)$ are defined in (10.4.3). Similarly, following the notations in Theorem 10.3.1, if $\Omega \Subset \mathbb{R}^3$ possesses a 3D edge corner $(B_h(\mathbf{x}_c') \times (-M, M)) \cap \Omega = S_h(\mathbf{x}_c') \times (-M, M)$, where $\mathbf{x}_c$ is the vertex of $S_h(\mathbf{x}_c)$ contained in the sector $W_{\mathbf{x}_c}(\theta_W)$ and the open angle θ_W of $W_{\mathbf{x}_c}(\theta_W)$ satisfies $\theta_W \in (0, 2\pi)\backslash\{\pi\}$, then $q \in C^\alpha(\overline{S_h}(\mathbf{x}_c') \times [-M, M])$, $\eta \equiv 0$ on $\partial\Omega$.
(c) The total wave field $\mathbf{u}$ is non-vanishing everywhere in the sense that for any $\mathbf{x} \in \mathbb{R}^n$,

$$\lim_{\rho \to +0} \frac{1}{m(B(\mathbf{x}, \rho))} \int_{B(\mathbf{x}, \rho)} |\mathbf{u}(\mathbf{x})| \mathrm{d}x \neq 0. \tag{10.4.4}$$

Remark 10.4.1. The assumption (10.4.4) can be fulfilled in certain generic scenario. For an illustration, let us consider a specific case by requiring the angular frequency $\omega \in \mathbb{R}_+$ sufficiently small. In the physical scenario, this is also equivalent to requiring that the size of the scatter, namely, $\operatorname{diam}(\Omega)$, is sufficiently small (compared to the operating wavelength). In such a case, from a physical point of view, the interruption of the incident field due to the scatter should be small, i.e., $\mathbf{u}^{sc}$ should be small compared to $\mathbf{u}^i$. Hence, if $\mathbf{u}^i$ is non-vanishing everywhere (say, $\mathbf{u}^i$ is a plane wave, namely, the Herglotz wave (10.2.3) or (10.3.2) with the densities being delta distributions), then $\mathbf{u} = \mathbf{u}^i + \mathbf{u}^{sc}$ should be non-vanishing everywhere. However, a rigorous justification of such a physical intuition will cost lengthy arguments, and we choose not to explore more about this point.

Next, we present a technical lemma concerning the regularity of the solution to the Lamé system around a corner (cf. [39, 40]) in $\mathbb{R}^2$. We would also like to refer interested readers to [21, 22, 27] on classical results of decomposing solutions to elliptic PDEs in corner domains.

Lemma 10.4.1 ([39, Theorem 2.3]). *Let Ω be a bounded open connected subset of $\mathbb{R}^2$, where the boundary $\partial\Omega$ of Ω is the union of a finite number of line segment $\overline{\Gamma}_\ell$, $\ell \in \Xi$. Fix a partition of Ξ into $\mathscr{D} \cup \mathscr{N}$, where $\mathscr{D}$ and $\mathscr{N}$ correspond to Dirichlet and Neumann boundary conditions, respectively. Given a vector field $\mathbf{f} \in L^2(\Omega)^2$ and $\mathbf{g}^{(\ell)} \in H^{1/2}(\Gamma_\ell)^2$ for all $\ell \in \mathscr{N}$, consider the weak solution $\mathbf{u} \in H^1(\Omega)^2$ of the Lamé system*

$$\mathscr{L}\mathbf{u} = \mathbf{f} \text{ in } \Omega, \tag{10.4.5}$$

with mixed boundary conditions

$$\begin{cases} \mathbf{u} = \mathbf{0} & \text{on } \Gamma_\ell, \quad \ell \in \mathscr{D}, \\ T_\nu \mathbf{u} = \mathbf{g}^{(\ell)} & \text{on } \Gamma_\ell, \quad \ell \in \mathscr{N}. \end{cases}$$

If Ω satisfies the assumption $\forall \ell_1, \ell_2 \in \Xi$ such that $\overline{\Gamma}_{\ell_1} \cap \overline{\Gamma}_{\ell_2} \neq \emptyset$, the interior angle $\angle(\Gamma_{\ell_1}, \Gamma_{\ell_2})$ fulfills $\angle(\Gamma_{\ell_1}, \Gamma_{\ell_2}) < 2\pi$, and moreover, if $\ell_1 \in \mathscr{D}$ and $\ell_2 \in \mathscr{N}$, $\angle(\Gamma_{\ell_1}, \Gamma_{\ell_2}) < \pi$, then a solution $\mathbf{u}$ of (10.4.5) with the data $\mathbf{f} \in L^2(\Omega)^2$ and $\mathbf{g}^{(\ell)} \in H^{1/2}(\Gamma_\ell)^2$, $\forall \ell \in \mathscr{N}$ satisfies $\mathbf{u} \in H^{3/2+\varepsilon}(\Omega)^2$ for some $\varepsilon > 0$.

Lemma 10.4.2. *Let $S_{2h} = W \cap B_{2h}$ and $\Gamma_{2h}^\pm = \partial S_{2h} \backslash \partial B_{2h}$, where W is the infinite sector defined in (10.2.1) with the opening angle $\theta_W \in (0, \pi)$. Suppose that $\mathbf{u} \in H^1(B_{2h})^2$ satisfies*

$$\begin{cases} \mathscr{L}\mathbf{u}^- + \omega^2 q_- \mathbf{u}^- = \mathbf{0} & \text{in } S_{2h}, \\ \mathscr{L}\mathbf{u}^+ + \omega^2 \mathbf{u}^+ = \mathbf{0} & \text{in } B_{2h}\backslash\overline{S_{2h}}, \\ \mathbf{u}^+ = \mathbf{u}^- & \text{on } \Gamma_{2h}^{\pm}, \end{cases} \tag{10.4.6}$$

where $\mathbf{u}^+ = \mathbf{u}|_{B_{2h}\backslash\overline{S_{2h}}}$, $\mathbf{u}^- = \mathbf{u}|_{S_{2h}}$, ω *is a positive constant, and* $q_- \in L^\infty(S_{2h})$. *Assume that* $\mathbf{u}^+$ *is real analytic in* $B_{2h}\backslash\overline{S_{2h}}$. *There exists* $\alpha \in (0,1)$ *such that* $\mathbf{u}^- \in C^\alpha(\overline{S_h})$.

Proof. Since $\mathbf{u}^+$ is real analytic in $B_{2h}\backslash\overline{S_{2h}}$, we let $\mathbf{w}$ be the analytic extension of $\mathbf{u}^+|_{B_h\backslash\overline{S_h}}$ in B_h. By using the transmission condition on $\Gamma_h^{\pm}$, one clearly has that $\mathbf{u}^- = \mathbf{u}^+ = \mathbf{w}$ on $\Gamma_h^{\pm}$. Set $\mathbf{v} = \mathbf{u}^- - \mathbf{w}$. Set l_h to denote the line segment with the starting point $h(\cos\theta_m, \sin\theta_m) \in \Gamma_h^-$ and the ending point $h(\cos\theta_M, \sin\theta_M) \in \Gamma_h^+$. It can be directly verified that

$$\mathscr{L}\mathbf{v} = \mathbf{f} \text{ in } \mathscr{T}_h; \;\; T_\nu \mathbf{v} = \mathbf{g} \text{ on } l_h; \;\; \mathbf{v} = \mathbf{0} \text{ on } \Gamma_h^{\pm},$$

where $\mathscr{T}_h$ is the open triangle formed by $\Gamma_h^{\pm}$, and l_h, $\mathbf{f} = -\omega^2(\mathbf{w} + q_-\mathbf{u}^-) \in L^2(S_h)^2$ and $\mathbf{g} \in C^\infty(\overline{l_h})^2$. By virtue of Lemma 10.4.1, one has that $\mathbf{v} \in H^{3/2+\varepsilon}(T_h)^2$ with $\varepsilon \in \mathbb{R}_+$. Therefore, by the Sobolev embedding theorem, it is clear that there exists $\alpha \in (0,1)$ such that $\mathbf{v} \in C^\alpha(\overline{S_h})^2$. Hence, we readily have that $\mathbf{u}^- \in C^\alpha(\overline{S_h})^2$.

The proof is complete. □

Remark 10.4.2. We would like to point out that the regularity result in Lemma 10.4.2 does not hold for the three-dimensional case in general. This is mainly due to the fact that the corresponding regularity result in Lemma 10.4.1 is generically not true around a general polyhedral corner in $\mathbb{R}^3$. Hence, we exclude the generalized transmission condition in (10.1.11) for an admissible elastic scatter in Definition 10.4.1. That is, we only consider the case $\eta \equiv 0$ on the boundary of an admissible elastic scatter in $\mathbb{R}^3$, which is different from the two-dimensional case.

Theorem 10.4.1. *Consider the elastic scattering problem* (10.4.1) *associated with the incident elastic wave field* $\mathbf{u}^i$ *and two elastic scatters* $(\Omega_j; q_j, \eta_j)$ $(j = 1,2)$ *in* $\mathbb{R}^n$ $(n = 2,3)$ *being admissible scattering configuration. Let* $\mathbf{u}_\beta^{j,\infty}(\hat{\mathbf{x}}; \mathbf{u}^i)$ *be the far-field pattern associated with the scatter* $(\Omega_j; q_j, \eta_j)$ *and the incident field* $\mathbf{u}^i$, $\beta = t, p$ *or* s. *If*

$$\mathbf{u}_\beta^{1,\infty}(\hat{\mathbf{x}}; \mathbf{u}^i) = \mathbf{u}_\beta^{2,\infty}(\hat{\mathbf{x}}; \mathbf{u}^i), \;\; \hat{\mathbf{x}} \in \mathbb{S}^{n-1}, \tag{10.4.7}$$

for a fixed incident wave $\mathbf{u}^i$, *then one has that*

$$\Omega_1 \Delta \Omega_2 := \left(\Omega_1 \backslash \Omega_2\right) \cup \left(\Omega_2 \backslash \Omega_1\right), \tag{10.4.8}$$

cannot possess a corner. Hence, if Ω_1 *and* Ω_2 *are convex polygons in* $\mathbb{R}^2$ *or convex polyhedron in* $\mathbb{R}^3$, *one must have*

$$\Omega_1 = \Omega_2. \tag{10.4.9}$$

Proof. We prove (10.4.8) by contradiction. Suppose that there is a corner contained in $\Omega_1 \Delta \Omega_2$. Without loss of generality, we may assume that the vertex O of the corner $\Omega_2 \cap W_{\mathbf{x}_c}(\theta_W)$ is such that $O \in \partial\Omega_2$ and $O \notin \overline{\Omega}_1$. Furthermore, one may assume that in two dimensions, O is the origin, whereas in three dimensions, the edge corner point $O = (\mathbf{x}_c', x_3^c)$ of the 3D edge corner $(B_h(\mathbf{x}_c') \times (-M,M)) \cap \Omega_2 = S_h(\mathbf{x}_c') \times (-M,M)$ fulfills that $\mathbf{x}_c'$ is the origin of $\mathbb{R}^2$.

Due to (10.4.7), applying Rellich's theorem (see [29,31]), we know that $\mathbf{u}_1^{sc} = \mathbf{u}_2^{sc}$ in $\mathbb{R}^n \backslash (\overline{\Omega}_1 \cup \overline{\Omega}_2)$. Thus,

$$\mathbf{u}_1(\mathbf{x}) = \mathbf{u}_2(\mathbf{x}), \tag{10.4.10}$$

for all $\mathbf{x} \in \mathbb{R}^n \backslash (\overline{\Omega}_1 \cup \overline{\Omega}_2)$. In what follows, we consider two separate cases.

Case 1 ($n=2$)**:** Following the notations in (10.2.2) and the setup of Theorem 10.2.2, we have from (10.4.10) that

$$\mathbf{u}_2^- = \mathbf{u}_2^+ = \mathbf{u}_1^+, \quad T_\nu \mathbf{u}_2^- = T_\nu \mathbf{u}_2^+ + \eta_2 \mathbf{u}_2^+ = T_\nu \mathbf{u}_1^+ + \eta_2 \mathbf{u}_1^+ \text{ on } \Gamma_h^\pm, \tag{10.4.11}$$

where the superscripts $(\cdot)^-, (\cdot)^+$ stand for the limits taken from Ω_2 and $\mathbb{R}^2\backslash\overline{\Omega_2}$, respectively. Moreover, we take $h \in \mathbb{R}_+$ sufficient small such that

$$\mathscr{L}\mathbf{u}_1^+ + \omega^2 \mathbf{u}_1^+ = \mathbf{0} \text{ in } B_h, \quad \mathscr{L}\mathbf{u}_2^- + \omega^2 q_2 \mathbf{u}_2^- = \mathbf{0} \text{ in } S_h. \tag{10.4.12}$$

Since $(\Omega_j; q_j, \eta_j)$, $j=1,2$, are admissible, we know that $q_2 \in C^\alpha(\overline{S_h})$ and $\eta_j \in C^\alpha(\overline{\Gamma_h^\pm})$. Clearly, $\mathbf{u}_2^- \in H^1(S_h)^2$ and $\mathbf{u}_1^+$ is real analytic in $B_{2h}\backslash\overline{S_{2h}}$. According to Lemma 10.4.2, we know that $\mathbf{u}_2^- \in C^\alpha(S_h)^2$, which implies that $q_2\mathbf{u}_2^- \in C^\alpha(\overline{S_h})$. Using the admissibility condition (b) in Definition 10.4.1, by (10.4.14) and applying Theorem 10.2.2, and also utilizing the fact that $\mathbf{u}_1$ is continuous at the vertex $\mathbf{0}$, we have

$$\mathbf{u}_1(\mathbf{0}) = \mathbf{0},$$

which contradicts to the admissibility condition (c) in Definition 10.4.1.

Case 2 ($n=3$)**:** Since $(\Omega_j; q_j, \eta_j)$, $j=1,2$, are admissible, we know that $\eta_2 \equiv 0$ on $\partial\Omega_2$. Therefore, from (10.4.10), following the setup of Theorem 10.3.1, it yields that

$$\mathbf{u}_2^- = \mathbf{u}_2^+ = \mathbf{u}_1^+, \quad T_\nu \mathbf{u}_2^- = T_\nu \mathbf{u}_2^+ = T_\nu \mathbf{u}_1^+ \text{ on } \Gamma_h^\pm \times (-M,M). \tag{10.4.13}$$

Moreover, we take $h \in \mathbb{R}_+$ sufficient small such that

$$\mathscr{L}\mathbf{u}_1^+ + \omega^2 \mathbf{u}_1^+ = \mathbf{0} \text{ in } B_h, \quad \mathscr{L}\mathbf{u}_2^- + \omega^2 q_2 \mathbf{u}_2^- = \mathbf{0} \text{ in } S_h \times (-M,M). \tag{10.4.14}$$

By the well-posedness of the direct problem (10.4.1) with $\eta \equiv 0$, we know that $\mathbf{u}_2 \in H^1(B_R)^3$ where B_R is a ball centered at the origin with the radius $R \in \mathbb{R}_+$ such that $\overline{S_h} \times [-M,M] \Subset B_R$ and $B_R \Subset \mathbb{R}^3\backslash\overline{\Omega_1}$. Let $\widetilde{q}_2 = q_2\chi_{\Omega_2} + 1\chi_{B_R\backslash\overline{\Omega_2}}$. Then

$$\mathscr{L}\mathbf{u}_1 + \omega^2 \mathbf{u}_1 = \mathbf{0} \text{ in } B_R, \quad \mathscr{L}\mathbf{u}_2 + \omega^2 \widetilde{q}_2 \mathbf{u}_2 = \mathbf{0} \text{ in } B_R.$$

Since $\widetilde{q}_2 \in L^\infty(B_R)$, by the interior elliptic regularity estimate [37], we have $\mathbf{u}_2^- \in H^2(B_{R'})^3$, where $B_{R'} \Subset B_R\backslash(\overline{S_h} \times [-M,M])$. Again using the interior elliptic regularity estimate, we have $\mathbf{u}_2^- \in W^{2,4}(\overline{S_h} \times [-M,M])^3$. Using the Sobolev embedding theorem, we have $\mathbf{u}_2^- \in C^{1,1/4}(\overline{S_h} \times [-M,M])^3$. Clearly, $\mathbf{u}_1^+$ is real analytic in $\overline{S_h} \times [-M,M]$. Therefore, one has $\mathbf{u}_1^+ - \mathbf{u}_1^+ \in C^{1,1/4}(\overline{S_h} \times [-M,M])^3$. Since $(\Omega_j; q_j, \eta_j)$, $j=1,2$, are admissible, we know that $q_2 \in C^\alpha(\overline{S_h} \times [-M,M])$. Using the admissibility condition (b) in Definition 10.4.1, by (10.4.14) and applying Corollary 10.3.2, we have

$$\mathbf{u}_1(\mathbf{0}) = \mathbf{0},$$

which contradicts to the admissibility condition (c) in Definition 10.4.1.

The conclusion (10.4.9) can be immediately obtained by using the contradiction argument and (10.4.8).

The proof is complete. □

Based on Definition 10.4.1, if we further assume that the surface parameter η is constant, we can recover η simultaneously in $\mathbb{R}^2$ once the shape of the scatter, namely, Ω is determined. However, in determining the surface conductive parameter, we need to assume that $q_1 = q_2 := q$ are known.

Theorem 10.4.2. *Consider the elastic scattering problem* (10.4.1) *in* $\mathbb{R}^2$ *associated with the incident elastic wave field* $\mathbf{u}^i$ *and two elastic scatters* $(\Omega_j; q, \eta_j)$ *being admissible scattering config-*

uration, where $\Omega_j = \Omega$ *for* $j = 1,2$ *and* $\eta_j \neq 0$, $j = 1,2$ *are two constants. Let* $\mathbf{u}_\beta^{j,\infty}(\hat{\mathbf{x}};\mathbf{u}^i)$ *be the far-field pattern associated with the scatter* $(\Omega_j; q_j, \eta_j)$ *and the incident field* $\mathbf{u}^i$, $\beta = t, p$, *or* s. *Suppose that*

$$\mathbf{u}_\beta^{1,\infty}(\hat{\mathbf{x}};\mathbf{u}^i) = \mathbf{u}_\beta^{2,\infty}(\hat{\mathbf{x}};\mathbf{u}^i), \quad \hat{\mathbf{x}} \in \mathbb{S}^1, \tag{10.4.15}$$

for a fixed incident wave $\mathbf{u}^i$. *Then if* ω *is not an eigenvalue of the partial differential operator* $\mathscr{L} + \omega^2 q$ *in* $H_0^1(\Omega)$, *we have* $\eta_1 = \eta_2$.

Proof. Due to (10.4.15), we have $\mathbf{u}_1^+ = \mathbf{u}_2^+$ for all $\mathbf{x} \in \mathbb{R}^2 \backslash \overline{\Omega}$ and thus $T_\nu \mathbf{u}_1^+ = T_\nu \mathbf{u}_2^+$ on $\partial\Omega$. Combining with the transmission condition in the scattering problem (10.4.1), we deduce that

$$\mathbf{u}_1^- = \mathbf{u}_1^+ = \mathbf{u}_2^+ = \mathbf{u}_2^- \quad \text{on } \partial\Omega.$$

Thus, we have

$$T_\nu(\mathbf{u}_1^- - \mathbf{u}_2^-) = T_\nu(\mathbf{u}_1^+ - \mathbf{u}_2^+) + \eta_1\mathbf{u}_1^+ - \eta_2\mathbf{u}_2^+ = (\eta_1 - \eta_2)\mathbf{u}_1^- \quad \text{on } \partial\Omega.$$

Set $\mathbf{v} := \mathbf{u}_1^- - \mathbf{u}_2^-$. Then $\mathbf{v}$ fulfills

$$\begin{cases} (\mathscr{L} + \omega^2 q)\mathbf{v} = \mathbf{0} & \text{in } \Omega, \\ \mathbf{v} = \mathbf{0} & \text{on } \partial\Omega, \\ T_\nu \mathbf{v} = (\eta_1 - \eta_2)\mathbf{u}_1^- & \text{on } \partial\Omega. \end{cases} \tag{10.4.16}$$

Since ω is not an eigenvalue of the operator $\mathscr{L} + \omega^2 q$ in $H_0^1(\Omega)$, one must have $\mathbf{v} = 0$ to (10.4.16). Substituting this into the Neumann boundary condition of (10.4.16), we know that $(\eta_1 - \eta_2)\mathbf{u}_1^- = T_\nu \mathbf{v} = \mathbf{0}$ on $\partial\Omega$.

Next, we prove the uniqueness of η by contradiction. Assume that $\eta_1 \neq \eta_2$. Since $(\eta_1 - \eta_2)\mathbf{u}_1^- = \mathbf{0}$ on $\partial\Omega$ and η_j, $j = 1,2$ are constants, we can deduce that $\mathbf{u}_1^- = \mathbf{0}$ on $\partial\Omega$. Then u_1^- satisfies

$$\begin{cases} (\mathscr{L} + \omega^2 q)\mathbf{u}_1^- = \mathbf{0} & \text{in } \Omega, \\ \mathbf{u}_1^- = \mathbf{0} & \text{on } \partial\Omega. \end{cases}$$

Similar to (10.4.16), this Dirichlet problem also only has a trivial solution $\mathbf{u}_1^- = \mathbf{0}$ in Ω due to that ω is not an eigenvalue of $\mathscr{L} + \omega^2 q$ in $H_0^1(\Omega)$. Hence, we can derive $\mathbf{u}_1^+ = \mathbf{u}_1^- = 0$ and

$$T_\nu \mathbf{u}_1^- = T_\nu \mathbf{u}_1^+ + \eta_1 \mathbf{u}_1^+ = T_\nu \mathbf{u}_1^+ = \mathbf{0} \quad \text{on } \partial\Omega,$$

which implies that $\mathbf{u}_1 \equiv \mathbf{0}$ in $\mathbb{R}^n$ and thus $\mathbf{u}_1^{sc} = -\mathbf{u}^i$. This contradicts to the fact that $\mathbf{u}_1^{sc}$ satisfies the Kupradze radiation condition.

The proof is complete. □

Remark 10.4.3. In Theorem 10.4.2, it is required that ω is not an eigenvalue of $\mathscr{L} + \omega^2 q$ in $H_0^1(\Omega)$. Clearly, if q is negative-valued in Ω, this condition is obviously fulfilled. On the other hand, if q is positive-valued in Ω, then this condition can be readily fulfilled when $\omega \in \mathbb{R}_+$ is sufficiently small.

References

1. M. Abramowitz and I. A. Stegun, *Handbook of mathematical functions: with formulas, graphs, and mathematical tables*, vol. 55, Courier Corporation, (1964). s
2. T. S. Angell and A. Kirsch, *The conductive boundary condition for Maxwells equations*, SIAM Journal on Applied Mathematics, **52(6)** (1992), 1597–1610.

3. T. Arens, *The scattering of plane elastic waves by a one-dimensional periodic surface,* Math. Meth. Appl. Sci., **22** (1999), 55–72.
4. C. Bellis, F. Cakoni and B. B. Guzina, *Nature of the transmission eigenvalue spectrum for elastic bodies*, IMA J. Appl. Math., **78**(2013), 895–923.
5. C. Bellis and B. B. Guzina, *On the existence and uniqueness of a solution to the interior transmission problem for piecewise-homogeneous solids*, J. Elasticity, **101** (2010), 29–57.
6. E. Blåsten, *Nonradiating sources and transmission eigenfunctions vanish at corners and edges*, SIAM J. Math. Anal., **50(6)** (2018), 6255–6270.
7. E. Blåsten, X. Li,H. Liu and Y. Wang, On vanishing and localization near cusps of transmission eigenfunctions: a numerical study, Inverse Problems, **33** (2017), 105001.
8. E. Blåsten and Y.-H. Lin, *Radiating and non-radiating sources in elasticity*, Inverse Problems, 35, 1 (2019) 015005.
9. E. Blåsten and H. Liu, *On vanishing near corners of transmission eigenfunctions*, J. Funct. Anal., **273(11)** (2017), 3616–3632. Addendum, arXiv:1710.08089
10. E. Blåsten and H. Liu, *Recovering piecewise constant refractive indices by a single far-field pattern*,Inverse Problems, **36** (2020), 085005.
11. E. Blåsten and H. Liu, *On corners scattering stably and stable shape determination by a single far-field pattern*, Indiana Univ. Math. J., **70** (2021), no. 3, 907–947.
12. E. Blåsten and H. Liu, *Scattering by curvatures, radiationless sources, transmission eigenfunctions and inverse scattering problems*, SIAM Journal on Mathematical Analysis, **53** (2021), no. 4, 3801–3837.
13. E. Blåsten, H. Liu and J. Xiao, *On an electromagnetic problem in a corner and its applications*, Anal. PDE, **14(7)** (2021), 2207–2224.
14. E. Blåsten, L. Päivärinta and J. Sylvester, *Corners always scatter*, Comm. Math. Phys., **331** (2014), 725–753.
15. F. Cakoni and H. Haddar, *Transmission eigenvalues in inverse scattering theory*, in "Inverse Problems and Applications: Inside Out II", Math. Sci. Res. Inst. Publ., Vol. 60, pp. 529–580, Cambridge Univ. Press., Cambridge, (2013).
16. F. Cakoni and J. Xiao, *On corner scattering for operators of divergence form and applications to inverse scattering*, Comm. Partial Differential Equations, in press, (2020).
17. X. Cao, H. Diao and H. Liu, *Determining a piecewise conductive medium body by a single far-field measurement*, CSIAM Trans. Appl. Math., **1** (2020), 740–765.
18. D. Colton and R. Kress, *Looking back on inverse scattering theory*, SIAM Review, **60(40)** (2018), 779–807.
19. D. Colton and R. Kress, *Inverse acoustic and electromagnetic scattering theory*, volume 93 of Applied Mathematical Sciences. Springer, Cham, 4th edition, 2019.
20. M. Costabel, *Boundary integral operators on Lipschitz domains: elementary results*, SIAM J. Math. Anal., **19(3)** (1988), 613–626.
21. M. Costabel, *A remark on the regularity of solutions of Maxwell's equations on Lipschitz domains*, Math. Methods Appl. Sci., **12** (1990), 365–368.
22. M. Dauge, *Elliptic Boundary Value Problems in Corner Domains-Smoothness and Asymptotics of Solutions*, Lecture Notes in Mathematics, **1341**. Springer-Verlag, Berlin (1988).
23. H. Diao, X. Cao and H. Liu, *On the geometric structures of conductive transmission eigenfunctions and their application*, Comm. Partial Differential Equations, **46(4)** (2021), 630–679.
24. H. Diao, H. Liu, and B. Sun, *On a local geometric property of the generalized elastic transmission eigenfunctions and application*, Inverse Problems, **37** (2021), no. 10, Paper No. 105015, 36 pp.
25. S. A. Faghidian, *A smoothed inverse eigenstrain method for reconstruction of the regularized residual fields*, International Journal of Solids and Structures, **51** (2014), 4427–4434.
26. S. A. Faghidian, *Inverse determination of the regularized residual stress and eigenstrain fields due to surface peening*, J. Strain Analysis, **50(2)** (2015), 84–91.
27. P. Grisvard, *Boundary Value Problems in Non-Smooth Domains*, Pitman, London (1985).
28. P. Hähner, *A uniqueness theorem in inverse scattering of elastic waves*, IMA J. Appl. Math., **51** (1993), 201–215.
29. P. Hähner, *On acoustic, electromagnetic, and elastic scattering problems in inhomogeneous media*, Universität Göttingen, Habilitation Thesis, (1998).
30. P. Hähner, *On uniqueness for an inverse problem in inhomogeneous elasticity*, IMA J. Appl. Math., **67** (2002), 127–143.
31. V. D. Kupradze, *Three-dimensional Problems of the Mathematical Theory of Elasticity and Thermoelasticity*, Amsterdam, North-Holland, (1979).
32. X. Ji and H. Liu, On isotropic cloaking and interior transmission eigenvalue problems, European J. Appl. Math., **29(2)** (2018), 253–280.
33. H. Liu, *On local and global structures of transmission eigenfunctions and beyond*, J. Inverse Ill-Posed Probl., **30(2)** (2022), 287–305.
34. H. Liu and C.-H. Tsou, *Stable determination of polygonal inclusions in Calderón's problem by a single partial boundary measurement*, Inverse Problems, **36(8)** (2020), 085010, 23 pp.

35. H. Liu, C.-H. Tsou and W. Yang, *On Calderón's inverse inclusion problem with smooth shapes by a single partial boundary measurement*, arXiv:2006.10586
36. H. Liu and J. Zou, *On uniqueness in inverse acoustic and electromagnetic obstacle scattering problems*, Journal of Physics: Conference Series, Applied Inverse Problems, Vancouver, Canada, (2007).
37. W. Mclean, *Strongly Elliptic Systems and Boundary Integral Equation*, Cambridge University Press, Cambridge, (2000).
38. Q. Meng, Z. Bai, H. Diao and H. Liu, *Effective medium theory for embedded obstacles in elasticity with applications to inverse problems*, SIAM J. Appl. Math., **82** (2022), No. 2, 720–749.
39. S. Nicaise, *About the Lamé system in a polygonal or a polyhedral domain and a coupled problem between the Lamé system and the plate equation, I: Regularity of the solutions*, Ann. Scuola Norm. Sup. Pisa Cl. Sci., **19**(4) (1992), 327–361.
40. A. Rössle, *Corner singularities and regularity of weak solutions for the two-dimensional Lamé equations on domains with angular corners*, Journal of Elasticity, **60** (2000), 57–75.

Chapter 11
Geometric Properties of Helmholtz's Transmission Eigenfunctions Induced by Curvatures and Applications

11.1 Overview

In this chapter, we consider two types of scenarios in inverse scattering problems, where we follow the treatment in [9]. The first one is concerned with radiationless or non-radiating monochromatic sources, and the other one is concerned with non-radiating waves that impinge against a certain given scatterer consisting of an inhomogeneous index of refraction. In this chapter, we show that such invisible objects have certain geometrical properties. This allows us to classify radiating sources and incident waves that are always radiating for scatterers. Moreover, they also help us to provide unique determination results for a longstanding inverse scattering problem in certain scenarios of practical importance. The second one is concerned with geometrical structure of transmission eigenfunctions at highly curved point. First, we provide a relationship among the value of the transmission eigenfunction, the diameter of the domain, and the underlying refractive index, which indicates that if the domain is sufficiently small, then the transmission eigenfunction is nearly vanishing. Then we further localize and geometrize this "smallness" result. Briefly, interior transmission eigenfunctions must be nearly vanishing at a high-curvature point on the boundary. Moreover, the higher the curvature, the smaller the eigenfunction must be at the high-curvature point. This nearly vanishing behavior readily implies that as long as the shape of a scatterer possesses a highly curved part, then it scatters every incident wave field nontrivially unless the wave is vanishingly small at the highly curved part. Furthermore, we give uniqueness results for Schiffer's problem in determining the support of an active source or an inhomogeneous medium with a single far-field pattern in certain scenarios of practical interest. These are the unique determination results for Schiffer's problem concerning scatterers with smooth shapes. The corresponding uniqueness result on Calderón's inverse inclusion problem with smooth shapes by a single partial boundary measurement can be found in [50].

The rest of the chapter is organized as follows. In Sect. 11.2, we consider the radiationless source and its geometric characterizations. In Sect. 11.3, we consider the non-radiating waves and transmission eigenfunctions and their geometric characterizations. Section 11.4 is devoted to the study of Schiffer's problem.

H. Diao, H. Liu, *Spectral Geometry and Inverse Scattering Theory*,
https://doi.org/10.1007/978-3-031-34615-6_11

11.2 Geometrical Characterizations of Radiationless Sources

11.2.1 Wave Scattering from an Active Source

Let $f : \mathbb{R}^n \to \mathbb{C}$ be a function having compact support, $f = \chi_\Omega \varphi$, where $\Omega \subset \mathbb{R}^n$ is a bounded domain* and $\varphi \in L^\infty(\mathbb{R}^n)$, $\varphi \neq 0$, in a neighborhood of $\partial\Omega$. The set Ω is the external *shape* of f, while φ describes the *intensity* of the source at various points in Ω. We assume that φ and Ω do not depend on the wavenumber k that's fixed. In other words, we are considering monochromatic scattering. The source f produces a radiating wave $u \in H^2_{loc}(\mathbb{R}^n)$ given by the unique solution to

$$(\Delta + k^2)u = f, \qquad \lim_{r\to\infty} r^{\frac{n-1}{2}}(\partial_r - \mathrm{i}k)u = 0, \tag{11.2.1}$$

where $r = |\mathbf{x}|$ for $x \in \mathbb{R}^n$. The limit in (11.2.1) is known as the Sommerfeld radiation condition which characterizes the outgoing nature of the radiating wave. By the limiting absorption principle (cf. [61]), the solution to (11.2.1) can be computed as follows:

$$\begin{aligned} u = (\Delta + k^2)^{-1} f &= \lim_{\varepsilon\to+0} \left(\Delta + (k - \mathrm{i}\varepsilon)\right)^{-1} f \\ &= -\lim_{\varepsilon\to+0} \int_{\mathbb{R}^n} \frac{e^{\mathrm{i}x\cdot\xi} \widehat{f}(\xi)}{|\xi|^2 - (k - \mathrm{i}\varepsilon)^2} \,\mathrm{d}\xi, \end{aligned} \tag{11.2.2}$$

where $\widehat{f}(\xi) := \mathscr{F}f(\xi) = (2\pi)^{-n} \int_{\mathbb{R}^n} f(x)e^{-\mathrm{i}\xi\cdot x}\,\mathrm{d}x$ signifies the Fourier transform of f. Inverting the Fourier transform in (11.2.2), one has the following integral representation:

$$u = (\Delta + k^2)^{-1} f := -\frac{\mathrm{i}}{4}\left(\frac{k}{2\pi}\right)^{\frac{n-2}{2}} \int_{\mathbb{R}^n} |\mathbf{x} - \mathbf{y}|^{\frac{2-n}{2}} H^{(1)}_{\frac{n-2}{2}}(k|\mathbf{x} - \mathbf{y}|) f(\mathbf{y})\,\mathrm{d}\mathbf{y}, \tag{11.2.3}$$

where $H^{(1)}_{(n-2)/2}$ is the first-kind Hankel function of order $(n-2)/2$. The stationary phase applied to (11.2.3) yields that

$$u(\mathbf{x}) = \frac{e^{\mathrm{i}k|\mathbf{x}|}}{|\mathbf{x}|^{(n-1)/2}} C_{n,k} \int_{\mathbb{R}^n} e^{-\mathrm{i}k\hat{\mathbf{x}}\cdot\mathbf{y}} f(\mathbf{y})\,\mathrm{d}\mathbf{y} + \mathscr{O}(|\mathbf{x}|^{\frac{n}{2}}) \quad \text{as} \quad |\mathbf{x}| \to \infty, \tag{11.2.4}$$

where $\hat{\mathbf{x}} := \mathbf{x}/|\mathbf{x}| \in \mathbb{S}^{n-1}$, $\mathbf{x} \in \mathbb{R}^n \backslash \{0\}$, and

$$C_{n,k} = \frac{-\mathrm{i}}{\sqrt{8\pi}} \left(\frac{k}{2\pi}\right)^{\frac{n-2}{2}} e^{-\frac{(n-1)\pi}{4}\mathrm{i}}.$$

The *far-field pattern* of u is given by

$$u_\infty(\hat{\mathbf{x}}) := C_{n,k} \int_{\mathbb{R}^n} e^{-\mathrm{i}k\hat{\mathbf{x}}\cdot\mathbf{y}} f(\mathbf{y})\,\mathrm{d}\mathbf{y} = (2\pi)^n C_{n,k}\, \mathscr{F}f(k\hat{\mathbf{x}}) \in L^2(\mathbb{S}^{n-1}). \tag{11.2.5}$$

As discussed earlier, we are particularly interested in the case where the source f does not radiate, and one has $u_\infty \equiv 0$. By the Rellich lemma (cf. [18]), which establishes the one-to-one correspondence between the wave field and its far-field pattern, one has $u = 0$ in the unbounded component of $\mathbb{R}^n \backslash \overline{\Omega}$. Hence, in such a case, the source is also referred to as non-radiating or radiationless. According to (11.2.5), one clearly has $\widehat{f}(k\hat{\mathbf{x}}) \equiv 0$ for $\hat{\mathbf{x}} \in \mathbb{S}^{n-1}$ for a radiationless source. Hence, characterizing radiationless sources, one actually characterizes functions with compact supports whose Fourier transforms vanish on the sphere of radius k.

* We shall consider only bounded domains Ω for which $H^2_0(\Omega) = \{u_{|\Omega} \mid u \in H^2(\mathbb{R}^n), u = 0 \text{ in } \mathbb{R}^n \setminus \overline{\Omega}\}$.

A classical example is given by a source of constant intensity supported on a ball, namely, the source is of the form $f = c_0\chi_{B_{r_0}}$, where $c_0 \in \mathbb{C}$, $c_0 \neq 0$ and $B_{r_0} := B(0, r_0)$ is a central ball of radius $r_0 \in \mathbb{R}_+$. By the properties of the Bessel functions (cf. [31, B.3]), one has $\mathscr{F}\chi_{B_{r_0}}(k\hat{\mathbf{x}}) = \gamma J_{n/2}(kr_0)$, where $\gamma > 0$ is a constant depending on the dimension n, wavenumber k, and radius r_0 of the ball. Here, $J_{n/2}$ is a *Bessel function* of order $n/2$. Hence, if the radius of the support of the source, measured in units of wavenumber, is a zero of the Bessel function, then the source is radiationless. In particular, by using the fact that there is a smallest positive zero of the Bessel function, a sufficiently small kr_0 implies that the source must be radiating. In what follows, we shall first generalize this classical example to a more general scenario. Indeed, we provide a quantitative relationship between the intensity of a generic radiating source and the diameter of its support (in units of wavenumber). This relationship readily implies that if a source's support is sufficiently small (compared to its intensity as well as the underlying wavelength), then it must be radiating.

We further localize and geometrize the result concerning sources of small support. By localizing, we mean that instead of considering a source with a small support, we consider a source whose support is "locally small." We know that for a domain with a small diameter, the (extrinsic) curvature of its boundary surface must be large. Hence, we naturally characterize the "locally small" domain as the existence of a boundary point where the boundary surface curvature is very large. This is referred to as the "geometrization" of the "local smallness." That is, we consider sources whose support contains a high-curvature point on its boundary. We basically extend the result on the source with a sufficiently small support to this "locally small" case. In fact, we give a certain quantitative relationship between the intensity of the source at the high-curvature point and the corresponding curvature. This relationship readily implies that if the support of a source contains a boundary point with a sufficiently high curvature, then it must be radiating. It also implies that the intensity of a radiationless source must be nearly vanishing near a high-curvature point on the boundary of its support.

11.2.2 Small-Support Sources Must Be Radiating

In this section, we provide that a source with a relatively small support compared to its intensity must be radiating in dimensions $n = 2$ and $n = 3$. Consider the scattering problem (11.2.1); we have the following:

Theorem 11.2.1. *There is a universal constant $C \in \mathbb{R}_+$ with the following property. Let $n \in \{2, 3\}$, $k \in \mathbb{R}_+$, and $\Omega \subset \mathbb{R}^n$ be a bounded Lipschitz domain whose complement is connected. Let $\varphi \in C^{1/2}(\mathbb{R}^n)$, and let $u \in H^2_{loc}(\mathbb{R}^n)$ be the unique outgoing solution to $(\Delta + k^2)u = \chi_\Omega\varphi$. Let $\delta = d(\Omega)k$ be the diameter of Ω in units*† *of k^{-1}, where and also in what follows, $d(\Omega)$ signifies the Euclidean diameter of Ω. If*

$$\frac{\sup_{\partial\Omega}|\varphi|}{\sup_\Omega|\varphi| + k^{-1/2}[\varphi]_{1/2,\Omega}} > C\big((1+\delta)\delta^{n/2} + 1\big)\delta^{1/2}, \tag{11.2.6}$$

then u_∞ cannot be identically zero.

We use the notation

$$[\varphi]_{\alpha,\Omega} := \sup_{\substack{\mathbf{x}\neq\mathbf{y}\\ \mathbf{x},\mathbf{y}\in\Omega}} \frac{|\varphi(\mathbf{x}) - \varphi(\mathbf{y})|}{|\mathbf{x}-\mathbf{y}|^\alpha}$$

for the Hölder space $C^\alpha(\overline{\Omega})$ seminorm.

† For example, if the wavelength λ is π meters and $d(\Omega) = 1\,\text{m}$, we have $k^{-1} = \lambda/(2\pi) = 0.5\,\text{m}$. Then $\delta = 2$, i.e., $d(\Omega)$ is twice the length of k^{-1}.

Remark. If φ is less smooth, for example, $\varphi \in C^\alpha(\mathbb{R}^n)$ with $0 < \alpha < 1/2$, the same conclusion holds if (11.2.6) is replaced by

$$\frac{\sup_{\partial\Omega}|\varphi|}{\sup_{\Omega}|\varphi| + k^{-\alpha}[\varphi]_{\alpha,\Omega}} > C\big((1+\delta)\delta^{(n+1)/2} + \delta^\alpha\big). \tag{11.2.7}$$

We note that the diameter of Ω is not the important quantity. Instead what's important is the diameter of *any* component Ω_c of Ω. In other words, if Ω has a component where φ and the diameter satisfy (11.2.6) or (11.2.7), then the far field cannot vanish identically. This is despite what shape or intensity the source might have elsewhere.

Theorem 11.2.2. *There is a universal constant $C \in \mathbb{R}_+$ with the following property. Let $n \in \{2,3\}$, $k \in \mathbb{R}_+$, and $\Omega \subset \mathbb{R}^n$ be a bounded Lipschitz domain whose complement is connected. Let $\varphi \in L^\infty(\mathbb{R}^n)$ with $\varphi_{|\Omega_c} \in C^\alpha(\overline{\Omega_c})$ for some $0 < \alpha \leq 1/2$ and some component Ω_c of Ω. Let $u \in H^2_{loc}(\mathbb{R}^n)$ be the unique outgoing solution to $(\Delta + k^2)u = \chi_\Omega\varphi$. Let $\delta = d(\Omega_c)k$ be the diameter of Ω_c in units of k^{-1}. If*

$$\frac{\sup_{\partial\Omega_c}|\varphi|}{\sup_{\Omega_c}|\varphi| + k^{-\alpha}[\varphi]_{\alpha,\Omega_c}} > C\big((1+\delta)\delta^{(n+1)/2} + \delta^\alpha\big), \tag{11.2.8}$$

then u_∞ cannot be identically zero.

An immediate consequence of Theorems 11.2.1 and 11.2.2 is that for a source with a given strength, namely, $\sup_{\mathbb{R}^n}|\varphi| + k^{-\alpha}[\varphi]_{\alpha,\mathbb{R}^n}$ fixed, if the size of (a component of) its support is sufficiently small in units of k^{-1} and $\sup_{\partial\Omega}|\varphi|$ has a positive lower bound, then it must be radiating; or in other words, if it is radiationless with a sufficiently small support, then its intensity must be much smaller on the boundary of its support than on the interior. It is particularly surprising to note that the intensity of a source does not matter if it is constant. No matter how small intensity it has, if it has a small size, it will prevent the total far field from being identically zero.

Corollary 11.2.1. *There is a universal $\varepsilon > 0$ such that if φ is constant and $d(\Omega) < \varepsilon/k$, then $u_\infty \not\equiv 0$. This ε can be chosen as the smallest positive solution to the equations*

$$C\big((1+\varepsilon)\varepsilon^{n/2} + 1\big)\varepsilon^{1/2} = 1 \tag{11.2.9}$$

with $n = 2$ and $n = 3$.

One can also infer that the size of a radiationless source must have a positive lower bound that depends on its intensity. More precisely, referring to formula (11.2.7), we have the following:

Corollary 11.2.2. *Let $\varphi \in C^\alpha(\mathbb{R}^n)$ with $0 < \alpha \leq 1/2$, and assume that the source $\chi_\Omega\varphi$ is radiationless. Then the diameter[‡] of Ω must be at least*

$$\min\left(1, \left(\frac{1}{3C}\frac{\sup_{\partial\Omega}|\varphi|}{\sup_{\Omega}|\varphi| + k^{-\alpha}[\varphi]_{\alpha,\Omega}}\right)^{1/\alpha}\right)\frac{1}{k}. \tag{11.2.10}$$

Here, C is as in Theorem 11.2.1.

Proof. If $d(\Omega) < k^{-1}$, then $\delta < 1$, and the right-hand side of (11.2.7) is smaller than $3C\delta^\alpha$. □

In order to prove Theorem 11.2.1, we first derive some auxiliary results. It is of particular importance for our results to know the exact dependence on k and Ω in the estimate.

[‡] This claim holds for the diameter of any component of Ω too by Theorem 11.2.2.

Lemma 11.2.1. *Let $\Omega \subset \mathbb{R}^n$, $n \in \{2,3\}$ be a bounded domain. Let $\varphi \in L^2(\mathbb{R}^n)$ and $u \in L^2_{loc}(\mathbb{R}^n)$ be the outgoing solution to $(\Delta + k^2)u = \chi_\Omega \varphi$. If $u = 0$ in $\mathbb{R}^n \setminus \overline{\Omega}$, then $u \in C^{1/2}(\mathbb{R}^n)$ and*

$$\|u\|_{L^\infty(\mathbb{R}^n)} + k^{-1/2}[u]_{1/2,\mathbb{R}^n} \leq C_n k^{n/2-1}\big(k^{-1} + d(\Omega)\big)\|\varphi\|_{L^2(\Omega)}$$

for some finite constant C_n depending only on the dimension n.

Proof. Denote

$$f(\mathbf{x}) = \chi_\Omega(\mathbf{x})\varphi(\mathbf{x}), \quad U(\mathbf{y}) = u(\mathbf{y}/k),$$
$$F(\mathbf{y}) = f(\mathbf{y}/k)/k^2, \quad \Omega_k = \{\mathbf{y} \in \mathbb{R}^n \mid \mathbf{y}/k \in \Omega\}.$$

Then $(\Delta + 1)U = F$ in $\mathbb{R}^n$ with $U \in L^2_{loc}$ and $F \in L^2(\mathbb{R}^n)$ being equal to zero outside of Ω_k.

Firstly note that $(-|\xi|^2 + 1)\hat{U}(\xi) = \hat{F}(\xi)$ and so $|\xi|^2\hat{U}(\xi) = \hat{U}(\xi) - \hat{F}(\xi)$ for $\xi \in \mathbb{R}^n$. Note also that $U = 0$ in $\mathbb{R}^n \setminus \overline{\Omega_k}$, and so actually $U \in L^2(\mathbb{R}^n)$ instead of being there locally only. We can thus deduce that $U \in H^2(\mathbb{R}^n)$ with the following estimate:

$$\|U\|_{H^2(\mathbb{R}^n)} = \big\|(1 + |\cdot|^2)\hat{U}\big\|_{L^2(\mathbb{R}^n)} = \|2\hat{U} - \hat{F}\|_{L^2(\mathbb{R}^n)} \leq \|F\|_{L^2(\mathbb{R}^n)} + 2\|U\|_{L^2(\mathbb{R}^n)}$$

by the Plancherel theorem.

By Sobolev embedding (e.g., Theorem 4.12 part II in [1]), there is a constant $C_n \in \mathbb{R}_+$ such that $H^2(\mathbb{R}^n) \hookrightarrow C^{1/2}(\mathbb{R}^n)$ with the estimate $\|U\|_{C^{1/2}(\mathbb{R}^n)} \leq C_n\|U\|_{H^2(\mathbb{R}^n)}$. If we combine this with the previous paragraph's estimate and recall that $F = U = 0$ in $\mathbb{R}^n \setminus \overline{\Omega_k}$, we have

$$\|U\|_{C^{1/2}(\mathbb{R}^n)} \leq C_n\big(\|F\|_{L^2(\Omega_k)} + 2\|U\|_{L^2(\Omega_k)}\big).$$

Let us return to the non-scaled variable x. Simple calculations show that the above estimate is equivalent to

$$\|u\|_{L^\infty(\mathbb{R}^n)} + k^{-1/2}[u]_{1/2,\mathbb{R}^n} \leq C_n k^{n/2}\big(k^{-2}\|\varphi\|_{L^2(\Omega)} + 2\|u\|_{L^2(\Omega)}\big).$$

We need an a priori estimate for $\|u\|_{L^2(\Omega)}$. The right type of estimate for our situation has been shown in [12, Section 2]. For a solution u to $(\Delta + k^2)u = f$ that satisfies a radiation condition and where the source f is zero outside a bounded domain Ω_s, they show that

$$\|u\|_{L^2(\Omega_r)} \leq C_n k^{-1}\sqrt{d(\Omega_r)d(\Omega_s)}\|f\|_{L^2(\Omega_s)}.$$

In our case, we have $\Omega_r = \Omega_s = \Omega$ and so the estimate $\|u\|_{L^2(\Omega)} \leq C_n k^{-1} d(\Omega)\|\varphi\|_{L^2(\Omega)}$. The proof follows by combining this with the previous section's final estimate. □

The next lemma shows that having zero far field is a linear property. A sum of waves has zero far field if and only if the individual summands do so too. This will be used to prove Theorem 11.2.2.

Lemma 11.2.2. *Let $\Omega \subset \mathbb{R}^n$ be a bounded Lipschitz domain whose complement is connected. Let $\varphi \in L^\infty(\mathbb{R}^n)$, and let $u \in H^2_{loc}(\mathbb{R}^n)$ satisfy $(\Delta + k^2)u = \chi_\Omega\varphi$ with Sommerfeld's radiation condition at infinity. Let Ω_c be a component of Ω, and let $u_c \in H^2_{loc}(\mathbb{R}^n)$ satisfy $(\Delta + k^2)u_c = \chi_{\Omega_c}\varphi$ and Sommerfeld's radiation condition at infinity.*

Then the total far field vanishes, $u_\infty = 0$, if and only if the individual far fields vanish, namely, $(u_c)_\infty = (u - u_c)_\infty = 0$.

Proof. The trivial direction follows by the linearity of the source to far-field map since $(\Delta + k^2)(u - u_c) = \chi_\Omega\varphi - \chi_{\Omega_c}\varphi$. Let us prove the nontrivial direction.

Assume that the far field of u vanishes. By Rellich's lemma and the connectedness of $\mathbb{R}^n \setminus \Omega$, we see that $u = 0$ in $\mathbb{R}^n \setminus \overline{\Omega}$. In particular, $u_{|\Omega_c} \in H_0^2(\Omega_c)$. Let

$$\tilde{u} = \begin{cases} u, & \Omega_c, \\ 0, & \mathbb{R}^n \setminus \Omega_c. \end{cases}$$

Because $u_{|\Omega_c} \in H_0^2(\Omega_c)$ and $(\Delta + k^2)u = \chi_{\Omega_c}\varphi$ in a neighborhood of $\overline{\Omega_c}$ (one that does not intersect $\Omega \setminus \Omega_c$), we see that

$$(\Delta + k^2)\tilde{u} = \chi_{\Omega_c}\varphi$$

in $\mathbb{R}^n$. Furthermore, we note that $\tilde{u}$ satisfies the Sommerfeld radiation condition at infinity trivially. We see that $\tilde{u} = u_c$, as both solve the same source scattering problem whose solution is known to be unique. This implies that $u_c = 0$ in $\mathbb{R}^n \setminus \overline{\Omega_c}$ and so $(u_c)_\infty = 0$, which readily yields the claim. □

We are in a position to present the proofs of Theorems 11.2.1 and 11.2.2.

Proof of 11.2.1. Let us prove the claim by contradiction. Assume that $u_\infty = 0$. By Rellich's theorem, the connectedness of $\mathbb{R}^n \setminus \overline{\Omega}$, and the unique continuation for $\Delta + k^2$, we see that $u = 0$ in $\mathbb{R}^n \setminus \overline{\Omega}$. Because $u \in H^2_{loc}$, we see that $u_{|\Omega} \in H_0^2(\Omega)$.

Let $g = \varphi_{|\Omega} - k^2 u_{|\Omega}$. By Lemma 11.2.1, we get that g is a continuous function. Let $\mathbf{p} \in \partial\Omega$. Since e $u_{|\partial\Omega} = 0$, we have

$$\varphi(\mathbf{p})m(\Omega) = g(\mathbf{p})m(\Omega) = \int_\Omega g(\mathbf{p})d\mathbf{x},$$

where and also in what follows, $m(\Omega)$ denotes the measure of Ω. On the other hand,

$$\int_\Omega g(\mathbf{x})d\mathbf{x} = \int_\Omega (\varphi - k^2 u)(x)d\mathbf{x} = \int_\Omega 1 \cdot \Delta u(\mathbf{x})d\mathbf{x} = \int_{\partial\Omega} \partial_\nu u(\mathbf{x})d\sigma(\mathbf{x}) = 0,$$

because $\partial_\nu u = 0$ follows from $u_{|\Omega} \in H_0^2(\Omega)$. Hence, $\varphi(\mathbf{p})m(\Omega) = \int_\Omega (g(\mathbf{p}) - g(\mathbf{x}))d\mathbf{x}$. Then

$$|\varphi(\mathbf{p})m(\Omega)| \leq [g]_{1/2,\Omega} \int_\Omega |\mathbf{p} - \mathbf{x}|^{1/2} d\mathbf{x} \leq [g]_{1/2,\Omega} m(\Omega)\sqrt{d(\Omega)}. \tag{11.2.11}$$

Next, let us estimate $[g]_{1/2,\Omega}$. Recall that $g = \varphi_{|\Omega} - k^2 u_{|\Omega}$. For $k^2[u]_{1/2,\Omega}$, we are going to use Lemma 11.2.1 again. Combining it with the fact that $\|\varphi\|_2 \leq \sqrt{m(\Omega)}\|\varphi\|_\infty$, we have

$$k^2[u]_{1/2,\Omega} = k^{5/2}k^{-1/2}[u]_{1/2,\Omega} \leq C_n k^{n/2+3/2}\sqrt{m(\Omega)}(k^{-1} + d(\Omega))\|\varphi\|_{L^\infty(\Omega)}.$$

With (11.2.11), we get

$$|\varphi(\mathbf{p})| \leq \sqrt{d(\Omega)}\Big([\varphi]_{1/2,\Omega} + C_n k^{n/2+3/2}\sqrt{m(\Omega)}(k^{-1} + d(\Omega))\|\varphi\|_{L^\infty(\Omega)}\Big).$$

Recall our definition of δ which gives $d(\Omega) = \delta k^{-1}$ and also $m(\Omega) \leq C_n(d(\Omega))^n = C_n\delta^n k^{-n}$ because Ω can be encapsulated into a sphere of radius $d(\Omega)$. This gives

$$\begin{aligned} |\varphi(\mathbf{p})| &\leq C_n\delta^{1/2}k^{-1/2}\big([\varphi]_{1/2,\Omega} + k^{3/2}\delta^{n/2}(k^{-1} + k^{-1}\delta)\|\varphi\|_{L^\infty(\Omega)}\big) \\ &\leq C_n\delta^{1/2}\big(k^{-1/2}[\varphi]_{1/2,\Omega} + \delta^{n/2}(1+\delta)\|\varphi\|_{L^\infty(\Omega)}\big) \\ &\leq C_n\delta^{1/2}\big(1 + \delta^{n/2}(1+\delta)\big)\big(k^{-1/2}[\varphi]_{1/2,\Omega} + \|\varphi\|_{L^\infty(\Omega)}\big) \end{aligned}$$

from which the claim follows by taking a supremum over $\mathbf{p} \in \partial\Omega$ and $n \in \{2,3\}$. □

Proof of Theorem 11.2.3. Let us assume the contrary, namely, $u_\infty = 0$. Let $u_c \in H^2_{loc}$ satisfy $(\Delta + k^2)u_c = \chi_{\Omega_c}\varphi$ and the Sommerfeld radiation condition at infinity. By Lemma 11.2.2, we see that $u_{c\infty} = 0$. By Theorem 11.2.1 when $\alpha = 1/2$ or the remark after it when $\alpha < 1/2$, we see that

$$\frac{\sup_{\partial\Omega_c}|\varphi|}{\sup_{\Omega_c}|\varphi| + k^{-\alpha}[\varphi]_{\alpha,\Omega}} \leq C\big((1+\delta)^{(n+1)/2} + \delta^\alpha\big),$$

which contradicts (11.2.8), so the claim is proved. □

Next, we localize and geometrize the result in Theorem 11.2.1. To that end, we first introduce the admissible K-curvature point in the next subsection for our study.

11.2.3 Admissible K-Curvature Boundary Points

In this section, we introduce admissible K-curvature boundary points that shall be used throughout the rest of the chapter. Let Ω be a bounded domain in $\mathbb{R}^n$ and $\mathbf{p} \in \partial\Omega$ be a fixed point. We next detail the conditions for $\mathbf{p}$ to be an admissible K-curvature point.

Definition 11.2.1. Let K, L, M, δ be positive constants and Ω be a bounded domain in $\mathbb{R}^n$, $n \geq 2$. A point $\mathbf{p} \in \partial\Omega$ is said to be an admissible K-curvature point with parameters L, M, δ if the following conditions are fulfilled; see Fig. 11.1 for a schematic illustration.

1. Up to a rigid motion, the point p is the origin $x = 0$ and $\mathbf{e}_n = (0, \ldots, 0, 1)$ is the interior unit normal vector to $\partial\Omega$ at $\mathbf{0}$.
2. Set $b = \sqrt{M}/K$ and $h = 1/K$. There is a C^3 function $\omega : B(0,b) \to \mathbb{R}_+ \cup \{0\}$ with $B(0,b) \subset \mathbb{R}^{n-1}$ such that if
$$\Omega_{b,h} = B(0,b) \times (-h,h) \cap \Omega, \tag{11.2.12}$$
then
$$\Omega_{b,h} = \{\mathbf{x} \in \mathbb{R}^n \mid |\mathbf{x}'| < b,\ -h < x_n < h,\ \omega(\mathbf{x}') < x_n < h\}, \tag{11.2.13}$$
where $\mathbf{x} = (\mathbf{x}', x_n)$ with $\mathbf{x}' \in \mathbb{R}^{n-1}$.
3. The function ω in Item 2 satisfies
$$\omega(\mathbf{x}') = K|\mathbf{x}'|^2 + \mathcal{O}(|\mathbf{x}'|^3) \quad \text{for} \quad \mathbf{x}' \in B(0,b). \tag{11.2.14}$$
4. We have $M \geq 1$, and there are $0 < K_- \leq K \leq K_+ < \infty$ such that
$$K_-|\mathbf{x}'|^2 \leq \omega(\mathbf{x}') \leq K_+|\mathbf{x}'|^2, \qquad |\mathbf{x}'| < b,$$
$$M^{-1} \leq \frac{K_\pm}{K} \leq M, \qquad K_+ - K_- \leq LK^{1-\delta}.$$
5. The intersection $V = \overline{\Omega_{b,h}} \cap \mathbb{R}^{n-1} \times \{h\}$ is a Lipschitz domain.

A simple example for an admissible K-curvature boundary point is that locally near $\mathbf{p}$, $\partial\Omega$ is the part of a paraboloid, namely, $\omega(\mathbf{x}') = K|\mathbf{x}'|^2$. In such a case, one can easily determine the values of parameters L, M, δ to fulfil the requirements in Definition 11.2.1. However, we allow the presence of more general geometries near p, and this can be guaranteed by the following lemma.

Lemma 11.2.3. *Assume that $\omega(\mathbf{x}') = K|\mathbf{x}'|^2 + \mathcal{O}(|\mathbf{x}'|^3)$ is a C^3 function. Let $L, \delta > 0$, $M \geq 1$, and*

$$c_n = \sup_{\mathbf{x}' \in \mathbb{R}^{n-1}} \frac{1}{|\mathbf{x}'|^3} \sum_{|\beta|=3} \frac{\mathbf{x}'^\beta}{\beta!}. \tag{11.2.15}$$

Let $f(K) = \max_{|\alpha|,|\beta|=3} \sup_{|\mathbf{x}'|<b} |\partial^\alpha \omega(\mathbf{x}')|/\beta!$, *where* $b = \sqrt{M}/K$. *Assume that*

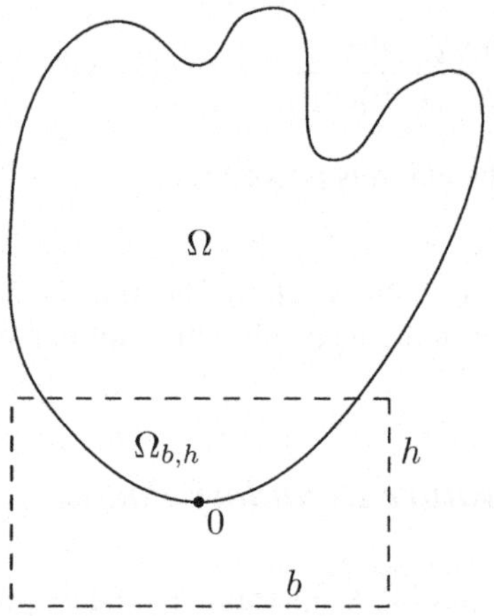

Fig. 11.1 The boundary neighborhood $\Omega_{b,h}$ of a high-curvature point

$$f(K) \leq \min\left(\frac{M-1}{c_n M^{3/2}} K^2, \frac{L}{2c_n\sqrt{M}} K^{2-\delta}\right).$$

Set

$$K_- = K - c_n f(K) b \text{ and } K_+ = K + c_n f(K) b. \tag{11.2.16}$$

Then one has

$$M^{-1} \leq \frac{K_-}{K} \leq K \leq \frac{K_+}{K} \leq M, \qquad K_+ - K_- \leq LK^{1-\delta}$$

and

$$K_- |\mathbf{x}'|^2 \leq \omega(\mathbf{x}') \leq K_+ |\mathbf{x}'|^2 \quad \text{when } |\mathbf{x}'| < b.$$

Proof. By Taylor's theorem, we first have

$$\omega(\mathbf{x}') = K|\mathbf{x}'|^2 + \sum_{|\beta|=3} R_\beta(\mathbf{x}') \mathbf{x}'^\beta,$$

where the functions R_β and $|\beta| = 3$ satisfy

$$|R_\beta(\mathbf{x}')| \leq \max_{|\alpha|=3} \sup_{|\mathbf{x}'|<b} |\partial^\alpha \omega(\mathbf{x}')|/\beta! = f(K).$$

Then one has

$$|\omega(\mathbf{x}') - K|\mathbf{x}'|^2| \leq c_n f(K) |\mathbf{x}'|^3, \tag{11.2.17}$$

where c_n is positive and finite because $|x_j x_k x_l| \leq |\mathbf{x}'||\mathbf{x}'||\mathbf{x}'| = |\mathbf{x}'|^3$. Let K_- and K_+ be defined in (11.2.16). Then if $|\mathbf{x}'| < b$, one can directly verify that

$$K_- |\mathbf{x}'|^2 \leq K|\mathbf{x}'|^2 - c_n f(K)|\mathbf{x}'|^3 \quad \text{and} \quad K|\mathbf{x}'|^2 + c_n f(K)|x'|^3 \leq K_+ |\mathbf{x}'|^2.$$

Hence, for $|\mathbf{x}'| < b$, there holds

$$K_- |\mathbf{x}'|^2 \leq K|\mathbf{x}'|^2 - c_n f(K)|\mathbf{x}'|^3 \leq \omega(\mathbf{x}').$$

The upper bound $\omega(\mathbf{x}') \leq K_+ |\mathbf{x}'|^2$ follows from a similar argument.

Next, we prove the bounds for $K_+ - K_-$ and $K_\pm/K$. Recall the assumed upper bound on $f(K)$ and $b = \sqrt{M}/K$. One has by straightforward calculations that

$$\frac{K_-}{K} = 1 - \frac{c_n\sqrt{M}f(K)}{K^2} \geq 1 - (1 - 1/M) = 1/M,$$
$$\frac{K_+}{K} = 1 + \frac{c_n\sqrt{B}f(K)}{K^2} \leq 1 + (M - 1) = M,$$

so

$$K_+ - K_- = 2c_n\sqrt{M}f(K)/K \leq LK^{1-\delta}.$$

The proof is complete. □

Remark 11.2.1. It can be directly verified that the usual curvature at the point p of the boundary surface is high if K is large in Definition 11.2.1. In fact, in the practically important three-dimensional case, it can be easily seen that for an admissible K-curvature point with a sufficiently large K, both the mean curvature and Gaussian curvature of the boundary surface at the point are high. In the following, we shall show that for a generic source if its support contains an admissible K-curvature point with a sufficiently large K (compared to the intensity of the source), then it must be radiating. This clearly elucidates the geometric viewpoint that the smallness is not the essential cause; instead, the high curvature is the essential cause for the radiating nature of a generic source. We believe that the result can be extended to a more general geometric setup by requiring that only the mean curvature is high. In fact, in a recent article by one of the authors [3] and from a reconstruction point of view, it is shown that the local shape of a scatterer around a boundary point with a high magnitude of mean curvature can be reconstructed more easily and stably due to the more significant scattering from the high-curvature point.

In the subsequent study, the concept of Rellich's lemma and unique continuation principle shall play an important role. They are the fundamental tools of bringing information from the far field to the near field and a fortiori to the boundary of the scatterer. In essence, if the far-field pattern is known, then the scattered wave is known in the component of the complement of the scatterer that is unbounded. Since the scatterer may be hollow and we are studying boundary behavior, we define for which boundary points the information from the far-field pattern can be propagated from infinity.

Definition 11.2.2. Let $U \subset \mathbb{R}^n$ be an open set and $\mathbf{p} \in \mathbb{R}^n$. We say that *$p$ is connected to infinity through U* if there is a continuous path $\gamma\colon \mathbb{R}_+ \to U$ such that $\lim_{s\to 0}\gamma(s) = \mathbf{p}$ and $\lim_{s\to\infty}|\gamma(s)| = \infty$.

Lemma 11.2.4. *Let $\Omega \subset \mathbb{R}^n$ be a bounded domain and $u^s, u'^s \in H^2_{loc}(\mathbb{R}^n) \cap C^0(\mathbb{R}^n)$ satisfy the Sommerfeld radiation condition and $(\Delta + k^2)u^s = (\Delta + k^2)u'^s = 0$ in $\mathbb{R}^n \setminus \overline{\Omega}$. If $u^s_\infty = u'^s_\infty$ and $\mathbf{p} \in \mathbb{R}^n$ is connected to infinity through $\mathbb{R}^n \setminus \overline{\Omega}$, then $u^s(\mathbf{p}) = u'^s(\mathbf{p})$.*

Proof. Elliptic regularity and Rellich's lemma (e.g., Lemma 2.11 in [18]) imply that $u^s = u'^s$ outside a large ball. Let $\gamma : \mathbb{R}_+ \to \mathbb{R}^n \setminus \overline{\Omega}$ be a path as in Definition 11.2.2. For each $s \in \mathbb{R}_+$, let $r(s) = d(\gamma(s), \Omega)$ be the distance from $\gamma(s)$ to Ω, and note that it is positive for $\mathbb{R}^n \setminus \overline{\Omega}$ is open. Let $U = \cup_{s>0} B(\gamma(s), r(s))$. Then U is a connected open set, $\mathbf{p} \in \overline{U}$, and U reaches infinity. The latter implies that $u^s = u'^s$ in an open ball in U, and by analyticity, we have $u^s = u'^s$ in U. Continuity implies the same conclusion at $\mathbf{p}$. The proof is complete. □

11.2.4 Geometric Structure of Radiationless Sources at Admissible K-Curvature Points

In this subsection, we derive the geometric characterization of a radiationless source at admissible K-curvature points on the boundary of its support.

Theorem 11.2.3. *Let $\Omega \subset \mathbb{R}^n$, $n \geq 2$ be a bounded domain with diameter at most D. Consider an active source of the form $\varphi\chi_\Omega$ with $\varphi \in C^\alpha(\overline{\Omega})$, $0 < \alpha < 1$. Assume that $\mathbf{p} \in \partial\Omega$ is an admissible K-curvature point with parameters L, M, δ and $K \geq e$. Assume further that $\mathbf{p}$ is connected to infinity through $\mathbb{R}^n \setminus \overline{\Omega}$. Then for any given wavenumber $k \in \mathbb{R}_+$ and Hölder smoothness index α, there exists a positive constant $\mathscr{E} = \mathscr{E}(\alpha, \delta, n, D, L, M, k) \in \mathbb{R}_+$ such that if*

$$\frac{|\varphi(\mathbf{p})|}{\max\left(1, \|\varphi\|_{C^\alpha(\overline{\Omega})}\right)} \geq \mathscr{E}(\ln K)^{(n+3)/2} K^{-\min(\alpha,\delta)/2}, \tag{11.2.18}$$

then the source $\chi_\Omega\varphi$ radiates a non-zero far-field pattern at wavenumber k.

Corollary 11.2.3. *Consider a source of the form $\chi_\Omega\varphi$, and let $\mathbf{p} \in \partial\Omega$ be an admissible K-curvature point as described in Theorem 11.2.3. Suppose the strength of the source is bounded, namely, $\|\varphi\|_{C^\alpha(\overline{\Omega})} \leq \mathscr{M}$ is bounded. If the source is radiationless, then there exists a constant $\mathscr{C} = \mathscr{C}(\alpha, \delta, n, D, L, M, k, \mathscr{M})$ such that*

$$|\varphi(p)| \leq \mathscr{C}(\ln K)^{(n+3)/2} K^{-\min(\alpha,\delta)/2}. \tag{11.2.19}$$

That is, if K is sufficiently large at $p \in \partial\Omega$, then the intensity of a radiationless source must be nearly vanishing at that high-curvature boundary point.

Remark 11.2.2. Comparing Theorems 11.2.1 and 11.2.3, one readily sees that the global geometrical parameter $\mathrm{diam}(\Omega)$ in (11.2.6) is replaced by the local geometrical parameter K in (11.2.18). Hence, Theorem 11.2.3 is a local and geometrized version of Theorem 11.2.1.

Remark 11.2.3. We would like to point out that according to our discussion made after (11.2.5), all the geometric properties considered in Theorems 11.2.1 and 11.2.3 and Corollaries 11.2.2 and 11.2.3 can be equally formulated for functions whose Fourier transforms vanish on a given sphere.

To prove Theorem 11.2.3, we need to derive the following auxiliary technical results.

Lemma 11.2.5. *Let $\Omega \subset \mathbb{R}^n$ be a bounded Lipschitz domain and $k \in \mathbb{R}_+$ be a fixed wavenumber. Let $u_0, u \in H^2(\Omega)$ and $\varphi \in L^\infty(\Omega)$ satisfy*

$$(\Delta + k^2)u = \varphi \quad \text{and} \quad \Delta u_0 = 0 \quad \text{in} \quad \Omega. \tag{11.2.20}$$

If $w = 0$ and $\partial_\nu w = 0$ on $\Gamma \subset \partial\Omega$, then

$$\int_\Omega (\varphi - k^2 u) u_0 \mathrm{d}\mathbf{x} = \int_{\partial\Omega\setminus\Gamma} (u_0 \partial_\nu u - u \partial_\nu u_0) \mathrm{d}\sigma, \tag{11.2.21}$$

where ν is the exterior unit normal vector on $\partial\Omega$.

Proof. By (11.2.20), we first have $\varphi - k^2 u = \Delta u$. From integration by parts, one can deduce

$$\int_\Omega (\varphi - k^2 u) u_0 \, \mathrm{d}\mathbf{x} = \int_\Omega (\Delta u u_0 - u \Delta u_0) \, dx = \int_{\partial\Omega} \left(u_0 \partial_\nu u - u \partial_\nu u_0\right) \mathrm{d}\sigma, \tag{11.2.22}$$

which completes the proof. □

Lemma 11.2.6. *Let $\Omega \subset \mathbb{R}^n$ be a domain and $0 \in \partial\Omega$ be an admissible K-curvature point with parameters L, M, δ. Let $\Omega_{b,h}$ be given in (11.2.12) in Definition 11.2.1 associated with $0 \in \partial\Omega$. Let $u_0(\mathbf{x}) = \exp(\rho\cdot x)$, where $\rho \in \mathbb{C}^n$ and $\rho\cdot\rho = 0$, and assume that $w \in H^2(\Omega_{b,h}) \cap C^0(\overline{\Omega_{b,h}})$ and $\varphi \in L^\infty(\Omega_{b,h})$ satisfy $(\Delta + k^2)w = \varphi$ for some $k > 0$ and $w = \partial_\nu w = 0$ on $\overline{\Omega_{b,h}} \cap \partial\Omega$. There holds*

$$\begin{aligned}\varphi(\mathbf{0})\int_{x_n>K|\mathbf{x}'|^2} e^{\rho\cdot\mathbf{x}}\mathrm{d}\mathbf{x} &= \varphi(0)\int_{x_n>\max(h,K|\mathbf{x}'|^2)} e^{\rho\cdot x}\mathrm{d}\mathbf{x}\\ &+\varphi(0)\left(\int_{K|\mathbf{x}'|^2<x_n<h} e^{\rho\cdot\mathbf{x}}\mathrm{d}\mathbf{x} - \int_{\Omega_{b,h}} e^{\rho\cdot\mathbf{x}}\mathrm{d}\mathbf{x}\right)\\ &-\int_{\Omega_{b,h}} e^{\rho\cdot\mathbf{x}}\big(\varphi(\mathbf{x})-\varphi(\mathbf{0})-k^2\big(w(\mathbf{x})-w(\mathbf{0})\big)\big)\mathrm{d}\mathbf{x}\\ &+\int_{\partial\Omega_{b,h}\setminus\partial\Omega}\left(e^{\rho\cdot\mathbf{x}}\partial_\nu w - w\partial_\nu e^{\rho\cdot\mathbf{x}}\right)\mathrm{d}\sigma.\end{aligned} \tag{11.2.23}$$

Proof. By straightforward calculations, one can first simplify the right-hand side of (11.2.23) to be

$$\begin{aligned}&\varphi(\mathbf{0})\int_{x_n>K|\mathbf{x}'|^2} e^{\rho\cdot\mathbf{x}}\mathrm{d}\mathbf{x} - \int_{\Omega_{b,h}} e^{\rho\cdot\mathbf{x}}\big(\varphi(\mathbf{x})-k^2\big(w(\mathbf{x})-w(\mathbf{0})\big)\big)\mathrm{d}\mathbf{x}\\ &+\int_{\partial\Omega_{b,h}\setminus\partial\Omega}\left(e^{\rho\cdot\mathbf{x}}\partial_\nu w - w\partial_\nu e^{\rho\cdot\mathbf{x}}\right)\mathrm{d}\sigma.\end{aligned} \tag{11.2.24}$$

Noting that $w(\mathbf{0}) = 0$, one has from (11.2.23) and (11.2.24) a similar identity to that in (11.2.21). Since $\rho\cdot\rho = 0$, it is clear that $\Delta u_0 = 0$, and hence, the claim follows from Lemma 11.2.5 because $\Omega_{b,h}$ is a Lipschitz domain. The proof is complete. □

Lemma 11.2.7. *Let $K \in \mathbb{R}_+$, $\rho \in \mathbb{C}^n$ with $\Re\rho_n < 0$, and $C_\infty = \{\mathbf{x} \in \mathbb{R}^n \mid x_n > K|\mathbf{x}'|^2\}$. Then one has*

$$\int_{C_\infty} e^{\rho\cdot\mathbf{x}}\mathrm{d}\mathbf{x} = \frac{1}{-\rho_n}\left(\frac{\pi}{-\rho_n K}\right)^{(n-1)/2}\exp\left(-\frac{\rho'\cdot\rho'}{4\rho_n K}\right), \tag{11.2.25}$$

where $\rho'\cdot\rho' = \rho_1^2 + \ldots + \rho_{n-1}^2$.

Proof. We have

$$\int_{C_\infty} e^{\rho\cdot\mathbf{x}}\mathrm{d}\mathbf{x} = \int_{\mathbb{R}^{n-1}} e^{\rho'\cdot\mathbf{x}'}\int_{K|\mathbf{x}'|^2}^{\infty} e^{\rho_n x_n}\mathrm{d}x_n\mathrm{d}\mathbf{x}' = -\frac{1}{\rho_n}\int_{\mathbb{R}^{n-1}} e^{\rho_n K|\mathbf{x}'|^2+\rho'\cdot\mathbf{x}'}\mathrm{d}\mathbf{x}'. \tag{11.2.26}$$

It is easily seen that the right-hand side term of (11.2.26) is the product of integrals of the form $\int_{-\infty}^{\infty}\exp(\rho_n K x_j^2 + \rho_j x_j)\mathrm{d}x_j$ with $j = 1,\ldots,n-1$. The integration formula for a complex Gaussian gives

$$\int_{-\infty}^{\infty} e^{At^2+Bt}\mathrm{d}t = \sqrt{-\frac{\pi}{A}}\exp\left(-\frac{B^2}{4A}\right)$$

when $\Re A < 0$. We have $\Re\rho_n K < 0$ and thus

$$\int_{-\infty}^{\infty} e^{\rho_n K x_j^2+\rho_j x_j}\mathrm{d}x_j = \sqrt{-\frac{\pi}{\rho_n K}}\exp\left(-\frac{\rho_j^2}{4\rho_n K}\right), \quad j = 1,\ldots,n-1. \tag{11.2.27}$$

The claim follows by plugging (11.2.27) into (11.2.26), along with some straightforward calculations. The proof is complete. □

Lemma 11.2.8. *Let $\tau, K, h \in \mathbb{R}_+$, $s \geq 0$ and $C_h = \{\mathbf{x} \in \mathbb{R}^n \mid h > x_n > K|\mathbf{x}'|^2\}$. Then we have*

$$\int_{C_h} e^{-\tau x_n}|\mathbf{x}|^s\mathrm{d}\mathbf{x} \leq C_{n,s}\left(h + K^{-1}\right)^{\frac{s}{2}} h^{\frac{n+s+1}{2}} K^{-\frac{n-1}{2}}, \tag{11.2.28}$$

where

$$C_{n,s} = \frac{\sigma(\mathbb{S}^{n-2})}{1+s/2}.$$

Proof. At first, we note that a horizontal slice of the paraboloid C_h has a radius $\sqrt{x_n/K}$ at the height x_n. Hence,

$$\int_{C_h} e^{-\tau x_n}|\mathbf{x}|^s \mathrm{d}\mathbf{x} = \int_0^h e^{-\tau x_n} \int_{B(0,\sqrt{x_n/K})} (x_n^2+|\mathbf{x}'|^2)^{s/2} \mathrm{d}\mathbf{x}' dx_n. \tag{11.2.29}$$

By using the polar coordinates $x' = r\theta$, $\theta \in \mathbb{S}^{n-2}$, $r = |\mathbf{x}|$, one can verify that

$$\begin{aligned} &\int_0^h e^{-\tau x_n} \int_{B(0,\sqrt{x_n/K})} (x_n^2+|\mathbf{x}'|^2)^{s/2} \mathrm{d}\mathbf{x}' dx_n \\ =&\sigma(\mathbb{S}^{n-2}) \int_0^h e^{-\tau x_n} \int_0^{\sqrt{x_n/K}} (x_n^2+r^2)^{s/2} r^{n-2} \mathrm{d}r \mathrm{d}x_n. \end{aligned} \tag{11.2.30}$$

Taking the upper bound of the integrands, once for r, and then for x_n, one can further show that

$$\begin{aligned} &\sigma(\mathbb{S}^{n-2}) \int_0^h e^{-\tau x_n} \int_0^{\sqrt{x_n/K}} (x_n^2+r^2)^{s/2} r^{n-2} \mathrm{d}r \mathrm{d}x_n \\ \leq&\sigma(\mathbb{S}^{n-2}) \int_0^h e^{-\tau x_n} \left(x_n^2+\frac{x_n}{K}\right)^{\frac{s}{2}} \left(\frac{x_n}{K}\right)^{\frac{n-1}{2}} \mathrm{d}x_n \\ \leq&\sigma(\mathbb{S}^{n-2}) \left(h+\frac{1}{K}\right)^{\frac{s}{2}} \left(\frac{h}{K}\right)^{\frac{n-1}{2}} \frac{h^{1+\frac{s}{2}}}{1+\frac{s}{2}}. \end{aligned} \tag{11.2.31}$$

Finally, by combining (11.2.29), (11.2.30), and (11.2.31), one readily has (11.2.28). The proof is complete. □

Lemma 11.2.9. *Let $\tau, K, h \in \mathbb{R}_+$ and $C_\ell = \{\mathbf{x} \in \mathbb{R}^n \mid \ell > x_n > K|\mathbf{x}'|^2\}$ for any $\ell \in \mathbb{R}_+ \cup \{\infty\}$. Then there holds*

$$\int_{C_\infty \setminus C_h} e^{-\tau x_n} \mathrm{d}\mathbf{x} \leq C_n \frac{1+(\tau h)^{\frac{n-1}{2}}}{\tau^{\frac{n+1}{2}} K^{\frac{n-1}{2}}} e^{-\tau h}, \tag{11.2.32}$$

where C_n depends only on n.

Proof. The proof proceeds as that of Lemma 11.2.8 but with $s = 0$ and $x_n \in (h,\infty)$ instead of $x_n \in (0,\infty)$. Changing to polar coordinates $\mathbf{x}' = r\theta$, $\theta \in \mathbb{S}^{n-2}$, $r = |\mathbf{x}'|$, and integrating $\int_0^{r_{max}} r^{n-2} \mathrm{d}r = r_{max}^{n-1}/(n-1)$ with $r_{max} = \sqrt{x_n/K}$, one has that

$$\begin{aligned} \int_{C_\infty \setminus C_h} e^{-\tau x_n} \mathrm{d}\mathbf{x} &= \int_h^\infty e^{-\tau x_n} \int_{B(0,\sqrt{x_n/K})} \mathrm{d}\mathbf{x}' \mathrm{d}x_n \\ &= \sigma(\mathbb{S}^{n-2}) \int_h^\infty e^{-\tau x_n} \int_0^{\sqrt{x_n/K}} r^{n-2} \mathrm{d}r \mathrm{d}x_n \\ &= \frac{\sigma(\mathbb{S}^{n-2})}{n-1} \int_h^\infty e^{-\tau x_n} \left(\frac{x_n}{K}\right)^{\frac{n-1}{2}} \mathrm{d}x_n. \end{aligned} \tag{11.2.33}$$

Next, by using the change of variables $t = \tau x_n$, one can further show that

$$\frac{\sigma(\mathbb{S}^{n-2})}{n-1} \int_h^\infty e^{-\tau x_n} \left(\frac{x_n}{K}\right)^{\frac{n-1}{2}} \mathrm{d}x_n = \frac{\sigma(\mathbb{S}^{n-2})}{n-1} \frac{1}{\tau(\tau K)^{\frac{n-1}{2}}} \int_{\tau h}^\infty e^{-t} t^{\frac{n+1}{2}-1} \mathrm{d}t. \tag{11.2.34}$$

Switching to $s = t - \tau h$ in the last integral in (11.2.34) allows us to estimate it as follows. Recall the definition of the Γ function $\Gamma(m) = \int_0^\infty e^{-s}s^{m-1}ds$ and also that $(A+B)^{a-1} \leq \max(1, 2^{a-2})(A^{a-1} + B^{a-1})$ for $a > 1$. We thus have

$$\begin{aligned}\int_{\tau h}^{\infty} e^{-t} t^{\frac{n-1}{2}} \mathrm{d}t &= e^{-\tau h} \int_0^\infty e^{-s}(s+\tau h)^{\frac{n+1}{2}-1}\mathrm{d}s \\ &\leq e^{-\tau h} \max\left(1, 2^{\frac{n+1}{2}-2}\right)\left(\Gamma\left(\frac{n+1}{2}\right) + (\tau h)^{\frac{n-1}{2}}\right).\end{aligned} \tag{11.2.35}$$

Finally, by combining (11.2.33), (11.2.34), and (11.2.35), one can readily verify (11.2.32). The proof is complete. □

Lemma 11.2.10. *Let $\tau, K_-, K_+, h \in \mathbb{R}_+$ with $K_+ > K_-$, and denote $C_\pm = \{\mathbf{x} \in \mathbb{R}^n \mid K_\pm |\mathbf{x}'|^2 < x_n < h\}$. Then there holds*

$$\int_{C_-\backslash C_+} e^{-\tau x_n} \mathrm{d}\mathbf{x} = \frac{\sigma(\mathbb{S}^{n-2})}{n-1}\left(\left(\frac{1}{K_-}\right)^{\frac{n-1}{2}} - \left(\frac{1}{K_+}\right)^{\frac{n-1}{2}}\right)\frac{1}{\tau^{\frac{n+1}{2}}}\gamma\left(\tau h, \frac{n+1}{2}\right), \tag{11.2.36}$$

where

$$\gamma(\mathbf{x}, a) := \int_0^{\mathbf{x}} \exp(-t)t^{a-1}dt, \quad a \in \mathbb{C}, \tag{11.2.37}$$

is the lower incomplete gamma function.

Proof. The proof can be proceeded as that of Lemma 11.2.9. We first note that the horizontal cut at the height x_h is an annulus of external radius $\sqrt{x_n/K_-}$ and internal radius $\sqrt{x_n/K_+}$. By using the polar coordinates in integrals and the fact that $\int_a^b r^{n-2}\mathrm{d}r = (b^{n-1} - a^{n-1})/(n-1)$, one can deduce as follows:

$$\begin{aligned}\int_{C_-\backslash C_+} e^{-\tau x_n} dx &= \int_0^h e^{-\tau x_n} \int_{\sqrt{x_n/K_+} < |\mathbf{x}'| < \sqrt{x_n/K_-}} \mathrm{d}\mathbf{x}'\mathrm{d}x_n \\ &= \sigma(\mathbb{S}^{n-2}) \int_0^h e^{-\tau x_n} \int_{\sqrt{x_n/K_+}}^{\sqrt{x_n/K_-}} r^{n-2}\mathrm{d}r\mathrm{d}x_n \\ &= \frac{\sigma(\mathbb{S}^{n-2})}{n-1}\left(\left(\frac{1}{K_-}\right)^{\frac{n-1}{2}} - \left(\frac{1}{K_+}\right)^{\frac{n-1}{2}}\right)\int_0^h e^{-\tau x_n} x_n^{\frac{n-1}{2}}\mathrm{d}x_n.\end{aligned} \tag{11.2.38}$$

Next, using again the change of variables $t = \tau x_n$ in the last integral in (11.2.38), along with the definition of the incomplete Γ function, one can further show that

$$\int_0^h e^{-\tau x_n} x_n^{\frac{n-1}{2}}\mathrm{d}x_n = \frac{1}{\tau^{\frac{n+1}{2}}}\int_0^{\tau h} e^{-t}t^{\frac{n+1}{2}-1}\mathrm{d}t. \tag{11.2.39}$$

Finally, by combining (11.2.38) and (11.2.39), one can readily show (11.2.36). The proof is complete. □

Proposition 11.2.1. *Let $\Omega \subset \mathbb{R}^n$, $n \geq 2$ be a domain and $p \in \partial\Omega$ be an admissible K-curvature point with parameters L, M, δ. Let $\Omega_{b,h}$ be introduced in (11.2.12) in Definition 11.2.1 associated with $p \in \partial\Omega$.*

Let $\mathbf{x}$ be the coordinates for which $p = 0$ in Definition 11.2.1. Let $u_0(\mathbf{x}) = \exp(\rho \cdot \mathbf{x})$, where $\rho = \mathrm{i}\tau e_1 - \tau e_n$ with $\tau \in \mathbb{R}_+$, and assume that $w \in H^2(\Omega_{b,h}) \cap C^{1,\beta}(\overline{\Omega_{b,h}})$, $0 < \beta \leq 1$, and $\varphi \in L^\infty(\Omega_{b,h})$ satisfy $(\Delta + k^2)w = \varphi$ in $\Omega_{b,h}$ for some $k > 0$ and $w = \partial_\nu w = 0$ on $\overline{\Omega_{b,h}} \cap \partial\Omega$. Then there holds

$$\begin{aligned} C_{n,\alpha}|\varphi(p)|\sqrt{\pi} \leq & (1+(\tau h)^{(n-1)/2})e^{\tau(\frac{1}{4K}-h)} + \left(\left(\frac{K}{K_-}\right)^{\frac{n-1}{2}} - \left(\frac{K}{K_+}\right)^{\frac{n-1}{2}}\right)e^{\frac{\tau}{4K}} \\ & + (\|\varphi\|_{C^\alpha} + k^2\|w\|_{C^{1,\beta}})(h+K_-^{-1})^{\alpha/2}h^{(n+\alpha+1)/2}(K/K_-)^{(n-1)/2}\tau^{3/2}e^{\frac{\tau}{4K}} \\ & + h^{\beta+(n-1)/2}(K/K_-)^{(n-1)/2}(1+\tau h)\tau^{(n+1)/2}e^{\tau(\frac{1}{4K}-h)}\|w\|_{C^{1,\beta}}, \end{aligned} \tag{11.2.40}$$

where $C_{n,\alpha}$ is a positive number.

Proof. In what follows, we make use of the coordinates $\mathbf{x}$ in Definition 11.2.1, and hence $\mathbf{p}$ is represented by $\mathbf{x} = \mathbf{0}$. First, by Lemma 11.2.6, we have

$$\begin{aligned} \varphi(\mathbf{0}) \int_{x_n > K|\mathbf{x}'|^2} e^{\rho\cdot\mathbf{x}}\mathrm{d}\mathbf{x} = & \varphi(\mathbf{0}) \int_{x_n > \max(h, K|\mathbf{x}'|^2)} e^{\rho\cdot\mathbf{x}}\mathrm{d}\mathbf{x} \\ & + \varphi(\mathbf{0}) \left(\int_{K|\mathbf{x}'|^2 < x_n < h} e^{\rho\cdot x}\mathrm{d}\mathbf{x} - \int_{\Omega_{b,h}} e^{\rho\cdot\mathbf{x}}\mathrm{d}\mathbf{x} \right) \\ & - \int_{\Omega_{b,h}} e^{\rho\cdot\mathbf{x}}(\varphi(\mathbf{x}) - \varphi(\mathbf{0}) - k^2(w(\mathbf{x}) - w(\mathbf{0})))\mathrm{d}\mathbf{x} \\ & + \int_{\partial\Omega_{b,h}\setminus\partial\Omega} (e^{\rho\cdot\mathbf{x}}\partial_\nu w - w\partial_\nu e^{\rho\cdot\mathbf{x}})\mathrm{d}\sigma \\ = & \varphi(\mathbf{0})\cdot I_1 + \varphi(\mathbf{0})\cdot I_2 + I_3 + I_4, \end{aligned} \tag{11.2.41}$$

where

$$I_1 := \int_{x_n > \max(h, K|\mathbf{x}'|^2)} e^{\rho\cdot x}\mathrm{d}\mathbf{x}, \tag{11.2.42}$$

$$I_2 := \int_{K|\mathbf{x}'|^2 < x_n < h} e^{\rho\cdot\mathbf{x}}\mathrm{d}\mathbf{x} - \int_{\Omega_{b,h}} e^{\rho\cdot\mathbf{x}}\mathrm{d}\mathbf{x}, \tag{11.2.43}$$

$$I_3 := -\int_{\Omega_{b,h}} e^{\rho\cdot\mathbf{x}}(\varphi(\mathbf{x}) - \varphi(\mathbf{0}) - k^2(w(\mathbf{x}) - w(\mathbf{0})))\mathrm{d}\mathbf{x}, \tag{11.2.44}$$

$$I_4 := \int_{\partial\Omega_{b,h}\setminus\partial\Omega} (e^{\rho\cdot\mathbf{x}}\partial_\nu w - w\partial_\nu e^{\rho\cdot\mathbf{x}})\mathrm{d}\sigma. \tag{11.2.45}$$

With the help of Lemmas 11.2.7–11.2.10, we next estimate the terms I_j, $j = 1, \ldots, 4$.
First of all, Lemma 11.2.7 implies that

$$\int_{x_n > K|\mathbf{x}'|^2} e^{\rho\cdot\mathbf{x}}\mathrm{d}\mathbf{x} = \left(\frac{\pi}{K}\right)^{(n-1)/2} \frac{1}{\tau^{(n+1)/2}} \exp\left(-\frac{\tau}{4K}\right), \tag{11.2.46}$$

which together with Lemma 11.2.9 gives

$$|I_1| = \left| \int_{x_n > \max(h, K|\mathbf{x}'|^2)} e^{\rho\cdot x}\mathrm{d}\mathbf{x} \right| \leq \int_{x_n > \max(h, K|\mathbf{x}'|^2)} e^{-\tau x_2}\mathrm{d}\mathbf{x} \leq C_n \frac{1+(\tau h)^{\frac{n-1}{2}}}{\tau^{\frac{n+1}{2}} K^{\frac{n-1}{2}}} e^{-\tau h}, \tag{11.2.47}$$

where C_n depends only on the dimension n.

We proceed with the estimates of the integral terms I_2 and I_3, respectively, in (11.2.43) and (11.2.44). Recall Definition 11.2.1, and let K_- and K_+ be as in Item 4 therein. In the definition, the distances $b, h > 0$ were chosen such that $h \leq K_- b^2$. Hence, the paraboloids $x_n = K_\pm|\mathbf{x}'|$ do not touch the sides of the cylinder $\{\mathbf{x} \mid |\mathbf{x}'| < b, -h < x_n < h\}$. Set

$$P_{b,h,\pm} = \{\mathbf{x} \in \mathbb{R}^n \mid K_\pm|\mathbf{x}'|^2 < x_n < h\}. \tag{11.2.48}$$

According to our discussion above and Item 4 of Definition 11.2.1, one can see that $P_{b,h,-} \subset \Omega_{b,h} \subset P_{b,h,+}$. Hence, one can show that

$$\begin{aligned} \left| \int_{K|\mathbf{x}'|^2<x_n<h} e^{\rho\cdot\mathbf{x}} \mathrm{d}\mathbf{x} - \int_{\Omega_{b,h}} e^{\rho\cdot\mathbf{x}} \mathrm{d}\mathbf{x} \right| &\leq \int_{\{K|\mathbf{x}'|^2<x_n<h\}\Delta\Omega_{b,h}} e^{-\tau x_n} \mathrm{d}\mathbf{x} \\ &\leq \int_{P_{b,h,-}\setminus P_{b,h,+}} e^{-\tau x_n} \mathrm{d}\mathbf{x} \end{aligned} \tag{11.2.49}$$

where and also in what follows, for two sets A and B, $A\Delta B := (A\cup B)\setminus(A\cap B)$ signifies the symmetric difference of the two sets. Next, by Lemma 11.2.10, we can further estimate that

$$\int_{P_{b,h,-}\setminus P_{b,h,+}} e^{-\tau x_n} \mathrm{d}\mathbf{x} = C_n\left(K_-^{-\frac{n-1}{2}} - K_+^{-\frac{n-1}{2}}\right)\tau^{-\frac{n+1}{2}}\gamma\left(\tau h, \frac{n+1}{2}\right), \tag{11.2.50}$$

where by the definition in (11.2.37), one clearly has

$$\gamma\left(\tau h, \frac{n+1}{2}\right) \leq \Gamma\left(\frac{n+1}{2}\right). \tag{11.2.51}$$

Finally, by combining (11.2.49), (11.2.50), and (11.2.51), one has

$$|I_2| \leq C_n\left(K_-^{-\frac{n-1}{2}} - K_+^{-\frac{n-1}{2}}\right)\tau^{-\frac{n+1}{2}}\Gamma\left(\frac{n+1}{2}\right) \tag{11.2.52}$$

For the third term I_3, we note that $w \in C^{1,\beta}$ and so is also in C^α. Hence, there holds

$$\left|\varphi(\mathbf{x}) - \varphi(\mathbf{0}) - k^2\big(w(\mathbf{x}) - w(\mathbf{0})\big)\right| \leq (\|\varphi\|_{C^\alpha} + k^2\|w\|_{C^\alpha})|\mathbf{x}|^\alpha. \tag{11.2.53}$$

On the other hand, we recall that $\Omega_{b,h} \subset P_{b,h,-}$. By applying Lemma 11.2.8 to estimate the integral on the second line below, one can deduce that

$$\begin{aligned} |I_3| &= \left| \int_{\Omega_{b,h}} e^{\rho\cdot\mathbf{x}} \big(\varphi(\mathbf{x}) - \varphi(\mathbf{0}) - k^2\big(w(\mathbf{x}) - w(\mathbf{0})\big)\big) \mathrm{d}\mathbf{x} \right| \\ &\leq \left(\|\varphi\|_{C^\alpha(\overline{\Omega_{b,h}})} + k^2\|w\|_{C^\alpha(\overline{\Omega_{b,h}})}\right) \int_{P_{b,h,-}} e^{-\tau x_2}|\mathbf{x}|^\alpha \mathrm{d}\mathbf{x} \\ &\leq C_{n,\alpha}\left(\|\varphi\|_{C^\alpha(\overline{\Omega_{b,h}})} + k^2\|w\|_{C^{1,\beta}(\overline{\Omega_{b,h}})}\right)\left(h + K_-^{-1}\right)^{\alpha/2} h^{(n+\alpha+1)/2} K_-^{-(n-1)/2}. \end{aligned} \tag{11.2.54}$$

For the last boundary integral term I_4 in (11.2.45), we first note that $V := \partial\Omega_{b,h} \setminus \partial\Omega$ is actually a horizontal slice because $h \leq K_- b^2$ and $P_{b,h,-} \supset \Omega$. Hence, $V = U \times \{h\}$ for some bounded domain $U \subset \mathbb{R}^{n-1}$. Its measure is at most the measure of a slice of $P_{b,h,-}$, so $\sigma(U) \leq \sigma(\mathbb{S}^{n-1})(h/K_-)^{(n-1)/2}$. On the other hand, we know that $w = 0$ and $\partial_\nu w = 0$ on $\partial\Omega$, so $\partial_n w = 0$ too, and any point of V has a distance at most h from $\partial\Omega$. The definition of the Hölder norm implies that

$$|\partial_n w(x', h)| \leq \|w\|_{C^{1,\beta}}(h - \omega(x'))^\beta, \tag{11.2.55}$$

where we recall that the graph of the function ω defines Ω and also that $\mathbf{x}' \in U$ with $\omega(\mathbf{x}') \leq h$. On the other hand, the graph stays above the zero line, $\omega(\mathbf{x}') \geq 0$, and hence, one obviously has from (11.2.55) that

$$|\partial_n w(\mathbf{x}', h)| \leq \|w\|_{C^{1,\beta}} h^\beta. \tag{11.2.56}$$

Next, the fundamental theorem of calculus implies that

$$w(\mathbf{x}', h) = \int_{\omega(\mathbf{x}')}^h \partial_n w(\mathbf{x}', s)\mathrm{d}s.$$

Thus, combining with (11.2.56) and recalling that $0 < \omega(\mathbf{x}') < h$, one has

$$|w(\mathbf{x}',h)| \leq \|w\|_{C^{1,\beta}} \int_{\omega(\mathbf{x}')}^{h} s^{\beta} \mathrm{d}s \leq \|w\|_{C^{1,\beta}} h^{1+\beta}/(1+\beta),$$

which readily implies with (11.2.56) and $\sigma(U) \leq C_n (h/K_-)^{(n-1)/2}$ that

$$\begin{aligned} |I_4| &= \left| \int_V (e^{\rho\cdot\mathbf{x}} \partial_\nu w - w \partial_\nu e^{\rho\cdot\mathbf{x}}) \mathrm{d}\sigma \right| \\ &\leq e^{-\tau h} \int_U \big(|\partial_n w(x',h)| + \tau |w(x',h)|\big) \mathrm{d}\mathbf{x}' \\ &\leq C_{n,\beta} h^{\beta+(n-1)/2} K_-^{-(n-1)/2} (1+\tau h) e^{-\tau h} \|w\|_{C^{1,\beta}(\overline{\Omega_{b,h}})}. \end{aligned} \tag{11.2.57}$$

Finally, by adding up (11.2.46), (11.2.47), (11.2.50), (11.2.54), and (11.2.57) and multiplying both sides by $K^{(n-1)/2}\tau^{(n+1)/2}\exp(\tau/4K)$, one can obtain (11.2.48). The proof is complete. □

We need one last lemma before attacking the problem of K-curvature scattering, namely, that non-scattering waves are $C^{1,\beta}$ smooth, after which we can apply Proposition 11.2.1.

Lemma 11.2.11. *Let $\Omega \subset \mathbb{R}^n$, $n \geq 2$ be a domain of diameter at most $D \in \mathbb{R}_+$. Let $u \in H_0^2(\Omega)$ satisfy*

$$(\Delta + k^2)u = \varphi$$

for some $\varphi \in L^\infty(\Omega)$ and $k \in \mathbb{R}_+$. Then

$$\|u\|_{C^{1,\beta}(\overline{\Omega})} \leq C \|\varphi\|_{L^\infty(\Omega)} \tag{11.2.58}$$

for any $0 \leq \beta < 1$ and some finite constant $C = C(D,k,n,\beta)$.

Proof. We first extend u by zero outside of Ω into a ball of radius $R_{D,k} > D$ such that k^2 is a *not* a Dirichlet eigenvalue for $-\Delta$ in balls of radius $R_{D,k}$. This is possible because $k = 0$ is the only common eigenvalue of $-\Delta$ among all large disks. Denote this ball by $B_{R,D,k}$. By a bit of abuse of notation, we still denote the extended function as u. Clearly, $u \in H^2(B_{R,D,k})$ satisfies

$$(\Delta + k^2)u = \chi_\Omega \varphi \quad \text{in } B_{R,D,k}, \quad u = 0 \quad \text{on } \partial B_{R,D,k}. \tag{11.2.59}$$

By Corollary 8.7 in [29], we first have from (11.2.59) that

$$\|u\|_{H^1(B_{R,D,k})} \leq C(R,D,k) \|\varphi\|_{L^2(\Omega)}. \tag{11.2.60}$$

Then by further applying Corollary 8.35 in [29], we have

$$\|u\|_{C^{1,\beta}(\overline{B_{R,D,k}})} \leq C(R,D,k,n,\beta)(\|u\|_{H^1(B_{R,D,k})} + \|\varphi\|_{L^\infty(\Omega)})$$

for any $0 \leq \beta < 1$, which together with (11.2.60) readily yields (11.2.58). The proof is complete. □

Next, we proceed to derive a critical inequality with the help of Proposition 11.2.1 by properly choosing the parameters appearing therein.

Proposition 11.2.2. *Let $\Omega \subset \mathbb{R}^n$, $n \geq 2$ be a bounded domain and $w \in H_0^2(\Omega)$ satisfy*

$$(\Delta + k^2)w = \varphi$$

for some $\varphi \in L^\infty(\Omega)$ and $k > 0$. Let $\mathbf{p} \in \partial\Omega$ be an admissible K-curvature point with parameters L, M, δ.

If φ *restricted to* $\overline{\Omega_{b,h}}$ *from Definition 11.2.1 is* C^α *smooth,* $0<\alpha<1$*, and* Ω *has a diameter at most D, then*

$$|\varphi(\mathbf{p})| \leq \mathscr{E} \max\left(1, \|\varphi\|_{C^\alpha}\right) (\ln K)^{(n+3)/2} K^{-\min(\alpha,\delta)/2} \tag{11.2.61}$$

for some $\mathscr{E}=\mathscr{E}(\alpha,\delta,n,D,L,M,k)\in\mathbb{R}_+$ *depending only on* α,δ,n,D,L,M,k.

Proof. First, we have by Lemma 11.2.11 that

$$\|w\|_{C^{1,\beta}(\overline{\Omega})} \leq C_{D,k,n,\beta}\|\varphi\|_{L^\infty(\Omega)}$$

for some finite constant $C_{D,k,n,\beta}$. This gives the required function regularity for applying Proposition 11.2.1. Hence, for any $\tau\in\mathbb{R}_+$, by Proposition 11.2.1, and assuming without loss of generality that $\mathbf{p}=\mathbf{0}$, we have

$$\begin{aligned} C_{n,k,\alpha}|\varphi(\mathbf{0})|\sqrt{\pi} &\leq (1+(\tau h)^{(n-1)/2})e^{\tau(\frac{1}{4K}-h)} + \left(\left(\frac{K}{K_-}\right)^{\frac{n-1}{2}} - \left(\frac{K}{K_+}\right)^{\frac{n-1}{2}}\right)e^{\frac{\tau}{4K}} \\ &+ \|\varphi\|_{C^\alpha}(h+K_-^{-1})^{\alpha/2}h^{(n+\alpha+1)/2}(K/K_-)^{(n-1)/2}\tau^{3/2}e^{\frac{\tau}{4K}} \\ &+ h^{\beta+(n-1)/2}(K/K_-)^{(n-1)/2}(1+\tau h)\tau^{(n+1)/2}e^{\tau(\frac{1}{4K}-h)}\|w\|_{C^{1,\beta}}. \end{aligned} \tag{11.2.62}$$

Let us start by estimating the difference of powers of K/K_- and K/K_+. Recall that $1/M \leq K_\pm/K \leq M$ by Item 4 of Definition 11.2.1. Consider the function $f(r)=r^{-s}$ with $f'(r)=-sr^{-s-1}$. By the mean value theorem,

$$|f(r_-)-f(r_+)| \leq \sup_{r_-<\xi<r_+} |f'(\xi)||r_+-r_-| = C_{s,M}|r_+-r_-|$$

when $1/M\leq r_-,r_+\leq M$. Recall also that $|K_+-K_-|\leq LK^{1-\delta}$ by Item 4. Hence,

$$\left|\left(\frac{K}{K_-}\right)^{\frac{n-1}{2}} - \left(\frac{K}{K_+}\right)^{\frac{n-1}{2}}\right| \leq C_{n,M}\left(\frac{K_+}{K}-\frac{K_-}{K}\right) = C_{n,L,M}K^{-\delta} \tag{11.2.63}$$

for some finite constant $C_{n,L,M}$.

Next, we recall that $h=1/K$ and $b=\sqrt{M}/K$ by Definition 11.2.1 and that $K/K_- < M$. Applying (11.2.63) to (11.2.62), estimating the constants, and then dividing them to the left-hand side give

$$\begin{aligned} C_{n,k,\alpha,L,M}|\varphi(\mathbf{0})|\sqrt{\pi} &\leq (1+(\tau/K)^{(n-1)/2})e^{-3\tau/4K} + K^{-\delta}e^{\tau/4K} \\ &+ \|\varphi\|_{C^\alpha}K^{-(n+2\alpha+1)/2}\tau^{3/2}e^{\tau/4K} \\ &+ K^{-\beta-(n-1)/2}(1+\tau/K)\tau^{(n+1)/2}e^{-3\tau/4K}\|w\|_{C^{1,\beta}}. \end{aligned} \tag{11.2.64}$$

Choose $\tau=4K\ln K^\gamma$ for some $\gamma\in\mathbb{R}_+$ to be specified in what follows. Since $K\geq e$, after dividing by the constants and $\max(1,\|\varphi\|_{C^\alpha})$, (11.2.64) can be further estimated above by

$$(\ln K)^{(n-1)/2}K^{-3\gamma} + K^{\gamma-\delta} + (\ln K)^{3/2}K^{1-n/2-\alpha+\gamma} + (\ln K)^{(n+3)/2}K^{1-\beta-3\gamma}. \tag{11.2.65}$$

One can directly verify that the quantity in (11.2.65) tends to zero as $K\to\infty$ if $0<\gamma<\min(\alpha,\delta)$ and $3\gamma>1-\beta$. These two conditions are fulfilled if one chooses $\beta=1-\min(\alpha,\delta)$ and $\gamma=\min(\alpha,\delta)/2$. For the final form of the upper bound, using the fact that $(\ln r)^{a_1}\leq a_1 r^{a_2}/a_2 e$ for any $a_1,a_2>0$, $r\geq e$, each of the terms in (11.2.65) can then be estimated above by

$$C_{n,L,M,\delta,\alpha}(\ln K)^{(n+3)/2}K^{-\min(\alpha,\delta)/2} \tag{11.2.66}$$

for some positive constant $C_{n,L,M,\delta,\alpha}$. By combining our discussion above, one arrives at (11.2.61). The proof is complete. □

Proof of Theorem 11.2.3. The proof follows from Proposition 11.2.2. Let $\tilde{\Omega}$ be the interior of the complement of the unbounded component of $\mathbb{R}^n \setminus \overline{\Omega}$; in other words, $\tilde{\Omega}$ is Ω with holes filled up. If $(\Delta + k^2)u = \chi_\Omega \varphi$ and $u \in H^2_{loc}(\mathbb{R}^n)$ radiates a zero far-field pattern, then by the Rellich lemma, $u_{|\tilde{\Omega}} \in H_0^2(\tilde{\Omega})$. Moreover, $\chi_\Omega \varphi = \varphi \in C^\alpha$ in the set $\overline{\Omega_{b,h}}$, where the latter is the notation introduced in Definition 11.2.1. Hence, one can readily show the claim in the theorem by Proposition 11.2.2. The proof is complete. □

11.3 Geometrical Characterizations of Non-radiating Waves and Transmission Eigenfunctions

In Sect. 11.2, we consider the wave scattering due to an active source that generates the wave propagation. In this section, we consider a different scattering scenario where one uses an incident field to generate the wave propagation in a uniform and homogeneous space. There is an inhomogeneous medium scatterer located in the space. The medium scatterer is passive and is characterized by its index of refraction which is different from that of the ambient space. The presence of the inhomogeneity interrupts the wave propagation and produces the wave scattering. Since we shall be considering scattering at a fixed wavenumber, one can also formulate such a scattering problem in the context of quantum scattering, where the refractive index is replaced by a potential function and the wavenumber is the energy level. However, in order to be more definite in our description, we stick to the former case of medium scattering in our subsequent discussion.

We are mainly concerned with the scenario that no wave scattering is generated; that is, the incident wave passes through the medium without being interrupted. In such a case, the incident field is referred to as a non-scattering wave. We aim to geometrically characterize the non-scattering waves associated with a given medium scatterer. The critical observation is that the incident wave interacting with the medium scatterer generates an active source which connects to our previous study on radiationless sources in Sect. 11.2. Nevertheless, due to the interaction of the incident wave and the medium scatterer, some physical phenomena manifest. Mathematically, we also need to introduce technical ingredients to deal with this situation. Furthermore, the study in Sect. 11.2 enables us to derive a certain elegant geometric structure of the so-called interior transmission eigenfunctions, which is of independent interest in spectral theory. In what follows, we first introduce medium scattering and non-scattering waves, invisibility cloaking, and transmission eigenvalue problems. Then we study the geometrical characterization of non-scattering waves and its implication to invisibility cloaking. Finally, we derive the intrinsic geometric structure of the interior transmission eigenfunctions.

11.3.1 Wave Scattering from an Inhomogeneous Medium

Let $V \in L^\infty(\mathbb{R}^n)$ be a complex-valued function such that $\Im V \geq 0$ and $\mathrm{supp}(V) \subset \Omega$. The function V signifies the index of refraction of an inhomogeneous medium supported in Ω. Let $u^i(\mathbf{x})$ be an incident field which is an entire solution to the Helmholtz equation

$$\Delta u^i + k^2 u^i = 0 \quad \text{in } \mathbb{R}^n. \tag{11.3.1}$$

For specific examples, one can take the incident field to be a plane wave $u^i(\mathbf{x}) = \exp(ik\theta \cdot \mathbf{x})$, where $\theta \in \mathbb{S}^{n-1}$ signifies an incident direction, or a Herglotz wave which is the superposition of plane waves of the form

$$u^i(\mathbf{x}) = \int_{\mathbb{S}^{n-1}} g(\theta)\exp(\mathrm{i}k\theta\cdot\mathbf{x})\,d\sigma(\theta), \quad g\in L^2(\mathbb{S}^{n-1}).$$

The presence of the inhomogeneity interrupts the propagation of the incident field. Let u^s signify the perturbation to the incident wave field. It is also called the scattered field. Set $u = u^i + u^s$ to be the total wave field. Medium scattering is governed by the following Helmholtz system :

$$\begin{aligned} &\big(\Delta + k^2(1+V)\big)u = 0, \quad u = u^i + u^s \quad \text{in } \mathbb{R}^n,\\ &\lim_{r\to\infty} r^{\frac{n-1}{2}}\big(\partial_r - \mathrm{i}k\big)u^s = 0. \end{aligned} \tag{11.3.2}$$

Here, $u\in H^2_{loc}(\mathbb{R}^n)$ and satisfies the following Lippmann-Schwinger equation:

$$u = u^i - k^2(\Delta+k^2)^{-1}(Vu), \tag{11.3.3}$$

where the integral operator $(\Delta+k^2)^{-1}$ is defined in (11.2.3). Similar to (11.2.5), one can have the far-field pattern of u^s from (11.3.3) by the stationary phase approximation

$$u^s_\infty(\hat{\mathbf{x}}) = -k^2 C_{n,k}\mathscr{F}(Vu)(k\hat{\mathbf{x}}) \in L^2(\mathbb{S}^{n-1}). \tag{11.3.4}$$

It is noted that in (11.3.4), u in the right-hand side is the unknown total wave field, which is in sharp difference to (11.2.5) for the source scattering and is responsible for the major technical difficulty of the study in the present section compared to that in Sect. 11.2.

Similar to the scattering by an active source, we are also particularly interested in the case that there is no scattering associated with the configuration consisting of the incident wave u^i and the inhomogeneous medium (Ω, V). If this occurs, then u^i is referred to as a non-scattering incident field. Suppose that $u^s_\infty\equiv 0$ and by the Rellich lemma , one immediately has that $u^s = 0$ in the unbounded component of $\mathbb{R}^n\backslash\overline{\Omega}$. If Ω is simply connected, by setting $w = u^i|_\Omega$, one can readily verify that there hold

$$(\Delta+k^2)w = 0 \qquad \text{in } \Omega, \tag{11.3.5}$$

$$(\Delta+k^2(1+V))u = 0 \qquad \text{in } \Omega, \tag{11.3.6}$$

$$w, u\in L^2(\Omega), \qquad u-w\in H^2_0(\Omega). \tag{11.3.7}$$

It is pointed out that the last condition means that $u = w$ and $\partial_\nu u = \partial_\nu w$ on $\partial\Omega$, which come from the standard transmission condition on the total wave field u in (11.3.2) across $\partial\Omega$, along with the fact that $u^s = 0$ in $\mathbb{R}^n\backslash\overline{\Omega}$.

Equations (11.3.5)–(11.3.7) are referred to as the *interior transmission problem* in the literature as discussed in the introduction. If, for some $k\in\mathbb{R}_+$, there exists a pair of nontrivial solutions to (11.3.5)–(11.3.7), then k is called a transmission eigenvalue and w, u are said to be the corresponding eigenfunctions. According to our discussion above, we know that if no scattering occurs for the Helmholtz system (11.3.2), i.e., invisibility , then the restrictions of the total wave u and the incident wave u^i form a pair of transmission eigenfunctions with the wavenumber being the transmission eigenvalue. On the other hand, for a pair of transmission eigenfunctions w and u, it is easily seen that if w can be (analytically) extended from Ω to the whole space $\mathbb{R}^n$ as an entire solution to the Helmholtz equation (11.3.1), which is still denoted by w, then w is a non-scattering incident wave field.

A well-known example is if Ω is a central ball and V is radially symmetric. Then there exist non-scattering incident waves, which in turn, by our discussion above, implies the existence of transmission eigenfunctions. Because of the radial symmetry, they can be analytically extended as entire solutions to the Helmholtz equation. It is widely believed in the literature that in general, transmission eigenfunctions cannot be analytically extended to the whole space as an entire solution to the Helmholtz equation associated with a generic (Ω, V). When none of them can be extended, this means that the inhomogeneous index of refraction scatters nontrivially every incident field.

Per our discussion in Sect. 11.1, transmission eigenfunctions cannot be analytically extended across a corner on $\partial\Omega$. That means corner singularities scatter every entire incident wave nontrivially [11, 36]. In what follows, we shall first provide a result applicable in many more situations by showing that if there is an admissible K-curvature point on $\partial\Omega$, then it must scatter every generic entire incident field nontrivially. Our study follows a similar spirit to that in Sect. 11.2 on the source scattering in a localized and geometrized manner. To that end, we first present some preliminary results on the direct scattering problem (11.3.2).

Definition 11.3.1. A nonempty subset

$$S \subset \{(k, \Omega, V) \mid k \in \mathbb{R}_+,\ \Omega \subset \mathbb{R}^n \text{ a bounded domain},\ V \in L^\infty(\mathbb{R}^n),\ V = 0 \text{ in } \mathbb{R}^n \setminus \overline{\Omega}\}$$

is called *a collection of scattering models*. It gives *uniformly well-posed scattering problems* if there is a map

$$\mathscr{U}_S : \{(d(\Omega)k, \|V\|_\infty) \mid (k, \Omega, V) \in S\} \to \mathbb{R},$$

which is non-decreasing in both of its arguments and such that for any incident wave $u^i \in L^2_{loc}$ and any triple $(k, \Omega, V) \in S$, we have a unique solution to (11.3.2) and the scattered wave satisfies

$$\|u^s\|_{L^2(\Omega)} \le \mathscr{U}_S(d(\Omega)k, \|V\|_\infty)\,\|u^i\|_{L^2(\Omega)}.$$

We have chosen this type of condition for the scattered wave because we are interested in how the shape (height and width) of the potential affects scattering. Furthermore, by dimensional analysis, one notices that the diameter $d(\Omega)$ and wavenumber k have units inverse to each other, while $\|V\|_\infty$ is unitless. Hence, $d(\Omega)k$ must appear as a combination in any norm estimate which is physically relevant. We are also using L^2 norms for simplicity although more general norms could apply in certain cases that are out of the scope of this topic. We will show in Proposition 11.3.1 that there are collections of scattering models that are physically interesting and give uniformly well-posed scattering problems. One such collection S is defined by $d(\Omega)k\|V\|_\infty < c_n$ for a given constant $c_n > 0$. In fact, we have the following:

Proposition 11.3.1. *For $n \ge 2$, there is a constant $c_n \in \mathbb{R}_+$ with the following property. Let $k \in \mathbb{R}_+$ and $V \in L^\infty(\mathbb{R}^n)$ with $V = 0$ outside of a bounded domain Ω. If $d(\Omega)k\|V\|_\infty < c_n$ and u^i is an incident wave, then the scattering system* (11.3.2) *has a unique solution u^s which also satisfies*

$$\|u^s\|_{L^2(\Omega)} \le \frac{d(\Omega)k\|V\|_\infty}{c_n - d(\Omega)k\|V\|_\infty}\|u^i\|_{L^2(\Omega)}.$$

Also

$$\|u\|_{L^2(\Omega)} \le \frac{c_n}{c_n - d(\Omega)k\|V\|_\infty}\|u^i\|_{L^2(\Omega)}$$

for the total wave.

Proof. By Section 2 in [12] (see the last paragraph in the proof of Lemma 11.2.1 for more details), we have

$$\left\|(\Delta + k^2)^{-1} f\right\|_{L^2(\Omega)} \le C_n d(\Omega) k^{-1} \|f\|_{L^2(\Omega)}$$

for any $f \in L^2(\mathbb{R}^n)$ with $f = 0$ outside Ω. Recall the Lippman-Schwinger equation (11.3.3), $u = u^i - k^2(\Delta + k^2)^{-1}(Vu)$. By the estimate right above, we have

$$\left\|-k^2(\Delta + k^2)^{-1}(Vu)\right\|_{L^2(\Omega)} \le C_n d(\Omega) k \|V\|_{L^\infty(\Omega)} \|u\|_{L^2(\Omega)} \tag{11.3.8}$$

because $V(x) = 0$ outside Ω.

The Neumann series for the Lippman-Schwinger equation converges in $L^2(\Omega)$ if the operator norm above is less than 1, i.e., if $C_n d(\Omega)k\|V\|_\infty < 1$. But this is one of our assumptions. Set

$$u = \sum_{j=0}^{\infty} \left(-k^2(\Delta + k^2)^{-1}V \right)^j u_i, \tag{11.3.9}$$

and we have convergence and norm estimate

$$\|u\|_{L^2(\Omega)} \leq \frac{1}{1 - C_n d(\Omega)k\|V\|_\infty} \|u^i\|_{L^2(\Omega)}. \tag{11.3.10}$$

The claim follows by applying (11.3.8) to $u^s = -k^2(\Delta + k^2)^{-1}Vu$. □

11.3.2 Geometric Characterizations of Non-scattering Incident Fields and Transmission Eigenfunctions

We are now in a position to geometrically characterize non-scattering configurations associated with the medium scattering system (11.3.2) and the interior transmission eigenfunctions. First, we show that a generic inhomogeneous medium with a sufficiently small support, compared to the underlying wavelength and the refractive index, then it must scatter any generic incident field nontrivially. In the following, we use the tilde notation in $\tilde{C}^{1/2}$ to denote dimensionless norms. In particular,

$$\|f\|_{\tilde{C}^{1/2}(\Omega)} = \|f\|_{L^\infty(\Omega)} + k^{-1/2} \sup_{\mathbf{x},\mathbf{y}\in\Omega} \frac{|f(\mathbf{x}) - f(\mathbf{y})|}{|\mathbf{x} - \mathbf{y}|^{1/2}}. \tag{11.3.11}$$

Theorem 11.3.1. *Let $n \in \{2,3\}$. Then there is a constant $C_n \in \mathbb{R}_+$ with the following property. Let S be a collection of scattering models that gives uniformly well-posed scattering problems. Assume furthermore that $\|V_{|\Omega}\|_{\tilde{C}^{1/2}} \leq M$ and $d(\Omega)k \leq \delta_M$ for all $(k,\Omega,V) \in S$. Next, let $(k,\Omega,V) \in S$ with Ω Lipschitz and $\mathbb{R}^n \setminus \Omega$ connected.*

If $u^i \in L^2_{loc}$ is an incident wave such that

$$\sup_{p\in\partial\Omega} \left| \frac{V(p)}{\|V\|_{\tilde{C}^{1/2}}} \frac{u^i(p)}{\|u^i\|_{\tilde{C}^{1/2}}} \right| > C_n\sqrt{d(\Omega)k}\left(1 + (1+M)(1+\delta_M)(1+\mathscr{U}_S(\delta_M,M))\delta_M^{n/2}\right), \tag{11.3.12}$$

then the far field u^s_∞ does not vanish identically.

Proof. We proceed by a reductio ad absurdum. Assume contrarily that $u^s_\infty \equiv 0$, and then by the Rellich lemma , we see that $u^s = 0$ in $\mathbb{R}^n \setminus \overline{\Omega}$. We are going to follow the ideas of the proof of Theorem 11.2.1. For this, note that $(\Delta + k^2)u^s = -k^2Vu$ and so we define the source term $f = -k^2Vu$. Set $g = f_{|\Omega} - k^2u^s_{|\Omega}$ so that $\Delta u^s = g$ and $g = -k^2Vu^i$ on $\partial\Omega$ when $u^s = 0$ outside of Ω.

Integrating by parts, we note that $\int_\Omega g(\mathbf{x})d\mathbf{x} = 0$, which further implies that

$$\begin{aligned} k^2|V(\mathbf{p})u^i(\mathbf{p})| = |g(p)| &\leq \frac{1}{m(\Omega)}\left|\int_\Omega \left(g(\mathbf{p}) - g(\mathbf{x})\right)d\mathbf{x}\right| \\ &\leq [g]_{1/2}\frac{1}{m(\Omega)}\int_\Omega |\mathbf{p} - \mathbf{x}|^{1/2}d\mathbf{x} \\ &\leq [g]_{1/2}\sqrt{d(\Omega)} \end{aligned} \tag{11.3.13}$$

for any $\mathbf{p} \in \partial\Omega$. Let us estimate $[g]_{1/2}$ next. Note that the seminorm $[g]_{1/2}$ is not multiplicative, but the norm $\|g\|_{\tilde{C}^{1/2}}$ is. Hence, we can switch to that norm first:

$$\begin{aligned}[g]_{1/2} &= k^{1/2}k^{-1/2}[g]_{1/2} \leq k^{1/2}\|g\|_{\tilde{C}^{1/2}} \\ &\leq k^{1/2}\|f\|_{\tilde{C}^{1/2}} + k^{5/2}\|u^s\|_{\tilde{C}^{1/2}} \\ &\leq k^{5/2}\|V\|_{\tilde{C}^{1/2}}\|u^i\|_{\tilde{C}^{1/2}} + k^{5/2}(1+M)\|u^s\|_{\tilde{C}^{1/2}}. \end{aligned} \tag{11.3.14}$$

By Lemma 11.2.1 and the trivial estimates $d(\Omega)k \leq \delta_M$, $\|V\|_\infty \leq \|V\|_{\tilde{C}^{1/2}}$, we have

$$\|u^s\|_{\tilde{C}^{1/2}} \leq C_n k^{n/2}(1+\delta_M)\|V\|_{\tilde{C}^{1/2}}\|u\|_{L^2(\Omega)}.$$

By the uniform well-posedness and $\|u^i\|_2 \leq \sqrt{m(\Omega)}\|u^i\|_\infty$, we also have

$$\begin{aligned}\|u\|_{L^2(\Omega)} &\leq \|u^i\|_{L^2(\Omega)} + \|u^s\|_{L^2(\Omega)} \leq \big(1+\mathscr{U}_S(d(\Omega)k, \|V\|_\infty)\big)\|u^i\|_{L^2(\Omega)} \\ &\leq C_n\big(1+\mathscr{U}_S(\delta_M, M)\big)\delta_M^{n/2}k^{-n/2}\|u^i\|_{\tilde{C}^{1/2}}.\end{aligned}$$

Estimating $\|u^s\|_{\tilde{C}^{1/2}}$ in (11.3.14) using the two previous estimates gives

$$\begin{aligned}[g]_{1/2} &\leq k^{5/2}\|V\|_{\tilde{C}^{1/2}}\|u^i\|_{\tilde{C}^{1/2}} \\ &\quad + C_n k^{5/2}(1+M)(1+\delta_M)\|V\|_{\tilde{C}^{1/2}}\big(1+\mathscr{U}_S(\delta_M, M)\big)\delta_M^{n/2}\|u^i\|_{\tilde{C}^{1/2}} \\ &\leq C_n k^{5/2}\|V\|_{\tilde{C}^{1/2}}\|u^i\|_{\tilde{C}^{1/2}}\Big(1+(1+M)(1+\delta_M)\big(1+\mathscr{U}_S(\delta_M, M)\big)\delta_M^{n/2}\Big). \end{aligned} \tag{11.3.15}$$

We recall that the condition of u_∞^s vanishing identically implies (11.3.13), which combined with (11.3.15) gives

$$\frac{|V(\mathbf{p})u^i(\mathbf{p})|}{\|V\|_{\tilde{C}^{1/2}}\|u^i\|_{\tilde{C}^{1/2}}} \leq C_n\sqrt{d(\Omega)k}\big(1+(1+M)(1+\delta_M)(1+\mathscr{U}_S(\delta_M, M))\delta_M^{n/2}\big).$$

The claim follows by taking the supremum over $p \in \partial\Omega$. □

Remark 11.3.1. According to (11.3.12), it can be easily inferred that if a generic inhomogeneous index of refraction possesses a sufficiently small support, compared to the underlying wavelength, then it scatters every generic incident wave nontrivially; that is, it cannot be identically invisible under the plane wave probing. As a simple illustration, one can consider a constant refractive index and the plane wave incidence of the form $\exp(ik\theta \cdot \mathbf{x})$ with $\theta \in \mathbb{S}^{n-1}$.

Next, we localize and geometrize the "smallness" result of Theorem 11.3.1.

Theorem 11.3.2. *Let $L, \delta, D, M_p, M_i, k \in \mathbb{R}_+$, $M > 1$, $0 < \alpha < 1$, $n \geq 2$ be the a priori constants. Let S be a collection of scattering models with wavenumber k that gives uniformly well-posed scattering problems. Then there exists a positive constant $\mathscr{C}$ satisfying the below.*

Let $(k, \Omega, V) \in S$, and we furthermore assume the following. The set Ω has diameter at most D, and there is $p \in \partial\Omega$ which is an admissible K-curvature point with parameters L, M, δ and $K \geq e$. Furthermore, p is connected to infinity through $\mathbb{R}^n \setminus \overline{\Omega}$. The potential V is of the form $V = \chi_\Omega \varphi$ with $\varphi \in C^\alpha(\mathbb{R}^n)$ and $\|\varphi\|_{C^\alpha(\overline{\Omega})} \leq M_p$.

Given any incident wave u^i satisfying $|u^i| \leq 1$ and $\|u^i\|_{C^{1,\alpha}(\mathbb{R}^n)} \leq M_i$, if

$$|\varphi(\mathbf{p})||u^i(\mathbf{p})| > \mathscr{C}(\ln K)^{(n+3)/2}K^{-\min(\alpha,\delta)/2}, \tag{11.3.16}$$

then $u_\infty^s \neq 0$.

In other words, an admissible high-curvature point on $\partial\Omega$ scatters every incident field nontrivially independent of the other parts of the inhomogeneous index of refraction.

Proof. Assume contrarily that $u^s_\infty \equiv 0$. Then by the Rellich lemma, we know that $u^s = 0$ in the unbounded component of $\mathbb{R}^n \setminus \overline{\Omega}$. Let B be a ball of diameter $2D$ that contains $\overline{\Omega}$. Then $u^s = \partial_\nu u^s = 0$ on its boundary.

We note that u is a solution to the following PDE: $(\Delta + k^2(1+V) - c)v = -cu$ in B and $v \in H^1(B_D)$ with $v = u^i$ on ∂B for any constant $c > 0$. By taking c large enough, Theorem 8.16 in [29] gives the unique solvability of the above PDE system, and so $u = v$, and moreover, it gives the integrability $u \in L^\infty(B)$ with a bound $\|u\|_{L^\infty(B)} \leq \|u^i\|_{L^\infty(B)} + C\|u\|_{L^2(B)}$ for some constant $C = C(n,k,D)$. The uniform well-posedness of the scattering problem given by Definition 11.3.1 implies $\|u\|_{L^2(B)} \leq C(D,M_p,k)$. Hence, u is bounded in B with a bound depending only on the a priori constants.

Set $f = -k^2 V u$. Then $\|f\|_{L^\infty(B)} \leq C(n,k,D,M_p)$ and

$$(\Delta + k^2)u^s = f \quad \text{for} \quad u^s \in H_0^2(B).$$

Lemma 11.2.11 implies that $u^s \in C^\alpha(\overline{B})$ with a norm bounded by the supremum of $|f|$, which is bounded by a priori constants according to the first part of the proof. Since $u = u^i + u^s$ and u^i is Hölder continuous, this further implies that $f \in C^\alpha(\overline{\Omega})$ with a norm bounded by the a priori constants.

Consider the domain Ω now, and let $\tilde{\Omega} \supset \Omega$ be open and simply connected such that $\partial\tilde{\Omega} \subset \partial\Omega$. We have $(\Delta + k^2)u^s = f$ and $u^s \in H_0^2(\tilde{\Omega})$ because $u^s = 0$ in the unbounded component of $\overline{\mathbb{R}^n \setminus \overline{\Omega}}$. The source f is Hölder continuous in Ω, and hence, it is obviously Hölder continuous in $\overline{\Omega_{b,h}}$ (cf. Definition 11.2.1 for the notation used here). By Proposition 11.2.2 and the estimate for f, we have

$$|f(\mathbf{p})| \leq C(\ln K)^{(n+3)/2} K^{-\min(\alpha,\delta)/2} \tag{11.3.17}$$

for some finite constant C depending on the a priori parameters. Set $\mathscr{C} = C/k^2$ and recall that $f = -k^2 V(u^i + u^s)$ with $u^s = 0$ at $\mathbf{x} = \mathbf{p}$. Thus, we have reached a contradiction with (11.3.16) and so the assumption of $u^s_\infty \equiv 0$ is false. The proof is complete. □

Corollary 11.3.1. *Consider the scattering configuration described in Theorem 11.3.2, and assume that there is no scattering, namely, $u^s_\infty \equiv 0$. Then*

$$|\varphi(\mathbf{p})||u^i(\mathbf{p})| \leq \psi(K), \quad \psi(K) := C(\ln K)^{(n+3)/2} K^{-\min(\alpha,\delta)/2}, \tag{11.3.18}$$

where C is a positive constant depending only on the a priori constants. It can be straightforwardly verified that $\lim_{K\to+\infty} \psi(K) = 0$.

Hence, if the medium's refractive index is not vanishing at a high-curvature point, then the incident field must be nearly vanishing at the high-curvature point.

The rest of the subsection is devoted to the study of the geometric structures of the interior transmission eigenfunctions in Eqs. (11.3.5)–(11.3.7). Before that, we would like to point out that if (w,u) is a pair of transmission eigenfunctions associated with the eigenvalue k, then it is obvious that $(\alpha w, \alpha u)$, $\alpha \in \mathbb{C}\backslash\{\mathbf{0}\}$, is a pair of transmission eigenfunctions associated with k as well. Hence, in what follows, we shall always normalize the transmission eigenfunctions in our study.

Theorem 11.3.3. *Let $n \in \{2,3\}$ and $k \in \mathbb{R}_+$. Let $\Omega \subset \mathbb{R}^n$ be a bounded Lipschitz domain of diameter δk^{-1} and $V \in \tilde{C}^{1/2}(\overline{\Omega})$ with $\inf_{\partial\Omega}|V| > 0$. Suppose that k is an interior transmission eigenvalue and $(u,w) \in L^2(\Omega) \times L^2(\Omega)$ is a pair of transmission eigenfunctions associated with k. If $u \in \tilde{C}^{1/2}(\overline{\Omega})$ with $\|u\|_{\tilde{C}^{1/2}(\overline{\Omega})} = 1$, then there holds*

$$\sup_{\partial\Omega}|u| \leq C\frac{\|V\|_{\tilde{C}^{1/2}}}{\inf_{\partial\Omega}|V|}\left((1+\delta)\delta^{n/2} + 1\right)\delta^{1/2}, \tag{11.3.19}$$

where $C \in \mathbb{R}_+$ is a universal constant independent of any other quantities here.

Proof. Let $f = -k^2 V u$ and extend both f and $(u-w)$ by zero to $\mathbb{R}^n \setminus \overline{\Omega}$. Then $(\Delta + k^2)(u-w) = \chi_\Omega f$ in $\mathbb{R}^n$, $u - w$ is trivially an outgoing solution. Furthermore, $u - w \in H^2_{loc}$ and $\|f\|_{\tilde{C}^{1/2}} \leq k^2 \|V\|_{\tilde{C}^{1/2}}$ because $\|u\|_{\tilde{C}^{1/2}} \leq 1$. Because $(u-w)_\infty \equiv 0$, Theorem 11.2.1 implies that

$$\sup_{\partial\Omega} |f| \leq C k^2 \|V\|_{\tilde{C}^{1/2}} \big((1+\delta)\delta^{n/2} + 1\big)\delta^{1/2}$$

for some universal constant $C \in \mathbb{R}_+$. The claim follows after dividing by $k^2 \inf_{\partial\Omega} |V|$. □

Equation (11.3.19) provides the relationship among the value of the transmission eigenfunction, the diameter of the domain, and the underlying refractive index. It indicates that if the domain is sufficiently small compared to the wavelength, then the transmission eigenfunction is nearly vanishing.

The following theorem localizes and geometrizes the "smallness" result of Theorem 11.3.3.

Theorem 11.1. *Let $L, \delta, D, k \in \mathbb{R}_+$, $M > 1$, $0 < \alpha < 1$, $n \geq 2$ be the a priori constants. Let $\Omega \subset \mathbb{R}^n$ be a bounded domain which has a diameter at most D. Assume that $p \in \partial\Omega$ is an admissible K-curvature point with parameters L, M, δ and $K \geq e$, and let $V \in C^\alpha(\overline{\Omega})$. Suppose that k is an interior transmission eigenvalue and $u, w \in L^2(\Omega)$ is a pair of transmission eigenfunctions associated with k. If $u \in C^\alpha(\overline{\Omega_{b,h}})$ with $\|u\|_{C^\alpha(\overline{\Omega_{b,h}})} = 1$, where $\Omega_{b,h}$ is defined in Definition 11.2.1 associated with the point p, then there holds*

$$|u(p)| \leq \mathscr{C} (\ln K)^{(n+3)/2} K^{-\min(\alpha,\delta)/2} \|V\|_{C^\alpha} / |V(p)|, \tag{11.3.20}$$

where $\mathscr{C}$ is a positive constant depending only on the a priori constants.

Proof. Set $f = -k^2 V u$ and note that $\|f\|_{C^\alpha(\overline{\Omega})} \leq k^2 \|V\|_{C^\alpha(\overline{\Omega})} \|u\|_{C^\alpha(\overline{\Omega})}$. Then $(\Delta + k^2)(u-w) = f$, $u - w \in H^2_0(\Omega)$. Proposition 11.2.2 immediately yields that

$$|f(p)| \leq C \|V\|_{C^\alpha} (\ln K)^{(n+3)/2} K^{-\min(\alpha,\delta)/2}$$

for some constant C depending on the a priori parameters. The claim follows after dividing by $k^2 |V(p)|$. □

11.4 Uniqueness Results for Inverse Scattering Problems

In this section, we consider the application of the results provided so far in the current article to the inverse scattering problem. The inverse problem associated with the source scattering system (11.2.1) can be described as identifying (Ω, φ) by knowledge of the corresponding far-field pattern $u_\infty(\hat{\mathbf{x}})$. By introducing an abstract operator $\mathscr{S}$ defined via (11.2.1) that sends the scatterer (Ω, φ) to the corresponding far-field pattern, the inverse problem can formulated as the following operator equation:

$$\mathscr{S}(\Omega, \varphi) = u_\infty(\hat{\mathbf{x}}), \quad \hat{\mathbf{x}} \in \mathbb{S}^{n-1}. \tag{11.4.1}$$

As discussed in the introduction, we are mainly concerned with the recovery of Ω, independent of the source density φ. It can be directly verified that in such a case, the inverse scattering problem (11.4.1) is nonlinear, and moreover, it is formally determined. We are mainly concerned with the uniqueness issue. That is, the sufficient condition to guarantee that for two scatterers (Ω, φ) and (Ω', φ') $\mathscr{S}(\Omega, \varphi) = \mathscr{S}(\Omega', \varphi')$ if and only if $\Omega = \Omega'$, without knowing φ and φ'. In what follows, we shall give such uniqueness results in two scenarios of practical importance, as long as φ and φ' are from certain generic classes.

Theorem 11.4.1. *Let $n \in \{2,3\}$, $0 < \alpha \leq 1/2$, $k \in \mathbb{R}_+$, and $\varphi, \varphi' \in C^\alpha(\Omega)$ with $\Omega, \Omega' \subset \mathbb{R}^n$ bounded Lipschitz domains. Assume that the complements of these sets are connected. Let $u, u' \in H^2_{loc}$ be outgoing solutions to the source problems*

$$
\begin{aligned}
(\Delta + k^2)u &= \chi_\Omega \varphi, \\
(\Delta + k^2)u' &= \chi_{\Omega'} \varphi'.
\end{aligned}
$$

If $u_\infty = u'_\infty$, then Ω cannot have a component Ω_c such that a) $\overline{\Omega_c} \cap \overline{\Omega'} = \emptyset$, b) it can be joined to infinity through $\mathbb{R}^n \setminus \overline{\Omega \cup \Omega'}$, and c)

$$
\frac{\sup_{\partial\Omega_c} |\varphi|}{\sup_{\Omega_c} |\varphi| + k^{-\alpha} [\varphi]_{\alpha,\Omega_c}} > C\big((1+\delta_c)\delta_c^{(n+1)/2} + \delta_c^\alpha\big) \tag{11.4.2}
$$

for the universal constant C of Theorem 11.2.2. Here, $\delta_c = d(\Omega_c)k$. A similar claim holds for the components of Ω'.

Proof. Assume that there would be such a component $\Omega_c \subset \Omega$. Then there is a bounded Lipschitz domain $W \subset \mathbb{R}^n$ such that $\Omega \cup \Omega' \subset W$, its complement is connected, and Ω_c is also a component of W. Set $w = u - u'$. Then

$$
(\Delta + k^2)w = \chi_W(\chi_\Omega \varphi - \chi'_\Omega \varphi'),
$$

it satisfies the Sommerfeld radiation condition , and $w_\infty = 0$. The source term above is equal to $-\chi_{\Omega_c}\varphi$ on Ω_c. By Theorem 11.2.2,

$$
\frac{\sup_{\partial\Omega_c} |\varphi|}{\sup_{\Omega_c} |\varphi| + k^{-\alpha} [\varphi]_{\alpha,\Omega_c}} \leq C\big((1+\delta_c)\delta_c^{(n+1)/2} + \delta_c^\alpha\big)
$$

where $\delta_c = d(\Omega_c)k$ as in this theorem's statement. But this is a contradiction with property c) of Ω_c whose existence was assumed. Hence, no such Ω_c exists. □

Corollary 11.4.1. *Under the situation of Theorem 11.4.1, let $\delta_0 > 0$ be the smallest positive solution to*

$$
C\big((1+\delta_0)\delta_0^{(n+1)/2} + \delta_0^\alpha\big) = \min\left(\frac{\sup_{\partial\Omega} |\varphi|}{\sup_\Omega |\varphi| + k^{-\alpha}[\varphi]_{\alpha,\Omega}}, \frac{\sup_{\partial\Omega'} |\varphi|}{\sup_{\Omega'} |\varphi| + k^{-\alpha}[\varphi]_{\alpha,\Omega'}}\right) \tag{11.4.3}
$$

or smaller. If $d(\Omega)k, d(\Omega')k < \delta_0$ and $\overline{\Omega} \cap \overline{\Omega'} = \emptyset$, then $u_\infty \neq u'_\infty$.

Proof. The condition $d(\Omega)k < \delta_0$ guarantees that (11.4.2) holds. □

Corollary 11.4.2. *Under the situation of Theorem 11.4.1, let δ_0 be as defined in Corollary 11.4.1, and assume further that Ω and Ω' are well-separated collections of small scatterers, namely,*

$$
\Omega = \bigcup_{j=1}^M \Omega_j, \qquad \Omega' = \bigcup_{l=1}^N \Omega'_l \tag{11.4.4}
$$

where Ω_j, Ω'_l each have a diameter at most $\delta_0 k^{-1}$, and $d(\Omega_{j_1}, \Omega_{j_2}), d(\Omega'_{l_1}, \Omega'_{l_2}) > 2\delta_0 k^{-1}$ for $j_1 \neq j_2$ and $l_1 \neq l_2$. Then if $u_\infty = u'_\infty$, we have $M = N$, and under a re-indexing, $\overline{\Omega_j} \cap \overline{\Omega'_j} \neq \emptyset$ for $j = 1, \ldots, M$.

Proof. Assume that Ω has a component Ω_{j_0} that does not touch[§] Ω'. The smallness from $d(\Omega_j)k < \delta_0$ guarantees the inequality (11.4.2). It remains to check that Ω_{j_0} can be joined to

[§] We say that A and B touch if $\overline{A} \cap \overline{B} \neq \emptyset$.

infinity through $\mathbb{R}^n \setminus \overline{\Omega \cup \Omega'}$. This follows because the distance between the various components of Ω is twice their maximal diameter, and the same holds for Ω': if a component Ω_j touches a component Ω_l', then they are encircled by an annulus of width at least $\delta_0 k^{-1}$ which doesn't intersect $\Omega \cup \Omega'$. The three conditions in Theorem 11.4.1 hold, so $u_\infty \neq u_\infty'$. Since the components are well-separated and small, a component of Ω' can only touch at most one component of Ω. Thus, $M = N$. □

Remark 11.4.1. Corollary 11.4.1 basically indicates that if two sources are of sufficiently small sizes (might be with different medium contents) and produce the same far-field pattern, then they must be very close to each other in the sense that their supports must have a nonempty intersection.

Remark 11.4.2. On the other hand, Corollary 11.4.2 implies that the exact number and approximate locations of well-separated scatterers are uniquely determined by a single far-field measurement, a question studied in [34]. This gives a proof for the numerical results in [32, 33].

Theorem 11.4.2. *Let $L, \delta, R_m, k, M_p, m_p \in \mathbb{R}_+$, $M > 1$, $0 < \alpha \leq 1/2$ be the a priori constants. Let $\Omega, \Omega' \subset B_{R_m} \subset \mathbb{R}^n$, $n \in \{2,3\}$ be bounded Lipschitz domains with connected complements, and let $\varphi, \varphi' \in C^\alpha(\overline{\Omega})$ such that*

$$m_p \leq |\varphi|, |\varphi'|, \qquad \|\varphi\|_{C^\alpha(\overline{\Omega})}, \|\varphi'\|_{C^\alpha(\overline{\Omega'})} \leq M_p. \tag{11.4.5}$$

Let $u, u' \in H^2_{loc}$ be outgoing solutions to $(\Delta + k^2)u = \chi_\Omega \varphi$ and $(\Delta + k^2)u' = \chi_{\Omega'} \varphi'$, respectively. Let u_∞ and u_∞' signify their far-field patterns.

Then there exist two positive constants C_1 and C_2, depending only on the a priori constants such that if $u_\infty = u_\infty'$, $k < C_2$, then $\Omega \setminus \Omega'$ cannot have an admissible K-curvature point p with parameters L, M, δ and $K > C_1$ and satisfying $d(p, \Omega') < \sqrt{1 + M/K}$ and connected to infinity through $\mathbb{R}^n \setminus \overline{\Omega \cup \Omega'}$.

Proof. Let $w = u - u'$. One has that $(\Delta + k^2)w = 0$ in $\mathbb{R}^n \setminus \overline{\Omega \cup \Omega'}$, and thus, by the Rellich lemma , $w = 0$ in Σ, where Σ is the unbounded connected component of $\mathbb{R}^n \setminus \overline{\Omega \cup \Omega'}$. Set $U = \mathbb{R}^n \setminus \Sigma$, then clearly $U \supset \Omega \cap \Omega'$. One easily sees that $w \in H_0^2(U)$ and $(\Delta + k^2)w = f$ in U where

$$f = \chi_\Omega \varphi - \chi_{\Omega'} \varphi'. \tag{11.4.6}$$

Let $p \in \partial\Omega \setminus \Omega'$ be the admissible K-curvature point as stated in the theorem, and consider the set $\Omega_{b,h}$ associated with p as specified in Definition 11.2.1. Because Ω' is of a distance $\sqrt{1 + M/K}$ from p, it does not intersect the rectangular neighborhood $B(0, b) \times (-h, h)$ that is used to define $\Omega_{b,h}$. Since p can be joined to infinity without passing through $\Omega \cup \Omega'$, we have $p \in \partial U$ and actually $\Omega_{b,h} = U_{b,h}$. Moreover, we have $f_{|U_{b,h}} = \varphi_{|U_{b,h}}$.

There holds

$$\|f\|_{C^\alpha(\overline{U_{b,h}})} = \|\varphi\|_{C^\alpha(\overline{U_{b,h}})} \leq M_p. \tag{11.4.7}$$

By (11.4.7), one can apply Proposition 11.2.2 to have

$$|f(p)| \leq \mathscr{E} (\ln K)^{(n+3)/2} K^{-\min(\alpha,\delta)/2}$$

for some constant $\mathscr{E} = \mathscr{E}(\alpha, \delta, n, R_m, L, M, k, M_p) \in \mathbb{R}_+$. Taking the lower bound $m_p \leq |\varphi|$ into account gives

$$m_p \leq \mathscr{E} (\ln K)^{(n+3)/2} K^{-\min(\alpha,\delta)/2}, \tag{11.4.8}$$

which is impossible when K is sufficiently small. This contradiction immediately yields that $\Omega \setminus \Omega'$ cannot have an admissible K-curvature point as stated in the theorem.

The proof is complete. □

Theorem 11.4.2 states a local uniqueness result for the inverse shape problem which basically indicates that if two sources produce the same far-field pattern, then the difference of the two scatterers cannot have a high-curvature point. On the other hand, if there is sufficient a priori knowledge about the shape of the underlying scatterer, one can also obtain approximate global uniqueness. As an illustrative example, one may consider an equilateral triangle in $\mathbb{R}^2$ with the three vertices being locally mollified to be admissible K-curvature points with sufficiently large K. Clearly, if two such kinds of scatterers produce the same far-field pattern, then by Theorem 11.4.2, they are approximately the same in the sense that their corresponding mollified vertices must be around distance K^{-1} from each other, respectively. Otherwise, the difference of the two scatterers would possess a high-curvature point.

Finally, we briefly remark the extension to the inverse medium scattering problem associated with (11.3.2). The inverse problem can be described as uniquely identifying Ω, independent of V, by knowledge of the corresponding far-field pattern $u^s_\infty(\hat{\mathbf{x}};u^i)$. Similar to the source scattering case, by introducing an abstract operator $\mathscr{T}$ defined via (11.3.2) that sends the scatterer (Ω,V) to the corresponding far-field pattern, the inverse problem can formulated as the following nonlinear equation:

$$\mathscr{T}(\Omega,V)=u^s_\infty(\hat{\mathbf{x}};u^i). \tag{11.4.9}$$

The corresponding uniqueness issue can be cast as deriving sufficient conditions to guarantee that $\mathscr{T}(\Omega,V)=\mathscr{T}(\Omega',V')$ if and only if $\Omega=\Omega'$, independent of V and V'. By noting that Helmholtz equation in (11.3.2) can be written as

$$(\Delta+k^2)u=\chi_\Omega\varphi, \quad \varphi:=-k^2Vu, \tag{11.4.10}$$

the study inverse medium problem (11.4.9) can obviously be reduced to that of the inverse source problem (11.4.1), and hence, the uniqueness results in Theorems 11.4.1 and 11.4.2 can be extended to the medium scattering case with proper modifications.

References

1. L. Alessandra and V. Felli, *Unique continuation from the edge of a crack*, arXiv:2004.11177
2. G. Alessandrini and L. Rondi, *Determining a sound-soft polyhedral scatterer by a single far-field measurement*, Proc. Amer. Math. Soc., **35** (2005), 1685–1691. s
3. H. Ammari, Y.-T. Chow and H. Liu, *Localized sensitivity analysis at high-curvature boundary points of reconstructing inclusions in transmission problems*, arXiv:1911.00820
4. E. Blåsten, *Nonradiating sources and transmission eigenfunctions vanish at corners and edges*, SIAM J. Math. Anal., **50(6)** (2018), 6255–6270.
5. E. Blåsten, X. Li,H. Liu and Y. Wang, On vanishing and localization near cusps of transmission eigenfunctions: a numerical study, Inverse Problems, **33** (2017), 105001.
6. E. Blåsten and H. Liu, *On vanishing near corners of transmission eigenfunctions*, J. Funct. Anal., **273(11)** (2017), 3616–3632. Addendum, arXiv:1710.08089
7. E. Blåsten and H. Liu, *Recovering piecewise constant refractive indices by a single far-field pattern*,Inverse Problems, **36** (2020), 085005.
8. E. Blåsten and H. Liu, *On corners scattering stably and stable shape determination by a single far-field pattern*, Indiana Univ. Math. J., **70** (2021), no. 3, 907–947.
9. E. Blåsten and H. Liu, E. Blåsten and H. Liu, *Scattering by curvatures, radiationless sources, transmission eigenfunctions and inverse scattering problems*, SIAM Journal on Mathematical Analysis, **53** (2021), no. 4, 3801–3837.
10. E. Blåsten and L. Päivärinta, *Completeness of generalized transmission eigenfunctions*, Inverse Problems, **29** (2013), 104002.
11. E. Blåsten, L. Päivärinta and J. Sylvester, *Corners always scatter*, Comm. Math. Phys., **331** (2014), 725–753.
12. E. Blåsten and J. Sylvester, *Translation-invariant estimates for operators with simple characteristics*, J. Differential Equations **263(9)** (2017), 5656–5695.
13. N. Bleistein and J.K. Cohen, *Nonuniqueness in the inverse source problem in acoustics and electromagnetics*, J. Math. Phys., **18** (1977), 194–201.
14. D. Bohm and M. Weinstein, *The self-oscillations of a charged particle*, Phys. Rev., 74;1789–1798, 1948.
15. F. Cakoni and H. Haddar, *Transmission eigenvalues in inverse scattering theory*, in "Inverse Problems and Applications: Inside Out II", Math. Sci. Res. Inst. Publ., Vol. 60, pp. 529–580, Cambridge Univ. Press., Cambridge, (2013).

16. J. Cheng and M. Yamamoto, *Uniqueness in an inverse scattering problem within non-trapping polygonal obstacles with at most two incoming waves*, Inverse Problems, **19** (2003), 1361–1384.
17. D. Colton, A. Kirsch and L. Päivärinta, *Far-field patterns for acoustic waves in an inhomogeneous medium*, SIAM J. Math. Anal., **20** (1989),1472–1483.
18. D. Colton and R. Kress, *Inverse acoustic and electromagnetic scattering theory*, volume 93 of Applied Mathematical Sciences. Springer, Cham, 4th edition, 2019.
19. D. Colton and P. Monk, *The inverse scattering problem for time-harmonic acoustic waves in an inhomogeneous medium*, Quart. J. Mech. Appl. Math., **41** (1988), 97–125.
20. D. Colton and B. Sleeman, *Uniqueness theorems for the inverse problem of acoustic scattering*, IMA J. Appl. Math., **31** (1983), 253–259.
21. A. Cossonniére and H. Haddar, *The electromagnetic interior transmission problem for regions with cavities*, SIAM J. Math. Anal., **43(4)** (2011), 1698–1715.
22. H. Diao, X. Cao and H. Liu, *On the geometric structures of conductive transmission eigenfunctions and their application*, Comm. Partial Differential Equations, **46(4)** (2021), 630–679.
23. A.J. Devaney and E. Wolf, *Non-radiating stochastic scalar sources*, In L. Mandel and E. Wolf, editors, Coherence and Quantum Optics V, 417–421, New York, (1984). Plenum Press.
24. P. Ehrenfest, *Ungleichförmige Elektrizitätsbewegungen ohne Magnet- und Strahlungsfeld*, Physik. Zeit., 11:708–709, (1910).
25. J. Elschner and M. Yamamoto, *Uniqueness in determining polyhedral sound-hard obstacles with a single incoming wave*, Inverse Problems, **24** (2008), 035004 (7pp).
26. F.G. Friedlander, *An inverse problem for radiation fields*, Proc. London Math. Soc., **2** (1973), 551–576.
27. A. Gamliel, K. Kim, A.I. Nachman, and E. Wolf, *A new method for specifying nonradiating monochromatic sources and their fields*, J. Opt. Soc. Am. A, **6** (1989), 1388–1393.
28. G. Gbur, *Nonradiating Sources and the Inverse Source Problem*, PhD Thesis, Univ. Rochester, (2001).
29. D. Gilbarg and N. S. Trudinger, *Elliptic partial differential equations of second order*, Vol. 224 of *Grundlehren der Mathematischen Wissenschaften [Fundamental Principles of Mathematical Sciences]*, Springer-Verlag, Berlin, 2nd edition, (1983).
30. G.H. Goedecke, *Classically radiationless motions and possible implications for quantum theory*, Phys. Rev., 135:B281–B288, (1964).
31. L. Grafakos, *Classical Fourier analysis*, second edition, Graduate Texts in Mathematics, 249, Springer, New York, (2008).
32. R. Griesmaier, M. Hanke and T. Raasch, *Inverse source problems for the Helmholtz equation and the windowed Fourier transform*, SIAM J. Sci. Comput. **34** (2012), no. 3, A1544–A1562.
33. R. Griesmaier, M. Hanke and T. Raasch, *Inverse source problems for the Helmholtz equation and the windowed Fourier transform II*, SIAM J. Sci. Comput. **35** (2013), no. 5, A2188–A2206.
34. R. Griesmaier, M. Hanke and J. Sylvester, *Far field splitting for the Helmholtz equation*, SIAM J. Numer. Anal. **52(1)** (2014), 343–362.
35. B.J. Hoenders and H.P. Baltes, *The scalar theory of nonradiating partially coherent sources*, Lettere al Nuovo Cimento, **23** (1979), 206–208.
36. G. Hu, M. Salo and E. Vesalainen, *Shape identification in inverse medium scattering problems with a single far-field pattern*, SIAM J. Math. Anal., **48** (2016),152–165.
37. M. Ikehata, *Reconstruction of a source domain from the Cauchy data*, Inverse Problems, **15** (1999), 637–645.
38. V. Isakov, *On uniqueness in the inverse transmission scattering problem*, Commun. Part. Diff. Eqns., **15** (1990), 1565–1587.
39. V. Isakov, *Inverse Problems for Partial Differential Equations*, 2nd edition, Springer-Verlag, New York, (2006).
40. K. Kim and E. Wolf, *Non-radiating monochromatic sources and their fields*, Opt. Commun., **59** (1986), 1–6.
41. A. Kirsch, *The denseness of the far field patterns for the transmission problem*, IMA J. Appl. Math., **37** (1986), 213–225.
42. A. Kirsch and R. Kress, *Uniqueness in inverse obstacle scattering*, Inverse Problems, **9** (1993), 285–299.
43. S. Kusiak and J. Sylvester, *The scattering support*, Comm. Pure Appl. Math., **56** (2003), 1525–1548.
44. S. Kusiak and J. Sylvester, *The convex scattering support in a background medium*, SIAM J. Math. Anal., **36** (2005), 1142–1158.
45. E. Lakshtanov and B. Vainberg. *Applications of elliptic operator theory to the isotropic interior transmission eigenvalue problem*, Inverse Problems, **29** (2013), 104003.
46. P. Lax and R. Phillips, *Scattering Theory*, Academic Press, New York and London, (1967).
47. J. Li, X. Li and H. Liu, *Reconstruction via the intrinsic geometric structures of interior transmission eigenfunctions*, arXiv:1706.04418, (2017).
48. H. Liu, M. Petrini, L. Rondi and J. Xiao, Stable determination of sound-hard polyhedral scatterers by a minimal number of scattering measurements, *J. Differential Equations*, **262(3)** (2017), 1631–1670.
49. H. Liu, L. Rondi and J. Xiao, Mosco convergence for $H(curl)$ spaces, higher integrability for Maxwell's equations, and stability in direct and inverse EM scattering problems, *J. Eur. Math. Soc.*, **21** (2019), no. 10, 2945–2993.

50. H. Liu, C.-H. Tsou and W. Yang, *On Calderón's inverse inclusion problem with smooth shapes by a single partial boundary measurement*, arXiv:2006.10586
51. H. Liu and J. Zou, *Uniqueness in an inverse acoustic obstacle scattering problem for both sound-hard and sound-soft polyhedral scatterers*, Inverse Problems, **22** (2006), 515–524.
52. E.A. Marengo and R.W. Ziolkowski, *On the radiating and nonradiating components of scalar, electromagnetic, and weak gravitational sources*, Phys. Rev. Lett., **83** (1999), 3345–3349.
53. L. Päivärinta, M. Salo and E. V. Vesalainen, *Strictly convex corners scatter*, Rev. Mat. Iberoam. **33(4)** (2017), 1369–1396.
54. L. Päivärinta and J. Sylvester, *Transmission eigenvalues*, SIAM J. Math. Anal., **40** (2008), 738–753.
55. L. Robbiano, *Spectral analysis of the interior transmission eigenvalue problem*, Inverse Problems, **29** (2013), 104001.
56. L. Rondi, *Stable determination of sound-soft polyhedral scatterers by a single measurement*, Indiana Univ. Math. J., **57** (2008), 1377–1408.
57. B. P. Rynne and B. D. Sleeman, *The interior transmission problem and inverse scattering from inhomogeneous media*, SIAM J. Math. Anal., **22** (1991), 1755–1762.
58. A. Sommerfeld, *Zur elektronentheorie. I. Allgemeine untersuchung des feldes eines beliebig bewegten elektrons*, Akad. der Wiss. (Gött.), Math. Phys. Klasse, Nach., 99–130, (1904).
59. A. Sommerfeld, *Zur elektronentheorie. II. Grundlagen für eine allgemeine dynamik des elektrons*, Akad. der Wiss. (Gött.), Math. Phys. Klasse, Nach., 363–439, (1904).
60. J. Sylvester and G. Uhlmann, *A global uniqueness theorem for an inverse boundary value problem*, Ann. of Math. (2) **125(1)** (1987), 153–169.
61. M. E. Taylor, *Partial Differential Equations II: Qualitative Studies of Linear Equations*, Appl. Math. Sci. 116, Springer-Verlag, New York, (1996).
62. G. Uhlmann, *Inside Out: Inverse Problems and Applications*, Cambridge University Press, Cambridge, (2003).

Chapter 12
Stable Determination of an Acoustic Medium Scatterer by a Single Far-Field Pattern

12.1 Overview

In this chapter, we are concerned with the stability analysis for inverse problems associated with time-harmonic acoustic scattering, where we follow the treatment in [3]. Let $k \in \mathbb{R}_+$ be a wavenumber of the acoustic wave, signifying the frequency of the wave propagation. Let $V \in L^\infty(\mathbb{R}^n)$, $n = 2, 3$, be a potential function, which signifies the material parameter of the medium at the point $\mathbf{x}$ and is related to the refractive index in our setting. We assume that $\mathrm{supp}(V) \subset B_R$, where B_R is a central ball of radius $R \in \mathbb{R}_+$ in $\mathbb{R}^n$. That is, the inhomogeneity of the medium is supported inside a given bounded domain of interest. The inhomogeneous medium is often referred to as a *scatterer*. An incident wave u^i is sent to interrogate the medium V, where the scattered wave u^s is generated. We let u denote the total wave field. The former is an entire solution to the Helmholtz equation $(\Delta + k^2)u^i = 0$. The aforementioned acoustic scattering problem can be formulate by

$$\big(\Delta + k^2(1+V)\big)u = 0 \quad \text{in} \quad \mathbb{R}^n. \tag{12.1.1}$$

Moreover, the scattered wave $u^s = u - u^i$ satisfies the Sommerfeld radiation condition

$$|\mathbf{x}|^{\frac{n-1}{2}}\big(\partial_r - \mathrm{i}k\big)u^s \to 0 \tag{12.1.2}$$

uniformly with respect to the angular variable $\theta := \mathbf{x}/|\mathbf{x}|$ as $r := |\mathbf{x}| \to \infty$. Here, ∂_r is the derivative along the radial direction from the origin. The radiation condition implies the existence of a far-field pattern. More precisely, there is a real analytic function on the unit sphere at infinity, namely, $A_{u^i} : \mathbb{S}^{n-1} \to \mathbb{C}$ satisfies

$$u(r\theta) = u^i(r\theta) + \frac{e^{ikr}}{r^{(n-1)/2}} A_{u^i}(\theta) + \mathscr{O}\Big(\frac{1}{r^{n/2}}\Big) \tag{12.1.3}$$

uniformly along the angular variable θ. This function is called the *far-field pattern* or *scattering amplitude* of u.

The inverse scattering problem that we are concerned with is to recover V or its shape, namely, the support, from the knowledge of $A_{u^i}(\theta)$. Furthermore, we consider the recovery in the formally determined case with a single far-field measurement, that is, the scattering amplitude produced from a single wave incidence. Based on the qualitative corner scattering result [4], it is shown in [9] that if two penetrable scatterers V and V' produce the same scattering amplitude for any single incident wave, namely, $A_{u^i} = A'_{u^i}$, then the difference of the supports of V and V', namely, $\mathrm{supp}(V) \triangle \mathrm{supp}(V') := \big(\mathrm{supp}(V) \setminus \mathrm{supp}(V')\big) \cup \big(\mathrm{supp}(V') \setminus \mathrm{supp}(V)\big)$, cannot have a corner of the type that appeared in the papers on corner scattering that shall be discussed in what follows. This means, in particular, that in the set of convex polygonal or cuboidal penetrable scatterers, the far-

H. Diao, H. Liu, *Spectral Geometry and Inverse Scattering Theory*,
https://doi.org/10.1007/978-3-031-34615-6_12

field pattern produced by sending any single incident wave uniquely determines the shape and location of the scatterer.

In this chapter, we sharply quantify the aforementioned uniqueness result on the shape determination by a single far-field pattern. More precisely, we provide logarithmic estimates in determining the shape of a medium scatterer supported in a 2D polygonal or 3D cuboidal domain. In essence, given two such penetrable mediums V and V' and a common incident wave u^i, if the far-field patterns of the scattered waves $u - u^i$ and $u' - u^i$ are ε-close to one another, then the supporting polytopes of V and V' are $\varphi(\varepsilon)$-close in the sense of Hausdorff distance. Here, φ is of double-logarithmic type. For precise statements, see Sect. 12.3.

The structure of the chapter is as follows. We define notation in the next section, which helps with stating the main theorems in Sect. 12.3. The proof idea is described in Sect. 12.4. The quantitative Rellich's theorem and propagation of smallness are proven in Sect. 12.5. The fundamental integral identity, along with estimates for its various terms, is shown in Sect. 12.6. The following one, Sect. 12.7, has the precise estimates for the complex geometric optics solutions. Finally, after all the ingredients have been prepared, the main theorems are proven in Sect. 12.8.

12.2 Notation

1. We use italic letters $P, Q, \ldots$ to denote polytopes; fraktura symbols $\mathfrak{P}, \mathfrak{Q}, \ldots$ for polyhedral cones; and calligraphic symbols $\mathscr{P}, \mathscr{Q}, \ldots$ for spherical cones. This is purely a stylistic choice: all symbols are defined in their context.
2. $B_R = B(0,R)$, $0 < R < \infty$: a priori domain of interest, where the scatterers are located in.
3. $P, P' \subset B_R$: the shape of the penetrable scatterers, which are open polytopes.
4. $d_H(P,P')$: the Hausdorff distance between the sets P and P', defined by

$$d_H(P,P') = \max\big(\sup_{\mathbf{x}\in P} d(\mathbf{x},P'),\ \sup_{\mathbf{x}'\in P'} d(\mathbf{x}',P')\big).$$

5. $\|P\|_{T(s,r)}$: a type of norm for the characteristic function χ_P. If it is finite, the latter is a multiplier in the Sobolev space $H_r^s(\mathbb{R}^n)$. See Definition 12.7.2.
6. u^i: incident wave.
7. u, u': corresponding total waves.

Definition 12.1 (Well-posed Scattering). A potential $V \in L^\infty(\mathbb{R}^n)$ is said to give a *well-posed scattering problem* if there is a finite $\mathscr{S}$ such that given any incident plane wave $u^i(\mathbf{x}) = \exp(ik\omega \cdot \mathbf{x})$, there is a unique $u \in H^2_{loc}$ such that

$$(\Delta + k^2(1+V))u = 0$$

and the scattered wave $u^s = u - u^i$ satisfies the Sommerfeld radiation condition. Moreover, it has to have the norm bound $\|u^s\|_{H^2(B_{2R})} \leq \mathscr{S}$.

Definition 12.2 (Admissible Shape). A polytope $P \subset B_R$ is *admissible* if

1. In 2D, it is a bounded open convex polygon
2. In 3D, it is a cuboid, i.e., there is a rigid motion taking P to $]0,a[\times]0,b[\times]0,c[$ for some $a,b,c > 0$

Definition 12.3 (Admissible Contrast). Given an admissible shape $P \subset B_R$, a function $\varphi : \mathbb{R}^n \to \mathbb{C}$ is *admissible* if

1. $\varphi \in C^\alpha$ for some $\alpha > 0$ in 2D and $\alpha > 1/4$ in 3D
2. $\varphi \neq 0$ at the vertices of P

If the wavenumber or the potential is small, $k^2 \|V\|_\infty < C_0$, then the Neumann series construction of the total wave shows directly that there is well-posed scattering. Unique continuation and Fredholm theory generalize this. For details, see Section 8.4 in [6]. An alternative approach is by Hörmander [8]; see, for example, the introduction in [9]. Note that if P and φ are admissible, then $V = \chi_P \varphi$ has well-posed scattering at any positive frequency $k > 0$.

Definition 12.4 (Non-vanishing Total Wave). We say that a potential $V \in L^\infty(B_R)$ produces a *non-vanishing total wave* if given any incident plane wave u^i, the total wave u vanishes nowhere in $B_R \setminus \operatorname{supp} V$.

We again emphasize that this condition is satisfied for k or $\|V\|_\infty$ small enough, but more general situations exist. It is well-known that the vanishing set (nodal set) of the total field cannot be too large; however, how it relates to a particular potential is an open problem.

12.3 Statement of the Stability Results

We assume the following a priori bounds on the potentials. Given any admissible shape P and function φ, it is possible to choose these parameters such that $V = \chi_P \varphi$ satisfies these bounds.

Definition 12.3.1 (A priori Bounds). The following two theorems have dimension $n \in \{2,3\}$, wavenumber $k > 0$, and radius of the domain of interest $R > 1$ fixed as a priori parameters. In addition,

1. The minimal distance from any vertex of P to a non-adjacent edge is at least ℓ which we assume at most 1 for technical reasons.
2. In 2D, P has angles at least $2\alpha_m > 0$ and at most $2\alpha_M < \pi$.
3. $\|P\|_{T(s,r)} \leq \mathscr{D}$; see Definition 12.7.2.
4. $\|\varphi\|_{C^\alpha} \leq \mathscr{M}$.
5. $|\varphi(\mathbf{x}_c)| \geq \mu$ for any vertex $\mathbf{x}_c$ of P.
6. If V is required to produce non-vanishing total waves, then assume that the infimum of the waves' absolute value in B_R is at least $c > 0$.

Theorem 12.3.1. *Let $V, V' \in L^\infty(B_R)$ be potentials of the form $V = \chi_P \varphi$ and $V' = \chi_{P'} \varphi'$ with P, P' and φ, φ' admissible by Definitions 12.2 and 12.3. Moreover, assume that V and V' produce non-vanishing total waves as in Definition 12.4.*

Let $\mathfrak{h} = d_H(P, P')$ be the Hausdorff distance of P and P'. Let $u^i(x) = \exp(ik\omega \cdot \mathbf{x})$ be any plane wave and u^s_∞, u'^s_∞ be the far-field patterns of the scattered waves produced by V and V', respectively.

There are constants $\varepsilon_{min}, C < \infty$, which depend on the a priori bounds of Definition 12.3.1 only and $\gamma = \gamma(\alpha, n, r, s) > 0$ such that if

$$\left\|u^s_\infty - u'^s_\infty\right\|_{L^2(\mathbb{S}^{n-1})} < \varepsilon_{min},$$

then

$$\mathfrak{h} \leq C \left(\ln\ln \frac{\mathscr{S}}{\|u^s_\infty - u'^s_\infty\|_{L^2(\mathbb{S}^{n-1})}} \right)^{-\gamma}. \tag{12.3.1}$$

We remark that in the following theorem, the refractive index function φ is allowed to vanish at the vertices. As long as there is one corner where it does not vanish, and the scatterer can fit inside the convex cone generated by that corner, then we can show a lower bound for the scattering amplitude.

Theorem 12.3.2. *Let $V \in L^\infty(B_R)$ be a potential of the form $V = \chi_P \varphi$ with P and φ admissible by Definition 12.2 and Definition 12.3.*

Recall that ℓ is a lower bound for the minimal vertex to non-adjacent edge distance of P. Let $u^i(x) = \exp(ik\omega \cdot x)$ be any plane wave and u^s_∞ be the far-field pattern of the scattered wave produced by V.

Then

$$\|u^s_\infty\|_{L^2(\mathbb{S}^{n-1})} \geq \min\left(\frac{\mathscr{S}}{\exp\exp(C\ell^{-2/\gamma}|\varphi(x_c)|^{-2-2/((n+5)\gamma)})}, \varepsilon_{min}\right), \tag{12.3.2}$$

where the constants $\varepsilon_{min}, C < \infty$ depend only on the a priori parameters of Definition 12.3.1 except for ℓ or μ and $\gamma = \gamma(\alpha, n, r, s) > 0$ is as in the previous theorem.

12.4 Idea of the Proofs

We start describing the proof of stability for scatterer support probing. After this, it is very convenient to show stability of corner scattering by having the second scatterer identically zero. Propagation of smallness is the first step.

Let $w = u - u'$ be the difference of the total (and hence scattered) waves from two potentials $V = \chi_P \varphi$ and $V' = \chi_{P'} \varphi'$. Its far-field pattern is the difference of the far-field patterns of u and u' and hence small when proving stability. We first propagate that smallness into the near field by an Isakov-type estimate. After that, we propagate it near the scatterers by a chain of balls argument and then into the scatterers by a delicate balancing argument using Hölder continuity.

Local issues are dealt with next. Focus on a vertex $\mathbf{x}_c \in \partial P$ which makes $d(\mathbf{x}_c, P')$ equal to the Hausdorff distance between P and P'. Let $P_h = P \cap B(\mathbf{x}_c, h)$ for some $h > 0$ small enough. We have two representations for the integral

$$\int_{P_h} V(\mathbf{x}) u_0(\mathbf{x}) u'(\mathbf{x}) \mathrm{d}\mathbf{x},$$

where u_0 is any (possibly nonphysical) solution to

$$\big(\Delta + k^2(1+V)\big) u_0 = 0 \tag{12.4.1}$$

and $u' : \mathbb{R}^n \to \mathbb{C}$ is the total wave satisfying $\big(\Delta + k^2(1+V')\big)u' = 0$ corresponding to the incident wave u^i. Near P_h, it is actually a solution to the constant coefficient equation

$$(\Delta + k^2) u' = 0, \tag{12.4.2}$$

because $V' = 0$ there.

For the first representation, we use (12.4.1) and Green's formula. The total wave u satisfies

$$\big(\Delta + k^2(1+V)\big) u = 0. \tag{12.4.3}$$

Integration by parts in a truncated cone Q_h slightly larger than P_h gives

$$k^2 \int_{P_h} V(x) u_0(x) u'(x) \mathrm{d}\mathbf{x} = -\int_{\partial Q_h} \big(u_0 \partial_\nu (u - u') - (u - u') \partial_\nu u_0\big) \mathrm{d}\sigma \tag{12.4.4}$$

by (12.4.2) and (12.4.3).

For the second representation, the a priori admissibility assumptions and the real analyticity of u' near P_h imply the splittings

$$V(\mathbf{x}) = \varphi(\mathbf{x}_c) + \varphi_\alpha(\mathbf{x}), \qquad |\varphi_\alpha(\mathbf{x})| \leq \|\varphi\|_{C^\alpha(C_h)} |\mathbf{x}-\mathbf{x}_c|^\alpha,$$
$$u'(\mathbf{x}) = u'(\mathbf{x}_c) + u'_1(\mathbf{x}), \qquad |u'_1(\mathbf{x})| \leq \mathscr{R}|\mathbf{x}-\mathbf{x}_c|.$$

Lastly, we choose $u_0 : \mathbb{R}^n \to \mathbb{C}$ to be a complex geometric optics solution

$$u_0(x) = e^{\rho\cdot(\mathbf{x}-\mathbf{x}_c)}(1+\psi(\mathbf{x}))$$

with $\rho \in \mathbb{C}^n$ such that $\exp(\rho \cdot \mathbf{x})$ decays exponentially in P_h as $|\rho| \to \infty$. We show that there are $p \geq 1$ and $\beta > 0$ such that

$$\|\psi\|_{L^p(\mathbb{R}^n)} \leq C|\Im\rho|^{-n/p-\beta}\|V\|$$

where C doesn't depend on ρ or V as long as $|\Im\rho|$ is large enough. Here, the norm $\|V\|$ is of new type and contains information about the geometry of the polytope P and a priori parameters related to φ.

Plug the above function splittings into $\int Vu_0u'\mathrm{d}\mathbf{x}$, and then estimate all of these integrals in terms of the norms of $u-u'$, $\varphi(x_c)$, $|\Re\rho|$, and h. After that, a choice of $|\Re\rho|$ proves an upper bound for $d_H(P,P')$ based on the smallness ε of the far-field pattern of $u-u'$.

12.5 From the Far Field to the Scatterer

The classical Rellich's theorem (Lemma 2.11 in [6]) says that if the far-field pattern of a scattered wave is zero, then the scattered wave is identically zero on the unbounded and connected component of space that's unperturbed by a potential or source term. In this section, we study what is the corresponding quantitative result: namely, having a penetrable scatterer and a far-field pattern whose norm is small but positive. This kind of question has been studied earlier for the easier case of impenetrable scatterers by Isakov [11, 12], and more recently by, for example, Rondi [18] and Chap. 6.

Our strategy in this section is as follows. We first generalize a far-field to near-field estimate in the style of Isakov [12] and Rondi and Sini [19] to the penetrable scatterer case. Then we use an L^∞ three-sphere inequality to propagate smallness from the boundary of B_{2R} to almost the support of the scatterer V. To proceed after that, use the Hölder continuity of $w = u-u'$. This allows the propagation to take the final step, crossing from outside the support of the potentials into the support. Lastly, we use an elliptic regularity estimate to see that the same operations can be done for $w = \nabla(u-u')$.

12.5.1 From the Far Field to the Near Field

We first show that if the far-field patterns A_{u^i} of u and A'_{u^i} of u' are close, then u and u' are close in $B_{2R}\setminus B_R$.

Lemma 12.5.1. *Let $A,\varepsilon,\mathscr{S} > 0$. Then there is a function $\ell : \mathbb{R}_+ \to \mathbb{R}_+$ such that for*

$$f(\varepsilon,\ell) = \left(\frac{\ell}{A}\right)^\ell \varepsilon^2 + \mathscr{S}^2,$$

we have

$$f(\varepsilon,\ell(\varepsilon)) \leq 2\max(\mathscr{S}^2,\varepsilon^2).$$

Moreover, when $\varepsilon < \mathscr{S}$, we may set $\ell(\varepsilon) = \sqrt{2A\ln\frac{\mathscr{S}}{\varepsilon}}$.

Proof. If $\varepsilon \geq \mathscr{S}$, choose $\ell(\varepsilon) = A$. Otherwise, $\ln(\mathscr{S}/\varepsilon) > 0$. We may set ℓ as in the statement, which implies that

$$\frac{\ell}{A}\ln\frac{\ell}{A} \leq \left(\frac{\ell}{A}\right)^2 = \frac{2}{A}\ln\frac{\mathscr{S}}{\varepsilon},$$

i.e., $(\ell/A)^\ell \leq \mathscr{S}^2/\varepsilon^2$, from which the claim follows. □

The following proposition generalizes Theorem 4.1 from Rondi and Sini [19] to the penetrable scatterer case.

Proposition 12.1. *Assume that $w^s \in H^2_{loc}(\mathbb{R}^n)$ satisfies $(\Delta + k^2)w^s = 0$ in $\mathbb{R}^n \setminus \overline{B}(0,R)$ and the Sommerfeld radiation condition. Let $B_0 > 1$, $\mathscr{S} \geq 0$, and assume the a priori bound $\|w^s\|_{L^2(B_{2R}\setminus B_R)} \leq \mathscr{S}$.*

Let $\varepsilon = \|w^s_\infty\|_{L^2(\mathbb{S}^{n-1})}$ where w^s_∞ is the far-field pattern of w^s. Then there is a constant $\mathscr{C} > 0$ depending only on k, R, B_0 such that if $\varepsilon < \mathscr{C}^{-1}\mathscr{S}$, then

$$\|w^s\|_{L^2(B_{2B_0R}\setminus B_{B_0R})} \leq \mathscr{C}\mathscr{S} B_0^{-\frac{1}{2}\sqrt{2ekR\ln(\mathscr{S}/\varepsilon)}}.$$

However, if not, then $\|w^s\|_{L^2(B_{2R}\setminus B_R)} \leq \mathscr{C}\varepsilon$.

Proof. By the assumptions on w^s, it is well-known that there is a sequence $b_j > 0$, $j = 0, 1, \ldots$ such that its far-field pattern w^s_∞ satisfies

$$\|w^s_\infty\|^2_{L^2(\mathbb{S}^{n-1})} = \sum_{j=0}^{\infty} b_j^2$$

and the function itself has

$$\|w^s\|^2_{L^2(S(0,r))} = \frac{\pi}{2}\sum_{j=0}^{\infty} b_j^2 kr \left|H^{(1)}_{j+(n-2)/2}(kr)\right|^2$$

for any $r > R$. Here, $H^{(1)}_\nu$ is a Hankel function of the first kind and order ν.

Let $j_0 \in \{0, 1, 2, \ldots\}$ and $B_0 > 1$. Then

$$\begin{aligned}
\|w^s\|^2_{L^2(S(0,r))} &= \frac{\pi}{2}\sum_{j=0}^{j_0} b_j^2 kr \left|H^{(1)}_{j+(n-2)/2}(kr)\right|^2 \\
&\quad + \frac{\pi}{2}\sum_{j=j_0+1}^{\infty} b_j^2 kr \frac{\left|H^{(1)}_{j+(n-2)/2}(kr)\right|^2}{\left|H^{(1)}_{j+(n-2)/2}(kr/B_0)\right|^2}\left|H^{(1)}_{j+(n-2)/2}(kr/B_0)\right|^2 \\
&\leq \frac{\pi}{2} kr \max_{0\leq j\leq j_0}\left|H^{(1)}_{j+(n-2)/2}(kr)\right|^2 \|w^s_\infty\|^2_{L^2(\mathbb{S}^{n-1})} \\
&\quad + B_0 \sup_{j>j_0}\frac{\left|H^{(1)}_{j+(n-2)/2}(kr)\right|^2}{\left|H^{(1)}_{j+(n-2)/2}(kr/B_0)\right|^2}\|w^s\|^2_{L^2(S(0,r/B_0))}
\end{aligned} \tag{12.5.1}$$

by the two formulas above. By Corollary 3.8 from Rondi and Sini [19], we see that if $0 < z_1 \leq z_2 < \infty$, then there is $C = C(z_1, z_2) < \infty$ such that

$$\left|H^{(1)}_0(z)\right|^2 \leq C^2 \leq C^2\frac{4}{\pi e z} \tag{12.5.2}$$

and

$$C^{-2}\frac{4}{\pi ez}\left(\frac{2\nu}{ez}\right)^{2\nu-1} \leq \left|H_\nu^{(1)}(z)\right|^2 \leq C^2\frac{4}{\pi ez}\left(\frac{2\nu}{ez}\right)^{2\nu-1} \tag{12.5.3}$$

for $z_1 \leq z \leq z_2$ and $\nu \in \{\frac{1}{2}, \frac{2}{2}, \frac{3}{2}, \ldots\}$. We will integrate the formula above for $\|w^s\|^2_{L^2(S(0,r))}$ along the segment $r \in [B_0 R, 2B_0 R]$, and so the minimal value of kr/B_0 will be $z_1 := kR > 0$, and the maximal value of the larger kr shall be $z_2 := 2B_0 kR < \infty$.

Write $\nu_0 = j_0 + (n-2)/2$, and assume that j_0 is large enough that $\nu_0 \geq ez_2/2 = eB_0 kR$ and $\nu_0 > 1$. These assumptions imply that $2\nu_0 \geq ez$ when $z_1 \leq z \leq z_2$ and thus also

$$\left|H_0^{(1)}(z)\right|^2 \leq \frac{4C^2}{\pi ez} \leq \frac{4C^2}{\pi ez}\left(\frac{2\nu_0}{ez}\right)^{2\nu_0-1}, \quad z_1 \leq z \leq z_2.$$

Next, if $1/2 \leq \nu \leq ez/2$ and it is a half-integer, we have

$$\left|H_\nu^{(1)}(z)\right|^2 \leq \frac{4C^2}{\pi ez}\left(\frac{2\nu}{ez}\right)^{2\nu-1} \leq \frac{4C^2}{\pi ez} \leq \frac{4C^2}{\pi ez}\left(\frac{2\nu_0}{ez}\right)^{2\nu_0-1}, \quad z_1 \leq z \leq z_2.$$

On the other hand, if $ez/2 \leq \nu \leq \nu_0$, then

$$\left|H_\nu^{(1)}(z)\right|^2 \leq \frac{4C^2}{\pi ez}\left(\frac{2\nu}{ez}\right)^{2\nu-1} \leq \frac{4C^2}{\pi ez}\left(\frac{2\nu_0}{ez}\right)^{2\nu_0-1}, \quad z_1 \leq z \leq z_2$$

because the function $\nu \mapsto (2\nu/(ez))^{2\nu-1}$ defined on $\mathbb{R}_+$ is increasing when $\ln 2\nu - \ln z - (2\nu)^{-1} \geq 0$. This is true when $2\nu \geq ez$ and $\nu \geq 1/2$. In conclusion, we can estimate

$$\left|H_{j+(n-2)/2}^{(1)}(kr)\right|^2 \leq \frac{4C^2}{\pi ekr}\left(\frac{2\nu_0}{ekr}\right)^{2\nu_0-1}, \quad B_0 R \leq r \leq 2B_0 R$$

in (12.5.1) when $0 \leq j \leq j_0$. Then, using the two Hankel function estimates (12.5.2) and (12.5.3) and recalling that $B_0^{1-2\nu} \leq B_0^{1-2\nu_0}$ when $\nu \geq \nu_0$ and $B_0 \geq 1$, we can continue estimating (12.5.1) with

$$\|w^s\|^2_{L^2(S(0,r))} \leq \frac{2C^2}{e}\left(\frac{2\nu_0}{ekr}\right)^{2\nu_0-1} \|w^s_\infty\|^2_{L^2(\mathbb{S}^{n-1})} + C^4 B_0^{1-2\nu_0}\|w^s\|^2_{L^2(S(0,r/B_0))} \tag{12.5.4}$$

whenever $B_0 R \leq r \leq 2B_0 R$.

Next, we integrate (12.5.4) by $\int_{B_0 R}^{2B_0 R} \ldots dr$ to get

$$\begin{aligned}
&\|w^s\|^2_{L^2(B_{2B_0R}\setminus B_{B_0R})} \\
&\quad \leq \frac{C^2 RB_0}{e}\frac{1}{\nu_0-1}\left(1-\frac{1}{2^{2\nu_0-2}}\right)\left(\frac{2\nu_0}{ekRB_0}\right)^{2\nu_0-1}\|w^s_\infty\|^2_{L^2(\mathbb{S}^{n-1})} \\
&\quad\quad + C^4 B_0^{2-2\nu_0}\|w^s\|^2_{L^2(B_{2R}\setminus B_R)}
\end{aligned}$$

where we have denoted 0-centered disks of radius ℓ by B_ℓ. Use the shorthand $\varepsilon = \|w^s_\infty\|_{L^2(\mathbb{S}^{n-1})}$, and recall from the proposition statement that $\mathscr{S} \geq \|w^s\|_{L^2(B_{2R}\setminus B_R)}$. Since $\nu_0 \in \frac{1}{2}\mathbb{N}$ and $\nu_0 > 1$, we have $|(1-e^{2-2\nu_0})/(\nu_0-1)| \leq 2$. Thus,

$$\|w^s\|^2_{L^2(B_{2B_0R}\setminus B_{B_0R})} \leq \max\left(\frac{2C^2 R}{e}, C^4\right)\frac{1}{B_0^{2\nu_0-2}}\left(\left(\frac{2\nu_0}{ekR}\right)^{2\nu_0-1}\varepsilon^2 + \mathscr{S}^2\right)$$

when $\nu_0 \in \frac{1}{2}\mathbb{N}$ with $\nu_0 > 1$ and $\nu_0 \geq eB_0 kR$.

We are now ready to fix ν_0. Let

$$\ell = \sqrt{2ekR\ln(\mathscr{S}/\varepsilon)}, \qquad \nu_0 = \lfloor \ell \rfloor /2. \tag{12.5.5}$$

If $\nu_0 < \max(3/2, eB_0kR)$, then

$$\max(3, 2eB_0kR) > 2\nu_0 = \lfloor \ell \rfloor > \ell - 1,$$

and so

$$\frac{\mathscr{S}}{\varepsilon} = \exp\left(\frac{\ell^2}{2ekR}\right) < \exp\left(\frac{(1+\max(3,2eB_0kR))^2}{2ekR}\right)$$

which implies $\|w^s\|_{L^2(B_{2R}\setminus B_R)} \leq \mathscr{C}\varepsilon$. On the other hand, this would follow even more directly if $\mathscr{S} < \varepsilon$. The other case, namely, $\nu_0 \geq \max(3/2, eB_0kR)$ and $\mathscr{S} \geq \varepsilon$, implies in particular that

$$\left(\frac{2\nu_0}{ekR}\right)^{2\nu_0-1} \leq \left(\frac{\ell}{ekR}\right)^{2\nu_0-1} \leq \left(\frac{\ell}{ekR}\right)^{\ell}$$

because $\ell \geq \lfloor \ell \rfloor = 2\nu_0$ and $2\nu_0 \geq 2eB_0kR \geq ekR$, as well as $2\nu_0 - 1 \geq 0$. Lemma 12.5.1 implies

$$\|w^s\|^2_{L^2(B_{2B_0R}\setminus B_0R)} \leq \max\left(\frac{2C^2R}{e}, C^4\right)\frac{2\mathscr{S}^2}{B_0^{2\nu_0-2}} \leq \max\left(\frac{2C^2R}{e}, C^4\right)\frac{2\mathscr{S}^2}{B_0^{\ell-3}}$$

because $2\nu_0 = \lfloor \ell \rfloor \geq \ell - 1$ and $\mathscr{S} \geq \varepsilon$ in this final case. The final claim follows from the choice of ℓ in (12.5.5). □

Corollary 12.5.1. *Let $w^s \in H^2_{loc}(\mathbb{R}^n)$ satisfy $(\Delta + k^2)w^s = 0$ in $\mathbb{R}^n \setminus \overline{B}(0,R)$ and the Sommerfeld radiation condition at infinity. Let w^s_∞ be its far-field pattern.*

Let $\mathscr{S} \geq 0$ and assume the a priori bound $\|w^s\|_{L^2(B_{2R}\setminus B_R)} \leq \mathscr{S}$. Denote $\varepsilon = \|w^s_\infty\|_{L^2(\mathbb{S}^{n-1})}$. Let A be a domain such that $\overline{A} \subset B_{2R}\setminus \overline{B}_R$. Then, for any smoothness index $r \in \mathbb{N}$, there are constants $c, C > 0$ depending only on k, r, R, A such that

$$\|w^s\|_{H^r(A)} \leq C\max\left(\varepsilon, \mathscr{S}e^{-c\sqrt{\ln(\mathscr{S}/\varepsilon)}}\right).$$

Proof. Elliptic interior regularity is the main tool to prove the claim. Firstly, if $f \in \mathscr{S}'(\mathbb{R}^n)$, then

$$\|f\|_{H^{s+2}(\mathbb{R}^n)} = \left\|(1+k^2)f - (\Delta+k^2)f\right\|_{H^s(\mathbb{R}^n)}$$

for any $s \in \mathbb{R}$. Let $B_0 > 1$ be such that $\overline{A} \subset \Omega := B_{2R}\setminus \overline{B}_{B_0R}$. If $\varphi \in C_0^\infty(\Omega)$, we have

$$\left\|(\Delta+k^2)(\varphi w^s)\right\|_{H^s(\mathbb{R}^n)} = \|2\nabla\varphi\cdot\nabla w^s + w^s\Delta\varphi\|_{H^s(\mathbb{R}^n)} \leq C_\varphi \|w^s\|_{H^{s+1}(\Omega)}.$$

Here, w^s was extended by zero outside of Ω. Let $\Omega' \subset \Omega$ be a subdomain a positive distance from the boundary of Ω. Now, if we have $\varphi \equiv 1$ on Ω', then

$$\begin{aligned}\|w^s\|_{H^{s+2}(\Omega')} &\leq \|\varphi w^s\|_{H^{s+2}(\mathbb{R}^n)} \leq (1+k^2)C_\varphi\|w^s\|_{H^s(\Omega)} + C_\varphi\|w^s\|_{H^{s+1}(\Omega)} \\ &\leq C_{k,\varphi}\|w^s\|_{H^{s+1}(\Omega)},\end{aligned}$$

by the two equations above.

Next, the proposition implies

$$\|w^s\|_{L^2(\Omega)} \leq \mathscr{C}\max\left(\varepsilon, \mathscr{S}B_0^{-\frac{1}{2}\sqrt{2ekR\ln(\mathscr{S}/\varepsilon)}}\right)$$

directly. Given $r \in \mathbb{N}$, take a sequence $A = \Omega_r \subset \Omega_{r-1} \subset \cdots \subset \Omega_0 = \Omega$ of sets whose boundaries are a positive distance apart. Also, take a sequence of smooth cutoff functions $\varphi_j \in C_0^\infty(\Omega_j)$ such that $\varphi_j \equiv 1$ on Ω_{j+1}. Then we use the last estimate of the previous paragraph inductively to get

$$\|w^s\|_{H^r(\Omega_r)} \leq C_{k,\varphi_0,\ldots,\varphi_{r-1}} \mathscr{C} \max\left(\varepsilon, \mathscr{S} B_0^{-\frac{1}{2}\sqrt{2ekR\ln(\mathscr{S}/\varepsilon)}}\right)$$

from the $L^2(\Omega)$ norm of w^s. □

12.5.2 A Three-Sphere Inequality and a Chain of Balls

We state an L^∞ three-ball inequality for solutions to the Helmholtz equation. It follows from Lemma 3.5 in [18] by suitable choices of parameters. After that, we prove a few lemmas and a proposition which allows us to propagate the smallness from outside a large ball along a straight line to near the scatterers V and V'.

Lemma 12.5.2. *There are positive constants R_m, C, c_1 such that $0 < c_1 < 1$, which depend only on k and satisfy the following: Let $x \in \mathbb{R}^n$ and $0 < 4r < R_m$. If w satisfies*

$$(\Delta + k^2)w = 0 \quad in \quad B_{4r} := B(x, 4r),$$

then

$$\|w\|_{B_{2r}} \leq C(2+\sqrt{2})^{\frac{3}{2}} \|w\|_{B_{4r}}^{1-\beta} \|w\|_{B_r}^{\beta}, \tag{12.5.6}$$

where the norms are L^∞ norms in the corresponding x-centered balls and β is a number that satisfies

$$\frac{c_1}{4} \leq \beta \leq 1 - \frac{3c_1}{4}.$$

Proof. Choose $\rho_1 = r$, $\rho = 2r$, $\rho_2 = 4r$, $\tilde{\rho}_0 = R_m$, and $s = 2^{3/2} r$ in Lemma 3.5 of [18]. Also choose $u(\cdot) = w(\cdot - x)$. □

Lemma 12.5.3. *Let $K \in \mathbb{N}$, $r > 0$ and $B_1, \ldots, B_K$ be a chain of balls with the following properties:*

1. *$4r < R_m$, the latter defined in Lemma 12.5.2.*
2. *The radius of each B_k is r.*
3. *The center-to-center distance of B_k to B_{k+1} is at most r.*

Let $U \subset \mathbb{R}^n$ be open and $w \in L^\infty(U)$ satisfy the Helmholtz equation $(\Delta + k^2)w = 0$ there and $\|w\|_{L^\infty(U)} \leq \mathscr{T}$ which we assume to be at least 1. Assume that each $B_k \subset U$ and moreover that $d(B_k, \partial U) \geq 3r$.

Then there are finite $C \geq 1$, $0 < c_2 < 1/4$ depending only on k such that

$$\|w\|_{B_K} \leq C\mathscr{T} \|w\|_{B_1}^{c_2^{K-1}}$$

if $\|w\|_{B_1} \leq 1$, where the norms are the L^∞ norms in the corresponding balls.

Proof. Lemma 12.5.2 and the fact that B_k is covered by the $2r$-radius ball with same center as B_{k-1} imply that

$$\|w\|_{B_k} \leq C(2+\sqrt{2})^{3/2} \mathscr{T}^{1-\beta} \|w\|_{B_{k-1}}^{\beta}.$$

Estimate $\|w\|_{B_K}$ as above, and continue telescopically to get

$$\|w\|_{B_K} \leq C^{1+\beta+\cdots+\beta^{K-2}} (2+\sqrt{2})^{\frac{3}{2}(1+\beta+\cdots+\beta^{K-2})} \mathscr{T}^{(1-\beta)(1+\beta+\cdots+\beta^{K-2})} \|w\|_{B_1}^{\beta^{K-1}}.$$

Note that $1+\cdots+\beta^{K-2} \leq 1/(1-\beta) \leq 4/(3c_1)$ and $\beta \geq c_1/4$. The claim follows by setting $c_2 = c_1/4$. □

Corollary 12.5.2. *Let $U \subset \mathbb{R}^n$ be open, $w \in L^\infty(U)$ such that $(\Delta + k^2)w = 0$. Let $\gamma \subset U$ be a rectifiable curve between two different points $\mathbf{x}, \mathbf{x}' \in U$ such that $B(\gamma, 4r) = \cup_{\mathbf{y}\in\gamma} B(\mathbf{y}, 4r) \subset U$ for some $r > 0$. Assume that the L^∞ norms satisfy $\|w\|_{B(\mathbf{x},r)} \leq 1$ and that $\|w\|_U \leq \mathscr{T}$ which is at least one.*

Then for any $\mathbf{y} \in \gamma$, we have

$$\|w\|_{B(\mathbf{y},r)} \leq C\mathscr{T}\, \|w\|_{B(\mathbf{x},r)}^{c_2^{d_\gamma(\mathbf{x},\mathbf{y})/r+1}} \leq C\mathscr{T}\, \|w\|_{B(\mathbf{x},r)}^{c_2^{d_\gamma(\mathbf{x},\mathbf{x}')/r+1}}$$

if $4r \leq R_m$ as in Lemma 12.5.2. Here, d_γ is the distance measured along γ.

Proof. Denote $l = d_\gamma(\mathbf{x},\mathbf{y})$. We build a sequence of balls, each of radius r and centers $x_1 = x, x_2, x_3, \ldots, x_{\lceil l/r \rceil}$. Finally, set $x_{\lceil l/r \rceil + 1} = y$. Choose them so that $d_\gamma(x_{k+1}, x_k) \leq r$. Hence, also $d(x_{k+1}, x_k) \leq r$. For example, if $l = 2r$, we would get the triple x, x_2, y with $2 = \lceil l/r \rceil$. For $l = (2+\frac{1}{2})r$, we would get the 4-tuple x, x_2, x_3, y with $3 = \lceil l/r \rceil$. Then use the previous lemma with $B_k = B(x_k, r)$ and $K = \lceil l/r \rceil + 1 \leq l/r + 2$. Since $\|w\|_{B(x,r)} \leq 1$ and $c_1/4 < 1$, both estimates follow. □

We are now ready to state and prove the propagation of smallness in the context of corner scattering. Recall that P and P' contain the supports of the potentials V and V' and both are contained in $B_R = B(0,R)$ for some fixed $R > 0$. Moreover, both are convex. This is important to ensure that $B_R \setminus (P \cup P')$ is simply connected.

Proposition 12.5.1. *Let $Q \subset B_R \subset \mathbb{R}^n$ be a convex polytope. Let w be a function such that $w \in L^\infty(B_{2R} \setminus Q)$ satisfies $(\Delta + k^2)w = 0$ in its domain, with L^∞ norm at most $\mathscr{T} \geq 1$. Let $4r \leq R_m$, the latter being from Lemma 12.5.2, and $2r < (1-2\lambda)R$ for some positive $\lambda < \frac{1}{2}$.*

Assume that $\|w\|_{L^\infty} \leq \delta \leq 1$ in $B_{(2-\lambda)R} \setminus B_{(1+\lambda)R}$. Then

$$\|w\|_{L^\infty(B_{2R} \setminus B(Q,4r))} \leq C\mathscr{T}\delta^{c_2^{(2+\lambda)R/r+2}}$$

where $C \geq 1$ and $0 < c_2 < 1/4$ are as in Lemma 12.5.3.

Proof. Let $x' \in B_{2R} \setminus B(Q,4r)$. Since Q is convex, there is a ray from x' into $B_{2R} \setminus B_{(1+\lambda)R}$ that's at least distance $4r$ from Q. It can be constructed as follows: consider the line from 0 to x' (if $x' = 0$ any line is fine). The point x' splits it into two rays. Take one of them not touching the convex set $B(Q,4r)$.

Cut a segment from the ray, starting at x' and ending distance r outside $B_{(1+\lambda)R}$ to make sure that $\|w\|_\infty \leq \delta$ in the first ball in the chain of balls we are about to use. This ball has radius r, and since $2r < (1-2\lambda)R$, it fits completely inside $B_{(2-\lambda)R} \setminus B_{(1+\lambda)R}$. The length of that segment is then at most $R + (1+\lambda)R + r$. Then use Corollary 12.5.2. □

12.5.3 Propagation of Smallness into the Perturbation

The purpose of the following proposition is to estimate $u - u'$ and $\nabla u - \nabla u'$ in Proposition 12.6.2. This is possible because these differences are Hölder continuous: the case of $u - u'$ follows directly from Sobolev embedding in $\mathbb{R}^2$ and $\mathbb{R}^3$ because $V, V' \in H^s(\mathbb{R}^n)$ for $s < 1/2$. The smoothness of the gradient follows from elliptic regularity estimates for boundary value problems with smooth boundary values. After all, $u - u'$ is real analytic outside of the supports of the potentials V and V'.

Proposition 12.5.2. *Let $Q \subset B_R \subset \mathbb{R}^n$ be a convex polytope. Let $w \in L^\infty(B_{2R})$ be such that $w \in C^\alpha(\overline{B}_{3R/2})$ with norm at most $\mathscr{T} \geq 1$ for some $0 < \alpha < 1$ and it satisfies $(\Delta + k^2)w = 0$ in $B_{2R} \setminus Q$. Assume that $|w(x)| \leq \delta$ in $B_{(2-\lambda)R} \setminus B_{(1+\lambda)R}$ for some positive $\lambda < \frac{1}{2}$, and let $A \geq 2 + \lambda$. If*

$$\delta < 1/\exp\exp\left(\frac{4AR\,|\ln c_2|/(1-\alpha)}{\min(R_m, R/2, 2(1-2\lambda)R)}\right), \tag{12.5.7}$$

where R_m is given in Lemma 12.5.2, then

$$|w(x)| \leq \frac{(8AR\,|\ln c_2|/(1-\alpha))^\alpha + C/c_2^2}{(\ln|\ln\delta|)^\alpha}\mathscr{T} \tag{12.5.8}$$

for $\mathbf{x} \in B_{3R/2}$ satisfying $d(\mathbf{x}, \partial Q) \leq 4AR\,|\ln c_2|/((1-\alpha)\ln|\ln\delta|)$. Here, C and c_2 are given by Lemma 12.5.3.

Proof. Choose

$$r = r(\delta) = \frac{AR\,|\ln c_2|}{(1-\alpha)\ln|\ln\delta|}$$

with c_2 from Lemma 12.5.3. Then $r > 0$. By the upper bound on δ, we have $4r < R_m$ and $2r < (1-2\lambda)R$ as required in Proposition 12.5.1. By that same proposition,

$$|w(\mathbf{x}')| \leq C\mathscr{T}\delta^{c_2^{(2+\lambda)R/r+2}}$$

when $\mathbf{x}' \in B_{2R}$, $d(\mathbf{x}', Q) \geq 4r$.

Let $d(\mathbf{x}, \partial Q) \leq 4r$ now. Then there is $\mathbf{y} \in \partial Q$ such that $|\mathbf{x} - \mathbf{y}| \leq 4r$. By the convexity of Q, there is $\mathbf{x}' \in \mathbb{R}^n$ with $d(\mathbf{x}', Q) = 4r$ and $|\mathbf{x}' - \mathbf{y}| = 4r$. The upper bound on δ implies $R + 4r < 3R/2$, and so $|\mathbf{x}'| \leq |\mathbf{x}' - \mathbf{y}| + |\mathbf{y}| \leq 4r + R \leq 3R/2$. Thus, $\mathbf{x}' \in B_{3R/2} \setminus B(Q, 4r)$ and $|\mathbf{x} - \mathbf{x}'| \leq |\mathbf{x} - \mathbf{y}| + |\mathbf{y} - \mathbf{x}'| \leq 8r$. Concluding, by the Hölder continuity of w, we have

$$|w(\mathbf{x})| \leq \|w\|_{C^\alpha(\overline{B}_{3R/2})}\left|\mathbf{x} - \mathbf{x}'\right|^\alpha + |w(\mathbf{x}')| \leq \mathscr{T}8^\alpha r^\alpha + C\mathscr{T}\delta^{c_2^{(2+\lambda)R/r+2}}$$

for $d(\mathbf{x}, \partial Q) \leq 4r$.

The choice of $r = r(\delta)$ implies that

$$r^\alpha = \frac{(AR\,|\ln c_2|/(1-\alpha))^\alpha}{(\ln|\ln\delta|)^\alpha}, \qquad \frac{(2+\lambda)R}{r} = -\frac{(2+\lambda)(1-\alpha)}{A}\frac{\ln|\ln\delta|}{\ln c_2},$$

and so

$$\delta^{c_2^{(2+\lambda)R/r+2}} = e^{-|\ln\delta| c_2^{(2+\lambda)R/r+2}} = e^{-c_2^2|\ln\delta|^{1-(2+\lambda)(1-\alpha)/A}}.$$

Now, since $2 + \lambda \leq A$ and $|\ln\delta| > 1$, we can continue the above with

$$\ldots \leq e^{-c_2^2|\ln\delta|^\alpha} \leq \frac{1}{c_2^2\,|\ln\delta|^\alpha} \leq \frac{1}{c_2^2(\ln|\ln\delta|)^\alpha}.$$

The claim follows. □

12.5.4 Quantitative Rellich's Theorem

Lemma 12.5.4. *Let $n \in \{2, 3\}$ and $q \in L^\infty(B_{2R})$ be supported in B_R for some $R > 0$. Let $w \in H^2(B_{2R})$, and assume that*

$$(\Delta + k^2(1+q))w = 0.$$

Then $w \in C^{1,\frac{1}{2}}(\overline{B}_{3R/2})$, *and there is* $C = C(R,k,n)$ *such that*

$$\|w\|_{C^{1,\frac{1}{2}}} \leq C\,(1+\|q\|_{L^\infty})\,\|w\|_{H^2}\,.$$

Proof. Interior elliptic regularity in the domain where $q \equiv 0$ (e.g., Theorem 8.10 by Gilbarg and Trudinger [7]) implies that $w \in H^s(B_{7R/4} \setminus \overline{B}_{5R/4})$ and a corresponding norm estimate for any $s \geq 0$ and in particular $s = (n+3)/2$. Adding Sobolev embedding gives then

$$\|w\|_{C^{1,\frac{1}{2}}(B_{7R/4}\setminus B_{5R/4})} \leq C\,\|w\|_{H^{\frac{n+3}{2}}(B_{7R/4}\setminus B_{5R/4})} \leq C\,\|w\|_{H^2(B_{2R}\setminus B_R)} \tag{12.5.9}$$

for some other constant $C = C(R,k,n)$. This implies that w has boundary values in $C^{1,1/2}(\partial B_{3R/2})$, i.e., more precisely that there is $\varphi \in C^{1,1/2}(\mathbb{R}^n)$ supported in $B_{7R/4} \setminus B_{5R/4}$ such that $w = \varphi$ on $\partial B_{3R/2}$.

Consider the Dirichlet problem for v

$$\Delta v = -k^2(1+q)w \quad \text{in} \quad B_{3R/2}, \quad v = \varphi \quad \text{on} \quad \partial B_{3R/2}. \tag{12.5.10}$$

We have $-k^2(1+q)w \in L^\infty$ and $\varphi \in C^{1,1/2}$. Theorem 8.34 in [7] gives unique solvability in the space of $C^{1,1/2}(\overline{B}_{3R/2})$ functions. However, to conclude that $w = v$ and a fortiori $w \in C^{1,1/2}$, we need something more. Consider equation (12.5.10) in $H^1(B_{3R/2})$. In this space, both v and w are solutions, and they satisfy

$$\Delta(v-w) = 0 \quad \text{in} \quad B_{3R/2}, \quad v - w = 0 \quad \text{on} \quad \partial B_{3R/2}.$$

By the H^1-maximum principle, $v = w$ in H^1. Hence, $w \in C^{1,1/2}$.

Finally, Theorem 8.33 in [7] gives an estimate for $\|v\|$ in $C^{1,1/2}(\overline{B}_{3R/2})$ based on the boundary and source data. Using that, the Sobolev embedding of $H^2 \hookrightarrow L^\infty$ in two and three dimensions, and (12.5.9) gives

$$\|w\|_{C^{1,\frac{1}{2}}(\overline{B}_{3R/2})} \leq C\left(\|w\|_{H^1(B_{2R})} + \left\|-k^2(1+q)w\right\|_{L^\infty(B_{2R})}\right)$$

for some constant $C = C(R,n)$ and the claim follows. □

Proposition 12.5.3. *Let* $R > 1$, $n \in \{2,3\}$ *and* $k > 0$. *Let* $u^i \in H^2_{loc}(\mathbb{R}^n)$ *be an incident wave such that* $(\Delta + k^2)u^i = 0$ *with* $\|u^i\|_{H^2(B_{2R})} \leq \mathscr{I}$.

Let $P, P' \subset B_R$ *be open convex polytopes and* $\varphi, \varphi' \in L^\infty(\mathbb{R}^n)$. *Let* $V = \chi_P\varphi$ *and* $V' = \chi_{P'}\varphi'$ *be two potentials with* $\|V\|_\infty, \|V'\|_\infty \leq \mathscr{M}$. *Also, let* $u, u' \in H^2_{loc}(\mathbb{R}^n)$ *be total waves satisfying*

$$(\Delta + k^2(1+V))u = (\Delta + k^2(1+V'))u' = 0$$

and whose scattered waves $u^s = u - u^i$ *and* $u'^s = u' - u^i$ *satisfy the Sommerfeld radiation condition. Let* $u^s_\infty, u'^s_\infty : \mathbb{S}^{n-1} \to \mathbb{C}$ *be their far-field patterns.*

Assume that $\|u^s\|, \|u'^s\| \leq \mathscr{S}$ *in* $H^2(B_{2R})$ *and* $\mathscr{S} \geq 1$. *Then there is* $\varepsilon_m = \varepsilon_m(\mathscr{S},k,R) > 0$ *such that if*

$$\|u^s_\infty - u'^s_\infty\|_{L^2(\mathbb{S}^{n-1})} \leq \varepsilon_m$$

and Q *is the convex hull of* P *and* P', *then* $u - u', \nabla u - \nabla u'$ *are continuous in* B_R *and*

$$\sup_{\partial Q}\left(|u-u'| + |\nabla u - \nabla u'|\right) \leq \mathscr{C}\left(\ln\ln(\mathscr{S}\,\|u^s_\infty - u'^s_\infty\|^{-1}_{L^2(\mathbb{S}^{n-1})})\right)^{-1/2}$$

for some $\mathscr{C} = C(k,R)(1+\mathscr{M})(\mathscr{I}+\mathscr{S})$.

Proof. Firstly, propagate smallness from the far field to the near field by using Corollary 12.5.1. Let w^s in that proposition be $u - u' = u^s - u'^s$, and denote $\varepsilon = \|u^s_\infty - u'^s_\infty\|_{L^2(\mathbb{S}^{n-1})}$. Note also that $\|w^s\|_{H^2(B_{2R})} \leq 2\mathscr{S}$ then. Choose the annulus $A = B_{(2-\lambda)R} \setminus \overline{B}_{(1+\lambda)R}$ for some positive $\lambda < \frac{1}{2}$. Corollary 12.5.1 implies that $w^s \in H^r(A)$ for any $r \in \mathbb{N}$. Moreover, in two and three dimensions, Sobolev embedding implies that $H^2(A) \hookrightarrow L^\infty(A)$. Hence, the estimate given by the corollary becomes

$$\|w^s\|_{L^\infty(A)}, \|\nabla w^s\|_{L^\infty(A)} \leq C \max\left(\varepsilon, \mathscr{S} e^{-c\sqrt{\ln(\mathscr{S}/\varepsilon)}}\right)$$

for $C > 1, c > 0$ depending on k, R, λ (we estimated $\ln(2\mathscr{S}/\varepsilon) \geq \ln(\mathscr{S}/\varepsilon)$). Our first requirement on ε_m is that the maximum picks the number on the right side. This happens if $\varepsilon \leq \mathscr{S} e^{-c^2}$ so let us require $\varepsilon_m \leq \mathscr{S} e^{-c^2}$.

The second step is to use the propagation of smallness by Proposition 12.5.2 for $w = w^s$ and also for $w = \partial_j w^s$, $j = 1, \ldots, n$. By Lemma 12.5.4, we have

$$\|u\|_{C^{1,1/2}} \leq C(1 + \mathscr{M})(\mathscr{I} + \mathscr{S})$$

in $\overline{B}_{3R/2}$ and similarly for u'. So $w^s, \partial_j w^s \in C^{1/2}$ for each j. Thus, the smoothness requirements of Proposition 12.5.2 are satisfied for each choice of w. Also $C = C(k, R)$. Set

$$\delta = C\mathscr{S} e^{-c\sqrt{\ln(\mathscr{S}/\varepsilon)}}.$$

We get a second upper bound on ε_m by requiring that δ satisfies (12.5.7). The right-hand side in that inequality depends only on $A = A(\lambda, R)$, k and R, so this second, updated, upper bound for ε_m still only depends on λ, k, R. Now Proposition 12.5.2 implies

$$|w(\mathbf{x})| \leq C(1 + \mathscr{M})(\mathscr{I} + \mathscr{S})(\ln|\ln\delta|)^{-1/2}$$

with $C = C(\lambda, k, R)$. The choice of δ implies

$$|\ln\delta| = c\sqrt{\ln(\mathscr{S}\varepsilon^{-1})} - \ln(C\mathscr{S}) \geq \frac{c}{2}\sqrt{\ln(\mathscr{S}\varepsilon^{-1})} \geq \left(\ln(\mathscr{S}\varepsilon^{-1})\right)^{1/4}$$

if ε_m is small enough (and again c, C depend only on k, λ, R). Thus,

$$(\ln|\ln\delta|)^{-1/2} \leq \left(\ln\left(\ln(\mathscr{S}\varepsilon^{-1})\right)^{1/4}\right)^{-1/2} = 2\left(\ln\ln(\mathscr{S}\varepsilon^{-1})\right)^{-1/2}$$

and the claim follows after choosing λ as a function of R, for example. □

12.6 From Boundary to Inside

We deal with particulars related to corner scattering in this section. More precisely, we prove the fundamental orthogonality identity which is the foundation upon which past results [4, 9, 17] were built on. Since we are proving stability instead of uniqueness, we have an extra boundary term here to deal with. Moreover, for future convenience, we do not assume that $u^i(\mathbf{x}_c) \neq 0$ in Proposition 12.6.2. This does not complicate the argument by much.

Proposition 12.6.1. *Let $Q_h \subset \mathbb{R}^n$ be a bounded Lipschitz domain, $V \in L^\infty(Q_h)$, $k > 0$, and $u^i, u, u_0 \in H^2(Q_h)$ satisfy*

$$(\Delta + k^2)u^i = 0,$$
$$(\Delta + k^2(1 + V))u = 0,$$
$$(\Delta + k^2(1 + V))u_0 = 0$$

in Q_h. Then

$$k^2 \int_{Q_h} V\, u_0\, u^i \,\mathrm{d}\mathbf{x} = \int_{\partial Q_h} \big(u_0\, \partial_\nu (u^i - u) - (u^i - u)\, \partial_\nu u_0\big)\mathrm{d}\sigma. \tag{12.6.1}$$

Proof. Use Green's formula after noting that

$$k^2 \int_{Q_h} V\, u_0\, u^i \,\mathrm{d}\mathbf{x} = \int_{Q_h} u_0 \big(\Delta + k^2(1+V)\big)(u^i - u)\,\mathrm{d}\mathbf{x}.$$

□

We consider only incident waves that do not vanish anywhere in this chapter. This means that in the following corollary, we would always have P_N is a constant and $N = 0$. The corollary is stated so that it applies also to the more general case where the incident wave can vanish up to a finite order N at x_c. This is for the convenience of future studies on the topic and also since the proof is not substantially more difficult in this case.

Proposition 12.6.2. *Let $\mathfrak{P}, \mathfrak{Q} \subset \mathbb{R}^n$ be open polyhedral cones with vertex x_c such that $\mathfrak{P} \subset \mathfrak{Q}$ and their boundaries are a subset of the union of at most $\mathscr{V}$ hyperplanes of codimension 1. Let $P_h = \mathfrak{P} \cap B(x_c, h)$ and $Q_h = \mathfrak{Q} \cap B(x_c, h)$ for $0 < h \leq 1$.*

Let $k > 0$ and $V, V' \in L^\infty(\mathbb{R}^n)$ be supported in $B_R \supset Q_h$ for some $R > 1$. Assume that $V = \chi_{\mathfrak{P}} \varphi$ and $V' = 0$ in Q_h for some measurable function $\varphi : P_h \to \mathbb{C}$. Let $u, u', u_0 \in H^2(B_{2R})$ satisfy

$$(\Delta + k^2(1+V))u = (\Delta + k^2(1+V))u_0 = 0, \quad (\Delta + k^2(1+V'))u' = 0.$$

If we have functions P_N, φ_α, u'_{N+1}, and ψ and a complex vector $\rho \in \mathbb{C}^n$ such that

$$\begin{aligned}
\varphi(\mathbf{x}) &= \varphi(\mathbf{x}_c) + \varphi_\alpha(\mathbf{x}), \\
u'(\mathbf{x}) &= P_N(\mathbf{x} - \mathbf{x}_c) + u'_{N+1}(\mathbf{x}), \\
u_0(\mathbf{x}) &= e^{\rho\cdot(\mathbf{x}-\mathbf{x}_c)}(1 + \psi(\mathbf{x}))
\end{aligned}$$

in P_h, then

$$\begin{aligned}
\varphi(\mathbf{x}_c) \int_{\mathfrak{P}} e^{\rho\cdot(\mathbf{x}-\mathbf{x}_c)} P_N(\mathbf{x}-\mathbf{x}_c)\mathrm{d}\mathbf{x} = \varphi(\mathbf{x}_c) &\int_{\mathfrak{P}\setminus P_h} e^{\rho\cdot(\mathbf{x}-\mathbf{x}_c)} P_N(\mathbf{x}-\mathbf{x}_c)\mathrm{d}\mathbf{x} \\
&- \int_{P_h} e^{\rho\cdot(\mathbf{x}-\mathbf{x}_c)} V(\mathbf{x}) u'_{N+1}(\mathbf{x})\mathrm{d}\mathbf{x} \\
&- \int_{P_h} e^{\rho\cdot(\mathbf{x}-\mathbf{x}_c)} \varphi_\alpha(\mathbf{x}) P_N(\mathbf{x}-\mathbf{x}_c)\mathrm{d}\mathbf{x} \\
&- \int_{P_h} e^{\rho\cdot(\mathbf{x}-\mathbf{x}_c)} V(\mathbf{x}) u'(\mathbf{x}) \psi(\mathbf{x})\mathrm{d}\mathbf{x} \\
&+ \frac{1}{k^2} \int_{\partial Q_h} \big(u_0 \partial_\nu (u - u') - (u - u')\partial_\nu u_0\big)\mathrm{d}\sigma.
\end{aligned} \tag{12.6.2}$$

Moreover, assume that $\psi \in L^p$ in Q_h, $p > 1$, and that

1. *$|\rho| \geq 1$ and $\Re\rho \cdot (\mathbf{x} - \mathbf{x}_c) \leq -\delta_0 |\mathbf{x} - \mathbf{x}_c| |\Re\rho|$ for some $\delta_0 > 0$ and any $\mathbf{x} \in Q_h$,*
2. *$|\varphi_\alpha(\mathbf{x})| \leq \mathscr{M} |\mathbf{x} - \mathbf{x}_c|^\alpha$, $|V(\mathbf{x})| \leq \mathscr{M}$ for $\mathbf{x} \in P_h$, and some $\alpha > 0$,*
3. *$|u'(\mathbf{x})| \leq \mathscr{F} |\mathbf{x} - \mathbf{x}_c|^N$ for $\mathbf{x} \in P_h$,*
4. *$|P_N(\mathbf{x} - \mathbf{x}_c)| \leq \mathscr{P} |\mathbf{x} - \mathbf{x}_c|^N$ for $\mathbf{x} \in P_h$,*
5. *$|u'_{N+1}(\mathbf{x})| \leq \mathscr{R} |\mathbf{x} - \mathbf{x}_c|^{N+1}$ for $\mathbf{x} \in P_h$,*

with $0 \leq N \leq \mathscr{N}$, then we have the norm estimate

$$
\begin{aligned}
C\left|\varphi(\mathbf{x}_c)\int_{\mathfrak{P}} e^{\rho\cdot(\mathbf{x}-\mathbf{x}_c)}P_N(\mathbf{x}-\mathbf{x}_c)\mathrm{d}\mathbf{x}\right| &\le |\Re\rho|^{-N-n}e^{-\delta_0|\Re\rho|h/2}+|\Re\rho|^{-N-n-\min(1,\alpha)}+|\Re\rho|^{-N-n/p'}\|\psi\|_{L^p(P_h)}\\
&\quad+h^{(n-1)/2}|\rho|(1+\|\psi\|_{H^2(B_{2R})})\sup_{\partial\Omega\cap B(\mathbf{x}_c,h)}\{|u-u'|,|\nabla u-\nabla u'|\} \qquad (12.6.3)\\
&\quad+h^{n/2-1}e^{-\delta_0|\Re\rho|h}|\rho|(1+\|\psi\|_{H^2(B_{2R})})(\|u\|_{H^2(B_{2R})}+\|u'\|_{H^2(B_{2R})}),
\end{aligned}
$$

where $1/p+1/p'=1$ *and* $C>0$ *depends on all the a priori parameters* $\mathscr{V},k,\mathscr{P},\mathscr{M},\mathscr{N},\mathscr{R},\mathscr{F},\alpha,\delta_0,n,p$.

Proof. The integral identity is a direct calculation using Proposition 12.6.1 with Q_h and $u^i=u'$ and then noting that $V=0$ on $Q_h\setminus P_h$. For the others, we use the incomplete gamma functions γ and $\Gamma:\mathbb{R}_+\times\mathbb{R}_+\to\mathbb{R}$

$$
\gamma(s,\mathbf{x})=\int_0^{\mathbf{x}}e^{-t}t^{s-1}\mathrm{d}t,\qquad \Gamma(s,\mathbf{x})=\int_{\mathbf{x}}^{\infty}e^{-t}t^{s-1}\mathrm{d}t,
$$

which satisfy $\gamma(s,\mathbf{x})\le\Gamma(s)\le\lceil s-1\rceil!$ and $\Gamma(s,\mathbf{x})\le 2^s\Gamma(s)e^{-\mathbf{x}/2}$, where $\Gamma(s)$ represents the ordinary, complete, gamma function. The latter estimate follows from splitting $e^{-t}\le e^{-t/2}e^{-x/2}$ in the integral, expanding the integration limits to $(0,\infty)$, and switching to the integration variable $t'=t/2$. By a radial change of coordinates, the first integral on the right has the upper bound

$$
\begin{aligned}
\left|\int_{\mathfrak{P}\setminus P_h} e^{\rho\cdot(\mathbf{x}-\mathbf{x}_c)}P_N(\mathbf{x}-\mathbf{x}_c)\mathrm{d}\mathbf{x}\right| &\le \int_{\mathfrak{P}\setminus P_h} e^{-\delta_0|\Re\rho||\mathbf{x}-\mathbf{x}_c|}\mathscr{P}|\mathbf{x}-\mathbf{x}_c|^N\mathrm{d}\mathbf{x}\\
&\le \mathscr{P}\sigma(\mathbb{S}^{n-1})\int_h^\infty e^{-\delta_0|\Re\rho|r}r^{N+n-1}\mathrm{d}r\\
&\le \left(\frac{2}{\delta_0}\right)^{N+n}(N+n)!\,\mathscr{P}\sigma(\mathbb{S}^{n-1})|\Re\rho|^{-N-n}e^{-\delta_0|\Re\rho|h/2}\\
&\le C_{\delta_0,\mathscr{N},n,\mathscr{P}}|\Re\rho|^{-N-n}e^{-\delta_0|\Re\rho|h/2}
\end{aligned}
$$

for the first integral on the right.

For the integral inside P_h, we note that

$$
\begin{aligned}
\int_{P_h} e^{\Re\rho\cdot(\mathbf{x}-\mathbf{x}_c)q'}|\mathbf{x}-\mathbf{x}_c|^{Bq'}\mathrm{d}\mathbf{x} &= \int_{\mathbb{S}^{n-1}\cap(P_h-\mathbf{x}_c)}\int_0^h e^{-\delta_0 q'|\Re\rho|r}r^{Bq'+n-1}\mathrm{d}r\mathrm{d}\sigma(\theta)\\
&\le \sigma(\mathbb{S}^{n-1})\int_0^{\delta_0 q'|\Re\rho|h} e^{-r'}r'^{Bq'+n-1}\frac{\mathrm{d}r'}{(\delta_0 q'|\Re\rho|)^{Bq'+n}}\\
&= \sigma(\mathbb{S}^{n-1})\gamma(Bq'+n,\delta_0 q'|\Re\rho|h)(\delta_0 q'|\Re\rho|)^{-Bq'-n}.
\end{aligned}
$$

Use this to prove the following estimate, each of which shall be applied to the next three integrals in (12.6.2). Let f,g be functions such that $|f(\mathbf{x})|\le A|\mathbf{x}-\mathbf{x}_c|^B$ with $A\le\mathscr{A}$, $B\le\mathscr{B}$, and $g\in L^q$. Then

$$
\begin{aligned}
\left|\int_{P_h} e^{\rho\cdot(\mathbf{x}-\mathbf{x}_c)}f(\mathbf{x})g(\mathbf{x})\mathrm{d}\mathbf{x}\right| &\le A\left(\int_{P_h} e^{\Re\rho\cdot(\mathbf{x}-\mathbf{x}_c)q'}|\mathbf{x}-\mathbf{x}_c|^{Bq'}\mathrm{d}\mathbf{x}\right)^{1/q'}\|g\|_{L^q(P_h)}\\
&\le A\left(\frac{\sigma(\mathbb{S}^{n-1})\gamma(Bq'+n,\delta_0 q'|\Re\rho|h)}{(\delta_0 q'|\Re\rho|)^{Bq'+n}}\right)^{1/q'}\|g\|_{L^q(P_h)}\\
&\le A\left(\frac{\sigma(\mathbb{S}^{n-1})\lceil Bq'+n\rceil!}{(\delta_0 q'|\Re\rho|)^{Bq'+n}}\right)^{1/q'}\|g\|_{L^q(P_h)}\\
&\le C_{\mathscr{A},\mathscr{B},n,\delta_0,q}|\Re\rho|^{-B-n/q'}\|g\|_{L^q(P_h)}
\end{aligned}
$$

where $1/q + 1/q' = 1$. Choosing

- $q = \infty$, $\mathscr{A} = \mathscr{P}\mathscr{M}$, $B = N + \alpha \leq \mathscr{N} + \alpha$,
- $q = \infty$, $\mathscr{A} = \mathscr{M}\mathscr{R}$, $B = N + 1 \leq \mathscr{N} + 1$, and
- $q = p$, $\mathscr{A} = \mathscr{M}\mathscr{F}$, $B = N \leq \mathscr{N}$

gives the three estimates

$$\left| \int_{P_h} e^{\rho\cdot(\mathbf{x}-\mathbf{x}_c)} \varphi_\alpha(\mathbf{x})\, P_N(\mathbf{x}-\mathbf{x}_c)\, \mathrm{d}\mathbf{x} \right| \leq C_{\mathscr{P},\mathscr{M},\mathscr{N},\alpha,n,\delta_0} |\Re\rho|^{-N-n-\alpha},$$
$$\left| \int_{P_h} e^{\rho\cdot(\mathbf{x}-\mathbf{x}_c)} V(\mathbf{x})\, u'_{N+1}(\mathbf{x})\, \mathrm{d}\mathbf{x} \right| \leq C_{\mathscr{M},\mathscr{R},\mathscr{N},n,\delta_0} |\Re\rho|^{-N-n-1},$$
$$\left| \int_{P_h} e^{\rho\cdot(\mathbf{x}-\mathbf{x}_c)} V(\mathbf{x})\, u'(\mathbf{x})\, \psi(\mathbf{x})\, dx \right| \leq C_{\mathscr{M},\mathscr{F},\mathscr{N},n,\delta_0,p} |\Re\rho|^{-N-n/p'} \|\psi\|_{L^p(P_h)}.$$

Only the boundary integral is left in (12.6.2). Let us split the boundary into two pieces: $\partial Q_h = (\partial\Omega \cap B(\mathbf{x}_c,h)) \cup (\Omega \cap S(\mathbf{x}_c,h))$. For the first piece, use the Cauchy-Schwartz inequality which gives

$$\left| \int_{\partial\Omega\cap B(\mathbf{x}_c,h)} \left(u_0\, \partial_\nu(u-u') - (u-u')\, \partial_\nu u_0\right) \mathrm{d}\sigma(\mathbf{x}) \right| \leq \sqrt{\sigma(\partial\Omega\cap B(\mathbf{x}_c,h))}\cdot$$
$$\cdot\left((1+|\rho|)(1+\|\psi\|_{L^2(\partial\Omega\cap B(\mathbf{x}_c,h))}) + \|\partial_\nu\psi\|_{L^2(\partial\Omega\cap B(\mathbf{x}_c,h))}\right) \|u-u'\|_{NF},$$

where $\|f\|_{NF}$ denotes the maximum of $|f|$ and $|\nabla f|$ on $\partial\Omega \cap B(\mathbf{x}_c,h)$. This estimate uses $|\exp(\rho\cdot(\mathbf{x}-\mathbf{x}_c))| \leq 1$ in Ω.

Both $\|\psi\|_2$ and $\|\partial_\nu\psi\|_2$ can be estimated by $C_{\mathscr{V},n}\|\psi\|_{H^2(B_{2R})}$ in the set $\partial\Omega\cap B(\mathbf{x}_c,h)$ since $h \leq 1$ and so $B(\mathbf{x}_c,h) \subset B_{2R}$. The constant depends on $\mathscr{V}$ instead of Ω because

$$\partial\Omega \subset \bigcap_{j=1}^{\mathscr{V}} \partial H_j \cap B(\mathbf{x}_c,1)$$

for some half-spaces H_j that pass through $\mathbf{x}_c$. The trace norm is identical in each of the sets $H_j \cap B(\mathbf{x}_c,1)$. By an easier argument, we see that $\sqrt{\sigma(\partial\Omega\cap B(\mathbf{x}_c,h))} \leq C_{\mathscr{V},n} h^{(n-1)/2}$ and the estimate for the first part of the boundary term in (12.6.2) follows.

For estimating the last integral, the one over $\Omega\cap S(x_c,h)$, the Cauchy-Schwartz inequality gives

$$\left| \int_{\Omega\cap S(\mathbf{x}_c,h)} \left(u_0\, \partial_\nu(u-u') - (u-u')\, \partial_\nu u_0\right) \mathrm{d}\sigma(\mathbf{x}) \right|$$
$$\leq \sqrt{\sigma(\Omega\cap S(\mathbf{x}_c,h))}\, \|u-u'\|_{C^1(\overline{B}(\mathbf{x}_c,h))}\, e^{-\delta_0|\Re\rho|h}$$
$$\cdot\left((1+|\rho|)(1+\|\psi\|_{L^2(\Omega\cap S(\mathbf{x}_c,h))}) + \|\partial_\nu\psi\|_{L^2(\Omega\cap S(\mathbf{x}_c,h))}\right).$$

We can estimate by C^1 norm by Lemma 12.5.4 which gives $\|u-u'\|_{C^{1,1/2}} \leq C(1+\mathscr{M})(\|u\|_2 + \|u'\|_2)$ where the $\|\cdot\|_2$ norm is the $H^2(B_{2R})$ norm.

For estimating ψ, let us consider how the trace norm depends on h when the trace operator maps $H^1(B(\mathbf{x}_c,h)) \to L^2(S(\mathbf{x}_c,h))$. We do this by scaling the variables, for example, by having $g(\mathbf{y}) = f(h(\mathbf{y}-\mathbf{x}_c)+\mathbf{x}_c)$ and $f(\mathbf{x}) = g((\mathbf{x}-\mathbf{x}_c)/h+\mathbf{x}_c)$. Now

$$\|f\|_{L^2(S(\mathbf{x}_c,h))} = h^{\frac{n-1}{2}} \|g\|_{L^2(S(\mathbf{x}_c,1))} \leq Ch^{\frac{n-1}{2}} \|g\|_{H^1(B(\mathbf{x}_c,1))}$$
$$\leq Ch^{-\frac{1}{2}}(1+h)\|f\|_{H^1(B(\mathbf{x}_c,h))} \leq Ch^{-\frac{1}{2}} \|f\|_{H^1(B(\mathbf{x}_c,h))}$$

because of $h \leq 1$. Hence, we see that $\|\psi\|_2$ and $\|\partial_\nu \psi\|_2$ can be estimated by $C_n h^{-1/2} \|\psi\|_{H^2(B_{2R})}$ in $L^2(\mathfrak{Q} \cap S(\mathbf{x}_c, h))$. However, note that

$$\sqrt{\sigma(\mathfrak{Q} \cap S(\mathbf{x}_c, h))} \leq Ch^{(n-1)/2}$$

so the final estimate (12.6.3) follows. □

To prove the final stability results, we need a lower bound on the left-hand side of (12.6.3). This is nontrivial. In [4] and [17], it is shown that the left-hand side does not vanish. We do need a quantitative version, for example, of the form: given a polynomial P_N satisfying some a priori conditions, the left-hand side is greater than C which does not depend on P_N. This turns out to require a too fine analysis in the context of support probing. However, we can avoid this because we assumed that $u'(\mathbf{x}_c) \neq 0$, which implies that $P_N(\mathbf{x}) \equiv u'(\mathbf{x}_c)$ is constant.

Lemma 12.6.1. *Let $n \in \{2,3\}$, $0 < 2\alpha_m < 2\alpha_M < 2\alpha' < \pi$, and $k > 0$. For $\mathscr{Q}, \mathfrak{P} \subset \mathbb{R}^n$, we say $(\mathscr{Q}, \mathfrak{P}) \in \mathscr{G}(\alpha_m, \alpha_M, \alpha', n)$ if the following are satisfied:*

1. *$\mathscr{Q}$ is an open spherical cone.*
2. *$\mathfrak{P}$ is an open convex polyhedral cone.*
3. *$\mathscr{Q}$ and $\mathfrak{P}$ have a common vertex $\mathbf{x}_c \in \mathbb{R}^n$.*
4. *$\mathfrak{P} \subset \mathscr{Q}$.*
5. *$\mathscr{Q}$ has opening angle at most $2\alpha'$.*
6. *In 2D, $\mathfrak{P}$ has opening angle in $]2\alpha_m, 2\alpha_M[$.*
7. *In 3D, $\mathfrak{P}$ can be transformed to $]0,\infty[^3$ by a rigid motion.*

If $(\mathscr{Q}, \mathfrak{P}) \in \mathscr{G}(\alpha_m, \alpha_M, \alpha', n)$, then there is $\tau_0 = kC(\alpha_m, \alpha_M, \alpha', n) > 0$, and $c = c(\alpha_m, \alpha_M, n) > 0$ with the following properties. There is a curve $\tau \mapsto \rho(\tau) \in \mathbb{C}^n$ (which depends on $\mathscr{Q}$) satisfying $\rho(\tau) \cdot \rho(\tau) + k^2 = 0$, $\tau = |\Re \rho(\tau)|$,

$$\Re \rho(\tau) \cdot (\mathbf{x} - \mathbf{x}_c) \leq -\cos\alpha' \, |\Re \rho(\tau)| \, |\mathbf{x} - \mathbf{x}_c|$$

for all $x \in \mathscr{Q}$ and such that if $\tau \geq \tau_0$, then

$$\left| \int_{\mathfrak{P}} e^{\rho(\tau)\cdot(\mathbf{x}-\mathbf{x}_c)} \mathrm{d}\mathbf{x} \right| \geq c\tau^{-n}.$$

Proof. We start by proving the claim for $\zeta \cdot \zeta = 0$ instead of $\rho \cdot \rho + k^2 = 0$. Consider the cases $n = 2$ and $n = 3$ separately. Let $(\mathscr{Q}, \mathfrak{P}) \in \mathscr{G}(\alpha_m, \alpha_M, \alpha', 2)$. Then there is a rigid motion $M_{\mathfrak{P}}$ and $\alpha \in [2\alpha_m, 2\alpha_M]$ such that $M_{\mathfrak{P}}$ takes $\mathfrak{P}$ to $\{x \in \mathbb{R}^2 \mid x_2 > 0, x_1 > ax_2\}$ where $a = 1/\tan\alpha$. We have $M_{\mathfrak{P}} x = R_{\mathfrak{P}}(\mathbf{x} - \mathbf{x}_c)$ for some rotation $R_{\mathfrak{P}}$. Denote $\xi = R_{\mathfrak{P}} \zeta$. Then

$$\int_{\mathfrak{P}} e^{\zeta\cdot(\mathbf{x}-\mathbf{x}_c)} \mathrm{d}\mathbf{x} = \int_0^\infty \int_{ay_2}^\infty e^{\xi\cdot y} \mathrm{d}y_1 \mathrm{d}y_2 = \frac{1}{\xi_1(\xi_2 + a\xi_1)}$$

if $\Re\xi_1 < 0$ and $\Re(\xi_2 + a\xi_1) < 0$. If $\zeta \cdot \zeta = 0$ and $|\Re\zeta| = 1$, then the same is true for its rotated version ξ, and so $|\xi_1| = |\xi_2| = 1$. This implies $|\xi_2/\xi_1 + a| \leq 1 + |a|$. Thus,

$$\left| \int_{\mathfrak{P}} e^{\zeta\cdot(\mathbf{x}-\mathbf{x}_c)} \mathrm{d}\mathbf{x} \right| \geq \frac{1}{1+|a|} > 0$$

because $|a|$ can be estimated above by $1/\min|\tan\alpha|$, where the minimum is taken over $2\alpha_m \leq \alpha \leq 2\alpha_M$, and the limits are away from 0 and π.

The conditions $\Re\xi_1 < 0$ and $\Re(\xi_2 + a\xi_1) < 0$ are implied at once if

$$\Re\zeta \cdot (\mathbf{x} - \mathbf{x}_c) \leq -\cos\alpha' \, |\mathbf{x} - \mathbf{x}_c|$$

for all $x \in \mathcal{Q}$ as this means that the map $\mathbf{x} \mapsto \exp(\Re\zeta \cdot (\mathbf{x}-\mathbf{x}_c))$ is exponentially decreasing in $\mathcal{Q}$ and a fortiori in $\mathfrak{P}$. We can now build ζ. Let $-\Re\zeta$ be the unit vector on the central axis of $\mathcal{Q}$ to make the above inequality valid. Next choose $\Im\zeta$ such that $\Im\zeta \perp \Re\zeta$, $|\Im\zeta| = 1 = |\Re\zeta|$. This implies $\zeta \cdot \zeta = 0$.

Consider the 3D case now. Let $(\mathcal{Q},\mathfrak{P}) \in \mathcal{G}(\alpha_m, \alpha_M, \alpha', 3)$. Then there is a rigid motion $M_{\mathfrak{P}}$ bringing $\mathfrak{P}$ to $]0,\infty[^3$. We have $M_{\mathfrak{P}}x = R_{\mathfrak{P}}(x-x_c)$ for some rotation $R_{\mathfrak{P}}$. Denote again $\xi = R_{\mathfrak{P}}\zeta$. Then

$$\int_{\mathfrak{P}} e^{\zeta\cdot(x-x_c)}dx = \int_{]0,\infty[^3} e^{\xi\cdot y}dy = \frac{-1}{\xi_1\xi_2\xi_3}$$

as long as $\xi_j < 0$ for all j. As before, $\zeta\cdot\zeta = 0$ and $|\Re\zeta| = 1$ imply $|\xi| \le \sqrt{2}$ and the lower bound of $2^{-3/2}$ for the integral. The conditions $\xi_j < 0$ follow from

$$\Re\zeta \cdot (\mathbf{x}-\mathbf{x}_c) \le -\cos\alpha' \, |\mathbf{x}-\mathbf{x}_c| \quad \text{in} \quad \mathcal{Q}.$$

The choice of ζ is made as in the 2D case.

To recap, in both 2D and 3D, for any $(\mathcal{Q},\mathfrak{P}) \in \mathcal{G}(\alpha_m, \alpha_M, \alpha', n)$, we found $\zeta \in \mathbb{C}^n$ satisfying $\zeta\cdot\zeta = 0$, $|\zeta| = 1$, $\Re\zeta\cdot(\mathbf{x}-\mathbf{x}_c) \le -\cos\alpha' \, |\mathbf{x}-\mathbf{x}_c|$ for all $x \in \mathcal{Q}$ with $\mathbf{x}_c$ the vertex, and finally

$$\left| \int_{\mathfrak{P}} e^{\zeta\cdot(\mathbf{x}-\mathbf{x}_c)}\mathrm{d}\mathbf{x} \right| \ge 2C_{\alpha_m,\alpha_M,n} > 0.$$

Let us build the curve $\rho(\tau)$ next. Set

$$\rho(\tau) = \tau\Re\zeta + i\sqrt{\tau^2+k^2}\,\Im\zeta.$$

It is easy to see that $\rho(\tau)/\tau \to \zeta$ as $\tau \to \infty$ and even easier to see that $\Re\rho(\tau)\cdot(\mathbf{x}-\mathbf{x}_c) \le -\cos\alpha' \, |\Re\rho(\tau)| \, |\mathbf{x}-\mathbf{x}_c|$ for $\mathbf{x} \in \mathcal{Q}$. Write $\mathcal{L}(\zeta) = \int_{\mathfrak{P}} \exp(\zeta\cdot(\mathbf{x}-\mathbf{x}_c))dx$ to conserve space. We quantify how far $\mathcal{L}(\rho(\tau)/\tau)$ is from $\mathcal{L}(\zeta)$ next. Ideally, we want an estimate that does not depend on $\mathcal{Q}$ or $\mathfrak{P}$.

If we set $f(r) = \exp((\Re\zeta + ir\Im\zeta)\cdot(\mathbf{x}-\mathbf{x}_c))$, then $f(1) = \exp(\zeta\cdot(\mathbf{x}-\mathbf{x}_c))$ and $f(\sqrt{1+k^2/\tau^2}) = \exp(\rho(\tau)/\tau\cdot(\mathbf{x}-\mathbf{x}_c))$. By the mean value theorem,

$$\left| f(1) - f\left(\sqrt{1+\tfrac{k^2}{\tau^2}}\right) \right| \le \sup_{1<r<\sqrt{1+\frac{k^2}{\tau^2}}} |f'(r)| \left| \sqrt{1+\frac{k^2}{\tau^2}} - 1 \right|.$$

Note that $\sqrt{1+k^2/\tau^2} - 1 \le k/\tau$. Also $f'(r) = i\Im\zeta\cdot(\mathbf{x}-\mathbf{x}_c)f(r)$, and because $|\Im\zeta| = |\Re\zeta| = 1$, we have $|f'(r)| \le |\mathbf{x}-\mathbf{x}_c|\exp(-\cos\alpha' \, |\mathbf{x}-\mathbf{x}_c|)$. Hence,

$$\left| f(1) - f\left(\sqrt{1+\tfrac{k^2}{\tau^2}}\right) \right| \le \frac{k}{\tau} |\mathbf{x}-\mathbf{x}_c| \, e^{-\cos\alpha' |\mathbf{x}-\mathbf{x}_c|}.$$

We see finally that

$$\begin{aligned}
\left| \mathcal{L}(\zeta) - \mathcal{L}\left(\frac{\rho(\tau)}{\tau}\right) \right| &= \left| \int_{\mathfrak{P}} (f(1) - f(\sqrt{1+k^2/\tau^2}))\mathrm{d}\mathbf{x} \right| \\
&\le \frac{k}{\tau} \int_{\mathfrak{P}} e^{-\cos\alpha' |\mathbf{x}-\mathbf{x}_c|} |\mathbf{x}-\mathbf{x}_c| \mathrm{d}\mathbf{x} \\
&\le \sigma(\mathfrak{P}\cap\mathbb{S}^{n-1}) \frac{k}{\tau} \int_0^\infty e^{-\cos\alpha' r} r^{1+n-1} dr \le C_{\alpha',n} k\tau^{-1}
\end{aligned}$$

because we can estimate $\sigma(\mathfrak{P}\cap\mathbb{S}^{n-1}) \le \sigma(\mathbb{S}^{n-1})$ and $\cos\alpha' > 0$ since $\alpha' < \pi/2$.

Now, it is easily seen that $\mathscr{L}(\rho(\tau)/\tau) = \tau^n \mathscr{L}(\rho(\tau))$. Recall that our choice of ζ implies that $|\mathscr{L}(\zeta)| \geq 2C_{\alpha_m,\alpha_M,n}$. By the triangle inequality, we have

$$|\tau^n \mathscr{L}(\rho(\tau))| = \left|\mathscr{L}\left(\frac{\rho(\tau)}{\tau}\right)\right| \geq |\mathscr{L}(\zeta)| - \left|\mathscr{L}(\zeta) - \mathscr{L}\left(\frac{\rho(\tau)}{\tau}\right)\right|$$
$$> 2C_{\alpha_m,\alpha_M,n} - C_{\alpha',n}k\tau^{-1} \geq C_{\alpha_m,\alpha_M,n} > 0$$

under the condition that $\tau \geq C_{\alpha',n}k/C_{\alpha_m,\alpha_M,n}$ which is finite and depends only on the a priori parameters. □

12.7 Complex Geometric Optics Solution

The construction of the CGO solutions for corner scattering was first shown in [4] and [17]. We do the analysis more precisely and keep track of what parameters the various bounds depend on. This involves defining a "norm" for polyhedral regions. We start by solving the Faddeev equation, then prove estimates for potentials supported on polytopes, and finally build the complex geometric optics solutions.

Lemma 12.7.1. *Let $s \geq 0$ and $1 < r < 2$ such that $1/r + 1/r' = 1$ and*

$$\frac{2}{n+1} \leq \frac{1}{r} - \frac{1}{r'} < \frac{2}{n}.$$

Let q be a measurable function such that the pointwise multiplier operator m_q maps $H^s_{r'}(\mathbb{R}^n) \to H^s_r(\mathbb{R}^n)$, and let $f \in H^s_r(\mathbb{R}^n)$.

Let $I_0 = (2M\|m_q\|_{H^s_{r'} \to H^s_r})^{2+n/r'-n/r}$, where $M = M(r,s,n) \geq 1$ is fixed in the proof. Then if $\rho \in \mathbb{C}^n$, $|\Im\rho| \geq I_0$, there is $\psi \in H^s_{r'}(\mathbb{R}^n)$ satisfying

$$(\Delta + 2\rho\cdot\nabla + q)\psi = f,$$

$$\|\psi\|_{H^s_{r'}(\mathbb{R}^n)} \leq 2M|\Im\rho|^{-(2+\frac{n}{r'}-\frac{n}{r})}\|f\|_{H^s_r(\mathbb{R}^n)}.$$

There are also $p \geq 2$ and a Sobolev embedding constant $E = E(s,n,r) \geq 1$ such that

$$\|\psi\|_{L^p(\mathbb{R}^n)} \leq EM|\Im\rho|^{-(2+\frac{n}{r'}-\frac{n}{r})}\|f\|_{H^s_r(\mathbb{R}^n)}.$$

We have the following observations about the choice of p and the decay rate of ψ compared to $|\Im\rho|^{-n/p}$.

1. *If $s > \frac{n}{r'}$, then $p = \infty$ and $2 + \frac{n}{r'} - \frac{n}{r} > \frac{n}{p}$.*
2. *If $s = \frac{n}{r'}$, then we may choose any finite p such that $\frac{1}{p} < \frac{2}{n} + \frac{1}{r'} - \frac{1}{r}$ which is positive, and then $2 + \frac{n}{r'} - \frac{n}{r} > \frac{n}{p}$.*
3. *If $\frac{n}{r} - 2 < s < \frac{n}{r'}$, then $s - \frac{n}{r'} = -\frac{n}{p}$ and $2 + \frac{n}{r'} - \frac{n}{r} > \frac{n}{p}$.*
4. *If $s \leq \frac{n}{r} - 2$, then $s - \frac{n}{r'} = -\frac{n}{p}$ but $2 + \frac{n}{r'} - \frac{n}{r} \leq \frac{n}{p}$.*

Lastly, if $f \in L^2_{loc}$ and $q \in L^\infty_{loc}$, then given any bounded domain, for example, B_{3R}, we have the elliptic regularity estimate

$$\|\psi\|_{H^2(B_{2R})} \leq C_R\big(\|f\|_{L^2(B_{3R})} + (1 + |\rho|^2 + \|q\|_{L^\infty(B_{3R})})\|\psi\|_{L^p(\mathbb{R}^n)}\big),$$

where C_R depends only on R.

Proof. Fix $M < \infty$ as the ρ-independent constant in the estimate

$$\|f\|_{L^{r'}(\mathbb{R}^n)} \le M|\Im\rho|^{n(1/r-1/r')-2}\|(\Delta+2\rho\cdot\nabla)f\|_{L^r(\mathbb{R}^n)}$$

by Kenig et al. [13] or in Theorem 5.4 in the notes [20]. By Proposition 3.3 in [17], the equation

$$(\Delta+2\rho\cdot\nabla+q)\psi = f$$

has a solution $\psi \in H^s_{r'}(\mathbb{R}^n)$ when $|\Im\rho| \ge I_0$. Moreover, it satisfies

$$\|\psi\|_{H^s_{r'}(\mathbb{R}^n)} \le 2M|\Im\rho|^{-(2+n/r'-n/r)}\|f\|_{H^s_r(\mathbb{R}^n)}.$$

Sobolev embedding implies the L^p estimates in the four cases of the statement. Note that in each case, we have $p \ge r' > 2$.

The elliptic regularity estimate needs some work. First assume that $G \in H^s(\mathbb{R}^n)$, $F \in H^s(\mathbb{R}^n)$, and $(\Delta+2\rho\cdot\nabla)G = F$. Then

$$\begin{aligned}\|G\|_{H^{s+2}(\mathbb{R}^n)} &= \left\|(1+|\xi|^2)^{s/2}(1+|\xi|^2)\hat{G}\right\|_{L^2(\mathbb{R}^n)}\\ &= \left\|(1+|\xi|^2)^{s/2}(\hat{G}+2i\rho\cdot\xi\hat{G}-\hat{F})\right\|_{L^2(\mathbb{R}^n)}\\ &\le \|G\|_{H^s(\mathbb{R}^n)} + \|F\|_{H^s(\mathbb{R}^n)} + 2\left\|(1+|\xi|^2)^{s/2}\rho\cdot\xi\hat{G}\right\|_{L^2(\mathbb{R}^n)}\end{aligned}$$

because $(-|\xi|^2+2i\rho\cdot\xi)\hat{G} = \hat{F}$. By looking at what happens when $|\xi|$ is larger or smaller than $3|\rho|$, we see that $|\rho\cdot\xi| \le |-|\xi|^2+2i\rho\cdot\xi|+3|\rho|^2$. Hence,

$$\|G\|_{H^{s+2}(\mathbb{R}^n)} \le 3\|F\|_{H^s(\mathbb{R}^n)} + (1+6|\rho|^2)\|G\|_{H^s(\mathbb{R}^n)}. \tag{12.7.1}$$

Now let $\chi, \tilde{\chi} \in C_0^\infty(\mathbb{R}^n)$ such that $\tilde{\chi} \equiv 1$ on $\operatorname{supp}\chi$. Assume that $f \in L^2_{loc}$ and $q \in L^\infty_{loc}$. Next, we get

$$(\Delta+2\rho\cdot\nabla)(\chi\psi) = \chi(f-q\psi)+2\nabla\chi\cdot\nabla(\tilde{\chi}\psi)+(\Delta\chi+2\rho\cdot\nabla\chi)\psi \tag{12.7.2}$$

in the distribution sense. We have $q\psi \in L^p_{loc}$, $p \ge 2$ so $\chi q\psi \in L^2(\mathbb{R}^n)$. Similarly, $\tilde{\chi}\psi \in L^2(\mathbb{R}^n)$ and so $\nabla\chi\cdot\nabla(\tilde{\chi}\psi) \in H^{-1}(\mathbb{R}^n)$. The last term on the right-hand side is in $L^2(\mathbb{R}^n)$. By absorbing all the norms of χ, $\tilde{\chi}$ into a constant, we get the estimate

$$C_{\chi,\tilde{\chi},p}\big(\|f\|_{L^2(\operatorname{supp}\chi)} + (1+|\rho|+\|q\|_{L^\infty(\operatorname{supp}\chi)})\|\psi\|_{L^p(\mathbb{R}^n)}\big)$$

for the $H^{-1}(\mathbb{R}^n)$ norm of the right-hand side. By (12.7.1) and since $\psi \in L^p$, $p \ge 2$,

$$\|\chi\psi\|_{H^1(\mathbb{R}^n)} \le \tilde{C}_{\chi,\tilde{\chi},p}\big(\|f\|_{L^2(\operatorname{supp}\chi)} + (1+|\rho|+|\rho|^2+\|q\|_{L^\infty(\operatorname{supp}\chi)})\|\psi\|_{L^p(\mathbb{R}^n)}\big)$$

and this is true no matter the choice of $\chi, \tilde{\chi} \in C_0^\infty(\mathbb{R}^n)$, $\tilde{\chi} \equiv 1$ on $\operatorname{supp}\chi$.

Consider the bounded domain B_{2R} now. Take a chain of cutoff functions $\chi, \tilde{\chi}, \overline{\chi} \in C_0^\infty(B_{3R})$ such that $\overline{\chi} \equiv 1$ on $\operatorname{supp}\tilde{\chi}$, $\tilde{\chi} \equiv 1$ on $\operatorname{supp}\chi$, and finally $\chi \equiv 1$ on B_{2R}. Then $\chi\psi \in H^2(\mathbb{R}^n)$ according to (12.7.1) if the right-hand side of (12.7.2) is in $L^2(\mathbb{R}^n)$. But this is indeed true by going through the previous paragraph while substituting $(\tilde{\chi}, \overline{\chi})$ for $(\chi, \tilde{\chi})$. This gives the final estimate

$$\begin{aligned}\|\psi\|_{H^2(B_{2R})} &\le \|\chi\psi\|_{H^2(\mathbb{R}^n)}\\ &\le \mathscr{C}_{\chi,\tilde{\chi},\overline{\chi},p}\big(\|f\|_{L^2(\operatorname{supp}\tilde{\chi})} + (1+|\rho|+|\rho|^2+\|q\|_{L^\infty(\operatorname{supp}\tilde{\chi})})\|\psi\|_{L^p(\mathbb{R}^n)}\big),\end{aligned}$$

which can be bounded above by the estimate of the statement. Note that the test functions can be chosen based exclusively on the set B_{2R} and their norms have a finite supremum, while p explores the whole set $[2,\infty]$. Hence, the constant can be made to depend only on R. □

The next estimate concerns a potential consisting of a Hölder-continuous function multiplying the characteristic function of a polytope. For a clearer notation, we define a multiplier norm for a polytope first.

Definition 12.7.1. A set $P \subset \mathbb{R}^n$ is a *bounded open polytope* if P is bounded and open and $\overline{P}$ is a finite union of finite intersections of closed half-spaces.

Definition 12.7.2. Let $P \subset \mathbb{R}^n$ be a bounded open polytope. We say a collection $\{H_{jl} \mid j = 1,\ldots,J, l = 1,\ldots,L_j\}$ of half-spaces is a *triangulation* of P if $J \in \mathbb{N}, L_1,\ldots,L_J \in \mathbb{N}, H \subset H_{jl} \subset \overline{H}$ for some open half-space $H \in \mathbb{R}^n$, the intersections $\bigcap_l H_{jl}$ are disjoint for different j, and

$$P = \bigcup_{j=1}^{J} \bigcap_{l=1}^{L_j} H_{jl}.$$

If $s \in \mathbb{R}$ and $1 \le r < \infty$, let $C_{s,r} \in \mathbb{R} \cup \{+\infty\}$ be the norm of the map $H_r^s(\mathbb{R}^n) \to H_r^s(\mathbb{R}^n)$, $f \mapsto \chi_H f$, where $H \subset \mathbb{R}^n$ is a half-space. Then by $\|P\|_{T(s,r)}$, we mean

$$\|P\|_{T(s,r)} = \inf\left\{ \sum_{j=1}^{J} C_{s,r}^{L_j} \,\middle|\, (H_{jl})_{j,l} \text{ is a triangulation of } P \right\}. \tag{12.7.3}$$

Lemma 12.7.2. *Let $P \subset \mathbb{R}^n$ be a bounded open polytope, $s \ge 0$, $r \ge 1$, and $sr < 1$. Then $\|P\|_{T(s,r)} < \infty$ and $\|\chi_P f\|_{H_r^s(\mathbb{R}^n)} \le \|P\|_{T(s,r)} \|f\|_{H_r^s(\mathbb{R}^n)}$. Moreover, we have $\|P\|_{T(s_0,r)} \le \|P\|_{T(s_1,r)}$ if $s_0 \le s_1$.*

Proof. By definition, P has a finite triangulation of let us say $m < \infty$ simplices. Each simplex in $\mathbb{R}^n$ is the intersection of $n+1$ half-spaces. By Triebel [21], Section 2.8.7, the map $f \mapsto \chi_H f$ is bounded in $H_r^s(\mathbb{R}^n)$ under the conditions for s and r given. Hence, $\|P\|_{T(s,r)} \le m C_{s,r}^{n+1} < \infty$. If $(H_{jl})_{j,l}$ is a triangulation, then the intersections $\bigcap_{l=1}^{L_j} H_{jl}$ are disjoint, so $\chi_P = \sum_{j=1}^{J} \prod_{l=1}^{L_j} \chi_{H_{jl}}$, and thus, $\|\chi_P f\|_{H_r^s} \le \sum_{j=1}^{J} C_{s,r}^{L_j} \|f\|_{H_r^s}$. The multiplier estimate follows by taking the infimum over all triangulations. The last claim follows since complex interpolation of Sobolev spaces implies that $C_{s_0,r} \le C_{s_1,r}$ if $s_0 \le s_1$. □

Lemma 12.7.3. *Let $V = \chi_P \varphi$ with $P \subset B_R$ an open polytope and $\varphi \in C^\alpha(\mathbb{R}^n)$ with $\alpha > 0$. Let $0 \le s < \alpha$, $1 \le r \le 2$, and $sr < 1$. Then $V \in H_r^s(\mathbb{R}^n)$,*

$$\|V\|_{H_r^s(\mathbb{R}^n)} \le C_{\alpha,s,r,R} \|P\|_{T(s,r)} \|\varphi\|_{C^\alpha(\mathbb{R}^n)},$$

and

$$\|Vf\|_{H_r^s(\mathbb{R}^n)} \le C_{\alpha,s,r,R} \|P\|_{T(s,r)} \|\varphi\|_{C^\alpha(\mathbb{R}^n)} \|f\|_{H_{r'}^s(\mathbb{R}^n)},$$

where $1/r + 1/r' = 1$ and $\|P\|_{T(s,r)}$ is defined in Definition 12.7.2.

Proof. Let $\Phi \in C_0^\infty$ be such that $\Phi = 1$ on B_R. Then we have the representation

$$V = \chi_P \varphi \Phi,$$

which helps us prove the estimates.

By the last corollary of Section 4.2.2 in [22], there is a finite upper bound $C_{\alpha,s,r}$ for the pointwise multiplier operator norm of any C^α function multiplying in $H_r^s(\mathbb{R}^n)$ when $s < \alpha$. Then the first claim

$$\|V\|_{H^s_r(\mathbb{R}^n)} \leq C_{\alpha,s,r} \|\varphi\|_{C^\alpha(\mathbb{R}^n)} \|\chi_P \Phi\|_{H^s_r(\mathbb{R}^n)} \leq C_{\alpha,s,r,\Phi} \|P\|_{T(s,r)} \|\varphi\|_{C^\alpha(\mathbb{R}^n)}$$

follows from Lemma 12.7.2 since $\|P\|_{T(s,r)} < \infty$ by $s \geq 0$, $r \geq 1$, and $sr < 1$.

By Päivärinta et al. [17] Proposition 3.5 or [2] Theorem 7.5, the product of a $H^s_{r'}(B_R)$ and $H^s_{r/(2-r)}(B_R)$ function is in $H^s_r(B_R)$ when $s \geq 0$ and $1 \leq r \leq 2$. According to [22], we know that C^α functions are pointwise multipliers for $H^s_{r/(2-r)}$ too. The last claim

$$\begin{aligned}
\|Vf\|_{H^s_r(\mathbb{R}^n)} &\leq \|P\|_{T(s,r)} \|\varphi\Phi f\|_{H^s_r(B_R)} \\
&\leq M_{s,r} \|P\|_{T(s,r)} \|\varphi\Phi\|_{H^s_{\frac{r}{2-r}}(B_R)} \|f\|_{H^s_{r'}(B_R)} \\
&\leq C_{\alpha,s,r,\Phi} \|P\|_{T(s,r)} \|\varphi\|_{C^\alpha(\mathbb{R}^n)} \|f\|_{H^s_{r'}(\mathbb{R}^n)}
\end{aligned}$$

follows then because V is supported in $\overline{B}_R$. □

We are now ready to specialize previous lemmas into proving the existence of the complex geometric optics solutions in the context of corner scattering in two and three dimensions.

The conditions on the Hölder smoothness index α of the following proposition follow from various requirements: For the half-space multipliers, we needed $sr < 1$ and $s < \alpha$. To have good enough error decay estimates for ψ from Lemma 12.7.1, we need $s > n/r - 2$. Combining these gives $n - 2r < sr < 1$, i.e., $r > (n-1)/2$. On the other hand, we must have $1/r - 1/r' \geq 2/(n+1)$, i.e., $r \leq 2(n+1)/(n+3)$ in Lemma 12.7.1. These two inequalities have solutions only when $n \in \{2,3\}$. The use of these solutions for corner scattering in higher dimensions requires the Fourier transforms of Besov spaces [4].

Since α is the parameter that ultimately decides which potentials are admissible, we want a largest possible range for it. This is achieved by making s, and thus $n/r - 2$, as small as possible. Hence, r must be largest, and a fortiori, we choose $r = 2(n+1)/(n+3)$.

Proposition 12.7.1. *Let $n \in \{2,3\}$ and $0 \leq s < 5/6$ in 2D or $1/4 < s < 3/4$ in 3D. Let $\varphi \in C^\alpha(\mathbb{R}^n)$ with $\alpha > s$ and $\|\varphi\|_{C^\alpha} \leq \mathcal{M}$. Let $P \subset B_R$ be a bounded open polytope, $r = 2(n+1)/(n+3)$, and assume that $\|P\|_{T(s,r)} \leq \mathcal{D}$.*

Let $k > 0$ and set $V = \chi_P \varphi$. Then there is $p \geq 2$ and $C_{\alpha,s,n,R} < \infty$ with the following properties. If $\rho \in \mathbb{C}^n$, $\rho \cdot \rho + k^2 = 0$, $|\Im\rho| \geq (C_{\alpha,s,n,R} k^2 \mathcal{D}\mathcal{M})^{(n+1)/2}$, then there is $\psi \in L^p(\mathbb{R}^n)$ such that $u_0(x) = \exp(\rho \cdot x)(1 + \psi(x))$ satisfies

$$(\Delta + k^2(1+V))u_0 = 0 \quad in \quad \mathbb{R}^n$$

and

$$\|\psi\|_{L^p(\mathbb{R}^n)} \leq C_{\alpha,s,n,R} k^2 \mathcal{D}\mathcal{M} |\Im\rho|^{-n/p-\beta}$$

with $\beta = \beta(s,n) > 0$. Moreover, $\psi \in H^2(B_{2R})$ with norm estimate

$$\|\psi\|_{H^2(2R)} \leq C_{\alpha,s,n,R}(1 + |\rho|^2 + (1+k^2)\mathcal{M}).$$

Proof. Set $q = k^2 V$ and $f = -k^2 V$. Now $0 \leq s < \alpha$, $1 \leq r \leq 2$, and $sr < 1$, so by Lemma 12.7.3, we have

$$\|f\|_{H^s_r(\mathbb{R}^n)}, \|m_q\|_{H^s_{r'} \to H^s_r} \leq C_{\alpha,s,n,R} k^2 \|P\|_{T(s,r)} \|\varphi\|_{C^\alpha(\mathbb{R}^n)},$$

where m_q is the pointwise multiplier operator.

We have $1/r - 1/r' = 2/(n+1)$, $r \leq 2$. The lower bound for $|\Im\rho|$ matches Lemma 12.7.1 so we have existence of ψ. The condition $s > n/r - 2$ that's required for the good enough error term decay is also satisfied by our a priori requirements on s.

For the H^2 norm estimate, note that $I_0 = (C_{\alpha,s,n,R} k^2 \mathcal{D}\mathcal{M})^{(n+1)/2}$ and the bound for $\|f\|_{H^s_r}$ imply that $\|\psi\|_p \leq C_{s,n}$. We also see that $\|f\|_{L^2} \leq C_{n,R}\mathcal{M}$ by its definition. □

12.8 Stability Proofs

We first give two auxiliary lemmas.

Lemma 12.8.1. *Let $P, P' \subset \mathbb{R}^2$ be two open bounded convex polygons. Let Q be the convex hull of $P \cup P'$. If $\mathbf{x}_c$ is a vertex of P such that $d(\mathbf{x}_c, P') = d_H(P, P')$, where d_H gives the Hausdorff distance,*

$$d_H(P,P') = \max\left(\sup_{\mathbf{x}\in P} d(\mathbf{x},P'),\ \sup_{\mathbf{x}'\in P'} d(P,\mathbf{x}')\right),$$

then $\mathbf{x}_c$ is a vertex of Q. If the angle of P at x_c is α, then the angle of Q at $\mathbf{x}_c$ is at most $(\alpha+\pi)/2 < \pi$.

Proof. Let a and b be the vertices of P on the adjacent edges to $\mathbf{x}_c$. Let $C \in \overline{P'}$ be any point such that $d(\mathbf{x}_c, C) = d_H(P,P')$, and let $h = d_H(P,P')$. Consider the circle $S(\mathbf{x}_c, h)$. Let H_a be an open half-plane tangent to $S(\mathbf{x}_c, h)$, parallel to the segment $\mathbf{x}_c a$ and such that it is on the opposite side of $\mathbf{x}_c a$ than b. Construct H_b similarly. See Fig. 12.1a. Let H_C be the closed half-space tangent to $S(\mathbf{x}_c, h)$ at C with $\mathbf{x}_c \notin H_C$.

Let $\mathbf{x}' \in P'$. If $\mathbf{x}' \in H_a$, then $d(\mathbf{x}', P) \geq d(\mathbf{x}', \ell_{\mathbf{x}_c,a}) > h$ where $\ell_{\mathbf{x}_c,a}$ is a line through $\mathbf{x}_c$ and a. This follows from the convexity of P: the polygon is contained in the cone with vertex $\mathbf{x}_c$ and edges defined by a and b. Thus, $d_H(P,P') \geq d(x', P) > h = d_H(P,P')$, a contradiction. The proof for $\mathbf{x}' \in H_b$ is similar. Consider H_C next: the convexity of P' implies that the segment $\mathbf{x}'C$ belongs to $\overline{P'}$. If $\mathbf{x}' \notin H_C$, then there is $\mathbf{y}' \in \mathbf{x}'C \cap B(x_c, h)$ by the non-tangency of $\mathbf{x}'C$. Then $\mathbf{y}' \in \overline{P'}$ and $d(\mathbf{x}_c, \mathbf{y}') < h$ so $d_H(P,P') < h$, a contradiction again. Thus, we see that $P' \subset H_a^\complement \cap H_b^\complement \cap H_C$.

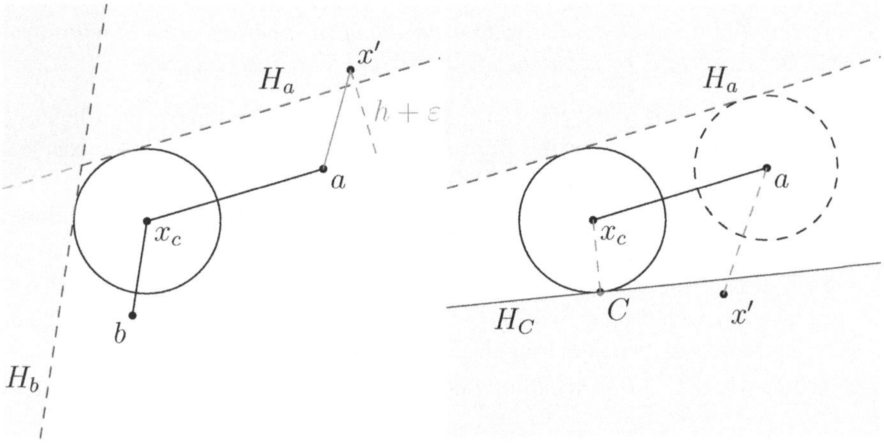

Fig. 12.1 **(a)** $P' \subset H_a^\complement \cap H_b^\complement \cap H_C$, **(b)** ray $\mathbf{x}_c$ to a must meet H_C

Next, H_C must be distance h from a: if it were not, then for any $\mathbf{x}' \in P'$, we have $d(a, \mathbf{x}') \geq d(a, H_C) > h$ since $P' \subset H_C$ as was shown above. Hence, ∂H_C and ∂H_a either are parallel (a case we skip in this proof) or meet at a point A', in which case the ray from $\mathbf{x}_c$ toward a intersects H_C. Do the same for b to get B'. See Fig. 12.1b. This means that $S(\mathbf{x}_c, h)$ is the incircle of the triangle formed by H_a, H_b, and H_C.

We can now see that $\mathbf{x}_c$ is a vertex of Q. First of all, $\mathbf{x}_c \in \overline{Q}$ since $\mathbf{x}_c \in \overline{P}$. Also, P is inside the angle $a\mathbf{x}_c b$ and P' inside the angle $A'\mathbf{x}_c B'$, which is obviously less than π. Thus, $\mathbf{x}_c$ is a vertex of Q. Moreover, its angle is at most $\angle A'\mathbf{x}_c B'$. See Fig. 12.2a.

Let X be the intersection of ∂H_a and ∂H_b. This is a well-defined point since $0 < \angle a\mathbf{x}_c b < \pi$. We have $\angle A'XB' = \angle a\mathbf{x}_c b = \alpha$ by parallel transport of $\mathbf{x}_c a$ to XA' and $\mathbf{x}_c b$ to XB'. Let the perpendiculars from $\mathbf{x}_c$ to XA', $A'B'$, and $B'X$ have base points B_h, C, and A_h, respectively. See

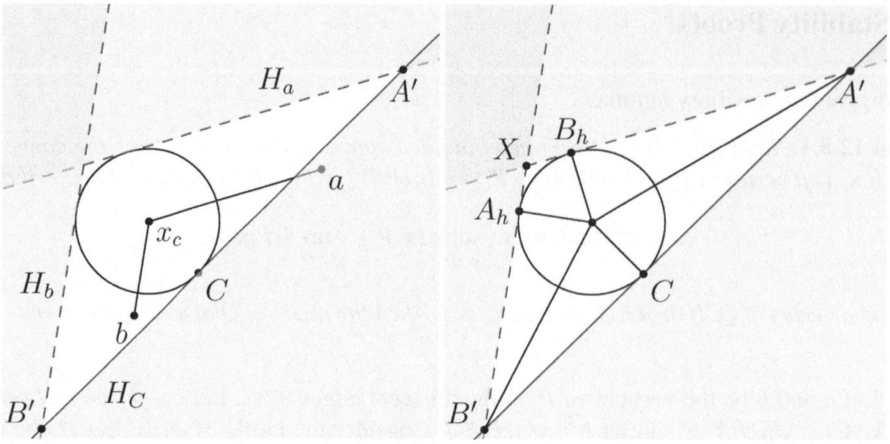

Fig. 12.2 (**a**) $S(\mathbf{x}_c, h)$ is an incircle, (**b**) solving $\angle A'\mathbf{x}_c B'$

Fig. 12.2b. Then $\angle A_h\mathbf{x}_c B_h = \pi - \alpha$, $\angle B_h\mathbf{x}_c A' = \angle A'\mathbf{x}_c C$, and $\angle C\mathbf{x}_c B' = \angle B'\mathbf{x}_c A_h$. This implies that $\angle A'\mathbf{x}_c B' = (\alpha + \pi)/2$ at once since the sum of all of these angles is 2π. $\square$

Lemma 12.8.2. *Let $P, P' \subset \mathbb{R}^3$ be two open cuboids. Let Q be the convex hull of $P \cup P'$. If $\mathbf{x}_c$ is a vertex of P such that $d(\mathbf{x}_c, P') = d_H(P, P')$, where d_H gives the Hausdorff distance,*

$$d_H(P,P') = \max\big(\sup_{\mathbf{x}\in P} d(\mathbf{x},P'),\ \sup_{\mathbf{x}'\in P'} d(P,\mathbf{x}')\big),$$

then $\mathbf{x}_c$ is a vertex of Q. The latter can also fit inside an open spherical cone $\mathscr{Q}$ with vertex $\mathbf{x}_c$ and opening angle $2\alpha' < \pi$. Here, α' is independent of P and P' or their location.

Proof. The proof proceeds as in the proof of Lemma 12.8.1. We can choose coordinates such that $\mathbf{x}_c = 0$ and the three edges of P starting from $\mathbf{x}_c$ lie on the positive coordinate axes having unit vectors e_1, e_1, and e_3. Let $h = d(\mathbf{x}_c, C) = d_H(P, P')$ for some $C \in \overline{P}'$.

If we set $H_j = \{\mathbf{x} \mid \mathbf{x} \cdot e_j < -h\}$, then as in the 2D proof, we see that $P' \subset H_j^{\complement}$. Similarly, if H_C is the closed half-space tangent to $S(\mathbf{x}_c, h)$ at C, we see that $P' \subset H_C$. Hence, $P' \subset H_1^{\complement} \cap H_2^{\complement} \cap H_3^{\complement} \cap H_C$.

If $C_3 < 0$, i.e., it is on the lower hemisphere of $S(\mathbf{x}_c, h)$, then there is $\mathbf{x} \in P$ with $d(\mathbf{x}, C) > h = d_H(P, P')$. Just take any $\mathbf{x}$ on the axis with $\mathbf{x}_3 > 0$. The contradiction, seen also if $C_1 < 0$ or $C_2 < 0$, forces C to be on the closed spherical triangle $T = \{x \mid |\mathbf{x}| = 1, x_j \geq 0\}$.

Now, no matter where $C \in T$ is, recalling that $P' \subset H_1^{\complement} \cap H_2^{\complement} \cap H_3^{\complement} \cap H_C$, it is easy to see that

$$\sup_{A,B \in P \cup P'} \angle A\mathbf{x}_c B < \pi$$

and hence that $H_1^{\complement} \cap H_2^{\complement} \cap H_3^{\complement} \cap H_C$ fits inside an spherical cone that does not contain a plane. Moreover, the minimal required angle of the spherical cone depends continuously on the location of $C \in T$. Compactness of the latter implies the claim. $\square$

We are ready to proof the final theorem whose statement is on page 339.

Proof of Theorem 12.3.1. By Lemma 12.8.1 and Lemma 12.8.2 and possibly switching the symbols P and P' (and their associated waves and potentials), we may assume that $\mathfrak{h} = d(\mathbf{x}_c, P')$ with $\mathbf{x}_c$ a vertex of P. We use the total wave u' of the second potential V' as a "local incident wave" in the neighborhood of $\mathbf{x}_c$. This is allowed since $(\Delta + k^2)u' = 0$ there because $V' = 0$ around $\mathbf{x}_c$.

The potentials V and V' give well-posed scattering. Denote the L^2 norm of the difference of the far-field patterns by $\varepsilon = \|u_\infty^s - u_\infty'^s\|_{L^2(\mathbb{S}^{n-1})}$. Use Proposition 12.5.3. If Q is the convex hull of

$P \cup P'$, then

$$\sup_{\partial Q} \left(\left| u^s - u'^s \right| + \left| \nabla (u^s - u'^s) \right| \right) \leq \frac{C}{\sqrt{\ln \ln \frac{\mathscr{S}}{\varepsilon}}} \tag{12.8.1}$$

when $\varepsilon < \varepsilon_m$. Here, C and ε_m depend only on the a priori parameters. Denote the right-hand side by $\delta(\varepsilon)$ to conserve space in formulas.

Let $\mathfrak{Q}$ be the polyhedral cone generated by the convex hull Q at x_c. By Lemma 12.8.1 and Lemma 12.8.2, there is an open spherical cone $\mathscr{Q} \supset \mathfrak{Q} \supset Q$ with vertex $\mathbf{x}_c$ having opening angle at most $2\alpha' = 2\alpha'(\alpha_m, \alpha_M) < \pi$. Let $\mathfrak{P}$ be the cone generated by P at its vertex $\mathbf{x}_c$. Remember for later that $(\mathscr{Q}, \mathfrak{P}) \in \mathscr{G}(\alpha_m, \alpha_M, \alpha', n)$ using the notation from Lemma 12.6.1.

Let $h = \min(\ell, \mathfrak{h})$, and it is enough to consider the case $h > 0$. We have $P \cap B(\mathbf{x}_c, h) = \mathfrak{P} \cap B(\mathbf{x}_c, h)$ and $Q \cap B(\mathbf{x}_c, h) = \mathfrak{Q} \cap B(\mathbf{x}_c, h)$. Denote the former by P_h and the latter by Q_h. We also have $P_h \cap P' = Q_h \cap P' = \emptyset$.

We want to use Proposition 12.6.2 next. The conditions of non-vanishing total waves of Definition 12.4 imply that we have $N = 0$, $P_N(x) \equiv u'(\mathbf{x}_c) \neq 0$. Moreover, as in the proof of Proposition 12.5.3, we see that u' is Lipschitz with norm at most $C(k, R, \mathscr{M}, \mathscr{S})$. The other conditions of Proposition 12.6.2 are also satisfied. Recall also $\delta(\varepsilon) = C/\sqrt{\ln \ln(\mathscr{S}/\varepsilon)}$ from (12.8.1) and that $\|u\|, \|u'\| \leq C_{k,R,\mathscr{S}}$ in $H^2(B_{2R})$. We can absorb this constant into the constants of the inequality. Hence, there is a constant C depending only on a priori parameters such that if $1/p + 1/p' = 1$, then

$$\begin{aligned} C \left| \varphi(\mathbf{x}_c) \int_{\mathfrak{P}} e^{\rho \cdot (\mathbf{x} - \mathbf{x}_c)} u'(\mathbf{x}_c) \mathrm{d}\mathbf{x} \right| &\leq |\Re \rho|^{-n} e^{-\delta_0 |\Re \rho| h/2} \\ &+ |\Re \rho|^{-n - \min(1,\alpha)} + |\Re \rho|^{-n/p'} \|\psi\|_{L^p(P_h)} \\ &+ h^{(n-1)/2} |\rho| (1 + \|\psi\|_{H^2(B_{2R})}) \delta(\varepsilon) \\ &+ h^{n/2-1} e^{-\delta_0 |\Re \rho| h} |\rho| (1 + \|\psi\|_{H^2(B_{2R})}) \end{aligned} \tag{12.8.2}$$

whenever $u_0 \in H^2(B_{2R})$ satisfies $(\Delta + k^2(1 + V))u_0 = 0$,

$$u_0(x) = e^{\rho \cdot (\mathbf{x} - \mathbf{x}_c)} (1 + \psi(\mathbf{x})),$$

$\psi \in L^p$ in Q_h with $\rho \in \mathbb{C}^n$, $|\rho| \geq 1$, and $\Re \rho \cdot (\mathbf{x} - \mathbf{x}_c) \leq -\delta_0 |\Re \rho| |\mathbf{x} - \mathbf{x}_c|$ for some $\delta_0 > 0$ and any $\mathbf{x} \in Q_h$.

Recall that $(\mathscr{Q}, \mathfrak{P}) \in \mathscr{G}(\alpha_m, \alpha_M, \alpha', n)$, and hence, we may use Lemma 12.6.1. It gives us constants $\tau_0 = \tau_0(k, \alpha_m, \alpha_M, \alpha', n)$ and $c = c(\alpha_m, \alpha_M, n) > 0$ and a curve $\tau \mapsto \rho(\tau) \in \mathbb{C}^n$, $\tau = |\Re \rho(\tau)|$ satisfying the conditions required of ρ above with $\delta_0 = \cos \alpha' > 0$, $\rho(\tau) \cdot \rho(\tau) + k^2 = 0$, and

$$\left| \int_{\mathfrak{P}} e^{\rho \cdot (\mathbf{x} - \mathbf{x}_c)} u'(x_c) \mathrm{d}\mathbf{x} \right| \geq c \left| u'(\mathbf{x}_c) \right| \tau^{-n} \tag{12.8.3}$$

whenever $\tau \geq \tau_0$.

If $\tau \geq \max(\tau_0, C_0)$, with the constant C_0 depending on a priori parameters and arising from Proposition 12.7.1, then the latter gives existence of u_0 and ψ required above. We may indeed use that proposition because the a priori bounds on the Hölder smoothness index α imply the existence of a suitable Sobolev smoothness index s used in there. Finally, it gives the estimates

$$\|\psi\|_{L^p(\mathbb{R}^n)} \leq C |\Im \rho|^{-n/p - \beta}$$

for some $\beta = \beta(s, n) > 0$ and

$$\|\psi\|_{H^2(B_{2R})} \leq C(1 + |\rho|^2)$$

where C again depends only on the a priori parameters.

We have all the fundamental estimates now. Let us apply them. We have $\exp(-\mathbf{x}) \leq \mathbf{x}^{-1}$ and $\exp(-\mathbf{x}) \leq (n+4)!x^{-n-4}$ for all $\mathbf{x} > 0$. Also, since $\rho(\tau)\cdot\rho(\tau)+k^2 = 0$, we get $|\rho(\tau)| = \sqrt{k^2+2\tau^2}$. By taking a new lower bound for τ, for example, $\tau \geq k$, we may assume that $|\rho(\tau)| \leq \sqrt{3}\tau$. Hence, we can estimate

$$\begin{aligned}
&|\Re\rho|^{-n} e^{-\delta_0|\Re\rho|h/2} \leq C|\Re\rho|^{-n-1}h^{-1},\\
&|\Re\rho|^{-n/p'}\|\psi\|_{L^p} \leq C|\Re\rho|^{-n-\beta},\\
&h^{(n-1)/2}|\rho|(1+\|\psi\|_{H^2(B_{2R})})\delta(\varepsilon) \leq Ch^{(n-1)/2}|\Re\rho|^3\delta(\varepsilon),\\
&h^{n/2-1}e^{-\delta_0|\Re\rho|h}|\rho|(1+\|\psi\|_{H^2(B_{2R})}) \leq Ch^{-n/2-5}|\Re\rho|^{-n-1}
\end{aligned}$$

in (12.8.2). Divide the new constants to the left-hand side, take the lower bound (12.8.3) into account, and use the a priori assumption $|u'(\mathbf{x})| \geq c > 0$ in $B_R \setminus P'$. Finally, using $h \leq 1$ and $\tau \geq 1$, we get

$$c|\varphi(\mathbf{x}_c)| \leq h^{(n-1)/2}(\delta(\varepsilon)\tau^{n+3} + h^{-n-9/2}\tau^{-m})$$

where $m = \min(1,\alpha,\beta)$. This holds as long as $\tau \geq \max(\tau_0, C_0, k)$ and $h = \min(\ell, \mathfrak{h})$. To make formulas simpler, we estimate the right-hand side above and get

$$c|\varphi(\mathbf{x}_c)| \leq \delta(\varepsilon)\tau^{n+5} + h^{-n-5}\tau^{-m}. \tag{12.8.4}$$

Setting $\tau = \tau_e$ with

$$\tau_e = \left(\frac{1}{h^{n+5}\delta(\varepsilon)}\right)^{\frac{1}{m+n+5}}$$

makes both terms on the right-hand side of (12.8.4) equal (which gives the minimum modulo constants), and the inequality becomes

$$c|\varphi(\mathbf{x}_c)| \leq 2h^{-\frac{(n+5)^2}{m+n+5}}\delta(\varepsilon)^{\frac{m}{m+n+5}}. \tag{12.8.5}$$

Note that if ε is small enough, then

$$\tau_e \geq \tau_e h^{\frac{n+5}{m+n+5}} = (\delta(\varepsilon))^{-\frac{1}{m+n+5}} \geq \max(\tau_0, C_0, k)$$

and so we can choose $\tau = \tau_e$ in (12.8.5). Solving for h in it gives

$$\min(\ell,\mathfrak{h}) = h \leq C\delta^{\frac{m}{(n+5)^2}}|\varphi(\mathbf{x}_c)|^{-\frac{m}{(n+5)^2}-\frac{1}{n+5}}.$$

By the a priori bounds of Definition 12.3.1, we have $|\varphi(x_c)| \geq \mu > 0$. Hence, if ε is again small enough (now also depending on μ and ℓ), then the right-hand side is smaller than ℓ, and so $\min(\ell,\mathfrak{h}) = \mathfrak{h}$. Writing out the definition of $\delta(\varepsilon)$ gives

$$\mathfrak{h} \leq C\left(\ln\ln\frac{S}{\varepsilon}\right)^{-\frac{m}{2(n+5)^2}}$$

and the claim is proven. □

Proof of Theorem 12.3.2. The proof uses the same lemmas and propositions as the proof of Theorem 12.3.1. Now instead of having two nontrivial potentials V and V', we choose the following: $P' = \emptyset$ and $V' \equiv 0$. This implies that $u' = u^i$, $u'^s \equiv 0$, and $u'^s_\infty \equiv 0$ among others. In particular, $V' \equiv 0$ is trivially admissible.

Proceed as in the proof of Theorem 12.3.1, except that choose $h = \ell$ instead of $h = \min(\ell, d_H(P,P'))$. Up to showing (12.8.5), none of the constants depend on μ or ℓ. Now, if ε is small enough, let's say at most ε_{min} which depends only on a priori parameters except for ℓ, $\varphi(x_c)$, then

$$(\delta(\varepsilon))^{-\frac{1}{m+n+5}} \geq \max(\tau_0, C_0, k)$$

and we can again let $\tau = \tau_e$ in (12.8.5). Solving for ε in it gives

$$\|u^s_\infty\|_{L^2(\mathbb{S}^{n-1})} = \varepsilon \geq \frac{\mathscr{S}}{\exp\exp(C\ell^{-2/\gamma}|\varphi(\mathbf{x}_c)|^{-2-2/((n+5)\gamma)})}$$

for $\gamma = \min(1, \alpha, \beta)/(n+5)^2$ as in the previous proof and a constant C depending on a priori data but not ℓ or $\varphi(\mathbf{x}_c)$. If on the other hand $\varepsilon > \varepsilon_{min}$, the claim is immediately true. □

References

1. G. Alessandrini and L. Rondi, *Determining a sound-soft polyhedral scatterer by a single far-field measurement*, Proc. Amer. Math. Soc., **35** (2005), 1685–1691.
2. A. Behzadan and M. Holst, Multiplication in Sobolev spaces, revisited, *Ark. Mat.* **59** (2021), no. 2, 275–306.
3. E. Blåsten and H. Liu, *On corners scattering stably and stable shape determination by a single far-field pattern*, Indiana Univ. Math. J., **70(3)** (2021), 907–947.
4. E. Blåsten, L. Päivärinta and J. Sylvester, *Corners always scatter*, Comm. Math. Phys., **331** (2014), 725–753.
5. J. Cheng and M. Yamamoto, *Uniqueness in an inverse scattering problem within non-trapping polygonal obstacles with at most two incoming waves*, Inverse Problems, **19** (2003), 1361–1384.
6. D. Colton and R. Kress, *Inverse acoustic and electromagnetic scattering theory*, volume 93 of Applied Mathematical Sciences. Springer, Cham, 4th edition, 2019.
7. D. Gilbarg and N. S. Trudinger, *Elliptic partial differential equations of second order*, Vol. 224 of *Grundlehren der Mathematischen Wissenschaften [Fundamental Principles of Mathematical Sciences]*, Springer-Verlag, Berlin, 2nd edition, (1983).
8. L. Hörmander, *The analysis of linear partial differential operators. II*, Grundlehren der mathematischen Wissenschaften, 257, Springer-Verlag, Berlin, (1983).
9. G. Hu, M. Salo and E. Vesalainen, *Shape identification in inverse medium scattering problems with a single far-field pattern*, SIAM J. Math. Anal., **48** (2016),152–165.
10. V. Isakov, *On uniqueness in the inverse transmission scattering problem*, Commun. Part. Diff. Eqns., **15** (1990), 1565–1587.
11. V. Isakov, *Stability estimates for obstacles in inverse scattering*, J. Comp. Appl. Math. **42** (1992) 79–88.
12. V. Isakov, *New stability results for soft obstacles in inverse scattering*, Inverse Problems **9** (1993) 535–543.
13. C. E. Kenig, A. Ruiz and C. D. Sogge, Uniform Sobolev inequalities and unique continuation for second order constant coefficient differential operators, *Duke Math. J.* **55(2)** (1987), 329–347.
14. H. Liu, M. Petrini, L. Rondi and J. Xiao, Stable determination of sound-hard polyhedral scatterers by a minimal number of scattering measurements, *J. Differential Equations*, **262(3)** (2017), 1631–1670.
15. H. Liu and J. Zou, *Uniqueness in an inverse acoustic obstacle scattering problem for both sound-hard and sound-soft polyhedral scatterers*, Inverse Problems, **22** (2006), 515–524.
16. A. I. Nachman, Reconstructions from boundary measurements, *Ann. of Math. (2)* **128(3)** (1988), 531–576.
17. L. Päivärinta, M. Salo and E. V. Vesalainen, *Strictly convex corners scatter*, Rev. Mat. Iberoam. **33(4)** (2017), 1369–1396.
18. L. Rondi, *Stable determination of sound-soft polyhedral scatterers by a single measurement*, Indiana Univ. Math. J., **57** (2008), 1377–1408.
19. L. Rondi and M. Sini, *Stable determination of a scattered wave from its far-field pattern: the high frequency asymptotics*, Arch. Ration. Mech. Anal. **218** (2015), 1–54.
20. A. Ruiz, *Harmonic Analysis and Inverse Problems, Lecture Notes*, (2002), accessed 24.01.2018.
21. H. Triebel, *Theory of function spaces*, Monographs in Mathematics, 78, Birkhäuser Verlag, Basel, (1983).
22. H. Triebel, *Theory of function spaces. II*, Monographs in Mathematics, 84, Birkhäuser Verlag, Basel, (1992).
23. G. Uhlmann, *Inside Out: Inverse Problems and Applications*, Cambridge University Press, Cambridge, (2003).
24. G. Uhlmann, Visibility and invisibility, in *ICIAM 07—6th International Congress on Industrial and Applied Mathematics*, 381–408, Eur. Math. Soc., Zürich, (2009).

Chapter 13
Stable Determination of an Elastic Medium Scatterer by a Single Far-Field Measurement and Beyond

13.1 Overview

13.1.1 Mathematical Setup

We are mainly concerned with the time-harmonic elastic wave scattering due to the impingement of an incident field on an inhomogeneous isotropic medium scatterer as well as the associated inverse problem of determining the scatterer from the corresponding far-field measurement. We follow the treatment in [2].

For clarity, we present the forward elastic medium scattering problem considered in this chapter again. Let Ω be a bounded Lipschitz domain in $\mathbb{R}^n$ with a connected complement $\mathbb{R}^n\backslash\overline{\Omega}$, $n = 2,3$. In the physical setup, Ω is the support of an inhomogeneous elastic scatterer embedded in a uniformly homogeneous background space. The elastic medium parameters are characterized by ρ and λ,μ, which are, respectively, referred to as the density and the bulk moduli. It is assumed that λ and μ are real constants satisfying the strong convexity conditions

$$\mu > 0 \quad \text{and} \quad n\lambda + 2\mu > 0.$$

It is also assumed that $\rho \in L^\infty(\mathbb{R}^n)$ and $\Omega = \mathrm{supp}(1-\rho)$. That is, by normalization, we assume that the density of the homogeneous background space is 1, whereas the scatterer is characterized by the inhomogeneous density ρ in Ω.

To introduce the Lamé system that describes the elastic scattering, we let $\omega \in \mathbb{R}_+$ signify the angular frequency of the time-harmonic elastic wave. Denote

$$\kappa_{\mathrm{s}} = \omega\sqrt{1/\mu} \text{ and } \kappa_{\mathrm{p}} = \omega\sqrt{1/(\lambda+2\mu)}$$

by the shear and compressional wavenumbers, respectively. Let $\mathbf{u}^i$ denote the time-harmonic plane incident wave of the following form:

$$\mathbf{u}^i = \alpha_1\,\mathbf{d}\,e^{\mathrm{i}\kappa_{\mathrm{p}}\mathbf{x}\cdot\mathbf{d}} + \alpha_2\,\mathbf{d}^\perp e^{\mathrm{i}\kappa_{\mathrm{s}}\mathbf{x}\cdot\mathbf{d}}, \quad \alpha_1,\alpha_2 \in \mathbb{C}, \quad |\alpha_1|+|\alpha_2| \neq 0, \tag{13.1.1}$$

where $\mathbf{d} \in \mathbb{S}^{n-1}$ is called the incident direction and $\mathbf{d}^\perp$ is a unit vector orthogonal to $\mathbf{d}$. Due to the interaction between the incident elastic wave $\mathbf{u}^i$ and the elastic scatter Ω, the scattered elastic wave $\mathbf{u}^s \in \mathbb{C}^n$ is generated. This induces the total elastic wave $\mathbf{u} \in \mathbb{C}^n$, which is the superposition of $\mathbf{u}^i$ and $\mathbf{u}^s$, namely, $\mathbf{u} := \mathbf{u}^i + \mathbf{u}^s$, and satisfies the Navier equation

$$\Delta^*\mathbf{u} + \rho\,\omega^2\mathbf{u} = \mathbf{0} \quad \text{in} \quad \mathbb{R}^n, \quad \Delta^*\mathbf{u} := \mu\Delta\mathbf{u} + (\lambda+\mu)\nabla(\nabla\cdot\mathbf{u}). \tag{13.1.2}$$

H. Diao, H. Liu, *Spectral Geometry and Inverse Scattering Theory*,
https://doi.org/10.1007/978-3-031-34615-6_13

Clearly, $\mathbf{u}^i$ is an entire solution to (13.1.2) with $\rho = 1$. By the Helmholtz decomposition, any solution $\mathbf{u} \in H^2_{loc}(\mathbb{R}^n)^n$ to (13.1.2) can be decomposed as follows:

$$\mathbf{u} = \mathbf{u}_\mathrm{p} + \mathbf{u}_\mathrm{s}, \tag{13.1.3}$$

where $\mathbf{u}_\mathrm{p}$ and $\mathbf{u}_\mathrm{s}$ satisfy the equations

$$\begin{cases} (\Delta + \kappa_\mathrm{p}^2)\mathbf{u}_\mathrm{p} = \mathbf{0}, & \nabla \times \mathbf{u}_\mathrm{p} = \mathbf{0}, \\ (\Delta + \kappa_\mathrm{s}^2)\mathbf{u}_\mathrm{s} = \mathbf{0}, & \nabla \cdot \mathbf{u}_\mathrm{s} = 0. \end{cases} \tag{13.1.4}$$

The scattered $\mathbf{u}^s$ satisfies the Kupradze radiation condition

$$\begin{cases} \lim\limits_{|\mathbf{x}|\to\infty} |\mathbf{x}|^{(n-1)/2}\left(\frac{\partial \mathbf{u}_\mathrm{p}^s}{\partial |\mathbf{x}|} - \mathrm{i}\kappa_\mathrm{p}\mathbf{u}_\mathrm{p}^s\right) = \mathbf{0}, \\ \lim\limits_{|\mathbf{x}|\to\infty} |\mathbf{x}|^{(n-1)/2}\left(\frac{\partial \mathbf{u}_\mathrm{s}^s}{\partial |\mathbf{x}|} - \mathrm{i}\kappa_\mathrm{s}\mathbf{u}_\mathrm{s}^s\right) = \mathbf{0}, \end{cases} \tag{13.1.5}$$

and admits the following asymptotic expansion (cf. [23]):

$$\mathbf{u}^s(\mathbf{x}) = \frac{\exp(\mathrm{i}\kappa_\mathrm{p}|\mathbf{x}|)}{|\mathbf{x}|^{(n-1)/2}}\mathbf{U}_\mathrm{p}(\hat{\mathbf{x}}) + \frac{\exp(\mathrm{i}\kappa_\mathrm{s}|\mathbf{x}|)}{|\mathbf{x}|^{(n-1)/2}}\mathbf{U}_\mathrm{s}(\hat{\mathbf{x}}) + \mathscr{O}(|\mathbf{x}|^{-\frac{n+1}{2}}) \quad \text{as} \quad |\mathbf{x}| \to \infty, \tag{13.1.6}$$

which holds uniformly in all directions $\hat{\mathbf{x}} \in \mathbb{S}^{n-1}$, where $\mathrm{i} := \sqrt{-1}$ is the imaginary unit. The vector fields $\mathbf{U}_\mathrm{p}$ and $\mathbf{U}_\mathrm{s}$ are referred to as the longitudinal and transversal far-field patterns, respectively, which, respectively, characterize the asymptotic behaviors of the normal part $\mathbf{u}_\mathrm{p}^s$ and of the tangential part $\mathbf{u}_\mathrm{s}^s$ of $\mathbf{u}^s$. Let $\mathbf{U} = (\mathbf{U}_\mathrm{p}, \mathbf{U}_\mathrm{s})$ denote the far-field pattern of $\mathbf{u}^s$, which is an element of the liner space $L^2(\mathbb{S}^{n-1})^n \times L^2(\mathbb{S}^{n-1})^n$ equipped with the following norm:

$$\|(\mathbf{w}_1, \mathbf{w}_2)\|^2_{L^2(\mathbb{S}^{n-1}, \mathbb{C}^n \times \mathbb{C}^n)} = \|\mathbf{w}_1\|^2_{L^2(\mathbb{S}^{n-1})^n} + \|\mathbf{w}_2\|^2_{L^2(\mathbb{S}^{n-1})^n},$$

where $(\mathbf{w}_1, \mathbf{w}_2) \in L^2(\mathbb{S}^{n-1})^n \times L^2(\mathbb{S}^{n-1})^n$.

The inverse problem of our study can be introduced as determining the support of the inhomogeneous scattering object, namely, Ω, independent of its material content (λ, μ, ρ), by knowledge of the far-field pattern $\mathbf{U}(\hat{\mathbf{x}})$. By introducing an operator $\mathscr{F}$ which is defined by the forward scattering system (13.1.1)–(13.1.6) and sends the scatterer $(\Omega; \lambda, \mu, \rho)$ to the corresponding far-field pattern associated with an incident field $\mathbf{u}^i$, the inverse problem can be abstractly given as

$$\mathscr{F}(\Omega) = \mathbf{U}(\hat{\mathbf{x}}; \mathbf{u}^i), \quad \hat{\mathbf{x}} \in \mathbb{S}^{n-1}. \tag{13.1.7}$$

It is emphasized that in our study of (13.1.7), the material content of Ω, namely, (λ, μ, ρ), is not required to be known in advance though it belongs to a certain general a priori class as shall be introduced in what follows. Moreover, we shall consider the case that only a fixed $\mathbf{u}^i$ is used, namely, $\mathbf{u}^i$ is given in (13.1.1) with fixed $\eta_1, \eta_2, \mathbf{d}$ and $\mathbf{d}^\perp$. In such a case, $\mathbf{U}(\hat{\mathbf{x}}; \mathbf{u}^i)$ is said to be a single far-field measurement. The inverse scattering problem with a single far-field measurement constitutes a longstanding problem in the literature. Finally, we remark that by direction verifications, the inverse problem (13.1.7) is nonlinear and it is formally determined with a single far-field measurement. Recently, several qualitative uniqueness results in determining convex polyhedral medium scatterers were established in [30] for electrostatics. In Chap. 12 and [31], quantitative stability estimates of double-logarithmic type were established in determining the convex polyhedral shape of an acoustic medium scatterer. The related stable determinations of obstacle scatterers by finite measurements were established in [1, 21, 35, 37, 45] for acoustic scattering, Chap. 7 for electromagnetic scattering, and [39, 40, 43] for elastic scattering. Recent developments in elastic and acoustic-elastic scattering problems can be founded in [18, 38].

The second focus of our study is under what conditions, $\mathbf{U} \equiv \mathbf{0}$. This is another aspect of the inverse scattering problem (13.1.7), and in such a case, the underlying object $(\Omega;\lambda,\mu,\rho)$ is invisible with respect to the far-field measurement.

13.1.2 Statement of the Main Stability Result and Its Implication to Invisibility

First, we introduce the a priori class of elastic scatterers for our study.

Definition 13.1.1. Let Ω be a bounded Lipschitz domain in $\mathbb{R}^n$ with a connected complement $\mathbb{R}^n\backslash\overline{\Omega}$, $n = 2,3$. In addition, $\Omega = \mathrm{supp}(1-\rho) \subset \overline{B_R}$, where B_R signifies a central ball of radius R. We say $(\Omega;\rho) \in \mathscr{K}$ if the following conditions are satisfied:

(a) $\Omega \subset \mathbb{R}^2$ is a convex polygon and the opening angle at each vertex of Ω is in $(2\alpha_m, 2\alpha_M)$, $\alpha_m > 0$, and $\alpha_M < \pi/2$.
(b) $\Omega \subset \mathbb{R}^3$ is a convex polyhedron.
(c) The distances of any vertex of Ω to its non-adjacent edges are at least l_0, $0 < l_0 \leq 1$.
(d) $\rho(\mathbf{x})$ is a uniformly θ-Hölder-continuous function in $\overline{\Omega}$, $0 < \theta \leq 1$. In addition, $|\rho(\mathbf{x}_0) - 1| \geq \varepsilon_0 > 0$ at any vertex $\mathbf{x}_0$ of Ω.

Definition 13.1.2. We say ρ is called an admissible density function if for a time-harmonic plane incident wave $\mathbf{u}^i$ of the form (13.1.1), the forward scattering system (13.1.1)–(13.1.5) admits a unique solution $\mathbf{u} \in H^2_{loc}(\mathbb{R}^n)^n$ such that the scattered wave $\mathbf{u}^s = \mathbf{u} - \mathbf{u}^i$ is the radiating solution which fulfils $\|\mathbf{u}^s\|_{H^2(B_{2R})^n} \leq \mathscr{N}$ and $|\mathbf{u}(\mathbf{x})| \geq \zeta_0$ for $\mathbf{x} \in S_\Omega$, where $\mathscr{N}$ and ζ_0 are a priori positive constants and S_Ω denotes the set of vertices of Ω.

Remark 13.1.1. The admissible conditions on $\rho(\mathbf{x})$ and $\mathbf{u}$ in Definitions 13.1.1 and 13.1.2 can be fulfilled in generic physical scenarios. For example, when $\rho(\mathbf{x})$ is given by two distinct constants in $\overline{\Omega}$ and $\mathbb{R}^n \setminus \overline{\Omega}$, say

$$\rho(\mathbf{x}) = \begin{cases} a \neq 1, & \mathbf{x} \in \overline{\Omega}, \\ 1, & \mathbf{x} \in \mathbb{R}^n\backslash\overline{\Omega}, \end{cases}$$

one readily sees that $\rho(\mathbf{x})$ satisfies the assumption (d) in Definition 13.1.1. Furthermore, by the well-posedness of the direct scattering problem (13.1.2), there exists a unique solution $\mathbf{u} \in H^1_{loc}(\mathbb{R}^n)^n$ to (13.1.2) (cf. [36, Chapter 4]). By further noting that the jump singularity is only attached to the lower-order term ρ and the standard elliptic interior regularity estimate (cf. [36, Theorem 4.16]), one has $\mathbf{u} \in H^2_{loc}(\mathbb{R}^n)^n$. Regarding the technical conditions $\|\mathbf{u}^s\|_{H^2(B_{2R})^n} \leq \mathscr{N}$ and $|\mathbf{u}(\mathbf{x})| \geq \zeta_0$ for $\mathbf{x} \in S_\Omega$, we can consider a specific case $\omega \cdot \mathrm{diam}(\Omega) \ll 1$, where ω is the angular frequency and $\mathrm{diam}(\Omega)$ is the diameter of the scatterer Ω. In the physical setup, this means the scatterer is much small in size compared to the operating wavelength. Under this situation, the incident wave $\mathbf{u}^i$ given by (13.1.1) is dominant in the total wave field $\mathbf{u}$, and the energy of the scattered wave field $\mathbf{u}^s$ is small (cf. [13, 25]). Hence, the assumption $\|\mathbf{u}^s\|_{H^2(B_{2R})^n} \leq \mathscr{N}$ can be satisfied for an a priori positive constant $\mathscr{N}$ and the total wave field $|\mathbf{u}(\mathbf{x})| \geq \zeta_0$ for $\mathbf{x} \in S_\Omega$, for a certain a priori positive constant ζ_0. In fact, $\|\mathbf{u}^s\|_{H^2(B_{2R})^n} \leq \mathscr{N}$ is clearly a generic physical condition by the well-posedness of the forward scattering problem, whereas for the technical condition $|\mathbf{u}(\mathbf{x})| \geq \zeta_0$ for $\mathbf{x} \in S_\Omega$, we also believe it holds generically.

In what follows, (Ω,ρ) is said to be an admissible polyhedral scatterer if it fulfils the conditions in Definitions 13.1.1 and 13.1.2. The parameters $\{R, \mathscr{N}, l_0, \varepsilon_0, \zeta_0, \alpha_m, \alpha_M\}$ in the above admissibility definitions are referred to as the a priori parameters. It is remarked that the admissibility requirements in Definitions 13.1.1 and 13.1.2 will be needed in our stability study and can be fulfilled in certain general scenarios, which shall become clearer in our subsequent discussion.

Next, we present the logarithmic stability estimate for the inverse problem (13.1.7) on the shape determination of the elastic medium scatterer by a single far-field pattern. Henceforth, we define the Hausdorff distance of two medium scatterers $(\Omega;\rho)$ and $(\Omega';\rho')$ as follows:

$$d_{\mathscr{H}}(\Omega,\Omega') = \max\left\{\sup_{\mathbf{x}\in\Omega} \operatorname{dist}(\mathbf{x},\,\Omega'),\ \sup_{\mathbf{x}\in\Omega'} \operatorname{dist}(\mathbf{x},\,\Omega)\right\}. \tag{13.1.8}$$

Theorem 13.1.1. *Let $(\Omega;\rho)$ and $(\Omega';\rho')$ be two admissible polyhedral scatterers. Let $\mathbf{u}^i$ be a common time-harmonic plane incident wave of the form* (13.1.1). *Assume that $\mathbf{U}$ and $\mathbf{U}'$ are the far-field patterns of the scattered waves $\mathbf{u}^s$ and $\mathbf{u}'^s$ by the medium scatterers $(\Omega;\rho)$ and $(\Omega';\rho')$, respectively. For sufficiently small $\varepsilon\in\mathbb{R}_+$, if*

$$\|\mathbf{U}-\mathbf{U}'\|_{L^2(\mathbb{S}^{n-1},\mathbb{C}^n\times\mathbb{C}^n)} \le \varepsilon,$$

then

$$d_{\mathscr{H}}(\Omega,\Omega') \le C(\ln\ln(\mathscr{N}/\varepsilon))^{-\gamma}, \tag{13.1.9}$$

where $\mathscr{N}$ is given in Definition 13.1.2 and C and γ are positive constants, depending only on the a priori parameters involved in Definitions 13.1.1 and 13.1.2 as well as the Lamé constants λ,μ.

Remark 13.1.2. In Theorem 13.1.1, the shape of the scattering inhomogeneity is defined by the discontinuity interface of the density parameter ρ, which appears in the lower-order term in the Lamé system (13.1.2). As also discussed in Remark 13.1.1, this can guarantee that the solution $\mathbf{u}^s\in H^2_{loc}(\mathbb{R}^n)^n$, which is a key technical ingredient in deriving the stability estimate (13.1.9). It would be interesting to consider the case that the material discontinuity, say, e.g., jump singularity, appears in λ and μ. In such a case, since the singularity is attached to the coefficients in the leading-order term of the Lamé system, one generically does not have H^2 regularity, which shall pose significant challenges.

Remark 13.1.3. In (13.1.9), the stability is of a double-logarithmic type, which indicates the ill-posedness nature of the inverse scattering problem (13.1.7). In our proof of Theorem 13.1.1 in what follows, the logarithmic estimates arise in two steps: first, the propagation of the data from the far field to the near field yields one logarithmic-type ill-posedness, and then the further propagation of the data from the near field to the boundary of the scatterer gives rise to another logarithmic-type ill-posedness. It is emphasized that the two-phase propagation is necessary since the analytic continuation, namely, the three-sphere inequalities in what follows, degenerates when approaching the boundary of the scatterer. Our argument on this issue follows the spirit of the one developed in [6], where a related inverse scattering problem associated with the Schrödinger operator $-\Delta+q$ was studied. In a different but related physical context where the elastic scatterer is an impenetrable obstacle, a different approach of continuing the data, especially from the near field to the boundary, was developed in [43]; see also [28, 29, 42] for related studies for the inverse acoustic and electromagnetic scattering problems. It seems plausible that one might be able to increase the stability estimate to a certain extent by adapting the data continuation strategy in [43] to the current setup.

The argument in proving Theorem 13.1.1 is localized around a corner of the underlying polyhedral scatterer. As an interesting byproduct, we can provide another stability result where the medium scatterer is not necessarily polyhedral as long as it possesses a corner. The full technical details of the result will be given in Theorems 13.5.1 and 13.5.2, and we only provide a rough summary in the following theorem.

Theorem 13.1.2. *Consider the scattering problem* (13.1.1)–(13.1.5) *associated with a general medium scatterer (Ω,ρ), which is not necessarily polyhedral. Suppose that $\partial\Omega$ possesses a corner as described in Theorem 13.5.2. Then it holds that*

$$\|\mathbf{u}\|_{L^2(\mathbb{S}^{n-1},\mathbb{C}^n\times\mathbb{C}^n)} \ge C, \tag{13.1.10}$$

where C is a positive constant depending on a certain set of a priori parameters.

Theorem 13.1.2 indicates that for a general medium scatterer, if it possesses a corner, then it generically scatters every incident field stably, i.e., invisibility phenomenon cannot occur.

The rest of the chapter is organized as follows. In Sect. 13.2, we derive a critical auxiliary result about the propagation of smallness from far field to the boundary of the scatterer. Section 13.3 is devoted to a micro-local analysis of the scattering solution around a corner. The full details of the proof of Theorem 13.1.1 is presented in Sect. 13.4. In Sect. 13.5, we present the full technical details of Theorem 13.1.2.

13.2 Propagation of Smallness from Far Field to Boundary

The main goal of this section is to show how the smallness from the far-field pattern propagates to the boundary of the elastic medium scatterer, which is a key ingredient in the stability proof of Theorem 13.1.1. Throughout the rest of the chapter, we let $\mathbf{u}$ and $\mathbf{u}'$, respectively, denote the total wave fields corresponding to the scatterers $(\Omega;\rho)$ and $(\Omega';\rho')$ in Theorem 13.1.1.

13.2.1 Stability Estimates: From Far Field to Near Field

In this subsection, our aim is to estimate the difference of $\mathbf{u}$ and $\mathbf{u}'$ in $B_{2R}\backslash B_R$ close to the convex hull of Ω and Ω'. To begin with, we need to generalize Theorem 4.1 in [44] and Proposition 5.2 in [6] to the elastic inhomogeneous medium scatterer case.

Lemma 13.2.1. *Fix $t \in (1,2)$ and $\mathscr{T} > 0$. Let $\mathbf{u} \in H^2_{loc}(\mathbb{R}^n)^n$ be a solution to the Navier equation (13.1.2) in $\mathbb{R}^n\backslash\overline{B_R}$ with $\rho = 1$, which satisfies the Kupradze radiation condition; see Definition 13.1.2 and Remark 13.1.1. Let $\mathbf{U}$ be the far-field pattern of $\mathbf{u}$ with the norm $\varepsilon = \|\mathbf{U}\|_{L^2(\mathbb{S}^{n-1},\mathbb{C}^n\times\mathbb{C}^n)}$. Assume that $\|\mathbf{u}_\alpha\|_{L^2(B_{2R}\backslash B_R)^n} \leq \mathscr{T}, \alpha = \mathrm{p,s}$, where $\mathbf{u}_\mathrm{p}$ and $\mathbf{u}_\mathrm{s}$ are the longitudinal and the transversal parts of $\mathbf{u}$, respectively. Then there exists a constant $C = C(\kappa_\mathrm{p},\kappa_\mathrm{s},\mathscr{T},n,R,t) > 0$ such that the following stability estimate holds for $\varepsilon \leq \frac{\mathscr{T}}{C}$:*

$$\|\mathbf{u}\|_{L^2(B_{2tR}\backslash B_{tR})^n} \leq Ce^{-\hat{c}\sqrt{\ln\frac{\mathscr{T}}{\varepsilon}}}, \tag{13.2.1}$$

where $\kappa = \min\{\kappa_\mathrm{p},\kappa_\mathrm{s}\}$ and $\hat{c} \leq \ln t\sqrt{\frac{e\kappa R}{2}}$.

Proof. Let u^j and u^j_α denote the j-th components of $\mathbf{u}$ and $\mathbf{u}_\alpha$, respectively, where $\alpha = \mathrm{p,s}$. From (13.1.3) and (13.1.4), we obtain that u^j_α satisfies the Helmholtz equation

$$(\Delta + \kappa_\alpha^2)u^j_\alpha = 0, \quad j = 1,2,\ldots,n.$$

Then by [6, Proposition 5.2], we know that if $\varepsilon \leq \frac{\mathscr{T}}{C_\alpha}$, then

$$\|u^j_\alpha\|_{L^2(B_{2tR}\backslash B_{tR})} \leq C_\alpha\mathscr{T}e^{-\hat{c}\sqrt{\ln\frac{\mathscr{T}}{\varepsilon}}},$$

where $C_\alpha = C_\alpha(\kappa_\alpha,n,R,t) > 0$, $\alpha = \mathrm{p,s}$, $\hat{c} \leq \ln t\sqrt{\frac{e\kappa R}{2}}$ and $\kappa = \min\{\kappa_\mathrm{p},\kappa_\mathrm{s}\}$.

When $\varepsilon \leq \frac{\mathscr{T}}{\sqrt{2n\,(C_\mathrm{p}^2+C_\mathrm{s}^2)}}$, we can derive

$$\|\mathbf{u}\|_{L^2(B_{2tR}\backslash B_{tR})^n} = \sqrt{\sum_{j=1}^{n} \|u^j\|^2_{L^2(B_{2tR}\backslash B_{tR})}} \leq \sqrt{\sum_{\alpha=\mathrm{p,s}} \sum_{j=1}^{n} \|u^j_\alpha\|^2_{L^2(B_{2tR}\backslash B_{tR})}}$$
$$\leq \sqrt{2n\,(C_{\mathrm{p}}^2 + C_{\mathrm{s}}^2)}\, e^{-\hat{c}\sqrt{\ln\frac{\mathcal{T}}{\varepsilon}}}.$$

The proof is complete. □

With the help of elliptic interior regularity, we can derive the following stability estimate of the near fields in $B_{2R}\backslash\overline{B_R}$.

Proposition 13.2.1. *Fix $t \in (1,2)$ and $\mathcal{T} > 0$. Assume that $\mathbf{u} \in H^2_{loc}(\mathbb{R}^n)^n$ satisfies the Navier equation $(\Delta^* + \omega^2)\mathbf{u} = \mathbf{0}$ in $\mathbb{R}^n\backslash\overline{B_R}$ and the Kupradze radiation condition; see Definition 13.1.2 and Remark 13.1.1. Let $\mathbf{U}$ be the far-field pattern of $\mathbf{u}$ with the norm $\varepsilon = \|\mathbf{U}\|_{L^2(\mathbb{S}^{n-1},\mathbb{C}^n\times\mathbb{C}^n)}$. Assume that $\|\mathbf{u}_\alpha\|_{L^2(B_{2R}\backslash B_R)^n} \leq \mathcal{T}, \alpha = \mathrm{p,s}$, where $\mathbf{u}_{\mathrm{p}}$ and $\mathbf{u}_{\mathrm{s}}$ are the longitudinal and the transversal parts of $\mathbf{u}$, respectively. Let $K \subset B_{2R}\backslash\overline{B_{tR}}$ be a domain. Then there exists a positive constant $C = C(\kappa_{\mathrm{p}}, \kappa_{\mathrm{s}}, n, R, r, t)$ such that the following stability estimate holds for $\varepsilon \leq \frac{\mathcal{T}}{C}$:*

$$\|\mathbf{u}\|_{H^r(K)^n} \leq C\mathcal{T} e^{-\hat{c}\sqrt{\ln\frac{\mathcal{T}}{\varepsilon}}}, \tag{13.2.2}$$

where $r \in \mathbb{N}$ is an order of the Sobolev space, $\kappa = \min\{\kappa_{\mathrm{p}}, \kappa_{\mathrm{s}}\}$, and $\hat{c} \leq \ln t\sqrt{\frac{e\kappa R}{2}}$.

Proof. Note that $\mathbf{u}$ is the superposition of $\mathbf{u}_{\mathrm{p}}$ and $\mathbf{u}_{\mathrm{s}}$, which are vector-valued weak solutions to the Helmholtz equation with wavenumbers κ_{p} and κ_{s}, respectively. We use u^j_α to denote the j-th component of $\mathbf{u}_\alpha$, where $\alpha = \mathrm{p,s}$. Assume that two domains $\tilde{K}$ and K' with $\tilde{K} \subset K' \subset B_{2R}\backslash\overline{B_{tR}}$ and $\mathrm{dist}(\partial\tilde{K}, \partial K') > 0$. For any $s \in \mathbb{R}$ and the smooth cutoff function $\tilde{\phi} \in C_0^\infty(K)$ with $\tilde{\phi} \equiv 1$ in $\tilde{K}$, it is not difficult to prove the two subsequent properties:

$$\|u^j_\alpha\|_{H^{s+2}(\mathbb{R}^n)} = \|(1+\kappa_\alpha^2)u^j_\alpha - (\Delta + \kappa_\alpha^2)u^j_\alpha\|_{H^s(\mathbb{R}^n)}$$

and

$$\|(\Delta + \kappa_\alpha^2)(\tilde{\phi}\, u^j_\alpha)\|_{H^{s+2}(\mathbb{R}^n)} = \|2\nabla\tilde{\phi}\cdot\nabla u^j_\alpha + u^j_\alpha\Delta\tilde{\phi}\|_{H^s(\mathbb{R}^n)} \leq C_\alpha\|u^j_\alpha\|_{H^{s+1}(\Omega)},$$

where C_α depends on $\tilde{\phi}$. Thus, we obtain

$$\|\mathbf{u}\|^2_{H^{s+2}(\tilde{K})^n} = \|\mathbf{u}_{\mathrm{p}} + \mathbf{u}_{\mathrm{s}}\|^2_{H^{s+2}(\tilde{K})^n} \leq \sum_{\alpha=\mathrm{p,s}} \sum_{j=1}^{n} \|u^j_\alpha\|^2_{H^{s+2}(\tilde{K})}$$
$$\leq 2\sum_{\alpha=\mathrm{p,s}} \sum_{j=1}^{n} \|\tilde{\phi}\, u^j_\alpha\|^2_{H^{s+2}(\mathbb{R}^n)} \leq \sum_{\alpha=\mathrm{p,s}} \sum_{j=1}^{n} C^2_{\kappa_\alpha,\tilde{\phi}} \|u^j_\alpha\|^2_{H^s(K')}$$
$$\leq C^2_{\kappa_{\mathrm{p}},\kappa_{\mathrm{s}},n,\tilde{\phi}} \|\mathbf{u}\|^2_{H^{s+1}(K')^n},$$

which implies that

$$\|\mathbf{u}\|_{H^{s+2}(\tilde{K})^n} \leq C_{\kappa_{\mathrm{p}},\kappa_{\mathrm{s}},n,\tilde{\phi}} \|\mathbf{u}\|_{H^{s+1}(K')^n}. \tag{13.2.3}$$

By fixing $r \in \mathbb{N}$, there exists a subdomain sequence $\{K_j, \phi_j\}_{j=0}^r$ such that

$$\begin{cases} K_j \subset K_{j-1},\ \mathrm{dist}(\partial K_j, \partial K_{j-1}) > 0, \\ \phi_j \in C_0^\infty(K_{j-1}), \quad \phi_j \equiv 1 \quad \text{in } K_j, \\ K_0 = B_{2R}\backslash\overline{B_{tR}}, \quad K_r = K. \end{cases}$$

By using (13.2.3) and Lemma 13.2.1 repeatedly, we have

$$\|\mathbf{u}\|_{H^r(K)^n} \leq C_{\kappa_{\mathrm{p}},\kappa_{\mathrm{s}},\phi_r} \|\mathbf{u}\|_{H^{r-1}(K)^n} \leq \cdots \leq C_{\kappa_{\mathrm{p}},\kappa_{\mathrm{s}},\phi_r,\ldots,\phi_1} \|\mathbf{u}\|_{L^2(K)^n}$$

$$\leq C_{\kappa_p,\kappa_s,\phi_r,\ldots,\phi_1} \mathscr{T} e^{-\hat{c}\sqrt{\ln(\mathscr{T}/\varepsilon)}}.$$

The proof is complete. □

13.2.2 Stability Estimates: From Near Field to Boundary

In this subsection, we give the propagation of smallness from the near field to the boundary of Q which is the convex hull of Ω and Ω'. We are mainly interested in estimating the sum of $|\mathbf{u}-\mathbf{u}'|$ and $|\nabla\mathbf{u}-\nabla\mathbf{u}'|$ on ∂Q. To begin with, we introduce the following *three-sphere inequalities* which may be found, for instance, in [11].

Lemma 13.2.2 ([42]). *There exist positive constants $\hat{R}$, C and c, $0<c<1$, depending on κ only. Let $0<r_1<r_2<r_3<\hat{R}$ and u be a solution to $\Delta u+\kappa^2 u=0$ in B_{r_3}. For any $s\in(r_2,r_3)$, we have*

$$\|u\|_{L^\infty(B_{r_2})} \leq C\big(1-\frac{r_2}{s}\big)^{-3/2}\|u\|^{1-\beta}_{L^\infty(B_{r_3})}\|u\|^{\beta}_{L^\infty(B_{r_1})},$$

where β satisfies the following inequality:

$$\frac{c\log\frac{r_3}{s}}{\log\frac{r_3}{r_1}} \leq \beta \leq 1-\frac{c\log\frac{s}{r_1}}{\log\frac{r_3}{r_1}}.$$

From now on, we fix $r\in\mathbb{R}_+$ and set $r_1=r$, $r_2=2r$, $r_3=4r$, and $s=2\sqrt{2}r$.

Lemma 13.2.3. *Let $V\subset\mathbb{R}^n$ be a bounded connected domain and $\Gamma\subset V$ be a non-self-intersected curve with endpoints $\mathbf{x}$ and $\mathbf{x}'$. Fix positive constants $\mathscr{T}$ and $\hat{R}$ such that $4r<\hat{R}$ and $\mathscr{T}\geq 1$. Let $B(\Gamma,4r)=\bigcup_{\mathbf{y}\in\Gamma}B_{4r}(\mathbf{y})\subset V$. Moreover, we assume that $\mathbf{u}\in L^\infty(V)^n$ satisfies*

$$\begin{cases} (\Delta^*+\omega^2)\mathbf{u}=\mathbf{0}, \quad \|\mathbf{u}_\alpha\|_{L^\infty(V)^n}\leq\mathscr{T},\ \alpha=\mathrm{p},\mathrm{s}, \\ \max\big\{\|\mathbf{u}_\mathrm{p}\|_{L^\infty(B_r(\mathbf{x}))^n}, \|\mathbf{u}_\mathrm{s}\|_{L^\infty(B_r(\mathbf{x}))^n}\big\}\leq 1, \end{cases}$$

where $\mathbf{u}_\mathrm{p}$ and $\mathbf{u}_\mathrm{s}$ represent the longitudinal and the transversal parts of $\mathbf{u}$, respectively. Then the following result holds:

$$\|\mathbf{u}_\alpha\|_{L^\infty(B_r(\mathbf{x}'))^n} \leq C\mathscr{T}\|\mathbf{u}_\alpha\|^{\beta^{\frac{d_\Gamma}{r}+1}}_{L^\infty(B_r(\mathbf{y}))^n}, \quad \alpha=\mathrm{p},\mathrm{s},$$

where d_Γ is the distance measured along Γ.

Proof. Let $N=\lceil d_\Gamma/r\rceil$. There exists a sequence of balls, each of radius r and centered, respectively, at $\mathbf{x}=\mathbf{x}_1,\mathbf{x}_2,\cdots,\mathbf{x}_{N+1}=\mathbf{x}'$, such that $\mathbf{x}_k\in\Gamma$ and $|\mathbf{x}_{k+1}-\mathbf{x}_k|\leq d_\Gamma(\mathbf{x}_{k+1},\mathbf{x}_k)\leq r$, where $d_\Gamma(\mathbf{x}_{k+1},\mathbf{x}_k)$ signifies the length along the curve Γ between the points $\mathbf{x}_k$ and $\mathbf{x}_{k+1}$. Clearly, $B_r(\mathbf{x}_k)\subset B_{2r}(\mathbf{x}_{k-1})$. Let u^j_α denote the j-th component of $\mathbf{u}_\alpha$, $\alpha=\mathrm{p},\mathrm{s}$. Then we have

$$\begin{aligned} \|\mathbf{u}_\alpha\|_{L^\infty(B_r(\mathbf{x}'))^n} &\leq \sum_{j=1}^n \|u^j_\alpha\|_{L^\infty(B_r(\mathbf{x}'))} \leq \sum_{j=1}^n \Big\{C_1\mathscr{T}^{1-\beta}\|u^j_\alpha\|_{L^\infty(B_r(\mathbf{x}_N))}\Big\} \\ &\leq \ldots \\ &\leq \sum_{j=1}^n \Big\{C_N\mathscr{T}^{(1-\beta)(1+\beta+\cdots+\beta^{N-1})}\|u^j_\alpha\|^{\beta^N}_{L^\infty(B_r(\mathbf{x}))}\Big\}. \end{aligned}$$

Note that $1+\beta+\cdots+\beta^N\leq 1/(1-\beta)$ and $N\leq d_\Gamma/r+1$; hence, we have the following claim:

$$\|\mathbf{u}_\alpha\|_{L^\infty(B_r(\mathbf{x}'))^n} \le C\mathscr{T}\|\mathbf{u}_\alpha\|^{\beta^N}_{L^\infty(B_r(\mathbf{x}))^n} \le C\mathscr{T}\|\mathbf{u}_\alpha\|^{\beta^{\frac{d_\gamma}{r}+1}}_{L^\infty(B_r(\mathbf{x}))^n}, \quad \alpha = \mathrm{p},\mathrm{s}.$$

The proof is complete. □

We now give the stability estimations of $\mathbf{u}_\mathrm{p}$ and $\mathbf{u}_\mathrm{s}$ near the scatterer $(\Omega;\rho)$.

Proposition 13.2.2. *For given positive constants $\mathscr{T},\hat{R}$ and $\theta,\beta \in (0,1)$ such that $\mathscr{T} \ge 1$. Let $Q \subset B_R$ be a convex polytope. Let $\mathbf{u} \in H^2_{loc}(B_{2R})^n$ satisfy the Navier equation $\Delta^*\mathbf{u} + \omega^2\mathbf{u} = \mathbf{0}$ in $B_{2R}\backslash\overline{Q}$. Moreover, we assume that*

$$\|\mathbf{u}_\alpha\|_{C^\theta\left(\overline{B_{3/2R}}\right)^n} \le \mathscr{T}, \quad \|\mathbf{u}_\alpha\|_{L^\infty(B_{7R/4}\backslash B_{5R/4})^n} \le \delta < 1,\ \alpha = \mathrm{p},\mathrm{s},$$

where $\mathbf{u}_\mathrm{p}$ and $\mathbf{u}_\mathrm{s}$ are the longitudinal and the transversal parts of $\mathbf{u}$, respectively. Let $r = \frac{9R|\ln\beta|}{4(1-\theta)\ln|\ln\delta|}$. If

$$\delta < \left[\exp\exp\left\{\frac{4r\ln|\ln\beta|}{\min\{\hat{R},R/2\}}\right\}\right]^{-1}, \tag{13.2.4}$$

then

$$\|\mathbf{u}_\alpha\|_{L^\infty(Q')^n} \le M(\ln|\ln\delta|)^{-\theta}\mathscr{T}, \quad \alpha = \mathrm{p},\mathrm{s}. \tag{13.2.5}$$

Here, $Q' = \left\{\mathbf{x} \in B_{3/2R}\,\middle|\,\mathrm{dist}(\mathbf{x},\partial Q) \le 4r\right\}$ and $M = M(\omega,R,\theta,\beta) > 1$.

Proof. Let $Q_1 = \left\{\mathbf{x} \in B_{2R}\,\middle|\,\mathrm{dist}(\mathbf{x},\partial Q) < 4r\right\}$. It is easy to reduce that $4r < \hat{R}$ and $2r < \frac{R}{4}$ from the upper bound on δ and the expression of r. For any point $\mathbf{x}' \in B_{2R}\backslash\overline{Q_1}$, there always exists a ray from $\mathbf{x}'$ into $B_{2R}\backslash B_{5/4R}$. We consider a special segment of the ray with endpoints $\mathbf{x}'$ and $\mathbf{y}$, where $\mathbf{y} \in B_{7/4R}\backslash B_{5/4R}$ and $\mathrm{dist}(\mathbf{y},\partial B_{5/4R}) = r$. One can see that $\|\mathbf{u}_\alpha\|_{L^\infty(B_r(\mathbf{y}))^n} \le \delta, \alpha = \mathrm{p},\mathrm{s}$. The length of that segment is smaller than $\frac{5R}{2} + r$. By Lemma 13.2.3, we have

$$\|\mathbf{u}_\alpha\|_{L^\infty(B_r(\mathbf{x}'))^n} \le C\mathscr{T}\|\mathbf{u}_\alpha\|^{\beta^{\frac{5R}{2r}+2}}_{L^\infty(B_r(\mathbf{y}))^n} \le C\mathscr{T}\delta^{\beta^{\frac{5R}{2r}+2}}, \quad \alpha = \mathrm{p},\mathrm{s}.$$

Then

$$\|\mathbf{u}_\alpha\|_{L^\infty(B_{2R}\backslash Q_1)^n} \le C'\mathscr{T}\delta^{\beta^{\frac{5R}{2r}+2}}, \quad \alpha = \mathrm{p},\mathrm{s}.$$

For any point $\mathbf{x}' \in Q_1$, there must be $\mathbf{y} \in \partial Q$ such that $|\mathbf{x}' - \mathbf{y}| \le 4r$. By the convexity of Q, there exists $\mathbf{x} \in \mathbb{R}^n\backslash Q$ such that $\mathrm{dist}(\mathbf{x},Q) = |\mathbf{x}-\mathbf{y}| = 4r$. The upper bound of δ implies $4r \le \frac{R}{2}$, and thus,

$$|\mathbf{x}| \le |\mathbf{x}-\mathbf{y}| + |\mathbf{y}| \le 4r + R \le \frac{3R}{2}.$$

Moreover, $|\mathbf{x}'-\mathbf{x}| \le |\mathbf{x}'-\mathbf{y}| + |\mathbf{y}-\mathbf{x}| \le 8r$. From the θ-Hölder continuity of $\mathbf{u}_\alpha$ and proposition 5.7 in [6], we can obtain that

$$\begin{aligned}\|\mathbf{u}_\alpha\|_{L^\infty(Q_1)^n} &\le \sum_{j=1}^n \|u^j_\alpha\|_{L^\infty(Q_1)} \le \sum_{j=1}^n \left(\|u^j_\alpha\|_{\mathbf{C}^\theta(B_{3/2R})}|\mathbf{x}-\mathbf{x}'|^\theta + \|u^j_\alpha\|_{L^\infty(Q_1)}\right)\\ &\le \sum_{j=1}^n C\mathscr{T}\left((8r)^\theta + \delta^{\beta^{\frac{5R}{2r}+2}}\right) \le \widetilde{C}\mathscr{T}\left((8r)^\theta + \delta^{\beta^{\frac{5R}{2r}+2}}\right),\end{aligned} \tag{13.2.6}$$

where $\widetilde{C}$ is a positive constant depending on κ_α and n.

Combining (13.2.4) with $|\ln\delta| > 1$, we have

$$r^\theta = \frac{(9R|\ln\beta|)^\theta}{((4-4\theta)\ln|\ln\delta|)^\theta}, \qquad \frac{5R}{2r} = \frac{10(1-\theta)\ln|\ln\delta|}{9|\ln\beta|}, \tag{13.2.7}$$

and

$$\delta^{\beta^{\frac{5R}{2r}+2}} = e^{-|\ln\delta|\beta^{\frac{5R}{2r}+2}} = e^{-\beta^2|\ln\delta|^{\frac{10\theta-1}{9}}} \leq e^{-\beta^2|\ln\delta|^\theta} \leq \beta^{-2}(|\ln\delta|)^{-\theta}. \tag{13.2.8}$$

Plugging (13.2.7) and (13.2.8) into (13.2.6), it yields that

$$\|\mathbf{u}_\alpha\|_{L^\infty(Q')^n} \leq \|\mathbf{u}_\alpha\|_{L^\infty(Q_1)^n} \leq M(\ln|\ln\delta|)^{-\theta}\mathscr{T}, \quad \alpha = \mathrm{p}, \mathrm{s},$$

where $M = \widetilde{C}\left[2^\theta R^\theta|\ln\beta|^\theta(1-\theta)^{-\theta} + \beta^{-2}\right]$.
The proof is complete. □

The next proposition states the propagation of smallness from the near field to the boundary of Q, which is the convex hull of Ω and Ω'. Firstly, we introduce the following lemma to derive Proposition 13.2.3.

Lemma 13.2.4. *Let $n = 2,3$ and $\rho \in L^\infty(B_{2R})$ be supported in the ball B_R with $R > 0$. Let $\mathbf{w} \in H^2(B_{2R})$ be the solution to (13.1.2) in B_{2R}. Then, $\mathbf{w}$ belongs to $C^{1,\frac{1}{2}}(\overline{B}_{3R/2})$ and has the following estimate:*

$$\|\mathbf{w}\|_{C^{1,\frac{1}{2}}(\overline{B}_{3R/2})^n} \leq C(1+\|\rho\|_{L^\infty(R^n)})\|\mathbf{w}\|_{H^2(B_{2R})^n}, \tag{13.2.9}$$

where C is a positive constant relying on R, n, $\kappa_{\mathrm{s}}, \kappa_{\mathrm{p}}, , \lambda, \mu$.

Proof. By virtue of the Helmholtz decomposition (13.1.3), each component of $\mathbf{w}_{\mathrm{p}}$ and $\mathbf{w}_{\mathrm{s}}$ satisfies the corresponding scalar Helmholtz equation associated with the shear and compressional wavenumbers, respectively, where $\mathbf{w}_{\mathrm{s}}$ and $\mathbf{w}_{\mathrm{p}}$ are the shear and compressional parts of $\mathbf{w}$, respectively. Therefore, using a similar argument to that in the proof of [6, Lemma 5.9], we can prove this lemma. □

Proposition 13.2.3. *Fix $n = 2,3$ and the a priori parameters $\mathscr{M}, \mathscr{P}$, and $\mathscr{N}$ greater than 1. Let $(\Omega;\rho)$ and $(\Omega';\rho')$ be two medium scatterers associated with two admissible density functions ρ and ρ', where $\max\{\|\rho\|_{L^\infty(R^n)}, \|\rho'\|_{L^\infty(R^n)}\} \leq \mathscr{M}$. Let $\mathbf{u}^i$ be an incident wave with the form (13.1.1) and $\|\mathbf{u}^i\|_{H^2(B_{2R})^n} \leq \mathscr{P}$. Assume that $\mathbf{U}$ and $\mathbf{U}'$ are the far-field patterns of the scattered waves $\mathbf{u}^s$ and $\mathbf{u}'^s$ by the medium scatterers $(\Omega;\rho)$ and $(\Omega';\rho')$, respectively, where $\mathbf{u}^s, \mathbf{u}'^s$ can be bounded by $\mathscr{N} > 1$ in $H^2(B_{2R})^n$. Let ε be a sufficiently small constant. If*

$$\|\mathbf{U}-\mathbf{U}'\|_{L^2(\mathbb{S}^{n-1},\mathbb{C}^n\times\mathbb{C}^n)} \leq \varepsilon,$$

then $\mathbf{u}-\mathbf{u}'$ and $\nabla\mathbf{u}-\nabla\mathbf{u}'$ are continuous in B_R, and moreover, we have

$$\sup_{\partial Q}\left(|\mathbf{u}-\mathbf{u}'| + |\nabla\mathbf{u}-\nabla\mathbf{u}'|\right) \leq C(\ln\ln(\mathscr{N}/\varepsilon))^{-1/2}, \tag{13.2.10}$$

where Q is the convex hull of Ω and Ω' and $C = C(\kappa_{\mathrm{p}}, \kappa_{\mathrm{s}}, n, R, \mathscr{P}, \mathscr{M}, \mathscr{N}) > 0$.

Proof. Denote $\varepsilon_0 = \|\mathbf{U}-\mathbf{U}'\|_{L^2(\mathbb{S}^{n-1},\mathbb{C}^n\times\mathbb{C}^n)}$ and $\mathbf{v} = \mathbf{u}-\mathbf{u}' = \mathbf{u}^s - \mathbf{u}'^s$. One can easily see that $\|\mathbf{v}\|_{H^2(B_{2R})} \leq 2\mathscr{P}$. Let $\Omega = B_{7/4R}\backslash B_{5/4R}$. From Proposition 13.2.1 and the Sobolev embedding $H^2(\Omega) \to L^\infty(\Omega)$, there exists a positive constant $C_1 = C_1(\kappa_{\mathrm{p}}, \kappa_{\mathrm{s}}, n, R)$ such that the following estimates hold for sufficiently small ε:

$$\|\mathbf{v}\|_{L^\infty(\Omega)^n} \leq \|\mathbf{v}\|_{H^2(\Omega)^n} \leq C_1\mathscr{N}e^{-c_0\sqrt{\ln\frac{\mathscr{N}}{\varepsilon_0}}} \leq C_1\mathscr{N}e^{-c_0\sqrt{\ln\frac{\mathscr{N}}{\varepsilon}}},$$

$$\|\nabla \mathbf{v}\|_{L^\infty(\Omega)^n} \le \|\mathbf{v}\|_{H^2(\Omega)^n} \le C_1 \mathscr{N} e^{-c_0\sqrt{\ln \frac{\mathscr{N}}{\varepsilon_0}}} \le C_1 \mathscr{N} e^{-c_0\sqrt{\ln \frac{\mathscr{N}}{\varepsilon}}},$$

where $\kappa = \min\{\kappa_p, \kappa_s\}$ and $c_0 = \ln(\frac{5}{4})\sqrt{\frac{e\kappa R}{2}}$. According to Lemma 13.2.4, $\|\rho\|_{L^\infty}(R^n)$, $\|\rho'\|_{L^\infty}(R^n) \le \mathscr{M}$, $\|\mathbf{u}^i\|_{H^2(B_{2R})^n} \le \mathscr{P}$, $\|\mathbf{u}^s\|_{H^2(B_{2R})^n} \le \mathscr{N}$, and $\|\mathbf{u}'^s\|_{H^2(B_{2R})^n} \le \mathscr{N}$, we obtain

$$\|\mathbf{u}_\alpha\|_{C^{1,\frac{1}{2}}(\overline{B_{3/2R}})^n} \le C_\alpha(1+\mathscr{M})(\mathscr{P}+\mathscr{N}),\ \alpha = \mathrm{p}, \mathrm{s},$$

$$\|\mathbf{u}'_\alpha\|_{C^{1,\frac{1}{2}}(\overline{B_{3/2R}})^n} \le C_\alpha(1+\mathscr{M})(\mathscr{P}+\mathscr{N}),\ \alpha = \mathrm{p}, \mathrm{s},$$

where $(\mathbf{u}_s, \mathbf{u}_p)$ and $(\mathbf{u}'_s, \mathbf{u}'_p)$ are transverse and longitudinal wave pairs of $\mathbf{u}$ and $\mathbf{u}'$, respectively. Thus, $\mathbf{v}$ and $\partial_j \mathbf{v} \in C^{\frac{1}{2}}(\overline{B_{3/2R}})^n$, $j = 1, 2, \ldots, n$. This shows that $\mathbf{u} - \mathbf{u}'$ and $\nabla\mathbf{u} - \nabla\mathbf{u}'$ are continuous in B_R.

Fix $\beta \in (0,1)$ and denote $Q' = \{\mathbf{x} \in B_{3/2R} \mid \mathrm{dist}(\mathbf{x}, \partial Q) \le 4 \cdot \frac{9R|\ln\beta|}{2\ln|\ln\delta|}\}$. We choose $\delta(\varepsilon) = C_2 \mathscr{N} e^{-c_0\sqrt{\ln(\mathscr{N}/\varepsilon)}}$ satisfying (13.2.4). According to Proposition 13.2.2, we know that

$$\|\mathbf{v}_\alpha\|_{L^\infty(Q')^n} \le C_\alpha(1+\mathscr{M})(2\mathscr{P}+\mathscr{N})(\ln|\ln\delta|)^{-\frac{1}{2}}, \quad \alpha = \mathrm{p}, \mathrm{s}.$$

Hence, one can deduce that

$$\begin{aligned}
\|\mathbf{v}\|_{L^\infty(Q')^n} &\le \max\{C_p, C_s\}(1+\mathscr{M})(2\mathscr{P}+\mathscr{N})(\ln\ln|\delta|)^{-\frac{1}{2}} \\
&\le \max\{C_p, C_s\}(1+\mathscr{M})(2\mathscr{P}+\mathscr{N})\Big[\ln\big(\frac{c_0}{2}\ln(\mathscr{N}/\varepsilon)\big)\Big]^{-\frac{1}{2}} \\
&\le \max\{C_p, C_s\}(1+\mathscr{M})(2\mathscr{P}+\mathscr{N})\big(\ln\ln(\mathscr{N}/\varepsilon)^{\frac{1}{4}}\big)^{-\frac{1}{2}} \\
&\le 2\max\{C_p, C_s\}(1+\mathscr{M})(2\mathscr{P}+\mathscr{N})\big(\ln\ln(\mathscr{N}/\varepsilon)\big)^{-\frac{1}{2}}.
\end{aligned} \tag{13.2.11}$$

Similar to the upper bound of $\|\mathbf{v}\|_{L^\infty(Q')^n}$ in (13.2.11), we can obtain the upper bound of $\partial_j \mathbf{v}$ by following a similar argument. Hence, we can prove (13.2.10).

The proof is complete. □

13.3 Micro-local Analysis of Corner Scattering

The present section is devoted to analyzing the quantitative behaviors of the scattered field locally around a corner. We first present two auxiliary lemmas.

Lemma 13.3.1 ([6, Lemma 8.2]). *Fix $n = 2$. Let Q be the convex hull of Ω and Ω', where Ω and Ω' are two open bounded convex polygons. If $\mathbf{x}_0$ is a vertex of Ω such that $\mathrm{dist}(\mathbf{x}_0, \Omega') = \mathrm{dist}_{\mathscr{H}}(\Omega, \Omega')$, then $\mathbf{x}_0$ is also a vertex of Q. Let σ denote the angle of K at $\mathbf{x}_0$; then the angle σ' of Q at $\mathbf{x}_0$ satisfies the inequality $\sigma < \sigma' \le (\sigma+\pi)/2 < \pi$.*

Lemma 13.3.2 ([6, Lemma 8.3]). *Fix $n = 3$. Let Q be the convex hull of Ω and Ω', where Ω and Ω' are open convex polyhedral cones. If $\mathbf{x}_0$ is a vertex of Ω such that $\mathrm{dist}(\mathbf{x}_0, \Omega') = \mathrm{dist}_{\mathscr{H}}(\Omega, \Omega')$, then $\mathbf{x}_0$ is also a vertex of Q. And Q can fit inside such an open spherical cone $\mathscr{A}$ that $\mathbf{x}_0$ is the vertex of $\mathscr{A}$ whose open angle σ at the vertex $\mathbf{x}_0$ is smaller than π, where σ does not depend on Ω, Ω' and their locations.*

Next, we provide an integral identity which follows from Betti's second formula.

Proposition 13.3.1. *Let $\omega > 0$ and $\rho(\mathbf{x}) \in C^\theta(S)$, where $S \subset \mathbb{R}^n$ is a bounded Lipschitz domain and $\theta \in (0,1)$. If $\mathbf{u}, \mathbf{u}'$ and $\mathbf{u}_0 \in H^2(S)^n$ satisfy the Navier equations*

$$\begin{cases} \Delta^* \mathbf{u} + \rho\,\omega^2 \mathbf{u} = \mathbf{0}, \\ \Delta^* \mathbf{u}' + \ \omega^2 \mathbf{u}' = \mathbf{0}, \\ \Delta^* \mathbf{u}_0 + \omega^2 \mathbf{u}_0 = \mathbf{0} \end{cases}$$

in S, then

$$\omega^2 \int_S (\rho - 1)\,\mathbf{u}\cdot\mathbf{u}_0\,\mathrm{d}\mathbf{x} = \int_{\partial S} \mathbf{u}_0 \cdot T_{\hat{\mathbf{v}}}(\mathbf{u}^i - \mathbf{u}) - (\mathbf{u}^i - \mathbf{u})\cdot T_{\hat{\mathbf{v}}}(\mathbf{u}_0)\,\mathrm{d}\sigma. \tag{13.3.1}$$

Here, $\hat{\mathbf{v}}$ denotes the outward unit normal to the boundary of S, and the conormal derivative $T_{\hat{\mathbf{v}}}(\mathbf{u})$ with u^j being the j-th component of $\mathbf{u}$ is defined by

$$T_{\hat{\mathbf{v}}}(\mathbf{u}) = \begin{cases} 2\mu\,\partial_{\hat{\mathbf{v}}}\mathbf{u} + \lambda\,\hat{\mathbf{v}}(\nabla\cdot\mathbf{u}) + \mu\,(\partial_2 u^1 - \partial_1 u^2)\hat{\mathbf{v}}^{\perp}, & n = 2, \\ 2\mu\,\partial_{\hat{\mathbf{v}}}\mathbf{u} + \lambda\,\hat{\mathbf{v}}\,(\nabla\cdot\mathbf{u}) + \mu\,\hat{\mathbf{v}}\times(\nabla\times\mathbf{u}), & n = 3. \end{cases} \tag{13.3.2}$$

The following complex geometric optics (CGO) solution $\mathbf{u}_0$ is introduced in [23]:

$$\mathbf{u}_0(\mathbf{x}) = e^{\xi\cdot(\mathbf{x}-\mathbf{x}_0)}\eta, \tag{13.3.3}$$

where

$$\begin{aligned} &\xi = \tau\mathbf{p} + \mathrm{i}\sqrt{\kappa_s^2 + \tau^2}\,\mathbf{p}^{\perp},\ \eta = \mathbf{p}^{\perp} - \mathrm{i}\sqrt{1 + \kappa_s^2/\tau^2}\,\mathbf{p}, \\ &\tau > \kappa_s, \quad \mathbf{p}\cdot\mathbf{p}^{\perp} = 0, \quad \mathbf{p}^{\perp}, \mathbf{p} \in \mathbb{S}^{n-1}. \end{aligned} \tag{13.3.4}$$

It is directly verified that

$$\Delta^* \mathbf{u}_0 + \omega^2 \mathbf{u}_0 = \mathbf{0} \ \text{ in } \ \mathbb{R}^n. \tag{13.3.5}$$

Note that

$$|\xi| = \sqrt{2\tau^2 + \kappa_s^2} \quad \text{and} \quad |\eta| = \sqrt{2 + \kappa_s^2/\tau^2} \le \sqrt{3}. \tag{13.3.6}$$

In Proposition 13.3.2, we shall present asymptotic analysis of the volume integral of the CGO solution $\mathbf{u}_0$ over a polyhedral cone with respect to τ as $\tau \to +\infty$. Before proving that, we now introduce a special cone pair $(\mathscr{A}, \mathscr{C})$ as follows.

Definition 13.3.1. Let $n \in \{2,3\}$, $0 < \alpha_m < \alpha_M < \frac{\pi}{2}$, $\omega > 0$. Assume that $\mathscr{A}, \mathscr{C} \subset \mathbb{R}^n$ are, respectively, an open polyhedral cone and an open spherical cone. We say $(\mathscr{A}, \mathscr{C}) \in \mathscr{F}(\alpha_m, \alpha_M, n)$ if the following conditions are fulfilled:

(a) $\mathscr{A} \subset \mathbb{R}^n$ is a convex polyhedral cone.
(b) $\mathscr{A}$ and $\mathscr{C}$ both take $\mathbf{x}_0 \in \mathbb{R}^n$ as the apex and $\mathscr{A} \subset \mathscr{C}$.
(c) The opening angle of $\mathscr{C}$ at the apex $\mathbf{x}_0$ is no more than π.
(d) The opening angle of $\mathscr{A} \subset \mathbb{R}^2$ at the apex $\mathbf{x}_0$ is in $(2\alpha_m, 2\alpha_M)$.

Proposition 13.3.2. *Fix $n \in \{2,3\}$, $0 < 2\alpha_m < 2\alpha_M < \pi$, $\omega > 0$. Let $\mathbf{u}$ be the solution to* (13.1.2) *with $\Omega = \mathrm{supp}(\rho - 1)$. In addition, Ω takes $\mathbf{x}_0$ as its apex, and $\mathbf{u}(\mathbf{x}_0)$ is a nontrivial constant complex vector. Let $\mathscr{A}$ and $\mathscr{C}$ be an open polyhedral cone and an open spherical cone, respectively; they both take $\mathbf{x}_0$ as their vertex. If $(\mathscr{A}, \mathscr{C}) \in \mathscr{F}(\alpha_m, \alpha_M, n)$, then there exists a vector ξ defined by* (13.3.4) *such that*

$$\mathbf{p}\cdot(\mathbf{x} - \mathbf{x}_0) \le -\delta_0|\mathbf{x} - \mathbf{x}_0| \ \textit{for any}\ \mathbf{x} \in \mathscr{A}, \tag{13.3.7}$$

where δ_0 is a positive constant only depending on $\mathbf{p}$, Ω, and $\mathscr{A}$. Moreover, we have

$$\left|\int_{\mathscr{A}} e^{\xi\cdot(\mathbf{x}-\mathbf{x}_0)}\mathrm{d}x\right| \ge C\tau^{-n}, \tag{13.3.8}$$

and

$$\left|\mathbf{u}(\mathbf{x}_0)\cdot\eta\right| \geq C_0 > 0, \tag{13.3.9}$$

where $\tau = \left|\Re\xi\right|$, η *is defined in* (13.3.4), $\mathbf{p}^{\perp} \in \mathbb{S}^{n-1}$ *satisfies* $\mathbf{p}\cdot\mathbf{p}^{\perp} = 0$, *and* C_0 *is a positive constant only depending on* ζ_0 *and* $\mathscr{A}$.

Proof. The above conclusions can be obtained directly from [6, Lemma 6.3] and [8, Lemma 2.2] except that we need to prove the inequality (13.3.9). Note that

$$\begin{aligned}\mathbf{u}(\mathbf{x}_0)\cdot\eta &= \mathrm{i}\big(\Im\mathbf{u}(\mathbf{x}_0)\cdot\mathbf{p}^{\perp} - \sqrt{1+\kappa_s^2/\tau^2}\,\Re\mathbf{u}(\mathbf{x}_0)\cdot\mathbf{p}\big)\\ &\quad + \Re\mathbf{u}(\mathbf{x}_0)\cdot\mathbf{p}^{\perp} + \sqrt{1+\kappa_s^2/\tau^2}\,\Im\mathbf{u}(\mathbf{x}_0)\cdot\mathbf{p},\end{aligned} \tag{13.3.10}$$

where $\Re\mathbf{u}(\mathbf{x}_0)$ and $\Im\mathbf{u}(\mathbf{x}_0)$ stand for the real and imaginary parts of $\mathbf{u}(\mathbf{x}_0)$, respectively. Since $\mathbf{u}(\mathbf{x}_0)$ is a nontrivial constant complex vector, without loss of generality, we assume that $\Im\mathbf{u}(\mathbf{x}_0) \neq \mathbf{0}$. For simplicity, we only consider three-dimensional scenarios, and the two-dimensional conclusion can be similarly proved. In fact, once $\mathbf{p}$ is fixed, $\mathbf{p}^{\perp}$ belongs to Π, where Π represents the plane defined by $\Pi = \{\mathbf{x}\in\mathbb{R}^3 \mid \mathbf{x}\cdot\mathbf{p} = 0\}$. We can easily prove (13.3.9) by virtue of choosing a special unit vector $\mathbf{p}^{\perp}$ as follows:

(I) Case 1: $\Im\mathbf{u}(\mathbf{x}_0)\cdot\mathbf{p} < 0$. In order to prove (13.3.9), we consider the lower bound for the real part of $\mathbf{u}(\mathbf{x}_0)\cdot\eta$.
Let us distinguish two separate situations. If $\Re\mathbf{u}(\mathbf{x}_0) \in \Pi$ (including $\Re\mathbf{u}(\mathbf{x}_0)//\Pi$), we can take $\mathbf{p}^{\perp} \in \Pi$ such that $\angle(\mathbf{p}^{\perp}, \Re\mathbf{u}(\mathbf{x}_0)) \in (\pi/2, \pi)$, where $\angle(\mathbf{p}^{\perp}, \Re\mathbf{u}(\mathbf{x}_0))$ is the angle between $\Re\mathbf{u}(\mathbf{x}_0)$ and $\mathbf{p}^{\perp}$. If $\Re\mathbf{u}(\mathbf{x}_0) \notin \Pi$, it is clear that there must exist a unit vector $\mathbf{p}^{\perp} \in \Pi$ such that $\angle(\mathbf{p}^{\perp}, \Re\mathbf{u}(\mathbf{x}_0)) \in (\pi/2, \pi)$. Hence, we always can choose the unit vector $\mathbf{p}^{\perp} \in \Pi$ fulfilling the following inequality:

$$\Re\mathbf{u}(\mathbf{x}_0)\cdot\mathbf{p}^{\perp} < 0.$$

It can be directly seen that the signs of $\Re\mathbf{u}(\mathbf{x}_0)\cdot\mathbf{p}^{\perp}$ and $\Im\mathbf{u}(\mathbf{x}_0)\cdot\mathbf{p}$ are the same. Therefore, we have

$$\begin{aligned}\left|\mathbf{u}(\mathbf{x}_0)\cdot\eta\right| &\geq \left|\Re\mathbf{u}(\mathbf{x}_0)\cdot\mathbf{p}^{\perp} + \sqrt{1+\kappa_s^2/\tau^2}\,\Im\mathbf{u}(\mathbf{x}_0)\cdot\mathbf{p}\right|\\ &\geq \left|\Re\mathbf{u}(\mathbf{x}_0)\cdot\mathbf{p}^{\perp} + \Im\mathbf{u}(\mathbf{x}_0)\cdot\mathbf{p}\right| := C_0 > 0.\end{aligned} \tag{13.3.11}$$

(II) Case 2: $\Im\mathbf{u}(\mathbf{x}_0)\cdot\mathbf{p} = 0$. In order to prove (13.3.9), we consider the lower bound for the imaginary part of $\mathbf{u}(\mathbf{x}_0)\cdot\eta$.
Let us investigate the following two situations. If $\Re\mathbf{u}(\mathbf{x}_0) \in \Pi$ (including $\Re\mathbf{u}(\mathbf{x}_0)//\Pi$), then $\Re\mathbf{u}(\mathbf{x}_0)\cdot\mathbf{p} = 0$. Thus, we can take $\mathbf{p}^{\perp} \in \Pi$ such that $\angle(\mathbf{p}^{\perp}, \Re\mathbf{u}(\mathbf{x}_0)) \neq \pi/2$. If $\Re\mathbf{u}(\mathbf{x}_0) \notin \Pi$ and $\Re\mathbf{u}(\mathbf{x}_0)\cdot\mathbf{p} > 0$, we can choose a unit vector $\mathbf{p}^{\perp}$ belonging to Π such that $\angle(\mathbf{p}^{\perp}, \Im\mathbf{u}(\mathbf{x}_0)) \in (\pi/2, \pi)$, which implies $\mathrm{sign}(\Im\mathbf{u}(\mathbf{x}_0)\cdot\mathbf{p}^{\perp}) = -\mathrm{sign}(\Re\mathbf{u}(\mathbf{x}_0)\cdot\mathbf{p})$. Here, $\mathrm{sign}(a)$ is the sign of a real number a. Therefore, we obtain

$$\begin{aligned}\left|\mathbf{u}(\mathbf{x}_0)\cdot\eta\right| &\geq \left|\Im\mathbf{u}(\mathbf{x}_0)\cdot\mathbf{p}^{\perp} - \sqrt{1+\kappa_s^2/\tau^2}\,\Re\mathbf{u}(\mathbf{x}_0)\cdot\mathbf{p}\right|\\ &\geq \left|\Im\mathbf{u}(\mathbf{x}_0)\cdot\mathbf{p}^{\perp} - \Re\mathbf{u}(\mathbf{x}_0)\cdot\mathbf{p}\right| := C_0 > 0.\end{aligned}$$

(III) Case 3: $\Im\mathbf{u}(\mathbf{x}_0)\cdot\mathbf{p} > 0$. Similar to Case 1, in order to prove (13.3.9), we can consider the lower bound for the real part of $\mathbf{u}(\mathbf{x}_0)\cdot\eta$.
Similar to the above two cases, let us consider the following two situations. If $\Re\mathbf{u}(\mathbf{x}_0) \in \Pi$ (including $\Re\mathbf{u}(\mathbf{x}_0)//\Pi$), we can take $\mathbf{p}^{\perp} \in \Pi$ such that $\angle(\mathbf{p}^{\perp}, \Re\mathbf{u}(\mathbf{x}_0)) \in [0, \pi/2)$. If $\Re\mathbf{u}(\mathbf{x}_0) \notin \Pi$, there must exist a unit vector $\mathbf{p}^{\perp} \in \Pi$ such that $\angle(\mathbf{p}^{\perp}, \Re\mathbf{u}(\mathbf{x}_0)) \in [0, \pi/2)$. Hence, for the above two situations, we can always choose a unit vector $\mathbf{p}^{\perp}$ fulfilling the following inequality as $\mathbf{p}^{\perp}$:

$$\Re\mathbf{u}(\mathbf{x}_0)\cdot\mathbf{p}^{\perp} \geq 0.$$

By the same deduction for Case 1, we can also obtain the inequality (13.3.11).

The proof is complete. □

A critical asymptotic estimate with respect to $\Re\xi$ for the volume integral associated with the CGO solution $\mathbf{u}_0$ and the total wave field $\mathbf{u}$ near a vertex of Ω is presented in the following proposition, which is the key ingredient to derive the stability result in Theorem 13.1.1.

Proposition 13.3.3. *Let* $\mathbf{u}$ *and* $\mathbf{u}'$ *be the solutions to* (13.1.2) *with* $\Omega, \Omega', \rho, \rho'$ *satisfying the assumptions in Theorem 13.1.1. Let* $\mathbf{x}_0$ *be a vertex of* Ω *such that* $\mathrm{dist}(\mathbf{x}_0, \Omega') = \mathrm{dist}_{\mathscr{H}}(\Omega, \Omega')$. *Fix* $\omega, \tau > 0$ *and* $h \in (0,1)$ *such that* $B_h(\mathbf{x}_0) \cap \Omega' = \emptyset$. *Denote* $S_h = B_h(\mathbf{x}_0) \cap \Omega$ *and* $D_h = B_h(\mathbf{x}_0) \cap Q$, *where* Q *is the convex hull of* Ω *and* Ω'. *Let* $\mathbf{u}(\mathbf{x}_0)$ *be a nontrivial constant complex vector. Assume that* $\mathscr{A}$ *and* $\mathscr{B}$ *represent the polyhedral cone generated by* K *and* Q *at the vertex* $\mathbf{x}_0$, *respectively. Let* $\rho \in C^{\theta_1}(S_h)$ *and* $\mathbf{u} \in C^{\theta_2}(S_h)^n$, $\theta_j \in (0,1)$, $j = 1,2$, *which yield that*

$$\rho(\mathbf{x}) = 1 + \rho(\mathbf{x}_0) + \rho_1(\mathbf{x}), \quad |\rho_1(\mathbf{x})| \le \mathscr{M}|\mathbf{x} - \mathbf{x}_0|^{\theta_1}, \ \theta_1 \in (0,1), \tag{13.3.12}$$

$$\mathbf{u}(\mathbf{x}) = \mathbf{u}(\mathbf{x}_0) + \mathbf{u}_1(\mathbf{x}), \qquad |\mathbf{u}_1(\mathbf{x})| \le \mathscr{R}|\mathbf{x} - \mathbf{x}_0|^{\theta_2}, \ \theta_2 \in (0,1), \tag{13.3.13}$$

where $\mathscr{M} = \|\rho_1\|_{C^{\theta_1}(\Omega)}$ *and* $\mathscr{R} = \|\mathbf{u}_1\|_{C^{\theta_2}(\Omega)}$.

Then we have the integral identity

$$\begin{aligned}(\rho(\mathbf{x}_0) - 1)\int_{\mathscr{A}} e^{\xi\cdot(\mathbf{x}-\mathbf{x}_0)}\eta \cdot \mathbf{u}(\mathbf{x}_0)\mathrm{d}x &= \frac{1}{\omega^2}\int_{\partial D_h} \mathbf{u}_0 \cdot T_{\hat{\nu}}(\mathbf{u}^i - \mathbf{u}) - (\mathbf{u}^i - \mathbf{u})T_{\hat{\nu}}(\mathbf{u}_0)\mathrm{d}\sigma \\ &+ (\rho(\mathbf{x}_0) - 1)\int_{\mathscr{A}\setminus S_h} e^{\xi\cdot(\mathbf{x}-\mathbf{x}_0)}\eta \cdot \mathbf{u}(\mathbf{x}_0)\mathrm{d}x \\ &- \int_{S_h} e^{\xi\cdot(\mathbf{x}-\mathbf{x}_0)}(\rho_1 + 1)\eta \cdot \mathbf{u}(\mathbf{x}_0)\mathrm{d}x \\ &- \int_{S_h} e^{\xi\cdot(\mathbf{x}-\mathbf{x}_0)}(\rho - 1)\eta \cdot \mathbf{u}_1\mathrm{d}x \end{aligned}\tag{13.3.14}$$

and the estimate

$$\begin{aligned}C\Big|(\rho(\mathbf{x}_0) - 1)\int_{\mathscr{A}} e^{\xi\cdot(\mathbf{x}-\mathbf{x}_0)}\eta \cdot \mathbf{u}(\mathbf{x}_0)\mathrm{d}x\Big| &\le |\Re\xi|^{-n}e^{-\delta_0|\Re\xi|h/2} + |\Re\xi|^{-n-\min\{\theta_1,\theta_2\}} \\ &+ h^{n-1}\sup_{\partial\mathscr{B}\cap B_h(\mathbf{x}_0)}\big\{|\nabla\mathbf{u} - \nabla\mathbf{u}'| + |\mathbf{u} - \mathbf{u}'|\big\} \\ &+ h^{n-1}e^{-\delta_0|\Re\xi|h}\big\{\|\mathbf{u}_{\mathrm{s}}\|_{H^2(B_{2R})^n} + \|\mathbf{u}_{\mathrm{p}}\|_{H^2(B_{2R})^n} \\ &+ \|\mathbf{u}'_{\mathrm{p}}\|_{H^2(B_{2R})^n} + \|\mathbf{u}'_{\mathrm{s}}\|_{H^2(B_{2R})^n}\big\},\end{aligned}\tag{13.3.15}$$

where δ_0 *coincides with the one in* (13.3.7) *and* C *is a positive constant depending on the a priori parameters.*

Proof. According to (13.3.3) and (13.3.5), the integral identity (13.3.14) follows directly from (13.3.1) by utilizing (13.3.12) and (13.3.13).

To estimate each term in the right-hand side of (13.3.14), let us recall the incomplete gamma functions $\gamma(\cdot,\cdot)$ and $\Gamma(\cdot,\cdot)$ (cf. [6]) which are defined by

$$\gamma : \mathbb{R}_+ \times \mathbb{R}_+ \to \mathbb{R} : (s,x) \mapsto \gamma(s,x), \quad \Gamma : \mathbb{R}_+ \times \mathbb{R}_+ \to \mathbb{R} : (s,x) \mapsto \Gamma(s,x),$$

where

$$\gamma(s,x) = \int_0^x e^{-t}t^{s-1}\mathrm{d}t, \quad \Gamma(s,x) = \int_x^\infty t^{s-1}e^{-t}\mathrm{d}t. \tag{13.3.16}$$

From the fact $e^{-t} \leq e^{-t/2}e^{-x/2}$ and a variable substitution $t' = t/2$, it is easy to get the estimation

$$\gamma(s,x) \leq \Gamma(s) \leq \lceil s-1 \rceil, \quad \Gamma(s,x) \leq 2^s\Gamma(s)e^{-x/2}, \tag{13.3.17}$$

where $\lceil s \rceil$ is the largest integer satisfying $\lceil s \rceil \leq s$ and $\Gamma(s)$ is the complete gamma function of s bounded by $\lceil s \rceil !$.

To estimate the first four terms in the right-hand side of (13.3.14), in the following, we derive some important asymptotic inequalities with respect to the parameter $|\Re\xi|$ in the CGO solution $\mathbf{u}_0$. By using the polar coordinate transformation, as well as (13.3.7), (13.3.16), and (13.3.17), we can obtain

$$\begin{aligned} \left| \int_{\mathscr{A}\setminus S_h} e^{\xi\cdot(\mathbf{x}-\mathbf{x}_0)}\mathrm{dx} \right| &\leq \left| \int_{\mathscr{A}\setminus S_h} e^{-\delta_0|\Re\xi||\mathbf{x}-\mathbf{x}_0|}\mathrm{dx} \right| \leq \sigma(\mathbb{S}^{n-1}) \int_h^{\infty} e^{-\delta_0|\Re\xi|r} r^{n-1}\mathrm{dr} \\ &\leq \sigma(\mathbb{S}^{n-1}) \int_{h\delta_0|\Re\xi|r}^{\infty} e^{-t}t^{n-1}(\delta_0|\Re\xi|)^{-n}\mathrm{dt} \\ &= \sigma(\mathbb{S}^{n-1})(\delta_0|\Re\xi|)^{-n}\Gamma(n,\delta_0|\Re\xi|h) \\ &\leq C_{\delta_0,n}|\Re\xi|^{-n}e^{-\delta_0|\Re\xi|\frac{h}{2}} \end{aligned} \tag{13.3.18}$$

as $|\Re\xi| \to +\infty$. Similarly, we have

$$\begin{aligned} \left| \int_{S_h} e^{\xi\cdot(\mathbf{x}-\mathbf{x}_0)}\mathrm{dx} \right| &\leq \int_{S_h} e^{-\delta_0|\Re\xi||\mathbf{x}-\mathbf{x}_0|}\mathrm{dx} \leq \sigma(\mathbb{S}^{n-1}) \int_0^h e^{-\delta_0|\Re\xi|r} r^{n-1}\mathrm{dr} \\ &\leq \sigma(\mathbb{S}^{n-1}) \int_0^{h\delta_0|\Re\xi|r} e^{-t}t^{n-1}(\delta_0|\Re\xi|)^{-n}\mathrm{dt} \\ &\leq \sigma(\mathbb{S}^{n-1})(\delta_0|\Re\xi|)^{-n}\gamma(n,\delta_0|\Re\xi|h) \\ &\leq C'_{\delta_0,n}|\Re\xi|^{-n} \end{aligned} \tag{13.3.19}$$

as $|\Re\xi| \to +\infty$, where $\sigma(\mathbb{S}^{n-1})$ is the measure of $\mathbb{S}^{n-1}$. For scalar functions f and g, which admit $|f| \leq A|\mathbf{x}-\mathbf{x}_0|^B$ with A and B being positive constants, and $g \in L^q(S_h)$ with $1/p+1/q = 1$. Combining the Hölder inequality, (13.3.7), and (13.3.16) with (13.3.17), we obtain

$$\begin{aligned} \left| \int_{S_h} e^{\xi\cdot(\mathbf{x}-\mathbf{x}_0)} f g\mathrm{dx} \right| &\leq A\Big(\int_{S_h} e^{\Re\xi\cdot(\mathbf{x}-\mathbf{x}_0)p}|\mathbf{x}-\mathbf{x}_0|^{Bp}\mathrm{dx} \Big)^{\frac{1}{p}} \|g\|_{L^q(S_h)} \\ &\leq A\sigma(\mathbb{S}^{n-1})^{\frac{1}{p}}(\delta_0 p|\Re\xi|)^{-B-n/p}\big(\gamma(Bp+n,\, \delta_0 p|\Re\xi|h)\big)^{\frac{1}{p}} \|g\|_{L^q(S_h)} \\ &\leq C_{A,B,\delta_0,p}|\Re\xi|^{-B-n/p}\|g\|_{L^q(S_h)} \end{aligned} \tag{13.3.20}$$

as $|\Re\xi| \to +\infty$. By similar arguments, we can derive the following estimations for vector-type functions $\mathbf{f}$ and $\mathbf{g}$ satisfying $|\mathbf{f}| \leq A|\mathbf{x}-\mathbf{x}_0|^B$ with A and B being positive constants and $\mathbf{g} \in L^q(S_h)^n$ with $1/p+1/q = 1$,

$$\begin{aligned} \left| \int_{S_h} e^{\xi\cdot(\mathbf{x}-\mathbf{x}_0)} \mathbf{f}\cdot\mathbf{g}\mathrm{dx} \right| &\leq A\Big(\int_{S_h} e^{\Re\xi\cdot(\mathbf{x}-\mathbf{x}_0)p}|\mathbf{x}-\mathbf{x}_0|^{Bp}\mathrm{dx} \Big)^{\frac{1}{p}} \|\mathbf{g}\|_{L^q(S_h)^n} \\ &\leq A\sigma(\mathbb{S}^{n-1})^{\frac{1}{p}}(\delta_0 p|\Re\xi|)^{-B-n/p}\gamma(Bp+n,\, \delta_0 p|\Re\xi|h)^{\frac{1}{p}} \|\mathbf{g}\|_{L^q(S_h)^n} \\ &\leq C'_{A,B,\delta_0,p}|\Re\xi|^{-B-n/p}\|\mathbf{g}\|_{L^q(S_h)^n} \end{aligned} \tag{13.3.21}$$

as $|\Re\xi| \to +\infty$.

By virtue of (13.3.12) and (13.3.13), from (13.3.18) to (13.3.21), we can deduce that

$$R_1 := \left| \int_{\mathscr{A}\setminus S_h} e^{\xi\cdot(\mathbf{x}-\mathbf{x}_0)}\eta\cdot\mathbf{u}(\mathbf{x}_0)\mathrm{dx} \right| \leq C_{\delta_0,n}|\mathbf{u}(\mathbf{x}_0)||\Re\xi|^{-n}e^{-\delta_0|\Re\xi|\frac{h}{2}}, \tag{13.3.22a}$$

$$R_2 := \left| \int_{S_h} e^{\xi\cdot(\mathbf{x}-\mathbf{x}_0)} \eta \cdot \mathbf{u}(\mathbf{x}_0) \mathrm{d}\mathbf{x} \right| \leq C'_{\delta_0,n} |\mathbf{u}(\mathbf{x}_0)| |\Re\xi|^{-n}, \tag{13.3.22b}$$

$$R_3 := \left| \int_{S_h} e^{\xi\cdot(\mathbf{x}-\mathbf{x}_0)} (\rho_1 + 1)\, \eta \cdot \mathbf{u}(\mathbf{x}_0) \mathrm{d}\mathbf{x} \right| \leq C_{\mathscr{R},n,\delta_0} |\mathbf{u}(\mathbf{x}_0)| \left|\Re\xi\right|^{-n-\theta_1}, \tag{13.3.22c}$$

$$R_4 := \left| \int_{S_h} e^{\xi\cdot(\mathbf{x}-\mathbf{x}_0)} (\rho - 1)\eta \cdot \mathbf{u}_1 \mathrm{d}\mathbf{x} \right| \leq C_{\mathscr{M},\mathscr{R},n,\delta_0} |\Re\xi|^{-n-\theta_2}. \tag{13.3.22d}$$

For the boundary integral term in the right-hand side of (13.3.14), we divide the integration area ∂D_h into two parts $\partial\mathscr{B}\cap B_h(\mathbf{x}_0)$ and $\mathscr{S} := \partial S_h(\mathbf{x}_0)\backslash(\partial\mathscr{B}\cap B_h(\mathbf{x}_0))$. For the boundary integral over $\partial\mathscr{B}\cap B_h(\mathbf{x}_0)$, the Cauchy-Schwarz inequality implies that

$$\begin{aligned} I_1 :=& \left| \int_{\partial\mathscr{B}\cap B_h(\mathbf{x}_0)} \mathbf{u}_0 \cdot T_{\hat{\nu}}(\mathbf{u}' - \mathbf{u}) - (\mathbf{u}' - \mathbf{u}) T_{\hat{\nu}}(\mathbf{u}_0) \mathrm{d}\sigma \right| \\ \leq& \sqrt{\sigma(\partial\mathscr{B}\cap B_h(\mathbf{x}_0))} \left(\int_{\partial\mathscr{B}\cap B_h(\mathbf{x}_0)} \left|\mathbf{u}_0 \cdot T_{\hat{\nu}}(\mathbf{u}' - \mathbf{u}) - (\mathbf{u}' - \mathbf{u}) T_{\hat{\nu}}(\mathbf{u}_0)\right|^2 \mathrm{d}\sigma \right)^{1/2} \\ \leq& \sqrt{\sigma(\partial\mathscr{B}\cap B_h(\mathbf{x}_0))} \left[\int_{\partial\mathscr{B}\cap B_h(\mathbf{x}_0)} \left(|\mathbf{u}_0| + |T_{\hat{\nu}}(\mathbf{u}_0)|\right)^2 \mathrm{d}\sigma \right]^{1/2} T_1, \end{aligned} \tag{13.3.23}$$

where $T_1 = \sup_{\partial\mathscr{B}\cap B_h(\mathbf{x}_0)} \left\{ |T_{\hat{\nu}}(\mathbf{u}' - \mathbf{u})|, |\mathbf{u} - \mathbf{u}'| \right\}$. By virtue of (13.3.2) and (13.3.3), one can directly derive that

$$T_{\hat{\nu}}(\mathbf{u}_0) = \mu e^{\xi\cdot(\mathbf{x}-\mathbf{x}_0)} \left[(\xi\cdot\hat{\nu})\eta + (\eta\cdot\hat{\nu})\xi \right] \quad \text{and} \quad \left|T_{\hat{\nu}}(\mathbf{u} - \mathbf{u}')\right| \leq C_{\lambda,\mu} |\nabla(\mathbf{u} - \mathbf{u}')|,$$

where ν is the outward unit normal vector to $\partial\mathscr{B}\cap B_h(\mathbf{x}_0)$ and $C_{\lambda,\mu}$ is a positive constant depending on λ and μ. From (13.3.23), we obtain that

$$\begin{aligned} I_1 \leq& \sqrt{\sigma(\partial\mathscr{B}\cap B_h(\mathbf{x}_0))} \left[\int_{\partial\mathscr{B}\cap B_h(\mathbf{x}_0)} \left(|e^{\xi\cdot(\mathbf{x}-\mathbf{x}_0)}| + \left|\mu e^{\xi\cdot(\mathbf{x}-\mathbf{x}_0)} (\xi\cdot\hat{\nu})\eta \right.\right.\right. \\ & \left.\left.\left. + (\eta\cdot\hat{\nu})\xi\right| \right)^2 \mathrm{d}\sigma \right]^{1/2} T_1 \\ \leq& \sqrt{\sigma(\partial\mathscr{B}\cap B_h(\mathbf{x}_0))} \left[\int_{\partial\mathscr{B}\cap B_h(\mathbf{x}_0)} \left(|\eta| + \mu|\eta(\xi\cdot\hat{\nu}) + \xi(\eta\cdot\hat{\nu})| \right)^2 \mathrm{d}\sigma \right]^{1/2} T_1 \\ \leq& C'_{\lambda,\mu}\ \sigma(\partial\mathscr{B}\cap B_h(\mathbf{x}_0)) \sup_{\partial\mathscr{B}\cap B_h(\mathbf{x}_0)} \left\{ |\nabla(\mathbf{u} - \mathbf{u}')|, |\mathbf{u} - \mathbf{u}'| \right\}, \end{aligned} \tag{13.3.24}$$

where $C'_{\lambda,\mu}$ is a positive constant depending on λ and μ. There exist finite half-spaces $\{H_j\}_{j=1}^N$ through $\mathbf{x}_0$ with a codimension of 1 such that

$$\partial\mathscr{B}\cap B_h(\mathbf{x}_0) \subset \bigcap_{j=1}^N \left(H_j\cap B_h(\mathbf{x}_0)\right).$$

Hence, we have

$$\sigma(\partial\mathscr{B}\cap B_h(\mathbf{x}_0)) \leq C_{N,n}\sigma(\partial H_j\cap B_h(\mathbf{x}_0)) \leq C_{N,n}\, h^{n-1}, \tag{13.3.25}$$

where $C_{N,n}$ is a positive constant depending only n and N. Substituting (13.3.25) into (13.3.24), one has

$$I_1 \leq Ch^{n-1} \sup_{\partial\mathscr{B}\cap B_h(\mathbf{x}_0)} \left\{ |\nabla(\mathbf{u} - \mathbf{u}')|, |\mathbf{u} - \mathbf{u}'| \right\}, \tag{13.3.26}$$

where C is an a priori positive constant depending on λ, μ, N, and n.

For the boundary integral over $\mathscr{S} = \partial S_h(\mathbf{x}_0)\backslash(\partial\mathscr{B}\cap B_h(\mathbf{x}_0))$, by the Cauchy-Schwarz inequality and the fact that $\sigma(\mathscr{B}\cap S_h(\mathbf{x}_0)) \leq Ch^{n-1}$, where C is a positive constant depending on n and π, one has

$$
\begin{aligned}
I_2 :=& \left|\int_{\mathscr{S}} \mathbf{u}_0\, T_{\hat{\nu}}(\mathbf{u}-\mathbf{u}') - (\mathbf{u}-\mathbf{u}')\, T_{\hat{\nu}}(\mathbf{u}_0)\mathrm{d}\sigma\right| \\
\leq& \sqrt{\sigma(\mathscr{S})}\,\Big(\int_{\mathscr{S}} \left|\mathbf{u}_0\cdot T_{\hat{\nu}}(\mathbf{u}'-\mathbf{u}) - (\mathbf{u}'-\mathbf{u})T_{\hat{\nu}}(\mathbf{u}_0)\right|^2 \mathrm{d}\sigma\Big)^{1/2} T_2 \\
\leq& \sqrt{\sigma(\mathscr{S})}\,\Big(\int_{\mathscr{S}} \left(|\mathbf{u}_0| + |T_{\hat{\nu}}(\mathbf{u}_0)|\right)^2 \mathrm{d}\sigma\Big)^{1/2} T_3 \\
\leq& C'_{\lambda,\mu,n}\, \sigma(\mathscr{S})\, e^{-\delta_0|\Re\xi|h}\, T_3 \\
\leq& C'_{\lambda,\mu,n}\, h^{n-1}\, e^{-\delta_0|\Re\xi|h}\, T_4,
\end{aligned}
\tag{13.3.27}
$$

where

$$
\begin{aligned}
T_2 &= \sup_{\mathscr{S}} |\nabla(\mathbf{u}-\mathbf{u}')| + \sup_{\mathscr{S}} |\mathbf{u}-\mathbf{u}'|, \\
T_3 &= \|\mathbf{u}_s - \mathbf{u}'_s\|_{C^1(\overline{B_h(\mathbf{x}_0)})^n} + \|\mathbf{u}_p - \mathbf{u}'_p\|_{C^1(\overline{B_h(\mathbf{x}_0)})^n}, \\
T_4 &= \|\mathbf{u}_s\|_{H^2(B_{2R})^n} + \|\mathbf{u}_p\|_{H^2(B_{2R})^n} + \|\mathbf{u}'_s\|_{H^2(B_{2R})^n} + \|\mathbf{u}'_p\|_{H^2(B_{2R})^n}.
\end{aligned}
$$

Finally, it is easy to see that

$$
C\left|(\rho(\mathbf{x}_0)-1)\int_{\mathscr{A}} e^{\xi\cdot(\mathbf{x}-\mathbf{x}_0)}\eta\cdot\mathbf{u}(\mathbf{x}_0)\mathrm{d}\mathbf{x}\right| \leq R_1 + R_2 + R_3 + I_1 + I_2, \tag{13.3.28}
$$

where $C = C\big(\kappa_p, \kappa_s, \mathscr{M}, \mathscr{R}, \theta_1, \theta_2, n, \delta_0\big) > 0$. Substituting (13.3.22a) to (13.3.22d), (13.3.26), and (13.3.27) into (13.3.28), we prove (13.3.15). □

13.4 Proof of Theorem 13.1.1

In this section, we shall give the proof of Theorem 13.1.1.

Proof of Theorem 13.1.1. Let $\mathbf{x}_0$ be a vertex of Ω such that $\mathrm{dist}(\mathbf{x}_0, \Omega') = \mathrm{dist}_{\mathscr{H}}(\Omega, \Omega')$ and Q be the convex hull of Ω and Ω'. From Lemma 13.3.1 and Lemma 13.3.2, we know $\mathbf{x}_0$ must be a vertex of Q. Let $\mathscr{B}$ be the open polyhedral cone generated by Q at $\mathbf{x}_0$. Then there must be an open spherical cone $\mathscr{D}$ whose opening angle at $\mathbf{x}_0$ is in $(2\alpha_M, \pi) \subset (2\alpha_m, \pi)$. Let $\mathscr{A}$ denote the open polyhedral cone generated by Ω at $\mathbf{x}_0$ and $\mathscr{C}$ be an open spherical cone such that $(\mathscr{A}, \mathscr{C}) \in \mathscr{F}(\alpha_m, \alpha_M, n)$ (cf. Definition 13.3.1). Denote $\hbar = \mathrm{dist}(\mathbf{x}_0, \Omega') < 1$, and choose $h \leq \min\{l_0, \hbar\}$ such that

$$
\begin{cases}
\Delta^*\mathbf{u}' + \omega^2\mathbf{u}' = \mathbf{0} \quad \text{in} \quad S_h \subset D_h, \\
S_h \cap \Omega' = D_h \cap \Omega' = \emptyset,
\end{cases}
$$

where $S_h = \Omega \cap B_h(\mathbf{x}_0) = \mathscr{A} \cap B_h(\mathbf{x}_0)$ and $D_h = Q \cap B_h(\mathbf{x}_0) = \mathscr{B} \cap B_h(\mathbf{x}_0)$. From the proof of Proposition 13.2.3, we can derive that

$$
\|\mathbf{u}_\alpha\|_{H^2(B_{2R})^n}, \|\mathbf{u}'_\alpha\|_{H^2(B_{2R})^n} \leq C_{\kappa_p, \kappa_s, R, \mathscr{S}}, \quad \alpha = \mathrm{p}, \mathrm{s}.
$$

Thus,

$$
\mathbf{u} \in H^2(B_{2R})^n \text{ and } \mathbf{u}' \in H^2(B_{2R})^n.
$$

The a priori admissibility assumptions of ρ and the local H^2 regularity of $\mathbf{u}$ near S_h imply the following splittings:

$$\rho(\mathbf{x}) = 1 + \rho(\mathbf{x}_0) + \rho_1(\mathbf{x}), \quad |\rho_1(\mathbf{x})| \le \|\rho_1\|_{C^{\theta_1}(K\cap B_h(\mathbf{x}_0))}|\mathbf{x}-\mathbf{x}_0|^{\theta_1},\ \theta_1 \in (0,1),$$
$$\mathbf{u}(\mathbf{x}) = \mathbf{u}(\mathbf{x}_0) + \mathbf{u}_1(\mathbf{x}), \quad |\mathbf{u}_1(\mathbf{x})| \le \|\mathbf{u}_1\|_{C^{\theta_2}(K\cap B_h(\mathbf{x}_0))^n}|\mathbf{x}-\mathbf{x}_0|^{\theta_2},\ \theta_2 \in (0,1),$$

where we use the Sobolev embedding theorem from H^2 to C^{θ_2}. So the conditions of Proposition 13.3.3 are all satisfied. We absorb all the constants into the left-hand side, which depend only on the a priori parameters, and then (13.3.15) reduces to

$$\begin{aligned} C\Big|(\rho(\mathbf{x}_0)-1)\int_{\mathscr{A}} e^{\xi\cdot(\mathbf{x}-\mathbf{x}_0)}\eta\cdot\mathbf{u}(\mathbf{x}_0)\mathrm{d}x\Big| \le{}& |\Re\xi|^{-n}e^{-\delta_0|\Re\xi|h/2} + |\Re\xi|^{-n-\min\{\theta_1,\theta_2\}} \\ &+ h^{n-1}\sup_{\partial\mathscr{B}\cap B_h(\mathbf{x}_0)}\Big(|\mathbf{u}-\mathbf{u}'| + |\nabla\mathbf{u}-\nabla\mathbf{u}'|\Big)^{-1/2} \\ &+ h^{n-1}e^{-\delta_0|\Re\xi|h}. \end{aligned} \tag{13.4.1}$$

From Proposition 13.2.3 and the fact that $\partial Q \supset \partial\mathscr{B}\cap B_h(\mathbf{x}_0)$, we have

$$\sup_{\partial\mathscr{B}\cap B_h(\mathbf{x}_0)}\Big(|\mathbf{u}-\mathbf{u}'| + |\nabla\mathbf{u}-\nabla\mathbf{u}'|\Big) \le C'\left(\ln\ln(\mathscr{N}/\varepsilon)\right)^{-1/2}.$$

Let $\delta(\varepsilon) = (\ln\ln(\mathscr{N}/\varepsilon))^{-1/2}$, where C' depends only on those a priori parameters.
On the other hand, note that $(\mathscr{A},\mathscr{C}) \in \mathscr{F}(\alpha_m,\alpha_M,n)$. We can apply Proposition 13.3.2 to obtain an estimate of the term on the left-hand side of (13.3.14) as follows:

$$\Big|\int_{\mathscr{A}} e^{\xi(\tau)\cdot(\mathbf{x}-\mathbf{x}_0)}\eta\cdot\mathbf{u}(\mathbf{x}_0)\mathrm{d}x\Big| \ge C'_{\mathscr{A}}\tau^{-n}, \tag{13.4.2}$$

where $C'_{\mathscr{A}}$ is a positive constant only depending on $\mathscr{A}$ and ζ_0.

From Definition 13.1.1, we have $|\rho(\mathbf{x}_0)-1| \ne 0$. Combining (13.4.1) with (13.4.2), we have

$$C_1\big|(\rho(\mathbf{x}_0)-1)\big| \le e^{-\delta_0|\Re\xi|h/2} + \tau^{-\min\{\theta_1,\theta_2\}} + h^{n-1}\tau^n\delta(\varepsilon) + h^{n-1}\tau^n e^{-\delta_0|\Re\xi|h},$$

where C_1 is a positive constant depending only on those a priori parameters and $\mathbf{u}(\mathbf{x}_0)$.

Using the fact that $e^{-\mathbf{x}} \le \frac{1}{\mathbf{x}}$ and $e^{-\mathbf{x}} \le \frac{(n+4)!}{\mathbf{x}^{n+4}}$ for all $\mathbf{x} > \mathbf{0}$, we get

$$\begin{aligned} C_2\big|(\rho(\mathbf{x}_0)-1)\big| &\le \tau^{-1}h^{-1} + \tau^{-\min\{\theta_1,\theta_2\}} + h^{n-1}\tau^{n+1}\delta(\varepsilon) + h^{-5}\tau^{-3} \\ &\le C_0(\tau^{-m}h^{-5} + \tau^{n+1}\delta(\varepsilon)h^{n-1}) \\ &\le C_0h^{n-1}[\tau^{-m}h^{-n-4} + \tau^{n+1}\delta(\varepsilon)]. \end{aligned}$$

Here, $m = \min\{1,\theta_1,\theta_2\}$. Note that $\tau^4 > 1$, $h^{-n} > 1$, and $0 < h^{n-1} < 1$. Dividing both sides of the above inequality by h^{n-1}, we obtain the following inequality:

$$C_3\big|(\rho(\mathbf{x}_0)-1)\big| \le \tau^{-m}h^{-n-5} + \tau^{n+5}\delta(\varepsilon). \tag{13.4.3}$$

For determining a minimum modulo constant of the right-hand side of (13.4.3), setting

$$\tau = \tau_e = \left[\frac{1}{h^{n+5}\delta(\varepsilon)}\right]^{\frac{1}{m+n+1}}, \tag{13.4.4}$$

we have

$$C_4\big|(\rho(\mathbf{x}_0)-1)\big| \le 2\,\delta(\varepsilon)^{\frac{m}{m+n+5}}h^{-\frac{(n+5)^2}{m+n+5}}. \tag{13.4.5}$$

Note that $\tau_e \ge \tau_e h^{\frac{n+5}{m+n+5}} = \delta(\varepsilon)^{-\frac{1}{m+n+5}} \ge \tau_0$ and $h \le \hbar \le l_0$ for sufficiently small ε. From (13.4.5), we have

$$\hbar \le C_5\,\delta(\varepsilon)^{\frac{m}{m+n+5}} \left|\rho(\mathbf{x}_0)\right|^{-\frac{m}{(n+5)^2}-\frac{1}{n+5}} \le C_5\,\delta(\varepsilon)^{\frac{m}{m+n+5}} \left|\varepsilon_0+1\right|^{-\frac{m}{(n+5)^2}-\frac{1}{n+5}}$$
$$\le C_6\left(\ln\ln(\mathscr{N}/\varepsilon)\right)^{-\frac{m}{2(n+5)^2}},$$

and the claim readily follows.

The proof is complete. □

13.5 Related Results About Invisibility

The present section is devoted to the technical details of Theorem 13.1.2.

Theorem 13.5.1. *Let $(\Omega;\rho)$ be an admissible polyhedral scatterer. The plane incident wave $\mathbf{U}^i$ of the form* (13.1.1) *is a nontrivial entire solution to* (13.1.2) *with $\rho = 1$. Let $\mathbf{U}$ be the far-field pattern of the scattered wave $\mathbf{u}^s$ by medium scatterer $(\Omega;\rho)$. Then there exists a positive constant C only depending on the a priori parameters such that the following estimate holds:*

$$\|\mathbf{U}\|_{L^2(\mathbb{S}^{n-1},\mathbb{C}^n\times\mathbb{C}^n)} \ge C > 0. \tag{13.5.1}$$

In fact, $\Omega \subset \mathbb{R}^n$, $n = 2,3$, can be relaxed to a general bounded Lipschitz domain, which admits a convex polygonal/polyhedral point on its boundary, namely, there exits $\mathbf{x}_c \in \partial\Omega$ such that $\Omega \cap B_l(\mathbf{x}_c) \subset \mathbb{R}^2$ is a plane sector for $l_0 > l > 0$ and $\Omega \cap B_l(\mathbf{x}_c) \subset \mathbb{R}^3$ is a polyhedral cone with a spherical bottom surface for $l_0 > l > 0$. The general result is stated as follows.

Theorem 13.5.2. *Let $\Omega \subset \mathbb{R}^n$, $n = 2,3$, be a bounded Lipschitz domain, with a convex polygonal/polyhedral point $\mathbf{x}_c \in \partial\Omega$. Let $\rho(\mathbf{x})$ be a uniformly θ-Hölder-continuous function in Ω, $0 < \theta \le 1$. The plane incident wave $\mathbf{u}^i$ of the form* (13.1.1) *is a nontrivial entire solution to* (13.1.2) *with $\rho = 1$. Let $\mathbf{U}$ be the far-field pattern of the scattered wave $\mathbf{u}^s$ by medium scatterer $(\Omega;\rho)$. Suppose that the following conditions are fulfilled:*

(1) $\Omega = \mathrm{supp}(1-\rho) \subset \overline{B_R}$ *for some* $R > l > 0$.
(2) $\|\mathbf{u}^s\|_{H^2(B_R)^n} \le \mathscr{N}$ *for* $\mathscr{N} > 0$.
(3) $|\rho(\mathbf{x}_c) - 1| \ge \varepsilon_0 > 0$ *and* $\|\rho\|_{C^\theta(\Omega)} \le \mathscr{M}$ *for some* $\mathscr{M} > 0$.
(4) *For* $n = 2$, $0 < 2\alpha_m \le \alpha \le 2\alpha_M \le \pi$, *where α signifies the opening angle at* $\mathbf{x}_c$.

Then the following estimate holds:

$$\|\mathbf{U}\|_{L^2(\mathbb{S}^{n-1},\mathbb{C}^n\times\mathbb{C}^n)} \ge C > 0, \tag{13.5.2}$$

where C is a positive constant, which depends on the a priori parameters $\theta, l, \mathscr{N}, \mathscr{M}, \varepsilon_0$, ζ_0, and R.

Proof. We follow similar arguments in the proof of Theorem 13.1.1 with some necessary modifications. We take $\Omega' = \emptyset$ and $\rho' \equiv 1$ in $\mathbb{R}^n$. Clearly, the incident wave $\mathbf{u}^i$ of the form (13.1.1) is an entire solution to $\Delta^*\mathbf{u}' + \omega^2\mathbf{u}' = \mathbf{0}$, which yields $\mathbf{u}'^s \equiv \mathbf{0}$ and $\mathbf{U}' \equiv \mathbf{0}$. Denote $\varepsilon = \|\mathbf{U}\|_{L^2(\mathbb{S}^{n-1},\mathbb{C}^n\times\mathbb{C}^n)}$. In fact, we only consider the effect caused by $\mathbf{x}_c$ in its micro-local area $\Omega \cap B_h(\mathbf{x}_c)$, and therefore, we can reduce the corresponding reasoning in the proof Theorem 13.1.1. Here, we take $h = l$. After a series of derivations similar to Theorem 13.1.1, we can obtain the following inequality:

$$C_1\left|(\rho(\mathbf{x}_c)-1)\right| \le \tau^{-m} l^{-5} + \tau^{n+1}\delta(\varepsilon)\, l^{n-1}, \tag{13.5.3}$$

where $\delta(\varepsilon) = (\ln\ln(\mathscr{N}/\varepsilon))^{-1/2}$ and $C_1 > 0$ does not depend on l.

Setting

$$\tau = \tau_e = \left[\frac{1}{l^{n+5}\delta(\varepsilon)}\right]^{\frac{1}{m+n+1}},$$

we have

$$C_2\big|(\rho(\mathbf{x}_c)-1)\big| \le 2\,\delta(\varepsilon)^{\frac{m}{m+n+5}}\, l^{-\frac{(n+5)^2}{m+n+5}}. \tag{13.5.4}$$

It is straightforward to verify that for sufficient small ε, one has

$$\tau_e \ge \tau_e\, l^{\frac{n+5}{m+n+5}} = \frac{C_3}{\delta(\varepsilon)^{m+n+5}} = C_3(\ln\ln(\mathscr{N}/\varepsilon))^{\frac{(n+5)^2}{2(m+n+5)}} \ge \tau_0.$$

Using $\tau = \tau_e$ in (13.5.4), $|\rho(\mathbf{x}_c)-1| \ge \varepsilon_0 > 0$, and $\delta(\varepsilon) = (\ln\ln(\mathscr{N}/\varepsilon))^{-1/2}$,

$$\begin{aligned} \varepsilon &\ge \mathscr{N}\Big[\exp\exp\big\{C_3\, l^{-\frac{(n+5)^2}{m}}|(\rho(\mathbf{x}_c)-1)|^{-\frac{m+n+5}{m}}\big\}\Big]^{-1} \\ &\ge \mathscr{N}\Big[\exp\exp\big\{C_3\, l^{-\frac{(n+5)^2}{m}}\varepsilon_0^{-\frac{m+n+5}{m}}\big\}\Big]^{-1}. \end{aligned} \tag{13.5.5}$$

That is,

$$\|\mathbf{U}\|_{L^2(\mathbb{S}^{n-1},\mathbb{C}^n\times\mathbb{C}^n)} \ge C > 0,$$

where $C = \mathscr{N}\Big[\exp\exp\big\{C_3\, l^{-\frac{(n+5)^2}{m}}\varepsilon_0^{-\frac{m+n+5}{m}}\big\}\Big]^{-1}$.

The proof is complete. □

References

1. G. Alessandrini and L. Rondi, *Determining a sound-soft polyhedral scatterer by a single far-field measurement*, Proc. Amer. Math. Soc., **35** (2005), 1685–1691.
2. Z. Bai, H. Diao, H. Liu and Q. Meng, *Stable determination of an elastic medium scatterer by a single far-field measurement and beyond*, Calc. Var. Partial Differential Equations, **61** (2022), no. 5, Paper No. 170, (23 pp).
3. E. Blåsten, *Nonradiating sources and transmission eigenfunctions vanish at corners and edges*, SIAM J. Math. Anal., **50(6)** (2018), 6255–6270.
4. E. Blåsten and Y.-H. Lin, *Radiating and non-radiating sources in elasticity*, Inverse Problems, 35, 1 (2019) 015005.
5. E. Blåsten and H. Liu, *Recovering piecewise constant refractive indices by a single far-field pattern*,Inverse Problems, **36** (2020), 085005.
6. E. Blåsten and H. Liu, *On corners scattering stably and stable shape determination by a single far-field pattern*, Indiana Univ. Math. J., **70(3)** (2021), 907–947.
7. E. Blåsten and H. Liu, E. Blåsten and H. Liu, *Scattering by curvatures, radiationless sources, transmission eigenfunctions and inverse scattering problems*, SIAM Journal on Mathematical Analysis, **53** (2021), no. 4, 3801–3837.
8. E. Blåsten, H. Liu and J. Xiao, *On an electromagnetic problem in a corner and its applications*, Anal. PDE, **14(7)** (2021), 2207–2224.
9. E. Blåsten, L. Päivärinta and J. Sylvester, *Corners always scatter*, Comm. Math. Phys., **331** (2014), 725–753.
10. F. Cakoni and J. Xiao, *On corner scattering for operators of divergence form and applications to inverse scattering*, Comm. Partial Differential Equations, in press, (2020).
11. R. Brummelhuis, *Three-spheres theorem for second order elliptic equations*, J. Analyse Math. **65** (1995) 179–206.
12. X. Cao, H. Diao and H. Liu, *Determining a piecewise conductive medium body by a single far-field measurement*, CSIAM Trans. Appl. Math., **1** (2020), 740–765.
13. D. P. Challa and M. Sini, *The Foldy-Lax approximation of the scattered waves by many small bodies for the Lamé system*, Math. Nachr., **288(16)** (2015), 1834–1872.
14. D. Colton and R. Kress, *Looking back on inverse scattering theory*, SIAM Review, **60(40)** (2018), 779–807.
15. D. Colton and P. Monk, *The inverse scattering problem for time-harmonic acoustic waves in an inhomogeneous medium*, Quart. J. Mech. Appl. Math., **41** (1988), 97–125,
16. H. Diao, X. Cao and H. Liu, *On the geometric structures of conductive transmission eigenfunctions and their application*, Comm. Partial Differential Equations, **46(4)** (2021), 630–679.

17. H. Diao, X. Fei, H. Liu and K.Yang, *Visibility, invisibility and unique recovery of inverse electromagnetic problems with conical singularities*, arxiv 2204.02835, (2022).
18. H. Diao, H. Li, H. Liu and J. Tang, *Spectral properties of an acoustic-elastic transmission eigenvalue problem with applications*, J. Differential Equations, **371** (2023), 629–659.
19. H. Diao, H. Liu, and B. Sun, *On a local geometric property of the generalized elastic transmission eigenfunctions and application*, Inverse Problems, **37** (2021), no. 10, Paper No. 105015, 36 pp.
20. H. Diao, H. Liu, X. Wang and K. Yang, *On vanishing and localizing around corners of electromagnetic transmission resonance*, Partial Differ. Equ., **2**(2021), 78.
21. M. Di Cristo and L. Rondi, *Example of exponential instability for inverse inclusion and scattering problems*, Inverse problems, **19(3)** (2003), 685–701.
22. P. Hähner, *A uniqueness theorem in inverse scattering of elastic waves*, IMA J. Appl. Math., **51** (1993), 201–215.
23. P. Hähner, *On acoustic, electromagnetic, and elastic scattering problems in inhomogeneous media*, Universität Göttingen, Habilitation Thesis, (1998).
24. P. Hähner, *On uniqueness for an inverse problem in inhomogeneous elasticity*, IMA J. Appl. Math., **67** (2002), 127–143.
25. G. Hu and H. Liu, *Nearly cloaking the elastic wave fields*, J. Math. Pures Appl., **104(6)** (2015), 1045–1074.
26. G. Hu, M. Salo and E. Vesalainen, *Shape identification in inverse medium scattering problems with a single far-field pattern*, SIAM J. Math. Anal., **48** (2016),152–165.
27. H. Liu, *On local and global structures of transmission eigenfunctions and beyond*, J. Inverse Ill-Posed Probl., **30(2)** (2022), 287–305.
28. H. Liu, L. Rondi and J. Xiao, Mosco convergence for $H(curl)$ spaces, higher integrability for Maxwell's equations, and stability in direct and inverse EM scattering problems, *J. Eur. Math. Soc.*, in press, (2017).
29. H. Liu, M. Petrini, L. Rondi and J. Xiao, Stable determination of sound-hard polyhedral scatterers by a minimal number of scattering measurements, *J. Differential Equations*, **262(3)** (2017), 1631–1670.
30. H. Liu and C.-H. Tsou, *Stable determination of polygonal inclusions in Calderón's problem by a single partial boundary measurement*, Inverse Problems, **36(8)** (2020), 085010, 23 pp.
31. H. Liu and C. H. Tsou, *Stable determination by a single measurement, scattering bound and regularity of transmission eigenfunction*, Calc. Var., **61**(2022), 91.
32. H. Liu, C.-H. Tsou and W. Yang, *On Calderón's inverse inclusion problem with smooth shapes by a single partial boundary measurement*, arXiv:2006.10586
33. H. Liu and J. Xiao, Decoupling elastic waves and its applications, *J. Differential Equations*, **265(8)** (2017), 4442–4480.
34. H. Liu and J. Zou, *On uniqueness in inverse acoustic and electromagnetic obstacle scattering problems*, Journal of Physics: Conference Series, Applied Inverse Problems, Vancouver, Canada, (2007).
35. N. Mandache, *Exponential instability in an inverse problem for the Schrö equation,* Inverse Problems, **17(5)**(2001), 1435–1444.
36. W. Mclean, *Strongly Elliptic Systems and Boundary Integral Equation*, Cambridge University Press, Cambridge, (2000).
37. G. Menegatti and L. Rondi, *Stability for the acoustic scattering problem for sound-hard scatterers*, Inverse Probl. Imaging **7** (2013) 1307–1329.
38. Q. Meng, Z. Bai, H. Diao and H. Liu, *Effective medium theory for embedded obstacles in elasticity with applications to inverse problems*, SIAM J. Appl. Math., **82** (2022), No. 2, 720–749.
39. A. Morassi and E. Rosset, *Stable determination of cavities in elastic bodies*, Invese Problems, **20(2)**(2004), 453–480.
40. A. Morassi and E. Rosset, *Uniqueness and stability in determining a rigid inclusion in an elastic body*, Mem. Amer. Math. Soc., **20(938)**(2009), viii+5888pp.
41. L. Päivärinta, M. Salo and E. V. Vesalainen, *Strictly convex corners scatter*, Rev. Mat. Iberoam. **33(4)** (2017), 1369–1396.
42. L. Rondi, *Stable determination of sound-soft polyhedral scatterers by a single measurement*, Indiana Univ. Math. J., **57** (2008), 1377–1408.
43. L. Rondi, E. Sincich and M. Sini, *Stable determination of a rigid scatterer in elastodynamics*, SIAM J. Math. Anal., **53** (2021), 2660–2689.
44. L. Rondi and M. Sini, *Stable determination of a scattered wave from its far-field pattern: the high frequency asymptotics*, Arch. Ration. Mech. Anal. **218** (2015), 1–54.
45. E. Sincich and M. Sini, *Local stability for soft obstacles by a single measurement*, Inverse Probl. Imaging, **2(2)** (2008), 301–315.

Index

H. Diao, H. Liu, *Spectral Geometry and Inverse Scattering Theory*,
https://doi.org/10.1007/978-3-031-34615-6

The manufacturer's authorised representative in the EU is Springer Nature Customer Service Centre GmbH, Europaplatz 3, 69115 Heidelberg, Germany. If you have any concerns regarding our products, please contact ProductSafety@springernature.com

Printed and bound by CPI Group (UK) Ltd, Croydon, CR0 4YY
05/10/2026
02228168-0002